CRC HANDBOOK OF RADIATION MEASUREMENT AND PROTECTION

Allen B. Brodsky
Editor

SECTION OUTLINE

SECTION A

General Scientific and Engineering Information

SECTION B

Medical Radiological Physics and Safety

SECTION C

Health Physics Methods in Industrial, Academic, and Research Facilities

CRC Handbook
of
Radiation Measurement
and
Protection

Section A
Volume I: Physical Science and Engineering Data

Editor

Allen B. Brodsky

Certified American Board of Health Physics
American Board of Radiology
American Board of Industrial Hygiene

CRC PRESS, Inc.
2255 Palm Beach Lakes Boulevard · West Palm Beach, Florida 33409

Library of Congress Cataloging in Publication Data

Main entry under title:

CRC handbook of radiation measurement &
    protection.

    Includes bibliographies and indexes.
    CONTENTS: Section A. General scientific &
engineering information.   2. v.
    1. Radiation dosimetry—Handbooks, manuals,
etc. 2. Radiation—Safety measures—Handbooks,
manuals, etc. I. Brodsky, Allen B.
QC795.32R3C17   614.8'39'0202   78-10558
ISBN 0-8493-3750-X (complete set)

This book represents information obtained from authentic and highly regarded sources. Reprinted material is quoted with permission, and sources are indicated. A wide variety of references are listed. Every reasonable effort has been made to give reliable data and information, but the author and the publisher cannot assume responsibility for the validity of all materials or for the consequences of their use.

International Standard Book Number 0-8493-3750-X
Section A, 0-8493-3755-0
Volume I, 0-8493-3756-9

Library of Congress Card Number 78-10558

Printed in the United States

PREFACE

The purpose of this handbook series is to provide data and methods for dealing with the design and evaluation of radiation measurement instruments; monitoring and survey methods; selection of protective facilities, equipment and procedures for handling radioactive materials; effluent control and monitoring; radioactive waste disposal; transportation of radioactive materials; and many other related subjects. Aspects of these subjects involve not only all of the sciences but also management, legal, insurance, and public information considerations. The sciences themselves include the entire range of basic scientific and engineering methods, as broad a range as in the entire field of public health itself. This series of volumes will be divided as far as possible into logical groupings that hopefully will tend to maximize the usefulness of each volume, as well as the series of volumes, to its users.

The primary aim of the handbook series will be to include as much useful data as possible for the specialist needing ready access for the solution of problems most likely to arise in the radiation protection professions. However, some selected review of fundamental concepts is also included to enable persons with a basic science or engineering background to acquire the necessary knowledge to solve a majority of problems in especially important aspects of radiation protection. Also, since the profession is broad in discipline, an attempt has been made to fulfill the frequent need of professionals for a refresher course in some of the more important fundamentals needed to utilize data included in the handbook. Principles of management, organization, and procedures related to radiation safety will also be summarized in later volumes, with attention to presentation of methods for establishing new radiation safety programs based on the accumulated experience of others.

This series of volumes should, therefore, be useful in industrial, academic, medical, and governmental institutions and to scientists carrying out both applied and basic research in subjects related to radiation protection, to engineers designing protective facilities and equipment, to applied health physicists establishing radiation safety programs, to physicians involved with caring for patients exposed to radiation or radioactive materials, to public health and other government officials concerned with regulating radiation safety and developing national standards, to professors and students studying and synthesizing a broad base of knowledge in radiation protection, and to practically every other type of professional involved in radiation protection problems. The editor and publisher hope that the many users will offer continued suggestions for improving these volumes and for including data that will make them more useful.

A general acknowledgment of gratitude must be made to the advisory board, to the contributors for all volumes, and to the authors and publishers who have granted permission to include material in this series. Contributors of the sections in this series have been selected for their professional knowledge in the respective subject areas, and in most cases have generously devoted their own personal time to this effort without monetary compensation. Material included in this handbook series was selected according to the professional opinions of authors, compilers, and editors and does not necessarily reflect any official sanction of the agencies or companies with which the contributors are employed. Addresses indicated for the contributors are for identification and correspondence purposes only. The collective responsibility for the total content rests, of course, with the editor, who is solely responsible for any omissions either from inadvertence or necessity.

On the other hand, the editor has purposely built in some redundancy of certain important types of data, since it is often advisable to cross-check values obtained from different literature sources or approaches. The editor has also, to some degree, allowed repetition of similar data in different sections of the handbook where it was deemed advisable for the continuity of individual sections or for the convenience of the user due to the need for frequent cross reference between text and data. In cases where the users intend to rely on a tabulated or calculated value for the protection of humans, it is, of course, necessary to arrive at values in different ways in order to not only avoid inadvertent mistakes but also to ensure better understanding of the degree of reliability or range of uncertainty of the end result. Frequently, in radiation protection design as in general engineering practice, factors of safety must be included that are based on knowledge of the ranges of uncertainty in estimated risks. A further purpose of the redundancy of sources is also to include older as well as newer approaches in cases where the changes themselves may be of interest in evaluating the need for protective actions or changes in regulations or policies.

The editor is grateful to CRC Press for generally providing an outstanding facility and staff for the publication of data and reference sources currently needed in science and technology. The publishers, Bernard, Florence, and Earl Starkoff have also given invaluable guidance in determining the philosophy, scope, and organization of this handbook series. Their continued enthusiasm and encouragement has been an inspiration to the editor. Gerald A. Becker, Editorial Director when the project began, also provided early inspirational leadership in developing the concepts and format of this series. Kay Harter was helpful in the initial administration of this series. A general acknowledgment is made to all the CRC editorial and technical staff who worked so diligently to produce these volumes, particularly Susan Cubar, Sandy Pearlman, Michael Ference, Georgene C. Smith, Paul Gottehrer, and Jodi Willoughby. The editor would also like to acknowledge the invitation of Professors Niel Wald and Yen Wang to contribute to the *CRC Handbook of Radioactive Nuclides*; reprinting of some of the editor's sections from the above handbook ultimately inspired the development of this handbook series. Special mention should be made of the able assistance of Eileen M. Haycraft in initiating and carrying through so much of the detailed work necessary in Bethesda in compiling, drafting, and editing many of the sections of this series.

Allen Brodsky
Editor
Bethesda, Maryland
November 1978

ALLEN BRODSKY, EDITOR
CRC HANDBOOK OF RADIATION MEASUREMENT
AND PROTECTION

Dr. Allen Brodsky has a broad professional background and interest in the scientific, engineering, and management aspects of radiation protection. After graduation with a B.E. in Chemical Engineering (Johns Hopkins University) and 1 year of postgraduate study in Radiological Physics at the Oak Ridge National Laboratory, his broad career in health physics and radiation measurements pinpointed the need for further study, which led to the degrees of M.A. in Physics (Johns Hopkins University) and Sc.D. in Biostatistics and Radiation Health (University of Pittsburgh), followed by board certifications from the American Board of Health Physics, the American Board of Industrial Hygiene, and the American Board of Radiology.

In addition to his education, positions as Physicist and Health Physicist, Naval Reserve Laboratory; Radiological Defense Officer, Federal Civil Defense Administration: Health Physicist, U.S. Atomic Energy Commission: Associate Professor of Health Physics, University of Pittsburgh: Radiation Therapy Physicist and Radiation Safety Officer, Mercy Hospital; and Adjunct Research Professor, School of Pharmacy, Duquesne University stimulated his interest in the need for readily available data to solve radiation measurement and protection problems, and to design facilities, equipment, and procedures for the safe handling of radiation sources and radioactive materials.

The Editor's research interests include the development of radiation measurement and dosimetry methods, hazard evaluation, facility design, methodology for health and mortality follow-up of radiation workers, and the theory of cancer induction by environmental agents. His present position at the U.S. Nuclear Regulatory Commission involves the development of guides for radiation protection in medical institutions, industry, and universities.

The editor welcomes suggestions or offers of further contributions of data for the handbook series at the following address:

Dr. Allen Brodsky
P. O. Box 34471
West Bethesda, MD 20034

PREFACE
SECTION A: GENERAL SCIENTIFIC AND ENGINEERING INFORMATION

This first volume of the *Handbook of Radiation Measurement and Protection* contains data, references, and some review material to provide much of the basic scientific and engineering information utilized in everyday work in professions related to radiation protection. Some reference is also made to related sections in future volumes where applications requiring the data in this volume are presented.

In addition to the acknowledgments given in the preface to the entire handbook series, the editor would like to acknowledge the excellent assistance of Michael Ference, Georgene C. Smith, and Paul Gottehrer of CRC Press in handling the final editing and styling of sections and management of other details of publication. Again, the outstanding assistance of Eileen M. Haycraft in preparing, typing, editing, and checking many of the sections is acknowledged.

ADVISORY BOARD MEMBERS

Dr. Walter S. Snyder
7017 Nubbin Ridge Road
Knoxville, Tennessee 37919

Dr. J. Newell Stannard
Department of Radiation Biology and
Biophysics (Emeritus)
The University of Rochester
School of Medicine and Industry
Rochester, New York 14642

Dr. Conrad P. Straub
2330 Chalet Drive
Minneapolis, Minnesota 55421

Dr. Niel Wald, Chairman
Department of Radiation Health
Graduate School of Public Health
University of Pittsburgh
130 DeSoto Street
Pittsburgh, Pennsylvania 15261

CONTRIBUTORS
SECTION A
VOLUME I: PHYSICAL SCIENCE DATA

Mitchell L. Borke
Professor
School of Pharmacy
Duquesne University
Pittsburgh, Pennsylvania 15219

Allen B. Brodsky
Health Physicist
U.S. Regulatory Commission
Washington, D.C.

Bernard L. Cohen
Professor of Physics
314 Nuclear Laboratory
Physics Department
University of Pittsburgh
Pittsburgh, Pennsylvania 15261

L. T. Dillman
Ohio Wesleyan University
Delaware, Ohio 43015

Ronald L. Kathren
Staff Scientist
Battelle Memorial Institute
Pacific Northwest Laboratories
Richland, Washington 99392

Alfred W. Klement, Jr.
Consultant
10105 Summit Avenue
Kensington, Maryland 20795

Edward B. Sanders
Senior Scientist and Project Leader of Organic
  Chemistry
Phillip Morris Research Center
Richmond, Virginia 23261

Tuvia Schlesinger
Health Physics Department
Soreq Nuclear Center
Yauneh, Israel

Imogene Sevin
17704 Millcrest Drive
Derwood, Maryland 20855

James E. Turner
Physicist
Health and Safety Research Division
Oak Ridge National Laboratory
Oak Ridge, Tennessee 37830

DEDICATION

This volume is dedicated to the many people whose scientific and professional contributions have produced a body of knowledge useful in protecting both workers and the public against unreasonable and unnecessary risks from radiation exposure. Some of those who have personally inspired the editor by a combination of unlimited devotion, uniqueness of contribution, and qualities of leadership are

Elda E. Anderson*
S. R. Bernard
Herman Cember
Charles L. Dunham*
Robley D. Evans
James C. Hart*
Wright H. Langham*
Karl Z. Morgan
Lester R. Rogers

Eugene Saenger
S. Marshall Sanders
Walter S. Snyder*
J. Newell Stannard
Lauriston S. Taylor
James E. Turner
John C. Villforth
Niel Wald
Forrest Western*

* Deceased.

TABLE OF CONTENTS

Introduction

1. INTRODUCTION

Allen Brodsky

1.1. GENERAL ORGANIZATION OF THE HANDBOOK SERIES

As stated in the Preface to the Handbook Series, the purpose of the series is to provide data and methods for dealing with many of the aspects of radiation protection encountered by professional scientists, engineers, physicians, administrators, and regulators in their daily activities. This introduction provides a brief guide to the information in the handbook series and other sources of information in the field of radiation protection.

Major section numbers in the series, i.e., Sections 1, 2, 3, etc., will continue consecutively throughout the series and be distributed among a number of volumes. An attempt will be made to use this consecutive section numbering to enable authors and/or compilers to cross reference other sections in the series that contain data or information of a nature related to or needed for their own subject matter. Sections will be divided into volumes in such a way that each volume will contain, as much as possible, a logical grouping of subjects that would make the volume by itself coherent and useful to as many users as possible. In addition, subject groupings of each volume will be selected with consideration of convenience to the user in learning the scope and logic of the handbook. Since the subject matter in radiation protection is still expanding rapidly and the material received from individual section authors has often been more generous in volume than expected, the exact content of each volume will not be specified in advance but will be optimized at the time of publication to accomplish the foregoing objectives.

This handbook series will initially focus on ionizing radiations, since the body of knowledge of these radiations is more well developed and of interest to defined segments of the profession. Also, there is some information on other radiations in the *CRC Handbook of Laboratory Safety* and other available summaries. There will also be some reference to regulatory requirements for nonionizing radiations in a later volume of the series. Section A contains the more basic scientific and engineering data that will be utilized in later sections of the series. Data in Section A are divided into sections on physical, chemical, biological, and mathematical data and methods most utilized in radiation measurement and protection. Only radiation measurement data useful in radiation dosimetry and protection are emphasized in this series. The *CRC Handbook of Radioactive Nuclides* contains information on the many industrial and medical applications of radioactive materials.

An index is provided at the end of each volume. After several volumes of the series have been published, a general index for the series is planned. The editor hopes that the general index will be improved as a result of experience with the use of individual volumes and feedback from the users.

The user should note that the equation, table, and figure numbering systems are such that each number will tell the user the major subsection, i.e., 1.2, 6.5, etc., to which the respective equation, table, or figure belongs. This will be helpful to authors as well as users in cross referencing useful information from other sections that are pertinent to the subject at hand.

No handbook can be completely current in such a rapidly changing field. An attempt has been made to incorporate the most recent literature references applicable to each volume up to the time of publication. However, there are many thousands of references that cannot be listed without making the handbook unwieldy.

Thus, a list of major scientific and technical periodicals that can be researched by the users is contained in Section 1.2 below.

1.2. LIST OF SCIENTIFIC AND TECHNICAL PERIODICALS

The following is a list of some of the major periodicals that present information and research findings in fields related to radiation protection and measurement. The professional, who needs summary data or texts not referenced in this handbook, may utilize the computer searching capabilities of modern libraries to find book titles and authors of interest.

AAAS Bulletin (American Association for the Advancement of Science)
ACS Cancer News (American Cancer Society)
Acta Radiologica, Therapy, Physics, Biology
Advances in Biological and Medical Physics
Advances in Cancer Research
Advances in Chemotherapy
Advances in Radiation Biology
American Journal of Roentgenology, Radium Therapy and Nuclear Medicine
Annales de Radiologie
*Applied Health Physics Abstracts**
Bio-Medical Engineering
Biomedical Sciences Instrumentation (see *National Biomedical Sciences Instrumentation Symposium*)
British Journal of Radiology
Bulletin du Cancer
Cancer Bulletin
Cancer Chemotherapy Abstracts
Cancer Research and supplements
Carcinogenesis Abstracts
Clinical Radiology
Computers and Biomedical Research
Dental Radiography and Photography
Energy Abstracts
Ergebnisse der Medizinischen Radiologie
European Journal of Cancer
Excerpta Medica: Section 14. Radiology
Excerpta Medica: Section 16. Cancer
Health Physics
IEEE Transactions on Bio-Medical Engineering (Institute of Electrical and Electronics Engineers)
International Journal of Applied Radiation and Isotopes
International Journal of Cancer
Journal of Biomedical Material Research
Journal of the National Cancer Institute
Journal of Nuclear Biology and Medicine
Journal of Nuclear Medicine
Journal de Radiologie, d'Electrologie et de Medicine Nucleaire
Journal of Theoretical Biology
Medical and Biological Engineering
Medical Electronics and Communications Abstracts
Medical Electronics News
Medical Physics
Medical Radiography and Photography
Monographs on Nuclear Medicine and Biology
National Cancer Institute (U.S.) *Monograph Series*
National Cancer Institute of Canada. Canada Cancer Conference. Proceedings
Nuclear Instruments and Methods

---

*   This quarterly journal is an excellent source of abstracts of publications broadly covering the subject of radiation measurement and protection. It is obtainable from *Applied Health Physics Abstracts,* Nuclear Technology Publishing, 180 Sandyhurst Lane, Ashford, Kent TN 25, 4nX England.

Nuclear News
Nuclear Safety
Nuclear Science Abstracts
Oncology; Journal of Clinical and Experimental Cancer Research
Physics in Medicine, Biology, and Medical Physics
Progress in Clinical Cancer
Radiation Research and Supplements
Radiobiologica Radiotherapia
Radiologia Clinica et Biologica
Radiologic Technology
Radiological Health Data and Reports
Radiology
Recent Results in Cancer Research
Science
Science News
Strahlentherapie and supplements
Symposium on Fundamental Cancer Research
U.S. National Cancer Institute. Monographs
Yearbook of Cancer
Yearbook of Nuclear Medicine

1.3. ORGANIZATIONS AND AGENCIES INVOLVED IN DEVELOPMENT OF RADIATION PROTECTION STANDARDS OR REGULATIONS

The following list of addresses will enable the user to contact major organizations and agencies that might be able to supply information not found in this handbook or the cited references. These organizations continually publish new documents on radiation protection and measurement; thus, the user who will continue to be active in this field should have himself placed on the mailing list of the organizations and agencies listed below. In addition, a nuclear information center is available at the Oak Ridge National Laboratory, Oak Ridge, Tenn. 37830, which will provide computer searches of certain categories of information in the radiation protection fields, including listings of literature referenced according to selected key words. Physical and biological data needed for internal dose calculations are available from data banks maintained in the Health and Safety Research Division of the Oak Ridge National Laboratory by Drs. S. R. Bernard and Mary R. Ford. Other reports and additional agencies publishing standards in the nuclear field related to radiation protection are given in the section by W. B. Cottrell in the *CRC Handbook of Laboratory Safety.*[1] However, the user should note the more current addresses in the list below that replace some of those in Cottrell's section.

The user developing radiation safety programs or practices should also consult current regulations of the federal agencies in the list below as well as state and local governments that might have radiation safety regulations covering activities in their jurisdiction. A list is given below of those state agencies that have an agreement with the Nuclear Regulatory Commission to carry out state licensing and regulatory functions for certain categories of uses of radioactive materials. Further guidance on regulations and standards will be given in a section of Volume II.

Major Organizations and Agencies

American Association of Physicists in Medicine (AAPM)
Radiation Protection Committee
335 East 45th Street
New York, N. Y 10017

(The AAPM Radiation Protection Committee has developed guidance for radiation protection and reducing exposures ALARA in various types of procedures in medical institutions using radioactive materials.)

American Conference of Governmental Industrial Hygienists
Committee on Industrial Ventilation
P.O. Box 453
Lansing, Mich. 48902

(This committee regularly updates its "Industrial Ventilation" manual, which provides engineering data and methods equally applicable to design of medical facilities handling radioactive materials.)

American Industrial Hygiene Association (AIHA)
66 S. Miller Road
Akron, Ohio 44313

(The AIHA has established a laboratory for accreditation of laboratories carrying out analyses related to the health and safety of employees and is proposing to extend this program to bioassay of radioactive materials for internal dose estimation. They may be contacted as well as HPS for expert assistance on planning bioassay services or for ventilation design and general laboratory safety considerations.)

American National Standards Institute, Inc. (ANSI)
IEEE Standards Office
345 East 47th Street
New York, N.Y. 10017

Assistant Administrator for Environment and Safety
Department of Energy
Washington, D.C. 20545

Bureau of Radiological Health (BRH)
Food and Drug Administration
Public Health Service
U.S. Department of Health, Education, and Welfare
Rockville, Md. 20852

Department of Transportation
Washington, D.C. 20590

Environmental Protection Agency
Washington, D.C. 20460
Federal Radiation Council (FRC)

(Functions of the Federal Radiation Council have now been transferred to the Environmental Protection Agency; documents are available from U.S. Government Printing Office, Washington, D.C. 20402)

International Atomic Energy Agency (IAEA)
Publishing Section
Kartner Ring 11
P. O. Box 590
A-1011 Vienna, Austria

(Publication orders from the U.S. should be addressed to: UNIPUB, Inc., P. O. Box 433, New York, N.Y. 10016.)

International Commission on Radiation Units and Measurements (ICRU)
7910 Woodmont Avenue
Washington, D.C. 20014

(Order publications from ICRU Publications, P.O. Box 30165, Washington, D.C. 20014.)

International Commission on Radiological Protection (ICRP)

(Order publications from Pergamon Press, Inc., Elmsford, N.Y. 10523, or through bookstores in the United States.)

International Labour Office (ILO)
Geneva, Switzerland

(Most publications of ILO useful in medical institutions are available from other agencies as joint publications.)

Joint Commission on Accreditation of Hospitals (JCAH)
John Hancock Building
875 N. Michigan Avenue
Chicago, Ill. 60611
Attention: Publications
   (The JCHA has published a manual of standards, including environmental aspects of radiation safety, which contains the basis for their hospital inspection program for accreditation. These standards will be issued in revised form as available.)

Medical Internal Radiation Dose Committee (MIRD)
Society of Nuclear Medicine
475 Park Avenue South
New York, N.Y. 10016
   (The MIRD committee has in recent years issued a number of excellent compilations of data and methods for calculating doses to humans from radionuclides used in medicine. These same data and methods are useful in monitoring and estimating both internal and external occupational exposures to hospital employees.)

National Academy of Sciences — National Research Council (NAS—NRC)
2101 Constitution Avenue, N.W.
Washington, D.C.

   National Council on Radiation Protection and Measurements (NCRP)
7910 Woodmont Avenue
Washington, D.C. 20014
   (Order publications from NCRP Publications, P.O. Box 30175, Washington, D.C. 20014.)

National Institute for Occupational Safety and Health and
National Institute for Environmental Health Sciences (NIOSH and NIEHS)
Department of Health, Education, and Welfare
Washington, D.C.
   (Publications are available from the National Technical Information
Service, Springfield, Va. 22151.)

Standards Committee
Health Physics Society (HPS)
4720 Montgomery Lane
Bethesda, Md. 20014

World Health Organization (WHO)
Distribution and Sales Service
1211 Geneva 27, Switzerland

U.N. Scientific Committee on the Effects of Atomic Radiation
   (Reports are available from U.N. Industrial Development Organization, Lerchenfelder Strasse 1, A-1070, Vienna, Austria)

U.S. Nuclear Regulatory Commission
Washington, D.C. 20555

AGREEMENT STATES
(AS OF APRIL 5, 1977)

Alabama   205-832-5992

Mr. Aubrey Godwin, Director
Division of Radiological Health
Environmental Health Administration
Room 314, State Office Building
Montgomery, Ala. 36130

Arizona   602-271-4845

Mr. Donald C. Gilbert, Executive Director
Arizona Atomic Energy Commission
First Floor — Commerce Building
1601 West Jefferson Street
Phoenix, Ariz. 85007

Arkansas   501-661-2307

Mr. David D. Snellings, Jr., Director
Division of Radiological Health
Arkansas Department of Health
4815 West Markham
Little Rock, Ark. 72201

California   916-445-0931 (License Inspection)

Mr. Joe Ward, Chief 916-322-2073
Radiologic Health Section
Department of Health
714 P Street, Rm. 498
Sacramento, Cal. 95814

Colorado   303-388-6111 Ext. 246/247

Mr. Albert J. Hazle, Director
Occupational and Radiological Health Division
Department of Public Health
4210 East 11th Avenue
Denver, Col. 80220

Florida   904-487-1004

Mr. Ulray Clark, Administrator
Radiological Health Program
Health Program Office
Department of Health and Rehabilitative Service
1323 Winewood Blvd.
Tallahassee, Fla. 32301

Georgia   404-894-5795

Mr. Richard H. Fetz, Director
Radiological Health Unit
Department of Human Resources
47 Trinity Avenue
Atlanta, Ga. 30334

Idaho   208-384-3335

Mr. Michael Christie, Supervisor
Radiation Control Section
Idaho Department of Health and Welfare
Statehouse
Boise, Id. 83720

Kansas   913-296-3821

Mr. Gerald W. Allen, Director
Bureau of Radiation Control
Division of Environment
Department of Health and Environment
Building 740
Forbes Field
Topeka, Kans. 66620

Kentucky   502-564-3700

Mr. Charles M. Hardin, Manager
Radiation Control Branch
Bureau for Health Services
Department for Human Resources
275 East Main Street
Frankfort, Ky. 40601

Louisiana   504-389-5963

Mr. B. Jim Porter, Administrator
Division of Radiation Control
Natural Resources and Energy
Department of Conservation
P.O. Box 14690
Baton Rouge, La. 70808

Maryland   301-383-2744/2735

Mr. Robert E. Corcoran, Chief
Division of Radiation Control
Department of Health and Mental Hygiene
201 W. Preston Street
Baltimore, Md. 21201

Mississippi   601-354-6657/6670

Mr. Eddie S. Fuente, Supervisor
Radiological Health Unit
State Board of Health
Jackson, Miss. 39205

Nebraska   402-471-2168

Mr. Ellis Simmons, Director
Division of Radiological Health
State Department of Health
301 Centennial Mall South
P.O. Box 05007
Lincoln, Nebr. 68509

Nevada   702-885-4750

Mr. William C. Horton, Supervisor
Division of Health
Department of Human Resources
Carson City, Nev. 89710

New Hampshire   603-271-2281

Mr. John R. Stanton, Director
Radiation Control Agency
Division of Public Health Services
State Department of Health and Welfare
State Laboratory Building
Hazen Drive
Concord, N.H. 03301

w Mexico   505-827-5271 ext. 240

. Russell Rhoades, Chief
cupational Health & Radiation
otection Division
vironmental Improvement Agency
te of New Mexico
). Box 2348
ta Fe, N.M. 87503

w York   518-474-2178

. T. K. DeBoer, Director
chnical Development Programs
w York State Energy Office
an St. Bldg., Core 1, 2nd Fl.
pire State Plaza
oany, N. Y. 12223

rth Carolina   919-733-4283

. Dayne H. Brown, Head
diation Protection Branch
vision of Facility Service
x 12200
eigh, N. C. 27605

rth Dakota   701-224-2374

. Gene A. Christianson, Director
vision of Environmental Engineering
diological Health Program
te Department of Health
)0 Missouri Avenue
marck, N. D. 58501

Oregon   503-229-5797

Dr. Marshall Parrott, D.Sc.
Radiation Control Service
Division of Health
Department of Human Resources
1400 South West Fifth Avenue
Portland, Ore. 97201

South Carolina   803-758-5548

Mr. Heyward Shealy, Chief
Bureau of Radiological Health
State Department of Health and   Environmental Control
J. Marion Sims Building
2600 Bull Street
Columbia, S. C. 29201

Tennessee   615-741-7812

Mr. Robert H. Wolle, Director
Division of Occupational and Radiological Health
Department of Public Health
727 Cordell Hull State Office Building
Nashville, Tenn. 37219

Texas   512-458-7341 or -7686

Mr. Martin C. Wukasch, P.E., Director
Division of Occupational Health and Radiation Control
Texas Department of Health Resources
Austin, Tex. 78756

Washington   206-753-3459

Mr. Robert C. Will, Supervisor
Radiation Control Unit
Division of Health
State Department of Social and Health Services
Olympia, Wash. 98501

1.4. RATE OF GROWTH OF STANDARDS AND LITERATURE IN RADIATION PROTECTION

In the 1971 Mid-Year Symposium of the Health Physics Society on "Radiation Protection Standards — Quo Vadis,"[2] Herbert Parker attempted to estimate the growth rate of standards in this field. Portions are quoted from his statement as follows.

"The topic, as you know, went high brow and asked, 'Quo Vadis?' Let me answer just one aspect of that quantitatively. I've long kept, as a matter of personal interest, a record of the number of applicable protection handbooks. In 1944 and even in 1954 they were few enough that you could show legible slides of them. ...you could actually read the title of these various things. And in 1944 there were six of those... really... only five, because I threw in one on protection of radium during air raids . . . In 1954 there were sixteen. In 1966, when the count read: 53, I gave up trying to make slides. You ended up taking a photograph of a bookcase by that time. In 1971, courtesy of Jack Selby, who did the counting, there are 114. If I were going to demonstrate it, I would have to do like this by saying there are 18 of NCRP types and 15 of ICRP types and so on... which includes a rather recent addition, 19 from ANSI, which are judged by Jack to be directly related to radiation protection out of the 35 published nuclear standards.''

A plot of Parker's count of standards in his book surveys yielded exponential formula $e^{0.148T}$, from which he made a projection of 9662 standards in the year 2001 but did not provide us with the base year of origin from which time T was measured. Further, he gave it to society members for homework, to ''... determine the year at which the sole occupation of a health physicist becomes the reading of protection handbooks.''

The editor carried out Parker's homework assignment 4 years later and obtained a slightly different exponential growth function, shown in Figure 1.4-1. Data, given by Parker, have been plotted and fitted with a standard eyeball regression line (see Section 6.2), which is compared with Parker's function. The editor's line, however, projects downward to pass through a very significant point — an index value of one solitary voluntary standard at the reference year 1928, the year of birth of the NCRP when Lauriston Taylor became the grandfather of radiation protection standards.[3,4] The fit obviously confirms the validity of the exponential model proposed by Parker, but the editor's equation is

$$N = e^{0.1085 \ (T-1928)}$$

rather than Parker's

$$N = e^{0.148 T'}$$

where T' is measured from Parker's secret date of origin.

The editor's projected eyeball line intersects a **very** significant point — the year 1928. This point is precisely known,[3,4] and its variance is zero. Therefore, the fit of any realistic quantitative exponential model of standards publications must be constrained to pass through the index point[1928.1], and only one degree of freedom can be allowed to adjust ''slope'' or the value of the exponential coefficient. New predictions of this now precise model of the exponential natural growth of voluntary radiation protection standards (not including all other types of literature sources) are summarized in Table 1.4-1. One finds that there would be only 2470 standards to read in the year 2001 A.D. — very comforting, indeed, compared to Parker's 9662. This observation is the editor's rational excuse for limiting this handbook's scope as well as his defense for its need. He hopes that the following sections will be invaluable to the user.

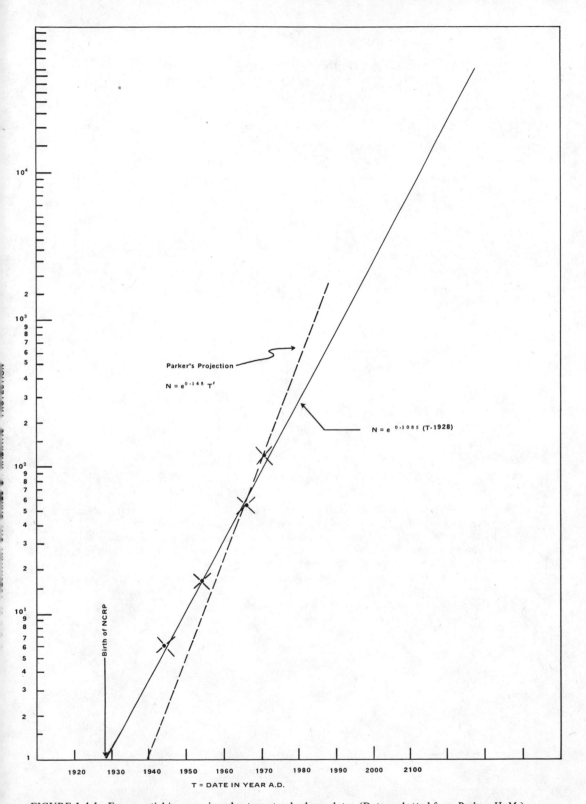

FIGURE 1.4-1. Exponential increase in voluntary standards vs. date. (Data replotted from Parker, H. M.)

Table 1.4-1

COMPARISON OF EXPONENTIAL MODELS
PREDICTING NUMBERS OF STANDARDS vs. TIME

Date (in year A.D.)	Parker's and Selby's data (no. of voluntary standards)	Predicted No. of Voluntary Standards	
		Parker's model (1971)	Brodsky's model (1975)
1928 (NCRP born)	—	—	1
1944	6	2	6
1954	16	9	17
1966	53	59	61
1971	114	114	105
Projections			
1981		432	314
1991		1897	930
2001		9662	2470

REFERENCES

1. **Cottrell, W. B.,** International organizations producing nuclear standards, in *CRC Handbook of Laboratory Safety,* 2nd ed., Steere, N. V., Ed., CRC Press, Cleveland, 1976, 503—508.
2. **Parker, H. M.,** Radiation Protection Standards — quo vadis, in *Proc. of the Midyear Symp. of the Health Physics Society, Richland, Wash.,* Columbia Chapter, Health Physics Society, Battelle Northwest Laboratories, Washington, D.C., 1971, 8—10.
3. **Taylor, L. S.,** Role of the NCRP in the development of radiation protection standards, in *Radiation Protection Standards — Quo Vadis,* Columbia Chapter, Health Physics Society, Battelle Northwest Laboratories, Washington, D.C., 1971, 16—32.
4. **Taylor, L. S.,** *Radiation Protection Standards,* CRC Press, Cleveland, 1971.

2. HISTORICAL DEVELOPMENT OF RADIATION MEASUREMENT AND PROTECTION

Ronald L. Kathren

2.1. X-RAY ANTECEDENTS

When Wilhelm Konrad Roentgen (1845 to 1923) made his insightful discovery of X-rays on November 8, 1895, he opened wide a door that had been previously closed to mankind and, by so doing, created whole new areas of study that would grow into professional disciplines in their own right. Roentgen had begun his researches with cathode rays the month before, following the efforts of the great Heinrich Hertz (1857 to 1894), who had died tragically of an infection the previous year at the early age of 36, and Philipp Lenard (1862 to 1947), Hertz's pupil and protege who had knocked loudly at the same door.

The stage had been set for the discovery of X-rays some 30 years before when Herman Sprengel, a German technician, devised the mercury air pump, a device that enabled a relatively high vacuum to be pulled in a short time. This led to the development of evacuated tubes filled with various rare gases at low pressures by H. Geissler (1815 to 1879), a glassblower at the University of Bonn. When subjected to an electric discharge, these tubes showed strange and wonderful color effects similar to those noted previously be Faraday and others who did not have the benefit of the highly evacuated tubes. Geissler, in collaboration with Julius Plucker (1801 to 1868), the Professor of Mathematics and Physics, studied the pretty green fluorescence that occurred in the wall of the tube opposite one of the electrodes as well as the spectrum of light emitted by various gases in the highly evacuated tubes. Together with his student, Wilhelm Hittorf (1824 to 1914), Plucker used Geissler's tubes to study the effects of a magnet on the strange radiations streaming from the cathode, demonstrating that the cathode emanation was in fact deflected by a magnetic field.

After being awarded the Ph.D. in 1852, Hittorf served for a short time as Privatdocent at Bonn, continuing his researches into electrical discharge phenomena. He soon moved to the Academy of Munster where, despite limited funding and facilities, he began to develop a satisfactory laboratory. He observed that the emanation or light from the cathode travelled in straight lines, producing heat and fluorescence in the glass walls of evacuated tubes. In 1969, he showed that if a solid object such as a metal cross was placed between the cathode and the tube wall then the shadow of the object would appear in the fluorescent glow in the tube wall. This study along with Hittorf's early work with Plucker led G. F. Varley, an English physicist, to theorize that negatively charged particles were emitted from the cathode.

There were, however, many as yet undiscovered facts and unanswered questions about the mysterious cathode emanations. In Berlin, physicist Eugen Goldstein (1850 to 1930) verified and expanded the work of his countryman Hittorf, coining the term "cathode rays" (kathodenstrahlen) to describe the strange emanation, that he considered a radiation rather than a stream of particles.

However, it was the great English physicist, Sir William Crookes (1832 to 1919) who ultimately established the nature of the cathode rays by a series of brilliantly conceived and executed experiments. On August 22, 1879, sixteen years before Roentgen's discovery, Crookes demonstrated to members of the Royal Institute that solid matter was actually emitted from the cathode, even possessing sufficient energy to drive a small wheel placed in its path within the nearly evacuated tube. This dramatic experiment, originally performed by Crookes a few months before, served to expand and explain the cathode ray observations of Hittorf, Goldstein, and others.

Crookes regarded the cathode emanation as a fourth state of matter — so-called "radiant matter" — differing fundamentally from and to be ranked along with the traditional liquids, solids, and gases. Most physicists of the time thought of the cathode rays as either some sort of ether radiation similar to light or as particles. Opinions were essentially divided along geographic lines with scientists on the continent (Goldstein and Hittorf) subscribing to the ether wave idea while the British (Varley and Crookes) tended to favor the particulate theory. Not until 1897, almost two decades later and more than a year after the discovery of both X-rays and radioactivity, was the matter more or less settled in favor of the English by Sir Joseph John Thomson (1856 to 1940) who conclusively demonstrated the corpuscular nature of the cathode rays, determining both the velocity and charge to mass ratio of the particles emitted from the cathode. For these and related researches into the nature and conduction of electricity in gases, Thomson was awarded the Nobel Prize in Physics in 1906.

Experimental work with cathode rays was intensified following the classic experiments of Crookes. The brilliant Hertz performed a series of important investigations on the properties of cathode rays, including studies of their penetrating power through various materials. In 1891, he was joined by Philipp Lenard, a young Hungarian physicist who had been invited to Bonn as his assistant. Lenard, following the suggestions of Hertz, continued the studies of cathode ray absorption, making a tube with a thin aluminum window through which the rays could be brought out into air. He noted the invisibility of the cathode rays and observed a darkening of photographic plates with rays brought outside the tube, a phenomenon seen earlier by Goldstein who had placed plates inside the tube. Lenard deduced that the cathode rays were scattered by air molecules and developed a nine-layer aluminum step wedge to measure the penetrating power of cathode rays from various types of tubes. All in all, Lenard made numerous investigations of the properties of the cathode rays and very nearly made the great discovery of X-rays before Roentgen.

2.2 THE DISCOVERY OF X-RAY

It was, perhaps, the tragic death of Hertz on New Year's Day 1894 that threw the discovery of X-rays to Roentgen rather than Lenard. The aluminum window tube that Lenard made at the suggestion of Hertz had produced X-rays, and these were mixed with the cathode rays passing through the thin aluminum window. It is possible Lenard would have soon recognized the situation, but soon after the death of Hertz, he left Bonn. In the space of two years, he moved from the University of Bonn to Breslau, then to Aachen, and finally to the University of Heidelberg where he filled the Chair of Physics. During these moves, his experimental work suffered because his laboratory was not available much of the time. Indeed, when Roentgen made the great discovery, Lenard was in the process of moving and his laboratory apparatus, being in transit, was not available to him nor had it been for some weeks.

It is also possible, however, that Lenard would have missed the discovery as did his many colleagues who were experimenting with cathode rays and hence generating X-rays in the process. Many scientists unknowingly produced X-rays and observed their effects quite possibly as far back as 1784 when William Morgan (1750 to 1833), a Welsh mathematician and actuary, conducted experiments to determine if electricity would pass through a 'perfect' vacuum. In his experiments, witnessed by Benjamin Franklin among others, Morgan demonstrated that electricity was not transmitted across the vacuum induced in glass tubes. During one experiment, an evacuated glass tube cracked, allowing air to slowly leak in while the tube was still electrified. As the vacuum dropped, Morgan noted a series of colors, yellow-green first, and then blue, purple, and red. In his report to the Royal Society, Morgan suggested a colorimetric test for vacuum strength, but not

Collaege. The patient was a young boy, Edward McCarthy, and the physician was Frost's brother Gilman (1864 to 1947). A 20-min exposure was required and the resulting plate clearly demonstrated the diagnosis, a fracture of the left ulna.

Within the week, the second successful diagnostic radiograph was made, this one in Canada by physics professor John Cox (1851 to 1923) of McGill University. After failing to get a picture of another patient on February 5, Cox detected a bullet in the left leg of Tolson Cunning, a patient of Dr. Robert C. Kirkpatrick (1863 to 1897). Kirkpatrick had surgically searched without success. The exposure required 45 minutes, and Cox noted that the plate was underexposed and probably should have been exposed for twice as long. The case was a first in another important regard, for the radiograph was used as evidence in court on behalf of Mr. Cunning, who successfully sued the man who shot him.

Also in February, Pupin made his first diagnostic radiograph. The exact date is uncertain, but the available evidence indicates that the exposure was made about mid-month, although it could have been as early as the latter part of the first week. While the date is obscure, the circumstances are not. Pupin was asked by a New York surgeon, William T. Bull (1849 to 1909), for assistance in locating numerous shotgun pellets in the left hand of Prescott Hall Butler, an attorney who had accidentally been shot. To reduce the exposure time, Pupin used a crude intensifying screen of cardboard covered with fluorescent powder in conjunction with the photographic plate; however, the exposure still required a full hour. The resultant plate had extraordinarily high quality, clearly showing 50 pellets scattered throughout the hand. Thus, Pupin apparently became the first to apply the principle of image intensification to reduction of exposure time.

Therapeutic application of X-rays may have preceded even the diagnostic applications. According to the memoirs of Emil H. Grubbe (1875 to 1960), published more than a half century later in 1949, the first suggestion of the therapeutic potential of the X-ray was made on January 27, 1896 by J. E. Gilman, one of a group of physicians at Hahnemann Medical College of Chicago who were examining the X-ray dermatitis on Grubbé's left hand. Apparently impressed by the damage to what had been healthy tissues, Gilman suggested the possibility of destroying cancer, lupus, and other pathological conditions by means of X-ray treatment.

Coincidentally, two days later, a lady named Rose Lee, suffering from breast cancer, came to Grubbé's laboratory seeking X-ray treatment. Mrs. Lee bore a simple three-line letter of recommendation from her physician R. Ludlam, Sr., who had also been present at Grubbé's examination. Grubbé began therapy that day using a Crooke's tube 3 in. from the breast and a one-hr exposure. The treatment plan called for daily sessions until the development of dermatitis, this being a signal that the treatment had produced appropriate effects. In all, Mrs. Lee received 18 treatments, and died within a month; according to Grubbé, ". . . . no dramatic results were obtained."

Grubbé also claimed to have simultaneously treated 80-year old A. Carr for lupus, beginning treatment the day after he began with Mrs. Lee. Like Mrs. Lee, Mr. Carr had been referred by a physician present at Grubbé's examination, A. C. Halphide of Chicago. Also like Mrs. Lee, Mr. Carr died within the month. Neither of the two cases was written up for the medical literature because of the apparent failure of the treatments, and the claims of Grubbé, supported by the two letters of referral now in the Smithsonian Institution, were not made until 1933, long after all involved in the events were dead. Although the letters appear authentic, there is some doubt as to this fact, and the lack of other corroboration has lead some to dispute the claims of Grubbé.

The honor may in fact belong to an obscure German physician, Dr. Voigt, who som[...] [y]ears later (1904) was credited by the pioneering Austrian radiologist Leopold Freu[nd] [1]868 to 1943) as having palliated a case of nasopharyngeal carcinoma with X-ray, t[...] [...]se having been orally reported to the Hamburg Society of Physicians on February [...]96. In Chicago, Dr. H. Preston Pratt (1860 to 1939) claimed to have treate[d]

until more than a century later could it be known that the yellow-green electrical discharge he observed was in reality caused by soft X-rays produced in the tube.

Others had better opportunity, and also failed to recognize what they saw. Crookes saw the pale yellow-green color in his tubes and thus was producing X-rays. Photographic plates stored near his laboratory workbench became fogged; in 1879 (and possibly earlier) he returned such plates to the manufacturer as defective. On February 22, 1890, nearly six years before the Roentgen discovery, Professor Arthur W. Goodspeed (1860 to 1943) of the University of Pennsylvania actually had in hand an inadvertent X-ray picture of some coins to be used for car fare that had been temporarily placed on a photographic plate in the laboratory by his assistant William Jennings (1860 to 1945). At the time, Jennings was helping Goodspeed photograph the spark gap of a Ruhmkorff coil. The curiously developed plate clearly showing the circular outlines of the two coins was placed in a drawer along with other photographic curiosa to be studied at some later time. The time came nearly six years later, when Goodspeed learned of Roentgen's discovery. He immediately dug out the strange but now understandable plate and realized what had occurred. To his credit, he made no claim for priority of discovery but accorded full recognition to Roentgen.

In actual fact, the existence of electromagnetic radiations with wave lengths much shorter than visible light, i.e., X-rays, was predicted by the Disperison Theory of the Spectrum put forth by Hermann von Helmholtz (1821 to 1894). This great German physicist drew heavily on the electromagnetic field theory of James Clerk Maxwell (1831—1879), which had been first published in 1865, and on the experimental work of Hertz who during the years 1886 to 1888 showed that electromagnetic waves with wavelengths longer than light — the so-called Hertzian or radio waves — not only travelled with the velocity of light but could be reflected, refracted, and in other ways show the same properties as light. Thus, the mathematical existence of X-rays was in fact demonstrated a few years before the discovery. Ironically, the aging von Helmholtz was to die of a shipboard accident in the same year as his youthful former pupil Hertz; thus, he never saw the experimental demonstration of the mathematical prognostication he developed from the work of Hertz and Maxwell.

It was in October 1895 that Wilhelm Konrad Roentgen began his studies of cathode rays, repeating first the experiments of Lenard. Roentgen was then 50 years old and a well-recognized and established physics professor at the University of Wurzburg. Earlier in the year, he had turned down a most attractive offer of a professorship and the directorship of a new physical institute from the University of Freiburg. He was a good teacher with consummate skill as an experimenter and with interests that spanned the breadth of physics. Although he had dabbled in experiments with cathode rays prior to October 1895, he began to devote his full research attention to the cathode rays in that month, forsaking his previous studies relating to the effects of pressure on dielectric constants of liquids.

It was thus on the afternoon of Friday, November 8 that Roentgen took a pear-shaped all-glass Hittorf-Crookes tube, covered it with opaque black cardboard, and electrified it with a Ruhmkorff coil. His plan was to see if this tube, without a window to bring the cathode rays out into air, might not also excite a fluorescent screen. As he concluded his initial observation that there was no leakage of the fluorescence within the tube through the cardboard wrapping, Roentgen noticed a weak glow on a small table about a meter from the tube. With the aid of a hastily struck match in the darkened laboratory, he was surprised to find the little barium platinocyanide screen he intended to use was already producing the glow.

The observation that chance favors the prepared mind has been attributed to Louis Pasteur. It was thus that Roentgen repeated his serendipitous experiment several times and recognized that what he was observing was contrary to what was known about cathode rays, for these never penetrated more than a few centimeters in air. To Roentgen, it was obvious that some other force — a new kind of ray — was causing the fluorescence. So rapt was he in his experiments that he was unaware of his surroundings. He failed to hear Herr Marstaller, the diener, who entered the laboratory to look for some apparatus after his knock went unanswered. He ignored several calls to dinner, finally eating silently and swiftly returning to the laboratory. He told no one — not even Frau Roentgen — of his observations.

Over the weekend, he continued and extended his observations, testing the penetrating ability of the new rays. Inadverently, while holding a disk of lead between the tube and the screen, he saw the outline of his thumb and forefinger and the darker shadow of the bones. During the following weeks, he made strange shadowy photographs using the new rays instead of light; on the evening of December 22, he photographed Frau Roentgen's hand with the rays, using a 15-min exposure to obtain a shadowy picture that showed the outlines of the bones as well as the two rings she wore.

The last weeks in December were spent in preparing his paper, entitled "On A New Kind of Rays, A Preliminary Communication" (*Ueber eine neue Art von Strahlen*) which was given to the secretary of the Wurzburg Physical-Medical Society on December 28, 1895 with a request for publication in the proceedings of that organization without the customary oral presentation. On New Year's Day, 1896, he sent reprints of this communication and copies of his X-ray pictures to several of his colleagues, including Franz Exner in Vienna, Henri Poincare in Paris, and Sir Arthur Schuster in Manchester. It was Exner who released the news to the world, communicating the discovery to the Vienna *Freie Presse* through the son of the publisher. Thus, on Sunday, January 5, 1896, the first newspaper account of the discovery appeared; on the following day, the London *Standard* broadcast the news via a world cable.

Roentgen published but three papers totalling 34 pages on X-rays. They are clear, concise, classic works describing his systematic and thorough observations, logic and conclusions, and recognizing the works of others, particularly Lenard and Hertz. Although his publications were few and brief, little of additional fundamental importance was discovered for 15 years, certainly a tribute to Roentgen's scientific skills. His first communication described not only the rays but discussed their absorption in various materials, his inability to refract them by a prism or deflect them by magnetic fields, and fluorescent and photographic effects. The potential value in medicine and industry was mentioned. Four X-ray pictures were included and Frau Roentgen's hand was among them. Roentgen concluded his first work by suggesting that the X-rays were longitudinal ether vibrations similar to light.

The second communication was submitted on March 9, 1896, and contained a brief discussion of the electrical effects, including the discharge of charged bodies and a scale for measuring intensity by both photographic and fluorescent means. The final publication, submitted a year later on March 10, 1897 was the longest of the three but still brief and included a detailed discussion of the 'hardness' of the tube and production of X-rays, exponential nature of absorption, and note of his unsuccessful attempts at X-ray diffraction.

For his discovery and for the interpretation of the phenomena he observed, Roentgen was showered with honors, including the Rumford Medal of the Royal Society (1900) which he shared with Lenard, the Elliot-Cress Medal of the Franklin Institute, and numerous honorary memberships in scientific and technical societies. He was awarded 65 honorary degrees, including one in medicine, and received many offers of high posts in universities. Several cities and states gave him honorary citizenship. In 1901, Roentgen fittingly received the first Nobel Prize in physics, one of the few accolades he chose to personally accept. He remained shy and modest, declining the offer of nobility and the honorific prefix *von* to his name. He eschewed public appearances and made but two for the purpose of describing his discovery, one a demonstration on January 13, 1896 before

Kaiser Wilhelm II and the other on January 23 before the Physical-Medical Society of Wurzburg.

Although the discovery clearly was Roentgen's, doubts were expressed by some regarding this fact. Lenard, with whom he had been on good terms, cooled noticeably after the awarding of the Nobel Prize in 1901 and later became embittered and hostile to Roentgen. Lenard never used the term "Roentgen rays" preferring instead "high-frequency rays." Although Lenard himself received many honors, including the Nobel Prize in physics for 1904, he refused to recognize that the discovery belonged to Roentgen. Instead, he first chose to ignore Roentgen, failing to mention his name in connection with the rays and then, as the anti-Semitic high priest of Hitlerian science, even attempted to rewrite history.

Most people recognized however, that Lenard's work, although fundamental to the discovery, did not include the discovery itself. Arthur Goodspeed, the Philadelphia physicist who might have made a claim of priority on the basis of his accidental and unrecognized radiograph taken in 1890, made no such claim, deferring instead to Roentgen, and his words of denial — "We can claim no merit for the discovery, for no discovery was made" — should serve for Lenard and other pretenders to the discovery.

2.3. EARLY X-RAY APPLICATIONS IN MEDICINE

Few events in the history of science provoked the interest and wonder of the world did the discovery of the X-ray. Both the scientific and lay world was caught up in feverish craze, and all sorts of bizarre stories and events were recorded. A claim that bone projected by X-ray onto a dog's brain caused salivation appeared in the scien literature, and a bill banning the use of X-rays in opera glasses was introduced in the Jersey legislature for the purpose of protecting the modesty of the ladies on the Other even more remarkable applications were suggested in the press, including undergarments for those who desired to preserve the Victorian ideals. (Such garments have enjoyed occasional vogue, an "in" joke among nuclear scientists in and 50s and actually in the 60s and 70s as gonadal shields for patients and underwear of dubious value for sale to radiologic technicians.) The news recognized the potential value of X-rays in medicine very realistically and ver the Vienna Presse of January 5, 1876, mention was made of the possibili X-rays as an aid to medical diagnosis, particularly to identify broken bones objects.

The answer to the question of who made the first intentional medic however, is not easily answered, for many conflicting priority claims ha Several experimenters made nonmedical X-ray plates in early January, th being Alan A. C. Swinton (1863 to 1930), a Scottish engineer who radi on January 7 and 8, and a hand (for demonstration, not diagnosis) on J plates were exhibited on January 16 and published in *Nature* on Janu intentional successful medical diagnostic radiographs were probably tak by Albert von Mosetig-Moorhof (1838 to 1907), a Viennese surgeon w cases at the meeting of the Imperial Society of Physicians on Janua Mosetig-Moorhof had radiographed a patient with a revolver bulle twenty-year old lady with a congenital supernumerary left great t radiographs facilitated surgical treatment.

The facts pertaining to the first intentional radiograph made i Professor Michael Pupin (1858 to 1935) of Columbia claimed to on January 2, 1896, " . . . two weeks after the discovery many . . ." Obviously, the January 2 date is erroneous mistranscription of January 12 or 22, possibly February 2.

The first intentional medical radiograph taken in the Am taken on February 3, 1896, by astronomer Edwin B. Frost (1

consumptive patient on April 14, 1896, substantiating his claims with newspaper articles from the Chicago *Tribune*. Pratt did establish a commercial X-ray laboratory, quite possibly the first in the United States, on February 7, 1896.

More certain than the above are the applications of V. Despeignes of Lyon, France, who began treatment of a case of stomach cancer in May 1896, reporting the patient showed great improvement and relief from pain in an article dated July 23. In the United States, the earliest undisputed therapeutic applications were made by James William White (1850 to 1916), a surgeon at the University of Pennsylvania. In the fall of 1896, White began clinical studies of the value of X-rays for treatment of inoperable cancers; months later when he made his first report, he could give no conclusions except that the study was to be continued.

Another therapeutic line pursued by some related to the germicidal value of X-rays. In the *Lancet* for February 1, 1896, an English physician, T. Glover Lyon, suggested such possibilities, but on February 17, he noted that this was not the case with the bacilli of diphtheria and tuberculosis. Similar results were reported by others, including W. H. Parke, head of the New York City Board of Health Bacteriology Laboratory. Some researchers notably Pratt of Chicago, found quite the opposite and, as noted above, claimed to have treated a tuberculosis patient. Controversy raged in this area, but by 1900, it was generally conceded that, while X-rays were indeed bactericidal, they were ineffective for treatment of infection because of the long exposures and serious secondary reactions they produced.

The turbulent first year of the X-ray, 1896, saw the birth and emergence of radiological sciences in their own right. By the end of the year, no less than three books and more than 100 technical articles has been published in the U.S., in addition to numerous articles in the popular lay press. Dental applications had been suggested in April by William J. Morton (1845 to 1920), the pioneer New York City radiologist, and in July, Charles Edmund Kells, Jr. (1856 to 1930), the noted New Orleans dentist, exhibited his skiagraphs before the Southern Dental Association meeting in Ashville, South Carolina. And, in that same month, William Herbert Rollins (1852 to 1929) of Boston described a simple film-holding device for taking dental radiographs. Rollins proposed using a stack of several films, such that "As each film has less exposure than the one in front of it, the appearances vary and one is sure to give the information sought." Rollins also had already produced a fluoroscope for intraoral use, and later was to connect a stethoscope to a fluoroscope for chest examinations, quaintly naming that device the "seehear."

Other developments of significance in that first year following the discovery included opthalmologic applications. In August, two reports appeared in literature describing identification of foreign bodies in the eye. Industrial applications were begun in February at Carnegie Steel Works in an attempt to detect casting flaws. At Trinity College, Oxford, X-rays were used to distinguish real diamonds from radio-opaque fakes.

The first year also saw the start of several contemporary instrumentation techniques and applications. The rotating anode tube was introduced by Robert W. Wood (1868 to 1955), Johns Hopkins University physicist, also noted for his whimsical writings. Photofluorography was discovered by New York laryngologist Julius Bleyer (1859 to 1915), who also helped to design his state's electric chair. Stereo techniques were pioneered by physicist Elihu Thomson (1853 to 1937), Boston physicist John Trowbridge (1843 to 1923) created an oil-immersed X-ray tube, and in Philadelphia H. Lyman Sayen devised a self-regulating tube. None of these three physicists are ordinarily associated with these efforts, for in each case, the ideas were rediscovered, pursued, and brought to fruition by others (Figures 2-1 and 2-2).

FIGURE 2-1.   An early hand-held fluoroscope; the handle is inserted into the wooden block. The eyepiece was surrounded by black velvet for the comfort of the observer. This unit was manufactured by Patterson Screen Company, which was later taken over by E. I. DuPont De Nemour, Inc.

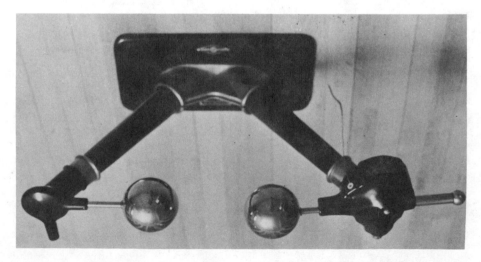

FIGURE 2-2.   A sphere gap device for monitoring KVP. This unit was suspended from the ceiling and provided a direot readout of voltage in a little meter to the right of the spheres; the right sphere could be adjusted as desired. Sphere gaps were used well into the 1930s.

2.4. THE DISCOVERY OF RADIOACTIVITY

Shortly after the discovery of X-rays, Henri Poincarei (1854 to 1912) suggested that X-rays, which were then thought to be produced by the action of cathode rays on fluorescent materials, might also be produced by ordinary phosphorescent substances. Responding to this suggestion, Becquerel undertook a systematic study of such substances, reasoning that the luminescence induced by sunlight might contain X-rays as well. On February 4, 1896, a scant month after the discovery of X-rays had excited the world, Antoine Henri Becquerel (1852 to 1908) communicated to the Paris Academy of Science his discovery of phosphorescent rays emitted from the compounds of uranium.

He began his studies with potassium uranyl sulfate, a material known to be highly fluorescent. After placing a quantity of the mineral on a photographic plate well wrapped in light-tight black paper, Becquerel then exposed the combination to sunlight for several days. When the plate was developed, the position of the mineral was clearly shown by dark spots. It was this observation that Becquerel communicated on that first Tuesday in February.

He continued his studies, and in late February, additional experiments were prepared. Because of cloudy weather, the exposure to sunlight was not made, and the plate and salts were instead stored in the dark interior of a drawer. When Becquerel developed the plate a few days later, he found the same kind of dark spots even greater in intensity than when he made the exposure in the sunlight. Becquerel had been expecting at most a weak effect, for the uranium had been exposed for only a short time and to diffuse light. The unexpected strong effect could be only interpreted as being caused by penetrating rays spontaneously emanating from the uranium salt, a fact that he communicated to the world through the Paris Academy on Monday, March 2, 1896.

Henri Becquerel was 44 years old and professor of physics at the Polytechnical School in Paris at the time of his discovery. He was the third in a family dynasty in physics that had begun with his grandfather, Antoine Cesar Becquerel (1788 to 1878) who had held the Chair of Physics at the Natural History Museum. His father Alexandre Edmond (1820 to 1891) held the same professorship that Henri held at the time of the discovery, and the uranium salt Henri used in his experiments has previously been used by his father for experiments on phosphorescence.

Like X-rays, the mysterious emanations from uranium had appeared before but had gone unrecognized. Nearly 30 years before the discovery in 1867, another French physicist, de St. Victor, had reported that uranium nitrate, after exposure to light, would affect a photographic plate causing an extraordinarily intense photographic effect. The cause was attributed to the light-induced flourescence which in some mysterious way was highly effective at provoking the photographic response, thus, the radiations from the uranium went undetected and unsuspected.

About the same time as Becquerel was carrying on his researches, Sylvanus P. Thompson (1856 to 1919) was performing similar studies and independently noted that uranium salts exposed to light could expose a photographic plate even through a thin metal sheet. Thompson wrote a letter communicating his results to Sir George Stokes, President of the Royal Society. However, Thompson failed to realize that the exposure to light was unnecessary; the rays were spontaneously and continuously given off by uranium. This finding was reserved for Becquerel alone.

Although the discovery of radioactivity (as the phenomenon was named by M. Curie in 1898) was extremely important, it went relatively unnoticed except in a small corner of the scientific world. Newspapers and magazines that had heralded (and still were doing so) the X-rays took little or no notice of the rays from uranium. Over the next two years, Becquerel studied the uranium emanations, determining that they did not diminish with time and hence were not stored light but a new and unexpected property of matter. He showed that the 'radiant activity,' as he called it, of uranium was similar to X-rays, and demonstrated the electrical and absorptive properties of the rays. In 1903, he was awarded the Nobel Prize, sharing the award in physics with the Curies.

While Becquerel's discovery received scant notice, the discovery of radium, a direct outgrowth of Becquerel's work, became one of the most publicized and popular stories in the history of science. Extension of Becquerel's work was undertaken by Pierre and Marie Curie, a husband and wife team at the School of Physics and Industrial Chemistry in Paris. Pierre Curie (1859 to 1906) was one of the most promising young physicists in France; when Becquerel discovered radioactivity, Curie was only 36 and had already made important contributions in crystal structure, magnetism and peizoelectricity. He earned his bachelor of science at sixteen and his licentiate in physical sciences at eighteen; the following year he began his academic career as demonstrator of physics at the Sorbonne. With his brother Jacques, he discovered and characterized the phenomenon of piezoelectricity before his 23rd birthday, an effort for which the brothers were awarded the prestigious Plante prize in 1895.

In 1894, Pierre Curie met Marie (nee Marja) Sklodowska (1867 to 1934) at the home of a Polish scientist. Marie, also Polish, was completing studies at the Sorbonne for the licentiate in mathematics. The following year the couple married, attracted in large measure by a common interest in physics. Pierre and Marie were excited by the discoveries of Becquerel, described in his many papers appearing in *Comptes Rendus* during 1896 and 1897. It was thus in 1897 that Marie Curie, by this time the mother of a daughter (Irene) who would grow into a great scientist in her own right, undertook the study of radioactivity as the topic of her doctoral thesis. She was in her studies aided by her husband Pierre; the first of their joint laboratory notebooks, simply titled "Curie 1897/1898 — Etude Uranium I," was started by Pierre on September 16, 1897. Marie's first entry — dealing with piezoelectric measurement of the radiation from uranium — appears on December 17, 1897, and most of the rest of the notebook is in her hand. The last entry, dated March 16, 1898, is a four page summary of Marie's 3 months of effort plus a table by Pierre.

On April 12, 1898, Marie published her first work on radioactivity, "Rayons émis par les composés de l'uranium et du thorium". The work included a description of the activity from thorium, a discovery independently made and reported by German physicist G. C. Schmidt a few weeks earlier. Marie also noted the high atomic weights of uranium and thorium, and most significantly reiterated in print what was apparently an idea of Pierre's — that certain minerals with high activity contain an element much more active than uranium or thorium. On July 13, 1898, Pierre recorded in the laboratory notebook the presence of a new element in the precipitate of bismuth. The new element was named polonium by the Curies in honor of Marie's native land, and in the publication describing it, the word 'radioactive' appeared for the first time in the literature of science.

The work of the Curies continued; late in 1898, a second new element — radium — was separated from the barium fraction with the aid of G. Bémont with whom their report was made on December 26, 1898. Bémont, one of the laboratory heads at the school, had assisted with the work, and although recognized as a coworker by the Curies, has been virtually forgotten except in an occasional obscure footnote to the discovery.

The tiny quantities of both radium and polonium were too small to permit isolation in elemental form. In the case of radium, identification was by radiological properties with spectrographic verification also made by M. Demarcay. Absolute verification would come from determination of the atomic weight, which would require a measurable quantity of the pure preparation. Therefore, in the same tiny, unheated shed that served as the laboratory in which radium and polonium were discovered, the Curies began the arduous task of obtaining a sufficient quantity of the purified product. By agreement, the chemical separation fell mainly to Marie, while Pierre would primarily concern himself with researches into the nature of radioactivity. In 1902, after four years of herculean effort, the task was done. From hundreds of kilograms of pitchblende, 100 mg of radium chloride was separated, and from this tiny product of many lengthy, arduous crystallizations, the existence of radium was unquestionably verified.

The Curies published many pioneering seminal works on radioactivity, among them a review presented before the International Congress of Physics in 1900 in which the atomic transformation hypothesis was discussed. In spite of her preoccupation with the separation of radium, Marie published six papers in 1899 — three of her own and three with Pierre — and eight in 1900, again half her own and half jointly with Pierre. In all, she was to publish 19 papers — 10 with Pierre prior to the separation of radium in 1902; Pierre was to publish 24.

Others, of course, were also active participants in the discovery of radioactivity and its nature. In 1899, Andre Debierne (b. 1874), a collaborator of the Curies and boyhood friend of Pierre, discovered actinium in the pitchblende residues. Rutherford, Ramsay, and Soddy took up the study of radioactivity, as did Hahn in Germany and others of lesser fame. By the end of 1904, Rutherford and Soddy had isolated some 20 radionuclides and proposed the theory of radioactive disintegration in 1902 as well as elucidating what they called the 'radioactive constant' (decay constant) and hence the fundamental decay law.

But it was in the efforts of the Curies in developing the discovery of Becquerel that the story of radioactivity began. For their labors, mostly carried out in a makeshift laboratory, the shy Curies received many honors, including the Nobel Prize in Physics for 1903, shared with Becquerel. Like Roentgen, they declined some honors, preferring instead to work for the sake of science. Most honors initially came from outside France, partially due to the fact that Marie was a woman. So it was that Pierre alone was awarded the Lacaze prize of the French Academy of Sciences in 1901. In 1903, Pierre declined the French Legion of Honor, expressing his thanks and noting ". . . I do not feel the slightest need to be decorated, but I am in the greatest need of a laboratory." Five years after the discovery of polonium and radium, the Curies were still working in an unheated shed, Marie with no official status, and under the sufferance of the head of the school!

Elemental radium was not obtained until 1910, years after the tragic accidental death of Pierre under the wheels of a truck on a Paris street. Fittingly, it was Marie who accomplished this by passing an electric current through molten radium chloride, separating the radium from the resulting amalgam with the mercury electrode. This great piece of work — done without Pierre — stilled many of those who criticized her because she was a woman and it earned her a second Nobel Prize, the first person to be so honored. In spite of her unquestionable scientific qualifications, she was denied membership in the Paris Academy of Sciences in 1911 solely because of her sex. The vote on that January day was close, but despite the sponsorship of the greatest scientists of France — Poincaré, Roux, Picard, and others — she was humiliated in what remains to this day an act of incomparable and blind bigotry. Ironically, among those leading the opposition was M. Amagat, who had been elected to the Academy over Pierre in 1902. Pierre himself was only barely elected to the Academy in 1905, perhaps the greatest scientific honor in France. Marie, with whom he shared so many magnificent discoveries, was never accorded membership, although she was so honored by many such academies in many other nations. Despite the rebuff by the French scientific community, she remained a loyal citizen and fruitful scientist and humanitarian.

2.5. COMMERCIAL DEVELOPMENT OF THE RADIOLOGICAL SCIENCES

Radiology and the radiological sciences grew out of the discovery of X-rays and radioactivity. As has already been shown, X-ray diagnosis and therapy began almost immediately and spurred efforts to improve apparatus. Initially, much effort was expended toward improving X-ray tubes and developing stable and reliable high voltage sources. Commercial production of X-ray tubes was begun in the U.S. in 1897 by the firm of E. Machlett and Son. The Machletts, father Ernst and son Robert (1872 to 1926), were glassblowers who struck out on their own and produced the first commercial X-ray tubes in the U. S. (Figure 2-3). They were followed by others, such as Green and Bauer of New York City which became one of the first large quantity X-ray tube producers, and General Electric Company.

General Electric Company was in fact the first firm in the U.S. to sell X-ray apparatus commercially. In the *Electrical World* dated August 22, 1896, an advertisement announced that the Edison Decorative and Miniature Lamp Department of General Electric was producing a complete line of X-ray apparatus. Other firms preceding GE offered specialized items: J. A. LeRoy of New York City sold a calcium tungstate-aluminum window fluoroscope as early as May, and L. E. Knott Apparatus Co., Boston, offered a new type of reliable coil a month later. The early days, often known as the 'gas tube era', saw many refinements and improvements in tubes, but in 1913, the entire radiological field was revolutionized by the hot cathode tube of William D. Coolidge (1873 to 1975), a physicist (and later executive) employed as a researcher by General Electric (Figure 2-4).

The Coolidge tube was made possible by the development of ductile tungsten and used a heated tungsten filament as the source of electrons. When coupled with a high vacuum,

FIGURE 2-3.   A gas X-ray tube manufactured by E. Machlett and Son in approximately 1906.

COOLIDGE X-RAY TUBE

JRE 2-4.   An early Coolidge X-ray tube (approximately 1912). This was one of the first universal type "hot cathode"

it was possible to rigidly and independently control both current and voltage, thus achieving reproducible output and permitting precise techniques to be developed. Other previously made technical advances were also vital, and included the 'Interrupterless Transformer' or rectifying switch invented by Homer C. Snook (1878 to 1942) in 1907. Snook's device greatly increased the output of X-ray tubes by achieving an output of more than 100 kv and 100 ma, indeed greater than could be safely carried by the forerunners of the Coolidge tube. The Snook transformer enjoyed considerable success and was a highly profitable item for the Roentgen Manufacturing Company of Philadelphia, which had been started by Snook and two associates in 1903 as the Radioelectric Comapny. Later, Snook's name was added, and the company became known as the Snook-Roentgen Manufacturing Co.

Picture quality was greatly enhanced by the production of improved plates and films. Films were used from the start in dentistry; however, plates were preferred for medical and other larger applications. By 1914, Eastman-Kodak offered large (11 X 17 in.) single coated film with sensitivity greater than any available plate. In 1916, double coated film appeared. Aside from the flammability of the cellulose nitrate base, film was clearly superior in all respects to plates, and conversion to film was rapid. By 1920, the use of plates for diagnostic radiology was becoming rare.

The problem of scattered radiation was partially solved by the use of a lead screen developed and patented by a German radiologist, Gustav Bucky (1800 to 1963), in 1913. The screen eliminated scatter and in so doing left its own image on the film, obscuring much of the radiograph. The grid was perfected in 1916 by Hollis Potter (1880 to 1964), an American radiologist in Chicago, who conceived of the idea of rapidly moving the grid back and forth in the beam, thus absorbing the scatter but not producing the image of the lead grid or screen on the film.

In the early years following the discovery of the X-ray, numerous firms entered the

X-ray business, supplying tubes and associated electrical apparatus for producing the high voltages required. Few offered a complete package, however, and hybrid apparatus was common. The lifetime of many of these firms was quite short, for the business was quite competitive and few of the newly formed companies could offer their customers the technical advice and skill they required.

By the early 1900s, the commercial sector had evolved toward stability and manufacturers could provide, in many cases, unique and patented products. Many of the smaller companies ultimately merged or were bought up by larger companies, thus consolidating their unique patented offerrings to the clear benefit of the customer. By 1920, what had begun as a chaotic and ever changing business had become a multimillion dollar industry in the U.S.

Credit must go to the commercial sector for developing (and sometimes appropriating development of others) the modern X-ray tube. Prior to the Coolidge tube, itself a product of industrial research, many companies brought X-ray tubes to market with various improvements, designed not only for the benefit of the customer but for the commercial advantage of the manufacturer. As early as 1897, Queen and Company of Philadelphia, which had started as a scientific optical firm in 1853, offered a self-regulating tube based on the automatic vacuum regulator of H. Lyman Sayen, an engineer employed in the Queen Laboratories. Sayen filed his patent (which was, of course, assigned to Queen) on April 27, 1897.

A tube with a cooled target — the so-called A-W-L X-Light Tube — was offered for a time by Oelling and Heinze Co. of Boston. This tube was developed in 1900 by William Rollins (1852 to 1929), the Boston dentist who contributed much to X-ray tube development as well as to X-ray protection. Rollins experimented with X-rays (or X-light as he called them) as a hobby and never sought to patent any of his numerous innovations, choosing instead to publish his discoveries and inventions in the open literature so they could be used by anyone. Rollins had been concerned with anode heating since the start of his researches, devising a water-cooled hollow platinum anode tube as early as 1897. He devised rotating anode tubes as well. The A-W-L tube featured a hollow, cooled anode which could be used continously with the then available currents without damage (or evaporation!) from heating. It also had a higher output (i.e. greater efficiency) than noncooled tubes and had a vacuum regulator (adjustable from outside the tube) that further improved its efficiency and stability. Although Rollins was given credit for the tube by Oelling and Heinze, he never received a penny in royalties or profit.

Another significant improvement in X-ray tubes that came from the commercial sector was the oil immersion principle developed by Henry F. Waite, Jr., (1874 to 1946) in 1918. Waite was the son of one of the founders of Waite and Bartlett Co. of New York City, a medical supply firm founded in 1879. The younger Waite graduated in medicine from New York University in 1897 and immediately began working for his father's firm. Waite and Bartlett was already a major manufacturer of a broad line of X-ray apparatus and was a well-recognized and respected company that maintained a leading position for nearly 4 decades when the company was sold to Picker X-Ray.

The first commercial oil-immersed unit was offered for sale in 1919 and featured a Coolidge hot cathode tube in addition to the oil filled tube encased in the 12 × 16 × 5¾-in. head with the transformer. The basic patent was applied for on January 31, 1919 and issued as Patent No. 1,334,936 on March 23, 1920; rights were later sold to General Electric in exchange for $1000 and a guarantee of 300 Coolidge tubes per year.

Waite was a highly imaginative and productive inventor of X-ray apparatus. In 1901, he patented a regulator for X-ray tubes, and in 1923, a vibrating Bucky diaphragm to better control scatter. His last patent was obtained in 1938 for an improved housing for X-ray apparatus.

There were, of course, numerous commercial contributions to all phases of X-ray work

— film, intensifying screens, transformers, high voltage, and current controls — and the list is seemingly without end. Many companies were to contribute, and one more — the Victor X-Ray Company — bears mention. Victor was actually formed in 1893 by Charles Samms (1868 to 1934) and Julius Wantz (1873 to 1952) in Chicago to provide electric dental apparatus. By 1896, Victor Electric Company was producing static machines for both X-ray and electrotherapy. Victor combined with or bought up many smaller companies in the early 1900s and was itself taken over by General Electric in 1920 and reorganized as the Victor X-Ray Company.

Victor pioneered education for X-ray technicians, largely through the efforts of Edward C. Jerman (1865 to 1936) who established an education department in 1917. In 1920, Jerman was to serve as the prime mover and founder of the American Society of X-Ray Technicians, which began with 14 members from nine states and the Canadian province of Manitoba. Jerman was the first president of the Society, which was originally known as the American Association of Radiological Technicians. He worked toward registration of technicians in conjunction with the Radiological Society of North America and the American Roentgen Ray Society, and in late 1922, the American Registry of X-Ray Technicians was born with Jerman serving as a member of the examining board. In 1933, he was to resign from the Registry Board, amidst rumors that he was using the organization to the advantage of his employer. Jerman, however, made innumerable contributions to the training of technicians and physicians through Victor X-Ray Corporation. In 1928, he published his experiences in teaching in the form of the classic book, *Modern X-Ray Technique.*

Victor also made numerous contributions in technical areas, both before and after the takeover by General Electric. From 1916 to 1926, Victor had the largest X-ray plant in the U.S. In 1926, Julius Grobe, a Victor engineer, patented the over-under table mount and brought out a line of completely shockproof X-ray apparatus 3 years later. The Victor name was discontinued by General Electric in 1930.

Radium was slow to start commercially, and not until 1913 had the first commercial production of radium begun in the United States, this being by the Radium Chemical Company of Pittsburgh. Within 8 years, Radium Chemical had produced over half the world supply — 71 g with a retail value of $8,500,000 ($120,000 per gram). Radium mining enterprises and the feasibility of commercial production had begun as early as 1902, as had quackery and nostrums based on radium.

There were several commercial manufacturers and suppliers of radium in the U. S. by 1920 including United States Radium Corporation, headquartered in New York City with factories in Orange, N.J. and mines in Colorado, The Radium Company of Colorado, Inc., and National Radium Products, Inc. By the mid-1920s, the virtual American monopoly on radium production was halted by rich ore discoveries in Africa and Canada and the price of radium fell to $35,000 per gram. Radium production in the U. S., based on low-yield Colorado ores, was no longer economically viable and fell to essentially zero.

2.6. EVOLUTION OF RADIOLOGICAL UNITS

Attempts were made from almost the beginning to accurately measure the output of X-rays or the emanations from radioactive materials. The earliest attempts were those of Friedrich Ernst Dorn (1848 to 1916), a German physicist who, in 1897, used an air thermometer to measure the heat produced by X-rays absorbed in sheets of metal. The great English physicist Ernest Rutherford (1871 to 1937), in collaboration with R. K. McLung, also tried to measure the heat produced by radiation with a bolometer. However, calorimetric methods were found unsatisfactory and abandoned in favor of other more promising techniques primarily because of the small amount of heat produced and the crude instruments available then.

Radiation-induced chemical changes provided that promise, particularly if the change was marked by a colorimetric effect. In 1902, Guido Holtzknecht (1872 to 1931) devised a "Chromoradiometer" consisting of fused HCl and Na_2CO_3 mixture which discolored when irradiated. The fundamental unit, H, was defined so that 3H produced a slight skin reaction.

Two French dermatologists, Raymond Sabourand (1864 to 1938) and Henri Noire (1878 to 1937), devised a similar method in 1904 based on colorimetric changes induced in a pastille of platino-barium cyanide. The method initially found wide acceptance, and from the change, a unit was developed and equated with the H unit. However, color changes would be produced by light or heat and the S-N method and unit were not fully reliable.

Another rather widely used dosimeter was devised by Robert Kienbock (1871 to 1954), an American radiologist who proposed exposing strips of silver bromide impregnated paper to the radiation and developing these according to a rigid protocol. The degree of blackening was determined in X units, with 10X = 5H = 1 S-N. Actually, photographic film had first been proposed in 1903 for dosimetry purposes by an American dermatologist, E. Stern, who suggested it be placed on the skin surface to determine the dose to the patient.

Although the chemical methods were highly energy dependent and subject to effects from light, humidity, heat, and other agents, they enjoyed a certain popularity among physicians and enjoyed wide use, particularly in Europe, well into the 1930s.

Other early dosimetry was based on measurement of the intensity of radiation-induced fluorescence by comparison with a standard lamp. The method was first proposed in 1902 by Gaston Countremoulins, a Belgian physicist, and improved in the late 1920s by two Germans, radiologist H. Wintz and physicist W. Rump. Although superior to chemical methods, even later ones based on production of free iodine from an iodoform-chloroform solution, the measurement of luminescence was cumbersome and required special apparatus.

In 1915, Robert Fuerstenau (b. 1887), a German physicist, developed an intensity meter that measured the resistance change induced by radiation in a selenium cell. An advantage of the cell was that it could be made water tight and, thus, used inside a water phantom. The cell, however, was energy dependent and also subject to fatigue which necessitated calibration before each use; response could not be equated in standard physical units.

Although ionization was among the first known effects of radiation, it was not until 1905 that the first applications were made to radiation dosimetry. In that year, Milton Franklin of Philadelphia put forth the possibility of using ionization as a means of quantification of X-rays, publishing his ideas in the *Philadelphia Medical Journal.* Two years later, a British physicist, Charles E. S. Phillips (1871 to 1945), advocated a unit based on ionization that could be used for both X-rays and emanations from radioactive materials. In the discussion following his paper, Charles Lester Leonard (1861 to 1913), one of the deans of early American radiology, spoke eloquently in support of Phillips' proposal, which came close to the modern definition of the roentgen. Phillips was an interesting if unorthodox physicist. He was self-taught having never attended college, but nonetheless, well accepted by the "establishment." Indeed, he was one of the initial editorial board members of *The Journal of the Röntgen Society* which began in England in 1904 — the only one of ten who had neither title nor degrees nor other initials following his name. Phillips was into X-rays very early, compiling in 1897 what was apparently the first bibliography of X-ray literature, listing over 1000 references.

In 1908, Paul Villard (1860 to 1934), the French physicist who had discovered gamma-rays in 1898, put forth as a standard unit the quantity of radiation that produced 1 esu/cm^3 of air at normal temperature and pressure. This proposal was essentially identical to the familiar roentgen unit, which was not adopted as such for 20 years when physicists and radiologists at the Second International Congress of Radiology meeting in Stockholm finally agreed. The roentgen unit has since undergone several revisions, the first a minor one in 1934 and the next in 1950.

In the interim, other physicists had devised similar units. William Duane (1872 to 1935) of Boston had slightly refined Villard's proposal and used the electrostatic unit-second which he measured with air ionization chambers. His chambers with low Z walls and electrodes, small size, and grounded guards are fundamentally identical to modern free-air chambers. Duane's unit was also virtually identical to the roentgen unit defined by the German physicist Hermann Behnken in 1924. In 1927, intercomparisons were made of the Duane and Behnken units using chambers of both experiments, and agreement was noted to within 1%, paving the way for the adoption by the international body the following year.

Over the years, radium became more available and more widely used. Its long half-life and the great penetrating power of its gamma rays made it suitable to radiation therapy. As early as January, 1902, William Rollins proposed the use of radium for therapy of lupus, skin cancer, and other dermatologic diseases, offering a source of radium to any Boston physician who wished to try such treatment. His offer received no response. In 1903, radium was first applied therapeutically by Margaret Cleaves (1848 to 1917) of New York who used it to treat cervical cancer by intrauterine placement. Two years earlier, Becquerel had been inadvertently burned by a vial carried in his vest pocket, and the Curies had experimentally demonstrated the tissue reaction to localized exposure. Radium was rapidly applied to medical use and a clinic devoted to radium therapy was opened in 1906 by two physicians, Louis-Frederic Wickham and Paul Degrais. Wickham improved dosage factors and was the first to apply filtration, which was fully developed by Henri Dominici (1867 to 1926). Wickham also was the first to use the "cross-fire" or multiple port techniques, but it was Dominici who, in the early days, scientifically established the techniques that came to be used.

Dosimetry of radium, even after the adoption of standard strengths and filtration, was at best haphazard well into the 1930s. Not until 1912 was an International Standard adopted, having been prepared by Marie Curie at the request of the 1910 Congress of Radiology and Electricity. Even so, the problem of reproducibly quantifiable tissue dose was unsolved. Over the next decade, at least a dozen separate units were proposed and in use, ranging from those based on various biological effects such as erythema to the simple and practical milligram-hour proposed by the great British radiologist Dawson Turner (1857 to 1928).

Other radium units were based, more or less, on activity. Thus in 1914, A. Debierne and Claude Regaud (1870 to 1940) proposed a millicurie unit based on the amount of radium destroyed by radioactive decay. In 1917, Gioacchino Failla defined a unit he called the "radon," which was equal to the amount of ionization produced by 1 g of radium at a distance of 1 cm. Under certain conditions specified by Failla, this unit could be used to quantify the dose delivered to tissues during radium therapy; clearly, this was an early expression of the need for and the means of an absorbed dose concept.

The following year, 1918, Sidney Russ a British physicist who was an early radiation pioneer, proposed a unit based on the lethal effect of radiation on mouse cancer. This clearly biologically based unit was named the "rad" by Russ. Others also proposed units with a biological basis, including Failla, who, in 1921, calculated the erythema dose in units of calories per unit area. Only Viktor Hess, an Austrian physicist, seems to have recommended a unit based on ionization; his unit, the "eve" was proposed in 1922 and was similar in many respects to that proposed by Villard for X-rays. However, the time was not yet ripe for establishing a unit; much basic measurement work, particularly with regard to absorbed dose, had yet to be done.

The formidable task of tissue dose was attacked by many. In 1912, Franz Theophil Christen (1873 to 1920), a Swiss mathematical physicist, contributed the concept of half-value layer, based on his studies of X-ray absorption, as a suitable method of characterizing the average energy or wavelength of an X-ray beam. This, coupled with ionization studies of intensity, — notably by William Duane, Bernard Szilard of Hungary, and Walther Friedrich of Germany — led to a reasonably accurate quantification of X-rays in air and paved the way for studies of X-ray dose in other absorbing media, specifically tissue.

In 1918, the German radiologists, Bernhard Kroenig (1863 to 1917) and Walther Friederich (1883 to 1969) of Freiburg, published their work on X-ray measurements, which included a discussion of the absorbed dose in tissue along the central axis of the beam. Their measurements were made with a horn wall ionization chamber, this material having been selected by Friederich as being similar to skin in absorptive properties and used in place of metal to minimize inaccuracies from scattered and characteristic radiations. Others produced similar absorbed dose plots, including the great German radiologist Hans G. D. Holfelder (1891 to 1944) in 1920 and physicists Friedrich Dessaner (1882 to 1963) and Friedrich Vierheller, who published a comprehensive study in the German journal *Strahlentherapie* the following year. About the same time, similar work was being carried on independently in the United States by physicists William Duane in Boston and Albert Bachem in Chicago. Thus, the isodose curve had its beginning about 1920, and this rather elegant technique provided a scientific basis for therapy dosimetry as well as serving to provide for reproducible systematic measurements.

To Otto Glasser (1895 to 1964), the German-American radiological physicist, credit must be given for coining the word "iosodose" and for refining and systematizing X-ray isodose technique. As early as 1919, Glasser, well known as the definitive biographer of Roentgen, had studied the distribution of both X-rays and radium gamma-rays in absorbing media, publishing his results in the German literature. He also worked with Friedrich on developing a series of equal intensity curves which he later named isodoses based on measurements of radiation intensity produced in water by an immersed source. Glasser was also instrumental in the development of the condenser R-meter, this work was done in the late 1920s in collaboration with Valentin B. Seitz, a Cleveland electrical engineer, and radiologist U. V. Portmann. The Glasser-Seitz chamber became famous as the Victoreen R-Meter, and manufacture was begun in 1929. Glasser had previously developed a popular X-ray dosimeter in conjunction with Robert E. Fricke, namely, the "thimble chamber" with an air equivalent wall.

Similar measurements were made of the dose associated with radium. Although Glasser and others had developed isodose plots for simple radium sources and applicators as early as 1919, the problem was complicated by the complex geometries of radium sources and human tissues in the therapeutic situation. Indeed, in 1917, Gioacchino Failla, the New York radiological physicist, had called attention to the fact that it was necessary to use the radiation reaching the tumor at any point as the criterion of dosage at that point, and obtained dose distributions for radium needles and tubes then in use. In the early 1920s, Edith Quimby enlarged upon this work and was instrumental in establishing a series of radium dose tables useful for uniformly distributed sources over areas and through volumes. Others, including the eminent Swedish physicist Rolf Sievert (b. 1896), took a more theoretical mathematical approach; Sievert's calculations of isodose curves for linear radium sources formed the basis for the Manchester system of radium dosimetry developed during the 1930s by radiologist James Ralston Paterson (b. 1897) and physicist Herbert M. Parker (b. 1910). This treatment system that relied heavily on Parker's original and seminal calculations came into widespread use and is still in use four decades later.

The depth dose and isodose work was important in that it not only set the stage for the establishment of an appropriate unit of exposure, namely the roentgen in 1928, but also refined and greatly improved both measurement and calculational techniques.

Radiological units have evolved continuously over the history or radiological science. The roentgen has gone through several changes from its original definition in 1928, the most recent being 1971. The discovery of the neutron in 1932 led to the development of the n-unit by L. H. Gray and J. Read in 1939; the roentgen was not only inappropriate vis-a-vis its definition that limited it to x- or gamma radiation but was unsuited for measurement of neutrons. The n-unit was defined as 1 esu/cm^3 produced in the sensitive volume of a special graphite chamber and was supposedly equivalent to the tissue dose produced by 1 R.

The inadequacies of the roentgen for mixed radiations led to the adoption of units based on the concept of energy absorption. In the mid-1930s, L. H. Gray had established the principle of energy deposition as the critical item with respect to interactions of radiation and matter and, more importantly, biological response. This helped (at least partially) to explain the variability of some biological units such as skin erythema dose, particularly when F. W. Spiers measured energy deposition in various tissues per roentgen of exposure a decade later.

Many pushed for the adoption of a dose unit based on energy absorption, with the feeling particularly strong among the physicists. British radiological physicist William V. Mayneord (b. 1902) introduced the gram-roentgen concept in 1940, and this unit became widely used in radiation therapy. In 1950, American physicists Ugo Fano and Lauriston Taylor proposed a unit based on energy absorption, and Herbert M. Parker suggested adoption of the rep (roentgen equivalent physical) which was used during the 1940s, applicable to all radiations, and originally defined as the quantity of radiation that liberates an amount of energy equal to the absorption of 1 R or x- or gamma-rays when absorbed.

The rep was an improvement but, unfortunately, not wholly satisfactory. First of all, it was tied to the roentgen, and as the value for W (energy to produce one ion pair in air) became better known, the value of the rep changed. Hence in 1953, only 3 years later, the International Commission on Radiological Units adopted the rad which was defined identically for every radiation and absorbing medium in units of energy absorbed per unit mass. One rad = 100 erg/g = 0.01 J/kg.

The origin of the word 'rad' is often given as an acronym derived from the phrase radiation (or roentgen) absorbed dose. This etymological origin, although quite common in otherwise accurate texts and articles, is incorrect. The rad was selected by the ICRU as a word by itself; the acronym idea, while perhaps useful as a mnemonic aid, is without merit and should not be perpetuated or given credence.

There also existed a need for a unit of biological dose, since different radiations may have different biological effectiveness even though the absorbed dose is the same. In the late 1940s, a unit called the rem, which considered both the absorbed energy and the relative biological effectiveness of the exposing radiation(s), came into use. Physicists R. D. Evans and, in particular, Parker proposed its use particularly for mixed radiation field dosimetry. The rem, which had been used in the Manhattan District radiation protection program in the early 1940s, fast found favor and is still used today although its definition has undergone several iterations.

The quantity of radium was originally simply expressed in units of mass of radium. However, the radioactive daughters of radium created complications, and the curie (equal to the number of disintegrations per unit time from 1 g of radium in equilibrium with daughters) was introduced in 1910 by the International Congress on Radiology and Electricity held in Brussels. The report on radium standards was presented to the Congress by no less a personage than Ernest Rutherford (1871 to 1937) who chaired the radium standardization committee. The curie unit was proposed in honor of Madame Curie who would prepare the first international standard of 21 mg of pure $RaCl_2$ sealed in a light glass ampule the following year. Indeed, the name was proposed before the unit was defined, and it was Marie Curie herself who dictatorially defined the unit; after all, it was named in her honor, and the Congress went along. In 1930, the International Commission on Radiological Units divorced the curie from the mass of radium, defining it instead in terms of absolute disintegrations per second, thereby paving the way for general applicability of the unit to any radioactive substance.

2.7. GROWTH IN THE USE OF RADIOACTIVE MATERIALS

An important area of radiological science, the use of artificially produced radio-nuclides, was made possible by the invention of the cyclotron in 1929 by Ernest O. Lawrence (1901 to 1958) and the discovery of artificial radioactivity by Frederic (1900 to 1958) and Irene Joliot-Curie (1897 to 1956) in 1933. It was in 1932 that James Chadwick (b. 1891) experimentally demonstrated at the Cavendish Laboratory at Cambridge the existence of the neutron which had been postulated 12 years earlier by Rutherford. Chadwick, who received the Nobel Prize in Physics in 1935 for his discovery, provided the explanation for the puzzling observations of the Joliot-Curies, and set the stage for their discovery of artificial radioactivity which was announced in *Comptes Rendus* in January 1934.

It was, however, the great Italian physicist Enrico Fermi (1901 to 1954) who expanded and developed the techniques of making artificial radionuclides along with his colleagues at the University of Rome. Fermi, who was awarded the 1938 Nobel Price in Physics for his work on radioactivity (including artificial radionuclides), filed for a patent on a "Process for the Production of Radioactive Substances" by neutron bombardment on October 3, 1934. His co-filers were four of his colleagues at the University of Rome: Eduardo Amaldi, Bruno Pontecorvo, Franco Rasetti, and Emilio Segre'. The American patent, Number 2,206,634, was issued on July 2, 1940 and assigned to G. M. Giannini and Co., Inc. of New York.

Although artificial radioactivity did not come upon the scene until the early 1930s, radioactive tracer work had begun two decades earlier. The first tracer study had been accomplished in 1913 by Georg von Hevesy (1886 to 1966) and Fritz Paneth (1887 to 1958) who used a naturally occurring radioisotope of lead to determine the solubilities of various lead compounds in water.

Ten years later, von Hevesy used ^{210}Pb, the 22-year half-life daughter of uranium, to study lead uptake by the broad bean *Vicia fabia*. The following year, he used the radioactive properties of bismuth to study its distribution in the rabbit. In 1934, after the discovery of deuterium by Harold C. Urey, von Hevesy obtained some heavy water and used it to measure water turnover time in the body and the exchange rate of water between a goldfish and its environment. Immediately after the discovery of artificial radioactivity, von Hevesy and his co-workers made ^{32}P by activation of sulfur with a RaBe source and began seminal studies of phosphorous metabolism in rats and humans the following year in collaboration with Otto Chiewitz. Still another pioneering first was to be accomplished by von Hevesy — activation analysis. In 1936, with the collaboration of H. Levi, impurities were detected in samples of rare earth compounds exposed to neutrons.

The remarkable studies of von Hevesy were continued and expanded by others. Within 5 years, several hundred papers had been published, including those on artificial isotopes of bromine, chlorine, carbon, cobalt, iron, and sodium. Von Hevesy himself remained active until shortly before his death at the age of 80 in 1966. He received many honors including the Nobel Prize in Medicine in 1943 and the U.S. Atomic Energy Commission Atoms for Peace Award in 1959.

In the middle 1930s, considerable thought was given to potential medical applications of cyclotrons, neutrons, and artificial radionuclides. Ernest O. Lawrence, the Nobel prize winning physicist who is noted for the development of the cyclotron, was one of the early advocates of using neutron beams for cancer therapy and also felt that various radionuclides might "hook onto cancer cells and radiate them internally." Along with radiologist Robert S. Stone (b. 1895), physician-scientist Joseph G. Hamilton (1907 to 1957), and Lawrence's physician brother John (b. 1904), the efforts of staff and facilities of the University of California Medical School Hospital in San Francisco and Radiation Laboratory in Berkeley were applied to the task. In the spring of 1936, Joseph Hamilton administered internal radiosodium to a leukemia patient for the first time, comparing its effectiveness with that of radium. The objective had been stated by Lawrence the previous year in a letter to John Cockcroft — using artificial radionuclides with superior chemical and biological properties as inexpensive substitutes for radium.

Throughout 1936 and 1937, cyclotron-produced radionuclides were studied intensively as potential therapeutic agents. John Lawrence, aided by biophysicist Hardin B. Jones, biologists K. G. Scott and E. Tuttle, and others at the University of California, established that ^{32}P was selectively localized in leukemic tissues. This led to the first therapeutic application of artificial radionuclides; on Christmas Eve 1937, John Lawrence began treatment of a patient suffering from chronic lymphatic leukemia with sodium radiophosphate. Within 2 years, the efficacy of ^{32}P for treating polycythemia vera was established, and such therapy is still current 4 decades later.

Perhaps the most well known of all biologically useful radionuclides is ^{131}I, which is not only a valuable tracer in plants and animals but has been of inestimable value in diagnosis and treatment of thyroid disease. In 1938, Joseph G. Hamilton was studying thyroid metabolism with ^{128}I. Finding the 25-min half-life too short, Hamilton asked Glenn T. Seaborg, who was awarded a Nobel Prize in Chemistry in 1951 for his work on heavy elements, if he could find an iodine isotope with a longer half-life. Seaborg asked what half-life would be most desirable, and Hamilton replied, "Oh, about a week." A short time later, Seaborg and his co-worker, J. J. Livingood, synthesized ^{131}I with a half-life of 8.08 days — almost precisely what Hamilton had ordered! Hyperthyroidism was being routinely treated with ^{131}I by 1941.

Tracers were also applied in industry. As early as 1941, ^{32}P was used to tag piston rings for lubrication and wear studies, and in 1944, copper alloys activated by deuterons were similarly used. Cold cathode electron tubes came into being using ^{60}Co or other beta emitters as the source of electrons.

The advent of the atomic age produced an explosive growth in the radiological sciences and marks an appropriate stopping point in the historical narrative. The discovery of fission in 1939 was followed by the first self-sustaining chain reaction on December 2, 1942 and led to the production of nuclear explosives in 1945. New and greatly expanded areas of radiological science were created — reactor technology, environmental studies, and weapons effects are but three. Nuclear medicine became a specialty in its own right, and new professions — nuclear engineering, health physics, and radiochemistry — came into being.

2.8. IN THE PUBLIC SECTOR: TALES OF TRAGEDY

From the very beginning, the public was fascinated by X-rays and radioactivity, a fascination that unfortunately gave rise to wildly exaggerated reports and unsupervised and uncontrolled uses, often with tragic results. Early publicity in newspapers and magazines (the mass media of the time) not only hailed the X-ray as a medical miracle but sometimes ascribed phenomenal and extraordinary physical and curative properties to it.

While most of the early reports in the lay press were fairly sober, New Yorkers read in one of their daily newspapers one February day of the use of X-rays to project anatomical figures into the brains of medical students at College of Physicians and

Surgeons enabling them to learn better and more quickly. A serious report indeed appeared not too long afterward in the scientific literature describing an experiment in which the image of a bone was projected via the X-ray onto the brain of a dog causing the beast to salivate Pavlovian fashion.

The X-rays were hailed in prose and verse. Humorous poems appeared in popular magazines. Among the earliest was one entitled "The New Photography," published in *Punch* January 25, 1896 and described X-ray portraiture including this quatrain of modesty directed to Roentgen:

> We do not wish, like Dr. Swift
> To take off our flesh and pose in
> Our bone, or show each little rift
> And joint for you to poke your nose in.

Life ran a cartoon in February, showing a farmer posing for a Roentgen photograph that appears as a skeleton and followed a month later with a poem by Lawrence K. Russel that described a cathode ray portrait of his love. *Photography* ran a poem that concluded with this Victorian rhyme: " . . . I hear they'll gaze thro' cloak and gown — and even stays, these naughty, naughty Roentgen Rays."

The public was completely smitten with X-rays, and almost overnight, a new entertainment was created. At museums, county and state fairs, and in private offices created for the purpose, the public was treated to a fluoroscopic view of a portion of their own anatomy, usually the hand, for a fee. Early in 1896, thousands viewed their bones at the New York Electrical Exposition held at Grand Central Palace using fluoroscopic apparatus made by none other than Thomas Edison. Some even wanted their brains examined. In a few people, latent bone diseases were discovered. Ironically, Edison, whose efforts introduced many to the wonders of the X-ray, was to halt his X-ray researches shortly thereafter because of his observations of eye troubles in his workers. In the short time he worked with X-rays, he devised a fluoroscope that was unpatented and left in the public domain because of its potential benefit to medicine; directly from his X-ray studies, Edison developed the first fluorescent light, applying for a patent on May 16, 1896.

For a few years, X-rays enjoyed great popularity with the general public who from time to time were treated to optimistic forecasts of yet-to-be wonders of X-rays. Among these were the use of X-ray "baths" for curing various diseases including tuberculosis and criminal behavior as well as X-ray identification of criminals to replace the cumbersome Bertillon measurements; the uniqueness of individual fingerprints had not yet been recognized. No less a personage than Frances E. Willard, president of the national Women's Christian Temperance Union stated that a superior use of X-rays would be to demonstrate the destruction of their own bodies to smokers and drinkers. Other reformers had even greater hopes, namely, that X-ray or radium treatments could cure the affliction of intemperance. In humorous rebuttal, a New York drinker named Joe Rogers noted that X-rays would never penetrate the smoky, alcohol-saturated atmosphere surrounding the intemperates; even were this not the case, the power of the intemperate person's breath would destroy the X-rays!

The wonderful dreams soon faded from the press replaced by the harsh reality of charlatans who promised miracles — for a fee. In addition to quack medical practitioners, X-ray and radium treatments were offered by nonmedical practitioners or as patent medicines, often with disastrous consequences. A favorite application of X-rays was as a depilatory. Thousands of young women were exposed to X-ray doses sufficient to produce epilation. Erythema, scarring, atrophy, and cancers were a common side effect of such treatments. Dr. Louis I. Harris, New York City Commissioner of Health, claimed to have seen "countless cases" of young women with disfigured faces caused by X-ray treatment for removal of superfluous hair. Even after the public became aware of the potential hazards of X-ray depilatories, the practice continued with the practitioners sometimes concealing the details of the treatment from their clients. This deplorable practice continued well into the 1930s and sporadically beyond.

Another once-common use of X-rays was in shoe fitting fluoroscopes that had their heyday in the 1930s and 1940s. Ironically, these devices were really unnecessary to the proper fitting of shoes; however, because of their novelty, they were of value as a sales tool. Some models had virtually unshielded tubes situated approximately opposite the pelvic area of a child using them; thus, the maximum exposure (other than to the feet) was to the gonads of the users. Doses of tens of rads might be given in a shoe-buying session, and at least one case of skin burns to a young customer has been reported.

Even more so than with X-rays, the lay public ascribed panacea properties to radium. Internal doses of radioactivity can be traced back to the early 1900s, when William J. Morton proposed the use of radioactive waters for treatment of a variety of ills. Although Morton's suggestions, originally published in reputable medical journals, did not find favor among his medical colleagues, the idea was adopted by patent medicine purveyors, operators of mineral springs, and charlatans of all types. To be sure, many of the "remedies" offered did not contain the claimed radium or other natural radioactivity; these could, in some cases, have been condemned as frauds under the Pure Food and Drugs Act of 1906.

One such patent medicine, named *Radol,* made its appearance in the early 1900s. The brainchild of a charlatan with an assumed name of Dr. Rupert Wells, *Radol* was billed as a radium impregnated cancer cure. It was, of course, neither radium containing nor cancer curing, being in fact an acidic solution of quinine sulfate with added alcohol. *Radol* was one of the targets of misbranding action under the early days of the Pure Food and Drug Act and was ultimately driven from the marketplace.

There were also numerous patent medicines containing radioactivity in high concentrations, the most famous being *Radiothor.* Since these were not mislabeled (i.e. they actually contained the radioactivity claimed on the label), they were safe from legal action so long as claims of cure were not made. However, the nostrum could be recommended for treatment (and indeed *Radiothor* was) for some 160 ailments, including cancer.

Radiothor was the brainchild of William J. A. Bailey, a notorious quack, who had once operated a fraudulent automobile manufacturing company. The nostrum flourished in the 1920s along with others, including *Linarium,* a claimed radioactive liniment; *Arium;* radium tablets; and *Thorone,* a liquid claimed to be 250 times as radioactive as radium and especially recommended for the treatment of impotence. The hazards of these patent medicines apart from the fact that they did not possess their claimed curative properties was that they could be fatal. In 1932, the tragic and horrible death of a wealthy young business man E. M. Byers focussed attention on the hazards of ingested radium. Byers had consumed several bottles of Radiothor daily for a few years, ultimately dying of osteogenic sarcoma of the jaw. Columbia University physicist Frederic Bonner Flinn calculated a bone burden of 36 μCi or approximately 900 times the current maximum permissible body burden for occupational exposure. The purveyors of *Radiothor,* however, were within the law and *Radiothor* was not banned from sale until the late 1930s by the Federal Trade Commission.

Radiothor was only one of many radium or thorium bearing nostrums, and the use of radioactivity was not confined to patent medicines. Hair tonics, tissue creams, and other cosmetics containing radium (or supposedly so) were sold well into the 1940s; even chocolate bars were made that supposedly contained radium. For those who wanted prevention rather than cure, radium or thorium could be added to water at home. The *Thoronator* was a small vial to which water could be added. Inside the vial was a porous cylinder that gave off thorium emanations and so charged the added water which, when drunk, would have tonic properties. Various cones and radioactive crocks appeared about 1905 and continued to be available well into the 1940s. In many instances, these contained concentrated radium or thorium bearing ores that would leach the natural radioactive materials into the water which was then drunk by the owner. Significant quantities of natural radioactivity could be ingested in this fashion; for example, the *Revigator,* billed as "A Genuine, Radium Ore, Patented Water Crock," would produce about 50 pCi/l of radium in water allowed to sit in it for 12 hr. The directions called for drinking 1.4 l per day of this water, which had a gross α-β concentration of about 1300 pCi/l. Hence, on an annual basis, a person following the instructions would ingest 0.026 μCi of ^{226}Ra per year.

There were other products that claimed special or super properties because of their radium content. In about 1920, a German toothpaste *"Radiogen"* was marketed. This toothpaste contained radium that would allegedly release a constant amount of radon as the teeth were brushed, thus reducing the decomposition of alimentary residues and eliminating "dental stone." The Germans were perhaps obsessed by radioactive toothpastes. During World War II, Project Alsos scientists (Alsos was the Allied mission concerned with German atomic bomb development) discovered that a German chemical firm had an extraordinary interest in thorium and had a virtual corner on the European supply. The Alsos scientists calculated that this amount of thorium constituted many times that required for use by ordinary industry for at least 20 years. Since thorium could enter into weapons research and development, Alsos was most interested. After extensive investigation and some unusual intelligence activity, the Alsos mission determined that the progressive German chemical firm, thinking ahead to postwar activities, planned to market a thoriated toothpaste! On a more down to earth note, *Radium Fertilizer,* promising "Eternal Sunshine" was marketed by National Radium Products Company of Grand Junction, Colorado in about 1920.

Although radium waters and similar preparations had virtually vanished by 1950, bathing in radium springs and visits to radioactive mines are still commonplace with Badgastein, Austria serving as a pilgrimage site for sufferers from many afflictions. Billed as "The Springs of Eternal Youth," the hot baths are acclaimed for their therapeutic effects which are largely attributed to dissolved radon. In the United States, the Free Enterprise Radon Mine in Boulder, Montana has been billed by the president of its owning company as providing beneficial effects for sufferers of arthritis. This mine operated for many years and may yet be open to the public.

The most well known essentially uncontrolled application of radium in the public sector was the use of radium for self luminous compounds. Around the time of World War I, radium became more generally available, and in the 1920s, the radium dial painting industry flourished. Luminous dials were first used in World War I in German submarines, having been suggested by the Austrian born physician Sabin A. von Sohocky (1882 to 1926) who emigrated to the U.S. after the war where he ultimately became president of U.S. Radium Corporation.

Dial painting was done by young women, who kept their brush tips pointed by moistening them with their lips. Thus, they ingested a considerable amount of radium, and in 1924, the syndrome prevalent among them was identified by Theodore Blum (1883 to 1970), a New York dentist, in a footnote to an article published in the *Journal of the American Dental Association.* This prompted Harrison Martland (1883 to 1954), then medical examiner for Essex County, N.J., to evaluate the dial painter mortality on the basis of radium exposure and led to a classic series of papers over a 20 year period as well as the recognition of the hazards of brush tipping. However, between 1917 and 1924, more than 2000 individuals had been employed in the dial painting industry and in addition to brush tipping were exposed by primitive working conditions as well as by such practices as cosmetic application of radioluminescent compounds to the hair and skin, a common practice among young dial painters.

2.9. RADIATION PROTECTION

With the immediate intensive and widespread use of the X-ray after its discovery, it was inevitable that X-ray injuries should soon appear. However, even after careful examination of the facts, the causal relationship between the use of the X-ray and subsequent injury was often overlooked; this was largely attributable to the belief that since X-rays could not be detected by the senses, they could therefore have no deleterious effects on the tissues. For several years, the controversy indeed raged about the hazardous nature of the X-rays, sparking acrimonious debates at meetings and harshly worded publications in the scientific literature. As late as 1905, it was not uncommon to find reports in the medical literature attesting to the benign or beneficial nature of the X-ray or, more commonly, explaining away skin damage as caused by other than X-ray exposure (Table 2–1).

The first reports of possible X-ray injury appeared on March 5, 1896, in the British journal *Science.* Two Americans, Thomas Edison and William J. Morton, independently described eye irritations associated with the use of X-rays and fluorescent materials. Edison indicated a possible causal relationship of these symptoms with the X-ray. It is quite possible that the eye troubles experienced by these two investigators were the result of fatigue or perhaps ultraviolet light associated with fluorescence, rather than of X-ray origin.

Reports of skin burns began to appear soon after in the American literature, creating much controversy prompting Elihu Thomson to attempt to settle the matter. In a classic experiment performed in late 1896, Thomson purposely exposed his left little finger over a period of several days to the direct beam of an X-ray tube for ½ hr each day at a target-to-skin distance of less than 1¼ in. No observable effects occurred for more than a week; then he noted pain, swelling, stiffness, and subsequent blistering, likening the damage done to that of "... high pitched waves of light in causing sunburn by penetrating more deeply into the tissues."

Thomson advanced the hypothesis that some sort of chemical action occurred within the tissues noting "... there is evidently a point beyond which exposure cannot go without causing serious trouble."

The turn of the century marked the great awakening with regard to X-ray protection. More and more scientists and physicians were accepting the evidence that pointed to X-ray dangers, the latter spurred by court awards for damages. The first such award was made in 1897 and granted $10,000 to a patient who incurred a necrosis form diagnostic X-ray procedures. The validity of the award was severely questioned with the medical journal maintaining that the idiosyncrasy of the patient was the major cause of the damage. Legal precedent was established 5 years later in an appeal case in which the court held that the same standards of care applicable to general medical practice also applied to the use of X-rays.

Public opinion was another important factor influencing the achievement of hazard control. Although the public, in general, seemed unaware of any problems associated with the use of the X-ray, some rather well versed vocal minorities exerted considerable pressure through the newspapers. A scathing indictment came from a crusading newspaper reporter, John Dennis. Dennis had made a study of the uses and abuses of X-rays, and on April 25, 1899, he unleashed a broadside at a meeting of the Seventh District Dental Society of the State of New York, recommending a state licensing commission consisting of two physicians and an electrician to limit X-ray usage to qualified practitioners. In no uncertain terms, he clearly stated that it was a criminal act for a radiographer to injure a subject. Unfortunately, the advanced ideas of Dennis have never been fully implemented. Less than two thirds of the states have comprehensive radiation control regulations and programs and fewer require licensure of technicians.

Recognition of the lethal effects of overexposure to the X-rays also helped to

Table 2-1

HISTORICAL MILESTONES IN RADIOLOGICAL PROTECTION AND MEASUREMENT 1784–1950

Year	Milestone
1784	W. Morgan unknowingly produces X-rays in experiment witnessed by B. Franklin
1839	M. Daguerre discovers photography
1865	H. Sprengel devises mercury air pump necessary for production of highly evacuated tubes; H. Geissler and J. Plucker observe fluorescence in evacuated tubes containing electrodes; J. Maxwell publishes electromagnetic field theory
1868	J. Plucker and W. Hittorf show cathode rays are deflected by magnetic field
1869	Hittorf shows cathode emanation stopped by solid object; E. Goldstein coins phrase "cathode rays"; W. Crookes notes fogging in photographic plates in his laboratory and complains of defective packaging
1879	W. Crookes shows cathode rays are solid matter with sufficient energy to drive a small wheel
1886	H. Hertz characterizes long wave electromagnetic radiation
1891	H. Hertz, assisted by P. Lenard, studies penetrating power of cathode rays
1895	W. Roentgen discovers X-rays (November 8)
1896	Roentgen's First Communication (January 3) dated December 26, 1895; Roentgen's Second Communication (March 9); First therapeutic applications of X-rays: Grubbé (January 27?) Voigt (January ?) Despeignes (May) First diagnostic X-ray in U.S. by E. Frost (February 3); H. Becquerel announces discovery of radioactivity (March 3); First book on X-rays published (April ?); First applications in dentistry (March) probably by C. Kells and W. Rollins; T. Edison reports eye injuries from X-rays (March); N. Tesla cautions experimenters not to get too close to X-ray tubes (June)
1897	J. J. Thomson demonstrates corpuscular nature of cathode rays; Roentgen's Third Communication (March 10) Queen & Co. introduces self regulating X-ray tube
1898	E. Thomson uses aluminum filter for X-ray protection; W. Rollins describes leaded X-ray tube housing; devises collimators; Discovery of radioactivity of thorium by G. Schmidt (March); M. and P. Curie discover polonium and coin word "radioactivity" (July 13); Curies discover radium (December 26); P. Villard discovers gamma rays
1900	Rollins develops tube with cooled target
1902	Rollins experimentally shows X-rays could kill higher life forms; Existence of radium verified by Curies by chemical methods
1904	Colorimetric dosimetry system devised by Saboroud and Noire
1905	Ionization unit proposed by M. Franklin
1907	Interrupterless transformer rectifying switch) devised by H. Snook
1911	International radium standard and Curie unit
1912	T. Christen puts forth concept of half value layer

Table 2-1 (continued)
HISTORICAL MILESTONES IN RADIOLOGICAL PROTECTION
AND MEASUREMENT 1784–1950

Year	Milestone
1913	G. Bucky introduces grid to eliminate effects of scatter; G. Von Hevesy and F. Paueth perform first tracer studies using lead isotopes to determine solubilities; W. D. Coolidge introduces tungsten target tubes, hot cathode tubes
1914	Curie unit proposed for radium by A. Debrierne and C. Regaud
1915	X-ray protection recommendations adopted by Roentgen Society (June)
1917	E. Jerman establishes X-ray education program at Victor X-ray Corporation
1918	H. Waite patents oil immersion principle
1922	American Roentgen Ray Society adopts radiation protection rules; American Registry of X-ray Technicians founded; G. Pfahler recommends personnel monitoring with film
1923	von Hevesy performs first radiotracer studies using ^{210}Pb
1925	A. Mutscheller puts forth first "tolerance dose" – ~0.2 R/day; (also, R. Sievert) ICRU founded
1926	E. Quimby devises film badge with energy compensating filters
1927	Dutch Board of Health recommends tolerance dose equivalent to 15 R/year; H. Muller shows genetic effects of radiation
1928	Roentgen unit adopted by Second International Congress on Radiology
1929	U.S. Advisory Committee on X-ray and Radium Protection (USACXRT) formed (forerunner of NCRP); L. Taylor makes first portable survey meter for X-ray monitoring
1931	USACXRT recommendations on radiation protection published as National Bureau of Standards Handbook 15; exposure limit of 0.2 R/day recommended
1932	G. Failla suggests limit of 0.1 R/day to whole body and 5 R/day to fingers, introducing concept of higher permissible dose to limited portions of body E. Chadwick discovers neutron
1934	ICRP recommends tolerance dose of 0.2 R/day Joliot–Curies discover artificial radioactivity; H. Urey discovers deuterium
1935	G. von Hevesy performs first artificial radiotracer studies using ^{32}P to measure water turnover rates in goldfish
1936	Bragg-Gray principle formulated
1937	First therapeutic application of artificial radioactivity by J. Lawrence who treated polycythemia vera with ^{32}P
1938	^{131}I made in cyclotron by G. T. Seaborg and J. Livingood
1939	O. Halin, F. Strassmann discover fission; E. Fermi patents first reactor (conceptual plan)
1941	ACXRP establishes first permissible body burden – 0.1 μCi for radium; First industrial tracer usage
1942–1945	Manhattan District – birth of Health Physics as a profession, and great developments in instrumentation and applied radiation protection
1944	First MPC's, initial MPC was for ^{239}Pu and was calculated by H. Parker; Rem and rep concepts used in Manhattan Project
1946	U.S. Atomic Energy Commission created; Civilian use of radioisotopes from fission process began

Table 2-1 (continued)
HISTORICAL MILESTONES IN RADIOLOGICAL PROTECTION
AND MEASUREMENT 1784–1950

Year	Milestone
1949	NCRP puts forth occupational exposure unit of 0.3 rem/week and introduces benefit-risk concept; First Tri-Partite Conference held to establish lung model and MPC's
1950	AEC incorporates NCRP recommendations into authorization and into regulation in 1957

emphasize the need for protective devices and techniques. In 1897, the defense attorney in a murder trial charged that the victim had died from X-ray exposure to the head, made in an attempt to locate the bullet. There is no record of whether this defense was successful.

In late 1900, a coroner's jury in England established the fact that a death had been caused from shock and exhaustion following an accident and also through the effects of the Roentgen rays on a weakened system. This decision was noted in the American literature without editorial comment. Shortly thereafter, William Rollins reported experimentally induced fatalities in guinea pigs exposed to both filtered and unfiltered X-rays. He also noted that a pregnant female aborted. Rollins' simple experiment was the first to demonstrate that X-rays could be lethal to the higher forms of animal life and, as such, marked a watershed in the beginnings of radiation studies.

A great boost to X-ray protection in the United States was provided by the American Roentgen Ray Society (ARRS). Formed in 1900 by a group that included physicians, engineers, and manufacturers, this organization provided an effective sounding board for those who advocated protection for its own sake. Persual of the *Proceedings* of the society over the years 1903 to 1907 indicates the great concern the membership had over the lack of protective measures for patient, operator, and experimenter. The ARRS was obviously a concerned and interested organization.

Pre-1900 recommendations for X-ray protection were few and based primarily upon erroneous assumptions regarding the nature of the biological effects. The very first recommendation appears to have been made by Nikola Tesla (1856 to 1943), the great Yugoslav-American inventor who in June 1896 advised experimenters not to get too close to the energized tube lest the eyes become inflamed.

The value of an enclosed tube for protection purposes was apparently first recognized by Elihu Thomson who noted in 1898 that aluminum sheeting would absorb the more hazardous "soft" rays but allow the more highly penetrating "harmless" rays to pass through. In that same year, Rollins described a leaded housing for the purpose of sharpening the radiographic image and later became one of the leading advocates of radio-opaque tube housings for protection purposes. In addition to his suggestions regarding "nonradiable" tube housings, he recommended radio-opaque glasses for fluoroscopists and lead shields for areas of the patient that were adjacent to those being X-rayed. The implications of collimation are strong throughout the writings of Rollins; in several of his papers, he discussed the theory and means of limiting the primary beam. Rollins even designed an X-ray tube with an internal metal collimator before 1900.

The ideas of Rollins with regard to the necessity for X-ray protection were challenged by Ernest A. Codman (1869 to 1940), who cited his experience of making 10,000 exposures with 4000 patients without a single case of skin burns. He acknowledged that some of the precautions advocated by Rollins would be needed if exposure were to be on a continuing basis; however, he thought that for practical use, it was necessary only to keep the hands away from the X-ray tube. The reply of Rollins to this attack was prompt and succinct: "An experiment can only be disproved by another experiment."

Of all the early scientists who attempted to further the art of X-ray protection, William Rollins is extraordinary. Trained at Harvard in both dentistry and medicine, he practiced dentistry in Boston and gained feeling for the problems of the medical practitioner. As a developer of new apparatus, he is seemingly without peer. Francis H. Williams (1852 to 1936), pioneer Boston radiologist and brother-in-law for whom Rollins created many new devices, paid him high tribute "for his unfailing aid . . . to obtain apparatus suitable for . . . examinations and particularly for the new and better forms of vacuum tubes he has devised."

It is unfortunate that the many warnings issued by Rollins went for the most part unheeded; his ideas for protective devices (unusually well designed) were not generally used. Perhaps the limited circulation of his writings may account for this, since most of his works appeared in what was essentially a regional medical journal or in engineering publications that did not reach many of the medical radiographers. Even his book, a compilation of his 181 papers published up to mid-1904, was privately printed and received only a very limited distribution. More likely, however, was that Rollins was far ahead of his time, and like so many other prophets in science, this shy and dedicated man went virtually unrecognized.

In 1902, Rollins published an excellent discussion of the theory and use of radio-opaque tube housings. Sensing that practicing radiographers failed to adopt leaded housings because of their bulk, he described a method whereby the unwanted radiation could be eliminated by painting the glass case of the tube itself with several layers of white lead in Japan. To test the coating, he recommended placing a photographic plate in contact with the painted surface of the tube housing. If the plate was not fogged by a 7 min exposure, the coating was assumed to be sufficiently thick. This simple test has been misinterpreted as a statement of "tolerance dose," and further extension credits Rollins with establishing the modern equivalent of a 10 R/day maximum permissible exposure. Although such a statement appears in several modern texts, Rollins clearly had no such intent.

Although filtration as a means of X-ray protection had been recognized by Rollins and others, it remained for George Pfahler (1874 to 1957) to strongly reiterate its value in 1905. In logical fashion, he explained the theory of selective absorption and reasoned that leather, because of its similarity to human skin, would make a highly satisfactory filter. His suggestions were widely adopted, partly because it was found that filtration would also improve radiograph quality. In addition, the hardened beam was found to be more effective in therapeutic work.

The deplorable practice of inserting the hand in the primary beam as a test of quality was common until about 1905. Francis H. Williams, otherwise notable for his efforts to achieve X-ray protection, wrote in 1901 that "a simple way of testing the quality of the light (X-ray) is to examine the hand with the fluorescent screen." As more reliable apparatus was developed, this practice fortunately died out, although pioneer New York radiologist and tube developer Henry Piffard (1842 to 1910) was impelled to issue another warning in 1906, and texts published as late as 1910 still referred to the method.

Many early radiologists concentrated on fluoroscopy, and many of the early protection schemes were built around protection of the fluoroscopist and his patients. Leaded garments were available and the prescient Rollins suggested several short bursts of X-ray rather than a single long exposure, thus taking advantage of the image-holding qualities of the fluorescent screen to reduce exposure to patient and physician alike.

Collimation in its strictest sense was a product of fluoroscopy, as was the development of leaded glass screens. At least one far-thinking physician advocated films rather than fluoroscopic examinations wherever possible, partly as a means of reducing patient and operator exposures.

In June 1915, at a meeting of the British Roentgen Society, a major step toward organized radiation protection was made with the adoption of a resolution put forth by C. C. Leyster calling for the adoption of stringent rules to protect X-ray personnel. However, for 5 years little was done. Then in 1920, interest was again expressed by both the British and American societies, sparked by the unsavory publicity in the press relating to aplastic anemia deaths among military roentgenologists. In September 1920, the ARRS established a standing radiation protection committee, and the following year, the British committee published the comprehensive organized radiation protection rules which were shortly afterward adopted verbatim by the ARRS, some 20 years after Rollins had done so as an individual, albeit in somewhat less comprehensive form.

In 1925, Arthur Mutscheller (1886 to 1972), a German-American physicist, proposed the first "tolerance" or maximum permissible dose — 72 R/year, whole body, the basis being one tenth the skin erythema dose. Others suggested similar "tolerance" doses in that same year, and in 1927, the Dutch Board of Health recommended the equivalent of 15 R/year.

The establishment of the roentgen unit in 1928 by the Second International Congress of Radiology was indeed important to protection efforts. Instrumentation, including personnel dosimeters, could now be calibrated in terms of the new unit and appropriate exposure limits and measurements made. The roentgen unit was arrived at after much labor and accepted by all as the International Unit. It was defined as "The unit of dose is that quantity of roentgen radiation which, when the secondary electrons are fully utilized and the wall effect of the chamber is avoided, produces in 1 cm^3 of atmospheric air at $0°C$ and 760 mm Hg pressure such a degree of conductivity that one electrostatic unit of charge is measured under saturation conditions." This somewhat cumbersome definition, revised in 1934 and again in 1950, was in essence the dose unit put forth a few years earlier by Herman Behnken, who had formulated a definition acceptable to the German Roentgen Society that adopted a similar unit as early as 1924. American physicist William Duane (1872 to 1935) had also formulated a dose unit in 1923, the esu-second, which was virtually identical to the roentgen. Both, however, were preceded and drew heavily on the original suggestion for an ionization-based unit put forth by Villard in 1908.

Personnel dosimetry, that vital core of radiation protection, had been suggested in 1922 by George Pfahler, who recommended that radiation workers routinely carry an unexposed dental X-ray film in their breast pockets. The film was to be developed at 2-week intervals, and the degree of blackening correlated in terms of skin erythema dose. Four years later, Edith Quimby (b. 1891), the New York radiological physicist whose career has spanned more than a half century, added filters to minimize energy dependence, and in 1928, Heinrich Franke (1881 to 1970) correlated the degree of film blackening with exposure in roentgens. By 1930, two Dutch scientists, Albert Bouwers and J. H. van der Tuuk, had increased the sensitivity of Franke's correlation, permitting film to be used for exposures down to 0.2 R/day (then the tolerance dose for x- and gamma radiation). However, the relative insensitivity of films available at that time limited their use for protection purposes as personnel dosimeters.

In the United States, a coordinated effort toward the establishment of radiation protection standards grew out of discussions in late 1928 among the major radiological societies who agreed that it was best to centralize American efforts in the National Bureau Standards. NBS had already established a long range program in radiation protection and was an independent agency with excellent facilities. Thus early in 1929, the Advisory Committee on X-Ray and Radium Protection (ACXRP) was formed; the Committee had two members each from the American Roentgen Ray Society, Radiological Society of North America, and X-Ray Equipment Manufacturers; a member from the American Medical Association; and as its chairman, Lauriston S. Taylor of the National Bureau of Standards.

Out of this committee grew the National Committee (now Council) on Radiation

Protection, formed originally in 1946 as a substantially enlarged body with much greater technical scope and interest in radiation protection. Lauriston Taylor, who had served as chairman of the ACXRP since its inception, was elected as chairman of the new body. Taylor ultimately would lead the ACXRP and its successor bodies for 47 years, retiring as President of the National Council on Radiation Protection and Measurements (NCRP) in 1977. All in all, Taylor had been active in radiation protection activities for 55 years at the time of his retirement from NCRP.

In 1931, the U.S. Advisory Committee on X-ray and Radium Protection (ACXRP) completed its 3-year effort on X-ray protection and its recommendations were published by the U.S. National Bureau of Standards as Handbook 15. This was the first in a long and still-growing series of NBS Handbooks dealing with radiation protection; H18, *Radium Protection,* would be the second and would follow a few years later leading a list that has grown to dozens.

Continuing attention was given to establishing an unequivocable generally accepted "safe" dose. In 1932, Failla proposed 36 R/year whole body and 5 R/day (1500 R/year) to the fingers introducing the concept of higher acceptable dose if exposure was limited to the extremities. Failla's whole-body recommendation was derived from existing data and based on the earlier (1925) work by Mutscheller. It was incorporated into H18 which appeared in 1934 and recommended 0.1 R/day (36 R/year). Internationally, the ICRP followed a few months later, but proposed 0.2 R/day as the recommended tolerance dose.

The problem of internal radium was also considered, receiving considerable impetus from the tragedy of the radium dial painters and imbibers of radium bearing patent medicines. The classic studies of Harrison Martland (1883 to 1954) the New Jersey pathologist, shed much light on radium toxicity, and in the late 1930's, in vivo determination of body burden was undertaken, most notably by Robley D. Evans (b. 1907), Professor of Physics at Massachusetts Institute of Technology. The early attempts at whole-body counting made use of a Geiger-Müller tube directed at appropriate portions of the anatomy. In 1941, the first permissible body burden was put forth by the ACXRT: 0.1 μCi of radium. This was based largely on the work of Evans and Martland.

2.10. THE MANHATTAN DISTRICT AND A LITTLE BEYOND

Sparked by the famous letter from Albert Einstein in the fall of 1939, President F. D. Roosevelt took steps that led to the establishment of the Manhattan Engineer District in August 1942. The Metallurgical Laboratory at the University of Chicago had already been operating for several months and from these organizations would come the atomic bomb. Also, tremendous advances in the radiological health sciences would accrue as spinoff benefits from the atomic bomb work.

At the start of World War II, the science and art of radiation protection and measurement was quite primitive. Measurement of radiation was generally accomplished with Geiger-Müller counters, and although a portable instrument was available, it was heavy, bulky, and temperamental. As plans were formulated for the first reactor, it was recognized that a new and powerful source of radiation would be created and that additional sources of radiation might also be produced as the project progressed. Arthur Holly Compton (1892 to 1962), the Nobel prize winning physicist who was director of the Metallurgical Project, and his senior associates had doubts about proceeding with the project, largely out of concern for the potential health hazards. Thus, in the summer of 1942, Ernest O. Wollan, then a cosmic ray physicist at the University of Chicago, was called upon to organize a group to study and control the hazards from radiation. His group was primarily concerned with shielding and, as such, was a part of the Engineering Division. The decision to form a Health Division was made in July, and on August 6, 1942, Robert S. Stone (b. 1895), a San Francisco radiologist with experience with artificial isotopes, was appointed Associate Director for Health of the Plutonium Projects. It was he who, in conjunction with Compton, coined the name Health Physics and incorporated Wollan's group into the Health Division.

When the first pile (the CP-1) achieved criticality, there were three health physicists — Wollan; Carl G. Gamertsfelder, a young physicist recently graduated; and Herbert M. Parker, the British medical physicist noted as codeveloper of the Paterson-Parker system of radium therapy. Gamertsfelder was present at the startup of the reactor that December day, measuring the radiation levels from it with an ionization chamber connected to an electrometer. Within the next several months, five more health physicists were added: Karl Z. Morgan, a cosmic ray physicist who was the physics department chairman at Lenoir-Rhyne College; James C. Hart (1917 to 1974), a chemical engineer from the DuPont Corporation (the project contractor); Robert R. Coveyou, a mathematician; O. G. Landsverk, a physicist who had long been concerned with accurate X-ray measurements and who was later to found the Landsverk Electrometer Company and manufacture hermetically sealed precision ion chambers; and L. A. Pardue, a cosmic ray physicist with nuclear emulsion experience. A ninth individual, John E. Rose, later headed the Chicago effort, but the above eight were the first cadre of professional health physicists.

The purpose of the Health Physics Section was, simply stated, to study penetrating radiations and devise the physical means for preventing damaging personnel doses. Accomplishment of this task was by no means simple, for not only were reactor radiations and plutonium new hazards, but only the very rudiments of radiation protection principles had been articulated, and available instrumentation was woefully inadequate to the task. Accordingly, one of the major efforts of the Manhattan District health physics efforts was directed toward the development of suitable monitoring instruments, many of which (such as the Juno and Cutie Pie) exist today in modern transistorized versions.

The efforts of the Health Physics Section were not limited to instrument development, much of which was done by other divisions to the specifications established by the Health Physics Division. Having evolved from a shielding group, it concerned itself with devising practical shields that would permit removal of radionuclides from the pile as well as experimental work with these materials in the laboratory setting. In addition to the physical aspects of protection, administrative procedures were devised, including the methods for area and personnel monitoring and the recording of associated data. Disposal of radioactive wastes was considered from the standpoint of release to air and water, and the activities of the section included evaluation of the results of physical measurements in terms of health hazards. Accordingly, studies were made of meteorological and aquatic dispersion, and various personnel protective measures and standards were devised.

The Health Physics Section was one of four administrative components of the Health Division headed by Stone, the others being the Medical, Biological Research, and Military Sections. The latter was short-lived and was formed to consider protective measures against military applications of radioactive materials by the enemy, a function that was soon taken over by the Army. Another function was to consider possible military use of radioactivity by the Allies, a study that was quickly abandoned.

The Biological Research Section was initially headed by Kenneth S. Cole, a biophysicist from the Columbia University College of Physicians and Surgeons. Cole had been consulted by Compton prior to the formation of the Health Division, and had been instrumental in the selection of Stone, a physician rather than a health physicist as director. The first staff members were Raymond E. Zirkle, a biophysicist, and Waldo E. Cohn, a biologist, who were joined shortly by H. J. Curtis and C. L. Prosser. The Biological Research Section was given the charter of obtaining data on the biological response to radiation; to establish tolerance doses; methods of diagnosis and detection of radiation injury on the basis of their studies and the discovery of treatment for personnel who might accidentally have been overexposed. The charter included both internal and external radiations, the former being perhaps more challenging because of the new nuclides and paucity of previous work. Underlying all studies was the goal of discovering the fundamental mechanism of radiation damage to biological systems, a goal which even today has not yet been realized.

The Medical Section was mainly concerned with the direct health of the project workforce. In this regard, industrial medical activities common at the time were instituted, but in addition, the section developed an extensive and comprehensive program of physiological monitoring, extending its efforts to devising appropriate special studies and tests to ascertain possible injury from relatively low levels of radiation and toxic materials. Emphasis was initially placed on hematology and kidney function, the latter studies to evaluate the possible chemotoxic effects of uranium. The studies were extended very quickly, however, to include liver function, enzyme behavior, urinalyses for radionuclides, and a variety of other tests. The Medical Section, with its studies on people, was complementary to the Biological Research Section in a sense, whose efforts were largely devoted to animal studies.

Selected to lead the Medical Section was Simeon T. Cantril, head of the radiology department at Swedish Hospital in Seattle. He was joined by Leon O. Jacobson, a hematologist from the University of Chicago who had been acting as liaison officer between the University and project. James J. Nickson, a physician associated with Memorial Hospital in New York City, served as assistant section head. Two industrial physicians, W. Daggett Norwood and Phillip A. Fuqua, both from the DuPont Corporation, joined as clinical staff. Norwood held not only a medical degree but one in electrical engineering and, hence, could effectively understand and reconcile the sometimes differing points of the medical and technical staff. In his later years, Norwood achieved a modicum of fame in a wholly different area, crossing the United States from coast to coast on foot.

The accomplishments of the Health Division during World War II were enormous and went into virtually every facet of radiological health. Even the environmental aspects were not ignored; for example, when the Hanford plant was being sited and designed, one of the major considerations was preventing radiological contamination of the environment. The Health Division was so successful in achieving its objective of ensuring that no one suffered serious injury from the radiological or toxicological substances peculiar to the effort, that some radiologists left the group, bored by the lack of injuries and unusual cases.

Some specific accomplishments included the universal adoption of preemployment and periodic physical examinations, particularly for those occupationally exposed to radiation. Hematological examination served as the principal criterion for evaluating possible acute and chronic overexposure, with monthly evaluations performed on radiation workers. Techniques were developed for analysis of activation products in blood and similar techniques were devised for ^{24}Na and ^{32}P in blood to evaluate neutron exposures. Monthly quantitative analyses for uranium and plutonium in urine were also done on radiation workers, and a comprehensive medical record form was devised and maintained for each worker. As a result of this intensive preventive program, many nonoccupational ailments were identified, a factor that permitted early treatment and improved the general health of the project workforce. Comparison of urological findings in various groups of exposed individuals was made, and no excess abnormalities were noted.

Other important areas of accomplishment were the setting of tolerance doses for both internal and external radiation and the development of monitoring instrumentation (Figure 2-5). The rep and the rem were both introduced to the Manhattan District by Herbert M. Parker, and the Maximum Permissible Concentration (MPC) concept was

FIGURE 2-5.   A "Fishpole" type survey meter used by radiation protection personnel in the Manhattan District.   This instrument measured β-γ radiations, weighed 5.5 lbs., and had a full scale range of 40 R/hr.   The health physicist using the instrument is wearing protective clothing and a respirator of the time.

applied to inhaled or ingested radionuclides. By 1943, some knowledge of the relative biological effectiveness (RBE) for both fast and slow neutrons had been obtained, and the animal research had revealed must important metabolic data, including lung clearance dynamics, important to the establishment of MPCs. It was indeed Parker again who made fundamental contributions in the MPC area, putting forth the first MPC for airborne plutonium in April 1944. Parker's proposed original MPC for Pu was 3.1×10^{-11} $\mu Ci/cm^3$ and was based on the dose to lung tissue. The current $(MPC)_a$ for insoluble ^{239}Pu (calculated by Walter S. Snyder and colleagues in the Health Physics Division, Oak Ridge National Laboratory (ORNL) for the NCRP and the International Commission of Radiological Protection) takes the lung as the critical organ and is 4×10^{-11} $\mu Ci/cm^3$ for a 40-hr week — virtually identical with Parker's original value.

FIGURE 2-6. A portable ion chamber survey meter, available just prior to World War II. The case is open showing the cylindrical chamber. The unit was reasonably light weight and stable, using an electrometer. The manufacturer was Victoreen Instrument Corporation.

In the instrumentation area, an early development was the pocket ionization chamber, used in pairs to measure the daily radiation exposure of personnel. Initially, these were not too rugged but, fortunately, the errors tended to be in the direction of overestimation of dose. Film badges were introduced somewhat later and were used in conjunction with the pocket chambers. Although the concept of personnel film dosimetry could be traced back to Pfahler and Quimby in the 1920s, the technique of using filters and correlating optical density with dose was greatly refined by Manhattan District personnel, and nuclear emulsions were also brought into use for personnel neutron monitoring, as were film rings for monitoring hand doses.

At the start of the war, there was apparently but a single commercial portable radiation survey meter available in the United States, a rather bulky instrument manufactured by Victoreen Instrument Company (Figure 2-6). There was a rather limited market for such instruments, since laboratory instruments were used at universities and other radiation research facilities, albeit in crude fashion, to monitor radiation levels. What was probably the first and no doubt for a long time the only portable survey meter was an ionization chamber developed by Lauriston S. Taylor of the National Bureau of Standard in 1929. The instrument featured three interchangable chambers and was specifically designed for X-ray monitoring.

Thus was the situation with respect to portable beta-gamma survey meters that faced the Manhattan District scientists at the start of the project. There were, of course, thimble chambers (Fricke-Glasser dosimeters) and R-meters, but these were not suited to everyday rugged use. Alpha and neutron monitoring instruments were unknown for all practical purposes. Instrument development for radiation monitoring purposes was carried out at both Clinton Laboratories (Oak Ridge) and Hanford. By 1945, no less than four portable instruments had been developed for alpha monitoring alone, including the Pluto, Poppy, and Zeus. For neutron monitoring, a compensated ion chamber instrument was devised. Since the device had two chambers, identical except that one was boron lined, it was named Chang and Eng after the famous Siamese twins (Figure 2-7). The Chang and Eng was the heaviest of all the so-called portable instruments, weighing 80 lb. Also out of the Clinton Laboratories (later renamed Oak Ridge National Laboratory and then Holifield National Laboratory) came the Cutie Pie, a pistol grip β-γ radiation survey meter that survives in modernized form to this day.

Maximum permissible exposures or tolerance doses were established for virtually all types of radiations at 0.1 rem/day, essentially the same as the recommendation given in H18 as put forth by the U.S. Advisory Committee on X-Ray and Radium Protection. However, internal doses were also considered, and for these, the MPC concept was developed. For airborne activity from the more common and more hazardous β-γ emitters, this level was taken as 10^{-7} μCi/cm^3, and for water, 5×10^{-4} μCi/cm^3. Permissible surface contamination levels were also established; for β-γ activity on surfaces, the maximum permissible level was set at 500 counts per minute as measured with an Eck and Krebs counter or other "standard" instrument.

The conclusion of World War II marked the end of a 4 year period of enormous growth and development in the radiological sciences. For a short time, there was a sort of consolidation of results and then, with the formation of the Atomic Energy Commission (AEC) in 1946, efforts began to pick up again with the formation of the various national laboratories. At Oak Ridge National Laboratory, training activities were expanded in the health physics and related radiological sciences areas, and 1947 saw the first civilian trainees enter the formal training program. In 1948, the AEC fellowship program was established at ORNL, Oak Ridge, Tennessee under Karl Z. Morgan and in 1949, Elda E. Anderson (1904 to 1961) was put in charge of the program at ORNL. Graduate degree programs were later established at two universities the following year with 20 fellowship students at each. In 1953, the University of Washington was added with field training provided at the Hanford Laboratories, and by 1960, the list included seven universities and several hundred individuals had received training.

Almost immediately after its establishment, the Manhattan District began to encourage the use of radionuclides from reactors. Paul C. Aebersold (1910 to 1964), a biophysicist who became known as "Mr. Isotope", had been involved since January 1946 as head of the Isotopes Division for the Manhattan District at Oak Ridge, had published the first catalog "Availability of Radioactive Isotopes" in *Science* on June 14, 1946. The first shipment was 1 mCi of ^{14}C as BaCO$_3$ to the Barnard Free Skin and Cancer Hospital in St. Louis on August 2, 1946 for research into cancer producing processes under the direction of two physicians, E. V. Cowdry and William L. Simpson. Although the Barnard clinic was the first, several others, including James Franck, recipient of the 1925 Nobel Prize in Chemistry, received shipments soon afterward.

During the first year of distribution, more than 1100 individual radioisotope shipments were made to 160 institutions in 31 states and Hawaii, which was a territory at that time. Of the 60 or so different nuclides, ^{131}I and ^{32}P were the most popular. Prior to making any shipments, a suitable shipping container was designed. This was the accomplishment of a young Army Corporal, Myron B. Hawkins. Hawkins, now a private consultant in California, was a graduate engineer assigned to the Special Engineering Detachment at Oak Ridge and also designed the familiar magenta and heliotrope (yellow) three bladed radiation warning symbol.

FIGURE 2-7.   The Chang and Eng neutron monitor was a compensated ion chamber instrument used for measuring fast neutrons. Readings were made through the microscope eyepiece. Several different sensitivity ranges were provided, permitting measurement of levels as high as about 300 mrem/hr. The instrument took 30 sec to respond and weighed 80 lb.

Following formation of the Atomic Energy Commission, the radiological health sciences were encouraged by a progressive policy and, for the time, munificent funding levels. In early 1947, Shields Warren, the noted Boston radiologist and pathologist who had served so well during the war, became the head of the Division of Biology and Medicine. This division paid special attention during its early days to the development of radiation protection programs and the training of personnel to staff them. Project Gabriel

— a program to evaluate the worldwide effects of ^{90}Sr fallout — was initiated, concluding in its unpublished report that the hazard at that time was not appreciable. The Gabriel study had been initially organized by Lauriston Taylor after conversations with Shields Warren of the AEC.

Almost immediately after the bombs were dropped over Japan, studies began on the survivors as well as on the physical effects. On August 11, 1945, 2 days after the bombing of Nagasaki, Colonel Stafford L. Warren received orders from General Groves to organize a survey team and proceed to Guam with the ultimate destination being Japan. Warren arrived in Hiroshima on September 8 to begin the studies of the Hiroshima and Nagasaki bombings.

The initial studies were concerned primarily with immediate clinical effects of radiation, as well as blast and other prompt effects. The limitations of these early studies performed by the Army under difficult conditions with limited funding and personnel were of concern to the National Academy of Sciences in 1946. Indeed, the Army advocated long-term studies, but was unable to fund them. Thus, in early 1947, the Academy formed a committee, and with AEC funding and other support, began a long-range study of nearly 100,000 survivors of the Hiroshima and Nagasaki bombings. This led to the formation of the Atomic Bomb Casualty Commission (ABCC) which has carried on the study of the survivors and their progeny for more than 30 years. The initial funding level of only $100,000 grew to $450,000 the next year, $1,400,000 in 1949, and $1,900,000 in 1950. About 300 full-time personnel, approximately half of them Japanese, were employed by the ABCC in 1950.

The ABCC had collected data on more than 150,000 persons by 1950. The data revealed a slight increase in leukemia among survivors and high incidence of cataracts among those receiving high doses. Data were obtained by the genetic group on 20,000 births, and this data would be of great value in setting population dose limits at some future time.

Attention had been directed to the potential genetic dangers by geneticist Herman J. Muller (1890 to 1967), whose discovery of the mutagenic effects of X-rays on fruit flies in the 1920s led to the Nobel Prize in 1946. Muller began a series of public lectures under the aegis of the honorary science fraternity, Society of Sigma Xi, during 1945 and 1946. Muller's task was to inform the public about radiation hazards and he was particularly sanguine in the area of low-level medical exposures. Muller's lectures undoubtedly led to recognition of the genetic hazards by many scientists who would have otherwise paid scant attention to effects other than somatic effects. Indeed, the population dose limits of the British Medical Research Council in 1956 and the NCRP in 1957 clearly reflect this awareness of the genetic effects, and the 1958 report of ICRP — its first publication — was even more definitive.

Concern for radiation protection, particularly internal emitters, led to the Tri-Partite Conferences, attended by representatives of the U.S., Canada, and United Kingdom. Only three conferences were held (October 1949, July 1950, and March 1953), but from these came a realistic and practical lung model and MPC's for most of the important radionuclides. The MPC's established at the 1950 meeting at Harwell were incorporated into the 1950 recommendations of the International Commission on Radiological Protection.

The tolerance dose of 0.1 R/day proposed in the 1930s and used throughout the Manhattan District was reexamined in the early postwar period. In 1949, the National Committee (now Council) on Radiation Protection and Measurement (NCRP and formerly Advisory Committee on X-Ray and Radium Protection, ACXRP) put forth a recommended occupational exposure limit of 15 rem/year (0.3 rem/week) whole body and also introduced the benefit vs. risk concept to radiation protection. The ICRP followed in 1950 as did the AEC, incorporating this level into their regulations. In 1953, the International Commission on Radiological Units and Measurements (ICRU) introduced the concept of absorbed dose, defining it as energy per unit mass imparted to matter by irradiation and providing an appropriate unit, the rad; 1 rad was defined as 100 ergs imparted to 1 g of absorbing medium, irrespective of the type or quality of the radiation in question.

Continued study and concern regarding exposure to radiation produced a joint recommendation by the National Academy of Sciences and ICRP in 1956 to lower the basic permissible dose to radiation workers to 5 rem/year. The following year, the NCRP introduced the age proration concept for occupational exposure and also recommended a limit of 0.5 rem/year for persons other than those occupationally exposed. Numerous studies of fallout effects were made, but perhaps the most significant study was published in 1958 by the United Nations Scientific Committee on the Effects of Atomic Radiation (UNSCEAR). The UNSCEAR Report tabulated sources of radiation exposure and evaluated the biological hazards on the basis of known doses and documented effects. This pioneering report was, in effect, a synthesis of knowledge accumulated to that time with regard to radiation effects and the biological bases for dose limits.

The Atomic Age had opened many new areas of science. Throughout the 1950s, the radiological health sciences grew as did the use of radiation, producing a maturation and sophistication of effort and results that may yet be too new to be treated with objectivity and historical perspective. Suffice it to say that the concern for radiological health has alway been great, and by 1950, an estimated 500 persons were involved on a professional level in full-time radiation protection work. Thus, the health physicist emerged as a new professional specialty, with the Health Physics Society forming in 1955 and growing to more than 3500 members by 1975. Other societies such as the Radiation Research Society (1952), Society of Nuclear Medicine (1958), and the American Association of Physicists in Medicine (1960) were also formed during the decade of the 1950s, signalling the arrival of the modern era of the radiological health sciences. A few years later, the International Radiation Protection Association came into being, and this body has promoted the common interest and research in the radiological health sciences without reference to national boundaries.

2.11. BIBLIOGRAPHIC EPILOGUE

More than 400 original sources were consulted in the preparation of this chapter with reference being made to original works wherever possible. To avoid a ponderous reference list of little interest to most readers, the bibliography was restricted to 20 listings selected to provide those desirous of greater in-depth study with a small library, which should satisfy all but the most demanding requirements for historical perspective in the science and professional practice of radiation protection and measurement. The works cited were consulted frequently and used freely in the preparation of the chapter. These references also provide hundreds of further entries from the literature for any reader desiring to examine original sources on specific topics of interest.

ACKNOWLEDGMENTS

The author is indebted to Allen Brodsky, Karl Z. Morgan, and especially Lauriston S. Taylor for their review of this chapter in draft form and for their helpful suggestions and comments. The author is also grateful for discussions with the late E. Dale Trout, who informally made many suggestions and permitted photographs of his historical collection.

REFERENCES

1. **Bizzell, O. M.,** Early history of radioisotopes from reactors, *Isot. Radiat. Technol.,* 4, 25–32, 1968.
2. **Bleich, A. R.,** *The Story of X-Rays from Röntgen to Isotopes,* Dover Publications, New York, 1960.
3. **Brecher, R. and Brecher, E.,** *The Rays,* Williams and Wilkins, Baltimore, 1969.
4. **Brown, P.,** *American Martyrs to Science Through the Roentgen Rays,* Charles C Thomas, Springfield, Ill., 1936.
5. **Bushong, S. C.,** *History of Protection in Diagnostic Radiology,* CRC Press, Cleveland, 1972.
6. **Curie, E.,** *Madame Curie,* Doubleday, New York, 1938.
7. **Curie, M.,** *Pierre Curie,* Macmillan, New York, 1923.
8. **Davis, K. S.,** History of radium, *Radiology,* 2, 334–342, 1924.
9. **Glasser, O., Ed.,** *The Science of Radiology,* Charles C Thomas, Springfield, Ill., 1933.
10. **Glasser, O.,** *William Conrad Roentgen and the Early History of the Roentgen Rays,* Charles C Thomas, Springfield, Ill., 1934.
11. **Glasser, O.,** The geneology of the roentgen rays, *Radiology,* 30, 180–200, 349–367, 1933.
12. **Grigg, E. R. N.,** *The Trail of the Invisible Light,* Charles C Thomas, Springfield, 1968.
13. **Hickey, P. M.,** "The First Decade of American Roentgenology," *Am. J. Roentgenol. Radium Ther.,* 201, 150–157, 249–356, 1928.
14. **Kathren, R. L.,** Early X-ray protection in the United States, *Health Phys.,* 8, 503–511, 1962.
15. **Kay, G. W. C.,** *Roentgenology,* Paul B. Hoeber, New York, 1928.
16. **Klickstein, H. S.,** *The Mallinckrodt Classics of Radiology,* Vol. 1 and 2, 1966.
17. **Stone, R. S.,** *Industrial Medicine on the Plutonium Project,* McGraw-Hill, New York, 1951.
18. **Taylor, L. S.,** "Brief History of the National Committee on Radiation Protection and Measurements (NCRP) Covering the Period 1929–1946," *Health Phys.,* 1, 3–10, 1958.
19. **Taylor, L. S.,** *Radiation Protection Standards,* CRC Press, Cleveland, 1971.
20. **Warren, S. L.,** "The Role of Radiology in the Development of the Atomic Bomb," in *Radiology in World War II,* Ahnfeldt, A. L., Ed., Office of the Surgeon General, U.S. Army, 1966, 831–935.

Physical Data

This section presents a selection of data and constants from the physical science handbooks and references indicated. These data are chosen for their usefulness in applications of radiation measurement and protection. More exhaustive data for specific research applications may be obtained from the references listed after each subsection. The subsection on nuclear cross sections by B. L. Cohen and the one on stopping power by J. E. Turner also review briefly the state of knowledge available on these parameters of frequent use in radiation measurement and dosimetry. The subsection by L. T. Dillman presents methods for converting nuclear decay scheme data into particle and quantum frequencies and energies useful for internal dose calculations. This section is primarily intended to contain, in a concise location, the more fundamental physics parameters that can be precisely measured. Additional related data will be found in the supplementary appendix to this handbook and in sections covering subject matter for which the particular data is most used.

3.1. FUNDAMENTAL CONSTANTS

Allen Brodsky

The following prefixes are frequently used in the field of radiation protection and are presented here for easy reference (see also Table 3.1-4).*

Prefix		Abbreviation
deci	$(= 10^{-1})$	d
centi	$(= 10^{-2})$	c
milli	$(= 10^{-3})$	m
micro	$(= 10^{-6})$	μ
nano	$(= 10^{-9})$	n
pico	$(= 10^{-12})$	p
femto	$(= 10^{-15})$	f
atto	$(= 10^{-18})$	a
deka	$(= 10)$	da
hecto	$(= 10^2)$	h
kilo	$(= 10^3)$	k
mega	$(= 10^6)$	M
giga	$(= 10^9)$	G
tera	$(= 10^{12})$	T

The following table (Table 3.1-1)† of fundamental constants is taken from the *Radiological Health Handbook*[1] and includes estimates of precision and units in both the meter-kilogram-second (MKS) and centimeter-gram-second (cgs) systems. An additional table of fundamental constants, including the new system internationale (SI) system of units as well as the cgs units, are given in Table 3.1-2 for comparison in checking with the values of Table 3.1-1.[2]

During the past few years, the SI system has been developed to include specific units of radioactivity and radiation dose measurement. These SI units of radiation dose and radioactivity measurement have not yet been put into practice in most installations but are currently in use in forthcoming publications of the International Commission of Radiological Protection (ICRP) and the International Commission on Radiological Units (ICRU). Therefore, most of the data in this handbook will be presented in the units that are used in daily practice; conversion factors are included in this section of the handbook for conversion of values from one system of units to another when needed.

Table 3.1-3 is reproduced from the *Federal Register* to summarize the current status of the SI system of units as presented by the National Bureau of Standards in the U.S.[3] Table 3.1-4 presents some conversion factors from the *Bureau of Radiological Health Handbook* that are useful in converting between units frequently used in radiation protection and dose estimation calculations. A number in units given in the left column is multiplied by the number in the center column to obtain the equivalent quantities in units in the right column.

---

*   *The following table of prefixes is taken from Reference 1 and used with their permission.*
†   *All tables appear at the end of the text.*

Densities and mean ionization potentials of the elements are presented in Table 3.1-5, and densities of common materials that may be useful in shielding calculations are given in Table 3.1-6. Densities of other materials, that are useful as phantoms in depth-dose determinations, are presented in Sections 7.1 and 7.3. Other relationships between constants used in radiation protection applications may be found in appropriate sections of the handbook. The index should be searched for conversion factors not found in this section.

CONSTANTS

Quantity	Value (±)	MKSA	CGS
Speed of light	$c = 2.997\ 925\ 3$	10^8 m sec^{-1}	10^{10} cm sec^{-1}
Boltzmann constant	$k = 1.380\ 54\ 18$	10^{-23} J° K^{-1}	10^{-16} erg° K^{-1}
Mass hydrogen atom	$m_H = 1.673\ 43\ 8$	10^{-27} kg	10^{-24} g
Proton mass	$m_p = 1.672\ 52\ 8$	10^{-27} kg	10^{-24} g
	$1.007\ 276\ 62\ 8$	u	u
Neutron	$m_n = 1.674\ 82\ 8$	10^{-27} kg	10^{-24} g
	$1.008\ 665\ 20\ 10$	u	u
Electron mass	$m_e = 9.109\ 1\ 4$	10^{-31} kg	10^{-28} g
	$5.485\ 97\ 3$	10^{-4} u	10^{-4} u
	$m_p m_e = 1.836\ 10\ 3$	10^3	10^3
Charge of positron	$e = 1.602\ 10\ 7$	10^{-19} C	
	$e = 4.802\ 98\ 20$		10^{-10} esu
	$e/c = 1.602\ 10\ 7$		10^{-20} emu
Charge to mass ratio	$e/m = 1.758\ 796\ 19$	10^{11} C kg^{-1}	
	$e/m = 5.272\ 74\ 6$		10^{17} esu g^{-1}
	$e/mc = 1.758\ 796\ 19$		10^7 emu g^{-1}
Electron radius	$r_e = 2.817\ 77\ 11$	10^{-16} m	10^{-13} cm
Thomson cross section	$(8\pi/3)r_e^2 = 6.651\ 6\ 5$	10^{-29} m^2	10^{-25} cm^2
Zeeman splitting constant	$e/4\pi mc = 4.668\ 58$	10^1 m^{-1} T^{-1}	

Table 3.1-1 (continued)
CONSTANTS

Quantity	Value (±)	MKSA	CGS
Planck constant	$e/4\pi mc^2$ = 4.668 58 / 4		10^{-5} cm^{-1} G^{-1}
	h = 6.625 6 / 5	10^{-34} J sec	10^{-27} erg sec
	$h/2\pi = \hbar$ = 1.054 50 / 7	10^{-34} J sec	10^{-27} erg sec
	h/e = 4.135 56 / 12	10^{-15} J sec C^{-1}	
	h/e = 1.397 47 / 4		10^{-17} erg sec esu^{-1}
	hc/e = 4.135 56 / 12		10^{-7} erg sec emu^{-1}
	h/k = 4.799 3 / 6	10^{-11} s K	10^{-11} sec K
1st radiation constant	$c_1 = 2\pi hc^2$ = 3.741 5 / 3	10^{-16} W m^2	10^{-5} erg cm^2 s^{-1}
2nd radiation constant	$c_2 = hc/k$ = 1.438 79 / 19	10^{-2} m K	cm K
Wien's radiation law	λ_{max} T = c_2/4.965 114 23 = 2.897 8 / 4	10^{-3} m K	10^{-1} cm K
Stefan-Boltzmann constant	σ = 5.669 7 / 2 9	10^{-8} W m^{-2} K^{-4}	10^{-5} erg cm^{-2} sec^{-1} K^{-4}
Fine structure constant	α = 7.297 20 / 10	10^{-3}	10^{-3}
	α^{-1} = 1.370 388 / 19	10^2	10^2
	α^2 = 5.324 92 / 14	10^{-5}	10^{-5}
Bohr radius	a_0 = 5.291 67 / 7	10^{-11} m	10^{-9} cm
Rydberg constant	$R\infty$ = 1.097 373 1 / 3	10^7 m^{-1}	10^5 cm^{-1}
	R_H = 1.096 775 8 / 3	10^7 m^{-1}	10^5 cm^{-1}

CONSTANTS

Quantity	Value (±)	MKSA	CGS
	$R_\infty hc = 2.179\ 72\ (17)$	10^{-18} J	10^{-11} erg
Bohr magneton	$\mu_B = 9.273\ 2\ (6)$	10^{-24} J T^{-1}	10^{-21} erg G^{-1}
Magnetic moment of electron	$\mu_e = 9.284\ 0\ (6)$	10^{-24} J T^{-1}	10^{-21} erg G^{-1}
	$\mu_e/\mu_B = 1.001\ 159\ 615\ (15)$		
Nuclear magneton	$\mu_N = 5.050\ 5\ (4)$	10^{-27} J T^{-1}	10^{-24} erg G^{-1}
Magnetic moment of proton	$\mu_p = 1.410\ 49\ (13)$	10^{-26} J T^{-1}	10^{-23} erg G^{-1}
	$\mu_p/\mu_N = 2.792\ 76\ (7)$		
Gyromagnetic ratio of proton	$\gamma_p = 2.675\ 19\ (2)$	10^{8} sec^{-1} T^{-1}	10^{4} sec^{-1} G^{-1}
Compton wave lengths Of electron	$\lambda_{Ce} = h/m_e = 2.426\ 21\ (6)$	10^{-12} m	10^{-10} cm
	$\lambda_C/2\pi = 3.861\ 44\ (9)$	10^{-13} m	10^{-11} cm
Of proton	$\lambda_{Cp} = h/m_pc = 1.321\ 40\ (4)$	10^{-15} m	10^{-13} cm
	$\lambda_{Cp}/2\pi = 2.103\ 07\ (6)$	10^{-16} m	10^{-14} cm
Of neutron	$\lambda_{Cn} = h/m_n c = 1.319\ 58\ (4)$	10^{-15} m	10^{-13} cm
	$\lambda_{Cn}/2\pi = 2.100\ 18$	10^{-16} m	10^{-14} cm
Avogadro constant	$N_A = 6.022\ 52\ (28)$	10^{23} mol^{-1}	10^{23} mol^{-1}
Molar volume of ideal gas at stp	$V_m = 2.241\ 36\ (30)$	10^{-2} m^3 mol^{-1}	10^{4} cm^3 mol^{-1}
Molar gas constant	$R = 8.314\ 3\ (12)$	J mol^{-1} K^{-1}	10^{7} erg mol^{-1} K^{-1}

Table 3.1-1 (continued)
CONSTANTS

Quantity	Value (±)	MKSA	CGS
Faraday constant	$F = N_A e = 9.648\ 70\ 16$	10^4 C mol^{-1}	10^{14} esu mol^{-1}
	$F = N_A e = 2.892\ 61\ 5$		10^3 emu mol^{-1}
	$F/c = N_A e/c = 9.648\ 70\ 16$		
Curie	$Ci = 3.7 \times 10^{10}$ dps		
Base of natural logarithm	$e = 2.718\ 281\ 828\ 4$		
Gravitational acceleration	$g = 9.806\ 65$	m sec^{-2}	10^2 cm sec^{-2}
Pi	$\pi = 3.141\ 592\ 653\ 59$		
Roentgen	$R = 2.58 \times 10^{-4}$ C kg^{-1}		
Energy equivalent of electron mass	$mc^2 =$	0.51 MeV	
Wave-length associated with 1 eV	$\lambda_o = 1.239\ 81$	10^{-6} m	10^{-4} cm
Ratio of chemical to unified mass scales	$r = M(O = 16)/M(^{12}C = 12) = 1.000\ 043\ 5$		
	$r = M(^{16}O = 16)/M(^{12}C = 12) = 1.000\ 317\ 92\ 2$		
Mass unit, unified mass scale	$u = 1/N_A = 1.660\ 43$	10^{-27} kg	10^{-24} g

From Bureau of Radiological Health, *Radiological Health Handbook*, U.S. Department of Health, Education, and Welfare, Public Health Service, Rockville, Md., 1970. With permission.

Table 3.1-2

FUNDAMENTAL PHYSICAL PARAMETERS

Fundamental Physical Constants

Quantity	Symbol	Value	Error (ppm)	Units (SI)	Units (cgs)
Velocity of light	c	2.9979250(10)	0.33	10^8 m sec^{-1}	10^{10} cm sec^{-1}
Fine-structure constant, $\{\mu_o c^2/4\pi\}(e^2/hc)$	α	7.297351(11)	1.5	10^{-3}	10^{-3}
	α^{-1}	137.03602(21)	1.5		
Electron charge	e	1.6021917(70)	4.4	10^{-19} C	10^{-20} emu
		4.803250(21)	4.4		10^{-10} esu
Planck's constant	h	6.626196(50)	7.6	10^{-34} J · sec	10^{-27} erg · sec
	$\hslash = h/2\pi$	1.0545919(80)	7.6	10^{-34} J · sec	10^{-27} erg · sec
Avogadro's number	N	6.022169(40)	6.6	10^{26} kmol^{-1}	10^{23} mol^{-1}
Atomic mass unit	amu	1.660531(11)	6.6	10^{-27} kg	10^{-24} g
Electron rest mass	m_e	9.109558(54)	6.0	10^{-31} kg	10^{-28} g
	m_e^*	5.485930(34)	6.2	10^{-4} amu	10^{-4} amu
Proton rest mass	M_p	1.672614(11)	6.6	10^{-27} kg	10^{-24} g
	M_p^*	1.00727661(8)	0.08	amu	amu
Neutron rest mass	M_n	1.674920(11)	6.6	10^{-27} kg	10^{-24} g
	M_n^*	1.0086520(10)	0.10	amu	amu
Ratio of proton mass to electron mass	M_p/m_e	1836.109(11)	6.2		
Electron charge to mass ratio	e/m_e	1.7588028(54)	3.1	10^{11} C kg^{-1}	10^7 emu g^{-1}
		5.272759(16)	3.1		10^{17} esu g^{-1}
Magnetic flux quantum, $[c]^{-1}(hc/2e)$	Φ_o	2.0678538(69)	3.3	10^{-15} T · m^2	10^{-7} G · cm^2
	h/e	4.135708(14)	3.3	10^{-15} J · sec C^{-1}	10^{-7} erg · sec emu^{-1}
		1.3795234(46)	3.3		10^{-17} erg · sec esu^{-1}
Quantum of circulation	$h/2m_e$	3.636947(11)	3.1	10^{-4} J · sec kg^{-1}	erg · sec g^{-1}
	h/m_e	7.273894(22)	3.1	10^{-4} J · sec kg^{-1}	erg · sec g^{-1}
Faraday constant, Ne	F	9.648670(54)	5.5	10^7 C kmol^{-1}	10^3 emu mol^{-1}
		2.892599(16)	5.5		10^{14} esu mol^{-1}
Rydberg constant $[\mu_o C^2/4\pi]^2(m_e e^4 h^3 c)$	$R\infty$	1.09737312(11)	0.10	10^7 m^{-1}	10^5 cm^{-1}
Bohr radius, $[\mu_o c^2/4\pi]^{-1} h^2/ m_e e^2) = \alpha/4\pi R\infty$	a_0	5.2917715(81)	1.5	10^{-11} m	10^{-9} cm
Classical electron radius $\{\mu_o c^2/4\pi\}(e^2/ m_e c^2) = \alpha^3/4\pi R\infty$	r_0	2.817939(13)	4.6	10^{-15} m	10^{-13} cm

Table 3.1-2 (continued)
FUNDAMENTAL PHYSICAL PARAMETERS

Fundamental Physical Constants

Quantity	Symbol	Value	Error (ppm)	SI	cgs
Electron magnetic moment in Bohr magnetons	μ_e/μ_B	1.0011596389(31)	0.0031		
Bohr magneton, {c}(eh/2m$_e$c)	μ_B	9.274096(65)	7.0	10^{-24} J T^{-1}	10^{-21} erg G^{-1}
Electron magnetic moment	μ_e	9.284851(65)	7.0	10^{-24} J T^{-1}	10^{-21} erg G^{-1}
Gyromagnetic ratio of protons in H$_2$O	γ'_p	2.6751270(82)	3.1	10^8 rad sec^{-1} · T^{-1}	10^4 rad sec^{-1} · G^{-1}
	$\gamma'_p/2\pi$	4.257597(13)	3.1	10^7 Hz T^{-1}	10^4 Hz G^{-1}
γ_p corrected for diamagnetism of H$_2$O	γ_p	2.6751965(82)	3.1	10^8 rad sec^{-1} · T^{-1}	10^4 rad sec^{-1} · G^{-1}
	$\gamma_p/2\pi$	4.257707(13)	3.1	10^7 Hz T^{-1}	10^3 Hz G^{-1}
Magnetic moment of protons in H$_2$O in Bohr magnetons	μ'_p/μ_B	1.52099312(10)	0.066	10^{-3}	10^{-3}
Proton magnetic moment in Bohr magnetons	μ_p/μ_B	1.52103264(46)	0.30	10^{-3}	10^{-3}
Proton magnetic moment	μ_p	1.4106203(99)	7.0	10^{-28} J T^{-1}	10^{-23} erg G^{-1}
Magnetic moment of protons in H$_2$O in nuclear magnetons	μ'_p/μ_n	2.792709(17)	6.2		
$\mu'_p \mu_n$ corrected for diamagnetism of H$_2$O	μ_p/μ_n	2.792782(17)	6.2		
Nuclear magneton [c] (eh2 M$_p$c)	μ_n	5.050951(50)	10	10^{-21} J T^{-1}	10^{-24} erg G^{-1}
Compton wavelength of the electron h/m$_e$c	λ_c	2.4263096(74)	3.1	10^{-12} m	10^{10} cm
	$\lambda_d/2\pi$	3.861592(12)	3.1	10^{-13} m	10^{-11} cm
Compton wavelength of the proton, h/M$_p$c	$\lambda_{c,p}$	1.3214409(90)	6.8	10^{-15} m	10^{-13} cm
	$\lambda_{c,p}/2\pi$	2.103139(14)	6.8	10^{-16} m	10^{-14} cm
Compton wavelength of the neutron, h/M$_n$c	$\lambda_{c,n}$	1.3196217(90)	6.8	10^{-15} m	10^{-13} cm
	$\lambda_{c,n}/2\pi$	2.100243(14)	6.8	10^{-16} m	10^{-14} cm
Gas constant	R_0	8.31434(35)	42	10^3 J kmol^{-1} · K^{-1}	10^7 erg mol^{-1} · K^{-1}
Boltzman's constant, R$_0$/N	k	1.380622(59)	43	10^{-23} J K^{-1}	10^{-16} erg K^{-1}
Stefan–Boltzman constant, $\pi^2 k^4/60\hbar^3 c^2$	σ	5.66961(96)	170	10^{-8} W m^{-2}K^4	10^{-5} erg sec^{-1} · cm^{-2} · K^{-4}
First radiation constant, $8\pi hc$	c_1	4.992579(38)	7.6	10^{-24} J · m	10^{-15} erg · cm
Second radiation constant, hc/k	c_2	1.438833(61)	43	10^{-2} m · K	cm · K
Gravitational constant	G	6.6732(31)	460	10^{-11} N · m^2 kg^{-2}	10^{-8} dyne · cm^2 g^{-2}

FUNDAMENTAL PHYSICAL PARAMETERS

Fundamental Physical Constants

Quantity	Symbol	Value	Error (ppm)	Units SI	cgs
kx-unit-to-angstrom conversion factor, $A = \lambda(A)/\lambda(kxu)$; $\lambda(CuK\alpha_1) \equiv 1.537400$ kxu	A	1.0020764(53)	5.3		
A*-to-angstrom conversion factor, $A = \lambda(A)/\lambda(A^*)$; $\lambda(WK\alpha_1) \equiv 0.2090100$ A*	A*	1.0000197(56)	5.6		

Energy Conversion Factors

Quantity	Value	Unit	Error (ppm)
1 kg	5.609538(24)	10^{29} MeV	4.4
1 amu	931.4812(52)	MeV	5.5
Electron mass	0.5110041(16)	MeV	3.1
Proton mass	938.2592(52)	MeV	5.5
Neutron mass	939.5527(52)	MeV	5.5
1 Electron volt	1.6021917(70)	10^{-19} J	4.4
		10^{-12} erg	
	2.4179659(81)	10^{14} Hz	3.3
	8.065465(27)	10^{3} m^{-1}	3.3
		10^{3} cm^{-1}	
	1.160485(49)	10^{4} K	42
Energy-wavelength conversion	1.239841(41)	10^{-6} eV · m	3.3
		10^{-4} eV · cm	
Rydberg constant, R∞	2.179914(17)	10^{-18} J	7.6
		10^{-11} erg	
	13.605826(45)	eV	3.3
	3.2898423(11)	10^{15} Hz	0.35

Table 3.1-2 (continued)
FUNDAMENTAL PHYSICAL PARAMETERS

Energy Conversion Factors

Quantity	Value	Unit	Error (ppm)
Bohr magneton, μ_B	1.578936(67)	10^5 K	43
	5.788381(18)	10^5 eV T^{-1}	3.1
	1.3996108(43)	10^{10} Hz T^{-1}	3.1
	46.68598(14)	m^{-1} · T^{-1}	3.1
		10^{-2} cm^{-1} · T^{-1}	
Nuclear magneton, μ_n	0.671733(29)	K T^{-1}	43
	3.152526(21)	10^{-8} eV T^{-1}	6.8
	7.622700(42)	10^6 Hz T^{-1}	5.5
	2.542659(14)	10^{-2} m^{-1} · T^{-1}	5.5
		10^{-4} cm^{-1} · T^{-1}	
	3.65846(16)	10^{-4} KT^{-1}	44
Gas constant, R_0	8.20562(35)	10^{-2} m^3 · atm kmol^{-1} · K^{-1}	42
Standard volume of ideal gas, V_0	22.4136	m^3 kmol^{-1}	

ᵃ Note that the unified atomic mass scale ^{12}C = 12 has been used throughout, that amu = atomic mass unit, C = coulomb, G = gauss, Hz = hertz = cycles per second, J = joule, K = kelvin (degrees kelvin), T = tesla (10^4 G), V = volt, and W = watt. In cases where formulas for constants are given (e.g., R), the relations are written as the product of two factors. The second factor, in parentheses, is the expression to be used when all quantities are expressed in cgs units, with the electron charge in electrostatic units. The first factor, in brackets, is to be included only if all quantities are expressed in SI units. The reader is reminded that with the exception of the auxiliary constants which have been taken to be exact, the uncertainties of these constants are correlated; therefore, the general law of error propagation must be used in calculating additional quantities requiring two or more of these constants.

Table 3.1-3*
METRIC SYSTEM OF WEIGHTS AND MEASURES

National Bureau of Standards

METRIC SYSTEM OF WEIGHTS AND MEASURES

Guidelines for Use

Section 403 of Pub. L. 93–380 states the policy of the United States to encourage educational agencies and institutions to prepare students to use the metric system of measurement as part of the regular education program and authorizes the U.S. Commissioner of Education to carry out a program of grants and contracts to fulfill this policy. Subsection 403 (a)(3) states, "For the purposes of this section, the term 'metric system of measurement' means the International System of Units as established by the General Conference of Weights and Measures in 1960 and interpreted or modified for the United States by the Secretary of Commerce." The National Bureau of Standards is responsible for "the custody, maintenance, and development of the national standards of measurement" (15 U.S.C. 272), and the Secretary has designated NBS to implement his responsibilities under subsection 403(a)(3). Pursuant to his authority under section 403, the U.S. Commissioner of Education has requested that NBS publish guidelines for use of the International System of Units, interpreted and modified for the United States. Accordingly, and in implementation of the Secretary's responsibilities under subsection 403(a)(3), the following tables and associated materials set forth guidelines for use of the International System of Units (hereinafter "SI"), as interpreted and modified for the United States by NBS on behalf of the Secretary of Commerce.

The SI is constructed from seven base units for independent quantities plus two supplementary units for plane angle and solid angle, listed in Table 1.

TABLE 1

Quantity	Name	Symbol
	SI BASE UNITS	
length	metre (meter)[1]	m
mass[2]	kilogram	kg
time	second	s
electric current	ampere	A
thermodynamic temperature[3]	kelvin	K
amount of substance	mole	mol
luminous intensity	candela	cd
	SI SUPPLEMENTARY UNITS	
plane angle	radian	rad
solid angle	steradian	sr

[1] Both spellings are acceptable.
[2] "Weight" is the commonly used term for "mass."
[3] It is acceptable to use the Celsius temperature (symbol t) defined by $t = T - T_0$ where T is the thermodynamic temperature, expressed in kelvins, and $T_0 = 273.15$ by definition. The unit "degree Celsius" is thus equal to the unit "kelvin" when used as an interval or difference of temperature. Celsius temperature is expressed in degrees Celsius (symbol °C).

Units for all other quantities are derived from these nine units. In Table 2 are listed 17 SI derived units with special names which were derived from the base and supplementary units in a coherent manner, which means in brief, that they are expressed as products and ratios of the nine base and supplementary units without numerical factors.

TABLE 2.—*SI derived units with special names*

Quantity	SI unit Name	Symbol	Expression in terms of other units
frequency	hertz	Hz	s⁻¹
force	newton	N	m·kg/s²
pressure, stress	pascal	Pa	N/m²
energy, work, quantity of heat.	joule	J	N·m
power, radiant flux	watt	W	J/s
quantity of electricity, electric charge.	coulomb	C	A·s
electric potential, potential difference, electromotive force.	volt	V	W/A
capacitance	farad	F	C/V
electric resistance	ohm	Ω	V/A
conductance	siemens	S	A/V
magnetic flux	weber	Wb	V·s
magnetic flux density.	tesla	T	Wb/m²
inductance	henry	H	Wb/A
luminous flux	lumen	lm	cd·sr
illuminance	lux	lx	lm/m²
activity (radioactive).	becquerel	Bq	s⁻¹
absorbed dose	gray	Gy	J/kg

All other SI derived units, such as those in tables 3 and 4, are similarly derived in a coherent manner from the 26 base, supplementary, and special-name SI units.

TABLE 3.—*Examples of SI derived units, expressed in terms of base units*

Quantity	SI unit	Unit symbol
area	square metre	m²
volume	cubic metre	m³
speed, velocity	metre per second	m/s
acceleration	metre per second squared.	m/s²
wave number	1 per metre	m⁻¹
density, mass density.	kilogram per cubic metre.	kg/m³
current density	ampere per square metre.	A/m²
magnetic field strength.	ampere per metre	A/m
concentration (of amount of substance).	mole per cubic metre.	mol/m³
specific volume	cubic metre per kilogram.	m³/kg
luminance	candela per square metre.	cd/m²

TABLE 4.—*Examples of SI derived units expressed by means of special names*

Quantity	Name	Unit symbol
dynamic viscosity	pascal second	Pa·s
moment of force	metre newton	N·m
surface tension	newton per metre	N/m
heat flux density, irradiance.	watt per square metre	W/m²
heat capacity, entropy.	joule per kelvin	J/K
specific heat capacity, specific entropy.	joule per kilogram kelvin.	J/(kg·K)
specific energy	joule per kilogram	J/kg
thermal conductivity.	watt per metre kelvin.	W/(m·K)
energy density	joule per cubic metre.	J/m³
electric field strength.	volt per metre	V/m
electric charge density.	coulomb per cubic metre.	C/m³
electric flux density.	coulomb per square metre.	C/m²
permittivity	farad per metre	F/m
permeability	henry per metre	H/m
molar energy	joule per mole	J/mol
molar entropy, molar heat capacity.	joule per mole kelvin.	J/(mol·K)

For use with the SI units there is a set of 16 prefixes (see table 5) to form multiples and submultiples of these units.

TABLE 5.—*SI prefixes*

Factor	Prefix	Symbol
10¹⁸	exa	E
10¹⁵	peta	P
10¹²	tera	T
10⁹	giga	G
10⁶	mega	M
10³	kilo	k
10²	hecto	h
10¹	deka	da
10⁻¹	deci	d
10⁻²	centi	c
10⁻³	milli	m
10⁻⁶	micro	μ
10⁻⁹	nano	n
10⁻¹²	pico	p
10⁻¹⁵	femto	f
10⁻¹⁸	atto	a

Certain units which are not part of the SI are used so widely that it is impractical to abandon them. The units that are accepted for continued use in the United States with the International System are listed in table 6.

TABLE 6.—*Units in use with the international system*

Name	Symbol	Value in SI unit
minute	min	1 min = 60 s
hour	h	1 h = 60 min = 3 600 s
day	d	1 d = 24 h = 86 400 s
degree	°	1° = (π/180) rad
minute	′	1′ = (1/60)° = (π/10 800) rad
second	″	1″ = (1/60)′ = (π/648 000) rad
litre (liter)[1]	l	1 l = 1 dm³ = 10⁻³ m³
metric ton or tonne	t	1 t = 10³ kg

[1] Both spellings are acceptable.

In those cases where their usage is already well established, the use, for a limited time, of the following units is accepted, subject to future review.

nautical mile	hectare	gal [1]
knot	barn	curie
angstrom	bar	rontgen
standard atmosphere	are	rad

[1] Not gallon.

Metric units and their symbols other than those enumerated above are not part of the International System of Units. Accordingly, the following units and terms listed in the table of metric units in section 2 of the act of July 28, 1866, that legalized the metric system of weights and measures in the United States, are no longer accepted for use in the United States:

myriameter
stere
millier or tonneau
quintal
myriagram
kilo (for kilogram)

For more information regarding the International System of Units, contact the Metric Information Office, National Bureau of Standards, U.S. Department of Commerce, Washington, D.C. 20234.

Dated: June 1, 1975.

RICHARD W. ROBERTS,
Director.

[FR Doc.75–15798 Filed 6–18–75;8:45 am]

Note: The kilogram is the only SI unit with a prefix. Because double prefixes are not to be used, the prefixes of Table 5, in the case of mass, are to be used with gram and not with kilogram.

From NBS Guidelines for Use of the Metric System, LC1056, U.S. Department of Commerce-National Bureau of Standards, Washington, D.C., 1975, as given in *Fed. Regist.*, 40(119), 1975. With permission.

Table 3.1-4
CONVERSION FACTORS

A	B	C

Area[a]

A	B	C
	10^{-24}	cm^2
circular mils	7.854×10^{-7}	in.^{2}
cm^2	10^{24}	
cm^2	0.1550	in.^{2}
cm^2	1.076×10^{-3}	ft^2
cm^2	10^{-4}	m^2
ft^2	929.0	cm^2
ft^2	144	in.^{2}
ft^2	9.290×10^{-2}	m^2
in.^{2}	6.452	cm^2
in.^{2}	6.944×10^{-3}	ft^2
in.^{2}	6.452×10^{-4}	m^2
m^2	1550	in.^{2}
m^2	10.76	ft^2
m^2	1.196	yd^2
m^2	3.861×10^{-7}	mi^2

Density[a]

A	B	C
cm^3	1.602×10^{-2}	ft^3/lb
ft^3/lb	62.43	cm^3/g
g/cm^3	62.43	lb/ft^3
lb/ft^3	1.602×10^{-2}	g/cm^3
$lb/in.^3$	27.68	g/cm^3
lb/gal	0.1198	g/cm^3

Electrical[a,b]

A	B	C
A	1	C
A	2.998×10^9	esu/sec
A	6.2418×10^{18}	electrons/sec
A-hr	3600.0	C
A-hr	0.03731	Faradays
C	2.998×10^9	statcoulombs
C	6.2418×10^{18}	electronic charges
C	1.036×10^{-5}	Faradays
Faradays/sec	9.650×10^4	A
Faradays	26.80	A-hr
Faradays	9.650×10^4	C
F	10^6	farads
International A	0.999835	A (absolute)
International V	1.00033	V (absolute)
International Ω	1.000495	Ω (absolute)
International V farady	9.654×10^4	J
μfarads	10^{-6}	F
μΩ	10^{-12}	MΩ
μΩ	10^{-6}	Ω
W	1	J/sec

Energy[a]

A	B	C
Btu	1.0548×10^3	J (absolute)
Btu	0.25198	kg-cal
Btu	1.0548×10^{10}	ergs
Btu	2.930×10^{-4}	kW-hr

Table 3.1-4 (continued)
CONVERSION FACTORS

A	B	C
Btu	0.556	g-cal/g
eV	1.6021×10^{-12}	ergs
eV	1.6021×10^{-19}	J (abs)
eV	10^{-3}	keV
eV	10^{-6}	MeV
ergs	10^{-7}	J (abs)
ergs	6.2418×10^{5}	MeV
ergs	6.2418×10^{11}	eV
ergs	1.0	dyne-cm
ergs	9.480×10^{-11}	Btu
ergs	7.375×10^{-8}	ft-lb
ergs	2.390×10^{-8}	g-cal
ergs	1.020×10^{-3}	g-cm
g-cal	3.968×10^{-3}	Btu
g-cal	4.186×10^{7}	ergs
J (abs)	10^{7}	ergs
J (abs)	0.7376	ft-lb
J (abs)	9.480×10^{-4}	Btu
g-cal/g	1.8	Btu/lb
kg-cal	3.968	Btu
kg-cal	3.087×10^{3}	ft-lb
ft-lb	1.356	J (abs)
ft-lb	3.239×10^{-4}	kg-cal
kw-hr	2.247×10^{19}	MeV
kW-hr	3.60×10^{-13}	ergs
MeV	1.6021×10^{-6}	ergs

Fission[a]

A	B	C
Btu	1.28×10^{-8}	^{235}U fissioned[c]
Btu	1.53×10^{-8}	^{235}U destroyed[c,d]
Btu	3.29×10^{13}	Fissions
Fission of 1 g ^{235}U	1	MW-days
Fissions	8.9058×10^{-18}	kWh
Fissions[c]	3.204×10^{-4}	ergs
kWh	2.7865×10^{17}	^{235}U fission neutrons[c]
kW/kg^{235}U	2.43×10^{10}	Average thermal neutron flux in fuel[c,e]
MW-days/ton U	1.174×10^{-4}	% U atoms fissioned[f]
MW/ton U	$2.68 \times 10^{10}/E^{g}$	Average thermal neutron flux in fuel[c,e]
Neutrons/kb	1×10^{21}	Neutrons/cm^2
W	3.121×10^{10}	Fissions/sec

Fluid Flow Rates[a]

A	B	C
cm^3/min	2.19×10^{-3}	ft^3/min
cm^3/sec	8.64×10^{-2}	m^3/day
cm^3/sec	1.585×10^{-2}	gal/min
cm^3/sec	3.60	l/hr
ft^3/min	4.72×10^{2}	cm^3/sec
ft^3/sec	4.488×10^{2}	gal/min
gal/min	2.228×10^{-3}	ft^3/sec
l/hr	0.278	cm^3/sec
l/min	15.851	gal/hr
m^3/day	11.57	cm^3/sec

Table 3.1-4 (continued)
CONVERSION FACTORS

A	B	C
yd³/min	0.450	ft³/sec
yd³/min	3.367	gal/sec
yd³/min	12.74	l/sec

Length*

Å	10^{-8}	cm
Å	10^{-10}	m
(μm)	10^{-3}	mm
μm	10^{-4}	cm
μm	10^{-6}	m
μm	3.937×10^{-5}	in.
mm	10^{-1}	cm
cm	0.3937	in.
cm	3.2808×10^{-2}	ft
cm	10^{-2}	m
m	39.370	in.
m	3.2808	ft
m	1.0936	yd
m	10^{-3}	km
m	6.2137×10^{-4}	mi
km	0.62137	mi
mils	10^{-3}	in.
mils	2.540×10^{-3}	cm
in.	10^3	mils
in.	2.5400	cm
ft	30.480	cm
rods	5.500	yd
mi	5280	ft
mi	1760	yd
mi	1.6094	km

Mass*

mg	10^{-3}	g
mg	3.527×10^{-5}	oz
mg	1.543×10^{-2}	gr
g	3.527×10^{-2}	oz
g	10^{-3}	kg
g	980.7	dynes
g	2.205×10^{-3}	lb
kg	2.205	lb
kg	p.0685	slugs
kg	9.807×10^5	dynes
lb	4.448×10^5	dynes
lb	453.592	g
lb	0.4536	kg
lb	16	oz
lb	0.0311	slugs
dynes	1.020×10^{-3}	g
dynes	2.248×10^{-6}	lb
u (unified-¹²C scale)	1.66043×10^{-27}	kg
amu (physical-¹⁶O scale)	1.65980×10^{-27}	kg
oz	28.35	g
oz	6.25×10^{-2}	lb

Table 3.1-4 (continued)
CONVERSION FACTORS

A	B	C

Miscellaneous*

Temperature
$$°C = (°F−32)/1.8 = (°F−32) 5/9$$
$$°F = 1.8°C + 32 = (9/5) °C + 32$$
$$°K = °C + 273.16$$
Wavelength to energy conversion
  keV = 12.40 per angstrom
  eV = $1.240×10^{-6}$ per meter

A	B	C
rad	57.296	degrees
eV	$1.78258×10^{-33}$	g
eV	$1.07356×10^{-9}$	u
erg	$1.11265×10^{-21}$	g
Proton masses	938.256	MeV
Neutron masses	939.550	MeV
Electron masses	511.006	keV
u (amu on ^{12}C scale)	931.478	MeV

Power*

A	B	C
Btu/hr	0.2162	ft-lb/sec
Btu/hr	0.0700	g-cal/sec
Btu/hr	$3.929×10^{-4}$	hp
Btu/hr	0.2930	W
Btu/min	12.97	ft-lb/sec
Btu/min	0.02357	hp
Btu/min	0.01758	kw
Btu/min	17.58	W
hp	42.42	Btu/min
hp	33,000	ft-lb/min
hp	550	ft-lb/sec
hp	10.69	kg-cal/min
hp	0.7457	kw
hp	$4.655×10^{15}$	MeV/sec
kg-cal/min	$9.356×10^{-2}$	hp
kw	14.33	kg-cal/min
kw	1.341	hp
kw	$6.243×10^{15}$	MeV/sec
W	10^7	ergs/sec
W	0.7376	ft-lb/sec
W	3.414	Btu/hr
W	0.05690	Btu/min
W	0.01433	kg-cal/min
ergs/sec	$5.688×10^{-9}$	Btu/min
ergs/sec	$4.425×10^{-6}$	ft-lb/min
ergs/sec	$1.433×10^{-9}$	kg-cal/min

Pressure*

A	B	C
atm	14.696	lb/in.²
atm	760	mm Hg (O°C)
atm	76.0	cm Hg (O° C)
atm	1.0133	bars
atm	$1.0332×10^3$	g/cm²
atm	29.921	in. Hg (O° C)
cm Hg	0.1934	lb/in.²
cm Hg	$1.316×10^{-2}$	atm
cm Hg	0.4465	ft of H_2O

Table 3.1-4 (continued)
CONVERSION FACTORS

A	B	C
in. Hg	0.4912	lb/in.^{2}
g/cm^2	1.4223×10^{-2}	lb/in.^{2}
bars	10^6	dynes/cm^2
bars	14.504	lb/in.^{2}
dynes/cm^2	1.4504×10^{-5}	lb/in.^{2}
dynes/cm^2	1.0197×10^{-3}	g/cm^2
lb/in.^{2}	27.673	in. of H$_2$O (4° C)
lb/in.^{2}	2.3066	ft of H$_2$O (4° C)
lb/in.^{2}	6.805×10^{-2}	atm
lb/in.^{2}	2.036	in. Hg (O° C)
lb/in.^{2}	5.1715	cm Hg
lb/in.^{2}	51.715	mm Hg
ft of H$_2$O	2.230	cm Hg
Ci	3.700×10^{10}	dis/sec

Radiological Units[b]

Ci	2.220×10^{12}	dis/min
Ci	10^3	mCi
Ci	10^6	μCi
Ci	10^{12}	pCi
Ci	10^{-3}	kCi
dis/min	4.505×10^{-10}	mCi
dis/min	4.505×10^{-7}	μCi
dis/sec	2.703×10^{-8}	mCi
dis/sec	2.703×10^{-5}	μCi
kCi	10^3	Ci
μCi	3.700×10^4	dis/sec
μCi	2.220×10^6	dis/min
mCi	3.700×10^7	dis/sec
mCi	2.220×10^9	dis/min
R	2.58×10^{-4}	C/kg of air
R	1	esu/cm^3 of air (stp)
R	2.082×10^9	ion prs/cm^3 of air (stp)
ft^3	1,728	in.^{3}
gal (U.S.)	231.0	in.^{3}
gal	0.13368	ft^3
l	33.8147	fluid oz
l	1.05671	qt
l	0.26418	gal
gm moles (gas)	22.4	l (stp)

[a] Multiply column A by column B to obtain column C. Divide column C by column B to obtain column A.

[b] Units are absolute unless otherwise noted.

[c] At 200 MeV per fission.

[d] Thermal neutron spectrum ($\alpha = 0.193$).

[e] σ (fission = 500 b).

[f] At 200 MeV per fission, in ^{235}U-^{238}U mixture of low ^{235} content.

[g] E = enrichment in grams ^{235}U per gram total. No other fissionable isotope present.

From Estabrook, G. M., *Nucleonics*, 18(11), 209, 1960, as given in Bureau of Radiological Health, *Radiological Health Handbook*, U. S. Department of Health, Education, and Welfare, Public Health Service, Rockville, Md., 1970. With permission.

Table 3.1-5
DENSITY OF ELEMENTS AND COMMON MATERIALS

Element	Atomic number	Atomic weight	Mean ionization potential	Density
H	1	1.00797	18.0	0.0586
He	2	4.0026	40.0	0.126
Li	3	6.939	39.032	0.534
Be	4	9.0122	56.0	1.8
B	5	10.811		2.34
C	6	12.01115	79.0	2.25
N	7	14.0067	92.0	0.808
O	8	15.9994	105.0	1.14
F	9	18.9984		1.11
Ne	10	20.183	130.016	1.2
Na	11	22.9898		0.971
Mg	12	24.312	156.4	1.74
Al	13	26.9815	163	2.699
Si	14	28.086		2.42
P	15	30.9738		1.82
S	16	32.064		2.07
Cl	17	35.453		1.56
Ar	18	39.948	240.0	1.40
K	19	39.102		0.87
Ca	20	40.08	200	1.55
Sc	21	44.956		3.02
Ti	22	47.90	225	4.5
V	23	50.942	254	5.96
Cr	24	51.996		7.1
Mn	25	54.9380		7.20
Fe	26	55.847	273	7.86
Co	27	58.9332	298	8.9
Ni	28	58.71	312	8.90
Cu	29	63.54	322	8.94
Zn	30	65.37	331	7.14
Ga	31	69.72		5.91
Ge	32	72.59		5.36
As	33	74.9216		5.73
Se	34	78.96		4.8
Br	35	79.909		3.12
Kr	36	83.80	493.68	2.6
Rb	37	85.47		1.53
Sr	38	87.62		2.54
Y	39	88.905		5.51
Zr	40	91.22		6.4
Nb	41	92.906	410	8.4
Mo	42	95.94	420	10.2
Tc	43	99		
Ru	44	101.07		12.2
Rh	45	102.905	450	12.5

Table 3.1-5 (continued)
DENSITY OF ELEMENTS AND COMMON MATERIALS

Element	Atomic number	Atomic weight	Mean ion-ization potential	Density
Pd	46	106.4	460	12.16
Ag	47	107.870	485	10.50
Cd	48	112.40	468.0	8.65
In	49	114.82	490	7.28
Sn	50	118.69	500	7.31
Sb	51	121.75		6.691
Te	52	127.60		6.24
I	53	126.9044		4.93
Xe	54	131.30	757.52	3.52
Cs	55	132.905		1.873
Ba	56	137.34		3.5
La	57	138.91		6.155
Ce	58	140.12		3.92
Pr	59	140.907		6.5
Nd	60	144.24		6.95
Pm	61	147		
Sm	62	150.35		7.8
Eu	63	151.96		5.24
Gd	64	157.25		
Tb	65	158.924		
Dy	66	162.50		8.56
Ho	67	164.930		
Er	68	167.26		4.77
Tm	69	168.934		
Yb	70	173.04		
Lu	71	174.97		
Hf	72	178.49		13.3
Ta	73	180.948	720	16.6
W	74	183.85	740	19.3
Re	75	186.2		20.53
Os	76	190.2		22.48
Ir	77	192.2	760	22.42
Pt	78	195.09	777	21.37
Au	79	196.967	786	19.32
Hg	80	200.59		13.546
Tl	81	204.37		11.85
Pb	82	207.19	818	11.35
Bi	83	208.980	826	9.747
Po	84	210		
At	85	210		
Rn	86	222		9.73
Fr	87	223		
Ra	88	226		
Ac	89	227		
Th	90	232.038		11.3

Table 3.1-5 (continued)
DENSITY OF ELEMENTS AND COMMON MATERIALS

Element	Atomic number	Atomic weight	Mean ion-ization potential	Density
Pa	91	231		
U	92	238.03	908	18.68
Np	93			
Pu	94			
Am	95			
Cm	96			
Bk	97			
Cf	98			
Es	99			
Fm	100			
Md	101			
No	102			
Lw	103			
Ku	104			

From Bureau of Radiological Health, *Radiological Health Handbook,* U.S. Department of Health, Education, and Welfare, Public Health Service, Rockville, Md., 1970. With permission.

Table 3.1-6
DENSITIES OF COMMON MATERIALS

Material	Density (gm/cm³)	Material	Density (gm/cm³)
Air	0.001293	Linoleum	1.18
Asbestos	2.0—2.8	Marble	2.47—2.86
Asphalt	1.1—1.5	Paraffin	0.87—0.91
Bone	1.7—2.0	Plaster, sand	1.54
Brick	1.4—2.5	Pressed wood:	
Cement	2.7—3.0	Pulp board	0.19
Clay	1.8—2.6	Sandstone	1.90
Concrete, siliceous	2.25—2.40	Slate	2.6—3.3
Ebonite	1.15	Tile	1.6—2.5
Gelatin	1.27	Water	1.000
Glass (common)	2.4—2.8	Water (heavy)	1.105
Glass (flint)	2.9—5.9	Wood:	
Granite	2.60—2.76	Oak	0.60—0.90
Graphite	2.30—2.72	White pine	0.35—0.50
Gypsum	2.31—2.33	Yellow pine	0.37—0.60
Limestone	1.87—2.76		

Data from medical X-ray protection up to three million volts, *Natl. Bur. Stand. Handb.,* 76, 1961; *Handbook of Chemistry and Physics,* 58th ed., Weast, R. C., Ed., CRC Press, Cleveland, 1977, F1; 1968 and Trout, E. D., Kelley, J. P., and Lucas, A. C., *Radiology,* 76,(2) 237—244, 1961; as given Bureau of Radiological Health, *Radiological Health Handbook,* U.S. Department of Health, Education, and Welfare, Public Health Service, Rockville, Md., 1970. With permission.

REFERENCES

1. **Bureau of Radiological Health,** *Radiological Health Handbook,* U.S. Department of Health, Education, and Welfare, Public Health Service, Rockville, Md., 1970.
2. **Gray, D. E., Ed.,** *American Institute of Physics Handbook,* 3rd ed., McGraw-Hill, New York, 1972, as adapted from Taylor, B. N., Parker, W. H., and Langenberg, D. N., *Rev. Mod. Phys.,* 41, 375, 1969.
3. *Fed. Regist.,* 40(119), 1975, as given in NBS Guidelines for Use of the Metric System, LC1056, U.S. Department of Commerce-National Bureau of Standards, Washington, D. C., 1975.

3.2. CHARTS OF NUCLIDES AND ELEMENTS

3.2.1 Chart of the Nuclides*

The following information on radioactive isotopes (nuclides) is reprinted here with the permission of the General Electric Company. Further details not included here and color-coded information on the nuclide chart may be obtained by ordering the complete "Chart of the Nuclides."† Additional nuclear data and radioactive decay schemes, particularly useful in dose calculations, may be found in Sections 3.3 to 3.5. Methods of utilizing this data in dose estimation and measurements are found in other sections. Some of the more important information for radiation protection and measurement is quoted below from the Chart of the Nuclides.††

---

* Holden, N. E. and Walker, F. W., Knolls Atomic Power Laboratory, Schenectady, N. Y., revised 1972.
† Copies may be obtained by contacting: Manager, Order Service, General Electric Company, Nuclear Energy Marketing Department, 175 Curtner Avenue, San Jose, California 95125; phone (408) 297-3000, Ext. 2251.
†† The Chart of the Nuclides can be found in the envelope at the end of the book.

NUCLIDES AND ISOTOPES *

INTRODUCTION

The earliest discussion of the atomic hypothesis is attributed to the ancient Greek philosophers who speculated about the mysteries of nature. In the fifth century B.C., Democritus believed that elementary substances (earth, water, fire and air) were formed by minute individual particles called atoms (the Greek word meaning indivisible). This vague philosophical speculation was given reality when John Dalton, between 1803 and 1808, showed how to determine the masses (atomic weight is the commonly used expression) of different atoms relative to one another.

In 1816, William Prout believed (based on the few atomic weights known) that all atomic weights were whole numbers and integral multiples of the atomic weight of hydrogen. He thought that all elements might be built up from hydrogen. His concept lost favor when elements such as chlorine were definitely shown to have noninteger atomic weights.

Periodic Properties of Elements

In 1869, Dmitri Mendeleev published a short note entitled, "The Correlation Between Properties of Elements and Their Atomic Weights." He arranged the elements in rows according to the magnitude of their atomic weights, beginning with the smallest weight. Elements that appeared in the same vertical column showed a remarkable similarity in their chemical properties. Mendeleev hypothesized that deviations from the expected periodicity were due to chemists' failure to discover some elements in nature. He predicted the properties of gallium, scandium, and germanium, which were subsequently discovered. Pairs of elements (for example, nickel and cobalt, iodine and tellurium) that did not fit the periodic properties of their columns were interchanged so that they would correspond. He argued that the atomic weight measurements for these elements must be in error. It is now known that the atomic number (see page 5), rather than the atomic weight, is the correct basis for the periodicity in the chemical properties of the elements. By coincidence, the list of elements ordered by atomic weight usually agrees with the list ordered by atomic number, except for the few cases observed by Mendeleev.

New Phenomena

Toward the end of the nineteenth century, the successes in chemistry, together with those of classical mechanics and electromagnetic theory, convinced some individuals that classical physics was a "closed book" and that workers in the field would henceforth merely advance existing knowledge to the next decimal place. This attitude changed in 1895 when Wilhelm Roentgen discovered X-rays and in 1896 when A. Henri Becquerel discovered natural radioactivity. Since such phenomena could not be explained by existing theories of matter, they created great interest.

In 1902, Ernest Rutherford and Frederick Soddy, in their theory of radioactive disintegration, proposed that radioactivity involves changes occurring within the atom. Their view met strong opposition because it was considered contrary to the established view on the permanency of the atom.

Early Models of Atomic Structure

Early experiments in the investigation of atomic structure disclosed three different types of radioactivity, called alpha, beta, and gamma radiation. Alpha rays were found to be positively charged helium ions; beta rays were found to be negatively charged electrons; and gamma rays were high-energy electromagnetic waves. In a magnetic field the alpha rays were deflected in one direction, the beta rays deflected in the opposite direction, and the gamma rays not deflected at all.

The discovery of radioactivity and Sir Joseph Thomson's proof of the independent existence of the electron were the starting points for theories of atomic structure. Thomson proposed one of the first models of the atom. His "plum pudding" model of internal structure depicted the atom as a homogeneous sphere of positive electric fluid (the pudding) in which were imbedded the negatively charged electrons (the plums). In this model the negatively charged electrons, which repel each other and which are attracted to the positive charge, assume certain stable positions inside the atom. If the electron distribution is disturbed by an external force, e.g., the violent collisions between atoms in a hot gas, the electrons vibrate about their equilibrium positions and emit electromagnetic radiation.

The homogeneous-atom concept was proved incorrect when Rutherford performed a series of experiments with a beam of high-speed alpha particles fired at a very thin metal foil. Most of the alpha particles passed straight through the foil or were scattered or deflected only slightly from their original paths. A small percentage of alpha particles were significantly deflected, however, with some alphas reversing their directions.

The Thomson model, in which the positive

*From Walker, F. W., Kirouac, G. J., and Rourke, F. M., *Nuclides & Isotopes*, 12th ed., General Electric Company, Knoll Atomic Power Laboratory, Schenectady, 1977. With permission.

harge was uniformly distributed throughout the ...om, would never permit a sufficiently large con...entration of this charge in one region to affect the ...pha particles significantly. Rutherford thought ...at "it (the experimental result) was about as ...edible as if you had fired a 15-inch shell at a piece ...f tissue paper and it came back and hit you."

To explain these results, Rutherford postu-...ted that the atom does not consist of a uniform ...phere of positive electrification, but that the posi-...ve charge is concentrated in a small region called ...e nucleus, at the center of the atom. In his dy-...amic planetary model, the nucleus plays the role ...f the sun and the electrons correspond to indi-...dual planets of the solar system revolving about ...e sun. This model, along with the classical phy-...cal laws of electricity and mechanics, provided ...n adequate explanation of the alpha particle's ...cattering. Subsequent experiments performed ...n seven different scattering materials and at ...ifferent alpha energies verified Rutherford's ...eory.

Electromagnetic theory demands that an ...scillating or revolving electric charge emit elec-...omagnetic waves. Such emission results in the ...ss of energy by the emitting particle. Applied to ...utherford's electrons, this energy loss would ...ause a steady contraction of the system, since ...e electrons would spiral into the central nucleus ...s their rotational energy was dissipated. This ...rocess would occur very rapidly and would di-...ctly contradict the permanent existence of ...oms. Also, if the radiation pattern produced by ...e atom were related to the energy radiated by its ...oving electron, this radiant energy would be ...hanging with the radius of curvature of the elec-...on's path. The pattern would consist of a contin-...ous range of wavelengths instead of the well-...efined discrete wavelengths that are character-...tic of each element.

ohr Atom

...nce the known stability of atomic systems could ...ot be reconciled with classical principles of ...echanics and electrodynamics, Niels Bohr in ...913 reasoned that classical physics laws must be ...rong when applied to the motion of the electron ...the atom. Max Planck revealed an essential ...mitation in the theories of classical physics in ...01 when he introduced the concept of discrete ...mounts of energy (the energy quantum) in his ...uantum theory of heat radiation. Albert Einstein ...d applied this concept to light in 1905, when he ...escribed the photoelectric effect. The quantum ...eory states that electromagnetic radiation (of ...hich light is one form) must be emitted or ab-...rbed in integral multiples of these energy quanta. ...hr coupled Rutherford's atom with the quantum ...eory to produce his quantum theory of atomic ...ructure.

Since a body that spins about its own axis or ...volves in an orbit about a central point possesses

angular momentum, Bohr assumed that the elec-tron's angular momentum was restricted to cer-tain values (he quantized the angular momentum). Each of the restricted values, which was described by a principal quantum number, n, would specify a particular circular orbit. An atomic system, whose electrons were in given orbits, would not emit electromagnetic radiation even though the particles were accelerating. The whole atom was said to be in a stationary state. Such an assump-tion is contrary to classical electrodynamics as mentioned earlier. Electromagnetic radiation would be emitted or absorbed only when an elec-tron changed from one allowed orbit to another allowed orbit. The energy difference between the two states would be emitted or absorbed in the form of a single quantum of radiant energy, pro-ducing a radiation pattern of a definite frequency ν, related to the energy E by the relation $E = h\nu$ already postulated by Planck and Einstein.

Quantum Numbers

The quantum theory was further refined in 1916 when Arnold Sommerfeld introduced an azimuthal quantum number, l, where $l \leq n - 1$, which per-mitted discrete elliptical orbits for electrons, in addition to the circular orbits. This change per-mitted the Bohr model to account for detailed structure in the pattern of radiation emitted by hydrogen and other atoms. To account for the change in the emitted radiation pattern when an atom is exposed to a magnetic field, a magnetic quantum number m (with permitted integral values from $-l$ to $+l$), was added. This quantum number designates different projections of the possible circular or elliptical orbits along the magnetic field direction in space. Finally, a spin quantum num-ber for the electron was postulated by Samuel Goudsmit and George Uhlenbeck to account for the close grouping of two or more spectral lines. An electron was considered to have an angular momentum about its own axis; in mechanical terms, this motion can be thought of as spin. In a magnetic field, the spin axis can have two direc-tions relative to the field.

The orbits in which the electrons move can be described by specifying a set of these four quan-tum numbers. All electrons with principal quan-tum number $n = 1$ are in the innermost orbit, called the K shell. All electrons with $n = 2$ fall into a sec-ond group, called the L shell. The total number of electrons in a shell is limited by the various possi-ble combinations of the other three quantum num-bers. When an electron shell is filled, the atom is in a stable configuration (the noble gas configura-tion) and does not easily undergo chemical reac-tions. If only one or two electrons are in the last unfilled shell, it is relatively easy for the atom to lose these electrons to another atom whose last unfilled shell has one or two vacancies. The first of these two atoms becomes positively charged (because of the loss of electrons); the second becomes negatively charged (because of the gain of electrons). These atoms can now attract each other and form a compound (ionic bonding).

The periodicity, or repetitive structure of the Mendeleev chart, is now understood to be due to the number of electrons in the atom. In a neutral atom the number of electrons is balanced by the equal number of protons (hydrogen nuclei with a positive charge and a mass of about 1836 electron masses) in the nucleus of the atom. Note that the atomic number of an element is equal to the number of unit positive charges carried by the nucleus and is not the same as the atomic weight. In 1913, Henry G. J. Moseley determined the magnitude of the nuclear charge by comparing the characteristic X-ray wavelengths of elements. The greater the atomic number of an atom, the shorter the waves of the characteristic X-rays associated with it. Identification of the atomic number of an element from its high-frequency spectrum provided a rule for fitting newly discovered elements into vacant places on the Mendeleev chart.

In 1923, Louis DeBroglie postulated that, in analogy with light having both a wave and a particle nature, matter should have a wave as well as a particle nature. The wavelength that he predicted for a particle was inversely proportional to the particle's momentum. Clinton J. Davisson and Lester H. Germer experimented with the scattering of electrons from a crystal. They showed that electrons definitely had wave properties with a wavelength corresponding to the value predicted by DeBroglie.

The mechanical picture offered for the classification of stationary states of atoms by the Bohr theory, and its subsequent modification, was handicapped by its reliance on many ad hoc postulates and by an inability to explain the intensities of radiation patterns emitted by atoms. A new departure was provided in 1926 by Erwin Schrödinger's establishment of wave-mechanics,* in which stationary states are conceived as proper solutions of a fundamental wave equation. In advanced theories, the mechanical models are no longer used.

Isotopes

Experimental investigations in nuclear physics began to require specialized instruments. One of the first of these instruments was the mass spectrograph developed by Francis W. Aston to measure the relative mass of the atoms of an element. If one or more of the negatively charged electrons are removed from an atom, the atom becomes positively charged and is called an ion, the Greek word meaning traveler. The mass spectrograph directed positive ions of an isolated (electrically charged) gas at a photographic plate. The ions were deflected by electric and magnetic fields, working at right angles so that all particles having

the same mass were brought to a focus at a fi. line. Heavier ions, having more inertia, were d flected less than were the lighter ions.

With the use of the mass spectrograph, it w discovered that some chemical elements have tw or more components, each with its own mas Natural chlorine, whose atomic weight is fraction (about 35.5), produced two lines on the phot graphic plate corresponding to masses very clo to 35 and 37. No particle was found with a fra tional mass (within the experimental error). Cor ponents of the same chemical element with diff ent mass numbers are called isotopes. Mo elements in their natural state consist of two more isotopes, although 20 elements have or one isotope: for example, aluminum, cobalt, a gold. Modifying Prout's hypothesis, Aston pr posed the whole-number rule which states that atomic masses are close to integers and that fra tional atomic weights of elements are due to t presence of two or more isotopes, each of whi has an approximately integral value. On the carbo 12 scale now used, where the atomic mass carbon-12 is exactly 12 units, all other isotop have atomic masses close to integers.

With the problem of fractional atomic weigh solved, physicists at first believed that nuclei co sisted of electrons and protons. A nucleus with atomic number Z and an atomic mass A wou consist of A protons, to account for the to mass, and A minus Z electrons to balance the e cess positive charge of the protons. This view the structure of the nucleus was altered in 19 when James Chadwick discovered the neutrc This particle has no electric charge and has a proximately the same mass as the proton.

It is now believed that neutral atoms cons of N neutrons, Z protons, and Z orbital electro with $A = N + Z$. Isotopes are nuclides with the sa Z but different N. For example, natural hydrog coneists almost entirely of atoms that contain o proton and one electron. However, a small amou (about 0.015%) of deuterium (heavy hydrogen) present in nature; deuterium consists of one p ton, one neutron, and one electron. In general, t situation becomes more complex as the heav elements are encountered. Natural tin, which h atomic number 50, consists of 10 isotopes masses 112, 114, 115, 116, 117, 118, 119, 120, 1 and 124. These isotopes differ from one anoth because, although each has 50 protons and electrons, each contains a different number neutrons (ranging from 62 to 74).

The nucleus is held together by attract forces between the neutrons and protons. The attractive forces are not completely understoc but it is known that they must be strong enough overcome the electrostatic repulsion between protons. Because of this repulsion, however, ratio of neutrons to protons increases for sta isotopes as the atomic number increases. Amo

---

* Wave-mechanics is equivalent to the matrix mechanics developed by Werner Heisenberg in 1925.

ght elements in nature, there is approximately
ne proton for every neutron. Among heavy stable
otopes, for every two protons there are approxi-
ately three neutrons.

As previously mentioned, Aston found that
e atomic masses were approximately integers.
ore accurate measurements indicate that the
tal mass of a nucleus is always less than the sum
the proton and neutron masses of which the
ucleus is composed. In 1905, Einstein had shown
at mass, m, was another form of energy, E, ex-
essed by his relationship $E=mc^2$, where c is the
locity of light. The mass deficiency of the nucleus
expressed as the nuclear binding energy. The
nding energy represents the amount of energy
quired to break the nucleus into its constituent
cleons. The ratio of the binding energy to the
mber of particles in the nucleus varies among
e stable elements. It is greater for elements with
ass numbers between 30 and 120 than it is for
ry light or very heavy stable elements.

tificial Radioactivity

1919, Rutherford's discovery of artificial radio-
ctivity achieved the feat vainly sought by the
cient alchemists, that is, changing one element
to another. Rutherford bombarded nitrogen gas
ith a stream of alpha particles. Some of the alpha
articles were absorbed by the nitrogen, protons
ere emitted, and a different element, oxygen,
as formed. The physicist uses symbolic language
represent the transformation as follows:

$$^{14}_{7}N + ^{4}_{2}He \rightarrow ^{17}_{8}O + ^{1}_{1}H.$$

he superscripts denote the total number of
cleons (number of protons plus neutrons), and
e subscripts denote the atomic number (number
protons) in each element. Note that the super-
ripts on one side of the arrow balance those on
e other side. The same is true for the subscripts.
e balance represents the conservation of the
mber of protons and neutrons separately.

This initial discovery has been followed by the
nstruction of large machines designed to accel-
ate charged particles such as protons and alpha
rticles to higher energies so that they may be
ed to bombard nuclei. Among these machines
e the Van de Graaff generator, the cyclotron,
e betatron, the linear accelerator (linac), and
hers. Beams of high-energy neutrons can also
produced. Since the neutron is electrically
utral, however, there is no electrostatic repul-
on between bombarding neutrons and the posi-
ely charged target nuclei. Even thermal neutrons
uld be used for nuclear reaction studies (thermal
utrons have energies that correspond to the
ost probable energy for a group of neutrons at
°F, that is, energies in the neighborhood of 0.025
), see Neutron Absorptive Properties.

utron Fission

ring the investigation of neutron-produced re-

actions in various target elements, Enrico Fermi
and his associates discovered different beta activities (distinguished by half-life) when uranium was
used as a target. They assumed that a transuran-
ium element had been produced (that is, an element whose atomic number was greater than 92).
In 1938, Otto Hahn and Fritz Strassman, repeating
the experiments, discovered part of the activity to
be due to barium (atomic number 56). Lise Meitner
and Otto Frisch suggested that the uranium nucleus had split into two roughly equal parts, barium
and krypton (the latter, atomic number 36), when
the uranium captured the incident neutron. This
reaction Frisch termed "fission," after the term
used to describe the division of cells in a living
organism. Since the mass defect (or binding energy) per particle is greater for the residual nuclei,
barium and krypton, than for the uranium, neutron
fission is accompanied by a large energy release.

For nuclear reactions other than fission, Fig. 1
illustrates the many combinations of incident (or
bombarding) and emitted particles, and how each
combination changes the original nucleus. This
figure is copied from the lower right corner of the
chart. A special type of shorthand is used on this
diagram to identify the data represented. An example is (p,n) which denotes a reaction in which
the nucleus absorbs a proton and emits a neutron.
The symbols used are:

n	neutron	t	triton (hydrogen-3
d	deuteron		nucleus)
p	proton	α	alpha particle
^{3}He	helium-3 nucleus	γ	gamma ray

Using these reactions, nuclear physicists have
produced far more artificially radioactive isotopes
than the stable or radioactive isotopes that occur
in nature. The term "nuclide" was proposed by

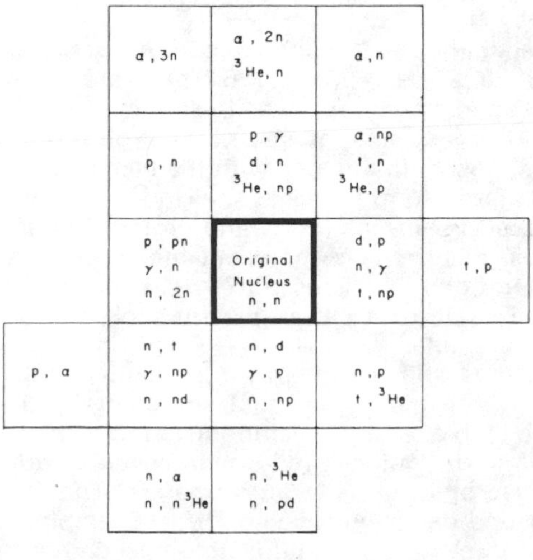

Figure 1. Changes Produced by Various Nuclear Reactions

Truman P. Kohman for a species of atom characterized by the number of neutrons and protons that the atom contains. The term is used in this booklet in this general sense to encompass both stable and radioactive species. The term isomer refers to a nuclide that possesses different radioactive properties in different long-lived energy states. At present, there are 2004 nuclides and 423 isomers known, of which 279 are stable or naturally occurring radioactive forms of the natural elements. Active nuclear research, which is conducted in many laboratories throughout the world, causes additions and changes in the list of nuclides. Since the last edition of the chart was published (1972) 7 nuclides and 16 isomers have been removed because they had been misassigned, and 158 isomers were removed to improve the clarity of the chart (see Metastable States, page 9). Also, 176 new nuclides and 64 new isomers have been added to the chart. In addition, there has been at least one change in more than 94% of the squares on the chart.

CHART OF THE NUCLIDES

(Data from major journal sources revised to July, 1977; data from all other sources revised to April, 1977.)

The general arrangement of the Chart is similar to that suggested by Emilio Segré and followed in previous editions. Because of its size, the Chart is presented in three overlapping sections. The numbers along the left-hand side, marking the horizontal rows, represent the atomic number Z (the number of protons in each nucleus of that row). Each horizontal row represents one element; the filled spaces indicate the known isotopes of that element. The numbers at the bottom of the vertical columns represent the number of neutrons in each nucleus of that column; the number is designated by N.

Heavy lines on the Chart occur for Z or N equal to 2, 8, 20, 28, 50, 82, and 126. These are the so-called "magic numbers," i.e., the numbers of neutrons (protons) present when a neutron (proton) shell is closed. In analogy with the electron shell model of the atom, a nuclear shell model has been developed for the neutrons and protons within a nucleus. Filled shells represent the most stable configurations. Nuclides having either a closed neutron shell, or a closed proton shell, or both, are most stable.

Spaces shaded in gray represent isotopes that occur in nature and that are generally considered stable. A black rectangular area at the top of a white or "colored" square indicates a radioactive isotope that is found in nature. The most recent one discovered being ^{244}Pu. Examples of such isotopes are (1) an unstable nuclide having a lifetime sufficiently long to have prevented the loss by disintegration of all atoms of that particu-

TABLE 1. Distribution of Stable Nuclides

A	Z	N	Number of Stable Nuclides
Even	Even	Even	168
Odd	Even	Odd	57
Odd	Odd	Even	50
Even	Odd	Odd	4
			279

lar nuclide that were available at the time the elements were formed, and (2) a short-lived nuclide that is a disintegration product of such a long-lived nuclide. Occasionally one nuclide has both the gray shading and the black top. This indicates a nuclide found in nature, such as rubidium-8 that is radioactive with a very long half-life. Squares with smaller black rectangular areas near the top represent members of one of the naturally radioactive decay chains (see page 16). The historic symbol is inserted in this smaller black area. White "colored" squares represent artificially produced radioactive nuclides.

The heavily bordered space at the left side of each horizontal row gives properties of the element as found in nature, including the chemical atomic weight (on a mass scale where the neutral atom of carbon-12=12.00000) and the thermal neutron absorption cross section (see Neutron Absorptive Properties.

Each of the other occupied spaces carries the chemical symbol (a list of these symbols is given on page 14 along with the atomic weights) and the mass number of the nuclide indicated. The mass number, designated by A, is the sum of the number of neutrons and protons in the nucleus. The number of neutrons N is equal to the difference between the mass number and the atomic number, that is A minus Z.

Stable Nuclides

Classifying the 279 stable nuclides by the evenness or oddness of Z and N gives four possible categories. The first category contains an even number of protons and an even number of neutrons (so called even-even nuclei). The other categories are even-odd, odd-even, and odd-odd. Table 1 shows the number of stable nuclides that fall in each category.

Table 1 shows that for the odd A nuclides there are approximately as many nuclides with an even number of protons (even Z) as with an even number of neutrons (even N). This is evidence that the nuclear force between two nucleons is independent of whether the nucleons are protons or neutrons. Odd-odd stable nuclides are scarce, and they are found only among the lightest nuclides. Their scarcity is due to a "pairing energy" between particles in the same shell. The condition of being in the same shell increases the binding energy hese particles, making them more stable than particles in different shells. An odd-odd nuclide contains at least one unpaired proton and one unpaired neutron which are usually in different shells and hence contribute weakly to the binding. For the lightest nuclei, however, the unpaired neutron and proton are in the same shell.

Diagonals running from upper left to lower right connect nuclides of different elements, which have the same mass numbers. For example, one line could connect calcium-40 which has 20 protons and 20 neutrons, with argon-40, which has 18 protons and 22 neutrons. Nuclides of the same mass number are called isobars; nuclides with the same number of neutrons are called isotones.

Data Display

The manner of displaying data is explained in the lower right corner of the chart. For stable nuclides, the first line contains the chemical symbol and mass number; the second line presents the atom percent of the natural element that this isotope represents (known as the absolute isotopic abundance, see page 10); the third line contains the thermal neutron cross section (see page 10); and the fourth line presents the isotopic mass of the neutral atom (the mass of the nucleus and its surrounding electrons). This mass is given in atomic mass units where carbon-12 is assigned a mass of 12.00000.

For long-lived, naturally occurring radioactive nuclides, the first line contains the chemical symbol and mass number, the second line presents the absolute isotopic abundance, and the third line contains the half-life. The half-life is the period of time in which half of the nuclei initially present in a given sample disintegrate. Additional lines present the decay modes (or types) and energies of decay and the isotopic mass of the nuclide. Energies are given in millions of electron volts (MeV). When more than one mode of decay occurs, the most prominent mode appears first (above, or to the left of, the other modes).

For radioactive nuclides that are not of the long-lived, naturally occurring type, the same information is presented except that the isotopic abundance is omitted and the last line of the pertinent square contains the beta-decay energy instead of the isotopic mass. For the heavy elements, where the major mode of decay is alpha-particle emission, the isotopic mass is retained in the last line. In many squares, a small black triangle appears in the lower right corner to indicate that the nuclide has been formed as a product in the thermal-neutron fission of uranium-235, uranium-233 or plutonium-239.

Metastable States

Note that certain squares are divided; for example, the square for aluminum-26. Such divisions occur when a nuclide has one or more isomeric states,

that is, when a nuclide has the same mass number and atomic number but possesses different radioactive properties in different long-lived energy states. An isomeric state is included on the chart if its half-life is one second or longer. Shorter half-lives are included only if they are fed in the decay of a parent nuclide or are part of a naturally occurring radioactive decay series. This arbitrary cutoff point of one second was chosen for convenience in presenting the data. Since many isomeric states were removed from the chart using this criterion, the reader is referred to IAEA report INDC(CCP)-89/N, April 1976, "Table of Lifetimes of Nuclear Levels," by Eh.E. Berlovich et. al. for a convenient listing of the half-lives of isomeric states (also includes references to earlier work). The presence on the chart of an isomeric state with half-life less than one second, but greater than one microsecond (one millionth of a second) may sometimes be inferred from the use of the symbol D, which means delayed radiation. When referring to isomeric states, the lower energy state is generally referred to as the ground state; the higher state, as the isomeric state. Frequently, the ground state is a stable nuclide. If one metastable state exists, it is shown on the left. If two exist, the higher energy state is shown on the left, the lower below it or to the right of it, and the ground state to the right of both or below them.

A mode of decay and usually also the decay energy shown in parentheses indicate that the decay results from a short-lived daughter that accompanies its parent. (In a radioactive decay, the original nuclide is called the parent or precursor; the resultant nuclide is called the daughter.) For example, nitrogen-17, with a half-life of 4.17 seconds, decays by negative beta emission (symbol β^-) into an exceedingly short-lived state of oxygen-17, which in turn emits a neutron. Thus, nitrogen-17 emits "delayed neutrons" with a half-life of 4.17 seconds.

Another example is 17.0-day palladium-103, which decays by K-electron capture mainly to the 56-minute rhodium-103 and, statistically less often, to stable rhodium-103. (K-electron capture occurs when the nucleus captures an electron from the K shell; the symbol is ϵ.) The 56-minute rhodium emits a gamma ray or an internal-conversion electron that corresponds to an isomeric transition of 0.03975 MeV. (An internal-conversion process involves the direct transfer of energy from the nucleus to one of the orbital electrons, and the electron is ejected from the atom; the symbol used is e⁻.) On the chart, the delayed gamma ray is assigned to the parent; inclusion of the energy in parentheses indicates that the gamma ray comes from the daughter, but continues to last as long as the disintegrating parent is still present.

A further example is provided by a standard laboratory radionuclide, 30.17-year cesium-137. This long-lived parent decays directly to a short-lived daughter, 2.552-minute barium-137, by negative beta emission. The 0.66164-MeV gamma ray which is emitted by the barium is included on the cesium square with a "D" to indicate delayed radiation proceeding through an isomeric level of the daughter nucleus. The symbol "D" is used most often to indicate decay through a level in the daughter nucleus of half-life greater than one microsecond.

Isotopic Abundances

The chemical atomic weights (relative atomic mass of an element) listed are those promulgated by the Commission on Atomic Weights—Inorganic Chemistry Division of the International Union of Pure and Applied Chemistry (IUPAC) in its 1975 Report and recently (1977) proposed changes by the Commission. Occasionally, a sample can display a significant difference from the accepted value because of artificial isotopic fractionation, artificial nuclear reaction, or a rare geological occurrence in small quantity and because of the radiogenic source or the extraterrestrial origin of the sample.

The isotopic abundance values, on an atom-percent basis, listed are from a 1977 evaluation performed for the meeting of the Commission on Atomic Weights. This list constitutes the first complete reevaluation of the data since 1971. The values do not always total exactly 100% for a given element because of the relative accuracy to which individual isotopic values are quoted.

Warning

The values listed are the current best estimate of the absolute isotopic abundances for the naturally occurring elements. However, samples have been observed for which the measured values differ significantly from those listed, especially lithium and boron. On the shelf, chemical reagents with a 3.75% ^{6}Li composition have been noted without any warning of the depletion in ^{6}Li content. Natural variation in boron from 19.8% ^{10}B to 20.1% have been measured. The sample atomic weight and the thermal neutron absorption cross section will then differ significantly from the value listed. It is highly recommended that samples be checked before the listed values are assumed to apply if these values are critical to the application.

Neutron Absorptive Properties

The Greek letter σ with various subscripts is used to identify the thermal neutron cross sections. The neutron cross section measures the probability of interaction of a neutron with matter. The cross section can be most easily visualized as a cross-sectional target area presented to the neutron by the nucleus. Cross sections are usually measured in units of barns per atom. A barn is the area of a square a millionth of a millionth of a centimeter on each side (10^{-24} square centimeters, cm^2).

Neutrons produced in the fission process have an average energy of approximately 2 MeV [2(10)^{6}eV]*. In a thermal reactor, these neutrons are slowed down by collisions with the moderator atoms until they are in thermal equilibrium with the moderator and have an average energy of 0.025 eV. Therefore, a spectrum of neutron energies exist in a reactor. An arbitrary division frequently used for this neutron energy spectrum is comprised of three regions:

1. The thermal region in which neutrons are in thermal equilibrium with the moderator.
2. The fast region in which the neutrons are produced from fission.
3. The intermediate or slowing-down region in which the neutrons have more or less been moderated but have not yet reached thermal equilibrium.

When neutrons reach thermal equilibrium with the moderator, their energies are determined by the thermal energy distribution of the moderator atoms, and the neutron energy spectrum becomes a Maxwellian distribution at the temperature (T in °K) of the moderator material. Setting the kinetic energy of neutron motion equal to the thermal energy of the moderator, one can obtain the most probable velocity of the neutron in thermal equilibrium. For a moderator at room temperature, the neutron velocity is 2200 meters per second. To obtain a reaction rate, one would use the density of neutrons, this most probable velocity of 2200 meter/sec, and the neutron cross section at the energy corresponding to a 2200 meter/sec velocity. The thermal neutron absorption cross section (symbol σ_a) is the sum of the cross section for all reactions except scattering of the neutron. For many materials, it is inversely proportional to the velocity of the neutron. These materials are called 1/v absorbers. A correction function, g(kT), is used to describe the departure of the cross section from the 1/v law for certain materials. The function g(kT) is defined as the ratio of the reaction rate in a Maxwellian flux at the temperature T to the reaction rate at 2200 meter/sec. For a 1/v absorber, g=1. Some of the materials with cross sections that are significantly non-1/v (g $\geq$ 1.1 or g $\leq$0.9) are listed in Table 2.

For a neutron slowing down in a moderating medium without absorption, the flux can be shown to be inversely proportional to the energy. Between 1 eV and a few keV (10^3 eV), especially for elements with intermediate and high mass numbers, there are often particular energies for which the rate of interaction is exceptionally high. A curve of the cross section versus energy in this

---

*1 electron volt (eV) is the energy produced, or the work done, when one electronic charge is moved through a potential difference of one volt.

region shows narrow peaks called resonances. The rate of occurrence of a particular reaction in this resonance region is proportional to the integral of the cross section as a function of neutron energy multiplied by the flux density, which is called a resonance integral. The resonance integral assumes that the flux density has an 1/E dependence that is not disturbed by the absorbing isotope; i.e., extreme dilution of the absorber in the moderator is assumed. Integrals of this type are called infinite dilution resonance integrals.

The measurement of the reaction rate in a reactor neutron spectrum produces contributions from intermediate neutrons with a 1/E shape and from thermal neutrons with a Maxwellian distribution of energies. To distinguish between these two contributions, a cadmium cover is often used. Cadmium filters do not curtail the intermediate neutron spectrum sharply, but a sharp "effective cadmium cutoff energy," E_{cd}, can be defined for a 1/v absorber. E_{cd} is the energy associated with a perfect filter* under which an irradiated material would have the same reaction rate as under a cadmium cover. Suitable cadmium filters terminate the intermediate neutron energy spectra at approximately 0.5 eV. This energy is sufficiently high to exclude most of the low-energy deviations of the flux from a 1/E shape.

The lower energy limit for the resonance integral will depend on the cadmium cutoff energy. The values listed on the chart assume a 0.5 eV cutoff energy. Contributions to the resonance integral will come from individual resonances in the cross section above 0.5 eV as well as from the 1/v smooth cross-section background. The 1/v contribution to the resonance integral will be a factor times the cross-section value at 2200 meter/sec. For a cadmium cutoff energy of 0.5 eV used here, the factor is 0.45.

The various cross sections are distinguished on the chart in the following manner. A 2200 meter/sec value for a particular reaction is listed as σ with a subscript appropriate for that reaction. A cross section for neutrons with energies below the cadmium cutoff is listed as $\bar{\sigma}$ with subscripts. All other cross sections including reactor spectrum values are shown as $\hat{\sigma}$ with subscripts.

At thermal energies, a number of reaction types (subscripts) are possible. The most probable reaction (the reaction with the largest cross section) is generally the neutron capture reaction symbol σ_γ) in which absorption of the neutron by the nucleus is accompanied by high-energy gamma ray emission. Occasionally, a proton or an alpha particle may be emitted, or a nucleus may fission upon neutron absorption (symbols σ_p, σ_a, and σ_f). Examples of these cross sections are found on the squares of beryllium-7, boron-10, and thorium-227, respectively. The symbol is followed by the cross-section value, followed by the

TABLE 2. Some Non-1/v Capture Cross Sections

Nuclide	g-Value	Nuclide	g-Value
^{113}Cd	1.34	^{157}Gd	0.85
^{135}Xe	1.16	^{176}Lu	1.75
^{149}Sm	1.64	^{182}Ta	1.64
^{151}Eu	0.89	^{239}Pu	1.13
^{155}Gd	0.84	^{243}Am	0.90

resonance integral value including the 1/v contribution.

A given nuclide might undergo two or more types of interactions, and its square would then contain two or more of these cross-section values. When neutron capture can lead to a metastable state as well as to the ground state, more than one value will appear beside the capture cross section for that nuclide. The cross-section value for metastable-state formation is listed on the left and that for direct-ground-state formation, on the right followed by similar values for the resonance integral. For two metastable states, the higher of the two states is on the left. For example, indium-115 has an indicated cross section and resonance integral σ_γ of $(87+75+41)$, $(2.6\times10^3+67\times10^1)$, which means that the cross section for formation of the 2.2 second state of indium-116 is 87 barns, the cross section for the direct formation of the 54.2 minute indium-116 is 75 barns and the resonance integral for direct plus indirect formation (by decay of the 2.2 second state) is 2.6×10^3 barns, and the cross section and resonance integral for formation of the 14.2 second ground state of indium-116 are 41 barns and 67×10^1 barns, respectively.

The designation mb or μb following the cross-section value indicates that the units of the cross section or resonance integral are millibarns per atom (10^{-27} cm^2/atom) or microbarns per atom (10^{-30} cm^2/atom), respectively. When no mb or μb appears on the chart square, the units of the cross section are barns per atom.

There is not enough space to indicate the one standard deviation uncertainty in the cross sections and resonance integrals on the chart. Therefore the following policy has been adopted: the values given have been rounded off so that the uncertainty in the last decimal place is less than or equal to five units. For example, a value of 6.78 ±0.08 barns would be quoted as 6.8 barns; similarly, a value of 270 ± 70 barns would be given as 3×10^2 barns.

Spins and Parities

In the upper right corner of the square for the ground state of a nuclide, and in the upper left corner of the isomeric state, are shown the spin and

---

* A perfect filter is one that has infinite absorption below the cutoff energy and zero absorption above the cutoff energy.

parity of the corresponding energy level. Each neutron and proton has an intrinsic angular momentum of ½ (in units of $h/2\pi$, where h is Planck's constant), similar to that of the electron, which combines with their orbital angular momentum to produce a resultant angular momentum called the nuclear spin. Since the orbital angular momentum is always zero or an integral multiple of $h/2\pi$, the nuclear spin (in units of $h/2\pi$) is always integer or half-odd-integer, depending upon whether the nucleus has an even or an odd number of nucleons. The concept of parity was introduced by the mathematical formalism of quantum theory and has no classical analogue. A system in a given state may have even parity (symbol +) or odd parity (symbol −). For aluminum-27, the spin and parity are shown as 5/+, where the 2 in the denominator of 5/2 has been removed to improve the readability of the chart. The ground states of all even-even nuclides are known to have spin and parity 0+; so 0+ has been omitted.

The arguments for the assignment of spin and parity to nuclear states can be divided into two classes: strong arguments such as measuring values directly and weak arguments such as inferring values indirectly. On the Chart, the absence of parentheses indicates spins and/or parities based on strong arguments; the presence of parentheses indicates spins and/or parities based on weak arguments. When the spins of both the ground state and an isomeric state are given for a particular nuclide, it is interesting to observe that these spins usually differ by two or more units of $h/2\pi$. The large angular momentum (spin) change is required for the gamma-ray transition between the states. Combining this spin change with the small energy differences (a few hundred keV) leads to a relatively long lifetime (mestastable state).

A useful reference on measured spins of nuclear states is "Nuclear Spins and Moments," by Gladys H. Fuller which appeared in Journal of Physical Chemistry Reference Data Vol. 5, No. 4, page 835 (1976).

Fission Yields

The binary fission process can occur in many different modes. A very large number of fission products are known, ranging from zinc (atomic number Z=30) to dysprosium (Z=66), and from mass number A=72 to A=163 in the thermal neutron fission of U-235. Fission into two equal fragments is by no means the most probable mode in thermal neutron fission. Asymmetric modes are much more favored, the maximum fission product yields occurring at A=95 and A=138. The asymmetry appears to become less pronounced with increasing bombarding energy of the neutron.

An enormous amount of radiochemical and mass spectrometry work was required to arrive at our present state of knowledge about fission products. It was necessary to develop chemical separation procedures, analyze radioactive decay and growth patterns, determine beta- and gamma-ray energies, establish mass assignments of many previously unknown nuclides, and to measure the fission yields.

Since the neutron-to-proton ratio is about 1.6 for ^{235}U and about 1.3 for the stable elements in the fission product region, the primary products of fission are generally on the neutron excess side of stability. Each such product decays by successive negatron decays to a stable isobar. Chains with as many as six negatron-decays have been established, and undoubtedly some fission products further removed from stability have escaped detection because of their very short half-lives. No neutron-deficient nuclides have been found among the products of thermal neutron fission; however, a few so-called shielded nuclides occur among the fission products. A shielded nuclide is one that has a stable isobar one unit lower in Z so that it is not formed as a daughter product in a beta-decay chain. Examples are Rb-86 and Cs-136. The fission yield of such a nuclide is presumably due entirely to its direct formation as a primary product. However, shielded nuclides are not indicated as fission products on the Chart by the presence of the triangle in the lower right-hand corner.

The known isobaric fission chains are indicated on the Chart by black (and sometimes orange) lines drawn diagonally off the corner of the first or last nuclide block for each mass number A. Those diagonal lines on the last nuclide (bottom) are total cumulative yields for U-235 thermal neutron fission; those on the first nuclide (top) are total cumulative yields for U-233 thermal neutron fission (above) and Pu-239 thermal neutron fission (below in parentheses). The total cumulative yield is the percentage of the total fissions that leads directly or indirectly to that nuclide. Since two fission products are emitted for every binary fission, the total cumulative yields of all the isobaric chains should add up to 200%. Direct or independent fission yields have been measured for a relatively few nuclides in some isobaric chains.

The independent yield of a particular fission product is the direct or instantaneous yield of that nuclide without any contribution from decay of preceding members of the isobaric chain. The measured total cumulative yield of a particular fission product represents the sum of its independent yield and of the independent yields of all its precursors.

The fission yields given in the Twelfth Edition of the Chart of the Nuclides are total cumulative chain yields (in percent) of the last isobaric chain member of U-235, U-233 and Pu-239 thermal neutron fission, and they were taken from NEDO-12154-2D, "Compilation of Fission Product Yields," by B. F. Rider and M. E. Meek, July (1977).

In addition to binary fission, ternary fission, or breakup into three products, also occurs but less frequently than binary fission. This process is a source of light-charged particles such as ^{3}H, ^{3}He, ^{4}He, ^{7}Be, etc. The long-lived, or stable, light nulides can build up in the nuclear fuel of an operating reactor. ^{3}H (tritium) is a ternary fission product important to reactor design engineers. Its yield in ternary thermal neutron fission of U-235 has been measured to be in the range of 0.0068 to 014%.

alf-Life Variability

here are a few nuclides for which measurable changes in the half-life, or disintegration constant, have been produced by chemical changes that alter the electron density near the nucleus. These effects are generally small, being a 0.27% greater half-life for Tc_2S_7 compared with $KTcO_4$ and a 0.08% greater half-life for BeF_2 compared with Be metal. The 6.02-hour Tc-99m decays by a weak isomeric transition, the decay taking place primarily by emission of M- and N- conversion electrons; the 53.28-day Be-7 decays by K-electron capture. Other nuclides for which measurable variability in half-life have been found are 18.8-

LIST OF ELEMENTS

ATOMIC NUMBER	SYMBOL	NAME	ATOMIC WEIGHT	ATOMIC NUMBER	SYMBOL	NAME	ATOMIC WEIGHT
0	n	neutron	......	52	Te	tellurium	127.60
1	H	hydrogen	1.0079	53	I	iodine	126.9045
2	He	helium	4.00260	54	Xe	xenon	131.30
3	Li	lithium	6.941	55	Cs	cesium	132.9054
4	Be	beryllium	9.01218	56	Ba	barium	137.33
5	B	boron	10.81	57	La	lanthanum	138.9055
6	C	carbon	12.011	58	Ce	cerium	140.12
7	N	nitrogen	14.0067	59	Pr	praseodymium	140.9077
8	O	oxygen	15.9994	60	Nd	neodymium	144.24
9	F	fluorine	18.998403	61	Pm	promethium	......
10	Ne	neon	20.179	62	Sm	samarium	150.4
11	Na	sodium	22.98977	63	Eu	europium	151.96
12	Mg	magnesium	24.305	64	Gd	gadolinium	157.25
13	Al	aluminum	26.98154	65	Tb	terbium	158.9254
14	Si	silicon	28.0855	66	Dy	dysprosium	162.50
15	P	phosphorus	30.97376	67	Ho	holmium	164.9304
16	S	sulfur	32.06	68	Er	erbium	167.26
17	Cl	chlorine	35.453	69	Tm	thulium	168.9342
18	Ar	argon	39.948	70	Yb	ytterbium	173.04
19	K	potassium	39.0983	71	Lu	lutetium	174.967
20	Ca	calcium	40.08	72	Hf	hafnium	178.49
21	Sc	scandium	44.9559	73	Ta	tantalum	180.9479
22	Ti	titanium	47.90	74	W	tungsten	183.85
23	V	vanadium	50.9415	75	Re	rhenium	186.207
24	Cr	chromium	51.996	76	Os	osmium	190.2
25	Mn	manganese	54.9380	77	Ir	iridium	192.22
26	Fe	iron	55.847	78	Pt	platinum	195.09
27	Co	cobalt	58.9332	79	Au	gold	196.9665
28	Ni	nickel	58.70	80	Hg	mercury	200.59
29	Cu	copper	63.546	81	Tl	thallium	204.37
30	Zn	zinc	65.38	82	Pb	lead	207.2
31	Ga	gallium	69.72	83	Bi	bismuth	208.9804
32	Ge	germanium	72.59	84	Po	polonium	......
33	As	arsenic	74.9216	85	At	astatine	......
34	Se	selenium	78.96	86	Rn	radon	......
35	Br	bromine	79.904	87	Fr	francium	......
36	Kr	krypton	83.80	88	Ra	radium	......
37	Rb	rubidium	85.4678	89	Ac	actinium	......
38	Sr	strontium	87.62	90	Th	thorium	232.0381
39	Y	yttrium	88.9059	91	Pa	protactinium	......
40	Zr	zirconium	91.22	92	U	uranium	238.029
41	Nb	niobium	92.9064	93	Np	neptunium	......
42	Mo	molybdenum	95.94	94	Pu	plutonium	......
43	Tc	technetium	......	95	Am	americium	......
44	Ru	ruthenium	101.07	96	Cm	curium	......
45	Rh	rhodium	102.9055	97	Bk	berkelium	......
46	Pd	palladium	106.4	98	Cf	californium	......
47	Ag	silver	107.868	99	Es	einsteinium	......
48	Cd	cadmium	112.41	100	Fm	fermium	......
49	In	indium	114.82	101	Md	mendelevium	......
50	Sn	tin	118.69	102	No	nobelium	......
51	Sb	antimony	121.75	103	Lr	lawrencium	......

second Nb-90m$_1$ and 26.1-minute U-235m. The effect found for Nb-90m$_1$ was a 3.6% greater decay rate for Nb metal compared with the niobium pentafluoride complex, the largest effect observed to date. A half-life variability claimed for 8.041-day I-131 of 0.5% greater in solution compared with crystals has been accounted for by escape of daughter Xe-131m and/or absorption of the radiations of Xe-131m. Since the decay of I-131 is by nuclear beta-decay, and since the chemical form was the iodide ion in both cases, one would not expect an alteration in electron density near the nucleus, and hence not an observable half-life variability.

A useful reference on this topic is "Survey on the Rate Perturbation of Nuclear Decay," by H. P. Hahn et. al., which appeared in Radiochimica Acta Vol. 23, pages 23–37 (1976).

Color

The color coding used on this edition of the Chart of the Nuclides emphasizes half-lives and cross sections. Two shades, or intensities, of each of four different colors—blue, green, yellow, and orange—are used for these identifications. The lower half of a nuclide block is used to identify the larger of thermal neutron cross section or resonance integral (neutron absorption properties, see page 10), with the darker or more intense shade of color designating these neutron absorption properties. The following ranges of neutron absorption properties are used:

> 10 to 100 barns — blue
> 100 to 500 barns — green
> 500 to 1000 barns — yellow
> Greater than 1000 barns — orange

The use of this color spectrum was chosen so that the hottest color, orange, coincided with the largest absorption values. The progression continued through yellow, and green so that the coolest color, blue, coincided with the lowest values.

The upper half of a nuclide block is used to identify half-lives, with the lighter, or less intense, shade of color designating these half-life properties. The following ranges of half-lives are used:

> 1 day to 10 days — orange
> 10 days to 100 days — yellow
> 100 days to 10 years — green
> 10 years to 10,000 years — blue

In this instance the hottest color, orange, was chosen for the lowest half-life values because these represented the highest values of radioactivity. The progression continued through yellow and green so that the coolest color, blue, was chosen for the highest half-life values, i.e., lowest activity values for the same number of atoms.

Radioisotopic Power and Production Data

In recent years the need has arisen to genera power in satellites orbiting Earth and to transm data from experiments left on the moon durir lunar landings. This power sometimes is suppli by radioisotopes. The kinetic energy possesse by the decay particles of a radioisotope is tran formed into thermal or radiant energy that can t used to produce electric power by some energ conversion technique. Thermocouples producir an electric voltage when the junctions of the tw dissimilar metals are kept at different temper tures are one form of an energy-conversion devic A thermionic converter in which electrons emitte by a hot surface are collected to produce an ele tric current is another form of an energy-conversic device.

The fuel for a radioisotopic power generat must be safe and reliable and low in both weig and cost. These requirements would result in rejection of radioisotopes with a half-life less tha 100 days because they would have to be replace too often. Half-lives greater than 100 years wou be rejected because the activity would be too lo to obtain an initial power density greater than 0 watts/gram. Radioisotopes with high-energ gamma ray emission would be rejected becaus this more penetrating radiation would requi additional shielding, which would add to the tot weight. Among the candidates that would quali as fuel sources for radioisotopic power gene tion are ^{3}H, ^{60}Co, ^{85}Kr, ^{90}Sr, ^{106}Ru, ^{137}Cs, ^{144}Ce, ^{147}P ^{170}Tm, ^{210}Po, ^{238}Pu, ^{241}Pu, ^{242}Am, and ^{244}Cm. T calculated initial power density in watts/gm f these sources are indicated by the values in t triangular box along the diagonal line through t appropriate mass chain.

In reactor technology, the term fuel burn refers to the amount of fissionable material that consumed (or the amount of power produced) t fore the fuel element is removed from the reac for processing. A common unit is the number megawatt-days of heat energy produced per met ton of fuel (MWD/MT), where a metric ton is 10 kilograms, or 2200 lb. For light water reactors, e riched fuel lifetimes would correspond to an op mum exposure of 30,000 MWD/MT. As a result the various nuclear reactions and decays th occur over this long exposure, various transura ium and fission product nuclides will be produc in quantity. Many of these nuclides are of intere for special power applications. The producti data for these nuclides are given on the Chart two ways:

1. Grams per metric ton of light water reac fuel exposed to 30,000 MWD/MT.
2. Kilograms per megawatt year of elect energy delivered from a reactor expos to 30,000 MWD/MT.

The grams per ton (metric) values are quot at discharge and are listed in the circle along t

agonal to the appropriate mass chain for both
w enrichment uranium fuel and plutonium mixed
xide fuel. The kilograms per 1000 megawatt years
  electric energy values assume a thermal-to-
ectric efficiency of 33% and are listed in the rec-
ngular box along the diagonal to the appropriate
ass chain.

adioactive Decay Chains

s nuclear processes occur, whether in natural
dioactivity or under artificially induced condi-
ons, the nuclides change in accordance with the
heme shown in Fig. 2. To understand the use of
is scheme more fully, consider the uranium-238
cay chain (one of three such chains found in
ture). On the Chart we start with the parent
anium-238 which emits an alpha particle. The
ughter nucleus is in the second space diagonally
wn to the left (see Fig. 2). This square represents
e isotope thorium-234. (This nuclide is also iden-
ed by the historical symbol uranium X₁, which
the name given to it before it was identified as
orium.)

Thorium-234 in turn emits a negative electron,
 the loss of mass is not appreciable. However,
ere is a loss of one negative charge, which means
at the atomic number Z increases by one. In
ect, one neutron has changed into a proton.
e move one space up and one space to the left
ee Fig. 2) leads to protactinium-234 which has
omeric states. Each of these states undergoes
gative beta emission, so another move diag-
ally upward to the left leads to uranium-234.

Uranium-234 emits an alpha particle ending at
orium-230. Another alpha decay yields radium-
6. Three further alpha decays result first in
don-222, then in polonium-218, and finally in
ad-214. However, this isotope of lead is unstable

and emits a negative electron producing bismuth-
214. A beta decay to polonium-214 is followed by
an alpha decay to lead-210. An alternate route
from bismuth-214 to lead-210 is taken in a small
fraction of the disintegrations since bismuth-214
can also emit an alpha particle and the resulting
thallium-210 beta-decays to lead-210.

In either case, lead-210 beta-decays to bis-
muth-210. Another beta decay produces polonium-
210 which alpha-decays to the stable isotope lead-
206. At this point the chain ends. Incidentally, in
many of the above steps, gamma rays and conversion electrons are also emitted.

Similarly, the two other natural radioactive
sequences may be traced. One is the actinium
series which starts with uranium-235 and ends
with lead-207. The other is the thorium series,
which goes from thorium-232 to lead-208. A precursor of the thorium series ²⁴⁴Pu has recently
been discovered in nature. A fourth, or neptunium,
series is also known. However, the half-life of the
parent, neptunium-237, is only about two million
years. Since the age of the earth is about 4.5 billion
years, most of the neptunium-237 present when
the earth was younger has already decayed, and
the series is not found in nature.

Pre-Fermi (Natural) Nuclear Reactor

The time interval between the formation of elements and the formation of Earth has been estimated to be 100 to 300 million years. There is
interest in some radionuclide that has a comparable half-life and that might have been present in
the early history of Earth to be used as a tracer for
studies of the formation of the solar system. ²⁴⁴Pu,
which has recently been found in nature, has the
proper decay characteristics, and its spontaneous
fission produces a large quantity of the heavy
xenon isotopes. Since the abundance of xenon is
low in terrestrial and meteorite samples, fissiogenic
xenon from ²⁴⁴Pu will produce noticeable anom-
alies in the xenon isotopic ratios as compared with
natural xenon. The measured isotopic values from
samples agree with the spectrum of xenon yields
from spontaneous fission of ²⁴⁴Pu. The presence
of ²⁴⁴Pu in the early solar system is considered to
be a crucial test in favor of the theory of the synthesis of chemical elements in stars.

Another source of atmospheric xenon is the
neutron-induced fission of ²³⁵U. In 1956, P. K.
Kuroda speculated that an assemblage of uranium
and water could easily become a self-supporting
chain reacting system in the early history of Earth.
These speculations have recently been borne out
by the French discovery at Oklo in the Gabon Re-
public in Africa of samples of natural uranium,
which have been found to be depleted in ²³⁵U.
(There is measurably less ²³⁵U in these samples
than in "ordinary" natural uranium.) Samples that
were slightly but definitely enriched in ²³⁵U were
also found. The amounts of various nuclides found
in the samples did not agree with the natural abun-

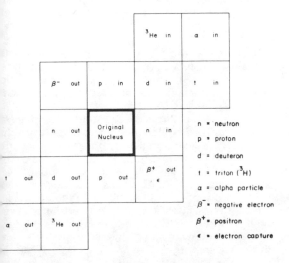

*igure 2.   Relative Locations of the Products of Various
Nuclear Processes*

dance of these nuclides. The distribution of isotopes of particular elements, after correction for neutron capture and natural background, agreed with the fission yield distribution from ^{235}U. From the Chart, it can be noted that neodymium has many stable isotopes at the end of short fission product chains which have large fission yields. The analysis of neodymium provided much of the information about the samples.

From radioactive dating methods, the samples are determined to be about 1.7 billion years old. The ^{235}U abundance in this natural reactor would have been 3% or more (prior to the ^{235}U decay over the past 1.7 billion years to its present 0.72% abundance). The geological evidence implies that large quantities of water were present to provide moderation of the fission neutrons. After fissioning, the samples would have been depleted in ^{235}U. Enriched samples would have resulted from the ^{238}U transforming into ^{239}Pu by neutron capture and decay and subsequent alpha decay of ^{239}Pu into ^{235}U. This conversion reaction would also have fed the reactor. Studies of the Oklo natural reactor indicate that about two-thirds of the total fission events were caused by the ^{235}U initially present in the ore and nearly one-third from ^{235}U converted from ^{238}U by neutron capture and subsequent decay.

By studying the amounts of the various nuclides in the ore, a wealth of information has been obtained. This information includes values of neutron flux, time-integrated neutron flux, conversion ratio, and fraction of epithermal neutrons. Finally, a determination has been made of the number of fissions that had occurred within the reactor and the percentage that had been caused by each of the fissioning nuclides ^{235}U, ^{238}U, and ^{239}Pu. It has been estimated that a total released energy of 15,000 megawatt years was generated by the Oklo reactor. This is approximately the same energy as generated by one large modern nuclear reactor in four years. It is interesting to note that this naturally occurring enriched converter reactor operated nearly two billion years before Fermi initiated the first man-made neutron chain reaction in 1942 with natural uranium fuel. Clearly, in this instance, man was the unknowing imitator of nature.

At least one other feature of the Oklo natural reactor has stimulated interest. This feature has been the remarkable stability of the Oklo ore deposit. The ^{235}U and ^{238}U fuels have been confined within the grains of the ore. Since almost half of the residual ^{235}U in the reactor zones is the daughter product of ^{239}Pu, this means that the plutonium has been "locked-up" for a time comparable to its 24,110 year half-life. Also, at least one-half of thirty or so fission product elements have remained immobilized in the ore. These findings have produced much interest from scientists working on the complex problem of the storage and disposal of radioactive fuel and wastes.

Elements Without Names

Conflicting claims by researchers at the Lawrence Berkeley Laboratory in California and at the Dubna Joint Institute for Nuclear Research in the Soviet Union have led to controversy in the priority of discovery and the naming of the elements 104 and 105. The Berkeley researchers have proposed the names Rutherfordium for 104 and Hahnium for 105, while the Soviet scientists have proposed the names Kurchatovium and Nielsbohrium. Discovery of element 106 has also been claimed by both groups, but so far no name has been proposed.

Because of the dispute over elements 104 and 105, the International Union of Pure and Applied Chemistry (IUPAC) and the International Union of Pure and Applied Physics (IUPAP), which are groups that help to mediate conflicting claims and determine who has the right to claim what, have jointly appointed a neutral committee to consider the matter. In addition to monitoring attempts by each of the research teams to reproduce experimental results obtained by the other, the committee is setting up ground rules or criteria for the discovery of chemical elements. Among these criteria are: proof, by some means, that the atomic number (Z) of the new element is different from the atomic number of all previously known elements. Acceptable means are chemical identification, identification of characteristic X-rays emitted by the new element itself or in connection with its decay, and proof of its genetic relationship with previously known element through an alpha particle decay chain. Another criterion is a suggestion that composite nuclear systems that have lifetimes less than about 10^{-14} seconds should not be considered new elements. It is hoped that with the application of reasonable ground rules to define what constitutes "discovery" of a chemical element and with impartial examination of the experimental evidence, the elements without a name will join the other named elements on the chart.

The Search for the Superheavies

The search for the superheavies started in the mid sixties when it was predicted that an "island of stability" of superheavy elements existed. Normally, as the atomic number is increased for heavy nuclides, shorter half-lives are expected. This is because of the large electrostatic repulsion between protons which overcomes the attractive nuclear forces and makes spontaneous fission of heavy nuclides more and more likely as their atomic numbers increase above 100. The probability of alpha decay also increases as coulomb forces become stronger. The shell model of the nucleus, in which the protons and neutrons are arranged in shells somewhat like atomic electrons, provides a way to circumvent this tendency to shorter half lives. When the proton and neutron shells are filled, the extra stability due to pairing leads to longer lifetimes. The magic numbers for which is closed shell condition would hold are thought to be 114 protons and 184 neutrons.

The longer predicted lifetimes for superheavy elements in or near the "island of stability" set off a flurry of activity as scientists in the United States, Europe, and the Soviet Union sought to make superheavy elements in accelerators or to detect them in nature. The search for the superheavies in nature is based on theories which suggested that they were created before the earth itself was formed, and are still present now provided that their half-lives are sufficiently long. Between 1968 and 1972, a Berkeley research team looked at over forty kinds of ores, minerals, and rocks including manganese nodules from the ocean floor, moon rocks, and gold and platinum in their natural states. Others have looked at meteorites, mineral-laden hot spring water, and newly erupted volcanic material. Soviet scientists even looked at medieval stained glass that was framed with lead to see whether or not the radiation characteristics of the superheavy elements had left "tracks" in the glass.

In 1976, a team of scientists from Oak Ridge National Laboratory, Florida State University, and the University of California at Davis announced that it had evidence for the existence of superheavy elements with atomic numbers 126 (strongest evidence), 116, 124, and 127. The scientists claimed to have detected the superheavy elements by bombarding small monazite inclusions (monazite is a rare earth-thorium phosphate mineral) in biotite (a mica mineral) with a finely focused proton beam to produce excited states. The energies of the X-ray signals observed agreed closely with those predicted for superheavy elements. Because the X-ray signals were weak, the evidence was not conclusive for the identification of the superheavy elements. Later it was found that the X-ray associated with element 126 was probably a gamma-ray resulting from proton interactions on the rare earth element cerium which is present in the monazite inclusions. This alternate explanation, and the lack of confirmatory evidence by more sensitive methods, has tended to negate the original findings. An interesting sidelight is that the monazite inclusions were selected as a likely material to contain the superheavy elements because of the presence of giant halos near the inclusions. These giant halos are damaged regions in the mineral caused by radioactive decay. The observed radii of the giant halos were thought to be too large to have been caused by alpha particles from any known naturally occurring alpha emitters. The monazite inclusions were at the center of these giant halos, and thus were assumed to contain the source of the high energy alpha particles. Later work by others suggested that the giant halos really resulted from the decay of naturally occurring alpha emitters and that the halos were enlarged by physical and chemical processes within the mineral.

If the half-lives of the superheavies are short enough so that they would have decayed away in the approximately four and a half billion years since the earth's creation, then they would no longer exist in nature. However, they might be detected by means of their decay products. Indirect evidence for the one-time existence of elements 113 to 115 in the solar system has been obtained by a group of University of Chicago scientists, who have reported possible decay products of superheavy elements 113 to 115 in the Allende meteorite. Attempts have also been made to synthesize superheavy elements by bombarding various heavy element targets with medium weight nuclei. For example, curium-248 bombarded with a beam of calcium-48 ions. However, all attempts at accelerator synthesis of superheavy elements have given negative results.

The existence of the superheavy elements has important implications for nuclear theory. If they are found they would provide data to test theoretical models of the nucleus such as the liquid drop plus shell model used to predict lifetimes in this region. Their existence in nature would also indicate whether or not superheavy elements were created by the process of nucleogenesis in stars. Because of these considerations, it is certain that there will be future attempts using other physical and chemical techniques involving separation, concentration, or nuclear bombardment to search for the superheavies.

Acknowledgments
The authors thank the large number of persons, who by correspondence and in discussion, contributed generously of their time and information. Special thanks are extended to Norman E. Holden of Brookhaven National Laboratory for his timely transmission of data, which otherwise might not have been available to us, and for many helpful discussions. Special thanks also are due to Professor A. H. Wapstra of the Technical University, Delft, Netherlands, who furnished us with the 1977 Atomic Mass Evaluation in advance of publication. Thanks also are due to Marianne Branagan, H. M. Eiland, C. R. Lubitz, J. E. MacDonald, W. E. Moore, J. T. Reynolds, and K. W. Seemann for their assistance in cross-section data evaluation and various other aspects of preparation of the accompanying Chart of the Nuclides. We are especially grateful to Claudette DesBois and F. Feiner without whose help this edition of the Chart would have been significantly delayed. We also were fortunate to have the help of the Technical and Document Library staffs at KAPL, in particular Helyn M. Walton, Elizabeth DeSimone, and Nora A. Balcer.

The Twelfth Edition of the Chart of the Nuclides is dedicated to Francis M. Rourke. "Frank" contributed greatly to the nuclear data evaluation before his untimely death in July of 1977.

3.3. NUCLEAR CROSS SECTIONS

Bernard L. Cohen

3.3.1. Definition, Measurement, and Use of Cross Sections

If one were to shoot a tiny bullet at an assemblage of objects, without aiming at any particular one of them, the probability of a hit, P, would be the fraction of the area as viewed by the bullet that is covered by the target objects, or

$$P = N\sigma \qquad\qquad (3.3\text{-}1)$$

where N is the number of objects per square centimeter of area perpendicular to the path of the bullet, and σ is the cross sectional area of each target object. In the world of atomic and nuclear particles, cross-sectional areas cannot be directly measured and even their meaning is not clear; however, when particles are shot at target nuclei, interesting events do occur and their probabilities are clearly proportional to N. Therefore, Equation 3.3-1 is still useful with σ serving as the proportionality constant. By analogy, it is called "cross section" and can be measured as the ratio P/N where N is now the number of target nuclei per square centimeter and P is the probability for any specific type of reaction or scattering event. For example, we can have $\sigma(n,\gamma)$ or $\sigma(n,p)$ which, in conjunction with Equation 3.3-1 give the probability for an (n,γ) or (n,p) reaction, respectively, when neutrons are incident on target nuclei. We can also have σ_T, the total cross section that gives the total probability for all reactions and scattering with incident neutrons.

If we are dealing with a stream of particles of intensity I (particles per second), and I is reduced by an amount dI by reactions and scatterings in travelling through the target material a distance dx, the probability of an interaction, P, is $-dI/I$, which from Equation 3.3-1 is

$$P = -\frac{dI}{I} = \sigma_T N = \sigma_T n\, dx \qquad\qquad (3.3\text{-}2)$$

where n is the number of target nuclei per cubic centimeter in the material, thus, N = ndx. Setting the second of these four equal quantities in Equation 3.3-2 equal to the fourth, integrating from x = 0 where I = I_0, and taking the exponential of both sides gives

$$I = I_0\, e^{-n\sigma_T x} \qquad\qquad (3.3\text{-}3)$$

By measuring the attenuation of a beam, I/I_0, as a function of thickness of material, x, one can measure σ_T by application of Equation 3.3-3.

One method for measuring the cross section for a nuclear reaction is by measuring the end product. For example, one can determine the cross section for

$$^{32}S\,(n,\,p)\,^{32}P \xrightarrow[\beta^- \text{ (14 day)}]{} S^{32}$$

by counting the emitted electrons from the beta decay at several times during its 14-day half-life decay and thereby calculate the total number of ^{32}P nuclei produced; this is just equal to the number of reactions that took place. If the number of incident neutrons is known, we then have the probability of a reaction per incident neutron, P; thus, $\sigma(n,p)$ can be calculated from Equation 3.3-1 by simply determining N, the number of sulfur nuclei per cubic centimeter. The number of sulfur nuclei may be determined by weighing the sulfur to find its mass and multiplying by Avogadro's number ($N_0 = 6.03 \times 10^{23}$ atoms/mol) divided by the grams per mole of sulfur, which in this case is just the atomic weight (32 g/mol); a measurement of the area (A_t) of the sulfur sample perpendicular to the direction of the incident beam then gives N (the number of nuclei/area) assuming that the thickness is uniform. Alternatively, N could be determined from N = nx, where x is the measured thickness and n can be calculated from the density of sulfur as obtained from published tables.

Instead of measuring the number of ^{32}P nuclei produced to determine $\sigma(n,p)$ for the above reaction, one could detect the emitted protons. This would have to be done at all angles of emission. A more practical procedure, however, is to measure separately the angular distribution of emitted protons by placing the detector sequentially at various angles. This gives $\frac{d\sigma}{d\Omega}$ (θ), defined analogously to σ in Equation 3.3-1 with P interpreted as the probability of an (n,p) reaction in which the proton is emitted within a unit solid angle at angle θ relative to the incident beam. The total cross section for the reaction is then

$$\sigma = \int \left(\frac{d\sigma}{d\Omega}\right) d\Omega = \int_0^\pi \frac{d\sigma}{d\Omega} (\theta) \cdot 2\pi \sin\theta \, d\theta \qquad (3.3\text{-}4)$$

The $d\sigma/d\Omega$ discussed here is called the "differential cross section"; it contains more detailed information about the reaction. Several other types of differential cross sections are sometimes used to display further detailed information. For example, if one is interested in the energy distribution of emitted protons, one can measure $\frac{d^2\sigma}{d\Omega dE}$ (θ, E) which is defined analogously to σ in Equation 3.3-1 with P interpreted as the probability for an (n,p) reaction in which the proton is emitted within a unit solid angle at θ and within a unit energy interval at energy E. The total (n,p) cross section is then

$$\sigma = \int_0^\infty \int_0^\pi \frac{d^2\sigma}{d\Omega dE} (\theta, E) \cdot 2\pi \sin\theta \, d\theta \cdot dE \qquad (3.3\text{-}5)$$

The unit for cross section is the barn where 1 barn = 10^{-24} cm^2. A millibarn, a microbarn, and a megabarn are 10^{-3}, 10^{-6}, and 10^6 barns, respectively. A typical unit for $(d\sigma/d\Omega)$ is mb/sr (millibarns/steradian).

If the cross section is known, the procedures outlined above can be used to calculate the probability for a reaction from Equation 3.3-1. For example, let the cross section for the ^{32}S (n,p) reaction be given as 300 mb, and seek the number of ^{32}P nuclei produced per second by 10^{13} neutrons per second passing through a sample 5 cm^2 in area that weighs 2 g. We may calculate N as

$$N = \frac{2\ g}{5\ cm^2} \times \frac{6 \times 10^{23}\ nuclei\ per\ mole}{32\ g/mol} = 7.5 \times 10^{21}\ nuclei\ per\ cm^2$$

From (1)

$$P = N\sigma = 7.5 \times 10^{21}\ \frac{nuclei}{cm^2} \times 300 \times 10^{-27}\ \frac{cm^2}{nucleus} = 2.2 \times 10^{-3}$$

The number of reactions per second is then the number of incident neutrons per second times this probability, $(1 \times 10^{13}) \times (2.2 \times 10^{-3}) = 2.1 \times 10^{10}$; this is then the number of ^{32}P nuclei produced/second by the beam of neutrons.

Expressing this procedure with symbols

$$R = I\sigma(M_t/A_t) \cdot (N_0\, ap/MW) \qquad (3.3\text{-}6)$$

where R = reactions per second, I = incident particles per second, M_t = mass of target, A_t = area of target, and $N_0 ap/MW$ = atoms of the target nucleus per gram of target material.

$$N_0 ap/MW = \frac{atoms}{molecule} \times \frac{molecules}{mol} \times \frac{mol}{gm} = ap \times N_0 \times 1/MW \qquad (3.3\text{-}7)$$

where p is the number of atoms per molecule of the element of interest (obtained from the chemical formula), and a is the isotopic abundance of the isotope of interest. For example, if we are interested in reactions in ^{37}Cl (a = 0.24) in a target of $CaCl_2$ (MW = 111, p = 2 atoms of Cl per molecule), we have MW/ap = 231.

As an illustration of the use of differential cross sections, assume that $d\sigma/d\Omega$ of the ^{32}S (n,p) reaction for protons emitted at $40°$ is 30 mb/sr and the beam and target are as above. How many protons per second strike a detector 1.5 cm^2 in area placed 5 cm from the target at $40°$? Using Equation 3.3-6 with σ and R changed to $d\sigma/d\Omega$ and $dR/d\Omega$ we find $\frac{dR}{d\Omega}$ = 2.2 × 10^9 reactions per sec-sr (in terms of reactions per second that emit protons per unit solid angle at $40°$ with respect to the incoming beam, at all azimuthal angles around the beam). The solid angle (in steradians) subtended by the detector equals its area divided by the square of the distance, $(1.5)/(5)^2$ = 0.06 sr. The number of protons striking the detector (ignoring absorption of protons by the target material) is 2.2 × 10^9 per second per steradian × 0.06 = 1.3 × 10^8 per second.

Our treatment to this point has been based on the assumption that the beam is smaller in area than the sample it intercepts. If the opposite is true, it is more appropriate to discuss the incident beam as a flux, ϕ, in neutrons per square centimeters per second. The effective beam then is ϕ times the area of the target, A_t, or

$$I = \phi A_t \qquad (3.3\text{-}7A)$$

When Equation 3.3-7A is inserted into Equation 3.3-6, A_t cancels, giving

$$R = \phi \sigma M_t \cdot \frac{N_0 ap}{MW} \qquad (3.3\text{-}8)$$

There is nothing in 3.3-8 that refers to the direction of the incident beam (as does A_T in Equation 3.3-6), so Equation 3.3-8 is valid where the flux is coming from diverse directions as in a reactor. For example, if a reactor flux is 10^{12} neutrons per square centimeter per second, the number of ^{32}P nuclei produced per second in a 1 g sample of sulfur is, from Equation 3.3-8

$$R = 1 \times 10^{12} \frac{neutrons}{cm^2\text{-}sec} \times 300 \times 10^{-27} \frac{cm^2}{nucleus} \times 1\,g \times \frac{6 \times 10^{23}\ nuclei\ per\ mole}{32\ g/mol} =$$

$$5.6 \times 10^9 \text{ per second}$$

When samples are thick, absorption of neutrons as given in Equation 3.3-3 must be taken into account. However, the situation is more complicated when the reaction is induced by charged particles such as protons since they lose energy rapidly as they pass through the target material, and σ is generally a sensitive function of this energy. For such cases Equation 3.3-6 must be converted to an integral

$$R = I \left(\frac{N_0 ap}{MW}\right) \int \sigma (E) [d(M_t/A_t)/dE]\ dE \qquad (3.3\text{-}9)$$

where the quantity in the square bracket is the thickness (in grams per cm^2) that reduces the energy by a unit amount or the inverse of what is commonly called the rate of energy loss, $-dE/d(\rho x)$ where ρ is the density of the material. The integral then becomes

$$R = I \left(\frac{N_0 ap}{MW}\right) \int_{E_i}^{0} \sigma (E) \left[\frac{dE}{d(\rho x)} (E)\right]^{-1} dE \qquad (3.3\text{-}10)$$

where E_i is the incident energy. The integral must be evaluated numerically. Graphs and tables of $[-dE/d(\rho x)]$ are widely available.[1] Methods and data for estimating this quantity, also called stopping power, are presented in Section 3.4.

3.3.2 Theoretical Estimates of Cross Sections[2]

3.3.2.1. Compound Nucleus Reactions in the MeV Region

The most common nuclear reaction mechanism below about 50 MeV is known as "compound nucleus". The incident particle strikes the target nucleus and is captured, forming a highly excited system called the compound nucleus in which the excitation energy is shared, as for molecules in a gas, among a considerable number of the neutrons and protons of which the nucleus is comprised. Eventually, in the course of the statistical fluctuations associated with this sharing, one of these particles (or some combination of them, such as an alpha particle) gets enough energy to escape. If it happened to be a proton, and if the incident particle was a neutron, the entire process would be described as an (n,p) reaction; however, except for the probability of forming the excited compound nucleus to begin with, the type of incident particle has nothing to do with the type of particle emitted. For example, if the compound nucleus were formed by an incident proton, alpha particle, or deuteron at the same excitation energy, its probability of decay by proton emission would be the same as if it were formed by an incident neutron.* Thus, the cross section for an (a,b) reaction (where a and b represent any particle) can be expressed as

$$\sigma (a, b) = \sigma_a \cdot \frac{f_b}{f_a + f_b + \ldots \ldots} \qquad (3.3\text{-}11)$$

where f_i is the probability per unit time for the compound nucleus to decay with emission of particle i, and the sum in the denominator is over all particles for which emission is energetically possible.

In Equation 3.3-11, σ_a is the cross section for formation of a compound nucleus; however, since this is by far the most probable result of particle *a* striking the nucleus, it is often taken to be the total reaction cross section, σ_r. Total reaction cross sections may be calculated with reasonable accuracy using widely-available computer programs based on quantum mechanical scattering theory with the nucleus represented by the "optical model". Such calculations are well developed and much used by nuclear physicists. For neutrons, the results may be reasonably well approximated by

$$\sigma_r = \pi (R + \lambda)^2 \qquad (3.3\text{-}12)$$

where $R = 1.5\ A^{1/3} \times 10^{-13}$ cm, often called the nuclear radius, A is the mass number of the target nucleus, and λ is $1/2\ \pi$ times the wavelength of the incident neutron = $(22\ \text{MeV}/E_n)^{1/2} \times 10^{-13}$ cm, where E_n is its energy. If we think of λ as the radius over which the neutron is spread out by its wave nature, Equation 3.3-12 is just the geometric cross section for the collision. For $E_n \gg 1$ MeV, $\lambda \ll R$; thus, σ_r is a slowly varying function of incident energy. For protons, alpha particles, and other charged particles, σ_r is reduced by the coulomb repulsion which prohibits any collisions at low energies and even at high energies deflects away some particles that would otherwise strike the nucleus. Plots of σ_r for protons and alpha particles are shown in Figure 3.3-1**. In summary, σ_a in Equation 3.3-11 is not difficult to obtain with a reasonable degree of accuracy.

---

* This is not quite true because the angular momentum brought in by the different particles is different; however, we ignore this complication.

** All figures appear at the end of the text.

The situation is more difficult for the f_i in Equation 3.3-11, but a statistical approach is available for estimates and we describe it briefly. Emission of particle i can occur with a series of energies corresponding to leaving the residual nucleus in its various states; we refer to each of these as a separate transition. In statistical theories, it is assumed that each of these transitions has an equal a priori probability; therefore, the more transitions available, the larger is f_i. The number of states increases exponentially with excitation energy; therefore, f_i is an exponentially increasing function of E_{0i}, the highest excitation energy of the residual nucleus that can be reached with emission of particle i. The other important factors affecting emission of particle i are the energy, E_i, with which it is emitted (the probability of each transition is proportional to E_i), and coulomb barriers that impede the emission of low-energy charged particles. As a consequence of these three factors, if emission of neutrons, protons, and alpha particles had the same E_{0i}, the energy distributions of emitted particles would be as in Figure 3.3-2. The curve for neutrons increases exponentially with decreasing E_n as this corresponds to leaving the residual nucleus with increasing excitation energy where there is an exponentially increasing number of states; for low energy neutrons, however, the energy spectrum is dominated by the proportionality to E_n. The spectrum is seen to resemble a Maxwell distribution for energies of molecules in a gas. The spectra for protons and alpha particles in Figure 3.3-2 are similar to that for neutrons except for the effects of the Coulomb barrier reducing low energy emissions; these effects are larger for alpha particles because they are doubly charged.

For the situation shown in Figure 3.3-2 where the three E_{0i} are equal, f_i would just be proportional to the areas under the three curves; hence, we see $f_n \gg f_p \gg f_\alpha$ and from Equation 3.3-11 neutron emission would predominate. However, since f_i is an exponentially increasing function of E_{0i}, if the energetics are such that $E_{0p} \gg E_{0n}$, proton emission can become important or even predominant.

Calculations of f_i depend on estimates of "level densities", the number of states as a function of excitation energy. Various estimates are available, and Figure 3.3-3 shows calculations of f_i based on one of them.[3]

To illustrate the methods of this section, let us calculate the cross section for a (p,α) reaction on ^{67}Zn induced by 12 MeV protons. From Equation 3.3-11

$$\sigma(p, \alpha) = \sigma_p \cdot \frac{f_\alpha}{f_n + f_p + f_\alpha} \qquad (3.3\text{-}11A)$$

From Figure 3.3-1 at 12 MeV for Z = 30 (the atomic number of Zn), σ_p = 0.82 barns. The f_i are a function of E_{0i} which are just

$$E_{0i} = E_a + Q(a, i) \qquad (3.3\text{-}13)$$

where E_a in this case is E_p = 12 MeV, and Q(a,i) is the energy release in an (a,i) reaction, readily available from tables.[4] For the reactions of interest here on Zn^{67}

$$Q(p, n) = -1.8, \quad E_{0n} = 10.2, \quad \log f_n = 5.5, \quad f_n = 3.2 \times 10^5$$
$$Q(p, p') - 0, \quad E_{0p} = 12.0, \quad \log f_p = 4.3, \quad f_p = 2.0 \times 10^4$$
$$Q(p, \alpha) = 2.4, \quad E_{0\alpha} = 14.4, \quad \log f_\alpha = 3.8, \quad f_\alpha = 6.3 \times 10^3$$

The Q-values are from Reference 4, the E_{0i} are from Equation 3.3-13 and with them the f_i are obtained from Figure 3.3-3. From Equation 3.3-11 then

$$\sigma(p, \alpha) = 0.82\, b \times \frac{6.3 \times 10^3}{(320 + 20 + 6.3) \times 10^3} = 0.015\, b$$

There are further refinements in this procedure, such as reducing E_{0i} by a pairing energy, Δ, if the final nucleus is odd-A and by 2Δ if it has even numbers of both neutrons and protons. Values of Δ are of the order of 1 MeV, but accurate values are available in the literature.[3] There are also other refinements available,[3] but in any case, the results of this type of calculation should be regarded only as rough estimates. There is an important exception to this statement in that we see from Figure 3.3-3 in nearly all realistic situations, f_n is much larger than f_p and f_α, and in these cases, Equation 3.3-11 reduces to

$$\sigma\,(a,\,n) \simeq \sigma_a \, .$$

which can be estimated rather accurately.

From Figure 3.3-2, we see that there is a strong tendency for particles to be emitted with much less than their maximum possible energy E_0; thus, the residual nucleus is left with considerable excitation energy. If this is sufficient to allow emission of another particle, that may happen in which case we get reactions like (p, αn), (n,2n), (α,2p), etc. As a result, there are a large number of complex reactions of these types occurring at higher energies. A relatively simple example is shown in Figure 3.3-4. The simplicity arises from the fact that the target has such a large charge (Z = 82) that particles other than neutrons are rarely emitted; therefore, the dominant reactions are (p,xn) with the cross section for the reaction with each value of x reaching its maximum several MeV above its energetic threshold and then decreasing at higher incident energies because that particular reaction is followed by further neutron emission.

From Figure 3.3-4, it is evident that at energies around 50 MeV, there may be about five neutrons emitted per reaction and this number increases with increasing bombarding energy. In the 10 to 20-MeV energy region, it is evident from our previous discussion that there is typically one neutron per reaction; so the cross section for production of neutrons is roughly equal to σ_R.

When particle emission is no longer energetically possible — this generally corresponds to excitation energies below about 8 MeV — the nucleus decays down to its ground state by successive gamma-ray emissions. These take much longer times (10^{-14} seconds or longer vs. 10^{-17}, typical for neutron emission with 1 MeV). In typical situations where reactions are induced by projectiles up to mass 4, about three gamma rays are emitted in the cascade; so there are typically three gamma rays per reaction for all energies above a few MeV. In reactions induced by energetic heavy ions, there is a great deal of angular momentum to be carried off, and since gamma rays rarely carry more than two units, there may be ten or more gamma rays per reaction.

A compound nucleus may also be formed when a nucleus is struck by an energetic gamma ray, and the manner in which it decays is no different than if it were formed at the same excitation energy by any other incident particle. The cross section for formation of a compound nucleus by gamma rays is dominated by effects of the photonuclear giant resonance that is centered at energies varying with mass from about 23 MeV in carbon to 12 MeV in heavy nuclei. The resonance is several megaelectronvolts wide and the cross section at its peak varies with target mass from less than 10 mb for carbon to over 500 mb in the heaviest nuclei.

3.3.2.2. Compound Nucleus Reactions Below 1 MeV

From Figure 3.3-2, we see that charged particles with less than 1-MeV incident energy cannot induce reactions except in the very lightest nuclei; therefore, low-energy reactions are induced principally by neutrons. On the other hand, many of the considerations to be discussed here apply up to several megaelectronvolts in light nuclei and include reactions induced by protons, deuterons, and alpha particles.

The distinction between reactions discussed here and in the previous section is that the energy levels in the compound nucleus at lower energies are well separated so a reaction can only occur if the incident energy is sufficient to excite one of them. The cross section as a function of incident energy is then characterized by narrow peaks called resonances. In the neighborhood of one of these resonances at neutron energy E_R, the cross section follows the well known Breit-Wigner formula

$$\sigma(n, b) = \pi \lambda^2 \frac{\Gamma_n \Gamma_b}{(E - E_R)^{2+} (\frac{\Gamma}{2})^2} \qquad (3.3\text{-}14)$$

where Γ_i are parameters called the "widths," which are characteristic of each individual resonance, and $\Gamma = \sum_i \Gamma_i$.

Another feature of low-energy reactions is that very few decay modes are available to the compound nucleus, and in most situations for neutrons below a few tenths megaelectronvolts, the only particle emission possible is neutron emission, leaving the residual nucleus in its lowest energy state. If this occurs, we have "elastic scattering" with the nucleus left unchanged. As the neutron energy is decreased, the time required for this neutron emission increases until, at about 100 eV in heavy nuclei, this time reaches 10^{-14} sec, which is roughly the time required for gamma-ray emission. Below this energy, gamma-ray emission usually occurs first; therefore, we have (n,γ) reactions that very often result in radioactive residual nuclei. For example, ^{23}Na (n,γ) produces ^{24}Na which is a beta-gamma emitter with a 15-hr half-life.

In elastic scattering, neutrons give up energy to the recoiling target nuclei until they reach thermal energies, about 0.025 eV, so the great majority of neutrons eventually become thermal neutrons. From our previous discussion, the dominant reaction at this energy is (n,γ). The cross section for thermal neutrons depends on the distance from the nearest resonance — (0.025 eV – E_R) in Equation 3.3-14 — and on Γ_n and Γ_γ for that resonance. Resonances are generally about 10 eV apart in heavy nuclei; thus, an average value for (0.025 eV – E_R) is about 5 eV. Typical values for widths in heavy nuclei are $\Gamma_\gamma \sim 0.15$ eV and $\Gamma_n \approx 10^{-3}$ $[E(eV)]^{1/2} = 2 \times 10^{-4}$ eV at E = 0.025 eV. Inserting these in Equation 3.3-14 gives a thermal neutron cross section $\sigma_{th} = 25$ b for typical heavy nuclei. However, locations of the closest resonance are subject to random variations between different nuclides, and if it should occur at 0.025 eV with these values for Γ_n and Γ_γ, Equation 3.3-14 gives $\sigma_{th} = 10^5$ b. Resonances near thermal energy occur by chance in several nuclei as in ^{113}Cd ($\sigma_{th} = 20,000$ b) and ^{135}Xe ($\sigma_{th} = 3 \times 10^6$ b). In light nuclei and in "closed-shell" heavy nuclei such as ^{208}Pb and ^{209}Bi, spacings between resonances are much larger and, thus, the closest resonance to thermal energy is usually very far away. This results in very small σ_{th}, e.g., 0.003 b in ^{12}C, 0.0002 b in ^{16}O, 0.0005 b in ^{208}Pb, etc.

In addition to (n,γ) reactions, low-energy neutrons can induce (n,p) and (n,α) reactions in a few nuclei light enough so coulomb barriers do not greatly inhibit emission of protons and alpha particles and where Q-values are especially favorable. Among these are ^{6}Li $(n,\alpha)^3$H, which is the source of tritium for hydrogen bombs and perhaps some day for controlled thermonuclear reactors; ^{10}B $(n,\alpha)^7$Li, which is the basis for a popular type of neutron detector and for control of reactors; and ^{14}N $(n,p)^{14}$C, which gives the basis for radiocarbon dating and is one of the more important reactions for consideration in interactions of thermal neutrons in tissue and in nuclear emulsions. Among heavy nuclei, the only possible reaction with low energy neutrons is (n,γ) except in the very few nuclei like ^{235}U, ^{239}Pu, etc., where low energy neutrons can induce fission.

3.3.2.3. Noncompound-nucleus Reactions

There are many situations in which a particle may strike a nucleus and induce a reaction without formation of a compound nucleus. These are sometimes called "direct reactions," and they have contributed greatly to our understanding of nuclear structure. We briefly discuss the most important types of these reactions.

3.3.2.3.1. Nucleon Transfer Reactions

A simple example of transfer reactions is (d,p) which may be pictured as a deuteron transferring its neutron into the nucleus as it passes and coming off as a proton. The neutron is said to be "stripped-off" and, thus, this is often called a stripping reaction. These reactions have been widely used for nuclear structure studies and, therefore, are very well understood. The energy spectrum of the protons is not unlike that from compound nucleus emission (see Figure 3.3-2) except that the coulomb barrier is less effective in inhibiting low-energy proton emission since the proton does not necessarily get very close to the nucleus. The reason for the preponderance of low-energy protons is that they correspond to inserting the neutron into a higher-energy orbit in the target nucleus; at higher energies, there are more of these orbits, and they are nearer to the surface of the nucleus. Typical total cross sections for (d,p) in the 10- to 20-MeV bombarding energy region are 100 mb. The (d,n) reaction that involves transfer of a proton rather than a neutron is essentially analogous except that the transfer of a proton is somewhat inhibited by coulomb barriers in heavy nuclei.

The inverse of a stripping reaction is "pick-up". For example, a proton passing a nucleus picks up a neutron from it and emerges as a deuteron. Deuterons are mostly emitted near the maximum available energy, as lower-energy deuterons correspond to picking up neutrons that are more strongly bound and further inside the nucleus. Total cross sections, therefore, are much smaller than for (d,p), typically of the order of 10 mb for incident protons in the 20-MeV region. The proton pick-up reaction (n,d) is analogous.

Many other one nucleon transfer reactions have been studied, such as (^{3}He,d), (^{4}He,t), (^{4}He,^{3}He), and their inverses and heavier ion reactions, such as (^{12}C,^{11}C), (^{16}O,^{15}O), (^{16}O,^{15}N), etc. There has also been extensive work on two nucleon transfer reactions, such as (p,t), (d,^{4}He), (^{3}He,p), and their inverses and many heavy ion reactions, such as (^{16}O,^{18}O), (^{14}N,^{12}C), etc. Other well known transfer reactions include 3-nucleon transfers, such as (p,α), (α,p), (^{16}O,^{13}N), etc. and alpha-transfers like (d,^{6}Li) and (^{16}O,^{12}C). In general, cross sections decrease with increasing complexity of the incident and outgoing particles and of that which is transferred, i.e., 2-nucleon transfers have smaller cross sections than one-nucleon transfers, and 3-nucleon transfer cross sections are still smaller. In pick-up reactions like (d,^{3}He), (p,α), etc., emitted particles tend to come off with nearly the maximum available energy while the opposite is true for stripping reactions like (^{3}He,d), (α,p), etc., and the latter have an order of magnitude larger total cross sections.

3.3.2.3.2. Direct Reaction Inelastic Scattering

Direct reaction inelastic scattering reactions, like (p,p$'$), (n,n$'$), (d,d$'$), (α,α'), and a few others, have proved very useful in nuclear structure studies and have thus been widely studied and are reasonably well understood. They are largely connected with excitation of rotational or vibrational motions in the nucleus, and their total cross sections are generally below 100 mb in the 20-MeV region.

3.3.2.3.3. Charge Exchange

Reactions like (p,n), (n,p), (^{3}He,t), etc. can proceed by a passing neutron changing into a proton or vice versa while the opposite occurs in the target nucleus. Cross sections are typically in the 1 mb region.

3.3.2.3.4. Particle-gamma Reactions

(p,γ) and (n,γ) reactions occur by a passing proton or neutron dropping into an orbit in the nucleus with the excess energy being carried off by a gamma ray. Typical cross sections in the 1- to 20-MeV region are about 1 mb.

3.3.2.3.5. Preequilibrium Reactions

In the formation of a compound nucleus, which is something of a statistical equilibrium energy sharing among the nucleons in the nucleus, there are large numbers of collisions between nucleons, whereas in most direct reactions, there are usually one or perhaps two collisions before the reaction is completed. Intermediate situations are called "preequilibrium reactions", although that term is often used to include direct reactions. In most situations, after two collisions, the energy available to any one nucleon is not sufficient for a quick escape; therefore, it is far more probable for further collisions to occur which means formation of a compound nucleus.

3.3.2.3.6. Knock-out Reactions

At energies above about 50 MeV, the initial stage of a nuclear reaction is best described as a simple nucleon-nucleon collision within the nucleus with a good chance for either or both colliding nucleons to emerge. At higher energies, there can be several collisions in which some of the principals emerge; after several such collisions, the energy available in single collisions is insufficient for particles to escape. This energy is shared among other nucleons in the nucleus in further collisions until equilibrium is reached; that is, we have a compound nucleus with an excitation energy in the 30- to 100-MeV-energy region. This then decays by further nucleon emission as described in our compound nucleus discussion, typically emitting perhaps three to six more neutrons. Note that the nucleons initially knocked-out are of high energy (tens of megaelectron-volts or more) and go predominantly in the direction of the incident particle, while those coming from the compound nucleus decay have energies of a few megaelectronvolts and come off with equal probability in all directions. A total of ten or more neutrons may be emitted from a heavy nucleus in a single reaction induced by a 1000-MeV particle.

There is little evidence for these knock-out reactions below 20 MeV, but above 100 MeV, their cross section approaches the geometric cross-sectional area of the nucleus which is of the order of 0.5 b.

3.3.3. Data Compilations

While theories are useful for gaining insight and understanding and for interpolating or extrapolating where experimental data are lacking, our knowledge of cross sections is largely based on experimental measurements. In general, these appear scattered through the scientific literature; therefore, compilations are necessary and, fortunately, are available. There are two principal sources for this information, the National Neutron Cross Section Center located at Brookhaven National Laboratory, and the journal *Nuclear Data Tables* (the name has recently been changed to *Atomic Data and Nuclear Data Tables*).

The Brookhaven Center is the U.S. segment of an international program with other segments operating in Saclay (France) representing Western Europe; Obninsk (USSR) representing the Soviet Union; and Vienna representing the International Atomic Energy Agency (IAEA), Eastern Europe, Asia, South America, Australia, and Africa. They operate an extensive computer data bank and occasionally issue publications. Among the more pertinent of these are: CINDA (Computer Index of Neutron Data), which gives a listing of literature references on neutron cross sections; Brookhaven National Laboratory report BNL-325 (3rd edition, 1976) entitled "Neutron Cross Sections," which gives data and derived parameters as a function of energy in graphical and tabular form; BNL-400 (3rd edition, 1975) entitled "Angular Distributions in Neutron Induced Reactions," which does likewise for angular distribution data; and BNL-17100 "ENDF (Evaluated nuclear data file) Cross Sections," which gives curves without data points for both cross sections and angular distributions. An alternative compilation of neutron data is the Lawrence Livermore Laboratory Experimental Cross Section Information Library

(ECSIL), a computer oriented system which in 1974 published a series of 14 thick volumes including separate plots of data from various measurements.

The journal *Nuclear Data Tables*, published approximately monthly by Academic Press, provides a wide variety of data from diverse sources including the U.S. Nuclear Data Compilation Center at Oak Ridge National Laboratory. The journal publishes up-to-date tables of nuclear masses, range-energy relationships, reaction Q-values, decay energies, etc., but we confine our discussion here to cross sections for nuclear reactions. One of the major services is an annual "Reaction List for Charged Particle Induced Reactions" which is an index to all papers in the scientific literature on that subject arranged so that sources of information on any reaction of interest can be rapidly located. Some other compilations of cross sections from this journal are listed below.

Reference 6 gives plots up to 15 MeV energy of cross sections for producing radioactive end products by (n,γ), (n,p), (n,a), $(n,2n)$, and (n,pn) reactions. Similar material is given in Reference 7.

Reference 8 gives these cross sections for neutron energies 13.1, 14.1, and 15.1 MeV and for a fission neutron spectrum.

Reference 9 gives plots vs. gamma-ray energy for all nuclei for which data are available of the total cross section for producing neutrons, and for producing one neutron, two neutrons, and three neutrons.

Plots and tables of cross sections and angular distributions for elastic scattering, (n,γ), (γ,n), and $(n,2n)$ reactions on H^1 and H^2 are given in References 10 and 11.

References 12 and 13 combine all available experimental information with theory to obtain total cross sections (σ_T); elastic scattering (σ_e) and reaction cross sections (σ_r); cross sections for (n,n'), $(n,2n)$, $(n,3n)$, and (n,γ); angular and energy distribution of emitted neutrons; gamma-ray production cross sections; and angular and energy distribution of emitted gamma rays for ^{206}Pb, ^{207}Pb, and ^{208}Pb.

Cross sections for producing monoenergetic neutrons and their angular distributions as a function of bombarding energy are given in References 14 and 15.

In References 16 to 18, a compilation of gamma ray energies and intensities from (n,γ) reactions is given. This is a joint effort between the Canadian and Soviet groups. Reference 19 gives gamma ray energies and intensities from neutron capture resonances and other energies in the range 5 to 300 keV.

Listed in Reference 20 are references on conversion electrons emitted in neutron capture reactions.

Reference 21 gives cross sections for production of gamma rays and their angular distributions from reactions induced by ^{14}N, ^{16}O, ^{18}O, and ^{19}F on ^{24}Mg, ^{26}Mg, ^{27}Al, and ^{28}Si with energies between 20 and 60 MeV.

Scattered compilations appear from time to time in various other publications and from other sources. The "Chart of the Nuclides," which hangs on walls of many nuclear installations and can be purchased from U.S. Government Printing Office (G.P.O.) Washington, D.C., gives thermal neutron capture cross sections and resonance integrals. The National Bureau of Standards publishes a compilation *Photonuclear Reaction Data* (NBS special publication 380 available from G.P.O.). The latest edition of the *Landoldt-Bornstein Tables*[22] (in section 3.3.4.1) gives excitation functions for charged-particle-induced reactions and combines them with theory to derive excitation functions and thick target yields for virtually all reactions on all target nuclei; thick target yields from these are given graphically below.

Many books on nuclear engineering have extensive tables, as do several popular handbooks of Chemistry, Physics, and Engineering.

3.3.4. Graphs and Tables
3.3.4.1. Thick Target Yields from Charged Particle Reactions

The plots given here are adapted from Landolt-Bornstein new data tables.[22] The results should be accurate within about a factor of three; more accurate results can be derived by using the more elaborate prescriptions given in the reference. Estimated yields are plotted as microcuries of product nucleus produced per microamp of beam assuming that the bombarding time, T, is much longer than the half-life, $0.69/\lambda$ where λ is the decay constant. For shorter bombarding times, the yield, Y, is

$$Y = Y_0 \ [1 - \exp(-\lambda T)]$$

where Y_0 is the yield given in the plots. The plots are based on the assumption that the target material is pure; if it is not, the yield is reduced roughly by the fraction of the target mass that is the target nucleus of interest.

For a very long bombarding time, the rate of decay, $3.7 \times 10^4 \ Y_0$ per second, is equal to the rate of production and, hence, to the rate at which reactions occur. If there are x neutrons produced per reaction (e.g., x = 2 in a (p,2n) reaction) the thick target neutron yield, Y_N, is then

$$Y_N = 3.7 \times 10^4 \ x \ Y_0 \ \text{neutrons per second per microamp}$$

The abscissas in the plots is the bombarding energy, E, minus the threshold, E_{th}. The latter is the negative of the reaction Q-value which is available from published tables,[4] or which can be calculated from mass tables.[5]

3.3.4.2. Neutron Cross Sections and Angular Distributions

Reproduced here in Figures 3.3-6.1 to 3.3-6.71 are curves for cross sections vs. energy up to 20 MeV for all neutron-induced reactions in selected elements and plots of elastic scattering cross sections for several energies and of the average scattering angle in the laboratory system, "mu bar". In a few cases, expanded plots are given for the resonance region. For fissile nuclei, data are also given for neutrons per fission and for the fission neutron energy spectra at a few energies. All data are from BNL-17100 except that data for ^{31}P and ^{32}S are not available there; hence, they were taken from BNL-325.

3.3.4.3. Activation Cross Sections vs. Energy

Plots are given in Figure 3.3-7 of cross section vs. energy for neutron reactions exciting long half-life radioactive products in selected nuclei. Cases where equivalent plots are available in section 3.3.4.2 are omitted. Data are from Reference 6.

3.3.4.4. Neutron Cross Sections

Table 3.3-1 from BNL-17100 gives thermal neutron cross sections (averaged over a Maxwell energy distribution for neutrons with 300°K temperature), cross sections for 14-MeV neutrons, and Resonance integrals (RI) defined as

$$RI = \int \frac{\sigma dE}{E}$$

where the limits in the integral are 0.5 eV to 20 MeV

3.3.4.5. Cross Sections for Photoneutron Reactions

Plots in Figure 3.3-8 are selections from a much more extensive group in Reference 9. Nuclear Data Tables, Vol. 15, p. 319 (1975). Cross sections shown are the sum of cross sections for all reactions that produce neutrons, including (γ,n), $(\gamma,2n)$, $(\gamma,3n)$, (γ,pn), etc. The original reference separates out some of these and they should be used to calculate neutron yields for gamma ray energies above the threshold for $(\gamma,2n)$ reactions. Note that the shape of the excitation function is dominated by the giant resonance which varies smoothly with nuclear mass. For nonspherical nuclei like Ta and U, the resonance is double-peaked and wider than for others.

Table 3.3-1
THERMAL AND 14 MeV NEUTRON CROSS SECTIONS
AND RESONANCE INTEGRALS

Material	Reaction	Thermal	14 MeV	Resonance integral
Hydrogen-1	Total	2.07975 + 1	6.92900 − 1	2.64600 + 2
	Elastic	2.04486 + 1	6.92900 − 1	2.64000 + 2
	(n,gamma)	2.94211 − 1	2.98300 − 5	1.49100 − 1
Hydrogen-2	Total	3.34985 + 0	7.97000 − 1	5.31700 + 1
	Elastic	3.35000 + 0	6.16000 − 1	5.29700 + 1
	Direct(n,2n)		1.61000 − 1	1.90400 − 1
	(n,gamma)	4.60714 − 4	9.17400 − 6	2.65400 − 4
Hydrogen-3	Total	1.29997 + 0	9.75000 − 1	2.44800 + 1
	Elastic	1.29997 + 0	9.25000 − 1	2.44400 + 1
	Direct(N,2N)		5.00000 − 2	3.69800 − 2
Helium-3	Total	4.72455 + 3	1.17000 + 0	2.42600 + 3
	Elastic	9.99965 − 1	9.71900 − 1	2.24500 + 1
	(n,p)	4.72039 + 3	1.22000 − 1	2.39000 + 3
	(n,d)		7.61000 − 2	8.85700 − 2
Helium-4	Total	7.59160 − 1	1.05900 + 0	2.20000 + 1
	Elastic	7.59160 − 1	1.05900 + 0	2.20000 + 1
Lithium-6	Total	8.33865 + 2	1.42000 + 0	4.44100 + 2
	Elastic	7.21238 − 1	8.80000 − 1	1.75800 + 1
	Inelastic		4.37800 − 1	1.09100 + 0
	(n,gamma)	3.41197 − 2	1.01700 − 5	1.73500 − 2
	(n,p)		7.20000 − 3	2.79900 − 2
	(n,alpha)	8.33085 + 2	2.60000 − 2	4.24900 + 2
Lithium-7	Total	1.08225 + 0	1.45400 + 0	2.26100 + 1
	Elastic	1.04986 + 0	9.74000 − 1	2.14900 + 1
	Inelastic		4.15000 − 1	1.12700 + 0
	Direct(n,2n)		2.20000 − 2	1.43100 − 2
	(n,gamma)	3.17493 − 2	1.00000 − 5	1.65300 − 2
	(n,d)		1.00000 − 2	5.73800 − 3
Beryllium-9	Total	6.00843 + 0	1.49100 − 0	8.90300 + 1
	Elastic	6.00000 + 0	9.40000 − 1	8.78500 + 1
	(n,gamma)	8.42580 − 3	1.00000 − 4	5.25100 − 3
	(n,p)		0.00000 + 0	3.86300 − 4
	(n,d)		0.00000 + 0	9.64700 − 4
	(n,t)		1.80000 − 2	1.85600 − 2
	(n,alpha)		1.10000 − 2	1.23600 − 1
Boron-10	Total	3.40246 + 3	1.45000 + 0	1.76200 + 3
	Elastic	2.10577 + 0	9.52800 − 1	3.78500 + 1
	Inelastic		2.85500 − 1	3.96500 − 1
	(n,p)	5.01593 − 4	3.28000 − 2	6.43900 − 2
	(n,d)		2.99700 − 2	3.96800 − 2
	(n,alpha)	3.40013 + 3	6.02200 − 2	1.72100 + 3
Boron-11	Total	5.04035 + 0	1.32000 + 0	7.50500 + 1
	Elastic	5.03580 + 0	6.02200 − 1	7.44600 + 1
	Inelastic		6.49800 − 1	5.55100 − 1
	Direct(N,2N)		2.00000 − 2	8.24500 − 3
	(n,gamma)	4.43136 − 3	2.15000 − 7	2.52800 − 3
	(n,p)		2.44000 − 3	1.41000 − 3

Table 3.3-1 (continued)
THERMAL AND 14 MeV NEUTRON CROSS SECTIONS
AND RESONANCE INTEGRALS

Material	Reaction	Thermal	14 MeV	Resonance integral
	(n,t)		1.50000 − 2	7.18300 − 3
	(n,alpha)		3.05500 − 2	1.78000 − 2
Carbon-12	Total	4.73130 + 0	1.27000 + 0	7.08600 + 1
	Elastic	4.72791 + 0	7.37900 − 1	7.02700 + 1
	Inelastic		4.50200 − 1	4.79500 − 1
	(n,gamma)	2.99174 − 3	0.00000 + 0	1.51000 − 3
	(n,alpha)		8.15100 − 2	1.07200 − 1
Nitrogen-14	Total	1.16716 + 1	1.56900 + 0	1.19800 + 2
	Elastic	9.95655 + 0	8.84900 − 1	1.17800 + 2
	Inelastic		4.30000 − 1	3.84500 − 1
	Direct(N,2N)		6.53000 − 3	3.89800 − 3
	(n,gamma)	6.64501 − 2	1.65000 − 5	3.38100 − 2
	(n,p)	1.61205 + 0	4.32500 − 2	9.65800 − 1
	(n,o)		4.11200 − 2	3.26400 − 2
	(n,t)		2.99500 − 2	2.74100 − 2
	(n,alpha)		1.01700 − 1	3.76500 − 1
	(n,2alpha)		3.15300 − 2	2.19200 − 2
Oxygen-16	Total	3.74834 + 0	1.61900 + 0	6.08500 + 1
	Elastic	3.74810 + 0	9.55900 − 1	6.02700 + 1
	Inelastic		4.74400 − 1	3.96500 − 1
	(n,gamma)	1.57745 − 4	7.56600 − 9	8.00700 − 5
	(n,p)		4.71200 − 2	2.00500 − 2
	(n,d)		1.53000 − 2	4.97600 − 3
	(n,alpha)		1.26300 − 1	1.53300 − 1
Fluorine-19	Total	4.00946 + 0	1.74800 + 0	7.23300 + 1
	Elastic	3.99730 + 0	8.26600 − 1	6.53000 + 1
	Inelastic		7.49100 − 1	6.73200 + 0
	Direct(N,2N)		4.90000 − 2	3.78300 − 2
	(n,gamma)	8.41934 − 3	1.00000 − 5	2.13500 − 2
	(n,p)		1.70000 − 2	3.31100 − 2
	(n,d)		1.20000 − 2	9.61000 − 3
	(n,t)		7.62200 − 3	5.18800 − 3
	(n,alpha)		2.50000 − 2	1.39300 − 1
Sodium-23	Total	3.79072 + 0	1.62800 + 0	1.37600 + 2
	Elastic	3.30728 + 0	7.81200 − 1	1.34800 + 2
	Inelastic		6.41000 − 1	2.33900 + 0
	Direct(N,2N)		2.00000 − 2	1.78500 − 2
	(n,gamma)	4.73265 − 1	2.12300 − 4	3.45900 − 1
	(n,p)		4.56000 − 2	6.43100 − 2
	(n,alpha)		1.40000 − 1	1.11900 − 1
Natural magnesium	Total	3.40298 + 0	1.78000 + 0	6.71200 + 1
	Elastic	3.33996 + 0	7.95700 − 1	6.51200 + 1
	Inelastic		5.87600 − 1	1.64300 + 0
	Direct(n,2n)		3.56000 − 2	2.80900 − 2
	(n,gamma)	5.58636 − 2	3.00000 − 4	3.34400 − 2
	(n,p)		1.71500 − 1	1.25900 − 1
	(n,alpha)		1.76500 − 1	1.44600 − 1

Table 3.3-1 (continued)
THERMAL AND 14 MeV NEUTRON CROSS SECTIONS
AND RESONANCE INTEGRALS

Material	Reaction	Thermal	14 MeV	Resonance integral
Aluminum-27	Total	1.58178 + 0	1.74400 + 0	4.07800 + 1
	Elastic	1.34767 + 0	6.51400 – 1	3.86000 + 1
	Inelastic		8.39400 – 1	1.72500 + 0
	Direct(n,2n)		2.40000 – 2	8.13200 – 2
	(n,gamma)	2.05764 – 1	4.99600 – 4	1.32600 – 1
	(n,p)		7.77000 – 2	1.03300 – 1
	(n,d)		2.63000 – 2	1.44700 – 2
	(n,t)		9.31200 – 6	2.25300 – 3
	(n,alpha)		1.24700 – 1	8.86400 – 2
Natural silicon	Total	2.30094 + 0	1.81400 + 0	4.11400 + 1
	Elastic	2.15000 + 0	7.30700 – 1	3.91300 + 1
	Inelastic		5.27600 – 1	1.31700 + 0
	Direct (n,2n)		4.66300 – 2	2.87400 – 2
	(n,gamma)	1.42226 – 1	5.00000 – 4	8.26400 – 2
	(n,p)		2.33300 – 1	2.64800 – 1
	(n,d)		1.88800 – 2	6.95800 – 3
	(n,alpha)		1.71100 – 1	1.63700 – 1
Natural chlorine	Total	4.63660 + 1	2.10000 + 0	1.20500 + 2
	Elastic	1.64611 + 1	1.03400 + 0	1.04400 + 2
	Inelastic		8.20000 – 1	1.67300 + 0
	Direct(n,2n)		2.20000 – 2	1.32200 – 2
	(n,gamma)	2.94927 + 1	1.00000 – 5	1.38200 + 1
	(n,p)	3.14630 – 1	9.80000 – 2	4.15800 – 1
	(n,alpha)		8.95000 – 2	1.59600 – 1
Natural potassium	Total	4.09847 + 0	2.04600 + 0	4.09600 + 1
	Elastic	2.17580 + 0	9.64000 – 1	3.77000 + 1
	Inelastic		3.99300 – 1	9.98600 – 1
	Direct(n,2n)		2.60000 – 3	7.40300 – 3
	(n,gamma)	1.86120 + 0	3.57100 – 5	1.21100 + 0
	(n,p)	4.52001 – 2	3.64000 – 1	6.65400 – 1
	(n,alpha)	4.07693 – 3	1.11800 – 1	2.27800 – 1
Natural calcium	Total	3.49483 + 0	2.14900 + 0	4.42700 + 1
	Elastic	2.98756 + 0	8.75800 – 1	4.17900 + 1
	Inelastic		2.43400 – 1	5.89600 – 1
	Direct(n,2n)		1.70000 – 3	4.95800 – 3
	(n,gamma)	3.81172 – 1	1.34200 – 5	2.22800 – 1
	(n,p)		2.15000 – 1	8.39100 – 1
	(n,d)		1.80000 – 2	1.27800 – 2
	(n,t)		4.26600 – 6	2.77000 – 3
	(n,helium-3)		7.30000 – 3	3.63600 – 3
	(n,alpha)	2.24080 – 3	1.40000 – 1	3.89700 – 1
	(n,2alpha)		7.00000 – 6	1.12500 – 3
Natural titanium	Total	9.80823 + 0	2.43900 + 0	1.53700 + 2
	Elastic	4.40000 + 0	1.20000 + 0	1.47700 + 2
	Inelastic		6.30000 – 1	2.81700 + 0
	Direct(n,2n)		5.00000 – 1	2.33800 – 1
	(n,gamma)	5.40828 + 0	2.62900 – 4	2.87400 + 0
	(n,p)		6.33300 – 2	3.96700 – 2
	(n,alpha)		4.50000 – 2	2.57400 – 2

Table 3.3-1 (continued)
THERMAL AND 14 MeV NEUTRON CROSS SECTIONS
AND RESONANCE INTEGRALS

Material	Reaction	Thermal	14 MeV	Resonance integral
Natural vanadium	Total	9.50047 + 0	2.32000 + 0	2.47400 + 2
	Elastic	4.99915 + 0	1.00600 + 0	2.41000 + 2
	Inelastic		5.63300 − 1	3.35900 + 0
	Direct(n,2n)		6.35000 − 1	3.64500 − 1
	(n,gamma)	4.48458 + 0	1.71400 − 4	2.53600 + 0
	(n,p)		3.61000 − 2	3.68000 − 2
	(n,d)		3.02000 − 2	1.99500 − 2
	(n,t)		1.23000 − 2	9.61600 − 3
	(n,alpha)		1.94000 − 2	1.29000 − 2
Natural chromium	Total	4.52048 + 0	2.42300 + 0	4.06000 + 1
	Elastic	3.14418 + 0	1.11200 + 0	3.67600 + 1
	Inelastic		8.07500 − 1	2.69700 + 0
	Direct(n,2n)		3.39700 − 1	2.49600 − 1
	(n,gamma)	1.35938 + 0	8.84000 − 4	7.38500 − 1
	(n,p)		3.22700 − 2	7.65200 − 2
	(n,d)		1.42900 − 2	9.66500 − 3
	(n,t)		8.61200 − 4	2.74200 − 3
	(n,helium-3)		3.49800 − 7	1.78000 − 4
	(n,alpha)		4.10000 − 2	2.41400 − 2
Manganese-55	Total	1.35487 + 1	2.58500 + 0	6.37600 + 2
	Elastic	1.75554 + 0	1.26500 + 0	6.17300 + 2
	Inelastic		3.93700 − 1	4.45300 + 0
	Direct(n,2n)		7.69000 − 1	4.27200 − 1
	(n,3n)		0.00000 + 0	8.53400 − 6
	(n,gamma)	1.17931 + 1	7.70000 − 4	1.53100 + 1
	(n,p)		4.70000 − 2	4.08800 − 2
	(n,d)		4.39800 − 3	2.85800 − 3
	(n,helium-3)		4.20000 − 4	1.57400 − 4
	(n,alpha)		3.26000 − 2	2.17200 − 2
Natural Iron	Total	1.36733 + 1	2.57200 + 0	1.40400 + 2
	Elastic	1.13997 + 1	1.17100 + 0	1.35400 + 2
	Inelastic		7.49200 − 1	3.09600 + 0
	Direct(n,2n)		4.32000 − 1	2.48100 − 1
	(n,gamma)	2.26883 + 0	2.10500 − 4	1.41600 + 0
	(n,p)		1.25000 − 1	1.21500 − 1
	(n,d)		1.90000 − 2	1.17500 − 2
	(n,t)		1.00000 − 4	3.99700 − 3
	(n,helium-3)		1.00000 − 4	1.51400 − 3
	(n,alpha)		3.99000 − 2	2.75500 − 2
Cobalt-59	Total	3.96299 + 1	2.68900 + 0	8.57700 + 2
	Elastic	6.62328 + 0	1.47800 + 0	7.77800 + 2
	Inelastic		4.56000 − 1	2.79300 + 0
	Direct(n,2n)		6.40000 − 1	3.57200 − 1
	(n,gamma)	3.30066 + 1	1.50000 − 3	7.66700 + 1
	(n,p)		8.15000 − 2	8.08700 − 2
	(n,d)		1.90000 − 3	1.71200 − 3
	(n,t)		9.00000 − 4	5.69100 − 4
	(n,alpha)		2.91000 − 2	1.90500 − 2
Natural nickel	Total	2.19574 + 1	2.71600 + 0	2.44900 + 2
	Elastic	1.78646 + 1	1.33000 + 0	2.39500 + 2

Table 3.3-1 (continued)
THERMAL AND 14 MeV NEUTRON CROSS SECTIONS
AND RESONANCE INTEGRALS

Material	Reaction	Thermal	14 MeV	Resonance integral
	Inelastic		7.49300 − 1	2.24300 + 0
	Direct(n,2n)		1.36000 − 1	8.30400 − 2
	(n,gamma)	4.05905 + 0	0.00000 + 0	2.25500 + 0
	(n,p)		2.97600 − 1	6.65200 − 1
	(n,alpha)		1.46400 − 1	1.52300 − 1
Natural copper	Total	1.23662 + 1	2.95700 + 0	1.50400 + 2
	Elastic	8.61142 + 0	1.45300 + 0	1.42000 + 2
	Inelastic		6.63800 − 1	3.29500 + 0
	Direct(n,2n)		6.01500 − 1	3.67400 − 1
	(n,3n)		0.00000 + 0	1.40200 − 3
	(n,gamma)	3.35733 + 0	2.30000 − 3	4.53400 + 0
	(n,p)		7.94700 − 2	1.42500 − 1
	(n,d)		5.38000 − 3	3.43000 − 3
	(n,helium-3)		2.27000 − 3	1.38400 − 3
	(n,alpha)		3.25700 − 2	2.73400 − 2
Krypton-78	Total	1.11571 + 1	3.64800 + 0	2.51600 + 2
	Elastic	6.86649 + 0	1.90100 + 0	1.88700 + 2
	Inelastic		1.36500 + 0	5.21000 + 0
	Direct(n,2n)		1.98500 − 1	1.29100 − 1
	(n,gamma)	4.29062 + 0	4.90000 − 3	5.73900 + 1
	(n,p)	1.89870 − 12	1.07000 − 1	8.70100 − 2
	(n,d)		2.25000 − 2	1.91100 − 2
	(n,t)		0.00000 + 0	3.63200 − 3
	(n,helium-3)		5.40000 − 4	1.65600 − 3
	(n,alpha)	4.18831 − 13	4.86000 − 2	3.17800 − 2
Krypton-80	Total	2.12512 + 1	3.69000 + 0	2.82100 + 2
	Elastic	8.56892 + 0	1.93800 + 0	2.15700 + 2
	Inelastic		8.99700 − 1	4.74100 + 0
	Direct(n,2n)		7.77000 − 1	4.49700 − 1
	(n,gamma)	1.26823 + 1	5.43000 − 3	6.11300 + 1
	(n,p)		4.98000 − 2	3.19400 − 2
	(n,d)		1.26000 − 2	1.31900 − 2
	(n,t)		0.00000 + 0	2.44000 − 3
	(n,helium-3)		4.02300 − 5	5.12800 − 4
	(n,alpha)	1.99443 − 13	7.14000 − 3	4.38000 − 3
Krypton-82	Total	3.68855 + 1	3.73000 + 0	3.94100 + 2
	Elastic	1.01419 + 1	1.97500 + 0	2.06000 + 2
	Inelastic		6.09200 − 1	4.44100 + 0
	Direct(n,2n)		1.11000 + 0	5.95900 − 1
	(n,gamma)	2.67436 + 1	4.26000 − 3	1.83100 + 2
	(n,p)		2.13000 − 2	1.12800 − 2
	(n,d)		6.10000 − 3	8.03000 − 3
	(n,t)		0.00000 + 0	1.60300 − 3
	(n,helium-3)		0.00000 + 0	9.69900 − 4
	(n,alpha)		4.45000 − 3	2.49600 − 3
Krypton-83	Total	1.93219 + 2	3.77700 + 0	3.71000 + 2
	Elastic	1.02049 + 1	2.00300 + 0	1.71300 + 2
	Inelastic		3.31000 − 1	6.88000 + 0
	Direct(n,2n)		1.41000 + 0	1.13600 + 0
	(n,3n)		0.00000 + 0	3.21400 − 5

Table 3.3-1 (continued)
THERMAL AND 14 MeV NEUTRON CROSS SECTIONS
AND RESONANCE INTEGRALS

Material	Reaction	Thermal	14 MeV	Resonance integral
	(n,gamma)	1.83014 + 2	5.95000 – 3	1.91700 + 2
	(n,p)		1.62000 – 2	1.19100 – 2
	(n,d)		5.90000 – 3	7.19300 – 3
	(n,t)		2.69000 – 3	4.84500 – 3
	(n,helium-3)		2.58700 – 5	5.59900 – 5
	(n,alpha)		2.53000 – 3	1.69700 – 3
Krypton-84	Total	6.78319 + 0	3.77000 + 0	1.37400 + 2
	Elastic	6.70771 + 0	2.01300 + 0	1.29100 + 2
	Inelastic		4.39100 – 1	4.08200 + 0
	Direct(n,2n)		1.30000 + 0	7.05400 – 1
	(n,gamma)	7.54778 – 2	3.90000 – 3	3.50100 + 0
	(n,p)		8.85000 – 3	4.26700 – 3
	(n,d)		2.97000 – 3	4.79800 – 3
	(n,t)		2.31000 – 5	1.06300 – 3
	(n,helium-3)		0.00000 + 0	5.81500 – 6
	(n,alpha)	6.38219 –13	2.16000 – 3	1.39300 – 3
Krypton-86	Total	6.06751 + 0	3.80800 + 0	1.29000 + 2
	Elastic	6.01149 + 0	2.05000 + 0	1.24200 + 2
	Inelastic		3.43100 – 1	3.83100 + 0
	Direct(n,2n)		1.41000 + 0	7.76300 – 1
	(n,3n)		0.00000 + 0	1.72500 – 2
	(n,gamma)	5.60244 – 2	7.00000 – 4	1.41600 – 1
	(n,p)		1.89000 – 3	1.59300 – 3
	(n,d)		1.00000 – 3	2.30000 – 3
	(n,t)		0.00000 + 0	5.33600 – 4
	(n,alpha)	3.98886 –13	1.65000 – 3	9.57700 – 4
Zircalloy-2	Total	6.72113 + 0	3.88300 + 0	1.28700 + 2
	Elastic	6.20000 + 0	2.21000 + 0	1.24000 + 2
	Inelastic		7.50000 – 1	3.15100 + 0
	Direct(n,2n)		8.84000 – 1	6.08800 – 1
	(n,gamma)	1.84348 – 1	2.80000 – 3	1.15500 + 0
	(n,p)		3.20000 – 2	2.39700 – 2
	(n,alpha)		4.00000 – 3	2.13900 – 3
Niobium-93	Total	6.73642 + 0	3.98000 + 0	1.22400 + 2
	Elastic	5.66393 + 0	2.29600 + 0	1.07600 + 2
	Inelastic		4.16000 – 1	4.60700 + 0
	Direct(n,2n)		1.21500 + 0	7.24000 – 1
	(n,3n)		0.00000 + 0	1.84200 – 2
	(n,gamma)	1.07250 + 0	1 8.00000 – 4	9.46100 + 0
	(n,p)		4.00000 – 2	3.45400 – 2
	(n,alpha)		9.20000 – 3	6.72800 – 3
Natural molybdenum	Total	7.41120 + 0	1 3.90100 + 0	1.70500 + 2
	Elastic	5.00000 + 0	1 2.05000 + 0	1.35600 + 2
	Inelastic		3.50000 – 1	4.29500 + 0
	Direct(n,2n)		1.50000 + 0	9.02500 – 1
	(n,3n)		0.00000 + 0	1.75200 – 1
	(n,gamma)	2.34866 + 0	1.31600 – 3	2.93300 + 1
Technetium-99	Total	2.19597 + 1	4.12500 + 0	4.81900 + 2
	Elastic	5.03235 + 0	2.10000 + 0	1.21600 + 2

Table 3.3-1 (continued)
THERMAL AND 14 MeV NEUTRON CROSS SECTIONS
AND RESONANCE INTEGRALS

Material	Reaction	Thermal	14 MeV	Resonance integral
	Inelastic		5.99400 − 1	5.88300 + 0
	Direct(n,2n)		1.42500 + 0	9.44400 − 1
	(n,gamma)	1.69273 + 1	4.40000 − 4	3.53400 + 2
Rhodium-103	Total	1.37792 + 2	4.25300 + 0	1.16500 + 3
	Elastic	3.46872 + 0	2.92500 + 0	1.09500 + 2
	Inelastic		5.32000 − 1	7.00600 + 0
	Direct(n,2n)		7.95500 − 1	4.44700 − 1
	(n,gamma)	1.34323 + 2	7.40000 − 4	1.04800 + 3
Silver-107	Total	3.83046 + 1	4.27000 + 0	2.46800 + 2
	Elastic	5.62203 + 0	2.71000 + 0	1.25600 + 2
	Inelastic		1.41900 − 2	5.74000 + 0
	Direct(n,2n)		1.49000 + 0	8.88400 − 1
	(n,gamma)	3.26825 + 1	1.69800 − 6	1.16300 + 2
	(n,p)		1.59200 − 2	1.08800 − 2
	(n,d)		1.43000 − 2	9.97300 − 3
	(n,t)		1.70000 − 3	1.47000 − 3
	(n,alpha)		2.38900 − 2	1.85900 − 2
Silver-109	Total	8.34426 + 1	4.26500 + 0	1.70500 + 3
	Elastic	1.67224 + 0	2.81000 + 0	2.41400 + 2
	Inelastic		4.81400 − 2	5.84000 + 0
	Direct(n,2n)		1.38000 + 0	8.47500 − 1
	(n,gamma)	8.17703 + 1	0.00000 + 0	1.45800 + 3
	(n,p)		1.66200 − 2	1.07400 − 2
	(n,alpha)		1.02400 − 2	8.38000 − 3
Natural cadmium	Total	2.90771 + 3	4.62000 + 0	1.87900 + 2
	Elastic	1.01219 + 1	2.65000 + 0	1.09400 + 2
	Inelastic		3.66000 − 1	5.00900 + 0
	Direct(n,2n)		1.58100 + 0	9.51600 − 1
	(n,gamma)	2.89759 + 3	2.30000 − 3	7.20100 + 1
	(n,p)		1.86000 − 2	1.29000 − 2
	(n,alpha)		2.60000 − 3	1.72500 − 3
Cadmium-113	Total	2.36268 + 4	4.62000 + 0	5.30600 + 2
	Elastic	4.74842 + 1	2.65000 + 0	1.16700 + 2
	Inelastic		2.16000 − 1	5.58200 + 0
	Direct(n,2n)		1.73700 + 0	1.22600 + 0
	(n,gamma)	2.35795 + 4	2.80000 − 3	4.05800 + 2
	(n,p)		1.27000 − 2	8.50600 − 3
	(n,alpha)		1.70000 − 3	1.06500 − 3
Xenon-124	Total	1.18337 + 2	4.80500 + 0	3.06400 + 3
	Elastic	4.26881 + 0	2.93200 + 0	5.75500 + 2
	Inelastic		7.57300 − 1	5.60000 + 0
	Direct(n,2n)		1.10000 + 0	6.00600 − 1
	(n,3n)		0.00000 + 0	3.37500 − 3
	(n,gamma)	1.14048 + 2	0.00000 + 0	2.48300 + 3
	(n,p)		1.16000 − 2	5.83900 − 3
	(n,d)		2.18000 − 3	5.12700 − 3
	(n,t)		2.05000 − 4	8.19400 − 4
	(n,helium-3)		1.09000 − 6	3.47300 − 4
	(n,alpha)		1.81000 − 3	1.54900 − 3

Table 3.3-1 (continued)
THERMAL AND 14 MeV NEUTRON CROSS SECTIONS
AND RESONANCE INTEGRALS

Material	Reaction	Thermal	14 MeV	Resonance integral
Xenon-126	Total	1.89716 + 1	4.84900 + 0	1.23100 + 2
	Elastic	4.30008 + 0	2.96800 + 0	1.03900 + 2
	Inelastic		5.34100 − 1	5.30000 + 0
	Direct(n,2n)		1.34000 + 0	7.47500 − 1
	(n,3n)		0.00000 + 0	1.27600 − 2
	(n,gamma)	1.46715 + 1	0.00000 + 0	1.37000 + 1
	(n,p)		4.25000 − 3	2.05700 − 3
	(n,d)		1.06000 − 3	2.99100 − 3
	(n,t)		3.55000 − 5	4.75600 − 4
	(n,helium-3)		9.13600 − 6	1.10600 − 4
	(n,alpha)		1.06000 − 3	7.65300 − 4
Xenon-128	Total	2.42084 + 1	4.88800 + 0	1.75500 + 2
	Elastic	4.29999 + 0	2.96500 + 0	1.45500 + 2
	Inelastic		4.09600 − 1	5.18400 + 0
	Direct(n,2n)		1.51000 + 0	8.74200 − 1
	(n,3n)		0.00000 + 0	5.35900 − 2
	(n,gamma)	3.11478 + 0	0.00000 + 0	1.13900 + 1
	(n,p)		2.07000 − 3	1.04200 − 3
	(n,d)		5.83000 − 4	1.83900 − 3
	(n,t)		2.12000 − 5	3.57000 − 4
	(n,helium-3)		5.48400 − 6	4.16500 − 5
	(n,alpha)		3.10000 − 4	2.46100 − 4
Xenon-129	Total	6.78343 + 1	4.90500 + 0	4.65400 + 2
	Elastic	4.64504 + 0	2.92500 + 0	1.65900 + 2
	Inelastic		7.50500 − 2	7.06100 + 0
	Direct(n,2n)		1.90000 + 0	1.60900 + 0
	(n,3n)		0.00000 + 0	6.74300 − 2
	(n,gamma)	1.58045 + 1	0.00000 + 0	2.55600 + 2
	(n,p)		3.95000 − 3	1.88100 − 3
	(n,d)		4.97000 − 4	1.58800 − 3
	(n,t)		3.43000 − 4	1.18400 − 3
	(n,alpha)		6.73000 − 4	6.50600 − 4
Xenon-130	Total	4.20722 + 1	4.92200 + 0	1.39800 + 2
	Elastic	4.30000 + 0	2.95300 + 0	1.05400 + 2
	Inelastic		5.28900 − 2	4.84500 + 0
	Direct(n,2n)		1.91000 + 0	1.08200 + 0
	(n,3n)		0.00000 + 0	1.05200 − 1
	(n,gamma)	5.63852 + 0	0.00000 + 0	4.27400 + 0
	(n,p)		5.85000 − 3	3.19500 − 3
	(n,d)		3.40000 − 4	1.15600 − 3
	(n,t)		9.11000 − 6	2.53400 − 4
	(n,alpha)		1.49000 − 4	1.19900 − 4
Xenon-131	Total	2.04904 + 2	4.93800 + 0	2.96400 + 3
	Elastic	4.30898 + 0	2.95000 + 0	1.98900 + 3
	Inelastic		3.20700 − 2	6.02800 + 0
	Direct(n,2n)		1.95000 + 0	1.72600 + 0
	(n,3n)		0.00000 + 0	1.17200 − 1
	(n,gamma)	7.98954 + 1	0.00000 + 0	8.76400 + 2
	(n,p)		4.86000 − 3	2.36600 − 3
	(n,d)		2.77000 − 4	9.62200 − 4
	(n,t)		2.43000 − 4	8.64900 − 4
	(n,alpha)		3.75000 − 4	3.15800 − 4

Table 3.3-1 (continued)
THERMAL AND 14 MeV NEUTRON CROSS SECTIONS AND RESONANCE INTEGRALS

Material	Reaction	Thermal	14 MeV	Resonance integral
Xenon-132	Total	6.78971 + 0	4.95200 + 0	1.16400 + 2
	Elastic	4.29999 + 0	2.98500 + 0	1.07300 + 2
	Inelastic		4.95600 − 3	4.52300 + 0
	Direct(n,2n)		1.96000 + 0	1.14300 + 0
	(n,3n)		0.00000 + 0	1.60900 − 1
	(n,gamma)	4.07290 − 1	0.00000 + 0	1.73200 + 0
	(n,p)		2.04000 − 3	1.37900 − 3
	(n,d)		2.03000 − 4	7.32500 − 4
	(n,t)		3.02000 − 6	1.74700 − 4
	(n,alpha)		6.84000 − 5	5.65000 − 5
Xenon-134	Total	5.83282 + 0	4.99200 + 0	1.13000 + 2
	Elastic	4.29951 + 0	2.98000 + 0	1.06100 + 2
	Inelastic		0.00000 + 0	3.96800 + 0
	Direct(n,2n)		2.01000 + 0	1.20000 + 0
	(n,3n)		0.00000 + 0	2.07000 − 1
	(n,gamma)	1.53331 + 0	0.00000 + 0	1.58200 + 0
	(n,p)		1.69000 − 3	1.61100 − 3
	(n,d)		1.24000 − 4	4.68600 − 4
	(n,t)		2.15000 − 6	1.38500 − 4
	(n,alpha)		5.18000 − 5	3.76400 − 5
Xenon-135	Total	3.09594 + 6	4.67200 + 0	1.28700 + 4
	Elastic	3.81965 + 5	3.45600 + 0	5.23000 + 3
	Inelastic		1.21400 + 0	3.82100 + 0
	(n,gamma)	2.71393 + 6	2.38400 − 3	7.64000 + 3
Xenon-136	Total	5.27753 + 0	4.99900 + 0	1.26600 + 2
	Elastic	4.30054 + 0	2.97600 + 0	1.21000 + 2
	Inelastic		3.23300 − 1	3.51100 + 0
	Direct(n,2n)		1.70000 + 0	1.06600 + 0
	(n,3n)		0.00000 + 0	2.60000 − 1
	(n,gamma)	1.41796 − 1	0.00000 + 0	1.23800 − 1
	(n,p)		2.82000 − 5	5.99300 − 5
	(n,d)		7.80000 − 5	3.03900 − 4
	(n,t)		1.58000 − 6	1.10500 − 4
	(n,alpha)		4.88000 − 5	3.14800 − 5
Cesium-133	Total	3.10666 + 1	4.86300 + 0	5.44300 + 2
	Elastic	4.85407 + 0	3.08300 + 0	1.56300 + 2
	Inelastic		1.59400 − 1	6.76400 + 0
	Direct(n,2n)		1.60100 + 0	1.05100 + 0
	(n,gamma)	2.62126 + 1	7.15700 − 3	3.80300 + 2
	(n,p)		1.15700 − 2	1.21200 − 2
	(n,alpha)		1.22000 − 3	9.52100 − 4
Samarium-149	Total	6.01469 + 4	5.20000 + 0	3.79600 + 3
	Elastic	3.47128 + 2	3.06700 + 0	6.01600 + 2
	Inelastic		3.00000 − 1	1.02200 + 1
	Direct(n,2n)		1.80000 + 0	1.36300 + 0
	(n,3n)		0.00000 + 0	1.82900 − 1
	(n,gamma)	5.97997 + 4	1.03000 − 3	3.18300 + 3
	(n,p)		1.60000 − 2	1.47200 − 2
	(n,alpha)		1.60000 − 2	1.47200 − 2

Table 3.3-1 (continued)
THERMAL AND 14 MeV NEUTRON CROSS SECTIONS
AND RESONANCE INTEGRALS

Material	Reaction	Thermal	14 MeV	Resonance integral
Europium-151	Total	7.39278 + 3	5.13000 + 0	3.56000 + 3
	Elastic	2.90261 + 0	2.84500 + 0	2.80100 + 2
	Inelastic		6.27300 − 2	9.13100 + 0
	Direct(n,2n)		2.17400 + 0	1.33400 + 0
	(n,3n)		0.00000 + 0	8.35100 − 2
	(n,gamma)	7.38987 + 3	4.10000 − 3	3.26400 + 3
	(n,p)	1.93904 −13	2.75500 − 2	1.71600 − 2
	(n,d)		5.46000 − 3	6.08300 − 3
	(n,t)		2.19800 − 3	4.15700 − 3
	(n,helium-3)		5.56900 − 6	9.21100 − 5
	(n,alpha)	1.95834 −12	9.13700 − 3	6.60300 − 3
Europium-152	Total	1.84644 + 3	4.93400 + 0	3.94600 + 3
	Elastic	5.27010 + 0	2.65000 + 0	2.42500 + 2
	Inelastic		2.17800 − 2	8.96400 + 0
	Direct(n,2n)		2.21500 + 0	1.62200 + 0
	(n,3n)		0.00000 + 0	1.92000 − 1
	(n,gamma)	1.84117 + 3	4.10000 − 3	3.69200 + 3
	(n,p)	1.50246 − 6	2.67500 − 2	1.88700 − 2
	(n,d)		3.34800 − 3	4.30900 − 3
	(n,t)		5.31500 − 3	6.26500 − 3
	(n,helium-3)		3.30800 − 6	4.84600 − 5
	(n,alpha)	2.16845 −12	7.73500 − 3	6.37600 − 3
Europium-153	Total	3.98403 + 2	5.17000 + 0	1.80800 + 3
	Elastic	4.66454 + 0	2.89800 + 0	2.27600 + 2
	Inelastic		1.43500 − 1	9.73600 + 0
	Direct(n,2n)		2.09400 + 0	1.35000 + 0
	(n,3n)		0.00000 + 0	1.67500 − 1
	(n,gamma)	3.93738 + 2	2.60000 − 3	1.56900 + 3
	(n,p)		7.96200 − 3	5.06100 − 3
	(n,d)		2.48700 − 3	3.85600 − 3
	(n,t)		8.17200 − 4	2.11900 − 3
	(n,helium-3)		2.28600 − 6	2.61100 − 5
	(n,alpha)	3.52148 −12	1.83800 − 2	1.15700 − 2
Europium-154	Total	1.20450 + 3	4.96800 + 0	2.78000 + 3
	Elastic	5.96904 + 0	2.61000 + 0	2.11700 + 2
	Inelastic		2.30000 − 2	9.42800 + 0
	Direct(n,2n)		2.21400 + 0	1.59700 + 0
	(n,3n)		0.00000 + 0	9.23700 − 2
	(n,gamma)	1.19853 + 3	2.60000 − 3	2.55700 + 3
	(n,p)		8.85100 − 2	5.89100 − 2
	(n,d)		1.47100 − 3	2.69500 − 3
	(n,t)		2.32200 − 3	3.77000 − 3
	(n,helium-3)		2.00100 − 6	1.45000 − 5
	(n,alpha)	3.58211 −12	2.56200 − 2	1.77200 − 2
Natural Gadolinium	Total	3.70006 + 4	5.47000 + 0	5.06900 + 2
	Elastic	1.39981 + 2	3.29800 + 0	1.27700 + 2
	Inelastic		3.99000 − 2	9.42300 + 0
	Direct(n,2n)		2.09700 + 0	1.47800 + 0
	(n,3n)		2.95000 − 2	1.47800 − 1
	(n,gamma)	3.68607 + 4	9.38500 − 4	3.67400 + 2

Table 3.3-1 (continued)
THERMAL AND 14 MeV NEUTRON CROSS SECTIONS
AND RESONANCE INTEGRALS

Material	Reaction	Thermal	14 MeV	Resonance integral
Dysprosium-164	Total	2.58610 + 3	5.26000 + 0	1.03600 + 3
	Elastic	3.80426 + 2	3.00600 + 0	6.97000 + 2
	Inelastic		4.00000 − 2	9.13200 + 0
	Direct(n,2n)		2.19800 + 0	1.22400 + 0
	(n,3n)		2.32300 − 3	4.36800 − 1
	(n,gamma)	2.20566 + 3	8.50000 − 3	3.28500 + 2
	(n,p)		1.00000 − 3	1.01300 − 3
	(n,alpha)		4.30000 − 3	4.53100 − 3
Lutetium-175	Total	2.76563 + 1	5.25700 + 0	9.12100 + 2
	Elastic	5.25654 + 0	3.02100 + 0	2.61600 + 2
	Inelastic		1.10000 − 1	7.89900 + 0
	Direct(n,2n)		2.12000 + 0	1.18600 + 0
	(n,3n)		0.00000 + 0	3.66600 − 1
	(n,gamma)	2.23990 + 1	2.00000 − 3	6.41000 + 2
	(n,p)		3.40000 − 3	4.87800 − 3
	(n,alpha)		1.00000 − 3	1.86300 − 3
Lutetium-176	Total	3.02823 + 3	5.26000 + 0	1.13000 + 3
	Elastic	4.15212 + 0	2.95200 + 0	2.04800 + 2
	Inelastic		1.10000 − 1	7.53800 + 0
	Direct(n,2n)		2.19000 + 0	1.61100 + 0
	(n,3n)		0.00000 + 0	3.98800 − 1
	(n,gamma)	3.02407 + 3	1 3.20000 − 3	9.15700 + 2
	(n,p)		3,40000 − 3	4.87800 − 3
	(n,alpha)		1.00000 − 3	1.86300 − 3
Tantalum-181	Total	2.49355 + 1	5.34900 + 0	1.00800 + 3
	Elastic	6.14233 + 0	2.89900 + 0	2.56300 + 2
	Inelastic		2.30000 − 1	1.04200 + 1
	Direct(n,2n)		2.21200 + 0	1.39800 + 0
	(n,3n)		0.00000 + 0	2.86800 − 1
	(n,gamma)	1.94848 + 1	4.50000 − 3	7.39200 + 2
	(n,p)		3.70000 − 3	3.27600 − 3
Tantalum-182	Total	1.20345 + 4	5.36400 + 0	1.39700 + 3
	Elastic	4.74674 + 1	2.74300 + 0	3.19100 + 2
	Inelastic		3.87000 − 1	1.12800 + 1
	Direct(n,2n)		2.15700 + 0	1.48200 + 0
	(n,3n)		7.10000 − 2	4.64800 − 1
	(n,gamma)	1.19870 + 4	4.20000 − 3	1.06500 + 3
	(n,alpha)		1.90000 − 3	2.25200 − 3
Tungsten-182	Total	3.00811 + 1	5.34200 + 0	1.09500 + 3
	Elastic	1.18111 + 1	2.74100 + 0	4.87200 + 2
	Inelastic		3.67200 − 1	8.64700 + 0
	Direct(n,2n)		2.22500 + 0	1.41900 + 0
	(n,3n)		0.00000 + 0	2.54700 − 1
	(n,gamma)	1.82700 + 1	3.60100 − 3	5.97100 + 2
	(n,p)		3.31800 − 3	3.67700 − 3
	(n,alpha)		1.10000 − 3	2.35800 − 3
Tungsten-183	Total	1.22689 + 1	5.34200 + 0	8.38500 + 2
	Elastic	3.40554 + 0	2.74200 + 0	4.71000 + 2
	Inelastic		3.26900 − 1	1.00300 + 1
	Direct(n,2n)		2.26300 + 0	1.92800 + 0

Table 3.3-1 (continued)
THERMAL AND 14 MeV NEUTRON CROSS SECTIONS
AND RESONANCE INTEGRALS

Material	Reaction	Thermal	14 MeV	Resonance integral
	(n,3n)		0.00000 + 0	3.08700 − 1
	(n,gamma)	8.86332 + 0	3.71300 − 3	3.55300 + 2
	(n,p)		4.00400 − 3	4.03100 − 3
	(n,alpha)		1.10000 − 3	2.35900 − 3
Tungsten-184	Total	5.83524 + 0	5.34200 + 0	3.34600 + 2
	Elastic	4.28459 + 0	2.74600 + 0	3.07700 + 2
	Inelastic		3.02300 − 1	8.56700 + 0
	Direct(n,2n)		2.28200 + 0	1.47900 + 0
	(n,3n)		3.82600 − 3	3.89000 − 1
	(n,gamma)	1.55065 + 0	3.85200 − 3	1.65200 + 1
	(n,p)		2.69300 − 3	3.34100 − 3
	(n,alpha)		1.10000 − 3	2.35800 − 3
Tungsten-186	Total	3.34896 + 1	5.34200 + 0	3.50600 + 3
	Elastic	2.42053 − 1	2.74400 + 0	2.97700 + 3
	Inelastic		2.50800 − 1	8.20700 + 0
	Direct(n,2n)		2.32300 + 0	1.48200 + 0
	(n,3n)		1.73300 − 2	4.99400 − 1
	(n,gamma)	3.32476 + 1	3.95800 − 3	5.18800 + 2
	(n,p)		1.45000 − 3	2.61300 − 3
	(n,alpha)		1.10000 − 3	2.35800 − 3
Rhenium-185	Total	1.22107 + 2	5.21900 + 0	2.18300 + 3
	Elastic	2.05020 + 1	2.70900 + 0	4.27500 + 2
	Inelastic		8.85700 − 1	7.90800 + 0
	Direct(n,2n)		1.62000 + 0	1.12300 + 0
	(n,3n)		0.00000 + 0	8.82900 − 2
	(n,gamma)	1.01590 + 2	4.34000 − 3	1.74600 + 3
Rhenium-187	Total	7.51494 + 1	5.22400 + 0	6.13600 + 2
	Elastic	1.00547 + 1	2.70100 + 0	3.18000 + 2
	Inelastic		8.71700 − 1	8.19700 + 0
	Direct(n,2n)		1.58800 + 0	1.06400 + 0
	(n,3n)		6.00000 − 2	2.93400 − 1
	(n,gamma)	6.51201 + 1	3.27000 − 3	2.86000 + 2
Natural lead	Total	1.13582 + 1	5.40000 + 0	1.71800 + 2
	Elastic	1.11940 + 1	2.89900 + 0	1.66500 + 2
	Inelastic		3.43900 − 1	3.31100 + 0
	Direct(n,2n)		2.15700 + 0	1.25400 + 0
	(n,3n)		0.00000 + 0	4.47500 − 1
	(n,gamma)	1.57748 − 1	1.91100 − 4	1.07000 − 1
Thorium-232	Total	1.83353 + 1	5.70000 + 0	3.12500 + 2
	Elastic	1.18064 + 1	2.90000 + 0	2.15300 + 2
	Inelastic		3.70000 − 1	8.88000 + 0
	Direct(n,2n)		1.56000 + 0	1.49100 + 0
	(n,3n)		5.00000 − 1	4.05500 − 1
	Fission		3.60000 − 1	5.92700 − 1
	(n,gamma)	6.52828 + 0	5.28800 − 3	8.55800 + 1
Protactinium-233	Total	4.43161 + 1	5.43000 + 0	1.02800 + 3
	Elastic	8.31582 + 0	3.01600 + 0	1.56800 + 2
	Inelastic		8.80000 − 2	1.21600 + 1

Table 3.3-1 (continued)
THERMAL AND 14 MeV NEUTRON CROSS SECTIONS
AND RESONANCE INTEGRALS

Material	Reaction	Thermal	14 Mev	Resonance integral
	Direct(n,2n)		5.20000 − 1	4.88900 − 1
	(n,3n)		2.10000 − 1	3.19100 − 1
	Fission		1.58900 + 0	2.94700 + 0
	(n,gamma)	3.60002 + 1	6.66000 − 3	8.55300 + 2
Uranium-233	Total	5.19713 + 2	6.02800 + 0	1.06800 + 3
	Elastic	1.41581 + 1	3.04200 + 0	1.66300 + 2
	Inelastic		8.80000 − 2	4.40200 + 0
	Direct(n,2n)		4.80000 − 1	5.12300 − 1
	(n,3n)		4.00000 − 2	3.70400 − 2
	Fission	4.63807 + 2	2.37000 + 0	7.62100 + 2
	(n,gamma)	4.17531 + 1	7.52800 − 3	1.34600 + 2
Uranium-234	Total	9.93135 + 1	5.38300 + 0	9.77000 + 2
	Elastic	1.54422 + 1	2.78000 + 0	3.32600 + 2
	Inelastic		1.50000 − 1	7.46200 + 0
	Direct(n,2n)		3.82000 − 1	2.57800 − 1
	(n,3n)		1.50000 − 1	8.60100 − 2
	Fission		1.91100 + 0	5.57800 + 0
	(n,gamma)	8.38714 + 1	1.03000 − 2	6.31000 + 2
Uranium-235	Total	6.07856 + 2	5.84000 + 0	5.99200 + 2
	Elastic	1.53744 + 1	3.13300 + 0	1.73600 + 2
	Inelastic		2.25000 − 1	5.56100 + 0
	Direct(n,2n)		2.96000 − 1	3.53700 − 1
	(n,3n)		3.20000 − 2	3.02700 − 2
	Fission	5.06695 + 2	2.15200 + 0	2.82000 + 2
	(n,gamma)	8.57816 + 1	2.44700 − 3	1.37700 + 2
Uranium-236	Total	1.40334 + 1	5.67200 + 0	7.47100 + 2
	Elastic	9.46212 + 0	2.78000 + 0	3.21900 + 2
	Inelastic		1.50000 − 1	7.88200 + 0
	Direct(n,2n)		5.50000 − 1	6.48200 − 1
	(n,3n)		5.50000 − 1	2.88600 − 1
	Fission		1.63200 + 0	3.42400 + 0
	(n,gamma)	4.57127 + 0	1.03000 − 2	4.13000 + 2
Uranium-238	Total	1.13472 + 1	5.79000 + 0	6.08500 + 2
	Elastic	8.94862 + 0	2.98600 + 0	3.19000 + 2
	Inelastic		3.13000 − 1	8.51700 + 0
	Direct(n,2n)		8.55000 − 1	1.01700 + 0
	(n,3n)		4.95000 − 1	3.58900 − 1
	Fission	4.13513 − 9	1.14000 + 0	2.05600 + 0
	(n,gamma)	2.39834 + 0	1.00000 − 3	2.77500 + 2
Neptunium-237	Total	1.63041 + 2	5.88300 + 0	8.34300 + 2
	Elastic	1.72917 + 1	2.97000 + 0	1.82600 + 2
	Inelastic		5.00000 − 3	6.62700 + 0
	Direct(n,2n)		4.87000 − 1	3.81200 − 1
	(n,3n)		8.06900 − 2	6.36100 − 2
	Fission	1.41962 − 2	2.33700 + 0	6.82000 + 0
	(n,gamma)	1.45733 + 2	3.50000 − 3	6.37900 + 2
Plutonium-238	Total	4.96101 + 2	5.92500 + 0	4.32300 + 2
	Elastic	2.01893 + 1	2.90400 + 0	2.53400 + 2

Table 3.3-1 (continued)
THERMAL AND 14 MeV NEUTRON CROSS SECTIONS
AND RESONANCE INTEGRALS

Material	Reaction	Thermal	14 MeV	Resonance integral
	Inelastic		5.05200 − 1	3.95200 + 0
	Direct(n,2n)		7.26300 − 2	7.29000 − 2
	(n,3n)		2.42300 − 2	1.56300 − 2
	Fission	1.37622 + 1	2.41400 + 0	3.07400 + 1
	(n,gamma)	4.61955 + 2	5.00000 − 3	1.43900 + 2
Plutonium-239	Total	9.73095 + 2	5.98800 + 0	6.77600 + 2
	Elastic	7.81556 + 0	2.99500 + 0	1.74400 + 2
	Inelastic		2.25000 − 1	4.98100 + 0
	Direct(n,2n)		2.05000 − 1	1.82600 − 1
	(n,3n)		2.20000 − 2	2.85900 − 2
	Fission	6.94244 + 2	2.53800 + 0	3.03900 + 2
	(n,gamma)	2.71036 + 2	2.55000 − 3	1.94100 + 2
Plutonium-240	Total	2.67413 + 2	6.11700 + 0	9.40400 + 3
	Elastic	3.57152 + 0	3.40700 + 0	9.41100 + 2
	Inelastic		1.85000 − 1	6.70500 + 0
	Direct(n,2n)		1.92000 − 1	1.50600 − 1
	(n,3n)		7.90000 − 2	4.87000 − 2
	Fission	5.25496 − 2	2.25200 + 0	9.54100 + 0
	(n,gamma)	2.63789 + 2	2.40000 − 3	8.44600 + 3
Plutonium-241	Total	1.27468 + 3	5.73400 + 0	8.90200 + 2
	Elastic	1.07363 + 1	2.62900 + 0	1.71800 + 2
	Inelastic		2.79000 − 1	7.59500 + 0
	Direct(n,2n)		1.05000 − 1	5.37600 − 1
	(n,3n)		1.28000 − 1	5.51700 − 2
	Fission	9.36651 + 2	2.57700 + 0	5.85600 + 2
	(n,gamma)	3.27278 + 2	1.58000 − 2	1.24700 + 2
Plutonium-242	Total	2.49255 + 1	6.00800 + 0	1.50800 + 3
	Elastic	8.35897 + 0	2.92200 + 0	3.68900 + 2
	Inelastic		5.47700 − 1	6.55700 + 0
	Direct(n,2n)		2.38300 − 1	4.20000 − 1
	(n,3n)		4.01500 − 1	2.15200 − 1
	Fission		1.89600 + 0	5.83800 + 0
	(n,gamma)	1.65462 + 1	2.54000 − 3	1.12600 + 3
Americium-241	Total	5.31467 + 2	5.85200 + 0	1.84400 + 3
	Elastic	9.98410 + 0	3.06500 + 0	1.74700 + 2
	Inelastic		1.33600 − 1	8.21700 + 0
	Fission	2.97021 + 0	2.65000 + 0	2.10800 + 1
	(n,gamma)	5.18373 + 2	1.00000 − 3	1.63600 + 3
Americium-243	Total	1.49841 + 2	4.60200 + 0	1.55600 + 3
	Elastic	7.52164 + 0	3.06500 + 0	1.83200 + 2
	Inelastic		1.33600 − 1	8.08700 + 0
	Fission		1.40000 + 0	4.36400 + 0
	(n,gamma)	1.42319 + 2	1.00000 − 3	1.36000 + 3
Curium-244	Total	2.17962 + 1	6.13800 + 0	1.02600 + 3
	Elastic	8.38818 + 0	2.96200 + 0	3.82600 + 2
	Inelastic		6.25800 − 1	5.38200 + 0
	Direct(n,2n)		9.58200 − 2	1.26900 − 1
	(n,3n)		8.24000 − 2	6.86200 − 2
	Fission	7.76911 − 1	2.36800 + 0	4.48600 + 1
	(n,gamma)	1.26111 + 1	4.79000 − 3	5.93200 + 2

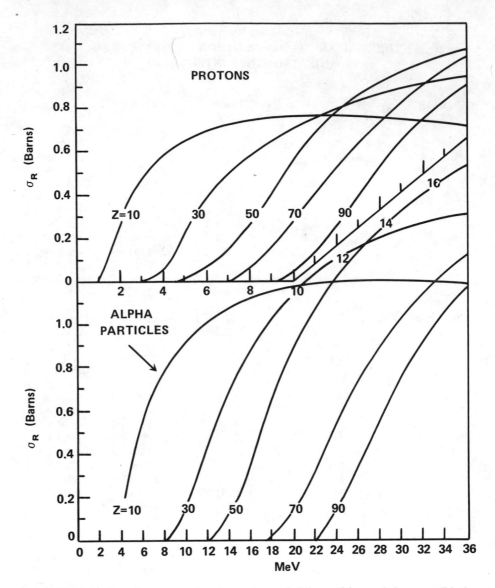

FIGURE 3.3-1.   Reaction cross sections for protons and alpha particles vs. their energy. Z is the atomic number of the target nucleus.

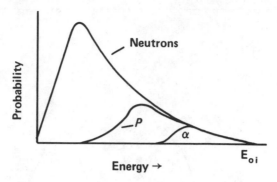

FIGURE 3.3-2.   Schematic energy spectra of neutrons, protons, and alpha particles emitted from a compound nucleus, assuming they all have the same maximum available energy E_{oi}.

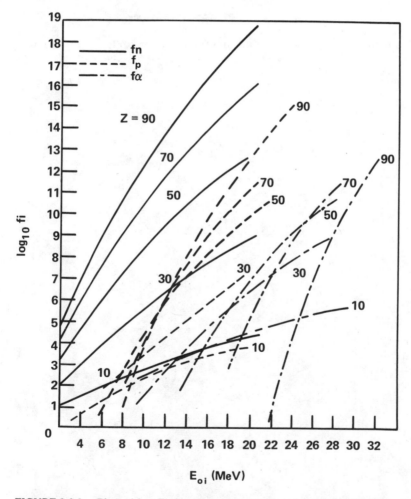

FIGURE 3.3-3.   Plots of f_i vs E_{oi} for neutrons, protons, and alpha particles. Numbers attached to curves are the atomic numbers of the target nuclei. These curves are calculated from statistical theory of nuclear reactions using Gilbert-Cameron level densities.[3]

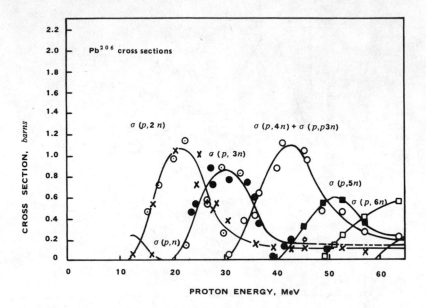

FIGURE 3.3-4.   Cross sections of (p,xn) reactions on ^{206}Pb vs. incident proton energy. Data are from Bell, R. E. and Skarsgard, H. M., *Can. J. Phys.,* 34, 745, 1956.

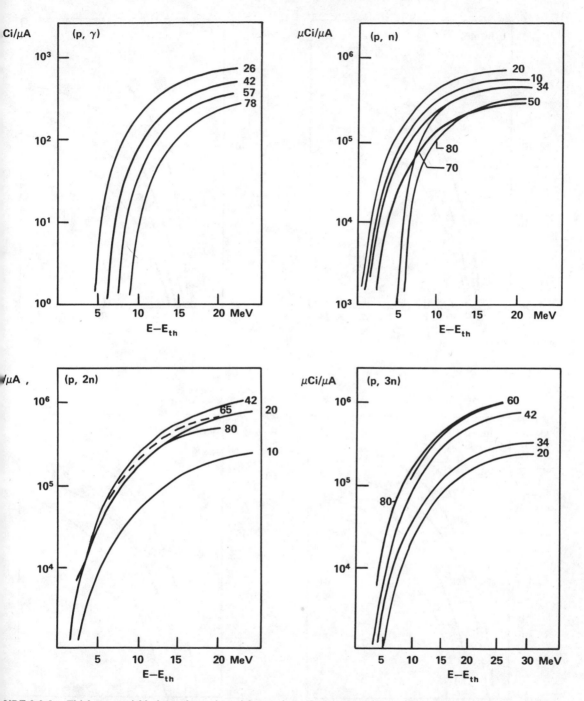

URE 3.3-5.   Thick target yields from charged particle reactions. Curves give microcuries per microamperes for elements
;iven atomic number, Z, as a function of particle energy minus the threshold energy for the reaction. Nuclides of the
e Z have about the same yield (order of magnitude) for the same $E - E_{th}$. See further description in Section 3.3-4.1.

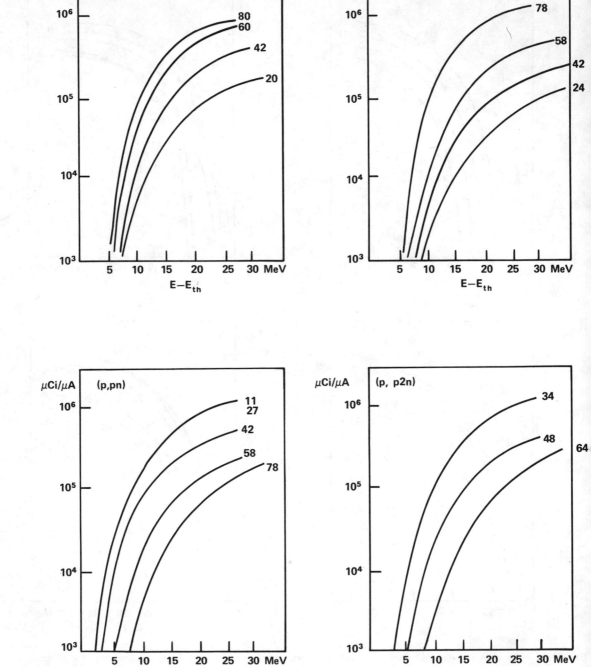

Figure 3.3-5 (continued)

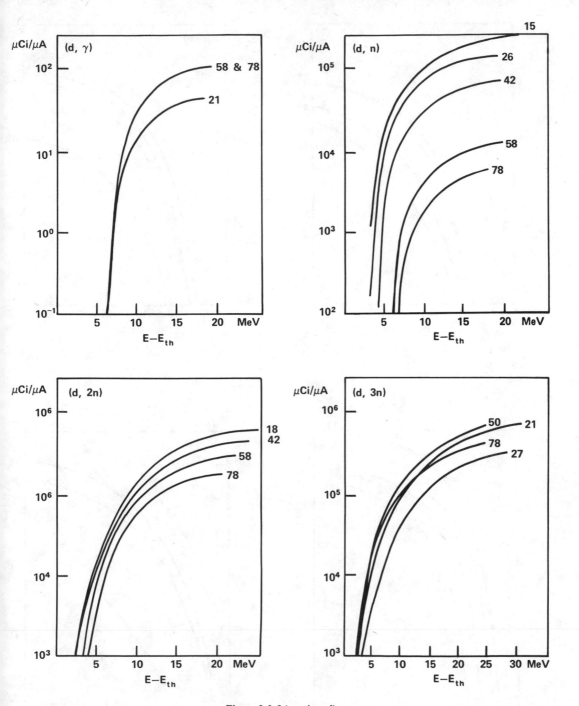

Figure 3.3-5 (continued)

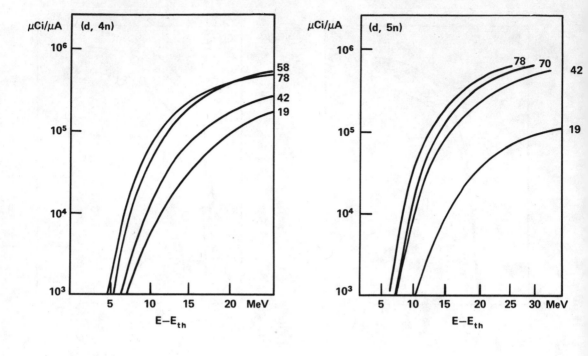

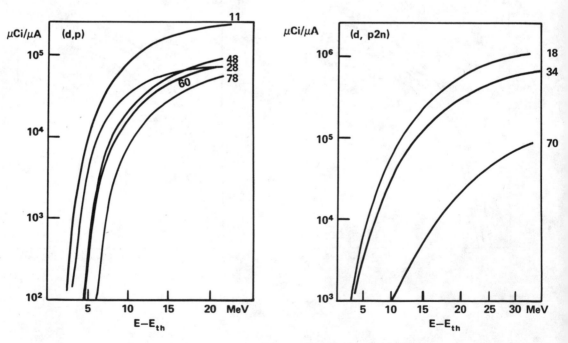

Figure 3.3-5 (continued)

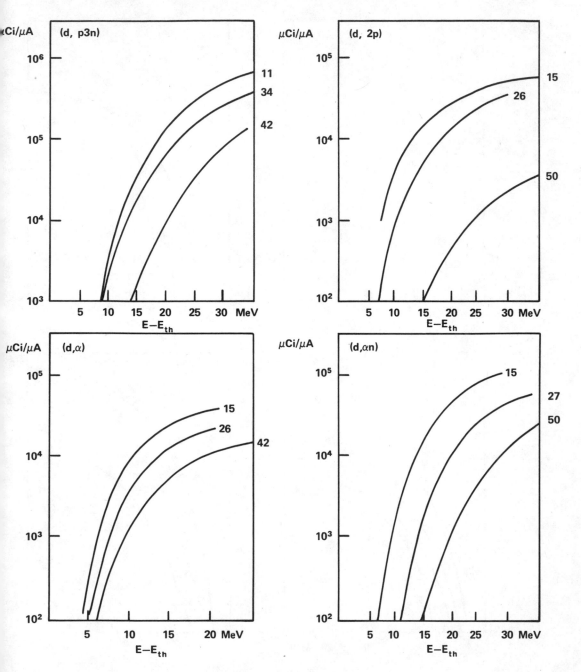

Figure 3.3-5 (continued)

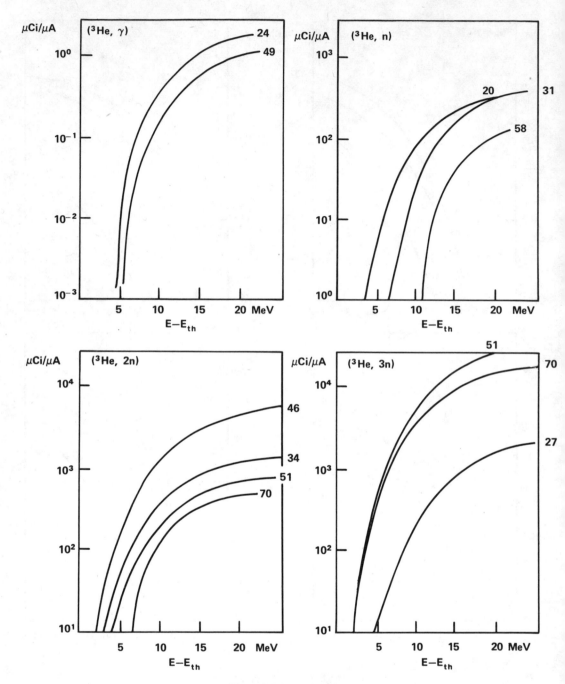

Figure 3.3-5 (continued)

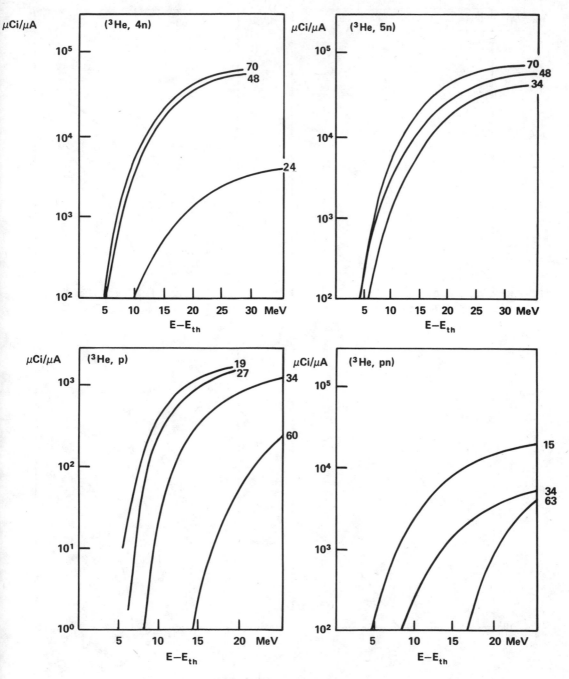

Figure 3.3-5 (continued)

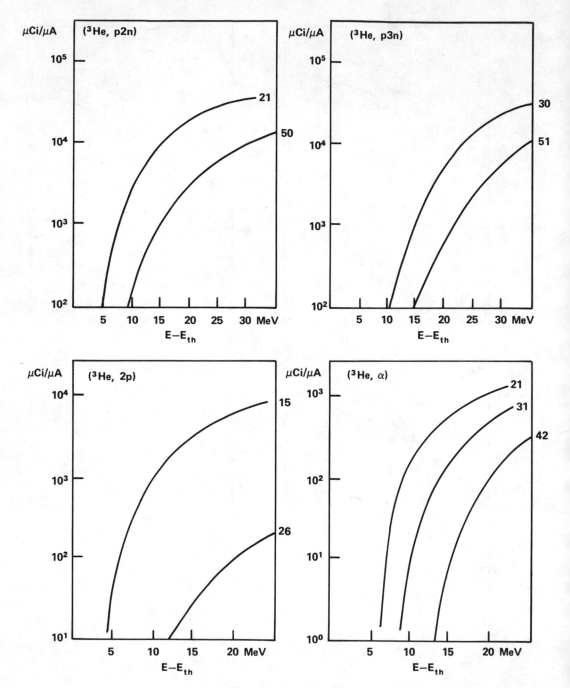

Figure 3.3-5 (continued)

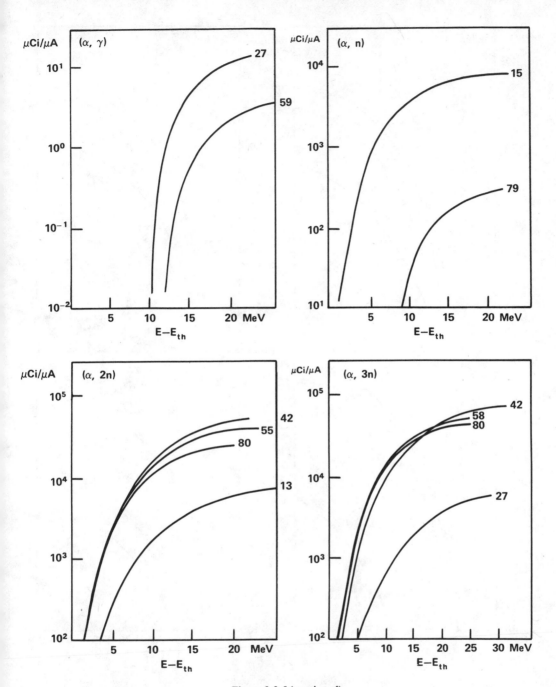

Figure 3.3-5 (continued)

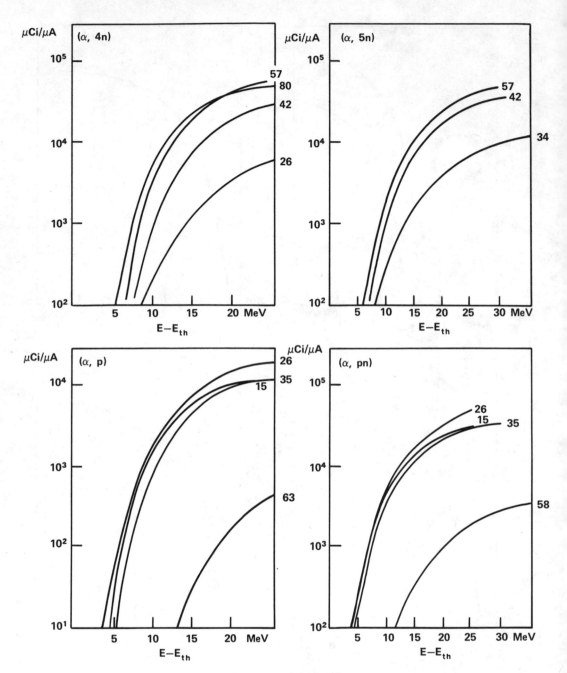

Figure 3.3-5 (continued)

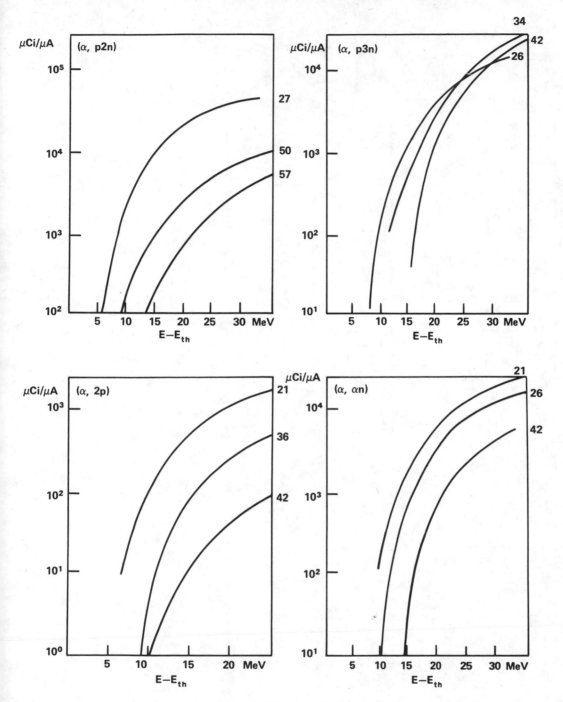

Figure 3.3-5 (continued)

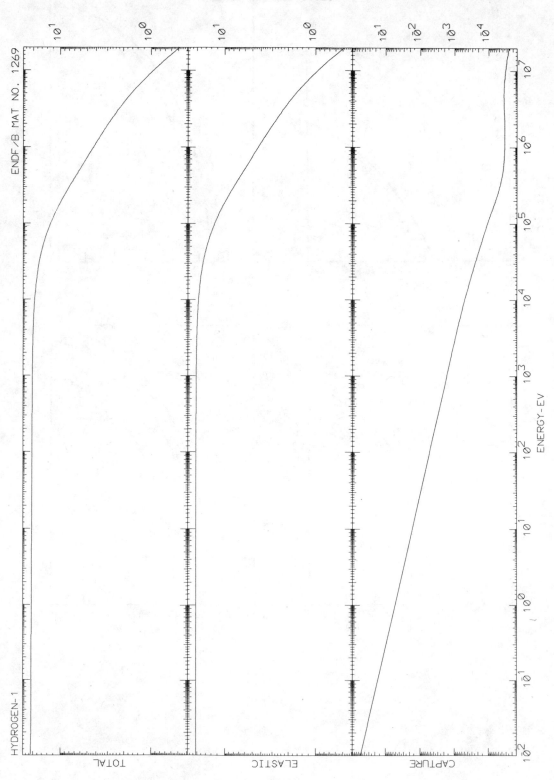

FIGURE 3.3-6.1 and 3.3-6.2.   Neutron cross section for $_1{}^1$H.

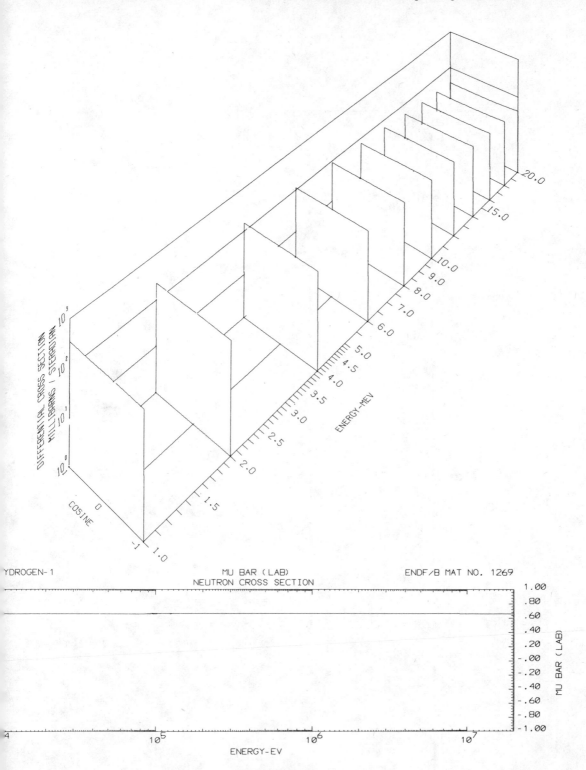

FIGURE 3.3-6.2.

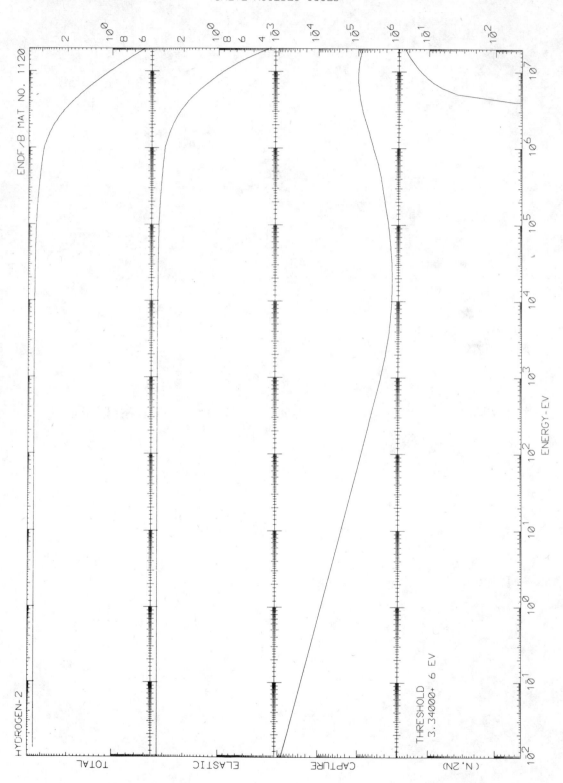

FIGURE 3.3-6.3 and 3.3-6.4.   Neutron cross section for $_1^2$H.

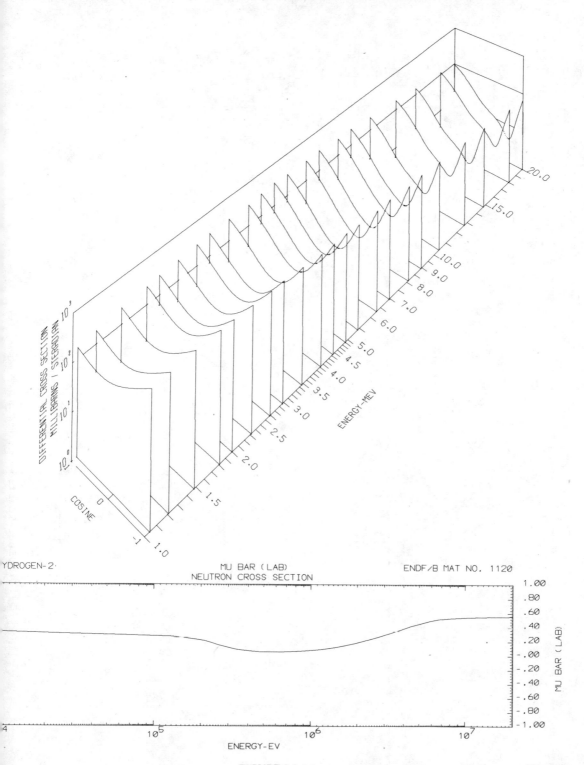

FIGURE 3.3-6.4.

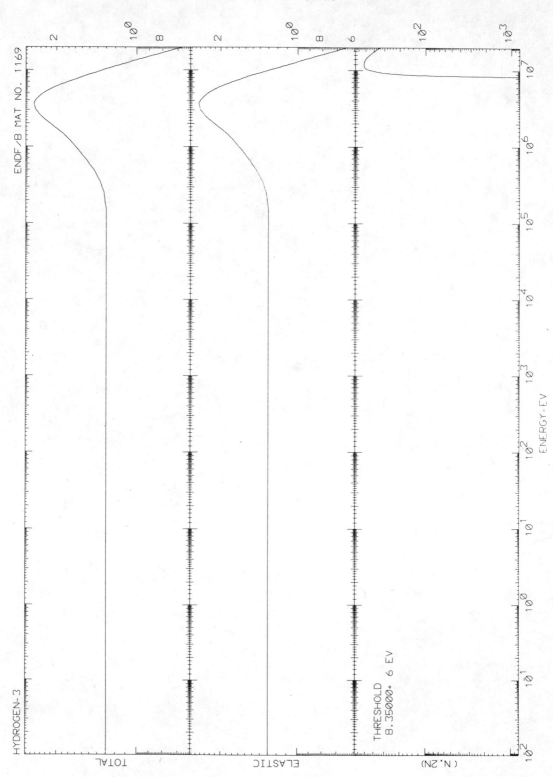

FIGURE 3.3-6.5 and 3.3-6.6. Neutron cross section for $_1{}^3$H.

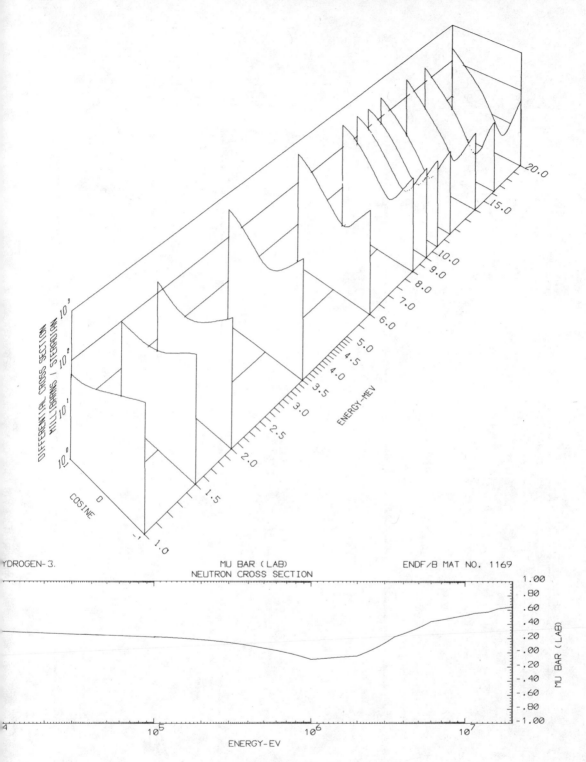

HYDROGEN-3.   MU BAR (LAB)   ENDF/B MAT NO. 1169
NEUTRON CROSS SECTION

FIGURE 3.3-6.6.

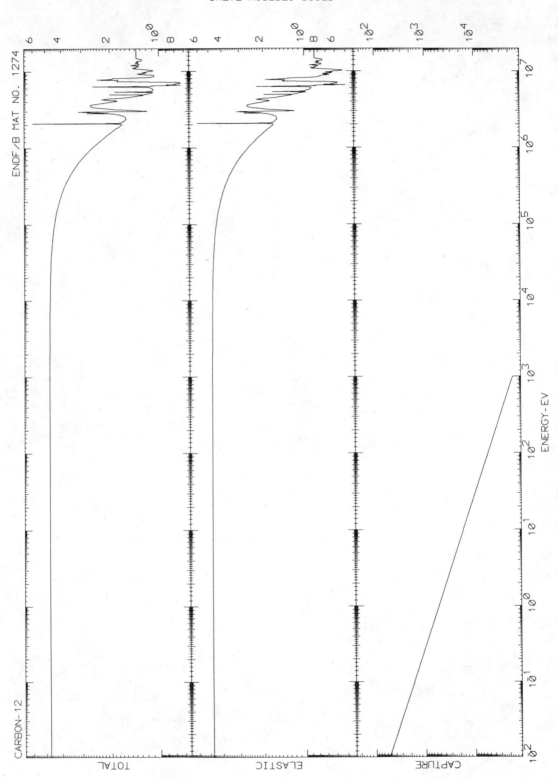

FIGURE 3.3-6.7 to 3.3-6.9.   Neutron cross section for $_6{}^{12}C$.

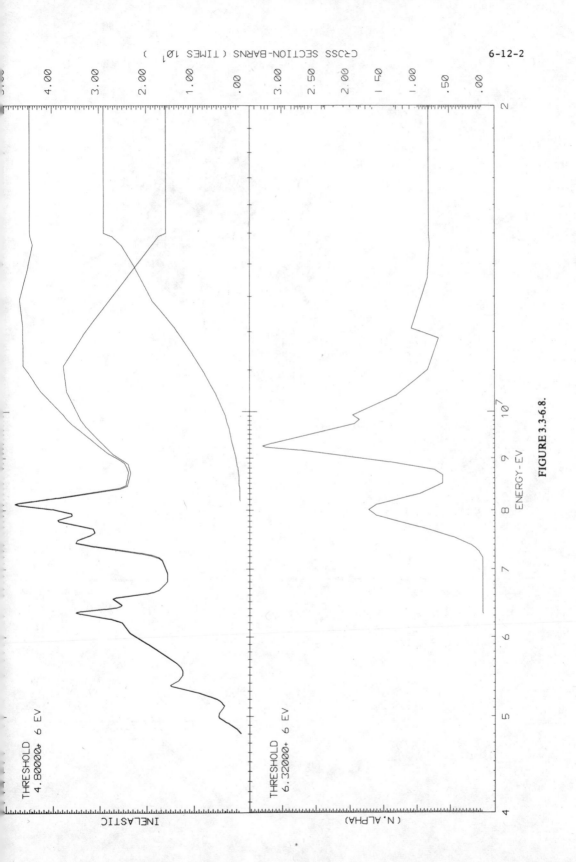

FIGURE 3.3-6.8.

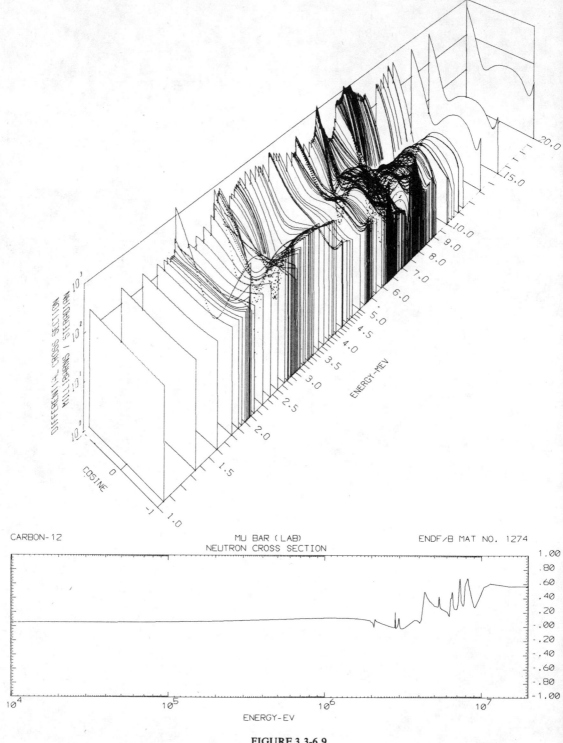

CARBON-12   MU BAR (LAB)   ENDF/B MAT NO. 1274
                                   NEUTRON CROSS SECTION

FIGURE 3.3-6.9.

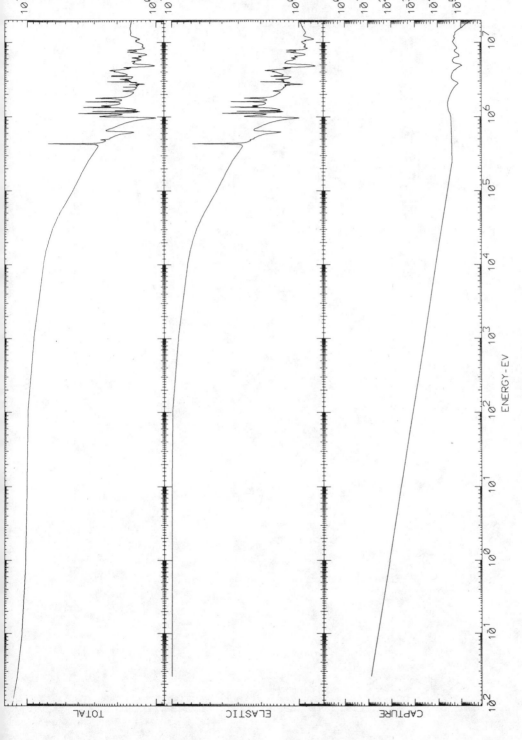

FIGURE 3.3-6.10 to 3.3-6.13.   Neutron cross section for $^{14}_7$N.

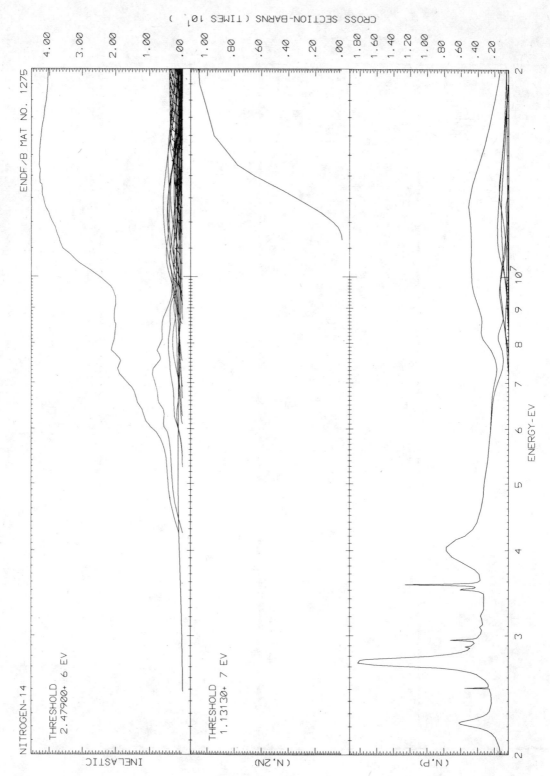

FIGURE 3.3-6.11.

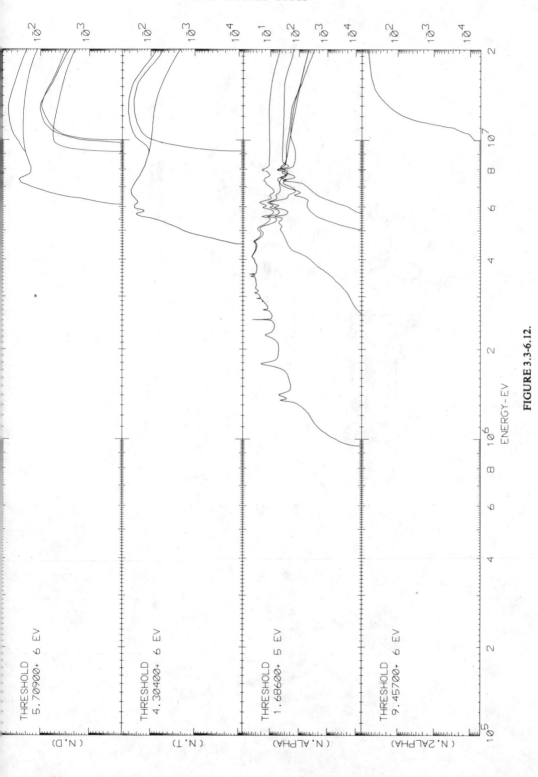

FIGURE 3.3-6.12.

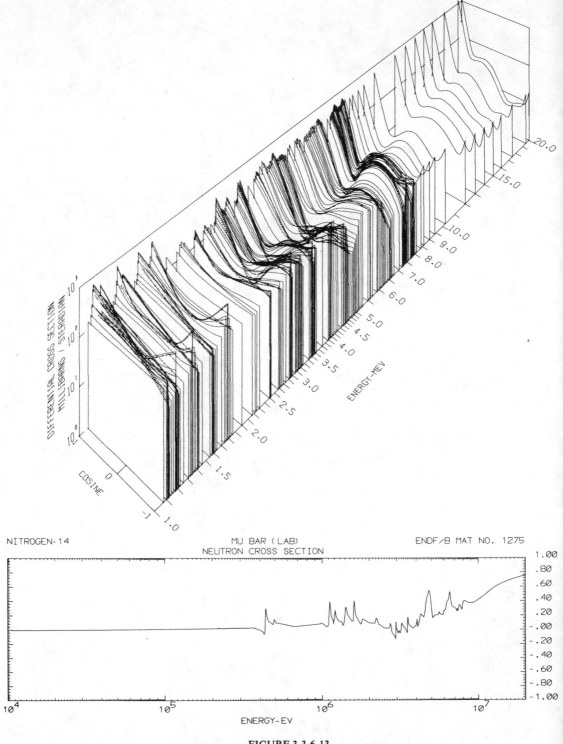

NITROGEN-14   MU BAR (LAB)   ENDF/B MAT NO. 1275
                          NEUTRON CROSS SECTION

FIGURE 3.3-6.13.

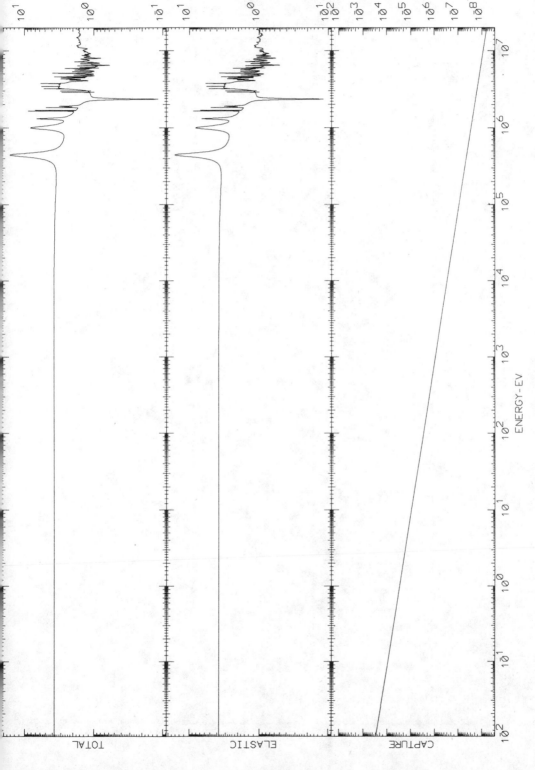

FIGURE 3.3-6.14 to 3.3-6.16.   Neutron cross section for $_8^{16}$O.

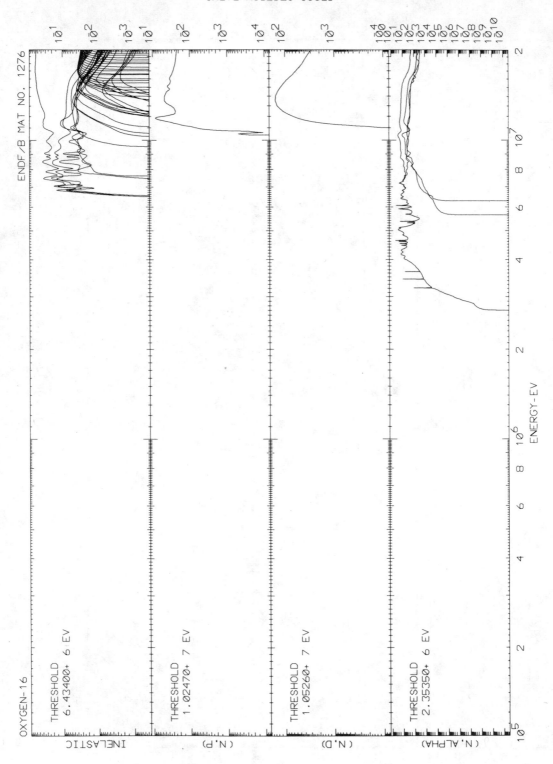

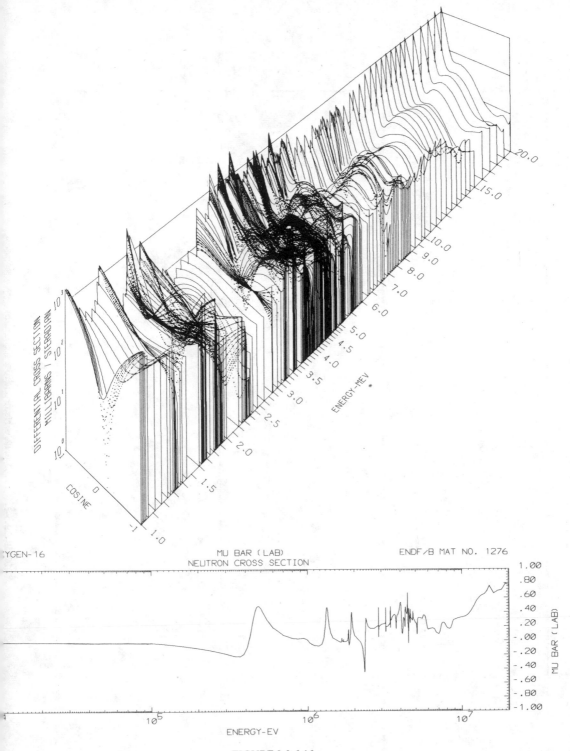

FIGURE 3.3-6.16.

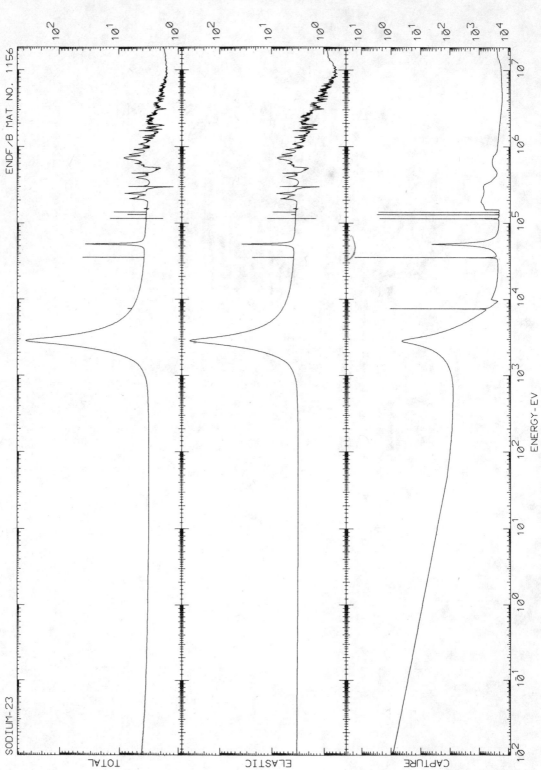

FIGURE 3.3-6.17 to 3.3-6.19. Neutron cross section for $_{11}^{23}$Na.

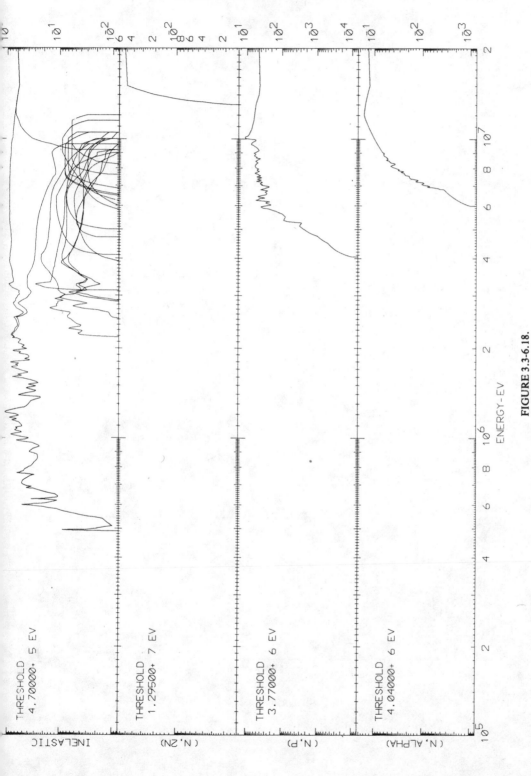

FIGURE 3.3-6.18.

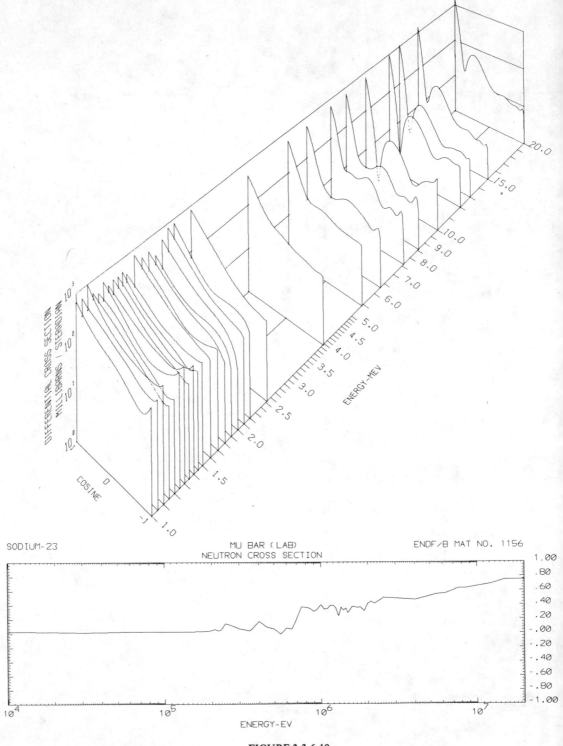

FIGURE 3.3-6.19.

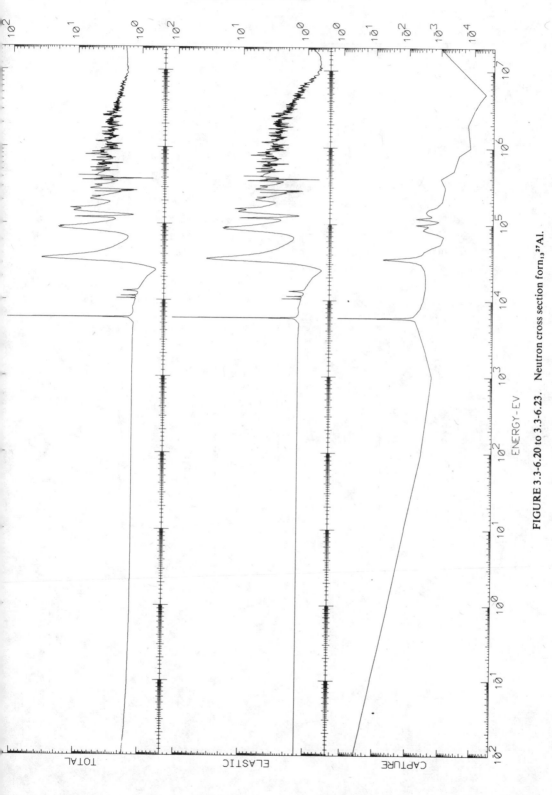

FIGURE 3.3-6.20 to 3.3-6.23.   Neutron cross section for $_{13}{}^{27}$Al.

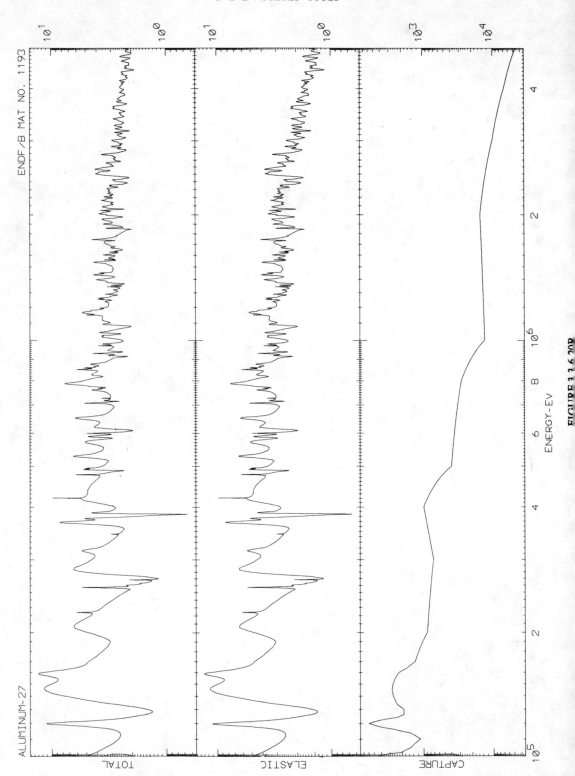

FIGURE 1.6.20B

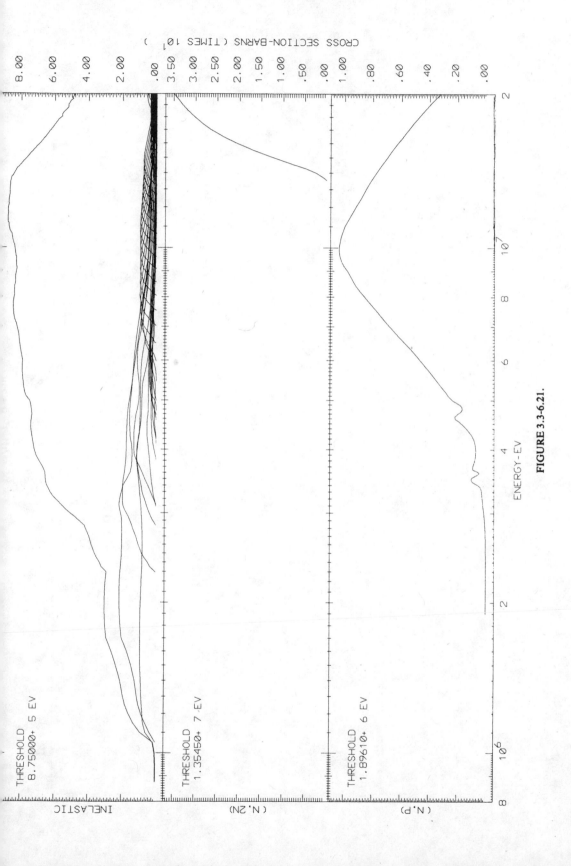

FIGURE 3.3-6.21.

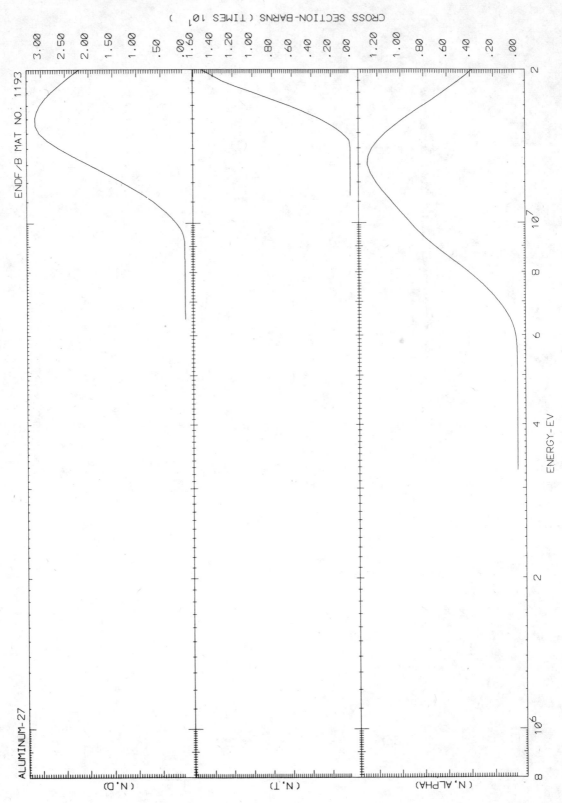

FIGURE 3.3-6.22.

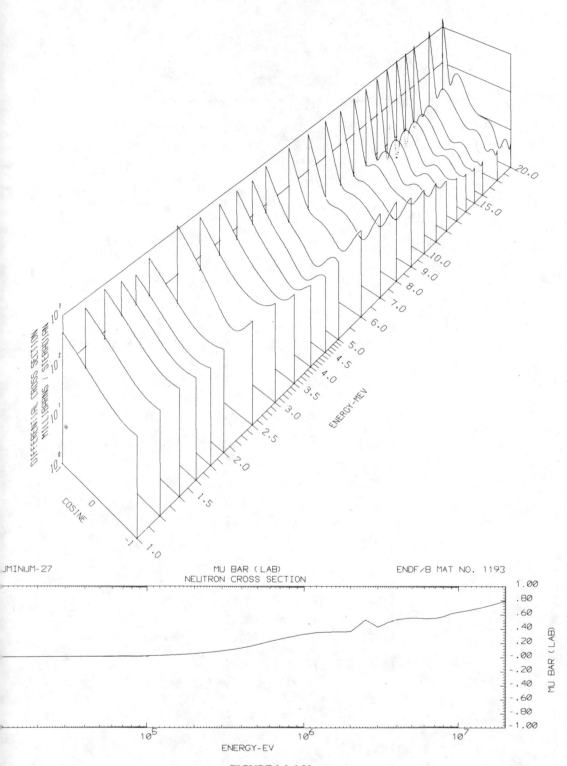

ALUMINUM-27   MU BAR (LAB)   ENDF/B MAT NO. 1193
NEUTRON CROSS SECTION

FIGURE 3.3-6.23.

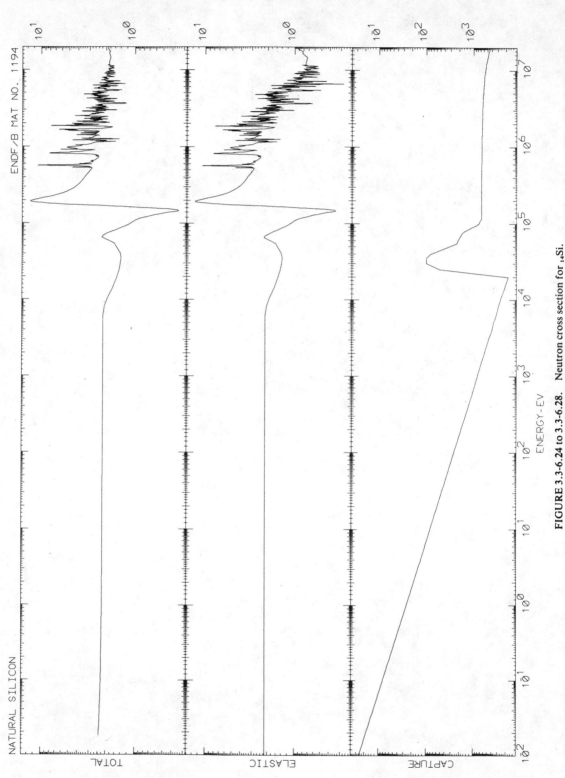

FIGURE 3.3-6.24 to 3.3-6.28. Neutron cross section for $_{14}$Si.

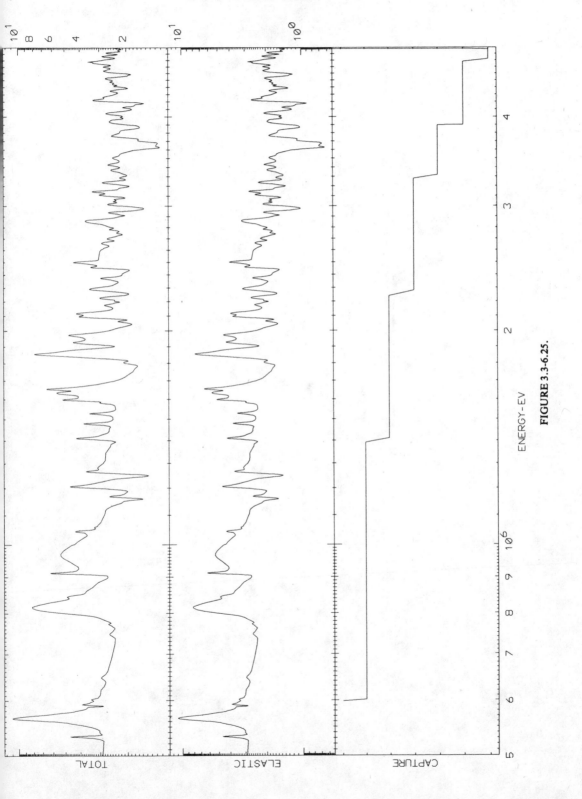

FIGURE 3.3-6.25.

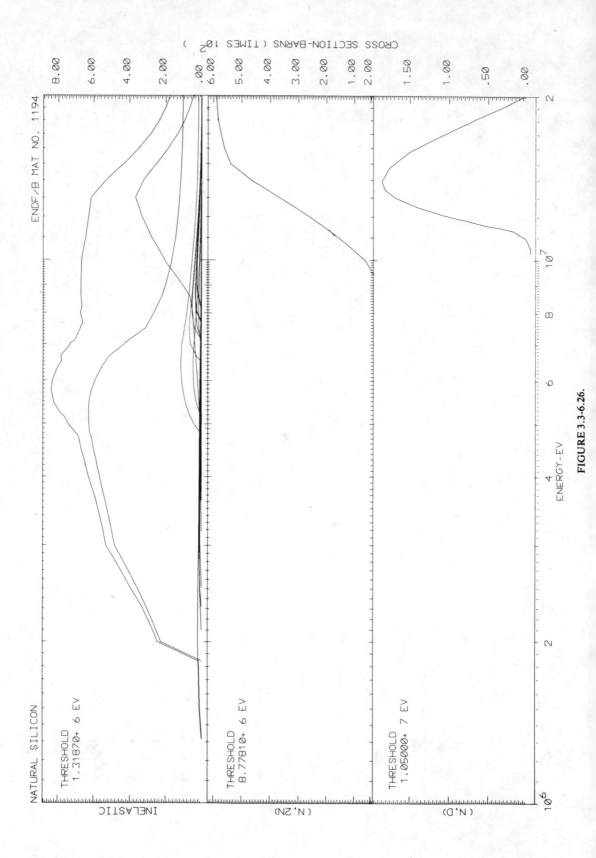

FIGURE 3.3-6.26.

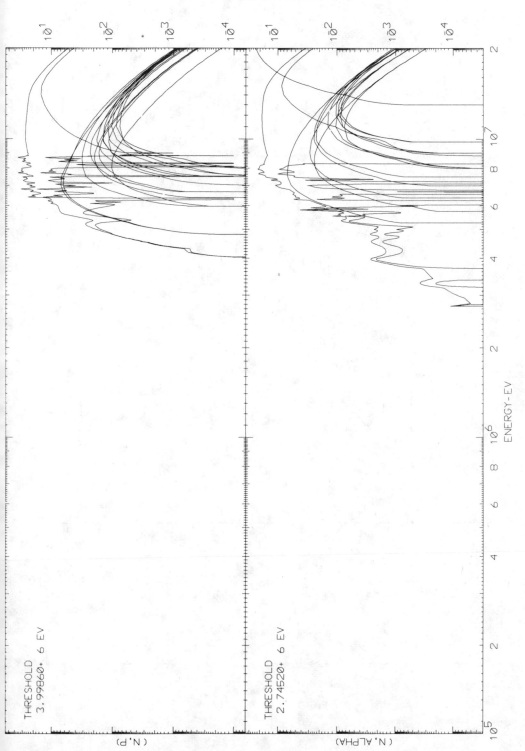

FIGURE 3.3-6.27.

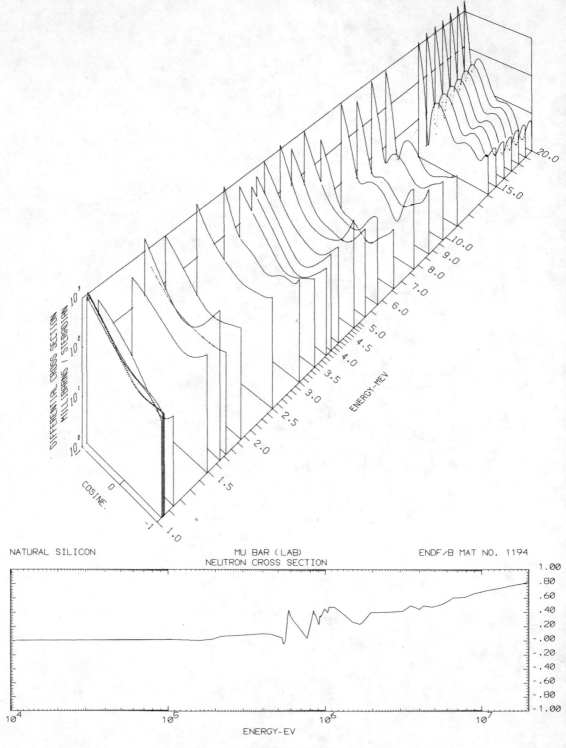

FIGURE 3.3-6.28.

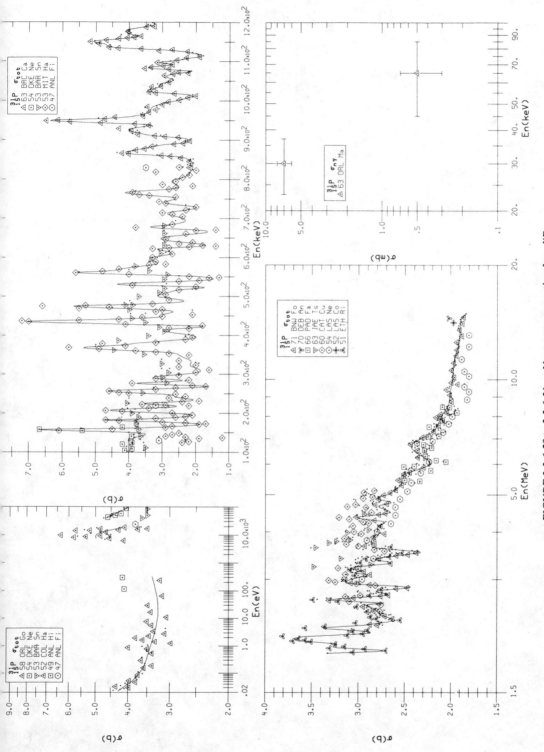

FIGURE 3.3-6.29 to 3.3-6.31.   Neutron cross section for $_{15}^{31}$P.

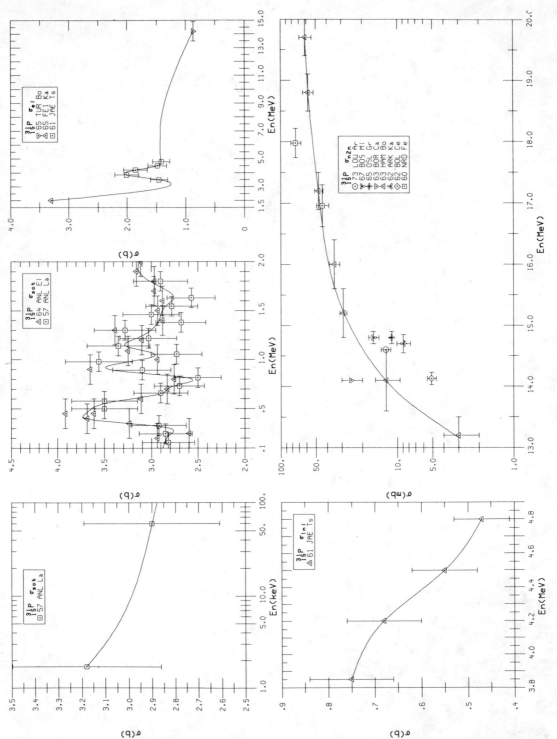

FIGURE 3.3-6.29.

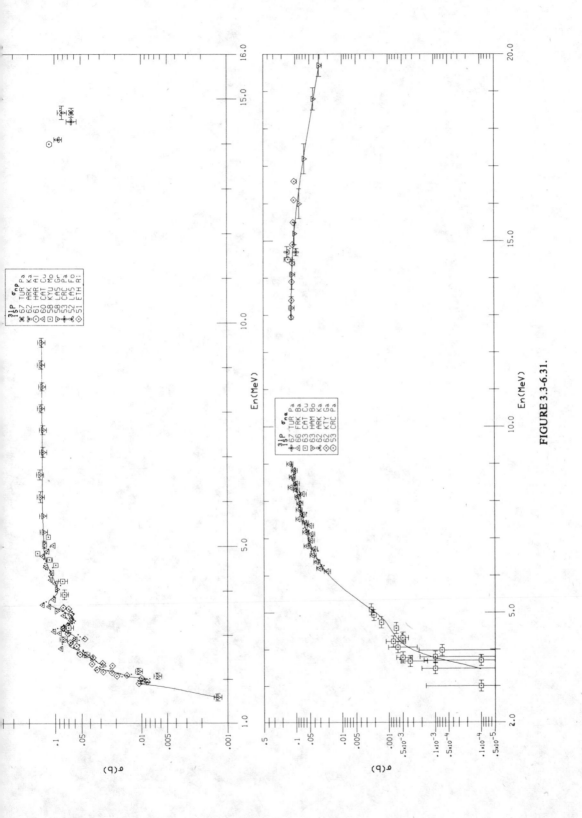

FIGURE 3.3-6.31.

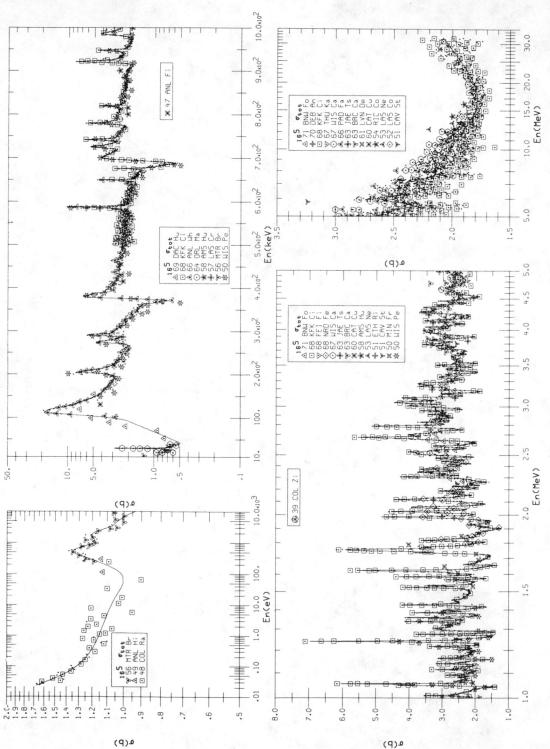

FIGURE 3.3-6.32 to 3.3-6.34.   Neutron cross sections for $_{16}^{32}S$.

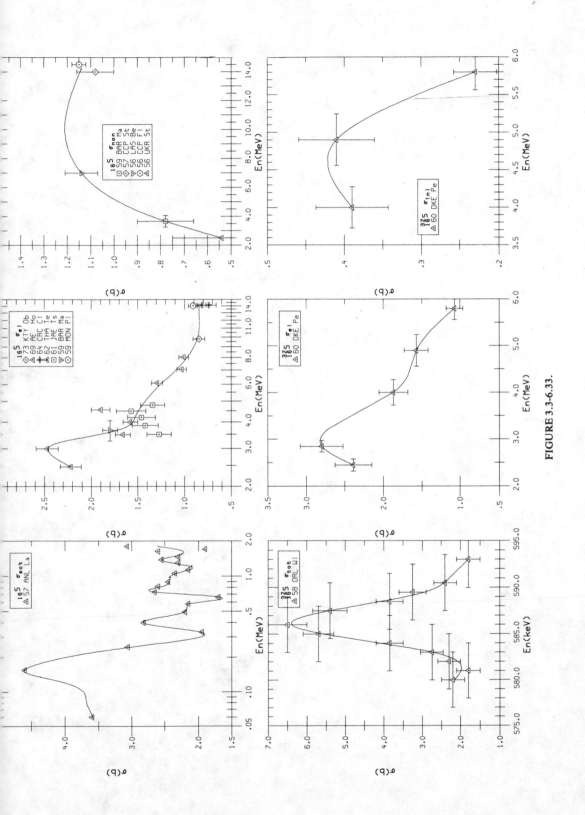

FIGURE 3.3-6.33.

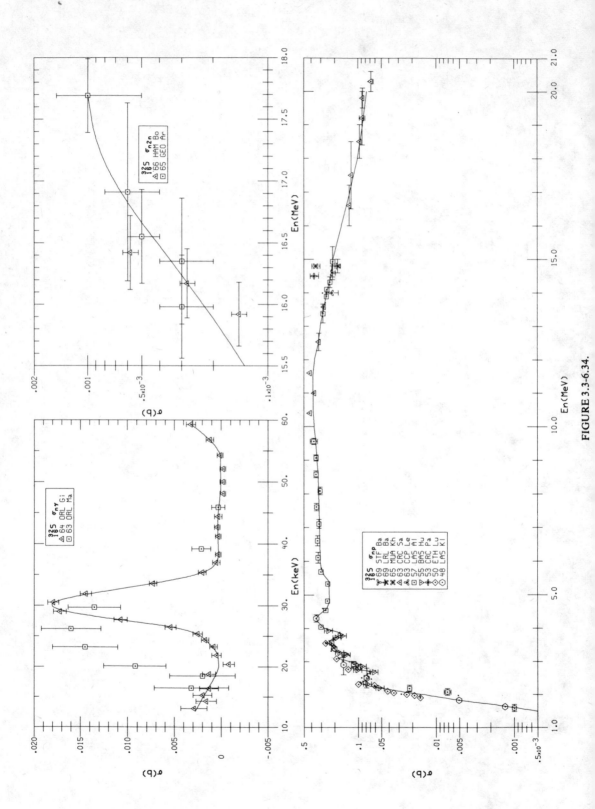

FIGURE 3.3-6.34.

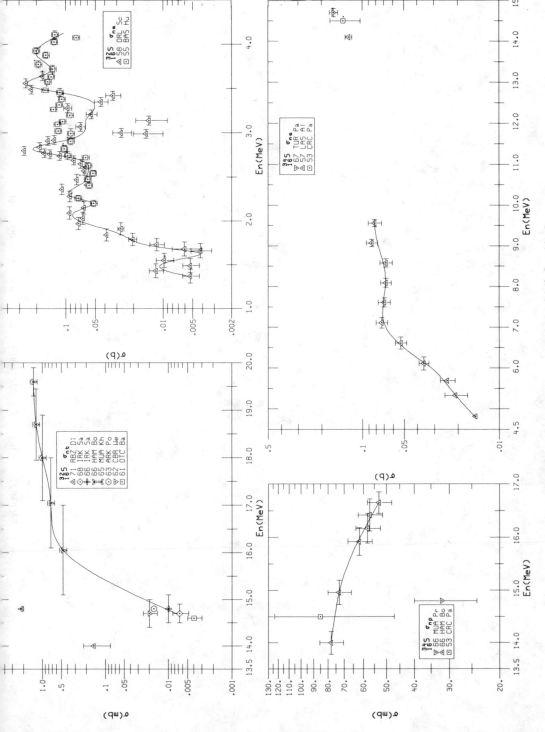

FIGURE 3.3-6.35. Neutron cross sections for $^{32}_{16}S$ and $^{34}_{16}S$.

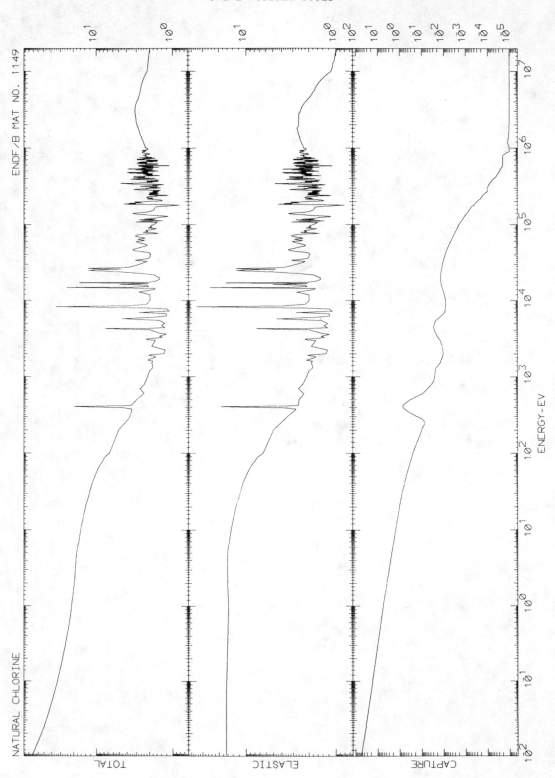

FIGURE 3.3-6.36 to 3.3-6.39. Neutron cross sections for natural chlorine.

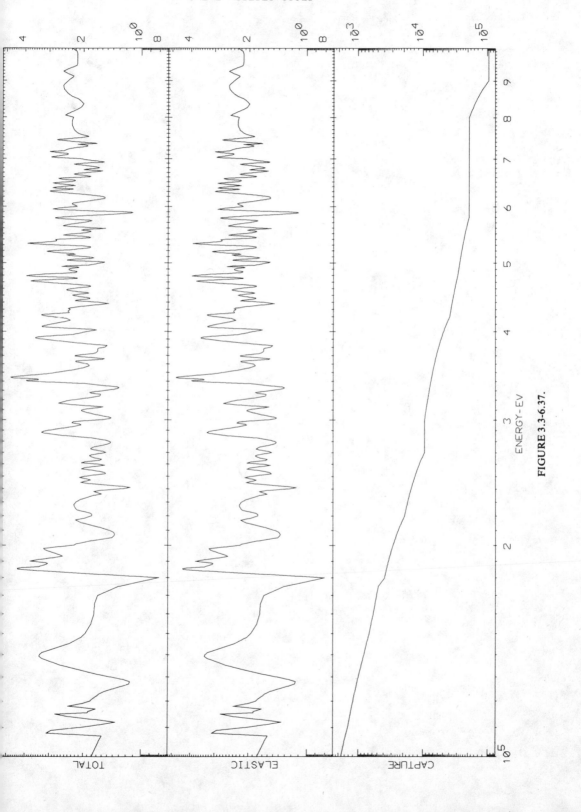

FIGURE 3.3-6.37.

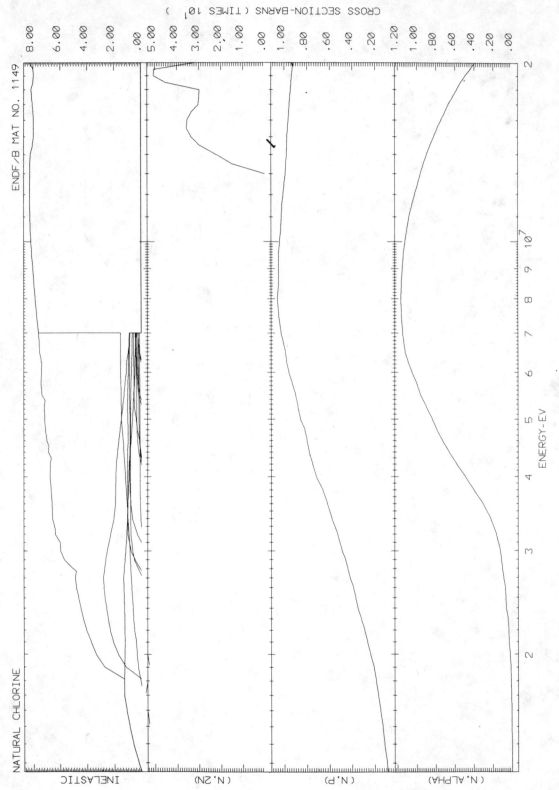

FIGURE 3.3-6.38.

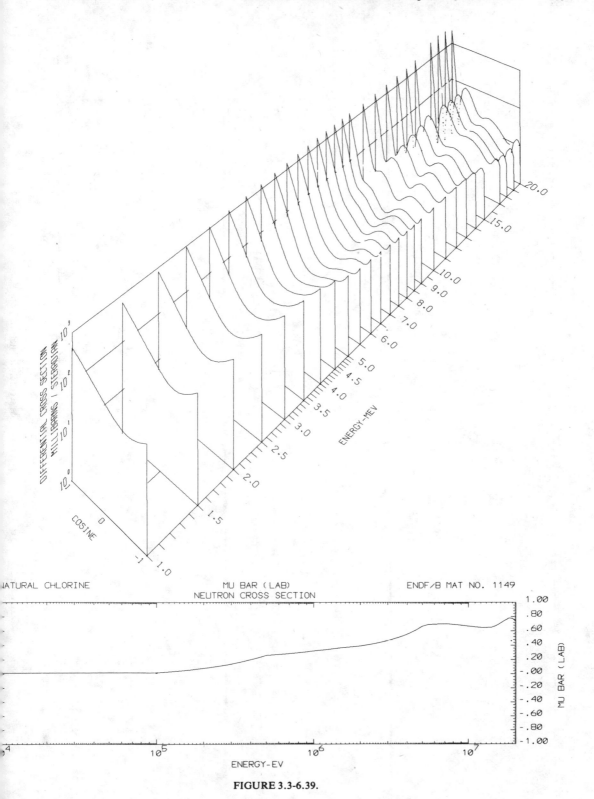

FIGURE 3.3-6.39.

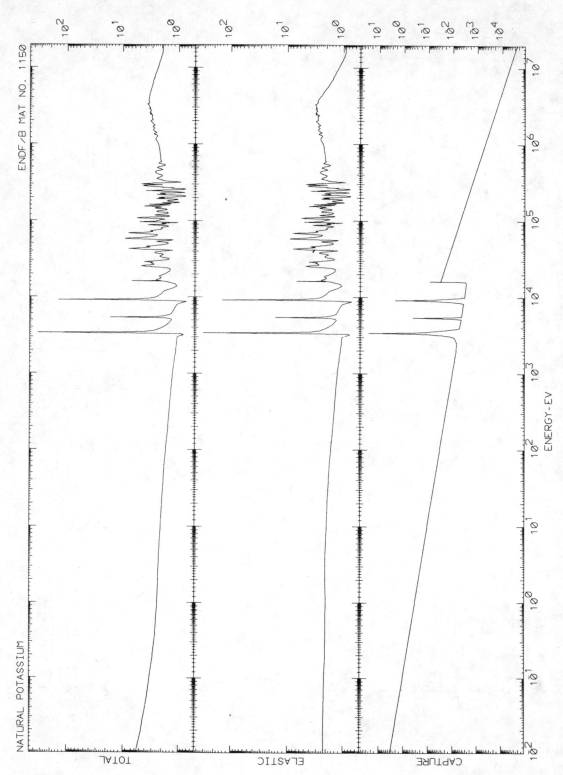

FIGURE 3.3-6.40 to 3.3-6.41.   Neutron cross sections for natural potassium.

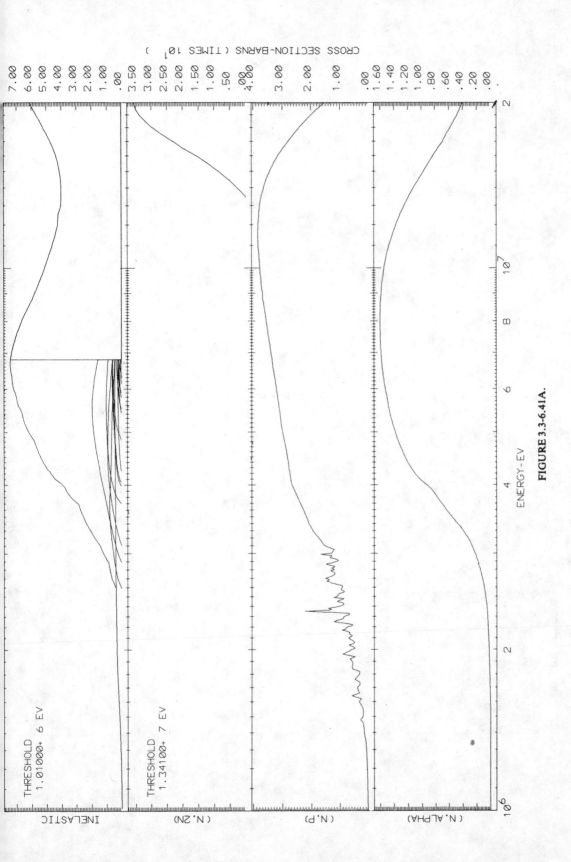

FIGURE 3.3-6.41A.

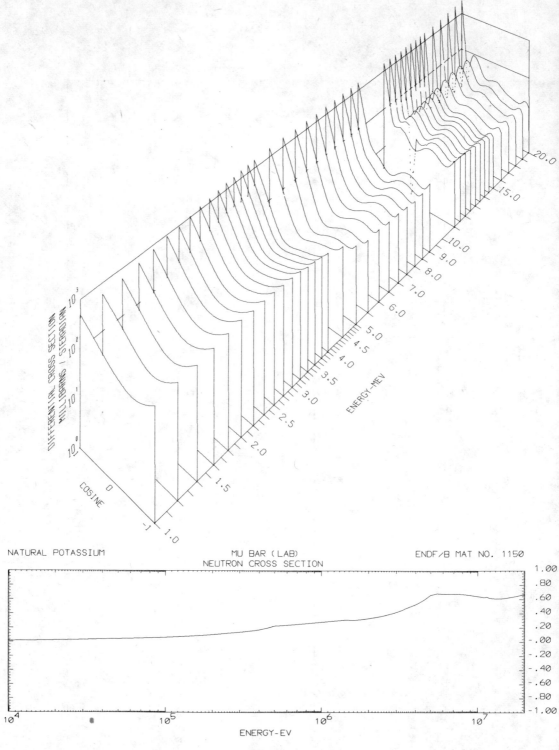

NATURAL POTASSIUM   MU BAR (LAB)   ENDF/B MAT NO. 1150
                                 NEUTRON CROSS SECTION

ENERGY-EV

FIGURE 3.3-6.41B.

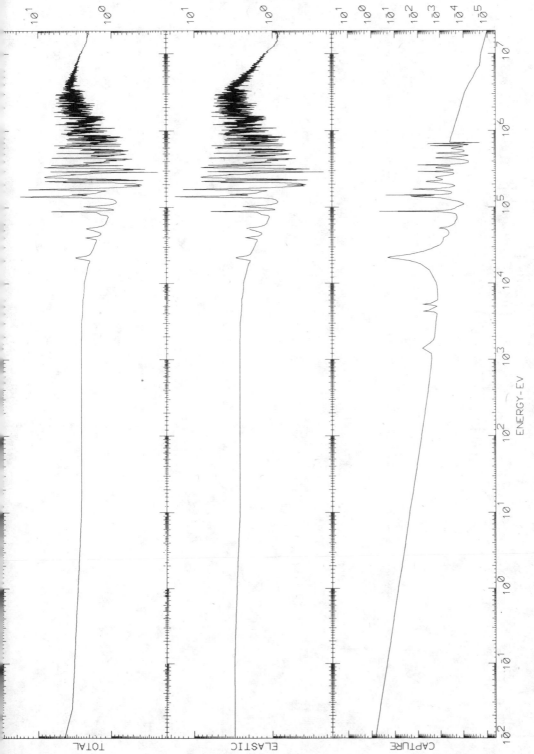

FIGURE 3.3-6.42 to 3.3-6.46.   Neutron cross sections for natural calcium.

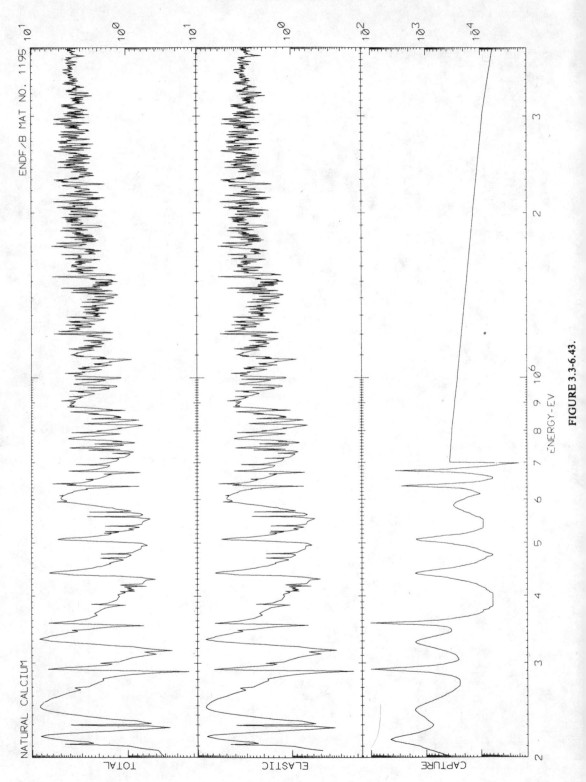

FIGURE 3.3-6.43.

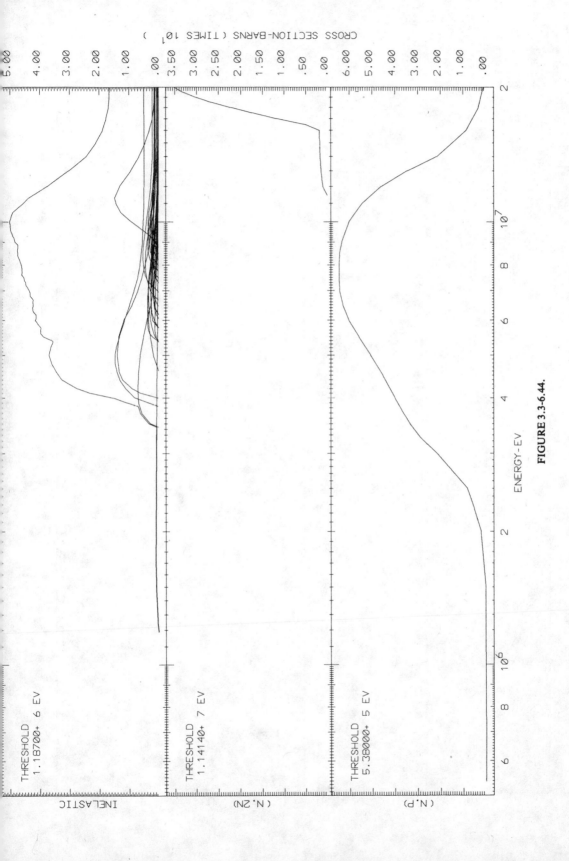

FIGURE 3.3-6.44.

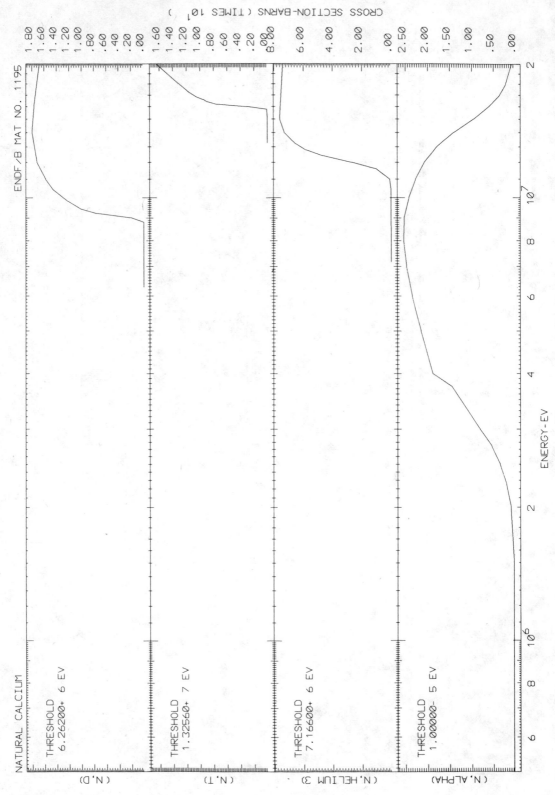

FIGURE 3.3-6.45.

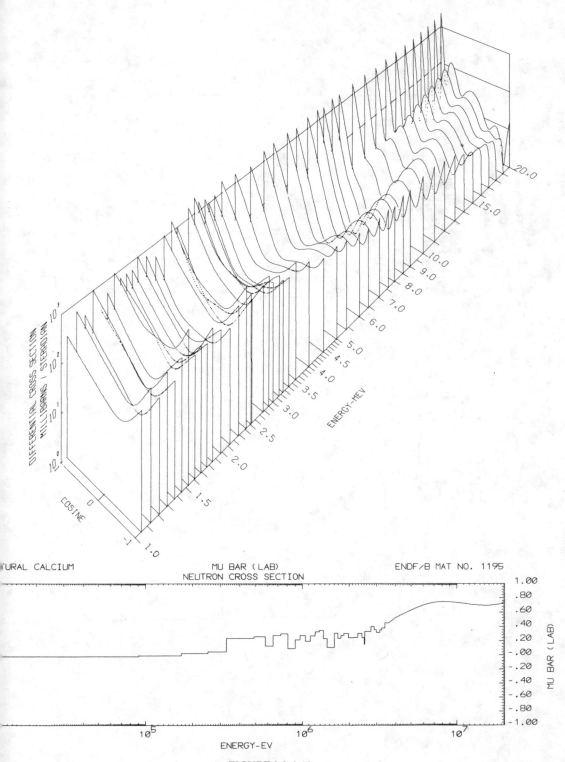

FIGURE 3.3-6.46.

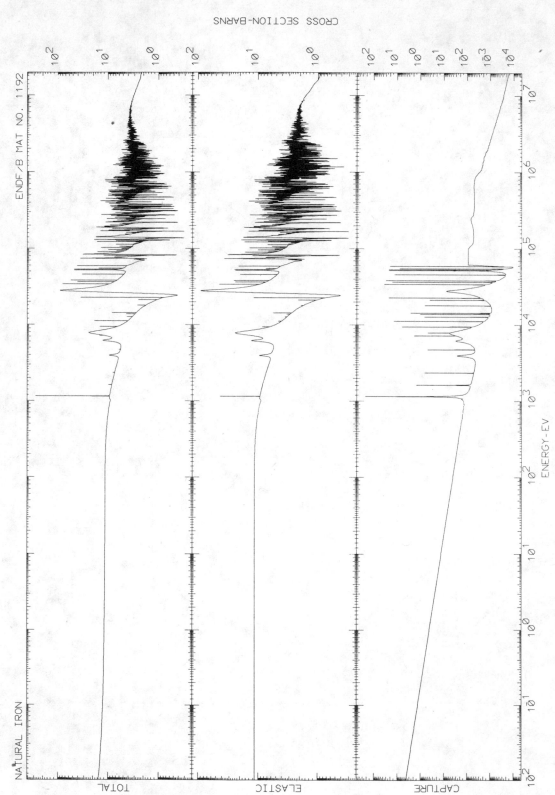

FIGURE 3.3-6.47 to 3.3-6.51. Neutron cross sections for natural iron.

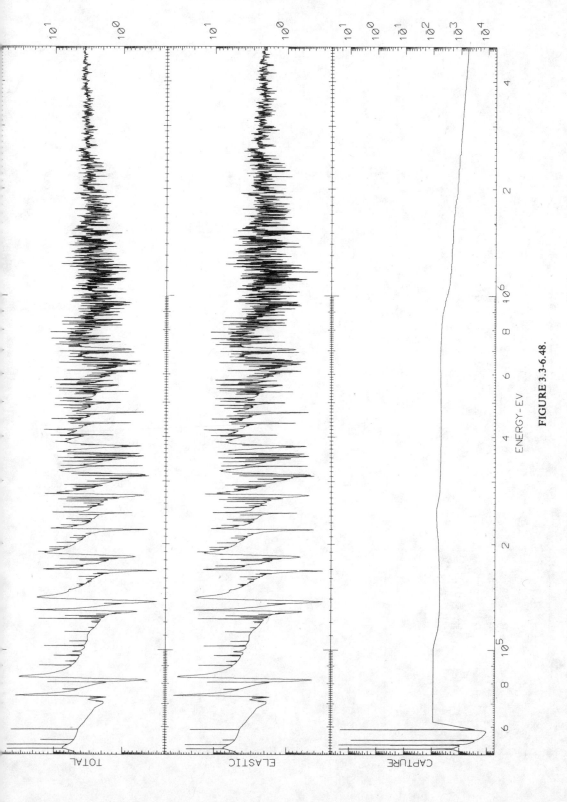

FIGURE 3.3-6.48.

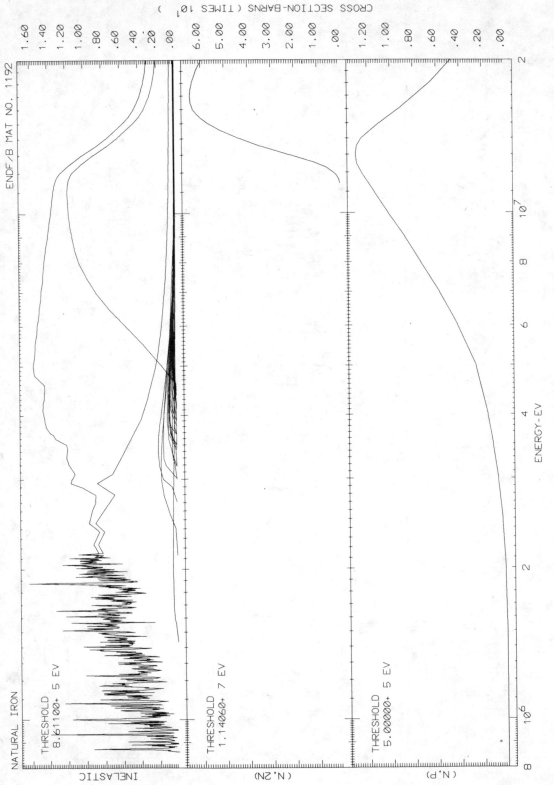

FIGURE 3.3-6.49.

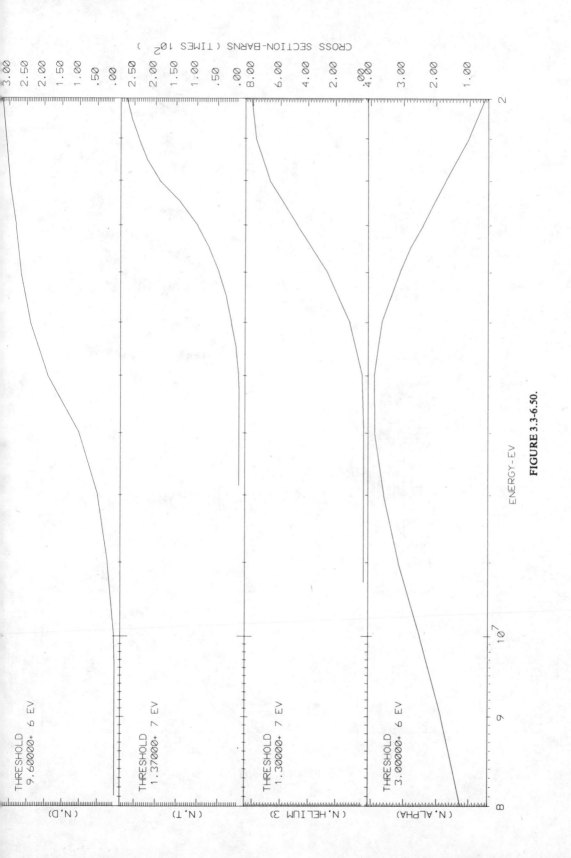

FIGURE 3.3-6.50.

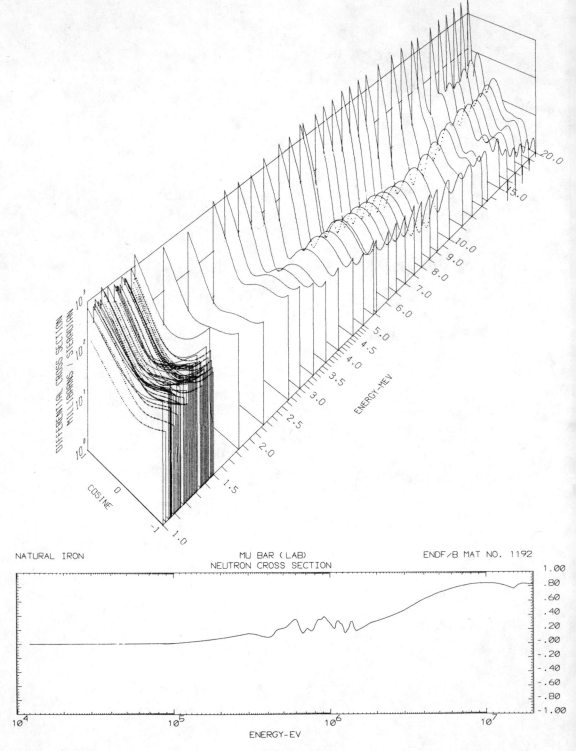

NATURAL IRON   MU BAR (LAB)   ENDF/B MAT NO. 1192
                            NEUTRON CROSS SECTION

FIGURE 3.3-6.51.

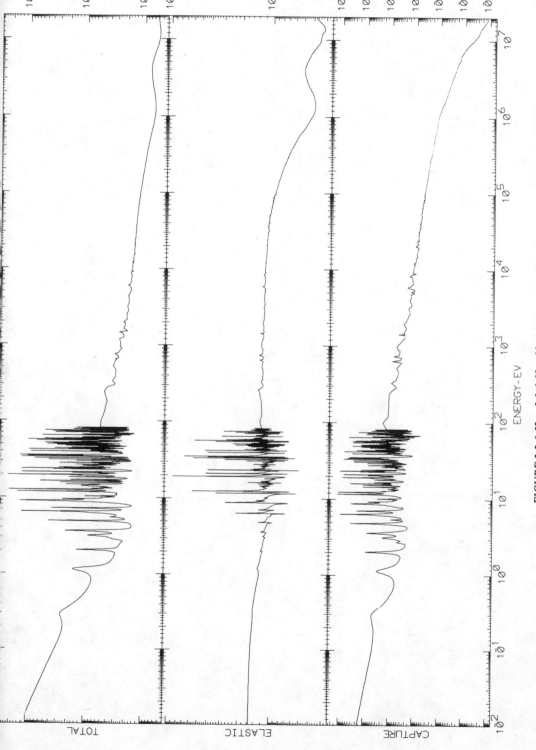

FIGURE 3.3-6.52 to 3.3-6.58.   Neutron cross sections for ^{235}U.

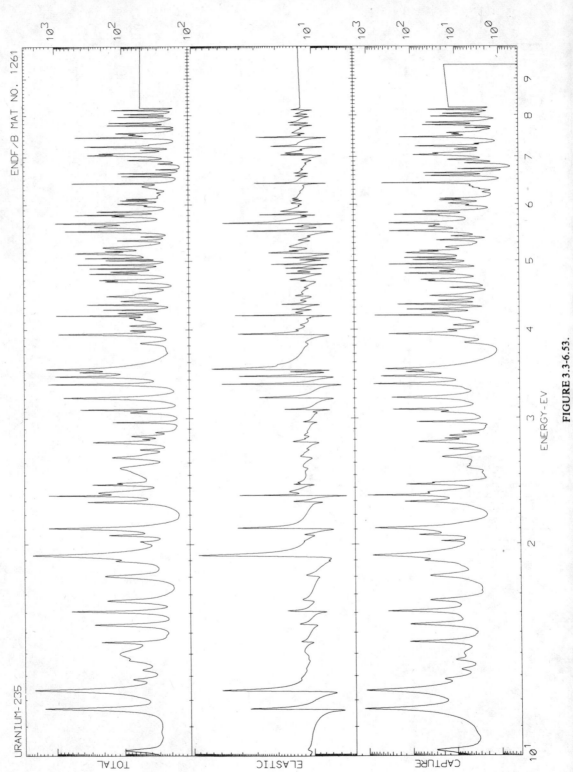

FIGURE 3.3-6.53.

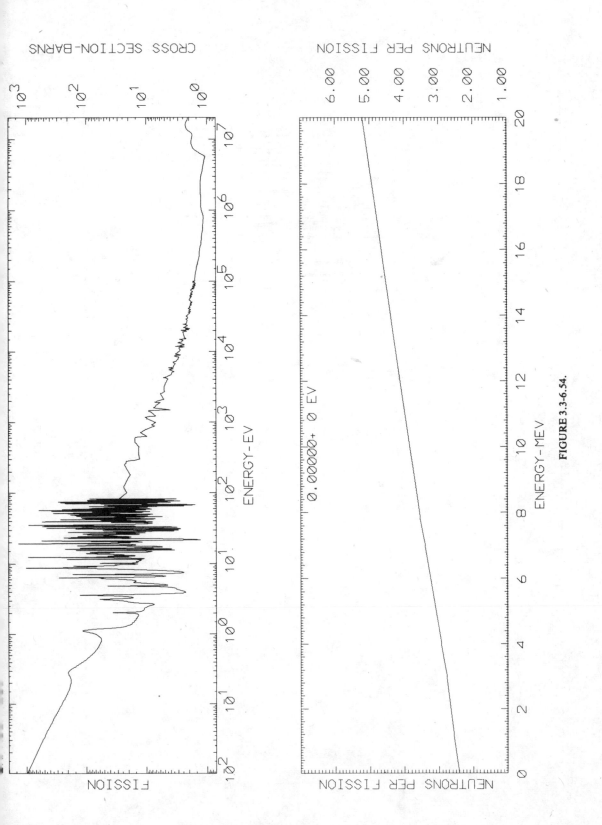

FIGURE 3.3-6.54.

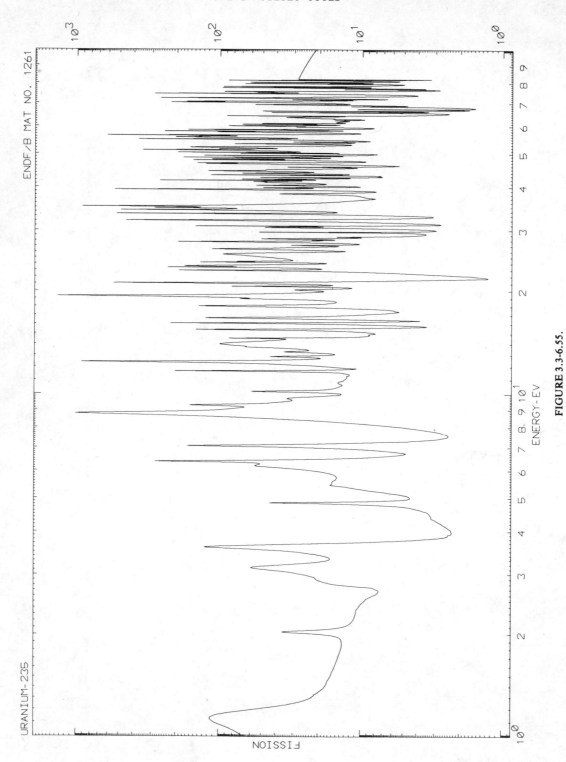

FIGURE 3.3-6.55.

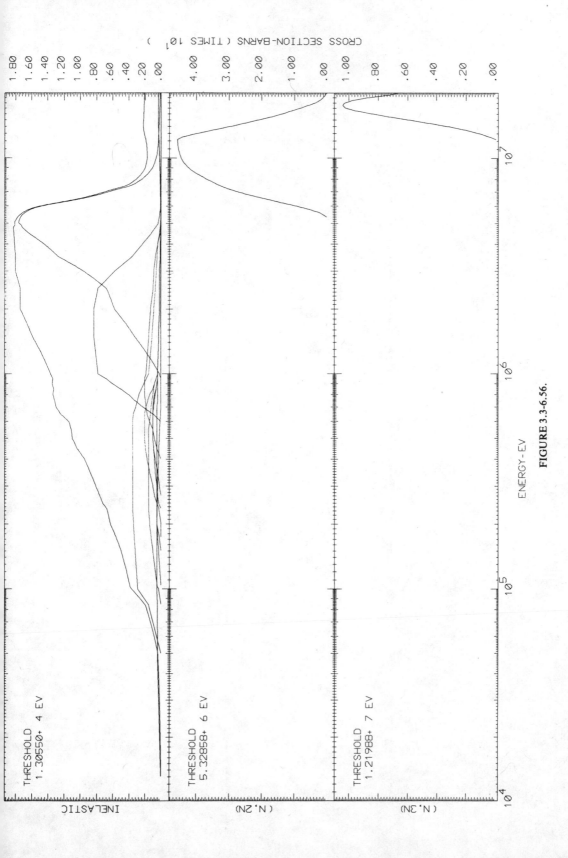

FIGURE 3.3-6.56.

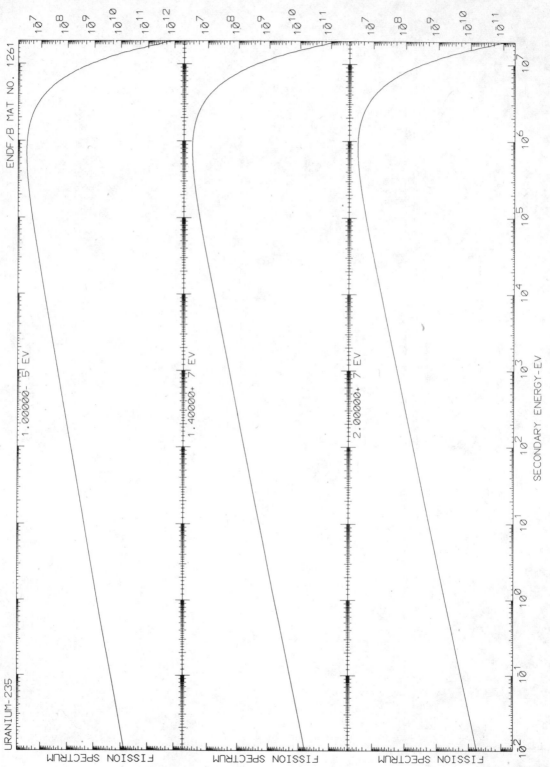

FIGURE 3.3-6.57.

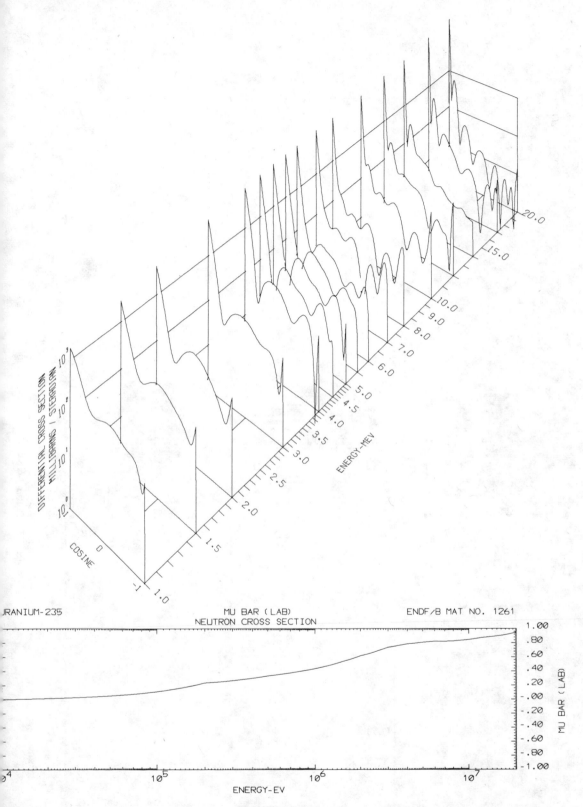

FIGURE 3.3-6.58.

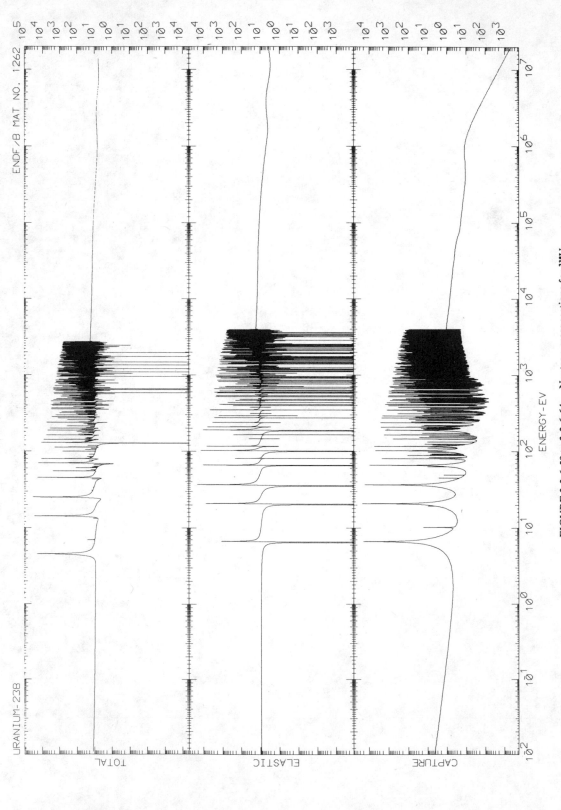

FIGURE 3.3-6.59 to 3.3-6.64.   Neutron cross sections for ^{238}U.

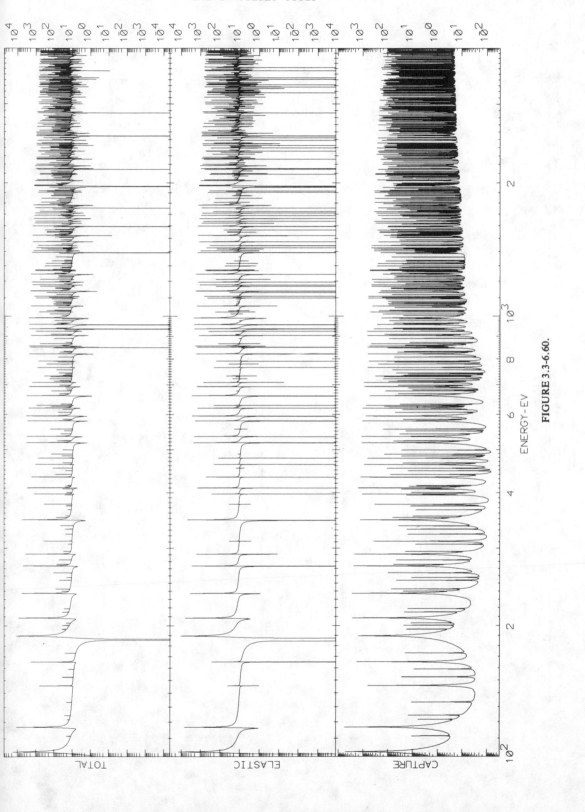

FIGURE 3.3-6.60.

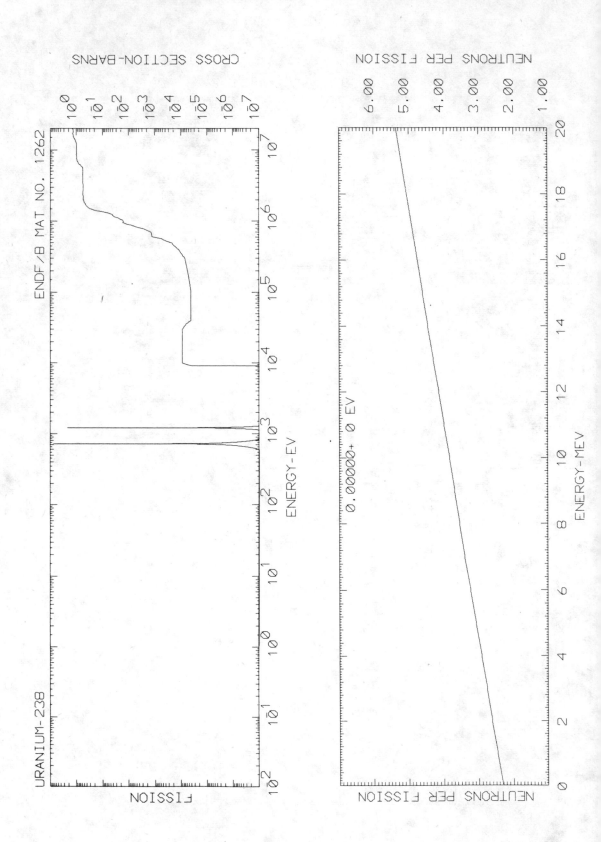

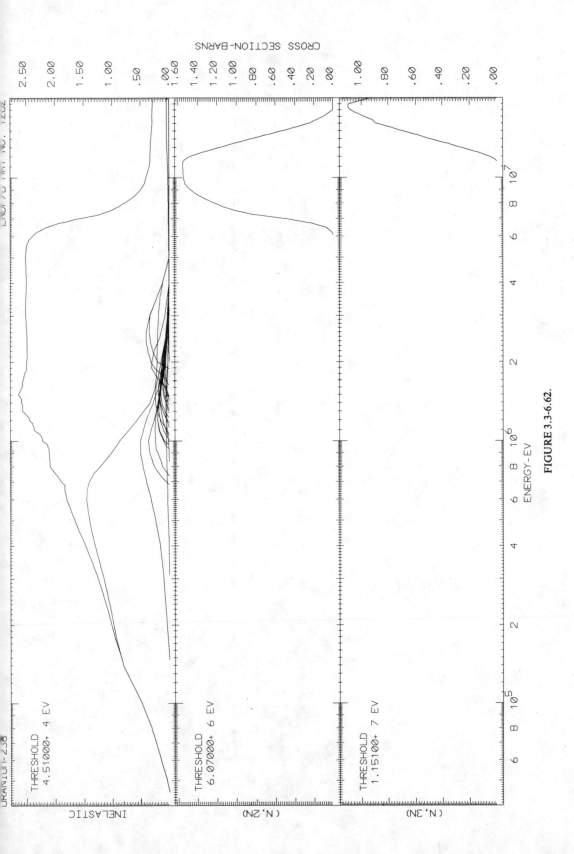

FIGURE 3.3-6.62.

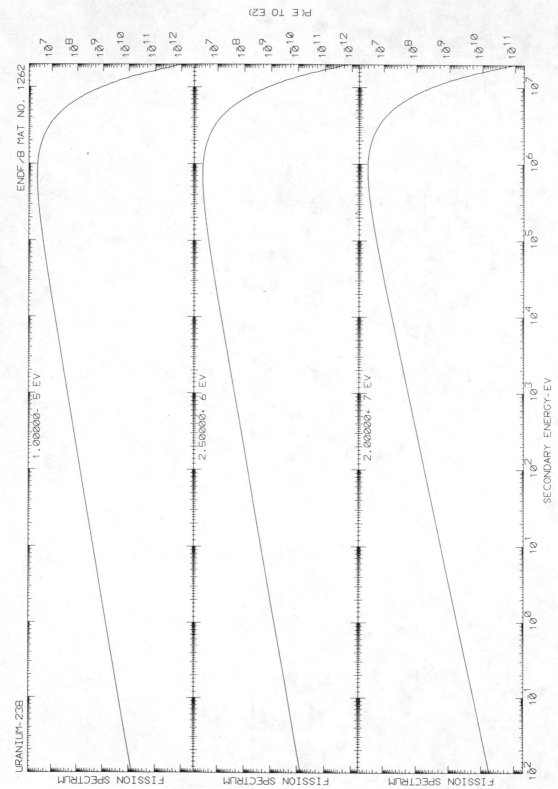

FIGURE 3.3-6.63.

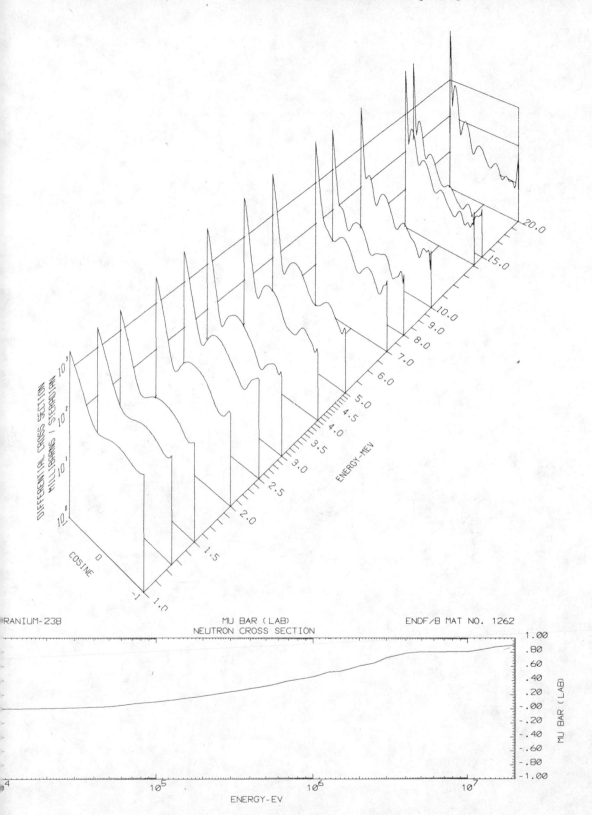

FIGURE 3.3-6.64.

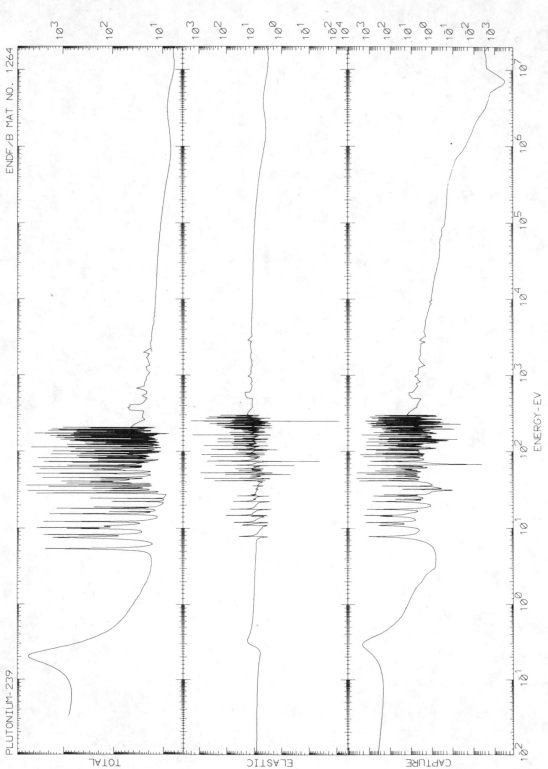

FIGURE 3.3-6.65 to 3.3-6.71. Neutron cross sections for ^{239}Pu.

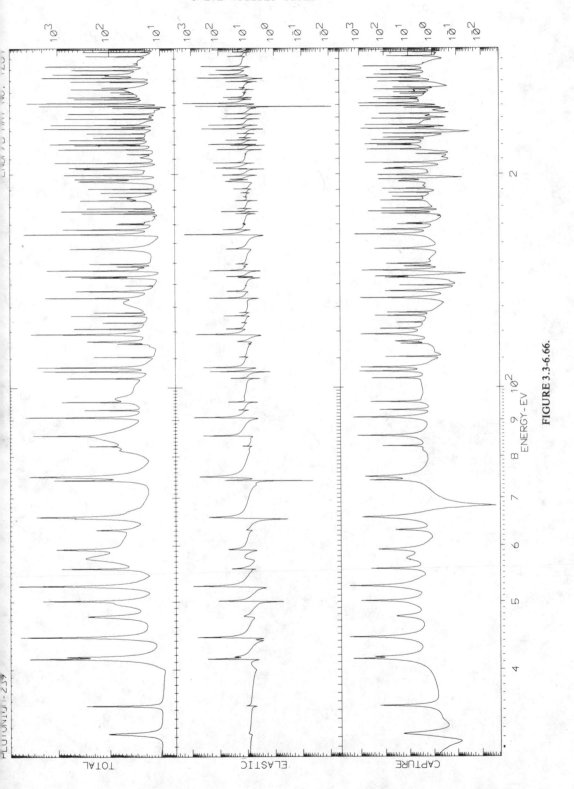

FIGURE 3.3-6.66.

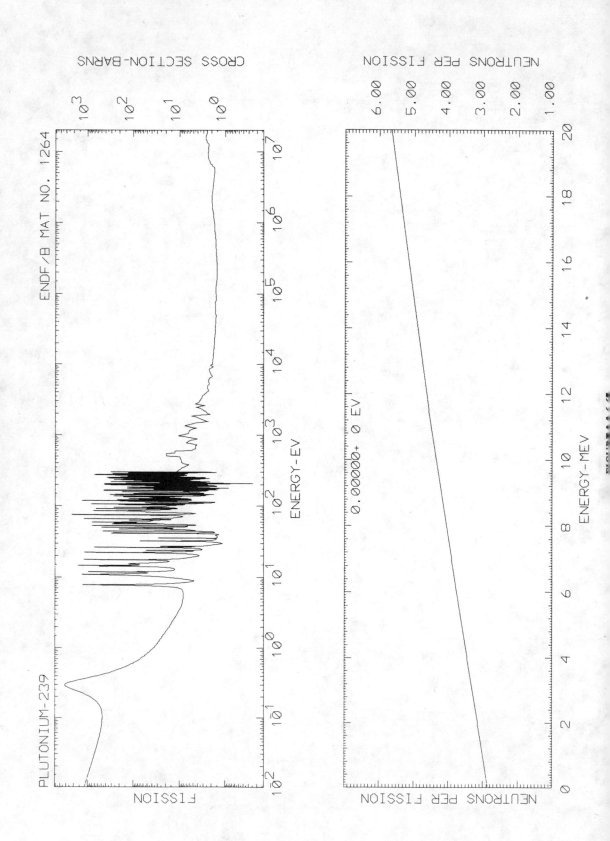

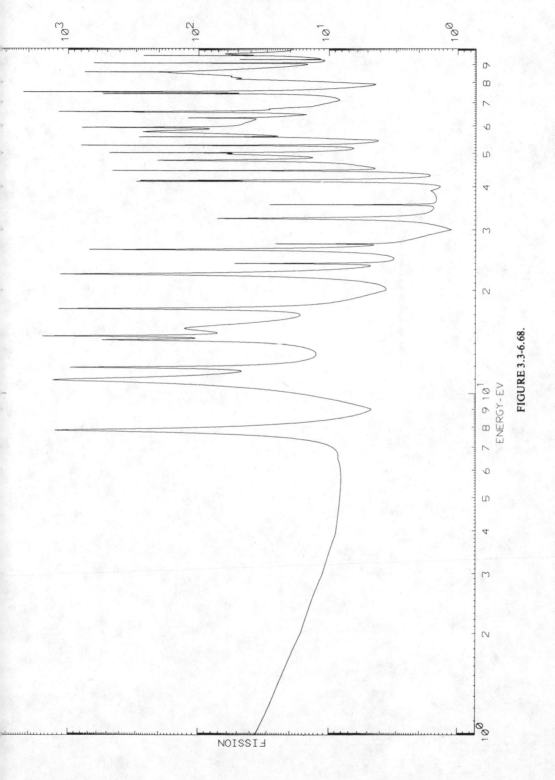

FIGURE 3.3-6.68.

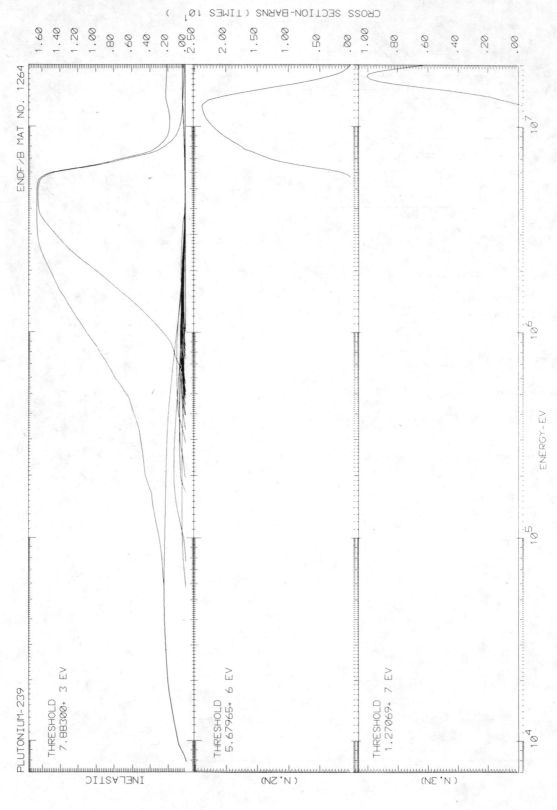

FIGURE 3.3-6.69.

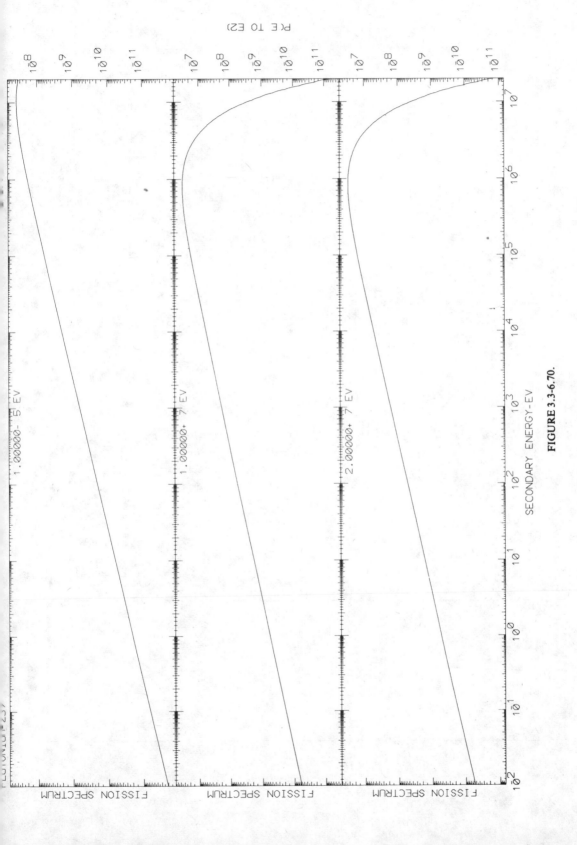

FIGURE 3.3-6.70.

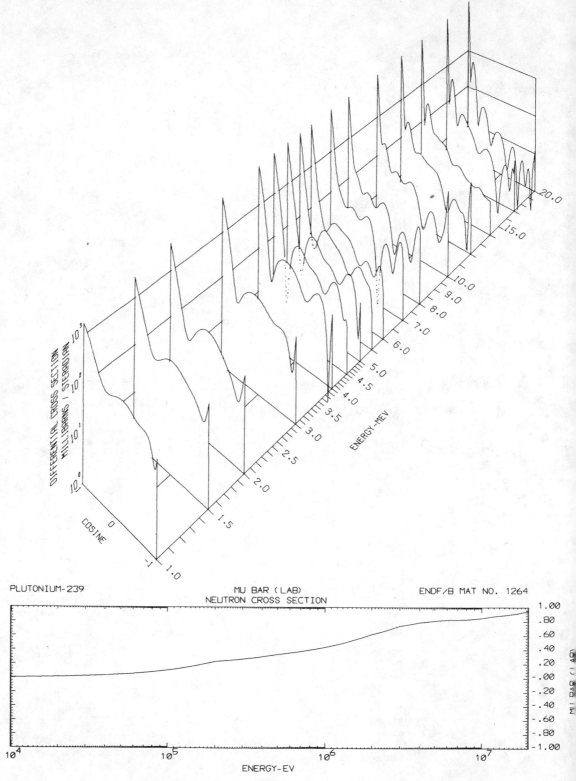

FIGURE 3.3-6.71.

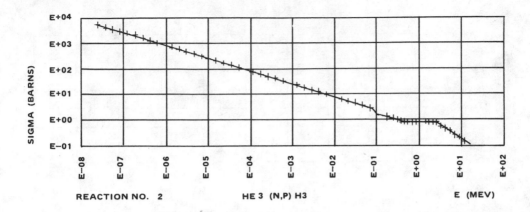

REACTION NO. 2   HE 3 (N,P) H3   E (MEV)

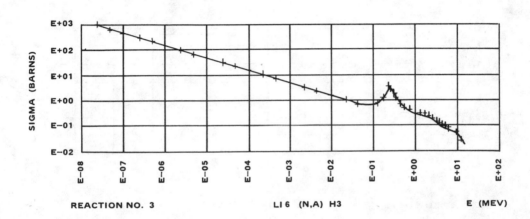

REACTION NO. 3   LI 6 (N,A) H3   E (MEV)

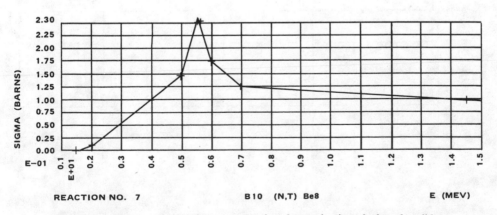

REACTION NO. 7   B 10 (N,T) Be8   E (MEV)

FIGURE 3.3-7.   Neutron activation cross sections for production of selected nuclides.

Cross-Section Curves

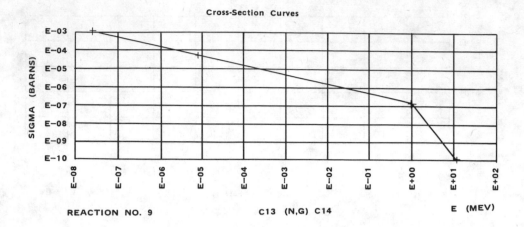

REACTION NO. 9   C13 (N,G) C14   E (MEV)

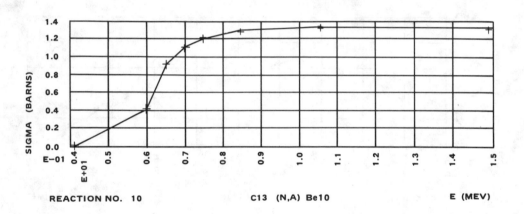

REACTION NO. 10   C13 (N,A) Be10   E (MEV)

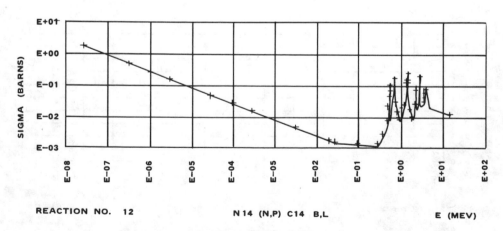

REACTION NO. 12   N 14 (N,P) C14   B,L   E (MEV)

Figure 3.3-7 (continued)

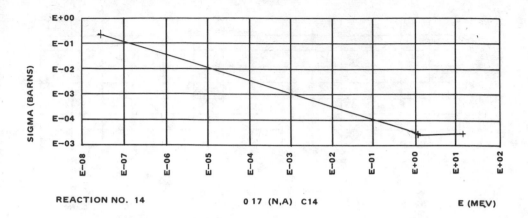

REACTION NO. 14   O 17 (N,A) C14   E (MEV)

REACTION NO. 25   SI 28 (N,P) AI 28   E (MEV)

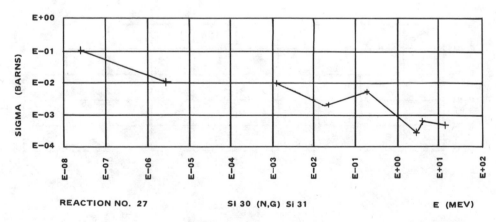

REACTION NO. 27   SI 30 (N,G) Si 31   E (MEV)

Figure 3.3-7 (continued)

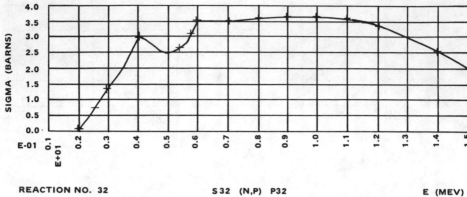

REACTION NO. 32   **S 32 (N,P) P32**   **E (MEV)**

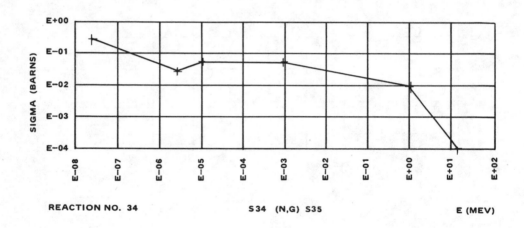

REACTION NO. 34   **S 34 (N,G) S35**   **E (MEV)**

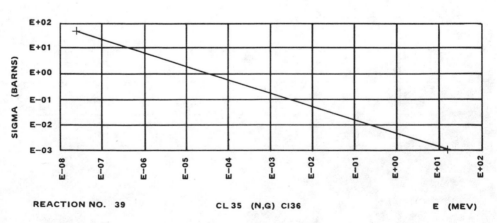

REACTION NO. 39   **CL 35 (N,G) Cl36**   **E (MEV)**

Figure 3.3-7 (continued)

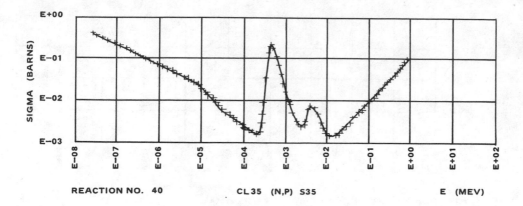

REACTION NO. 40   CL 35 (N,P) S35   E (MEV)

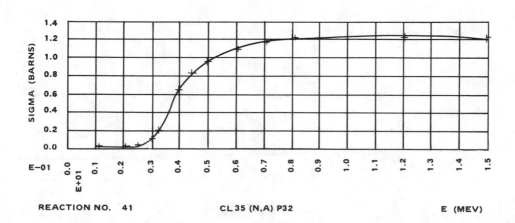

REACTION NO. 41   CL 35 (N,A) P32   E (MEV)

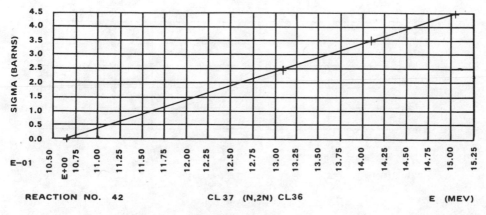

REACTION NO. 42   CL 37 (N,2N) CL36   E (MEV)

Figure 3.3-7 (continued)

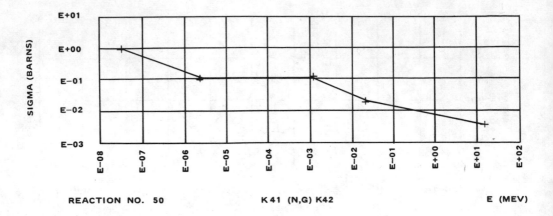

REACTION NO. 50   K 41 (N,G) K42   E (MEV)

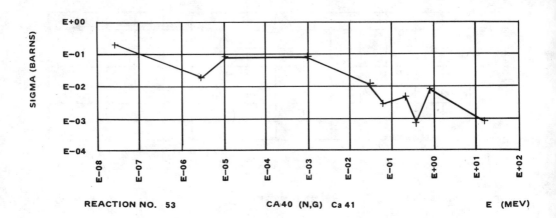

REACTION NO. 53   CA40 (N,G) Ca 41   E (MEV)

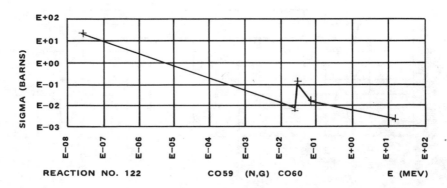

REACTION NO. 122   CO59 (N,G) CO60   E (MEV)

Figure 3.3-7 (continued)

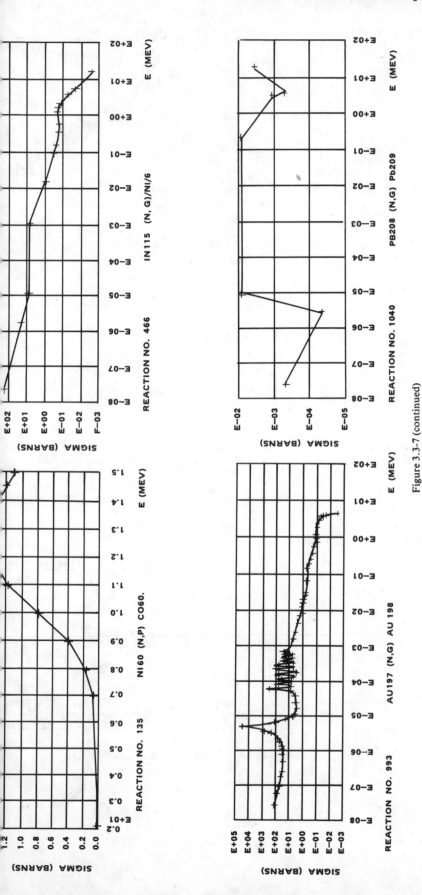

Figure 3.3-7 (continued)

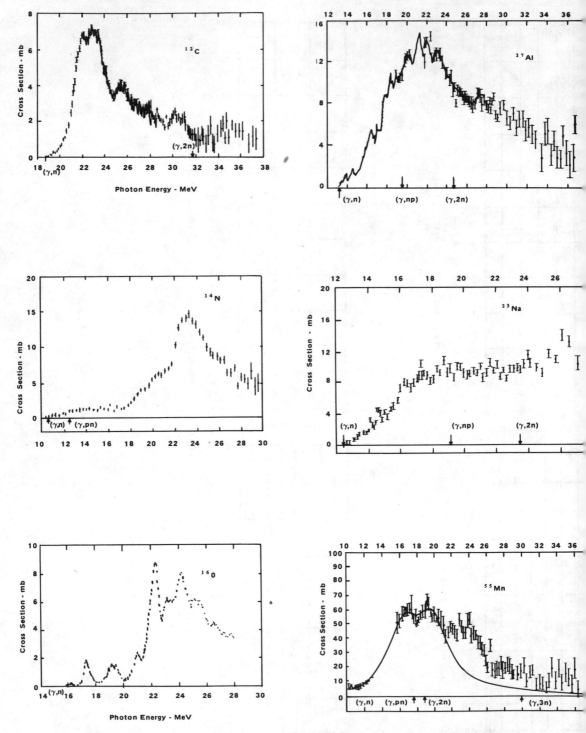

FIGURE 3.3-8.   Cross sections for neutron production by gamma rays (see Section 3.3.4.5).

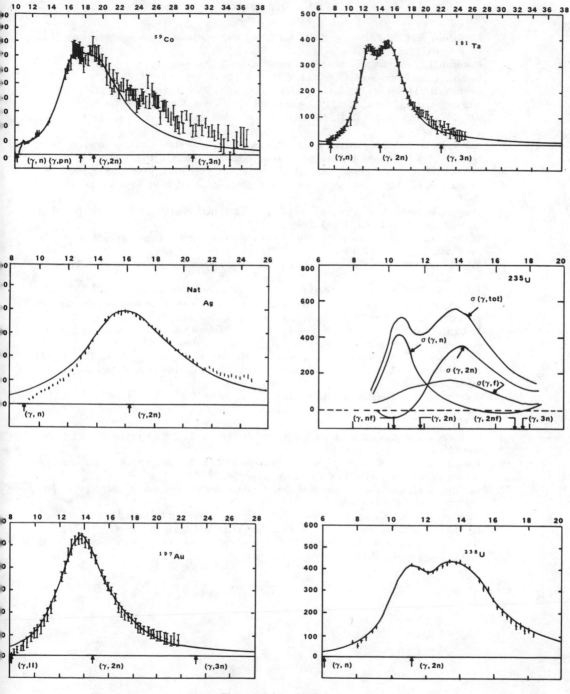

Figure 3.3-8 (continued)

REFERENCES

1. **Northcliffe, L. C. and Schilling, R. M.,** Range and stopping-power tables for heavy ions, *Nucl. Data Tables,* 7, 233, 1970.
2. **Cohen, B. L.,** *Concepts of Nuclear Physics,* McGraw-Hill, New York, 1971.
3. **Gilbert, A. and Cameron, A. G. W.,** *Can. J. Phys.,* 43, 1446, 1965.
4. **Gove, N. B. and Wapstra, A. H.,** Nuclear reaction Q-values, *Nucl. Data Tables,* 11, 127, 1972.
5. **Wapstra, A. H. and Gove, N. B.,** 1971 Atomic mass evaluation, *Nucl. Data Tables,* 9, 265, 1971.
6. **Alley, W. E. and Lessler, R. M.,** Neutron activation cross sections, *Nucl. Data Tables,* 11, 622, 1973.
7. **Jessen, P., Bormann, M., Dreyer, F., and Neuert, H.,** Experimental Excitation Functions (n,p), (n,t), (n,a), (n,2n), (n,np), and (n,na) Reactions, *Nucl. Data Tables,* 1, 103, 1966.
8. **Pearlstein, S.,** Calculated (n,2n) and (n,3n) cross sections, *Nucl. Data Tables,* 3, 327, 1967.
9. **Berman, B. L.,** Atlas of photoneutron cross sections, *At. Data Nucl. Data Tables,* 15, 319, 1975.
10. **Horsley, A.,** Neutron cross sections of hydrogen and deuterium in the energy range 10^{-4} eV to 20 MeV, *Nucl. Data Tables,* 2, 243, 1966.
11. **Horsley, A.,** Neutron cross sections of deuterium in the energy range 0.001 eV to 20 MeV, *Nucl. Data Tables,* 4, 321, 1968.
12. **Fu, C. Y. and Perey, F. G.,** Neutron cross section data for lead, *At. Data Nucl. Data Tables,* 16, 409, 1975.
13. **Fu, C. Y.,** Neutron cross section data for ^{40}Ca, *At. Data Nucl. Data Tables,* 17, 127, 1976.
14. **Liskien, H. and Paulsen, A.,** Neutron production cross sections for reactions that provide monoenergetic neutrons for T(p,n), D(d,n), and T(d,n), *Nucl. Data Tables,* 11, 569, 1973.
15. **Liskien, H. and Paulsen, A.,** Neutron production cross sections for reactions that provide monoenergetic neutrons for ^{7}Li(p,n), *At. Data Nucl. Data Tables,* 15, 57, 1975.
16. **Bartholomew, G. A., Doveika, A., Eastwood, K. M., Monaro, S., Groshev, L. V., Demidor, A. M., Pelekhov, V. I., and Sokolovskii, L. L.,** Compendium of thermal neutron capture γ-ray measurements, *Nucl. Data Tables,* 3, 367, 1966.
17. **Groshev, L. V., Demidor, A. M., Pelekhov, V. I., and Sokolovskii, L. L.,** Compendium of thermal neutron capture y-ray measurements, Z = 47 to Z = 67 (Ag to Ho), *Nucl. Data Tables,* 5, 1, 1968.
18. Compendium of thermal neutron capture y-ray measurements, Z = 68 to Z = 94 (ER to Pu), *Nucl. Data Tables,* 5, 243, 1968.
19. **Bird, J. R., Allen, B. J., Bergqvist, I., and Biggerstaff, J. A.,** Compilation of KeV-neutron-capture gamma ray spectra, *Nucl. Data Tables,* 11, 434, 1973.
20. **von Egidy, T.,** References for neutron capture conversion electron measurements, *Nucl. Data Tables,* 7, 465, 1970.
21. **Warburton, E. K., Kolata, J. J., Olness, J. W., Poletti, A. R., and Gorodetzky, P. H.,** Gamma rays from some heavy ion reactions, *At. Data Nucl. Data Tables,* 14, 147, 1974.
22. **Keller, K. A., Lange, J., and Munzel, H.,** *Landolt-Bornstein New Data Tables,* Group 1, Vol. 5, Part C, Shopper, H., Ed., Springer-Verlag, Berlin, 1974.

3.4. STOPPING POWERS AND RANGES OF CHARGED PARTICLES

James E. Turner

3.4.1. Introduction and Definitions

Understanding the interactions of charged particles with matter has been fundamental to the development of physics in this century. Charged-particle radiation provides an essential tool for probing and unravelling the structure of matter. In addition, the study of tracks in cloud chambers, emulsions, and other materials furnishes data on which much of our knowledge of the properties of the elementary particles is based. Understanding charged-particle interactions also forms the cornerstone for many applied disciplines, such as radiation dosimetry, which is a major subject of this series. In the present section, we will summarize some basic formulas and numerical material that are applicable to practical needs in dosimetry. A number of references are cited where additional detailed information can be found.

The passage of a charged particle through matter is characterized by the mean rate of energy loss along the particle's path through interactions with atomic electrons. This quantity, usually written $-dE/dx$ and expressed in units of energy per unit length (e.g., MeV/cm), is called the stopping power of the material. When stopping power is divided by the density ρ of the material, the mass stopping power, $-dE/\rho dx$ (MeV·cm^2/g), is obtained.

The total distance traveled by a charged particle in a material is called the range of the particle. Range can be expressed, therefore, in units of length (e.g., cm); it is also sometimes expressed in units of mass per unit area (g/cm^2), this number being obtained by multiplying the distance traveled by the density of the material. The range of a charged particle is inversely related to the stopping power of the material. Mathematically, the range R of a charged particle of a given kinetic energy T is the integral of the reciprocal of the stopping power from that energy down to zero energy:

$$R = \int_0^T \left(-\frac{dE}{dx} \right)^{-1} dE \qquad (3.4\text{-}1)$$

Implications of this formula are discussed below in Section 3.4.2.5.2.

Based on the known electromagnetic interactions of charged particles with atomic electrons, formulas describing energy-loss processes have been derived. The theoretical results are in good numerical agreement with experimental data. The description of the slowing-down process naturally divides into two physically different cases: "heavy" charged particles (i.e., more massive than the electron) and electrons and positrons. These cases are treated in the next two sections, after which some numerical stopping-power and range data are presented.

3.4.2. Heavy Charged Particles

The formula for the stopping power for heavy charged particles was derived by Bethe.[1] Its derivation and a review of various aspects of its applicability are given by Fano.[2] The Bethe formula for stopping power can be written:

$$-\frac{dE}{dx} = \frac{4\pi z^2 e^4 NZ}{mv^2} \left(\ln \frac{2mv^2}{I(1-\beta^2)} - \beta^2 - \frac{C}{Z} - \frac{1}{2}\delta \right) \qquad (3.4\text{-}2)$$

where $-e$ and m = charge and mass of electron; ze and v = charge and velocity of heavy particle; β = velocity of particle relative to velocity of light; Z, N, and I = atomic number, number of atoms per unit volume, and mean excitation energy of material traversed, respectively; and $\frac{C}{Z}$ and δ = shell and density-effect corrections. As written, Equation

3.4-2 is applicable to an elemental medium consisting of N atoms per unit volume with atomic number Z. For most applications in radiation dosimetry, if more than one element is present, the individual contributions can be computed from Equation 3.4-2 and added to find the total stopping power. The stopping powers of chemical compounds are within a few percent of the sum of the stopping powers of their elemental constituents.

Substitution of the numerical values for the universal constants into Equation 3.4-2 gives the stopping power

$$-\frac{dE}{dx} = \frac{5.10 \times 10^{-25} \, z^2 \, NZ}{\beta^2} \left(\ln \frac{1.022 \times 10^6 \, \beta^2}{I_{ev} \, (1 - \beta^2)} - \beta^2 - \frac{C}{Z} - \frac{1}{2} \delta \right) \text{ MeV/cm} \tag{3.4-3}$$

where N is the number of atoms per cubic centimeter of the medium and I_{ev} is expressed in electron volts. In terms of the density, ρ, of the medium, its atomic weight, A, and Avogadro's number, N_o, (= 6.023×10^{23} atoms/g atomic weight), we can write N = $N_0 \rho / A$. Equation 3.4-3 thus gives for the mass stopping power

$$-\frac{dE}{\rho dx} = \frac{0.307 \, z^2}{\beta^2} \cdot \frac{Z}{A} \left(\ln \frac{1.022 \times 10^6 \, \beta^2}{I_{eV} (1 - \beta^2)} - \beta^2 - \frac{C}{Z} - \frac{1}{2} \delta \right) \text{ MeV·cm}^2/\text{g} \tag{3.4-4}$$

We will next discuss the numerical evaluation of the quantities that enter Equations 3.4-3 and 3.4-4.

3.4.2.1. Calculation of β^2

In most applications, stopping powers and ranges are needed as functions of particle kinetic energy, T, rather than the velocity β. The relationship between these two quantities for a particle of rest mass, M, is given by

$$\beta^2 = \frac{T (T + 2Mc^2)}{(T + Mc^2)} \tag{3.4-5}$$

where c is the velocity of light and Mc^2 is the rest energy of the particle. If $T \ll Mc^2$ then Equation 3.4-5 reduces to the expression from classical mechanics, $v^2 = 2T/M$. Equation 3.4-5 is exact for all energies. Tables of T, as well as the expression $\ln[1.022 \times 10^6 \, \beta^2/(1 - \beta^2)] - \beta^2$, which enters Equations 3.4-3 and 3.4-4, are given by Bichsel[3] for protons, alpha particles, pions, muons, and electrons between the velocities $\beta = 0.032634$ and $\beta = 0.875028$, corresponding to protons with kinetic energies between 0.50 and 1000 MeV.

3.4.2.2. Mean Excitation Energy

The mean excitation energy, I, depends only on the stopping material. It has been calculated from first principles from its definition in Bethe's theory for only the simplest atoms. In practice, one relies on a combination of experimental data and theoretical arguments to determine I. Several surveys and compilations are available.[2-7] Since the logarithm of I enters the stopping-power formula, and because some uncertainty often exists in the evaluation of the shell-correction term, C/Z, measurement of stopping power does not determine I precisely. An analysis[6,7] of shell corrections and available experimental data led to the values of I given in Table 3.4-1. In the absence of specific data to the contrary, a linear interpolation over the atomic number, Z, can be used to obtain I for the elements not given. As can be seen from the table, a rough rule of thumb is $I_{eV} \sim 10$. The numerical values of I in electron volts from Table 3.4-1 can be used directly in Equations 3.4-3 and 3.4-4.

The adjusted mean excitation energy, I_{adj}, is sometimes used. It is related to I, and its logarithm differs from that of I by the magnitude of $(C/Z)_{\beta \to 1}$, the shell correction in the high-energy limit:

$$\ln I_{adj} = \ln I + (C/Z)_{\beta \to 1} \qquad\qquad (3.4\text{-}6)$$

As seen from Equation 3.4-2, I may be replaced by I_{adj} if shell corrections are assumed to vanish at high energies. This replacement is sometimes convenient in that I_{adj} is then determined experimentally by measurement of the stopping power.

3.4.2.3. Shell Corrections

Shell corrections C/Z, are not needed when the speed of the incident particle is much greater than that of the electrons bound in the atomic shells of the absorbing material. Such corrections are negligible in light, tissue-like materials at high particle velocities and are never large, even at low velocities. For example, shell corrections for protons in aluminum are no more than a few percent at energies of several megaelectronvolts, where they are largest. Shell corrections are not normally needed for work in radiation dosimetry.

On the other hand, the magnitude of shell corrections increases with increasing atomic number. Depending on the particular problem, it may be necessary to include them in shielding studies or in analyzing the interaction of heavy charged particles with solid-state detectors. Detailed calculations of the corrections for the K and L shells have been made by Walske.[8,9] Fano[2] and Bichsel[3] have given plots of the total C/Z for several atoms spanning the periodic table. Other sources of shell-correction data are cited in these two references. In addition, Barkas and Berger[10] give an empirical formula for calculating shell corrections.

3.4.2.4. Density Effect

The density-effect term, δ, describes the reduction in stopping power due to polarization of the medium in which the particle slows down. It is important only at high energies (e.g., $\sim$1000 MeV for protons in metals). The density effect has been reviewed

Table 3.4-1
VALUES OF I FOR AIR, WATER,
AND SOME CHEMICAL ELEMENTS

Atomic Number Z	Substance	I (eV)	Atomic Number Z	Substance	I (eV)
–	Air	92.9	30	Zn	319
–	Water	71.3	36	Kr	350
1	H	18.2	41	Nb	407
2	He	44.3	42	Mo	422
3	Li	37.4	45	Rh	440
4	Be	61.7	46	Pd	456
6	C	81.2	47	Ag	466
7	N	89.6	48	Cd	462
8	O	101	49	In	481
10	Ne	132	50	Sn	486
13	Al	163	54	Xe	480
17	Cl	176	73	Ta	692
18	Ar	189	74	W	704
20	Ca	187	77	Ir	730
22	Ti	224	78	Pt	711
23	V	250	79	Au	760
26	Fe	277	82	Pb	767
27	Co	290	90	Th	698
28	Ni	312	92	U	856
29	Cu	316			

by Crispin and Fowler.[11] Sternheimer[12,13] has given numerical formulas for calculating the density correction in a number of materials. Sternheimer's formulas are also given by Fano[2] and Bichsel.[3]

3.4.2.5. Additional Aspects of Particle Penetration
3.4.2.5.1. Fluctuation of Ion Charge at Low Velocities

Bethe's formula (Equation 3.4-2) applies to an ion with a fixed charge, ze. As it slows down in matter, a heavy particle will begin to capture atomic electrons with high probability at sufficiently low velocity. Captured electrons may be lost in subsequent collisions, and, thus, the charge of the ion fluctuates. The parameter that indicates when electron capture and loss become important is the ratio of the velocities of the captured electron and the ion, $ze^2/\hbar v$, where $\hbar = 1.05 \times 10^{-27}$ erg sec is Planck's constant divided by 2π.

Whaling[14] has given a number of empirical formulas for use at low energies where charge fluctuation is important, particularly for slow protons and alpha particles. Bichsel[3] also gives a procedure for taking into account the reduced charge of a slow ion. An extensive review and compilation of numerical data has been made by Northcliffe.[15,16]

3.4.2.5.2. Statistical Nature of Energy Loss

Energy losses to atomic electrons occur as discrete statistical events when a charged particle penetrates matter. Particles under identical conditions will experience different kinds and numbers of collisions, a phenomenon called straggling. As a result of straggling, all particles do not travel the same distance in losing a given amount of energy and they do not experience the same energy loss after traveling a given distance.

The Bethe formula (Equation 3.4-2) gives the average rate of energy loss per unit pathlength for an ion of specified charge and velocity. Equation 3.4-1 gives the range in the "continuous slowing-down approximation" calculated as though all particles under identical conditions slowed down continuously in the same fashion. The ranges of individual ions of a given type with the same initial energy will fluctuate about a mean value. If the fluctuations are symmetric about the mean then Equation 3.4-1 gives the most probable pathlength traveled. It should also be noted that the ions are deflected slightly by electronic collisions;* thus, the depth of penetration is somewhat shorter than the pathlength traveled.

These aspects of the energy loss process, not embodied in Equations 3.4-1 and 3.4-2, need not be considered in many applications for radiation dosimetry. On the other hand, they form the basis for some studies in track structure, microdosimetry, and detector response. Discussions can be found in the literature.[2,3,10,17,18]

3.4.2.5.3. Multiple Coulomb Scattering

In addition to collisions with atomic electrons, a charged particle will be deflected by the coulomb force it experiences from atomic nuclei. Typically, a heavy charged particle in condensed matter experiences a large number of small-angle elastic collisions with nuclei. Such multiple coulomb scattering will cause an initially parallel beam of ions to spread in penetrating a target. In calculating the absorbed dose from a broad beam of heavy ions, the effects of multiple coulomb scattering are negligible. On the other hand, the effects are large with relatively narrow beams, such as those obtained from most particle accelerators.

Multiple coulomb scattering is discussed in a number of texts.[19] Berger and Seltzer[17] summarize a number of studies of both multiple coulomb scattering and straggling effects discussed in the last section. Bichsel[3] gives an empirical representation for calculating

---

* Also by multiple coulomb scattering from atomic nuclei (see Section 3.4.2.5.3).

multiple coulomb scattering. Different methods for dose computations for negative pion beams have been studied.[20]

3.4.2.5.4. Other Effects

The foregoing sections have summarized Bethe's formulation of the stopping power of matter for heavy charged particles and indicated how various refinements can be made within the framework of the theory. Several extensions of the basic theory may be relevant under certain circumstances. First, extending Bethe's theory, which is calculated in the first Born approximation, to the second Born approximation introduces in the stopping-power formula terms proportional to $(ze)^3$, the cube of the charge of the ion. This extension thus leads to differences in stopping powers and ranges for positive and negative particles that are otherwise identical. Second, the Bethe formula rests on the assumption that the ratio of the electron and ion masses satisfies the inequality $m/M \ll \sqrt{1 - \beta^2}$. This assumption does not hold at extreme relativistic energies for which $\beta \sim 1$ (e.g., ~ 100 GeV for protons). The work on these extensions of the Bethe theory by a number of investigators is summarized in References 2, 21, and 22. Third, radiative corrections to the theory may be important at very high energies.[2,21]

3.4.2.6. Scaling of Stopping Powers and Ranges

As seen from Equation 3.4-2, the stopping power depends mainly on the velocity and charge of an ion. Apart from the often small corrections, C/Z and δ, and the effects discussed in Section 3.4.2.5, the stopping power of a material is the same function of v and z for heavy particles. To this approximation, we may write for a given material,

$$-\frac{dE}{dx} = z^2 f(\beta) \qquad\qquad (3.4\text{-}7)$$

Therefore, knowledge of the stopping power or range as a function of energy for one type of ion can often be used to infer these quantities for another ion. One needs only to properly scale the velocity and charge.

3.4.2.6.1. Stopping Power

An example will illustrate the scaling of stopping powers. We obtain the stopping power of water for a 4-MeV alpha particle based on that for protons (Table 3.4-3, Section 3.4.2.7). To do this, we find the square of the velocity, β^2, of the alpha particle, determine $f(\beta)$ from the proton table, and then multiply by $z^2 = 4$ in accordance with Equation 3.4-7. The rest energy of the alpha particle is $M_\alpha c^2 = 3727.3$ MeV. With $T = 4$ MeV, Equation 3.4-5 gives

$$\beta^2 = \frac{4(4 + 2 \times 3727.3)}{(4 + 3727.3)^2} = 0.0021429$$

With this value of β^2, linear interpolation in Table 3.4-3 gives $-dE/\rho dx = 270$ MeV cm^2/g for the proton mass stopping power, which is numerically equal to $f(\beta)$. Since $z^2 = 4$ for the alpha particle, the mass stopping power of water for the 4-MeV alpha particle is $-dE/\rho dx = 4 \times 270 = 1080$ MeV cm^2/g. Since $\rho = 1$ g/cm^3 for water, the stopping power is $-dE/dx = 1080$ MeV/cm.

Frequently, tables give the proton energy but do not include β^2. We would then need to find the kinetic energy of the proton with $\beta^2 = 0.0021429$. Since the rest energy of the proton is $M_p c^2 = 938.259$ MeV, its kinetic energy is

$$T = 938.259\left(\frac{1}{\sqrt{1 - \beta^2}} - 1\right) = 1.00692 \text{ MeV}$$

Again, linear interpolation in Table 3.4-3 gives $-dE/\rho dx = 270$ MeV cm^2/g for f(β).*

3.4.2.6.2. Ranges

Equation 3.4-1 expresses the range as a function of the kinetic energy, T, of a particle. We obtain a universal formula in terms of the velocity and charge. Since the total energy of a particle of rest mass, M, is given by $E = Mc^2/(1 - \beta^2)^{1/2}$, the differential in energy can be expressed by writing $dE = Mg(\beta)d\beta$, where g is a function that depends only on β. Combining this with Equation 3.4-7, we may write

$$R(\beta) = \frac{M}{z^2} R_o (\beta) \qquad (3.4\text{-}8)$$

where $R_o(\beta) = \int_o^\beta [g(\beta)/f(\beta)]\, d\beta$ is the same function of β for all heavy charged particles. Equation 3.4-8 implies that, if we know the range as a function of velocity for one type of particle, the range for another type scales directly as the ratio of the particle masses and inversely on the ratio of the squares of the charges. In comparing alpha particles and protons, for example, $M_\alpha \cong 4M_p$

$$\frac{R_\alpha (\beta)}{R_p (\beta)} = \frac{M_\alpha}{M_p} \cdot \frac{z_p^2}{z_\alpha^2} = 1$$

and, thus, alpha particles and protons of the same velocity have the same range in a given material. For deuterons, on the other hand, $M_d \cong 2M_p$ and $R_d(\beta) = 2R_p(\beta)$; therefore, deuterons of a given velocity have about twice the range of protons with the same velocity.

Table 3.4-2 lists the charge z and the ratio M/z^2 for a number of heavy particles relative to that for protons. This table can be used with the proton stopping-power and range tables in the next section to obtain these quantities for the other particles.

3.4.2.7. Stopping-Power and Range Tables

While the above formulas and information can be used for calculating stopping powers and ranges in a particular case of interest, these quantities can often be found in published tables. Section 3.4.4 describes some reference sources that contain numerical stopping-power and range tables that are of particular importance in many health-physics applications. Some of these references also contain useful empirical formulas.

We include here tables of stopping powers and ranges for protons in water, air, and lead. The data for soft tissue of unit density and for water are, for most purposes, identical.

Table 3.4-3 gives the mass stopping powers (MeV cm^2/g) of water, air, and lead for protons with energies from 0.10 to 5000 MeV. The stopping power in megaelectronvolts per centimeter is obtained by multiplying the values in Table 3.4-3 by the density of the material in grams per cubic centimeter. The mass stopping powers of soft tissue and bone are close to that of water. At 100 MeV, for example, the calculated mass stopping power of bone is 6.7% less than that of muscle, which is 1.1% less than that of water.[10] Proton stopping powers for other materials can be estimated crudely from the trends in the table based on the value of I (Table 3.4-1) for the material. Values of β^2 are included in Table 3.4-3 to assist in scaling to other heavy charged particles.

Table 3.4-4 gives the mean ranges (g/cm^2) of protons with energies from 1 to 5000 MeV in water, air, and lead. The range, in centimeters, can be obtained by dividing the

---

* Note that since β^2 is small in this example, the simpler classical expression $T = \frac{1}{2}Mv^2$ gives the same result as the relativistic Equation 3.4-5. Classically, for a given v (or β), the ratio of the kinetic energies of two particles is equal to the ratio of the masses of the particles.

Table 3.4-2
z AND M/z² RELATIVE
TO PROTON FOR HEAVY
CHARGED PARTICLES

Particle	z	M/z^2
Muon	±1	0.113
Pion	±1	0.149
Kaon	±1	0.526
Proton	+1	1.000
Deuteron	+1	1.999
Triton	+1	2.994
^{3}He	+2	0.748
Alpha particle	+2	0.993

Table 3.4-3
MASS STOPPING POWER FOR PROTONS
IN WATER, AIR, AND LEAD

Mass stopping power (MeV cm²/g)

Proton energy (MeV)	β^2	Water $\rho = 1.00$ g/cm³	Air $\rho = 0.00129$ g/cm³	Lead $\rho = 11.3$ g/cm³
0.1	0.000213	910	730	122
0.2	0.000426	740	580	127
0.3	0.000639	600	480	113
0.4	0.000852	500	410	100
0.5	0.001065	430	350	90
0.6	0.001278	381	310	83
0.7	0.001490	343	280	77
0.8	0.001703	317	260	71
0.9	0.001916	299	240	67
1.0	0.002128	271	220	62.9
1.2	0.002553	238	198	57.7
1.4	0.002978	213	177	53.5
1.6	0.003402	194	160	50.2
1.8	0.003826	179	148	47.7
2.0	0.004250	168	139	45.6
4.0	0.008472	100	84.2	31.4
6.0	0.012668	71.4	60.2	24.2
8.0	0.016837	55.8	47.3	20.0
10.0	0.020980	46.8	39.9	17.5
14	0.029188	35.6	30.5	13.7
18	0.037292	29.0	24.9	11.5
22	0.045296	24.7	21.2	9.98
26	0.053200	21.5	18.5	8.90
30	0.061007	19.2	16.5	8.06
34	0.068717	17.3	14.9	7.40
38	0.076333	15.8	13.6	6.86
42	0.083856	14.6	12.6	6.40
46	0.091287	13.6	11.7	6.01
50	0.098628	12.7	11.0	5.67
60	0.116597	11.0	9.50	5.00

Table 3.4-3 (continued)
MASS STOPPING POWER FOR PROTONS
IN WATER, AIR, AND LEAD

Mass stopping power (MeV cm^2/g)

Proton energy (MeV)	β^2	Water $\rho = 1.00$ g/cm^3	Air $\rho = 0.00129$ g/cm^3	Lead $\rho = 11.3$ g/cm^3
70	0.134033	9.74	8.43	4.50
80	0.150958	8.79	7.61	4.11
90	0.167392	8.03	6.96	3.79
100	0.183354	7.42	6.43	3.53
110	0.198860	6.92	6.00	3.31
120	0.213929	6.49	5.63	3.13
130	0.228577	6.13	5.32	2.97
140	0.242820	5.81	5.05	2.83
150	0.256671	5.54	4.81	2.71
160	0.270146	5.30	4.60	2.60
170	0.283258	5.08	4.42	2.51
180	0.296019	4.89	4.25	2.42
190	0.308443	4.72	4.10	2.34
200	0.320514	4.57	3.97	2.27
220	0.343803	4.30	3.74	2.15
240	0.365891	4.07	3.54	2.05
260	0.386882	3.88	3.38	1.96
280	0.406848	3.72	3.24	1.88
300	0.425854	3.58	3.11	1.81
320	0.443961	3.45	3.01	1.76
340	0.461225	3.34	2.91	1.71
360	0.477697	3.25	2.83	1.66
380	0.493425	3.16	2.75	1.62
400	0.508453	3.08	2.68	1.58
440	0.536570	2.95	2.57	1.52
480	0.562342	2.83	2.47	1.47
520	0.586023	2.74	2.39	1.42
560	0.607832	2.66	2.32	1.39
600	0.627963	2.59	2.26	1.35
640	0.646582	2.53	2.21	1.33
680	0.663837	2.48	2.17	1.30
720	0.679859	2.44	2.13	1.28
760	0.694763	2.40	2.10	1.27
800	0.708649	2.36	2.07	1.25
840	0.721609	2.33	2.04	1.24
880	0.733723	2.30	2.02	1.22
920	0.745063	2.28	2.00	1.21
960	0.755694	2.26	1.98	1.20
1000	0.765673	2.24	1.96	1.20
1200	0.807458	2.16	1.90	1.17
1400	0.838987	2.11	1.86	1.15
1600	0.863361	2.08	1.84	1.14
2000	0.898032	2.05	1.82	1.13
2400	0.921004	2.04	1.82	1.14

Table 3.4-3 (continued)
MASS STOPPING POWER FOR PROTONS
IN WATER, AIR, AND LEAD

Mass stopping power (Me V cm² /g)

Proton energy (Me V)	β^2	Water $\rho = 1.00\ g/cm^2$	Air $\rho = 0.00129\ g/cm^3$	Lead $\rho = 11.3\ g/cm^3$
2800	0.937005	2.03	1.82	1.14
3200	0.948594	2.03	1.83	1.15
3600	0.957257	2.04	1.85	1.17
4000	0.963901	2.05	1.86	1.18
5000	0.975035	2.07	1.90	1.20

values in the table by the density of the material. Equation 3.4-8 and Table 3.4-2 can be used to find the ranges of other heavy charged particles by scaling from Table 3.4-4. These tables are based on compilations by Barkas and Berger[10] and by Bichsel.[3]

3.4.3. Electrons and Positrons
3.4.3.1. Energy Loss by Collisions and Radiation

The theory of energy loss by low-energy electrons parallels that for heavy charged particles. The stopping-power formula is modified, however, because the reduced mass for the collision of one electron with another is m/2 and the two electrons that emerge from an ionizing collision are indistinguishable. By convention, the electron emerging with the higher energy is defined as the primary. In addition to these modifications, moreover, high-energy electrons also lose energy by emitting electromagnetic radiation, bremsstrahlung, when accelerated in the field of an atomic nucleus. The kinetic energy at which the rate of energy loss by radiation, $(-dE/dx)_{rad}$, is equal to that due to collisions with atomic electrons, $(-dE/dx)_{col}$, is given in approximation by $T_c \cong 800/Z$ MeV, where Z is the atomic number of the material. In water, for example, $T_c \cong 92$ MeV. The total rate of energy loss is the sum

$$\left(-\frac{dE}{dx}\right)_{tot} = \left(-\frac{dE}{dx}\right)_{col} + \left(-\frac{dE}{dx}\right)_{rad} \tag{3.4-9}$$

The collision stopping-power formula for electrons of kinetic energy T is

$$\left(-\frac{dE}{dx}\right)_{col} = \frac{2\pi e^4 NZ}{mv^2}\left[\ln\frac{Tmv^2}{2I^2\,(1-\beta^2)} + 1 - \beta^2 + \frac{1}{8}\,(1-\sqrt{1-\beta^2}) - (2\sqrt{1-\beta^2} - 1 + \beta^2)\ln 2 - \delta\right] \tag{3.4-10}$$

where all of the symbols have the same meaning as in Equation 3.4-2. No convenient single formula is available to represent $(-dE/dx)_{rad}$, which is calculated from the theory of Bethe and Heitler[23] with various modifications. For positrons, the collision stopping-power formula is

$$\left(\frac{-dE}{dx}\right)_{col}^{+} = \frac{2\pi e^4 NZ}{mv^2}\left\{\ln\frac{Tmv^2}{2I^2\,(1-\beta^2)} + 2\ln 2 - \frac{\beta^2}{12}\left[23 + \frac{14\sqrt{1-\beta^2}}{1+\sqrt{1-\beta^2}} + \frac{10\,(1-\beta^2)}{(1+\sqrt{1-\beta^2})^2} + \right.\right.$$

$$\left.\left.\frac{4\,(1-\beta^2)^{3/2}}{(1+\sqrt{1-\beta^2})^3}\right] - \delta\right\} \tag{3.4-11}$$

Table 3.4-4
MEAN RANGES OF PROTONS
IN WATER, AIR, AND LEAD

Proton energy (MeV)	Mean range (g/cm²)		
	Water $\rho = 1$ g/cm³	Air $\rho = 0.00129$ g/cm³	Lead $\rho = 11.3$ g/cm³
1.0	.0031	0.0039	.0116
1.2	.0037	0.0046	.0149
1.4	.0044	0.0055	.0185
1.6	.0052	0.0065	.0224
1.8	.0061	0.0076	.0265
2.0	.0071	0.0087	.0308
4.0	.0230	0.0277	.0844
6.0	.0471	0.0562	.158
8.0	.0791	0.0941	.250
10.0	.118	0.140	.356
14	.217	0.256	.617
18	.342	0.402	.937
22	.493	0.577	1.31
26	.667	0.779	1.74
30	.864	1.01	2.21
34	1.08	1.26	2.73
38	1.33	1.54	3.29
42	1.59	1.85	3.90
46	1.87	2.18	4.54
50	2.18	2.53	5.23
60	3.03	3.52	7.11
70	4.00	4.64	9.22
80	5.08	5.89	11.5
90	6.27	7.26	14.1
100	7.57	8.76	16.8
110	8.97	10.4	19.7
120	10.5	12.9	22.9
130	12.0	13.9	26.1
140	13.7	15.9	29.6
150	15.5	17.9	33.2
160	17.3	20.0	37.0
170	19.3	22.2	40.9
180	21.2	24.5	44.9
190	23.3	26.9	49.1
200	25.5	29.4	53.5
220	30.0	34.6	62.5
240	34.8	40.1	72.1
260	39.8	45.9	82.1
280	45.1	51.9	92.5
300	50.6	58.2	103
320	56.3	64.8	115
340	62.2	71.5	126
360	68.2	78.5	138
380	74.5	85.7	150

Table 3.4-4 (continued)
MEAN RANGES OF PROTONS
IN WATER, AIR, AND LEAD

Mean range (g/cm²)

Proton energy (MeV)	Water $\rho = 1$ g/cm³	Air $\rho = 0.00129$ g/cm³	Lead $\rho = 11.3$ g/cm³
400	80.9	93.1	163
440	94.2	108	189
480	108	124	215
520	122	141	243
560	137	158	272
600	152	175	301
640	168	193	331
680	184	211	361
720	200	230	392
760	217	249	423
800	234	268	455
840	251	288	487
880	268	307	520
920	285	327	553
960	303	347	586
1000	321	368	619
1200	412	471	789
1400	506	578	962
1600	601	686	1137
2000	795	905	1490
2400	991	1125	1843
2800	1188	1345	2194
3200	1384	1564	2542
3600	1581	1782	2887
4000	1776	1998	3229
5000	2262	2530	4069

The total rate of energy loss for electrons and positrons as a function of energy is almost the same over most of the energy scale. At low energies (T $\leq$ 0.1 MeV), $(-dE/dx)_{tot}$ is somewhat greater for positrons. Berger and Seltzer[24] and Nelms[25,26] provide detailed comparisons of electron and positron stopping powers.

3.4.3.2. Stopping-Power and Range Tables

Table 3.4-5 gives the collision and total stopping powers, $(-dE/dx)_{col}$ and $(-dE/dx)_{tot}$, of water for electrons with energies from 0.01 to 1000 MeV. The radiative energy loss, $(-dE/dx)_{rad}$, is the difference of these two quantities, as described by Equation 3.4-9. Ranges are given in Table 3.4-6 as calculated in the continuous slowing-down approximation (Section 3.4.2.5.2). These tables are taken from the work of Berger and Seltzer.[24]

3.4.4. Tabulations of Stopping Powers and Ranges

A number of tabulations of ranges and stopping powers are available in the literature. The following list describes some of those which are useful for work in radiation dosimetry.

Table 3.4-5

MASS STOPPING POWERS (MeV cm² /g) FOR ELECTRONS IN WATER, AIR, AND LEAD

Electron energy (MeV)	Water, $\rho = 1$ g/cm³		Air, $\rho = 0.00129$ g/cm³		Lead, $\rho = 11.3$ g/cm³	
	Collision $\left(-\dfrac{dE}{dx}\right)_{col}$	Total $\left(-\dfrac{dE}{dx}\right)_{tot}$	Collision $\left(-\dfrac{dE}{dx}\right)_{col}$	Total $\left(-\dfrac{dE}{dx}\right)_{tot}$	Collision $\left(-\dfrac{dE}{dx}\right)_{col}$	Total $\left(-\dfrac{dE}{dx}\right)_{tot}$
0.010	23.2	23.2	19.7	19.7	8.42	8.46
0.015	16.9	16.9	14.4	14.4	6.56	6.60
0.020	13.5	13.5	11.6	11.6	5.45	5.50
0.025	11.4	11.4	9.73	9.73	4.71	4.76
0.030	9.88	9.88	8.48	8.48	4.18	4.23
0.035	8.79	8.79	7.55	7.55	3.78	3.83
0.040	7.95	7.95	6.84	6.84	3.46	3.51
0.045	7.29	7.29	6.27	6.27	3.21	3.26
0.050	6.75	6.75	5.81	5.81	3.00	3.05
0.055	6.30	6.30	5.43	5.43	2.82	2.87
0.060	5.92	5.92	5.10	5.11	2.67	2.72
0.065	5.60	5.60	4.82	4.83	2.54	2.59
0.070	5.32	5.32	4.58	4.59	2.43	2.48
0.075	5.07	5.08	4.37	4.38	2.33	2.38
0.080	4.85	4.86	4.19	4.20	2.24	2.29
0.085	4.66	4.67	4.02	4.03	2.16	2.22
0.090	4.49	4.50	3.88	3.89	2.09	2.15
0.095	4.33	4.34	3.74	3.75	2.02	2.08
0.10	4.19	4.20	3.62	3.63	1.96	2.02
0.15	3.30	3.31	2.85	2.86	1.58	1.65
0.20	2.84	2.85	2.46	2.47	1.39	1.46
0.25	2.57	2.58	2.23	2.24	1.27	1.35
0.30	2.39	2.40	2.08	2.09	1.20	1.28
0.35	2.27	2.28	1.97	1.98	1.14	1.23
0.40	2.18	2.19	1.90	1.91	1.11	1.20
0.45	2.11	2.12	1.84	1.85	1.08	1.18
0.50	2.06	2.07	1.80	1.81	1.06	1.17
0.60	1.99	2.00	1.74	1.75	1.03	1.15
0.70	1.94	1.96	1.70	1.72	1.02	1.15
0.80	1.91	1.93	1.68	1.70	1.01	1.15
0.90	1.89	1.91	1.66	1.68	1.00	1.16
1.00	1.87	1.89	1.66	1.68	1.00	1.17
1.5	1.85	1.88	1.66	1.68	1.02	1.24
2.0	1.86	1.89	1.68	1.71	1.04	1.32
3.0	1.88	1.93	1.74	1.79	1.08	1.48
4.0	1.91	1.97	1.79	1.85	1.11	1.63
5.0	1.93	2.01	1.83	1.91	1.14	1.78
6.0	1.95	2.05	1.87	1.97	1.16	1.93
7.0	1.96	2.09	1.90	2.02	1.18	2.07
8.0	1.98	2.12	1.93	2.07	1.19	2.21
9.0	1.99	2.15	1.96	2.12	1.21	2.35
10	2.00	2.18	1.98	2.16	1.22	2.49
20	2.06	2.47	2.13	2.53	1.29	3.91

Table 3.4-5 (continued)
MASS STOPPING POWERS (MeV cm² /g) FOR
ELECTRONS IN WATER, AIR, AND LEAD

Electron energy (MeV)	Water, $\rho = 1$ g/cm³		Air, $\rho = 0.00129$ g/cm³		Lead, $\rho = 11.3$ g/cm³	
	Collision $\left(-\dfrac{dE}{dx}\right)_{col}$	Total $\left(-\dfrac{dE}{dx}\right)_{tot}$	Collision $\left(-\dfrac{dE}{dx}\right)_{col}$	Total $\left(-\dfrac{dE}{dx}\right)_{tot}$	Collision $\left(-\dfrac{dE}{dx}\right)_{col}$	Total $\left(-\dfrac{dE}{dx}\right)_{tot}$
30	2.10	2.74	2.23	2.86	1.33	5.34
40	2.13	3.01	2.28	3.16	1.36	6.78
50	2.14	3.28	2.32	3.44	1.38	8.30
60	2.16	3.54	2.36	3.72	1.40	9.83
80	2.19	4.08	2.40	4.27	1.42	12.9
100	2.20	4.61	2.43	4.81	1.44	16.0
200	2.26	7.27	2.52	7.47	1.49	31.6
300	2.30	9.95	2.56	10.1	1.51	47.2
400	2.32	12.6	2.59	12.8	1.53	62.9
500	2.34	15.3	2.61	15.4	1.54	78.6
600	2.36	18.0	2.63	18.1	1.55	94.3
800	2.38	23.3	2.66	23.3	1.57	126
1000	2.40	28.7	2.67	28.6	1.59	152

Table 3.4-6
ELECTRON RANGES IN
WATER, AIR, AND LEAD

Range (g/cm²)

Electron energy (MeV)	Water $\rho = 1$ g/cm³	Air $\rho = 0.00129$ g/cm³	Lead $\rho = 11.3$ g/cm³
0.010	0.00024	0.00029	0.00083
0.015	0.00050	0.00059	0.00150
0.020	0.00083	0.00098	0.00234
0.025	0.00124	0.00145	0.00332
0.030	0.00171	0.00201	0.00443
0.035	0.00225	0.00263	0.00568
0.040	0.00285	0.00333	0.00704
0.045	0.00351	0.00409	0.00852
0.050	0.00422	0.00492	0.0101
0.055	0.00499	0.00581	0.0118
0.060	0.00580	0.00676	0.0136
0.065	0.00667	0.00777	0.0155
0.070	0.00759	0.00883	0.0174
0.075	0.00855	0.00995	0.0195
0.080	0.00956	0.0111	0.0216
0.085	0.0106	0.0123	0.0239
0.090	0.0117	0.0136	0.0262
0.095	0.0128	0.0149	0.0285
0.10	0.0140	0.0163	0.0310
0.15	0.0276	0.0320	0.0586

Table 3.4-6 (continued)
ELECTRON RANGES IN
WATER, AIR, AND LEAD

Range (g/cm^2)

Electron energy (MeV)	Water $\rho = 1$ g/cm^3	Air $\rho = 0.00129$ g/cm^3	Lead $\rho = 11.3$ g/cm^3
0.20	0.0440	0.0509	0.0910
0.25	0.0625	0.0722	0.127
0.30	0.0826	0.0954	0.165
0.35	0.104	0.120	0.205
0.40	0.126	0.146	0.246
0.45	0.150	0.172	0.288
0.50	0.174	0.200	0.330
0.60	0.223	0.256	0.416
0.70	0.273	0.314	0.504
0.80	0.325	0.372	0.591
0.90	0.377	0.431	0.677
1.00	0.430	0.491	0.763
1.5	0.696	0.789	1.18
2.0	0.961	1.08	1.57
3.0	1.49	1.66	2.29
4.0	2.00	2.21	2.93
5.0	2.50	2.74	3.52
6.0	2.99	3.25	4.06
7.0	3.47	3.75	4.56
8.0	3.95	4.24	5.03
9.0	4.42	4.72	5.47
10	4.88	5.19	5.88
20	9.18	9.45	9.06
30	13.0	13.2	11.2
40	16.5	16.5	12.9
50	19.7	19.5	14.2
60	22.6	22.3	15.3
80	27.9	27.3	17.1
100	32.5	31.7	18.5
200	49.6	48.3	22.9
300	61.3	59.7	25.4
400	70.2	68.5	27.3
500	77.4	75.6	28.7
600	83.4	81.6	29.9
800	93.2	91.4	31.7
1000	101.	99.1	33.1

Armstrong and Chandler[27] describe a method of calculation and input required for a program to compute stopping powers and ranges for ions from low energies to several hundred gigaelectronvolts. They present stopping-power and range curves for protons, alpha particles, and O ions in water and for muons, pions, protons, deuterons, tritons, ^{3}He, alpha particles, Li, Be, B, C, N, and O in tissue.

Reference 23 contains a very extensive treatment of the basic theories and corrections for penetration of heavy charged particles and beta and gamma rays as well as many numerical data and graphs. There are nine pages of references, with only a few from later than 1951.

Reference 28 contains some of the tables for heavy charged particles that are included in Reference 3. More extensive treatment is given to some theoretical aspects. A table of mass stopping powers of C, Al, Cu, Ag, Pb, H_2O, and air for electrons from 0.01 to 100 MeV appears in this work, followed by a bibliography.

Bicshel[3] presents a concise and thorough collection of formulas, tables, and graphs, including subjects covered in Section 3.4.2.5. There are tables of β, β^2, and $\ln[2mv^2/(1 - \beta^2)] - \beta^2$ for protons, alpha particles, pions, muons, and electrons from $\beta = 0.033$ to 0.875 (proton energies from 0.5 to 1000 MeV). He includes tables of mass stopping powers for protons from 10 to 1000 MeV in Be, graphite, H_2O, Al, Cu, Ag, and Pb and proton ranges from 1 to 1000 MeV in these materials as well as proton stopping power from 0.01 to 12 MeV in 19 elements from Be through Au. Also included in this work are tables for straggling and for multiple coulomb scattering and some stopping-power and range information for electrons. An extensive bibliography is contained as well.

The review by Fano[2] is the most recent "complete" treatment of the theory of particle penetration and associated phenomena. In it, he summarizes (Section 8) tabulations available through early 1963.

Reference 29 is a collection of state-of-the-art reports with extensive numerical data. Included are stopping-power and range tables for protons from 1 to 5000 MeV for 36 values of I and mass stopping-power and range tables for protons, kaons, pions, and muons from 2 to 5000 MeV for 36 elements and compounds. Also covered are mass stopping-power and range tables for protons from 2 to 1000 MeV in nine substances, including concrete, muscle, and bone as well as range-energy curves for heavy ions. There are tables of mean energy loss by collisions with atomic electrons and by bremsstrahlung, mean range, and radiation yield for electrons (with a prescription for scaling to positrons) from 0.01 to 1 MeV in about 40 materials. This volume also contains a reprint of References 2 and 15.

Nelms[25,26] gives stopping-power and range tables and graphs for electrons and positrons in a number of materials. The supplement[26] includes the density effect.

Northcliffe[15] presents numerous curves for stopping powers for heavy ions at low energies and data on equilibrium charge distributions at low velocities. Also included is coverage of "nuclear" stopping power from elastic collisions with atomic nuclei. In a later work, Northcliffe[16] gives some graphs of range-energy relations for a number of ions with energies up to 10 MeV/AMU in a variety of materials.

Steward[30] presents a calculational method and several curves (e.g., Figures 11 and 19) for low-energy heavy ions (H through Rn) in water that are particularly useful in applications of track structure to effects modeling.

Finally, Whaling[14] describes a survey of experimental data on ranges and stopping powers of nuclear particles, primarily protons and alpha particles with energies less than 10 MeV. His work contains useful empirical formulas and a number of tables and graphs.

REFERENCES

1. **Bethe, H. A.,** Quantenmechanik der Ein- und Zwei-Elektronenprobleme, in *Handbuch der Physik,* Vol. 24, No. 1, Geiger, H. and Scheel, K., Eds., Springer-Verlag, Berlin, 1933, 491.
2. **Fano, U.,** Penetration of protons, alpha particles, and mesons, *Annu. Rev. Nucl. Sci.,* 13, 1–66, 1963.
3. **Bichsel, H.,** Passage of charged particles through matter, in *American Institute of Physics Handbook,* 3rd ed., Gray, D. E., Ed., McGraw-Hill, New York, 1972, 8-142–8-189.
4. Stopping powers for use with cavity chambers, *Handbook 79,* National Bureau of Standards, Washington, D.C., 1961.

5.   **Turner, J. E.,** Values of I and I_{adj} suggested by the subcommittee, in *Studies in Penetration of Charged Particles in Matter,* Fano, U., Ed., Publ. No. 1133, National Academy of Sciences, National Research Council, Washington, D.C., 1964, 99–101.

6.   **Dalton, P. and Turner, J. E.,** New evaluation of mean excitation energies for use in radiation dosimetry, *Health Phys.,* 15, 257–262, 1968.

7.   **Turner, J. E., Roecklein, P. D., and Vora, R. B.,** Mean excitation energies for chemical elements, *Health Phys.,* 18, 159–160, 1970.

8.   **Walske, M. C.,** The stopping power of K electrons, *Phys. Rev.,* 88, 1283–1289, 1952.

9.   **Walske, M. C.,** Stopping power of L electrons, *Phys. Rev.,* 101, 940–944, 1956.

10. **Barkas, W. H. and Berger, M. J.,** Tables of energy losses and ranges of heavy charged particles, in *Studies in Penetration of Charged Particles in Matter,* Fano, U., Ed., Publ. No. 1133, National Academy of Sciences, National Research Council, Washington, D.C., 1964, 103–172.

11. **Crispin, A. and Fowler, G. N.,** Density effect in the ionization energy loss of fast charged particles in matter, *Rev. Mod. Phys.,* 42, 290–316, 1970.

12. **Sternheimer, R. M.,** Density effect for the ionization loss in various materials, *Phys. Rev.,* 103, 511–515, 1956.

13. **Sternheimer, R. M.,** Density effect for the ionization loss of charged particles. II, *Phys. Rev.,* 164, 349–351, 1967.

14. **Whaling, W.,** The energy loss of charged particles in matter, in *Encyclopedia of Physics,* Vol. 34, No. 2, Fluegge, S., Ed., Springer-Verlag, Berlin, 1958, 193–217.

15. **Northcliffe, L. C.,** Passage of heavy ions through matter, *Annu. Rev. Nucl. Sci.,* 13, 67–102, 1963.

16. **Northcliffe, L. C.,** Passage of heavy ions through matter. II. Range-energy curves, in *Studies in Penetration of Charged Particles in Matter,* Fano, U., Ed., Publ. No. 1133, National Academy of Sciences, National Research Council, Washington, D.C., 1964, 173–186.

17. **Berger, M. J. and Seltzer, S. M.,** Multiple-scattering corrections for proton range measurements, in *Studies in Penetration of Charged Particles in Matter,* Fano, U., Ed., Publ. No. 1133, National Academy of Sciences, National Research Council, Washington, D.C., 1964, 69–98.

18. **Seltzer, S. M. and Berger, M. J.,** Energy-loss straggling of protons and mesons: tabulation of the Vavilov Distribution, in *Studies in Penetration of Charged Particles in Matter,* Fano, U., Ed., Publ. No. 1133, National Academy of Sciences, National Research Council, Washington, D.C., 1964, 187–203.

19. **Rossi, B.,** *High-Energy Particles,* Prentice-Hall, Englewood Cliffs, N.J., 1952.

20. **Hamm, R. N., Wright, H. A., and Turner, J. E.,** Monte Carlo treatment of multiple coulomb scattering in pion-beam dose calculations, *J. Appl. Phys.,* 46, 4445–4452, 1975.

21. **Fano, U.,** A list of currently unsolved problems, in *Studies in Penetration of Charged Particles in Matter,* Fano, U., Ed., Publ. No. 1133, National Academy of Sciences, National Research Council, Washington, D.C., 1964, 281–284.

22. **Inokuti, M., Itikawa, Y., and Turner, J. E.,** Addenda: Inelastic collissions of Fast charged particles with atoms and molecules – the be the theory revisted [*Rev. Mod. Phys.,* 43, 297, 1972]. *Rev. Mod. Phys.,* 50, 23–35, 1978.

23. **Bethe, H. A. and Ashkin, J.,** Passage of radiations through matter, in *Experimental Nuclear Physics,* Vol. 1, Segrè, E., Ed., John Wiley & Sons, New York, 1953, 166–357.

24. **Berger, M. J. and Seltzer, S. M.,** Tables of energy losses and ranges of electrons and positrons, in *Studies in Penetration of Charged Particles in Matter,* Fano, U., Ed., Publ. No. 1133, National Academy of Sciences, National Research Council, Washington, D.C., 1964, 205–268.

25. **Nelms, A. T.,** *Energy Loss and Range of Electrons and Positrons,* NBS Circular 577, National Bureau of Standards, Washington, D.C., 1956, 1–30.

26. **Nelms, A. T.,** *Energy Loss and Range of Electrons and Positrons,* Supplement to NBS Circular 577, National Bureau of Standards, Washington, D.C., 1958, 1–31.

27. **Armstrong, T. W. and Chandler, K. C.,** SPAR, A Fortran Program for Computing Stopping Powers and Ranges for Muons, Charged Pions, Protons, and Heavy Ions, ORNL-4869, Oak Ridge National Laboratory, Oak Ridge, Tenn., 1973, 1–31.

28. **Biehsel, H.,** Charged particle interactions, in *Radiation Dosimetry,* Vol. 1, Attix, F. H. and Roesch, W. C., Eds., Academic Press, New York, 1968, 157–228.

29. **Fano, U., Ed.,** *Studies in Penetration of Charged Particles in Matter,* Publ. No. 1133, National Academy of Sciences, National Research Council, Washington, D.C., 1964, 1–388.

30. **Steward, P. G.,** Stopping Power and Range for any Nucleus in the Specific Energy Interval 0.01 to 500 MeV/AMU in any Nongaseous Material, Report UCRL-18127, University of California, Berkeley, 1968.

3.5. X- AND GAMMA-RAY ABSORPTION
AND SCATTERING COEFFICIENTS

Allen Brodsky

The X- and gamma-ray coefficients in this section are chosen from those most useful for attenuation and dosimetry calculations in the field of radiation protection. Only basic data are given in this section. Methods of using this data in dose calculations are presented in other sections of the handbook.

Table 3.5-1* consists of a series of tables presenting mass energy-transfer coefficients and mass energy-absorption coefficients taken from Hubbell.[1] These data are recommended for the more accurate attenuation and energy absorption calculations for materials given in the table. The mass energy-transfer coefficient is used to calculate energy transferred to secondary electrons per unit mass thickness traversed by photons. The mass energy-absorption coefficient is used to calculate energy absorption per unit mass thickness in grams per square centimeter, absorbed doses, or in dosimetry calculations, as discussed in Section 7.1. An excellent review of the derivation and use of attenuation and absorption coefficients is given by Evans.[2]

Table 3.5-2 presents mass energy-absorption coefficients for air,[3] including attenuation, energy transfer, energy absorption, and conversion to energy and photon fluence. Also, the specific gamma-ray constant is given in this table in Roentgens per hour per millicurie at 1 cm for the case where exactly one gamma photon of energy is emitted per disintegration. Thus, the appropriate constant would need to be corrected by the gamma abundance for each gamma energy, in order to calculate a specific gamma-ray dose rate constant for a given nuclide. Section 3.6 presents or gives reference to data on specific gamma constants already calculated from decay scheme data for many nuclides of interest.

Table 3.5-3, taken from the 56th edition of the *Handbook of Chemistry and Physics,*[4] presents very low-energy X-ray total attenuation coefficients (which in most cases also approximate the mass energy coefficients) for various materials of interest. For materials not given in this table, coefficients from higher energy data may be plotted on a log-log plot and extrapolated to lower energies in regions not interrupted by any of the X-ray absorption peaks (see Section 3.7). Except for internal emitters, however, the X- or gamma-ray energies represented in this table are below the photon energies of usual interest for radiation protection. Doses from externally incident photons of energies less than 12 keV may be evaluated radiologically as similar to superficial exposure to beta-rays.[5] Values from Table 3.5-3 in the 2 to 4 Å region may, however, be used for estimating the position of the energy absorption coefficient curves in the 10 to 20 keV region. Additional X-ray cross sections are given in Tables 3.5-4 and 3.5-5 in the 4.5 to 25 keV region.

A visual picture of the variations in absorption cross sections with energy and with atomic number are often helpful. Curves showing both total attenuation and mass energy absorption coefficients vs. energy are shown in Figures 3.5-1 to 3.5-3 from Evans.[2] Additional curves from the *Radiological Health Handbook*[6] are given in Figures 3.5-4 to 3.5-7. Figures 3.5-8 through 3.5-11 from Snyder and Powell[7] present some early theoretical and experimental graphs of absorption coefficients. The curves for higher Z in Snyder and Powell have in some cases been shifted to agree with early experimental data which have since been modified. However, a comparison of these curves with the data in Table 3.5-1 will show that the curves are accurate for the most part and provide a good visual comparison of relative absorption coefficients as a function of energy for gases and materials of interest in dosimetry and personnel monitoring.

Another very useful set of curves in scattering calculations for radiation protection are

*All tables and figures appear at the end of the text.

those shown in Figure 3.5-12 from Hine and Brownell.[5] These curves have been calculated from the Klein-Nishina equations for differential cross sections of Compton scattering as a function of scattering angle. The dashed curves were obtained by multiplying the values in the solid curves, for each respective angle, by the fractional amount of energy retained by Compton photons scattered at that angle. Thus, the dashed curves give the fractional amount of incident gamma energy (per square centimeter) scattered per steradian at angle ϕ, per electron (per square centimeter). The reader is cautioned to notice that numbers on the vertical axis are to be multiplied by 10^{-26}. Additonal data basic to radiation protection and dosimetry calculations is presented in Tables 3.5-6 through 3.5-8.

Table 3.5-1

VALUES OF THE (a) MASS ENERGY-TRANSFER COEFFICIENT μ_K/ρ (cm^2/g), AND (b) MASS ENERGY-ABSORPTION COEFFICIENT, μ_{en}/ρ (cm^2/g)

Hydrogen through Sodium

$h\nu$(MeV)	$_1$H (a)	(b)	$_4$Be (a)	(b)	$_5$B (a)	(b)	$_6$C (a)	(b)	$_7$N (a)	(b)	$_8$O (a)	(b)	$_{11}$Na (a)	(b)
0.01	0.00986	0.00986	0.368	0.368	0.911	0.911	1.97	1.97	3.38	3.38	5.39	5.39	14.9	14.9
0.015	0.0110	0.0110	0.104	0.104	0.248	0.248	0.536	0.536	0.908	0.908	1.44	1.44	4.20	4.20
0.02	0.0135	0.0135	0.0469	0.0469	0.0998	0.0998	0.208	0.208	0.362	0.362	0.575	0.575	1.70	1.70
0.03	0.0185	0.0185	0.0195	0.0195	0.0338	0.0338	0.0594	0.0594	0.105	0.105	0.165	0.165	0.475	0.475
0.04	0.0231	0.0231	0.0146	0.0146	0.0210	0.0210	0.0306	0.0306	0.0493	0.0493	0.0733	0.0733	0.199	0.199
0.05	0.0271	0.0271	0.0142	0.0142	0.0175	0.0175	0.0233	0.0233	0.0319	0.0319	0.0437	0.0437	0.106	0.106
0.06	0.0306	0.0306	0.0147	0.0147	0.0170	0.0170	0.0211	0.0211	0.0256	0.0256	0.0322	0.0322	0.0668	0.0668
0.08	0.0362	0.0362	0.0166	0.0165	0.0179	0.0179	0.0205	0.0205	0.0223	0.0223	0.0249	0.0249	0.0382	0.0382
0.10	0.0406	0.0406	0.0184	0.0184	0.0194	0.0194	0.0215	0.0215	0.0224	0.0224	0.0237	0.0237	0.0297	0.0297
0.15	0.0481	0.0481	0.0216	0.0216	0.0226	0.0226	0.0245	0.0245	0.0247	0.0247	0.0251	0.0251	0.0260	0.0260
0.2	0.0525	0.0525	0.0235	0.0235	0.0245	0.0245	0.0265	0.0265	0.0267	0.0267	0.0268	0.0268	0.0264	0.0264
0.3	0.0569	0.0569	0.0255	0.0255	0.0266	0.0266	0.0287	0.0287	0.0287	0.0287	0.0288	0.0288	0.0277	0.0277
0.4	0.0586	0.0586	0.0262	0.0262	0.0273	0.0273	0.0295	0.0295	0.0295	0.0295	0.0295	0.0295	0.0284	0.0284
0.5	0.0593	0.0593	0.0265	0.0265	0.0276	0.0276	0.0297	0.0297	0.0297	0.0296	0.0297	0.0297	0.0285	0.0285
0.6	0.0587	0.0587	0.0263	0.0263	0.0274	0.0273	0.0296	0.0295	0.0296	0.0295	0.0296	0.0296	0.0284	0.0284
0.8	0.0574	0.0574	0.0257	0.0256	0.0268	0.0267	0.0289	0.0288	0.0289	0.0289	0.0289	0.0288	0.0277	0.0277
1.0	0.0555	0.0555	0.0248	0.0248	0.0259	0.0258	0.0279	0.0279	0.0280	0.0279	0.0280	0.0278	0.0268	0.0266
1.5	0.0507	0.0507	0.0227	0.0227	0.0237	0.0236	0.0256	0.0255	0.0256	0.0255	0.0256	0.0254	0.0245	0.0243
2	0.0465	0.0464	0.0208	0.0208	0.0218	0.0217	0.0235	0.0234	0.0236	0.0234	0.0236	0.0234	0.0227	0.0225
3	0.0399	0.0398	0.0181	0.0180	0.0190	0.0188	0.0206	0.0204	0.0207	0.0205	0.0208	0.0206	0.0202	0.0199
4	0.0353	0.0352	0.0163	0.0161	0.0172	0.0170	0.0187	0.0185	0.0189	0.0186	0.0191	0.0188	0.0188	0.0184
5	0.0319	0.0317	0.0149	0.0148	0.0158	0.0156	0.0174	0.0171	0.0177	0.0173	0.0179	0.0175	0.0179	0.0174
6	0.0292	0.0290	0.0140	0.0138	0.0149	0.0146	0.0164	0.0161	0.0167	0.0163	0.0171	0.0166	0.0173	0.0161
8	0.0253	0.0252	0.0126	0.0123	0.0135	0.0132	0.0151	0.0147	0.0156	0.0151	0.0160	0.0155	0.0167	0.0159
10	0.0227	0.0225	0.0117	0.0114	0.0127	0.0123	0.0143	0.0138	0.0149	0.0143	0.0154	0.0148	0.0161	0.0155

Table 3.5-1 (continued)
VALUES OF THE (a) MASS ENERGY-TRANSFER COEFFICIENT, μ_K/ρ (cm^2/g), AND (b) MASS ENERGY-ABSORPTION COEFFICIENT, μ_{en}/ρ (cm^2/g)

Magnesium through Potassium

$h\nu$(MeV)	$_{12}$Mg (a)	(b)	$_{13}$Al (a)	(b)	$_{14}$Si (a)	(b)	$_{15}$P (a)	(b)	$_{16}$S (a)	(b)	$_{18}$Ar (a)	(b)	$_{19}$K (a)	(b)
0.01	20.1	20.1	25.5	25.5	33.3	33.3	39.8	39.8	49.7	49.7	62.3	62.3	77.6	77.6
0.015	5.80	5.80	7.47	7.47	9.75	9.75	11.8	11.8	14.9	14.9	19.1	19.1	23.9	23.9
0.02	2.38	2.38	3.06	3.06	4.01	4.01	4.91	4.91	6.21	6.21	8.02	8.02	10.2	10.2
0.03	0.671	0.671	0.868	0.868	1.14	1.14	1.39	1.39	1.77	1.77	2.31	2.31	2.94	2.94
0.04	0.276	0.276	0.357	0.357	0.472	0.472	0.572	0.572	0.727	0.727	0.962	0.962	1.23	1.23
0.05	0.144	0.144	0.184	0.184	0.241	0.241	0.293	0.293	0.372	0.372	0.488	0.488	0.623	0.623
0.06	0.0888	0.0888	0.111	0.111	0.144	0.144	0.173	0.173	0.218	0.218	0.284	0.284	0.366	0.366
0.08	0.0475	0.0475	0.0562	0.0562	0.0700	0.0700	0.0820	0.0820	0.101	0.101	0.128	0.128	0.162	0.162
0.10	0.0346	0.0346	0.0386	0.0386	0.0459	0.0459	0.0511	0.0511	0.0609	0.0609	0.0735	0.0735	0.0913	0.0913
0.15	0.0279	0.0279	0.0285	0.0285	0.0312	0.0312	0.0322	0.0322	0.0357	0.0357	0.0377	0.0377	0.0442	0.0442
0.2	0.0277	0.0277	0.0276	0.0276	0.0292	0.0292	0.0293	0.0293	0.0311	0.0311	0.0304	0.0304	0.0343	0.0343
0.3	0.0288	0.0288	0.0282	0.0282	0.0294	0.0294	0.0288	0.0288	0.0299	0.0299	0.0278	0.0278	0.0304	0.0304
0.4	0.0294	0.0294	0.0287	0.0287	0.0298	0.0298	0.0291	0.0291	0.0301	0.0301	0.0275	0.0275	0.0298	0.0298
0.5	0.0294	0.0294	0.0287	0.0286	0.0298	0.0298	0.0291	0.0291	0.0300	0.0300	0.0272	0.0272	0.0294	0.0293
0.6	0.0293	0.0293	0.0286	0.0286	0.0296	0.0295	0.0288	0.0288	0.0297	0.0297	0.0296	0.0269	0.0291	0.0290
0.8	0.0286	0.0285	0.0279	0.0277	0.0289	0.0288	0.0280	0.0278	0.0290	0.0288	0.0262	0.0261	0.0283	0.0282
1.0	0.0276	0.0275	0.0270	0.0269	0.0279	0.0277	0.0272	0.0270	0.0280	0.0278	0.0253	0.0251	0.0273	0.0270
1.5	0.0253	0.0251	0.0247	0.0245	0.0255	0.0253	0.0248	0.0246	0.0256	0.0253	0.0232	0.0229	0.0250	0.0247
2	0.0234	0.0232	0.0229	0.0226	0.0237	0.0234	0.0231	0.0228	0.0238	0.0235	0.0215	0.0212	0.0233	0.0229
3	0.0210	0.0206	0.0206	0.0202	0.0214	0.0210	0.0209	0.0204	0.0216	0.0211	0.0198	0.0192	0.0214	0.0208
4	0.0196	0.0191	0.0193	0.0188	0.0202	0.0196	0.0198	0.0192	0.0205	0.0199	0.0189	0.0182	0.0206	0.0198
5	0.0187	0.0181	0.0185	0.0179	0.0194	0.0187	0.0191	0.0184	0.0200	0.0192	0.0185	0.0177	0.0202	0.0193
6	0.0182	0.0175	0.0181	0.0172	0.0191	0.0182	0.0188	0.0179	0.0197	0.0188	0.0184	0.0174	0.0202	0.0190
8	0.0177	0.0168	0.0177	0.0168	0.0187	0.0177	0.0187	0.0175	0.0197	0.0184	0.0186	0.0173	0.0205	0.0190
10	0.0175	0.0164	0.0176	0.0165	0.0188	0.0175	0.0188	0.0174	0.0200	0.0184	0.0190	0.0174	0.0210	0.0191

Table 3.5-1 (continued)

VALUES OF THE (a) MASS ENERGY-TRANSFER COEFFICIENT, μ_K/ρ (cm^2/g), AND (b) MASS ENERGY-ABSORPTION COEFFICIENT, μ_{en}/ρ (cm^2/g)

Calcium through Copper

$h\nu$(MeV)	$_{20}$Ca (a)	(b)	$_{26}$Fe (a)	(b)	$_{27}$Co (a)	(b)	$_{28}$Ni (a)	(b)	$_{29}$Cu (a)	(b)
0.01	91.6	91.6	142.	142.	148.6	148.6	161.	161.	160.	160.
0.015	28.6	28.6	49.3	49.3	52.5	52.5	58.2	58.2	59.4	59.4
0.02	12.2	12.2	22.8	22.8	24.4	24.4	27.4	27.4	28.2	28.2
0.03	3.60	3.60	7.28	7.28	7.97	7.97	9.06	9.06	9.50	9.50
0.04	1.50	1.50	3.17	3.17	3.49	3.49	4.03	4.03	4.24	4.24
0.05	0.764	0.764	1.64	1.64	1.64	1.64	2.15	2.15	2.22	2.22
0.06	0.444	0.444	0.961	0.961	1.08	1.08	1.25	1.25	1.32	1.32
0.08	0.196	0.196	0.414	0.414	0.454	0.454	0.528	0.528	0.573	0.573
0.10	0.109	0.109	0.219	0.219	0.243	0.243	0.284	0.284	0.302	0.302
0.15	0.0497	0.0497	0.0814	0.0814	0.0887	0.0887	0.101	0.101	0.106	0.106
0.2	0.0371	0.0371	0.0495	0.0495	0.0523	0.0523	0.0582	0.0582	0.0597	0.0597
0.3	0.0318	0.0318	0.0335	0.0335	0.0347	0.0347	0.0374	0.0374	0.0370	0.0370
0.4	0.0309	0.0309	0.0308	0.0308	0.0308	0.0308	0.0326	0.0326	0.0318	0.0318
0.5	0.0304	0.0304	0.0295	0.0295	0.0294	0.0294	0.0309	0.0309	0.0298	0.0298
0.6	0.0300	0.0299	0.0287	0.0286	0.0284	0.0283	0.0298	0.0297	0.0287	0.0286
0.8	0.0291	0.0289	0.0275	0.0273	0.0272	0.0270	0.0284	0.0282	0.0272	0.0271
1.0	0.0280	0.0278	0.0264	0.0262	0.0260	0.0258	0.0272	0.0269	0.0261	0.0258
1.5	0.0257	0.0254	0.0241	0.0237	0.0237	0.0234	0.0247	0.0243	0.0237	0.0233
2	0.0240	0.0236	0.0225	0.0220	0.0222	0.0217	0.0232	0.0227	0.0222	0.0217
3	0.0220	0.0214	0.0212	0.0204	0.0202	0.0202	0.0219	0.0211	0.0211	0.0202
4	0.0213	0.0205	0.0209	0.0199	0.0208	0.0197	0.0218	0.0207	0.0211	0.0200
5	0.0211	0.0200	0.0211	0.0198	0.0210	0.0196	0.0222	0.0207	0.0214	0.0200
6	0.0211	0.0198	0.0215	0.0199	0.0215	0.0198	0.0227	0.0209	0.0220	0.0202
8	0.0215	0.0198	0.0226	0.0204	0.0226	0.0203	0.0239	0.0215	0.0234	0.0209
10	0.0222	0.0201	0.0238	0.0209	0.0239	0.0210	0.0254	0.0222	0.0248	0.0215

Table 3.5-1 (continued)
VALUES OF THE (a) MASS ENERGY-TRANSFER
COEFFICIENT, μ_K/ρ (cm^2/g), AND (b) MASS
ENERGY-ABSORPTION COEFFICIENT, μ_{en}/ρ
(cm^2/g)

Tin

$h\nu$(MeV)	$_{50}$Sn (a)	(b)
0.0010	11110	11110
0.0015	3950	3950
0.0020	1954	1954
0.0030	705	705
0.0039288	360	360
L$_3$ edge		
0.0039288	1067	1067
0.0040	1019	1019
0.0041573	930	930
L$_2$ edge		
0.0041573	1187	1187
0.0044648	971	971
L$_1$ edge		
0.0044648	1207	1207
0.005	880	880
0.006	540	539
0.008	250	249
0.010	136.5	136.4
0.015	43.7	43.6
0.020	19.83	19.81
0.0291947	6.83	6.82
K edge		
0.0291947	16.70	16.69
0.030	16.18	16.17
0.04	9.97	9.96
0.05	6.25	6.24
0.06	4.20	4.19
0.08	2.19	2.18
0.10	1.257	1.250
0.15	0.446	0.442
0.20	0.211	0.209
0.30	0.0853	0.0843
0.4	0.0536	0.0530
0.5	0.0423	0.0416
0.6	0.0358	0.0353
0.8	0.0301	0.0294
1.0	0.0270	0.0264
1.5	0.0233	0.0226
2.0	0.0220	0.0210
3.0	0.0219	0.0205
4	0.0232	0.0212
5	0.0247	0.0221
6	0.0262	0.0230
8	0.0292	0.0245
10	0.0319	0.0258

Table 3.5-1 (continued)
VALUES OF THE (a) MASS ENERGY-TRANSFER COEFFICIENT, μ_K/ρ (cm^2/g), AND (b) MASS ENERGY-ABSORPTION COEFFICIENT, μ_{en}/ρ (cm^2/g)

Lead

hv(MeV)	$_{82}$Pb (a)	(b)
M$_1$ edge		
0.003854	1454	1453
0.004	1298	1297
0.005	747	747
0.006	479	479
0.008	230	230
0.010	131.0	130.7
0.0130406	66.2	66.0
L$_3$ edge		
0.0130406	128.8	128.8
0.015	91.7	91.7
0.0152053	89.6	89.6
L$_2$ edge		
0.0152053	113.0	113.0
0.015855	101.7	101.6
L$_1$ edge		
0.015855	123.0	123.0
0.02	69.2	69.1
0.03	24.6	24.6
0.04	11.83	11.78
0.05	6.57	6.54
0.06	4.11	4.08
0.08	1.924	1.908
0.088005	1.494	1.481
K edge		
0.088005	2.47	2.47
0.10	2.28	2.28
0.15	1.164	1.154
0.2	0.637	0.629
0.3	0.265	0.259
0.4	0.1474	0.1432
0.5	0.0984	0.0951
0.6	0.0737	0.0710
0.8	0.0503	0.0481
1.0	0.0396	0.0377
1.5	0.0288	0.0271
2	0.0259	0.0240
3	0.0260	0.0234
4	0.0281	0.0245
5	0.0306	0.0259
6	0.0331	0.0272
8	0.0378	0.0294
10	0.0419	0.0310

Table 3.5-1 (continued)
VALUES OF THE (a) MASS ENERGY-TRANSFER COEFFICIENT, μ_K/ρ (cm^2/g), AND (b) MASS ENERGY-ABSORPTION COEFFICIENT, μ_{en}/ρ (cm^2/g)

Uranium

$h\nu$(MeV)	$_{92}$U (a)	$_{92}$U (b)	$h\nu$(MeV)	$_{92}$U (a)	$_{92}$U (b)
N$_1$ edge					
0.001439	4460	4460	L$_2$ edge		
0.0015	4050	4050	0.020945	66.9	66.9
0.002	2110	2110	0.021771	61.2	61.1
0.003	818	818	L$_1$ edge		
0.003545	554	554	0.021771	69.6	69.6
M$_5$ edge			0.03	33.0	33.0
0.003545	1191	1191	0.04	16.33	16.27
0.003720	1055	1055	0.05	9.30	9.25
M$_4$ edge			0.06	5.82	5.78
0.03720	1520	1520	0.08	2.76	2.73
0.004	1258	1258	0.10	1.535	1.51
0.004299	1046	1046	0.011562	1.049	1.03
M$_3$ edge			K edge		
0.004299	1242	1242	0.11562	1.660	1.65
0.005	851	850	0.15	1.120	1.19
0.005179	781	780	0.20	0.712	0.70
M$_2$ edge			0.30	0.322	0.31
0.005179	839	839	0.4	0.1835	0.17
0.005546	707	706	0.5	0.1222	0.11
M$_1$ edge			0.6	0.0902	0.08
0.005546	739	739	0.8	0.0599	0.05
0.006	612	611	1.0	0.0458	0.04
0.008	300	299	1.5	0.0317	0.02
0.010	170.5	170.2	2	0.0278	0.02
0.015	59.6	59.4	3	0.0276	0.02
0.017165	42.2	42.1	4	0.0298	0.02
L$_3$ edge			5	0.0324	0.02
0.017165	80.1	80.1	6	0.0348	0.02
0.020	55.5	55.5	8	0.0392	0.02
0.020945	49.4	49.4	10	0.0432	0.03

Table 3.5-1 (continued)

VALUES OF THE (a) MASS ENERGY-TRANSFER COEFFICIENT, μ_K/ρ (cm²/g), AND (b) MASS ENERGY-ABSORPTION COEFFICIENT, μ_{en}/ρ (cm²/g)

Some Mixtures and Compounds

hν(MeV)	Air		Water		0.8N(0.4M) H₂SO₄ solution		Compact bone (femur)		Muscle (striated)		Polystyrene, (C₈H₈)ₙ	
	(a)	(b)	(a)	(b)	(a)	(b)	(a)	(b)	(a)	(b)	(a)	(b)
0.01	4.61	4.61	4.79	4.79	5.36	5.36	19.2	19.2	4.87	4.87	1.82	1.82
0.015	1.27	1.27	1.28	1.28	1.45	1.45	5.84	5.84	1.32	1.32	0.495	0.495
0.02	0.511	0.511	0.512	0.512	0.585	0.585	2.46	2.46	0.533	0.533	0.193	0.193
0.03	0.148	0.148	0.149	0.149	0.169	0.169	0.720	0.720	0.154	0.154	0.0562	0.0562
0.04	0.0668	0.0668	0.0677	0.0677	0.0761	0.0761	0.304	0.304	0.0701	0.0701	0.0300	0.0300
0.05	0.0406	0.0406	0.0418	0.0418	0.0460	0.0460	0.161	0.161	0.0431	0.0431	0.0236	0.0236
0.06	0.0305	0.0305	0.0320	0.0320	0.0344	0.0344	0.0998	0.0998	0.0328	0.0328	0.0218	0.0218
0.08	0.0243	0.0243	0.0262	0.0262	0.0271	0.0271	0.0537	0.0537	0.0264	0.0264	0.0217	0.0217
0.10	0.0234	0.0234	0.0256	0.0256	0.0260	0.0260	0.0387	0.0387	0.0256	0.0256	0.0231	0.0231
0.15	0.0250	0.0250	0.0277	0.0277	0.0277	0.0277	0.0305	0.0305	0.0275	0.0275	0.0263	0.0263
0.2	0.0268	0.0268	0.0297	0.0297	0.0296	0.0296	0.0301	0.0301	0.0294	0.0294	0.0286	0.0286
0.3	0.0287	0.0287	0.0319	0.0319	0.0319	0.0319	0.0310	0.0310	0.0317	0.0317	0.0309	0.0309
0.4	0.0295	0.0295	0.0328	0.0328	0.0327	0.0327	0.0315	0.0315	0.0325	0.0325	0.0318	0.0318
0.5	0.0297	0.0296	0.0330	0.0330	0.0330	0.0330	0.0317	0.0317	0.0328	0.0328	0.0321	0.0321
0.6	0.0296	0.0295	0.0329	0.0329	0.0328	0.0328	0.0315	0.0314	0.0326	0.0325	0.0319	0.0318
0.8	0.0289	0.0289	0.0321	0.0321	0.0320	0.0320	0.0307	0.0306	0.0318	0.0318	0.0311	0.0310
1.0	0.0280	0.0278	0.0311	0.0309	0.0310	0.0308	0.0297	0.0295	0.0308	0.0306	0.0300	0.0300
1.5	0.0256	0.0254	0.0284	0.0282	0.0283	0.0281	0.0272	0.0270	0.0282	0.0280	0.0275	0.0275
2	0.0236	0.0234	0.0262	0.0260	0.0261	0.0259	0.0251	0.0249	0.0259	0.0257	0.0253	0.0252
3	0.0207	0.0205	0.0229	0.0227	0.0229	0.0227	0.0221	0.0219	0.0227	0.0225	0.0221	0.0219
4	0.0189	0.0186	0.0209	0.0206	0.0209	0.0206	0.0204	0.0200	0.0207	0.0204	0.0200	0.0198
5	0.0178	0.0174	0.0195	0.0191	0.0194	0.0191	0.0192	0.0187	0.0193	0.0189	0.0185	0.0182
6	0.0168	0.0164	0.0185	0.0180	0.0184	0.0180	0.0184	0.0178	0.0183	0.0178	0.0174	0.0171
8	0.0157	0.0152	0.0170	0.0166	0.0171	0.0166	0.0173	0.0167	0.0169	0.0164	0.0159	0.0155
10	0.0151	0.0145	0.0162	0.0157	0.0162	0.0157	0.0168	0.0159	0.0160	0.0155	0.0150	0.0145

Table 3.5-1 (continued)

VALUES OF THE (a) MASS ENERGY-TRANSFER COEFFICIENT, μ_K/ρ (cm^2/g), AND (b) MASS ENERGY-ABSORPTION COEFFICIENT, μ_{en}/ρ (cm^2/g)

Some Mixtures and Compounds (continued)

$h\nu$(MeV)	Methyl methacrylate (Perspex®, plexiglass, Lucite®), $C_5H_8O_2$ (a)	(b)	Polyethylene, $(CH_2)_n$ (a)	(b)	Bakelite® (typical composition), $C_{45}H_{38}O_7$ (a)	(b)	Pyrex® glass (Corning #7740) (a)	(b)	Concrete (typical composition) (a)	(b)
0.01	2.91	2.91	1.69	1.69	2.43	2.43	16.5	16.5	25.5	25.5
0.015	0.783	0.783	0.461	0.461	0.658	0.0658	4.75	4.75	7.66	7.66
0.02	0.310	0.310	0.180	0.180	0.258	0.258	1.94	1.94	3.22	3.22
0.03	0.0899	0.0899	0.0535	0.0535	0.0748	0.0748	0.554	0.554	0.936	0.936
0.04	0.0437	0.0437	0.0295	0.0295	0.0374	0.0374	0.232	0.232	0.393	0.393
0.05	0.0301	0.0301	0.0238	0.0238	0.0269	0.0269	0.122	0.122	0.204	0.204
0.06	0.0254	0.0254	0.0225	0.0225	0.0235	0.0235	0.0768	0.0768	0.124	0.124
0.08	0.0232	0.0232	0.0228	0.0228	0.0221	0.0221	0.0428	0.0428	0.0625	0.0625
0.10	0.0238	0.0238	0.0243	0.0243	0.0230	0.0230	0.0325	0.0325	0.0424	0.0424
0.15	0.0266	0.0266	0.0279	0.0279	0.0260	0.0260	0.0274	0.0274	0.0290	0.0290
0.2	0.0287	0.0287	0.0303	0.0303	0.0281	0.0281	0.0276	0.0276	0.0290	0.0290
0.3	0.0310	0.0310	0.0328	0.0328	0.0303	0.0303	0.0289	0.0289	0.0295	0.0295
0.4	0.0318	0.0318	0.0337	0.0337	0.0312	0.0312	0.0295	0.0295	0.0298	0.0298
0.5	0.0322	0.0322	0.0340	0.0340	0.0315	0.0315	0.0297	0.0297	0.0300	0.0300
0.6	0.0319	0.0319	0.0338	0.0337	0.0313	0.0312	0.0295	0.0294	0.0297	0.0297
0.8	0.0312	0.0311	0.0330	0.0329	0.0305	0.0305	0.0288	0.0287	0.0290	0.0289
1.0	0.0302	0.0301	0.0319	0.0319	0.0295	0.0295	0.0279	0.0277	0.0281	0.0279
1.5	0.0276	0.0275	0.0292	0.0291	0.0270	0.0269	0.0254	0.0252	0.0256	0.0254
2	0.0254	0.0253	0.0268	0.0267	0.0248	0.0247	0.0235	0.0233	0.0237	0.0235
3	0.0222	0.0220	0.0234	0.0232	0.0217	0.0215	0.0209	0.0207	0.0212	0.0209
4	0.0202	0.0199	0.0211	0.0209	0.0197	0.0195	0.0194	0.0190	0.0198	0.0193
5	0.0187	0.0184	0.0195	0.0192	0.0183	0.0180	0.0184	0.0179	0.0188	0.0182
6	0.0177	0.0173	0.0182	0.0180	0.0173	0.0169	0.0178	0.0171	0.0183	0.0176
8	0.0162	0.0158	0.0166	0.0162	0.0158	0.0154	0.0170	0.0163	0.0176	0.0168
10	0.0153	0.0148	0.0155	0.0151	0.0150	0.0145	0.0166	0.0157	0.0174	0.0163

Note: See original source for data references.

From Hubbell, J. H., *Photon Cross Sections, Attenuation Coefficients, and Energy Absorption Coefficients from 10 KeV to 100 GeV,* NSRDS-NBS 29, U.S. Government Printing Office, Washington, D.C., August 1969.

MASS ABSORPTION COEFFICIENTS FOR AIR

Energy Fluence per Roentgen, Photon Fluence per Roentgen, and Exposure per Millicurie as a Function of Photon Energy

| Photon energy MeV | Mass coefficients for air, cm^2 g | | | | Energy fluence per roentgen, erg/cm^2-R | Photon fluence per roentgen, N/cm^2-R | Specific gamma ray constant (roentgen per hour per millicurie[a] at 1 cm) |
| | Attenuation | | Energy transfer $\dfrac{\mu_k}{\rho}$ | Energy absorption $\dfrac{\mu_{en}}{\rho}$ | | | |
	$\dfrac{\mu'}{\rho}$	$\dfrac{\mu}{\rho}$					
.010	4.82	5.04	4.61	4.61	18.8	11.8×10^8	9.00
.015	1.45	1.56	1.27	1.27	68.4	28.5	3.72
.02	.691	.758	.511	.511	170	53.1	2.00
.03	.318	.350	.148	.148	587	122	.866
.04	.229	.248	.0668	.0668	1301	203	.521
.05	.196	.206	.0406	.0406	2140	267	.396
.06	.179	.187	.0305	.0305	2849	297	.357
.08	.162	.167	.0243	.0243	3576	279	.379
.10	.151	.155	.0234	.0234	3714	232	.457
.15	.134	.136	.0250	.0250	3476	145	.732
.2	.123	.124	.0268	.0268	3243	101	1.05
.3	.106	.107	.0287	.0287	3028	63.1	1.68
.4	.0954	.0954	.0295	.0295	2946	46.0	2.30
.5	.0868		.0298	.0296	2936	36.7	2.89
.6	.0804		.0296	.0295	2946	30.7	3.45
^{137}Cs .661[b]	.0772		.0294	.0294	2956	27.9	4.51
.80	.0706		.0289	.0289	3007	23.5	5.43
1.0	.0635		.0280	.0278	3126	19.5	
^{60}Co 1.25[b]	.0572		.0268	.0266	3267	16.3	7.44
1.5	.0517		.0256	.0254	3421	14.3	9.13
2	.0444		.0236	.0234	3714	11.6	12.0
3	.0358		.0207	.0205	4239	8.83	14.5
4	.0308		.0189	.0186	4672	7.30	17.0
5	.0276		.0178	.0174	4994	6.24	19.2
6	.0252		.0168	.0164	5299	5.52	23.7
8	.0223		.0157	.0152	5717	4.47	28.3
10	.0204		.0151	.0145	5993	3.75	

Table 3.5-2 (continued)
MASS ABSORPTION COEFFICIENTS FOR AIR

Energy Fluence per Roentgen, Photon Fluence per Roentgen, and Exposure per Millicurie as a Function of Photon Energy (continued)

Equations:

$$\text{Energy fluence per roentgen} = \frac{86.9}{(\mu_{en}/\rho)_{air}} \cdot \frac{ergs}{cm^2 R}$$

$$\text{Photon fluence per roentgen} = \frac{86.9}{(\mu_{en}/\rho)_{air}} \cdot \frac{1}{1.6 \times 10^{-6}} \cdot \frac{1}{h\nu \text{ (in MeV)}} \; \frac{photons}{cm^2 R}$$

$$r = \frac{3.700 \times 10^7 \times 3600 \times (\mu_{en}/\rho)_{air}}{4\pi} \cdot \frac{h\nu \text{ (in MeV)}}{86.9} \times 1.6 \times 10^{-6} \; \frac{R \; cm^2}{hr \cdot mCi}$$

$$W_{air} = 33.7 \text{ eV/ion pair}$$

[a]Calculated on the basis of 3.700×10^7 gamma rays emitted per sec/mCi.
[b]To calculate the value for ^{137}Cs or ^{60}Co, one also requires information concerning the disintegration scheme.

From Johns, H. E. and Cunningham, J. R., *The Physics of Radiology*, 3rd ed., 1971, 735. Courtesy of Charles C Thomas, Publisher, Springfield, Ill.

Table 3.5-3
MASS ATTENUATION COEFFICIENTS*
FOR LOW-ENERGY X-RAYS

Wavelength Å	Formvar $(C_5H_7O_2)_x$	Collodion $(C_{12}H_{11}O_{22}N_6)_x$	Polypropylene $(CH_2)_x$	Cellulose acetate $(C_{10}H_{21}O_{15})_x$	Mylar $(C_{10}H_8O_4)_x$	Teflon $(CF_2)_x$	Energy (eV)
2.0	14	20	8	19	14	28	6199.0
4.0	113	156	69	150	116	220	3099.5
6.0	372	510	234	489	384	700	2066.3
8.0	850	1140	550	1100	870	1540	1549.8
10.0	1580	2110	1040	2020	1630	2800	1239.8
12.0	2600	3450	1740	3310	2680	4520	1033.2
14.0	3920	5200	2660	4950	4040	6700	885.6
16.0	5600	7300	3830	7000	5800	9400	774.9
18.0	7500	9800	5200	9400	7800	12600	688.8
20.0	9900	12800	6900	12300	10200	2780	619.9
22.0	12500	16200	8800	15500	12900	3540	563.5
24.0	8200	6400	11000	4850	8500	4430	516.6
26.0	10100	7900	13500	5900	10400	5400	476.8
28.0	12000	9400	16200	7100	12400	6500	442.8
30.0	14300	11100	19200	8400	14700	7800	413.3
32.0	16700	8200	22400	9900	17200	9100	387.4
34.0	19300	9500	25900	11400	19900	10600	364.6
36.0	22100	10800	29600	13100	22800	12100	344.4
38.0	25000	12300	33600	14900	25800	13800	326.3
40.0	28200	13900	37800	16800	29100	15600	309.9
42.0	31500	15500	42200	18700	32500	17500	295.2
44.0	3250	4540	1940	4370	3350	6900	281.8
46.0	3640	5100	2180	4900	3760	7800	269.5
48.0	4050	5600	2430	5400	4170	8600	258.3
50.0	4450	6200	2690	6000	4590	9500	248.0
52.0	4910	6800	2960	6600	5100	10500	238.4
54.0	5400	7500	3240	7200	5600	11500	229.6
56.0	5900	8200	3540	7900	6100	12500	221.4
58.0	6400	8900	3860	8600	6600	13600	213.8
60.0	7000	9700	4190	9300	7200	14800	206.6
62.0	7500	10500	4540	10100	7800	16000	200.0
64.0	8100	11300	4880	10900	8400	17300	193.7
66.0	8700	12100	5200	11700	9000	18600	187.8
68.0	9400	13100	5700	12600	9700	19900	182.3
70.0	10000	14000	6000	13500	10300	21300	177.1
72.0	10700	14900	6400	14400	11100	22700	172.2
74.0	11400	15900	6800	15400	11800	24200	167.5
76.0	12200	17000	7300	16300	12500	25700	163.1
78.0	12900	18000	7700	17300	13300	27200	158.9
80.0	13600	19100	8100	18400	14100	28800	155.0
82.0	14500	20200	8600	19500	14900	30500	151.2
84.0	15300	21400	9100	20600	15800	32100	147.6
86.0	16100	22600	9600	21700	16600	33800	144.2
88.0	17000	23700	10100	22900	17500	35500	140.9
90.0	17800	25000	10600	24100	18400	37300	137.8
92.0	18800	26300	11100	25300	19400	39100	134.8
94.0	19700	27600	11700	26500	20300	40900	131.9
96.0	20600	28900	12200	27800	21300	43000	129.1
98.0	21600	30300	12800	29200	22300	44600	126.5
100.0	22600	31700	13400	30500	23300	46300	124.0
105.0	25200	35300	15000	34000	26000	51000	118.1
110.0	27900	39100	16600	37600	28800	56000	112.7
115.0	30700	43000	18200	41300	31600	61000	107.8
120.0	33600	47000	20000	45100	34600	67000	103.3
125.0	36800	52000	21900	49600	38000	72000	99.2
130.0	39800	56000	23900	53000	41100	78000	95.4
135.0	43200	60000	25900	58000	44600	83000	91.8
140.0	46700	65000	28100	63000	48200	89000	88.6
145.0	50000	70000	30300	67000	52000	95000	85.5
150.0	54000	75000	32600	72000	55000	101000	82.7
155.0	57000	80000	35000	76000	59000	106000	80.0
160.0	61000	85000	37600	82000	63000	112000	77.5
165.0	65000	90000	40200	87000	67000	118000	75.1
170.0	69000	96000	42800	92000	72000	124000	72.9
175.0	73000	101000	45400	97000	75000	130000	70.8
180.0	77000	107000	47900	102000	80000	136000	68.9
185.0	82000	113000	51000	108000	84000	141000	67.0
190.0	86000	118000	54000	112000	88000	147000	65.3
195.0	90000	124000	57000	118000	93000	153000	63.6
200.0	95000	130000	61000	124000	98000	159000	62.0

ote: The wavelengths in this table are given in the left-hand column in angstrom units; the energies in electronvolts
e on the right.

om *Handbook of Chemistry and Physics,* 56th ed., Weast, R. C., Ed., CRC Press, Cleveland, 1975, E-136.

Table 3.5-3 (continued)
MASS ATTENUATION COEFFICIENTS
FOR LOW-ENERGY X-RAYS

Wavelength Å	Polystyrene (CH)$_x$	Nylon (C$_{12}$H$_{22}$O$_3$N$_2$)$_x$	Vyns (C$_{22}$H$_{33}$O$_2$Cl$_9$)$_x$	Saran (C$_2$H$_2$Cl$_2$)$_x$	Aluminum oxide Al$_2$O$_3$	Quartz (SiO$_2$)$_x$	Energy (eV)
2.0	9	12	114	164	68	77	6199.0
4.0	74	96	750	1070	480	530	3099.5
6.0	252	318	350	375	1440	1600	2066.3
8.0	590	730	780	820	860	980	1549.8
10.0	1120	1370	1440	1520	1580	1810	1239.8
12.0	1870	2270	2370	2490	2570	2950	1033.2
14.0	2870	3440	3590	3760	3840	4400	885.6
16.0	4120	4910	5100	5400	5500	6200	774.9
18.0	5600	6700	6900	7200	7300	8400	688.8
20.0	7500	8800	9000	9300	9600	11000	619.9
22.0	9500	11100	11400	11800	12100	13800	563.5
24.0	11900	10600	12800	14400	4290	4920	516.6
26.0	14600	13000	15500	17300	5200	6000	476.8
28.0	17400	15500	18400	20400	6300	7200	442.8
30.0	20700	18400	21400	23700	7500	8500	413.3
32.0	24200	17300	24700	27100	8700	9900	387.4
34.0	28000	20000	28100	30800	10100	11300	364.6
36.0	31900	22800	31600	34400	11500	12900	344.4
38.0	36200	25800	35400	38200	13100	14500	326.3
40.0	40700	29100	39100	42100	14700	16200	309.9
42.0	45500	32500	43100	46000	16300	18000	295.2
44.0	2090	2730	25700	37000	18100	19800	281.8
46.0	2350	3060	23600	33600	19900	21700	269.5
48.0	2620	3400	25300	36100	21800	23600	258.3
50.0	2900	3750	27300	38800	23700	25500	248.0
52.0	3190	4130	28900	41100	25700	27500	238.4
54.0	3500	4540	30600	43400	27700	29400	229.6
56.0	3820	4950	32200	45700	29800	31800	221.4
58.0	4160	5400	33900	47900	31800	33700	213.8
60.0	4510	5800	35100	49500	33900	35600	206.6
62.0	4890	6300			35900	37500	200.0
64.0	5300	6800			38000	39900	193.7
66.0	5600	7300			40600	41900	187.8
68.0	6100	7900			42800	44000	182.3
70.0	6500	8400			44900	46000	177.1
72.0	6900	9000			47100	48000	172.2
74.0	7400	9600			49300	50000	167.5
76.0	7800	10200			51000	52000	163.1
78.0	8300	10800			54000	54000	158.9
80.0	8800	11400			56000	56000	155.0
82.0	9300	12100			58000	52000	151.8
84.0	9800	12800			61000	53000	147.6
86.0	10300	13500			62000	56000	144.2
88.0	10900	14200			65000	57000	140.9
90.0	11400	15000			67000	59000	137.8
92.0	12000	15700			69000	61000	134.8
94.0	12600	16500			71000	63000	131.9
96.0	13200	17300			73000	65000	129.1
98.0	13800	18100			75000	67000	126.5
100.0	14500	19000			77000	69000	124.0
105.0	16100	21100			83000	73000	118.1
110.0	17900	23400			79000	78000	112.7
115.0	19600	25800			84000	82000	107.8
120.0	21600	28300			88000	86000	103.3
125.0	23600	31000			93000		99.2
130.0	25700	33600			97000		95.4
135.0	28000	36500			101000		91.8
140.0	30300	39400			106000		88.6
145.0	32700	42400			110000		85.5
150.0	35100	45600			114000		82.7
155.0	37700	48900			118000		80.0
160.0	40500	52000			122000		77.5
165.0	43300	56000			125000		75.1
170.0	46100	59000					72.9
175.0	48900	63000					70.8
180.0	52000	66000					68.9
185.0	55000	70000					67.0
190.0	58000	74000					65.3
195.0	62000	78000					63.6
200.0	65000	82000					62.0

Table 3.5-3 (continued)
MASS ATTENUATION COEFFICIENTS
FOR LOW-ENERGY X-RAYS

Wavelength	Stearate $CH_3(CH_2)_{16}COO^-$	Animal proteins C = 52.5% H = 7% S = 1.5% O = 22.5% N = 16.5%	Air O = 21% N = 78% Ar = 1%	P 10 CH_4 = 10% Ar = 90%	Methane CH_4	Q Gas C_4H_{10} = 1.3% He = 98.7%	Energy (eV)
2.0	10	16	21	230	7	1	6199.0
4.0	84	126	148	162	60	12	3099.5
6.0	281	361	481	467	205	41	2066.3
8.0	650	820	1090	1020	479	97	1549.8
10.0	1220	1530	2020	1850	910	185	1239.8
12.0	2030	2530	3310	3010	1520	313	1033.2
14.0	3080	3830	4980	4500	2330	484	885.6
16.0	4410	5500	7100	6400	3350	700	774.9
18.0	6000	7400	9500	8400	4570	970	688.8
20.0	7900	9700	12400	10900	6100	1300	619.9
22.0	10100	12300	15700	13500	7700	1670	563.5
24.0	10000	10300	14100	16400	9700	2110	516.6
26.0	12200	12600	17100	19600	11800	2620	476.8
28.0	14600	15100	20400	22800	14100	3160	442.8
30.0	17300	17800	24000	26300	16800	3780	413.3
32.0	20300	15500	2290	29700	19600	4460	387.4
34.0	23400	17900	2650	33300	22700	5200	364.6
36.0	26800	20400	3040	36900	25900	6000	344.4
38.0	30300	23100	3460	40500	29300	6900	326.3
40.0	34100	26000	3810	37600	33000	7800	309.9
42.0	38200	29100	4270	40900	36900	8800	295.2
44.0	2390	3750	4780	42600	1700	2960	281.8
46.0	2680	4180	5300	45600	1910	3380	269.5
48.0	2980	4610	5900	48900	2130	3840	258.3
50.0	3290	5000	6400	52000	2350	4340	248.0
52.0	3620	5500	6300		2590	4910	238.4
54.0	3970	5900	7000		2840	5500	229.6
56.0	4340	6400	7600		3100	6100	221.4
58.0	4730	6900	8200		3380	6700	213.8
60.0	5100	7500	8900		3660	7400	206.6
62.0	5600	8100	9700		3970	8200	200.0
64.0	6000	8700	10400		4270	8900	193.7
66.0	6400	9300	11200		4570	9700	187.8
68.0	6900	10000	12100		4940	10600	182.3
70.0	7400	10600	12900		5200	11500	177.1
72.0	7900	11300	13800		5600	12500	172.2
74.0	8400	12000	14700		6000	13500	167.5
76.0	8900	11600	15600		6400	14600	163.1
78.0	9500	12300	16600		6700	15600	158.9
80.0	10000	13000	17600		7100	16800	155.0
82.0	10600	13800	18600		7600	18000	151.2
84.0	11200	14600	19700		7900	19300	147.6
86.0	11800	15400	20800		8400	20500	144.2
88.0	12400	16200	21900		8800	21900	140.9
90.0	13100	17000	23100		9300	23300	137.8
92.0	13700	17900	24200		9700	24700	134.8
94.0	14400	18800	25400		10300	26200	131.9
96.0	15100	19700	26700		10700	27800	129.1
98.0	15800	20700	28000		11200	29400	126.5
100.0	16600	21600	29200		11800	31000	124.0
105.0	18400	24100	32700		13100	35400	118.1
110.0	20400	26700	36200		14500	40100	112.7
115.0	22500	29300	39900		15900	44800	107.8
120.0	24600	32200	43800		17500	50000	103.3
125.0	27000	35300	48000		19200	56000	99.2
130.0	29300	38200	52000		20900	62000	95.4
135.0	31800	41500	57000		22700	68000	91.8
140.0	34400	44800	61000		24600	75000	88.6
145.0	37000	48200	65000		26500	82000	85.5
150.0	39800	52000	71000		28500	89000	82.7
155.0	42600	55000	76000		30600	96000	80.0
160.0	45700	59000	80000		32900	104000	77.5
165.0	48600	63000	86000		35100	112000	75.1
170.0	52000	67000	91000		37400	121000	72.9
175.0	55000	71000	96000		39700	130000	70.8
180.0	58000	75000	102000		41900	138000	68.9
185.0	62000	79000	108000		44900	147000	67.0
190.0	65000	83000	114000		47200	157000	65.3
195.0	68000	88000	119000		50000	167000	63.6
200.0	72000	92000	125000		53000	177000	62.0

Tables 3.5-4 and 3.5-5
INTRODUCTION TO X-RAY CROSS SECTIONS

Alex F. Burr

These tables are part of an extensive report published by W. H. McMaster, et al. as UCRL 50174 and available from the National Technical Information Service, Springfield. Va. 22151. Section I of UCRL 50174 describes the data base and the treatment given it, Section II contains the total cross sections between 1 and 1000 keV for all the elements, Section III contains results used in producing Section II, and Section IV contains total cross sections for selected energies and is reproduced in part here. To obtain these values existing experimental x-ray total cross section data and theoretical cross section calculations were surveyed. The coherent (Rayleigh) scattering cross sections and the incoherent (Compton) scattering cross sections were computed. The photo-electric cross sections were obtained by least squares fitting of experimental data, theory, and interpolation of experiment and theory. The following table contains cross sections interpolated from the basic compilation at those wavelengths of most use to x-ray crystallographers. The wavelengths chosen were selected to correspond to those given in the International Tables for X-Ray Crystallography. The energy-to-wavelength conversion is given below.

Table 3.5-4
Energy-to-wavelength conversion

Target radiation	Å	keV	Target radiation	Å	keV
Ag K$\bar{\alpha}$	0.5608	22.105	Ni K$\bar{\alpha}$	1.6591	7.472
Kβ_1	0.4970	24.942	Kβ_1	1.5001	8.265
Pd K$\bar{\alpha}$	0.5869	21.125	Co K$\bar{\alpha}$	1.7902	6.925
Kβ_1	0.5205	23.819	Kβ_1	1.6208	7.649
Rh K$\bar{\alpha}$	0.6147	20.169	Fe K$\bar{\alpha}$	1.9373	6.400
Kβ_1	0.5456	22.724	Kβ_1	1.7565	7.058
Mo K$\bar{\alpha}$	0.7107	17.444	Mn K$\bar{\alpha}$	2.1031	5.895
Kβ_1	0.6323	19.608	Kβ_1	1.9102	6.490
Zn K$\bar{\alpha}$	1.4364	8.631	Cr K$\bar{\alpha}$	2.2909	5.412
Kβ_1	1.2952	9.572	Kβ_1	2.0848	5.947
Cu K$\bar{\alpha}$	1.5418	8.041	Ti K$\bar{\alpha}$	2.7496	4.509
Kβ_1	1.3922	8.905	Kβ_1	2.5138	4.932

Table 3.5-5
Total Cross Section in cm^2/g

Z KEV	1 H	2 He	3 Li	4 Be	5 B	6 C	7 N	8 O	9 F	10 Ne
4.51	.432	.661	2.10	5.63	12.9	25.6	43.0	64.8	91.8	125
4.93	.421	.550	1.62	4.25	9.69	19.4	32.6	49.4	70.3	95.9
5.41	.412	.465	1.24	3.18	7.23	14.5	24.4	37.2	53.1	72.7
5.90	.405	.405	.986	2.45	5.53	11.1	18.7	28.6	41.1	56.4
5.95	.405	.400	.964	2.39	5.39	10.8	18.2	27.9	40.0	54.9
6.40	.400	.362	.798	1.92	4.28	8.55	14.5	22.2	32.0	44.0
6.49	.400	.355	.770	1.84	4.10	8.18	13.9	21.3	30.6	42.2
6.93	.397	.329	.659	1.52	3.36	6.68	11.3	17.4	25.1	34.7
7.06	.396	.322	.631	1.44	3.17	6.30	10.7	16.5	23.7	32.8
7.47	.394	.303	.555	1.23	2.67	5.28	8.96	13.8	19.9	27.6
7.65	.393	.297	.528	1.15	2.49	4.92	8.33	12.9	18.6	25.7
8.04	.391	.284	.477	1.01	2.14	4.22	7.14	11.0	16.0	22.1
8.27	.390	.277	.452	.936	1.98	3.88	6.56	10.1	14.7	20.4
8.63	.389	.268	.417	.837	1.74	3.40	5.74	8.87	12.8	17.9
8.91	.388	.262	.394	.774	1.59	3.09	5.22	8.06	11.7	16.2
9.57	.386	.250	.349	.651	1.30	2.49	4.19	6.47	9.36	13.1
17.44	.373	.202	.197	.245	.345	.535	.790	1.15	1.58	2.21
19.61	.370	.197	.187	.222	.293	.429	.605	.855	1.15	1.60
20.17	.369	.196	.185	.217	.283	.408	.570	.799	1.07	1.48
21.13	.368	.195	.182	.210	.268	.379	.519	.717	.952	1.31
22.11	.366	.193	.179	.205	.256	.354	.476	.648	.851	1.16
22.72	.366	.192	.177	.201	.249	.340	.452	.610	.795	1.08
23.82	.364	.191	.175	.196	.239	.319	.416	.553	.711	.963
24.94	.363	.189	.173	.192	.229	.301	.385	.504	.640	.861

X-RAY CROSS SECTIONS (Continued)

Table 3.5-5 (continued)
Total Cross Section in cm²/g

Z KEV	11 Na	12 Mg	13 Al	14 Si	15 P	16 S	17 Cl	18 Ar	19 K	20 Ca
4.51	168	220	264	336	386	464	518	577	679	805
4.93	129	171	206	263	304	364	411	456	542	638
5.41	98.5	131	158	203	236	282	322	356	427	500
5.90	76.6	102	124	160	186	223	256	282	342	398
5.95	74.7	99.6	121	156	182	217	250	276	334	389
6.40	59.9	80.2	97.5	126	148	177	205	225	275	319
6.49	57.5	77	93.7	121	142	170	197	217	264	307
6.93	47.3	63.5	77.5	100	118	141	165	181	222	257
7.06	44.7	60.1	73.4	95.1	112	134	156	172	211	244
7.47	37.7	50.8	62.2	80.7	95.2	114	134	147	181	209
7.65	35.2	47.4	58.1	75.4	89.1	107	125	137	170	196
8.04	30.3	40.9	50.2	65.3	77.3	92.5	109	120	148	171
8.27	27.9	37.6	46.3	60.2	71.3	85.5	101	111	138	159
8.63	24.4	33.0	40.7	53.0	62.9	75.4	89.4	97.7	122	141
8.91	22.2	30.1	37.1	48.4	57.4	68.9	81.8	89.3	112	129
9.57	17.9	24.2	30.0	39.1	46.6	55.9	66.7	72.7	91.4	106
17.44	2.94	3.98	5.04	6.53	7.87	9.63	11.6	12.6	16.2	19
19.61	2.10	2.83	3.59	4.62	5.57	6.84	8.26	8.95	11.5	13.6
20.17	1.94	2.60	3.30	4.26	5.13	6.30	7.61	8.24	10.6	12.5
21.13	1.70	2.28	2.89	3.71	4.47	5.50	6.63	7.18	9.24	10.9
22.11	1.50	2.00	2.54	3.25	3.91	4.82	5.80	6.28	8.08	9.57
22.72	1.39	1.86	2.35	3.00	3.61	4.45	5.35	5.79	7.45	8.84
23.82	1.23	1.63	2.06	2.62	3.15	3.88	4.66	5.04	6.49	7.71
24.94	1.09	1.44	1.81	2.30	2.76	3.40	4.08	4.41	5.67	6.75

Z KEV	21 Sc	22 Ti	23 V	24 Cr	25 Mn	26 Fe	27 Co	28 Ni	29 Cu	30 Zn
4.51	819	111	125	143	160	188	206	240	257	280
4.93	658	86.8	97.3	111	125	147	161	188	201	220
5.41	521	571	75.1	85.7	96 1	113	125	146	155	172
5.90	420	459	513	67.4	75.6	88.9	98.4	115	123	137
5.95	411	449	501	65.8	73.8	86.8	96.1	113	120	134
6.40	339	370	411	462	59.9	70.4	78.3	91.8	97.4	110
6.49	327	357	396	445	57.6	67.7	75.3	88.3	93.6	106
6.93	276	301	333	375	405	56.3	62.9	73.8	78.1	88.7
7.06	262	286	316	357	385	53.3	59.6	70.0	74.1	84.3
7.47	226	246	271	307	331	367	50.9	59.8	63.2	72.4
7.65	212	231	255	288	311	346	47.7	56.1	59.2	68.0
8.04	186	202	223	252	273	304	339	48.8	51.5	59.5
8.27	173	188	206	234	253	283	315	45.2	47.7	55.3
8.63	153	167	183	208	225	253	281	306	42.3	49.2
8.91	141	153	168	191	207	234	259	283	38.7	45.3
9.57	116	126	138	157	170	194	214	236	245	37.4
17.44	21.0	23.3	25.2	29.3	31.9	37.7	41.0	47.2	49.3	55.5
19.61	15.0	16.7	18.1	21.0	22.9	27.2	29.5	34.2	35.8	40.3
20.17	13.8	15.4	16.7	19.4	21.1	25.1	27.2	31.6	33.1	37.2
21.13	12.1	13.4	14.6	17.0	18.5	22.0	23.8	27.7	29.0	32.7
22.11	10.5	11.8	12.8	14.9	16.2	19.3	20.9	24.3	25.5	28.7
22.72	9.72	10.9	11.8	13.7	15.0	17.8	19.3	22.5	23.6	26.6
23.82	8.47	9.48	10.3	12.0	13.1	15.6	16.9	19.7	20.7	23.3
24.94	7.40	8.30	9.02	10.5	11.5	13.7	14.8	17.3	18.2	20.4

X-RAY CROSS SECTIONS (Continued)

Table 3.5-5 (continued)
Total Cross Section in cm²/g

Z KEV	31 Ga	32 Ge	33 As	34 Se	35 Br	36 Kr	37 Rb	38 Sr	39 Y	40 Zr
4.51	309	329	368	403	435	464	508	552	599	648
4.93	242	258	288	317	342	366	400	436	473	511
5.41	187	200	224	246	266	285	312	339	369	399
5.90	148	158	178	195	211	226	248	270	294	317
5.95	144	155	173	190	206	221	242	263	287	310
6.40	117	126	142	156	169	181	198	215	235	254
6.49	113	122	136	150	162	174	191	207	227	244
6.93	94.2	102	114	125	136	146	160	174	190	205
7.06	89.3	96.8	108	119	129	138	152	165	181	195
7.47	76.2	82.9	92.5	101	110	119	130	141	155	167
7.65	71.4	77.8	86.8	95.1	104	111	122	132	145	157
8.04	62.1	67.9	75.7	82.9	90.3	97	106	115	127	137
8.27	57.5	63.0	70.1	77.8	83.7	89.9	98.5	107	118	127
8.63	50.9	56.0	62.2	68.0	74.2	79.8	87.4	94.7	105	113
8.91	46.7	51.4	57.0	62.3	68.1	73.2	80.2	86.8	96.2	103
9.57	38.1	42.3	46.7	51.0	55.8	60.0	65.7	71.0	79.0	84.8
17.44	56.9	60.5	66.0	68.8	74.7	79.1	83.0	88.0	97.6	16.1
19.61	41.7	44.3	48.6	51.2	55.6	58.6	62.1	65.6	72.6	75.2
20.17	38.6	41.0	45.1	47.6	51.7	54.5	57.8	61.0	67.5	70.1
21.13	34.0	36.1	39.7	42.1	45.6	48.1	51.1	54.0	59.7	62.1
22.11	29.9	31.8	35.0	37.2	40.4	42.5	45.3	47.9	53.0	55.2
22.72	27.7	29.4	32.4	34.6	37.5	39.5	42.1	44.5	49.2	51.4
23.82	24.3	25.8	28.5	30.5	33.1	34.8	37.2	39.3	43.5	45.5
24.94	21.4	22.7	25.1	26.9	29.2	30.7	32.9	34.8	38.5	40.4

Z KEV	41 Nb	42 Mo	43 Te	44 Ru	45 Rh	46 Pd	47 Ag	48 Cd	49 In	50 Sn
4.51	697	738	786	832	892	928	987	1064	1151	1128
4.93	552	585	621	660	708	739	785	842	906	899
5.41	432	457	486	518	555	581	617	659	706	709
5.90	344	365	387	414	444	466	495	526	561	569
5.95	336	357	378	404	434	455	484	514	548	557
6.40	276	293	310	333	357	375	398	422	449	460
6.49	266	282	299	320	344	361	384	407	433	443
6.93	223	237	251	269	289	304	324	342	363	374
7.06	212	225	238	256	275	289	308	325	345	356
7.47	182	193	204	220	236	249	265	279	295	307
7.65	170	181	192	207	222	234	249	262	277	289
8.04	149	158	168	181	194	205	218	229	242	253
8.27	138	147	156	168	180	190	203	213	225	236
8.63	122	130	138	149	160	169	180	189	200	210
8.91	112	120	127	137	147	156	166	174	183	193
9.57	92.0	98.2	104	113	121	128	137	143	151	159
17.44	17.0	18.4	19.8	21.3	23.1	24.4	26.4	27.7	29.1	31.2
19.61	81.2	13.3	14.3	15.4	16.7	17.6	19.1	20.1	21.2	22.6
20.17	75.6	79.3	13.2	14.2	15.4	16.3	17.7	18.6	19.6	20.9
21.13	67.1	70.3	73.0	12.5	13.5	14.3	15.5	16.4	17.3	18.4
22.11	59.6	62.5	65.0	11.0	11.9	12.6	13.7	14.5	15.3	16.2
22.72	55.5	58.2	60.6	63.8	11.0	11.6	12.7	13.4	14.1	15.0
23.82	49.1	51.5	53.7	56.6	58.3	10.2	11.1	11.8	12.4	13.2
24.94	43.6	45.7	47.7	50.3	52.0	57.0	97.8	10.4	11.0	11.6

X-RAY CROSS SECTIONS (Continued)

Table 3.5-5 (continued)
Total Cross Section in cm²/g

Z KEV	51 Sb	52 Te	53 I	54 Xe	55 Cs	56 Ba	57 La	58 Ce	59 Pr	60 Nd
4.51	997	753	293	300	324	334	355	383	414	433
4.93	926	843	921	683	259	266	282	304	330	344
5.41	733	769	835	755	803	587	223	240	261	271
5.90	592	617	666	701	742	660	677	521	210	218
5.95	579	603	651	685	725	645	662	509	205	213
6.40	479	497	535	565	597	615	636	592	450	464
6.49	462	479	516	545	575	593	613	571	610	447
6.93	391	404	434	459	484	499	519	559	596	532
7.06	373	385	413	437	460	475	494	532	567	506
7.47	322	331	355	376	395	408	427	459	488	506
7.65	303	312	333	353	372	384	402	432	459	476
8.04	267	273	292	310	325	336	354	379	402	418
8.27	248	254	271	288	302	312	329	352	374	389
8.63	221	226	241	256	269	278	294	314	333	347
8.91	204	208	222	236	248	256	271	290	307	320
9.57	168	172	183	195	204	211	224	240	253	265
17.44	33.0	33.9	36.3	38.3	40.4	42.4	45.3	48.6	50.8	53.3
19.61	23.9	24.7	26.5	27.9	29.5	31.0	33.1	35.5	37.1	38.9
20.17	22.1	22.8	24.6	25.8	27.3	28.7	30.7	33.0	34.4	36.0
21.13	19.4	20.1	21.7	22.7	24.1	25.4	27.0	29.1	30.3	31.7
22.11	17.1	17.7	19.2	20.0	21.3	22.5	23.9	25.7	26.8	28.0
22.72	15.8	16.4	17.8	18.6	19.8	20.9	22.1	23.9	24.9	26.0
23.82	13.8	14.4	15.7	16.3	17.4	18.4	19.5	21.0	21.9	22.9
24.94	12.1	12.7	13.9	14.4	15.4	16.3	17.2	18.6	19.3	20.2

Z KEV	61 Pm	62 Sm	63 Eu	64 Gd	65 Tb	66 Dy	67 Ho	68 Er	69 Tm	70 Yb
4.51	455	473	503	510	546	568	589	615	644	664
4.93	361	375	398	405	432	449	465	485	508	524
5.41	285	295	313	319	339	352	363	380	397	410
5.90	229	237	251	256	271	281	290	303	317	327
5.95	224	232	245	250	265	275	283	296	310	319
6.40	186	192	203	207	219	227	234	244	255	263
6.49	476	185	195	200	211	219	225	235	246	254
6.93	401	412	165	170	179	185	190	198	207	214
7.06	536	392	420	161	170	176	181	189	197	204
7.47	535	475	361	369	147	152	156	163	170	175
7.65	503	446	477	347	367	143	146	153	160	165
8.04	441	454	418	427	322	337	128	134	140	145
8.27	410	422	450	397	420	313	333	125	131	135
8.63	366	376	401	410	375	392	296	318	117	120
8.91	336	347	369	377	400	360	272	292	289	111
9.57	278	287	305	312	330	344	364	336	239	232
17.44	55.5	58.0	61.2	62.8	66.8	68.9	72.1	75.6	79.0	80.2
19.61	40.5	42.4	44.7	46.0	48.9	50.4	52.8	55.1	57.9	59.2
20.17	37.6	39.3	41.5	42.6	45.3	46.7	48.9	51.0	53.8	55.0
21.13	33.1	34.7	36.6	37.6	40.0	41.2	43.2	45.0	47.5	48.7
22.11	29.3	30.7	32.4	33.3	35.4	36.5	38.3	39.8	42.1	43.2
22.72	27.1	28.5	30.1	30.9	32.9	33.9	35.6	37.0	39.1	40.2
23.82	23.9	25.1	26.5	27.3	29.0	29.9	31.4	32.6	34.5	35.5
24.94	21.1	22.2	23.4	24.1	25.6	26.4	27.8	28.8	30.6	31.5

X-RAY CROSS SECTIONS (Continued)

Table 3.5-5 (continued)
Total Cross Section in cm² /g

Z KEV	71 Lu	72 Hf	73 Ta	74 W	75 Re	76 Os	77 Ir	78 Pt	79 Au	80 Hg
4.51	696	720	736	753	796	824	871	934	906	958
4.93	549	568	581	598	631	653	688	734	720	760
5.41	430	445	455	470	496	512	540	572	568	598
5.90	343	355	363	378	397	410	431	455	457	480
5.95	335	347	355	369	388	401	422	444	447	469
6.40	276	286	293	306	321	332	348	365	371	388
6.49	266	276	282	295	310	320	336	351	358	375
6.93	225	233	238	250	262	270	283	295	303	317
7.06	214	222	227	238	249	257	270	281	289	302
7.47	184	191	195	206	215	222	233	241	250	261
7.65	173	180	184	194	203	209	219	226	236	246
8.04	152	158	162	171	178	184	192	198	208	216
8.27	142	147	150	159	166	171	179	184	194	202
8.63	126	131	134	142	149	153	160	164	174	180
8.91	117	121	124	132	137	141	148	151	161	167
9.57	247	236	103	110	114	118	123	125	134	139
17.44	84.2	86.3	89.5	95.8	98.7	100	103	109	111	115
19.61	62.0	64.2	66.1	70.6	72.5	74.1	77.2	80.2	82.3	85.3
20.17	57.6	59.7	61.4	65.5	67.3	68.9	71.9	74.6	76.5	79.4
21.13	51.0	52.9	54.4	58.0	59.5	61.1	63.8	66.0	67.8	70.4
22.11	45.3	46.9	48.3	51.4	52.8	54.2	56.8	58.6	60.2	62.6
22.72	42.1	43.7	44.9	47.8	49.1	50.5	52.9	54.5	56.0	58.3
23.82	37.2	38.6	39.7	42.2	43.3	44.6	46.8	48.2	49.5	51.7
24.94	33.0	34.2	35.2	37.4	38.4	39.6	41.6	42.7	43.9	45.9

Z KEV	81 Tl	82 Pb	83 Bi	86 Rn	90 Th	92 U	95 Pu
4.51	991	1035	1066	1174	1098	1084	960
4.93	785	820	847	930	993	862	1023
5.41	617	645	667	731	844	774	803
5.90	494	517	536	586	678	672	731
5.95	483	505	524	573	663	657	772
6.40	400	418	435	474	549	545	638
6.49	386	403	419	458	530	526	615
6.93	326	341	355	387	449	446	520
7.06	311	325	338	369	428	425	495
7.47	268	280	293	318	370	368	427
7.65	253	264	276	300	348	347	402
8.04	222	232	243	264	307	306	353
8.27	207	216	227	246	286	285	329
8.63	185	194	203	220	256	256	294
8.91	171	179	188	203	237	236	271
9.57	142	149	156	169	197	197	225
17.44	119	123	126	117	99.5	96.7	48.8
19.61	88.3	90.6	93.5	101	73.3	72.6	79.0
20.17	82.0	84.1	87.0	93.8	95.0	67.8	73.7
21.13	72.7	74.5	77.2	83.3	97.2	84.2	65.7
22.11	64.6	66.1	68.7	74.1	86.3	86.9	58.8
22.72	60.1	61.5	64.0	69.1	80.3	81.1	76.4
23.82	53.2	54.4	56.7	61.2	71.0	72.1	78.3
24.94	47.2	48.2	50.4	54.4	62.9	64.2	69.8

Table 3.5-6
VALUES OF THE ATOMIC WEIGHT M, THE FACTOR $(N_A/M) \times 10^{-24}$
FOR CONVERTING ATTENUATION DATA FROM b/ATOM TO $cm^2\ g^{-1}$,
AND TYPICAL DENSITIES ρ FOR CONVERTING FROM $cm^2\ g^{-1}$ TO
cm^{-1}

Element		$M,^a$ atomic		
Z	Symbol	weight, g/g-atom	$\frac{N_A}{M} \times 10^{-24}, ^b \frac{cm^2}{g} / \frac{b}{atom}$	$\rho, ^c\ g/cm^3$
1	H	1.00797	0.5975	0.00008988 g, (H$_2$)
2	He	4.0026	.1505	.0001785 g
3	Li	6.939	.08679	.534
4	Be	9.0122	.06683	1.85
5	B	10.811	.05571	2.535
6	C	12.01115	.05014	2.25 (graphite)^d
7	N	14.0067	.04300	.001250 g, (N$_2$)
8	O	15.9994	.03764	.001429 g, (O$_2$)
9	F	18.9984	.03170	.001696 g, (F$_2$)
10	Ne	20.183	.02984	.0008999 g
11	Na	22.9898	.02620	.971
12	Mg	24.312	.02477	1.74
13	Al	26.9815	.02232	2.70
14	Si	28.086	.02144	2.42
15	P	30.9738	.01944	1.8−2.7
16	S	32.064	.01878	1.96−2.07
17	Cl	35.453	.01699	.003214 g, (Cl$_2$)
18	Ar	39.948	.01508	.001784 g
19	K	39.102	.01540	.87
20	Ca	40.08	.01503	1.55
21	Sc	44.956	.01340	3.02
22	Ti	47.90	.01257	4.5
23	V	50.942	.01182	5.87
24	Cr	51.996	.01158	7.14
25	Mn	54.9380	.01096	7.3
26	Fe	55.847	.01078	7.86
27	Co	58.9332	.01022	8.71
28	Ni	58.71	.01026	8.8
29	Cu	63.54	.009478	8.93
30	Zn	65.37	.009213	6.92
31	Ga	69.72	.008638	5.93
32	Ge	72.59	.008297	5.46
33	As	74.9216	.008038	5.73
34	Se	78.96	.007627	4.82
35	Br	79.909	.007537	3.12 l
36	Kr	83.80	.007187	.003743 g
37	Rb	85.47	.007046	1.53
38	Sr	87.62	.006873	2.6
39	Y	88.905	.006774	3.8
40	Zr	91.22	.006602	6.44

Table 3.5-6 (continued)
VALUES OF THE ATOMIC WEIGHT M, THE FACTOR $(N_A/M) \times 10^{-24}$ FOR CONVERTING ATTENUATION DATA FROM b/ATOM TO $cm^2 \ g^{-1}$, AND TYPICAL DENSITIES ρ FOR CONVERTING FROM $cm^2 \ g^{-1}$ TO cm^{-1}

Element		$M,^a$ atomic weight, g/g-atom	$\dfrac{N_A}{M} \times 10^{-24},^b \dfrac{cm^2}{g} / \dfrac{b}{atom}$	$\rho,^c \ g/cm^3$
Z	Symbol			
41	Nb	92.906	.006482	8.4
42	Mo	95.94	.006277	9.01
43	Tc	(99)	(.006083)	[11.50]
44	Ru	101.07	.005959	12.1
45	Rh	102.905	.005853	12.44
46	Pd	106.4	.005660	12.25
47	Ag	107.870	.005583	10.49
48	Cd	112.40	.005358	8.65
49	In	114.82	.005245	7.43
50	Sn	118.69	.005074	5.75−7.29
51	Sb	121.75	.004947	6.62
52	Te	127.60	.004720	6.25
53	I	126.9044	.004746	4.94
54	Xe	131.30	.004587	.005896 g
55	Cs	132.905	.004531	1.873
56	Ba	137.34	.004385	3.5
57	La	138.91	.004336	6.15
58	Ce	140.12	.004298	6.90
59	Pr	140.907	.004274	6.48
60	Nd	144.24	.004175	7.00
61	Pm	(145)	(.004153)	[7.22]e
62	Sm	150.35	.004006	7.7−7.8
63	Eu	151.96	.003963	[5.259]
64	Gd	157.25	.003830	[7.948]
65	Tb	158.924	.003790	[8.272]
66	Dy	162.50	.003706	[8.536]
67	Ho	164.930	.003652	[8.803]
68	Er	167.26	.003601	[9.051]f
69	Tm	168.934	.003565	[9.332]
70	Yb	173.04	.003480	[6.977]
71	Lu	174.97	.003442	[9.872]
72	Hf	178.49	.003374	13.3
73	Ta	180.948	.003328	17.1
74	W	183.85	.003276	19.3
75	Re	186.2	.003234	20.53
76	Os	190.2	.003166	22.8
77	Ir	192.2	.003133	22.42−22.8
78	Pt	195.09	.003087	21.4
79	Au	196.967	.003058	19.3
80	Hg	200.59	.003002	13.55 l

Table 3.5-6 (continued)
VALUES OF THE ATOMIC WEIGHT M, THE FACTOR $(N_A/M) \times 10^{-24}$ FOR CONVERTING ATTENUATION DATA FROM b/ATOM TO $cm^2 \ g^{-1}$, AND TYPICAL DENSITIES ρ FOR CONVERTING FROM $cm^2 \ g^{-1}$ TO cm^{-1}

Element		M,[a] atomic weight, g/g-atom	$\frac{N_A}{M} \times 10^{-24}$,[b] $\frac{cm^2}{g} \Big/ \frac{b}{atom}$	ρ,[c] g/cm³
Z	Symbol			
81	Tl	204.37	.002947	11.86
82	Pb	207.19	.002907	11.34
83	Bi	208.980	.002882	9.78
84	Po	(210)	(.002868)	[9.32]
85	At	(210)	(.002868)	–
86	Rn	(222)	(.002713)	0.00996 g
87	Fr	(223)	(.002701)	–
88	Ra	(226)	(.002665)	5(?)
89	Ac	(227)	(.002653)	[10.07]
90	Th	232.038	.002595	11.0
91	Pa	(231)	(.002607)	[15.37]
92	U	238.03	.002530	18.7
93	Np	(237)	(.002541)	[19.36]
94	Pu	(242)	(.002489)	[19.84]
95	Am	(243)	(.002478)	[11.7]
96	Cm	(247)	(.002438)	[~7]
97	Bk	(249)	(.002419)	–
98	Cf	(251)	(.002399)	–
99	Es	(254)	(.002371)	–
100	Fm	(253)	(.002380)	–

[a]Atomic weights are those recommended in 1961 by the International Union of Pure and Applied Chemistry, based on M = 12.0000 for C^{12}. These values are based on average isotropic abundances and ignore natural and artificial variations. In practice, for example, M can vary from 6.01513 for pure $_3Li^7$, with a corresponding variation in the conversion factor from 0.1001 to 0.08584.

[b]N_A = Avogadro's Number = 6.02252×10^{23} atoms/g-atom, C^{12} scale.

[c]Densities are for common solids, except as denoted liquid (l) or gas (g) at 20°C, and in square brackets ([ ]). Gas densities are at S.T.P.: 0°C, 76 cm Hg.

[d]Graphite theoretical density, based on X-ray diffraction data. Commercial-grade pile graphite has a density range of 1.5 to 1.9 g/cm³.

[e]The density of Pm^{147} is 7.22 g/cm³.

[f]Density values of erbium in the literature vary from 4.77 to the values 9.05 and 9.16.

From Hubbell, J. H., *Photon Cross Sections, Attenuation Coefficients, and Energy Absorption Coefficients, from 10 KeV to 100 GeV*, NSRDS-NBS-29, U.S. Government Printing Office, Washington, D.C., August 1969, 6.

Table 3.5-7

CONVERSION FACTORS AND DENSITIES FOR A
FEW COMPOUNDS AND MIXTURES

Substance	Conversion factor, $\dfrac{cm^2}{g} \Big/ \dfrac{b}{molecule}$	ρ, g/cm^3
H_2O	0.03344	1.00 1, 0.917 (ice)
SiO_2	.01002	2.32
NaI	.004019	3.667
Air ($20°$C, 76 cm Hg)	–	0.001205 g
Concrete	–	2.2 – 2.4
0.8 N H_2SO_4 solution	–	1.0494
Bone	–	1.7 – 2.0
Muscle	–	~ 1
Polystyrene, $(C_8H_8)_n$	–	1.05 – 1.07
Polyethylene, $(CH_2)_n$	–	0.92
Polymethyl methacrylate (Lucite®), $(C_5H_8O_2)_n$	–	1.19
Bakelite® (typical), $C_{43}H_{38}O_7$	–	1.20 – 1.70
Pyrex® glass (Corning No. 7740)	–	2.23

From Hubbell, J. H., *Photon Cross Sections, Attenuation Coefficients, and Energy Absorption Coefficients, from 10 KeV to 100 GeV*, NSRDS-NBS-29, U.S. Government Printing Office, Washington, D.C., August 1969, 7.

Table 3.5-8

FRACTIONS BY WEIGHT, w_j, OF ELEMENTS IN SOME MIXTURES AND COMPOUNDS

Element		H_2O	SiO_2	NaI	Air	Concrete	0.8 N H_2SO_4 solution[a]	Bone, compact	Muscle, striated	Polystyrene, $(C_8H_8)_n$	Lucite, $(C_5H_8O_2)_n$	Polyethylene, $(CH_2)_n$	Bakelite, $C_{45}H_{38}O_7$	Pyrex glass[b]
Z	Symbol													
1	H	0.1119	—	—	—	0.0056	0.1084	0.064	0.102	0.0774	0.0805	0.1437	0.0574	—
5	B	—	—	—	—	—	—	—	—	—	—	—	—	0.0401
6	C	—	—	—	—	—	—	0.278	0.123	0.9226	0.5999	0.8563	0.7746	—
7	N	—	—	—	0.755	—	—	0.027	0.035	—	—	—	—	—
8	O	0.8881	0.5326	—	0.232	0.4983	0.8791	0.410	0.72893	—	0.3196	—	0.1680	0.5396
11	Na	—	—	0.1534	—	0.0171	—	—	0.0008	—	—	—	—	0.0282
12	Mg	—	—	—	—	0.0024	—	0.002	0.0002	—	—	—	—	—
13	Al	—	—	—	—	0.0456	—	—	—	—	—	—	—	0.0116
14	Si	—	0.4674	—	—	0.3158	—	—	—	—	—	—	—	0.3772
15	P	—	—	—	—	—	—	0.070	0.002	—	—	—	—	—
16	S	—	—	—	—	0.0012	0.0125	0.002	0.005	—	—	—	—	—
18	Ar	—	—	—	0.013	—	—	—	—	—	—	—	—	—
19	K	—	—	—	—	0.0192	—	—	0.003	—	—	—	—	0.0033
20	Ca	—	—	—	—	0.0826	—	0.147	0.00007	—	—	—	—	—
26	Fe	—	—	—	—	0.0122	—	—	—	—	—	—	—	—
53	I	—	—	0.8466	—	—	—	—	—	—	—	—	—	—

[a] 3.832% H_2SO_4, 96.168% H_2O by weight.

[b] SiO_2, 80.7%; B_2O_3, 12.9%; Na_2O_3, 3.8%; Al_2O_3, 2.2%; K_2O, 0.4% by weight. 0.2% has been added to SiO_2, to make total = 100%. Otherwise, these percentages, with 80.5% SiO_2, are those given for Pyrex glass (Corning 7740).

From Hubbell, J. H., *Photon Cross Sections, Attenuation Coefficients, and Energy Absorption Coefficients, from 10 KeV to 100 GeV*, NSRDS-NBS-29, U.S. Government Printing Office, Washington, D.C., August 1969, 8.

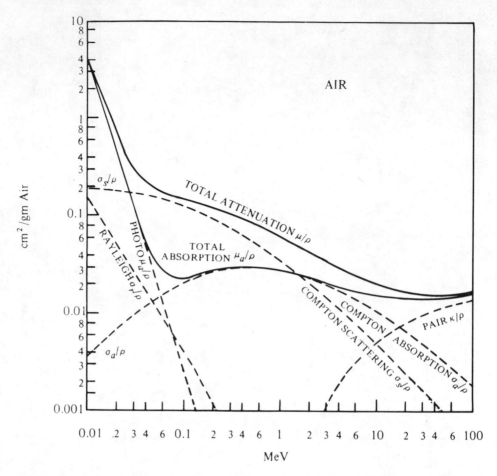

FIGURE 3.5-1.   Mass attenuation coefficients for photons in ''air'' taken as 78.04 vol % nitrogen, 21.2 vol % oxygen, and 0.94 vol % argon. At 0°C and 760 mm Hg pressure, the density of air is ϱ = 0.001293 g/cm³. (From Evans, R. D., in *Radiation Dosimetry,* 3rd ed.,Attix, F. H. and Roesch, W. C., Eds., Academic Press, New York, 1968. With permission.)

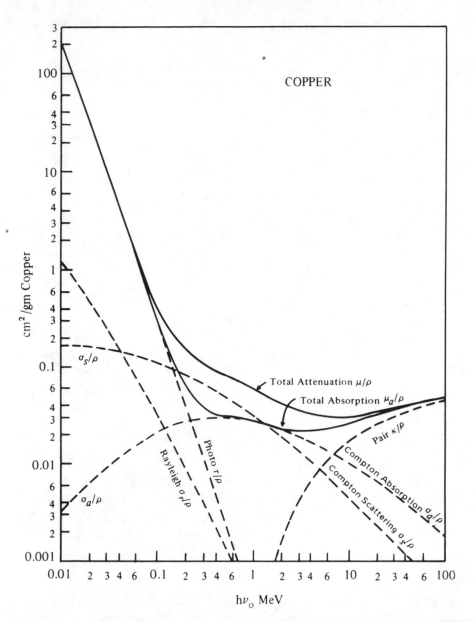

FIGURE 3.5-2.   Mass attenuation coefficients for photons in copper ($Z = 29$). The dashed branch on the μ_a/ϱ curve shows the effect of excluding the annihilation photons. The corresponding linear coefficients for copper may be obtained by multiplying all curves by $p = 8.92$ g/cm³ Cu. (From Evans, R. D., in *Radiation Dosimetry,* 3rd ed., Attix, F. H. and Roesch, W. C., Eds., Academic Press, New York, 1968. With permission.)

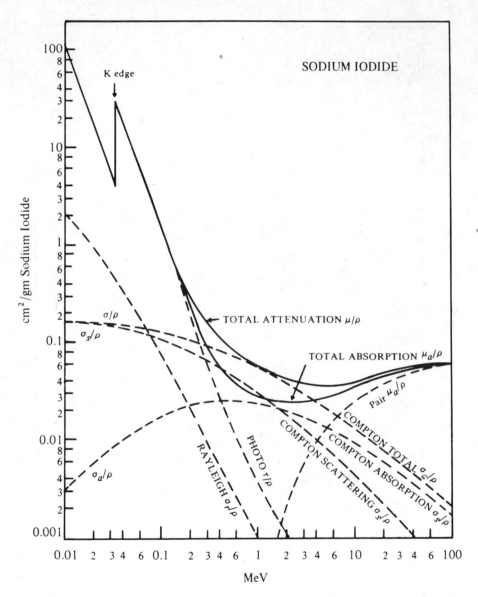

FIGURE 3.5-3. Mass attenuation coefficients for photons in pure NaI. The "Compton total" attenuation coefficient $(\sigma/\varrho) = (\sigma_a/\varrho) + (\sigma_s/\varrho)$ is shown explicitly, because of its usefulness in predicting the behavior of NaI(Tl) scintillators. The 0.1 to 0.2% thallium activator in NaI(Tl) scintillators is ignored here. The dashed branch on the μ_a/ϱ curve shows the effect of excluding annihilation photons. Linear attenuation coefficients for NaI may be obtained using $\varrho = 3.67$ g/cm³ NaI. (From Evans, R. D., in *Radiation Dosimetry,* 3rd ed., Attix, F. H. and Roesch, W. C., Eds., Academic Press, New York, 1968. With permission.

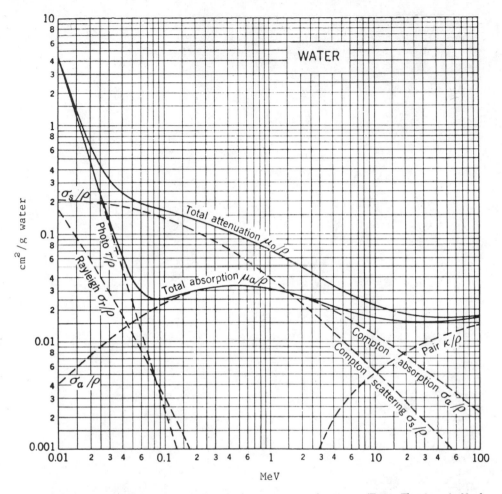

FIGURE 3.5-4. Mass attenuation coefficients for gamma-rays in water. (From *The Atomic Nucleus,* by R. D. Evans, Copyright 1955, by permission of McGraw-Hill Book Co., Inc., as reproduced from Radiological Health Handbook.)

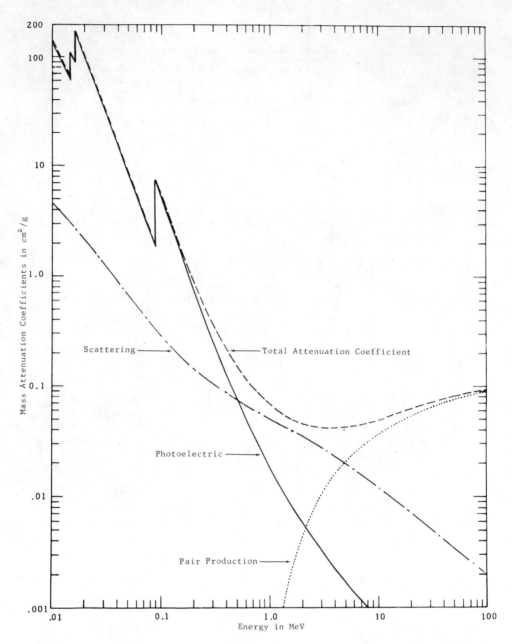

FIGURE 3.5-5.   Mass attenuation coefficients for lead. (From White, G. R., NBS Report No. 1003, National Bureau of Standards, Washington, D.C., 1952.)

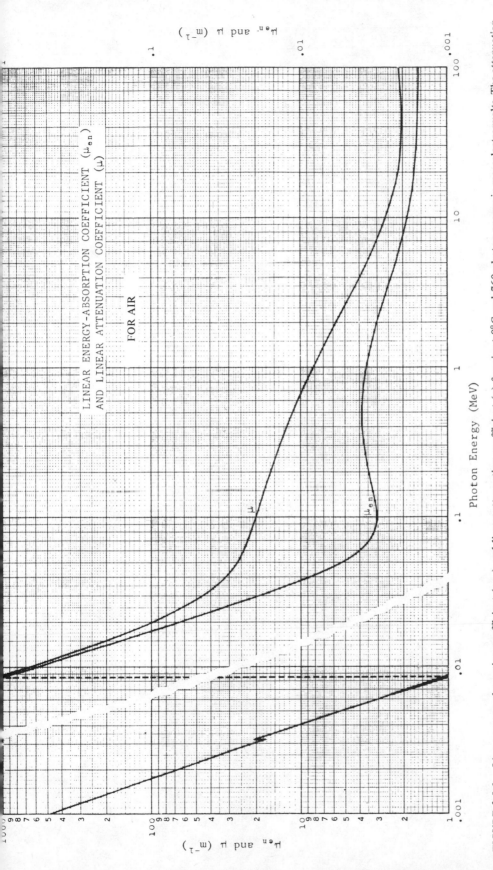

FIGURE 3.5-6. Linear energy-absorption coefficient (μ_{en}) and linear attenuation coefficient (μ) for air at 0°C and 760 photon energy in megaelectronvolts. The attenuation coefficients from 0.003 to 100 MeV were derived from mass attenuation coefficients (with coherent) given in NBS Circular 583, 1957 and its supplement, 1959. The energy-absorption coefficients for 0.01 to 100 MeV were derived from data published in *Engineering Compendium on Radiation Shielding*[8] The μ_{en} for the range 0.003 to 0.01 are based on the μ values adjusted for Compton and coherent scattering. The range below 0.003 is extrapolated and involves an uncertainty of about ±15 at 0.002. (From Radiological Health Handbook, U.S. Department of Health, Education, and Welfare, Public Health Service, Rockville, Md., January 1970.)

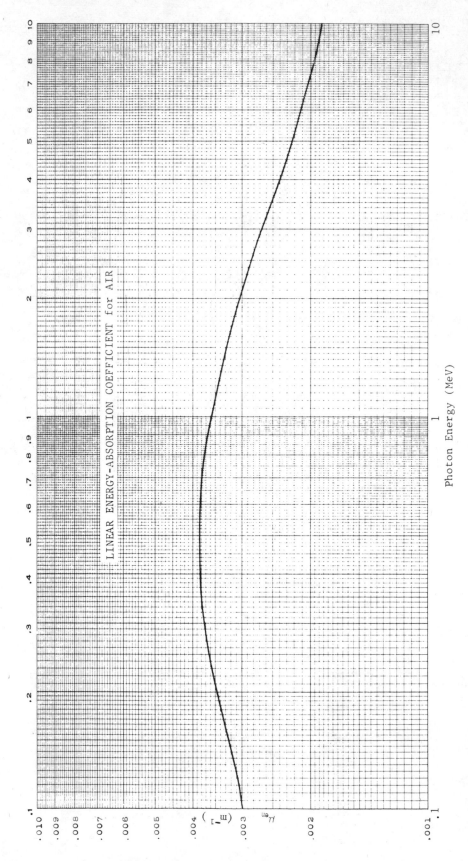

FIGURE 3.5-7.   Linear energy-absorption coefficient for air. (From Radiological Health Handbook, U.S. Department of Health, Education, and Welfare, Public Health Service, Rockville, Md., January 1970.

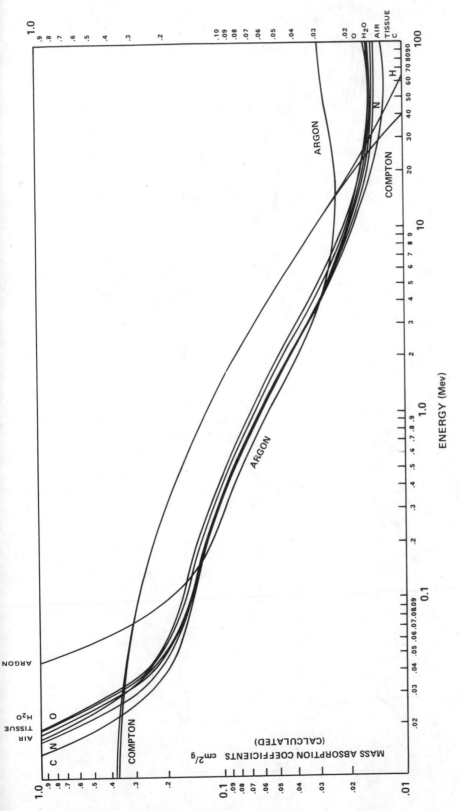

FIGURE 3.5-8.   Mass energy-absorption coefficients vs. photon energy.

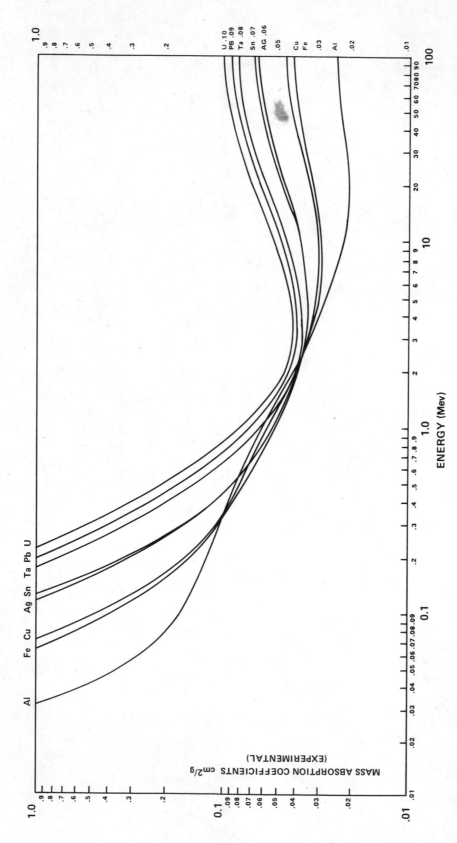

FIGURE 3.5-9. Curves for the heavier elements were shifted slightly to fit early data. The reader is cautioned to use the tables of more recent data for accurate calculations. These curves, however, present trends vs. Z and E for easy comparison and have been found to agree with most of the later experimental data. (from Snyder, W. S. and Powell, J. L., Absorption of Gamma Rays, ORNL-421, Series C, Suppl. 2 and 3, Oak Ridge National Laboratory, Oak Ridge, Tenn., 1948.)

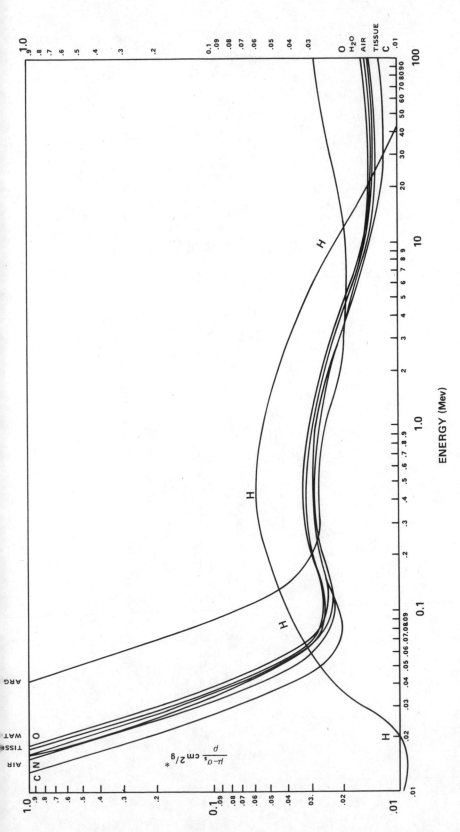

FIGURE 3.5-10. Curves for the heavier elements were shifted slightly to fit early data. The reader is cautioned to use the tables of more recent data for accurate calculations. These curves, however, present trends vs. Z and E for easy comparison and have been found to agree with most of the later experimental data. Note that this symbol (*) represents what is now commonly termed the mass-energy absorption coefficient. (From Snyder, W. S. and Powell, J. L., Absorption of Gamma Rays, ORNL-421, Series C, Suppl. 2 and 3, Oak Ridge National Laboratory, Oak Ridge, Tenn., 1948.)

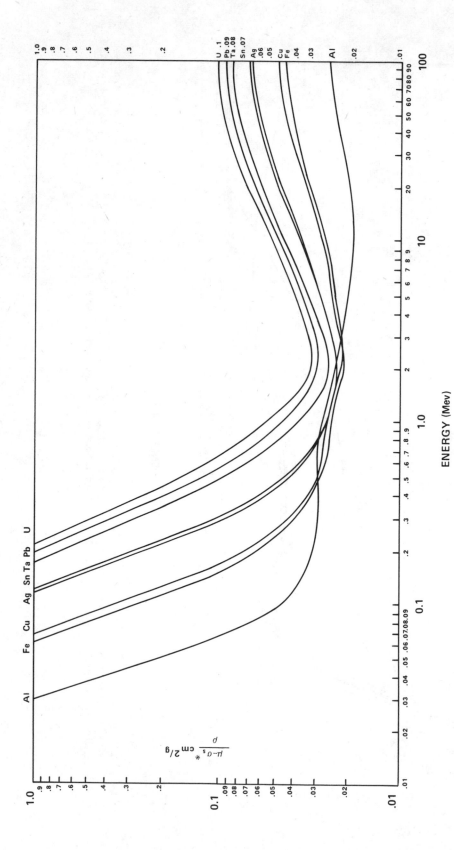

FIGURE 3.5-11.   Mass energy-absorption coefficients *vs.* photon energy.   Note that this symbol (*) represents what is now commonly termed the mass- energy absorption coefficient.

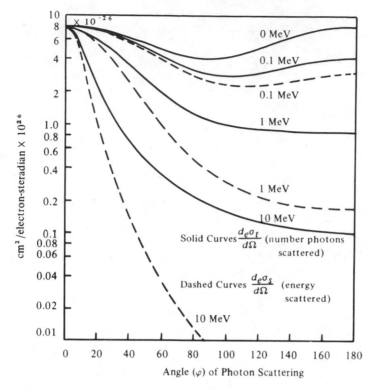

FIGURE 3.5-12. Graphs showing $d_e\sigma_t/d\Omega$ (solid curves) and $d_e\sigma_s/d\Omega$ (dashed curves) as a function of the angle of photon scattering, φ, for photon energies of 0, 0.1, 1.0, and 10 MeV. $d_e\sigma_t/d\Omega$ is the differential cross section per electron per unit solid angle for the number of photons scattered, and $d_e\sigma_s/d\Omega$ is the same cross section for the fraction of the energy scattered. (From Klein-Nishina equations as given in Hine, G. J. and Brownell, G. L., Eds., *Radiation Dosimetry*, 1st ed., Academic Press, New York, 1956. With permission.)

REFERENCES

1. **Hubbell, J. H.,** *Photon Cross Sections, Attenuation Coefficients, and Energy Absorption Coefficients from 10 keV to 100 GeV,* NSRDS-NBS-29, U.S. Government Printing Office, Washington, D.C., August 1969.

2. **Evans, R. D.,** *Radiation Dosimetry,* 3rd ed., Attix, F. H. and Roesch, W. C., Eds., Academic Press, New York, 1968.

3. **Johns, H. E. and Cunningham, J. R.,** *The Physics of Radiology,* 3rd ed., Charles C Thomas, Springfield, Ill., 1971, 735.

4. *Handbook of Chemistry and Physics,* 56th ed., Weast, R. C., Ed., CRC Press, Cleveland, 1975.

5. *Radiation Dosimetry,* 1st ed., Hine, G. J. and Brownell, G. L., Eds., Academic Press, New York, 1956.

6. *Radiological Health Handbook,* U.S. Department of Health, Education, and Welfare, Public Health Service, Rockville, Md., 1970.

7. **Snyder, W. S. and Powell, J. L.,** Absorption of Gamma Rays, ORNL-421, Series C, Suppl. 2 and 3, Oak Ridge National Laboratory, Oak Ridge, Tenn., 1948.

8. *Engineering Compendium on Radiation Shielding,* Vol. 1, Jaeger, R. G. et al., Eds., Springer-Verlag, New York, 1968, 183–184.

9. **Evans, R. D.,** *The Atomic Nucleus,* McGraw-Hill, New York, 1955.

10. **White, G. R.,** NBS Report No. 1003, National Bureau of Standards, Washington, D.C., 1952.

3.6. DECAY SCHEME DATA AND DOSE CONSTANTS FOR RADIONUCLIDES OF IMPORTANCE IN RADIOLOGICAL SCIENCE

L. T. Dillman

3.6.1. Decay Scheme Summary Data and Sources of Information on Nuclear Decay

3.6.1.1 Summary Decay Data

In this section some decay scheme data for nuclides of primary interest in radiation protection are summarized from several reviews. (The summary decay scheme data from the review of Lederer, Hollander, and Perlman,[1] as presented in the *Radiological Health Handbook*,[2] are not presented here. A new edition of the work of Lederer et al.[1] will be published in the near future. The reader should take note that there may be significant changes in some decay schemes from those presented in Reference 2, when the newer schemes become available.)

The most comprehensive and detailed decay scheme data compilation is found in *Nuclear Data Sheets*;[3] however, for purposes of radiation protection much of the information found here is superfluous. Furthermore, as explained below, certain vital information is missing; hence, data from MIRD pamphlet Number 10[4] and from the work of M. J. Martin,[5] where decay data in a form particularly useful for radiation protection are available, have been selected for inclusion in this section. Further information on original sources of these data and review data for nuclides not listed here may be located in the selected references given at the end of this section.

In making radiation dose calculations, it is necessary to determine the energy per disintegration deposited in a target region for a specified distribution of a radionuclide in a source region. Unfortunately, it is often the case that neither well-known decay scheme compilations such as References 1, 2, and 3 nor the original literature present decay scheme data in a form directly usable for radiation dose calculations. Relative intensities rather than absolute intensities are often given. Intensities and energies of secondary radiations, such as X-rays and Auger electrons, are often omitted. Internal conversion coefficient data are often incomplete. Several compilations that correct this difficulty have become recently available.

Pamphlet No. 10 of the Medical Internal Radiation Dose (MIRD) Committee of the Society of Nuclear Medicine[4] is one such compilation which includes energies and intensities of all photons and particles needed for internal dose calculations. Table 3.6-1* lists decay scheme data for selected radionuclides taken from Reference 4; for additional nuclides and a discussion of the method of deriving output data, the original source should be consulted.[4] The data in Table 3.6-1 briefly may be described as follows. Data from basic decay scheme compilations or from original literature are used as input in a computer program. The computer calculates the energies and intensities of all emitted radiations. This includes calculation of the average energy of β^+ or β^- particles, the energies and intensities of internal conversion electrons, and the energies and intensities of X-ray and Auger electrons. The nuclides, presented in Reference 4, having been selected by the MIRD Committee, primarily represent nuclides of interest in nuclear medicine. Calculations have since been carried out for additional nuclides by Dillman and others for estimating internal dose; thus, continuing reports of this group should be consulted for data on additional nuclides.

---

* All tables appear at the end of the text.

Summary data on nuclear decay schemes as well as equilibrium dose constants have also been provided for 194 radioactive nuclides by Martin.[5,*] Data for selected nuclides of interest in radiation protection have been included in Table 3.6-2. Some of these nuclides are the same as those selected for Table 3.6-1, so that a comparison of results from the two methods can be made for nuclides of greater importance. Although the original source should again be consulted if a more detailed description of the method of calculation is required, the following concise review of decay processes, abstracted from Martin,[5] is for the user's convenience.

3.6.1.2. Review of Radiation Processes
3.6.1.2.1. Decay by Alpha Emission (Alpha (α-) Decay)

In α-decay, a nucleus with atomic number Z and mass number A emits an α-particle (^{4}He nucleus, Z = 2, A = 4) and decays to a nucleus with atomic number (Z − 2) and mass number (A − 4).

For decay to a particular level in the daughter nucleus, the maximum energy available for the α-particle is

$$E_\alpha = Q_\alpha - E_R - E(\text{level}) = \frac{M}{M + M_\alpha} [Q_\alpha - E(\text{level})] = \frac{M}{M + M_\alpha} Q'_\alpha \qquad (3.6\text{-}1)$$

where Q_a is the atomic mass difference between ground states of parent and daughter, M_a and M are the masses of the α-particle and daughter atom, respectively, and $E_R = M_a Q'/(M_a + M)$ is the energy carried off by the recoil of the daughter atom.

3.6.1.2.2. Beta-minus (β⁻) Decay

In β⁻-decay, an antineutrino (ν) and a negative electron (β⁻) are emitted from the nucleus as a result of the process n→ p + β⁻ + ν. This decay increases the nuclear charge by one unit.

The energy released in a single β-transition is divided between the β-particle and the antineutrino in a statistical manner, so that when a large number of transitions are considered, both the antineutrinos and β-particles are found to have energy distributions extending from zero up to some maximum value. For decay to a particular level in the daughter nucleus, the maximum energy available is $E_{max} = Q^- - E(\text{level})$, where Q^- is the energy equivalent of the atomic mass difference between ground states of parent and daughter. The average energy of a β-particle in this transition is given by:

$$E_{av} = \int_0^{E_{max}} E \times N(E)dE / \int_0^{E_{max}} N(E)dE \qquad (3.6\text{-}2)$$

where N(E) is the number of β-particles with energy between E and E + dE. Thus the β⁻-particles from a particular transition constitute a "group" having a continuous distribution of energies with a definite E_{max} and E_{av}.

3.6.1.2.3. Beta-plus (β⁺-) Decay

In β⁺-decay, a neutrino (ν) and a positive electron (β⁺) (positron) are emitted from the nucleus as a result of the process p→ n + β⁺ + ν. This decay decreases nuclear charge by one unit.

As in β⁻-decay (see above), the β⁺-particles emitted in a transition to a particular level in the daughter nucleus have a continuous distribution of energies with a defi-

---

* The data in Reference 5 were prepared from the Evaluated Nuclear Structure Data File of the Nuclear Data Project, W. B. Ewbank, Director. Data given in Table 3.6-2 were prepared by Martin in June 1977; thus, some of the data may vary from those given in ORNL-5114 (Reference 5) due to recent updates of data.

nite E_{max} and E_{av}. For positron decay, $E_{max} = Q^+ - 2mc^2 - E(level)$, where Q^+ is the energy equivalent of the atomic mass difference between ground states of parent and daughter. Note that β^+-decay to a particular level cannot occur unless $Q^+ - E(level) > 2mc^2$ ($2mc^2 = 1022$ keV).

3.6.1.2.4. Electron-capture (ε-) Decay

In electron-capture decay, an atomic electron is captured by a nucleus and a neutrino is emitted as a result of the process $p + e^- \rightarrow n + \nu$. This decay decreases nuclear charge by one unit and leaves the daughter nucleus with a vacancy in one of its atomic shells. K-shell electron capture, for example, refers to a capture process in which the final-state vacancy is in the K-shell. For a K-shell electron-capture branch to a particular level, the energy released is given by $Q^+ - E(level) - E_K$, where E_K is the K-shell binding energy in the daughter atom.

The electron-capture process always competes with β^+-decay, but also can occur when the transition energy is too small to permit β^+-emission, that is, when $Q^+ - E(level) < 1022$ keV.

3.6.1.2.5. Gamma (γ-) Decay

Electromagnetic radiation is emitted by a nucleus in transition from a higher to a lower energy state. A numeric index associated with γ, as in $\gamma 1$, denotes a γ-ray emitted in the transition between a particular pair of nuclear levels. The energy of this gamma ray $E(\gamma 1)$ is equal to the energy difference between the two levels (except for a negligible amount of nuclear recoil energy, $E_{recoil}(keV) \simeq 5.4 \times 10^{-7} E^2(\gamma)/A$, where A is the mass number and $E(\gamma)$ is in keV).

3.6.1.2.6. Internal-conversion Electron Emission (ce)

An atomic electron can be emitted as an alternative to γ-ray emission in the transition of a nucleus from a higher to a lower energy state. In the internal-conversion process, the energy difference between these states is transferred directly to a bound atomic electron which is then ejected from the atom. An alphabetic index refers to the shell from which the atomic electron is ejected; thus, in Table 3.6-2, ce-K denotes K-shell conversion and ce-MNO denotes conversion in the M-, N-, and O-shells combined. A numeric index, as in ce-K-1, has a meaning analogous to that used above for γ-rays. For a transition with energy $E(\gamma 1)$, the K-shell internal-conversion electron is emitted with energy $E(ce-K-1) = E(\gamma 1) - E_K$, where E_K is the K-shell binding energy. The energy of an electron converted in one of the other shells is given by the same expression with the replacement of E_K by the binding energy appropriate to that shell. Similar notation is used in Table 3.6-1. For example, K INT CON ELECT denote K-shell internal-conversion. In this table all conversion electron entries immediately follow the γ-transitions with which they are associated; therefore, no numeric index is used for correlation purposes.

For a particular transition, the ratio of probability for emission of a K-conversion electron to that for emission of a γ-ray, $I(ce-K)/I(\gamma)$, is called the K-shell conversion coefficient for that transition. Conversion coefficients for the other shells are defined in an analogous manner.

3.6.1.2.7. X-Ray and Auger-electron Emission (X,e_A)

Whenever a vacancy is produced in an inner electron shell of an atom, the filling of this vacancy is accompanied by either X-ray or Auger-electron emission. Vacancies created by the filling of the initial vacancy will in turn produce further X-rays or Auger electrons. This cascade of radiation continues until all vacancies have been transferred to the outermost electron shell. Inner-shell vacancies are always produced in two types of nuclear decay, electron capture and internal conversion. Other processes which produce electron vacancies following a nuclear decay, such as electron shakeoff (ejection of one or more atomic electrons due to a sudden change in nuclear charge) or ejection of atomic electrons by escaping α- or β-particles, are not discussed here because of the low probability of their occurrence.

Fluorescence yield (ω) — The fluorescence yield for a particular atomic shell ($\omega_K, -\omega_L$, etc.) is defined as the probability that a vacancy in that shell will result in the emission of an X-ray. If n_K is the number of K-shell vacancies per disintegration, the number of K-X-rays will be $n_K \omega_K$ and the number of K-Auger electrons will be $n_K(1-\omega_K)$.

The most recent and thorough review of fluorescence yields is that of Bambynek et al.[6] K-shell fluorescence yields ω_K, adopted in computing results given in Tables 3.6-1 and 3.6-2 are the "fitted values" from Table III-V of Reference 6 above. Average L-shell fluorescence yields, $-\omega_L$, with uncertainties, have been estimated from Figure 4-34 of this same work. It should be noted that use of average L-shell fluorescence yields involves basic approximations that can conceivably lead to rather large errors in certain circumstances. Namely, average L-shell fluorescence yield depends upon the initial vacancy distribution among the L-subshells, which, in turn, varies considerably depending upon whether one is considering electron-capture or internal-conversion processes. The problem is further compounded by the existence of Coster-Kronig transitions between L-subshells; thus, average L-shell fluorescence yield, an average of a limited number of experimental points, may be considerably in error in some cases. However, at the time the data of Tables 3.6-1 and 3.6-2 were calculated, it was the best that could be done. More recent data on L-subshell fluorescence yields and Coster-Kronig yields are correcting this situation.

X-Rays An X-ray emitted as a result of the filling of a K-shell vacancy by an electron from a higher shell, for example, the Y-shell, has an energy $E_K - E_Y$, where E_K and E_Y are the electron-binding energies in the K- and Y- shells, respectively. This transition can be denoted by K − Y. In order of decreasing intensity, the most important transitions are

$$\begin{aligned}
K_{\alpha 1} &= K - L_3 & K_{\beta 2} &= K - N_3 \\
K_{\alpha 2} &= K - L_2 & K_{\beta 4} &= K - N_2 \\
K_{\beta 1} &= K - M_3 & K_{\beta 5} &= K - M_4 \\
K_{\beta 3} &= K - M_2 & &+ K - M_5
\end{aligned}$$

Energies and intensities for $K_{\alpha 1}$ and $K_{\alpha 2}$ lines and the $K_\beta = \Sigma K_{\beta i}$ group are given in Table 3.6-2. Individual K_β components are given in Table 3.6-1. X-Ray energies for both tables are taken from Bearden and Burr.[7] Intensity ratios, K_β/K_α and $K_{\alpha 2}/K_{\alpha 1}$ are taken from Rao et al.[8] for Table 3.6-2 and from Wapstra et al.[9] for Table 3.6-1. As already mentioned, the number of K-X-rays per disintegration is $n_K \omega_K$; thus,

$$\text{Number of } K_{\alpha 1}\text{-X-ray} = \frac{n_K \omega_K}{(1 + K_\beta/K_\alpha)(1 + K_{\alpha 2}/K_{\alpha 1})} \tag{3.6-3}$$

with similar expressions for $K_{\alpha 2}$ and K_β. The number of K-vacancies per disintegration is the sum of the vacancies produced by K-capture and those produced by internal conversion in the K-shell; thus,

$$n_K = \epsilon_K + I(\text{ce-K}), \tag{3.6-4}$$

where ϵ_K is the total number of K-captures per disintegration, and $I(\text{ce-K})$ is the total number of K-shell internal-conversion electrons per disintegration.

As in the case of the K-shell, many transitions contribute to the L-X-ray (X_L) spectrum. However, since the relative intensities of these transitions are not known for all Z-values and since the energy differences between the strong transitions are small ($\leqslant$ 3 keV), the total L-X-ray radiation is treated in Table 3.6-2 as a single group with the energy of the strongest transition, $L_{a1} = L_3 - M_5$. For Z < 37, the M_5-transitions have not been resolved experimentally from the M_4-transitions; in these cases, the energy given is that for the transition $L_{a1,a2} = L_3 - M_4,M_5$. In Table 3.6-1, for Z $\geqslant$ 70, there is an additional breakdown of L-X-rays into L_a, L_β, and L_γ components. Calculation of the number of L-X-rays per disintegration, $n_L\omega_L$, is similar to that for K-X-rays, except that in addition to vacancies produced by direct L-shell capture and by L-shell conversion, those created by the transfer of L-shell electrons to K-shell vacancies must also be included. Thus,

$$n_L = \epsilon_L + I(ce\text{-}L) + n_{KL} \times n_K \qquad (3.6\text{-}5)$$

where n_{KL} is the number of L-shell vacancies created per K-shell vacancy, and the other symbols have meanings analogous to those used above for the K-shell. For Table 3.6-2, values of n_{KL} are taken from Reference 8. At the time the data in Table 3.6-1 were calculated, tabulated values of n_{KL} were not available, so a slightly different approach was used. The quantity $n_{KL} n_K$ was determined by using information on ω_K and K_a/K_β X-ray ratios from Wapstra et al.[9] Information on KLL/KLX/KXY Auger electron ratios was determined from the data of Figure 8 in Reference 10.

Auger Electrons (e_A) — The Auger process competes with emission of X-rays as a means of carrying off energy released in the filling of an inner-shell vacancy by an electron from an outer shell. In this process, the filling of an inner-shell vacancy is accompanied by the simultaneous ejection of an outer-shell electron to the continuum. The resulting atom is left with two vacancies. The Auger-electron yield per disintegration of a particular atomic shell is $n_K(1 - \omega_K)$, $n_L(1 - -\omega_L)$ etc. If the original vacancy is in the K-shell and if this vacancy is filled by an electron from shell X with the ejection of an electron from shell Y, the transition is denoted by KXY. The energy of the ejected electron is $E_K - E_X - E_{Y}'$, where E_K and E_X are neutral-atom K- and X-shell binding energies, respectively, and E_{Y}', is the binding energy of a Y-shell electron in an atom containing a vacancy in the X-shell. The most intense K-Auger transitions are of the type KLL. Since the relative intensities of electrons in the KLL group are not accurately known for all Z-values and since the energy difference between transitions is small ($\leqslant$5 keV), the K-Auger electrons are treated in Table 3.6-2 as a single group with the energy of the strongest transitions, KL_2L_3. The energy of this transition as a function of atomic number is taken from Table 1 of Bergstrom and Nordling.[10] In Table 3.6-1, a weighted average energy, based on approximate relative intensities of the six major KLL subgroups, is given for the KLL-Auger electron group. Figure 3-5 of Reference 6 provides an example of a K-Auger electron spectrum. In the case of the L-Auger process, very little is known about the energies or relative intensities of the individual L-Auger electrons. Consequently, L-Auger electrons are treated as a single group with the energy appropriate to an $L_3M_4M_5$-transition. Figure 4-13 of Reference 6 is an example of an L-Auger electron spectrum.

3.6.1.2.8. Bremsstrahlung

In addition to any monoenergetic X-rays or γ-rays that may be present in radioactive decay, every β-decay produces continuous electromagnetic radiation called bremsstrahlung. Two processes contribute to this continuous spectrum.

"**External**" **bremsstrahlung** — This type of radiation is produced by collisions between βs (and conversion electrons) and the atoms of the material surrounding the radiating atoms. Since the intensity of the external bremsstrahlung depends on the atomic number Z of the surrounding material, this radiation is not included in the present tables. In many cases, however, this radiation cannot be neglected in dose calculations. An approximate value for the average energy of the external bremsstrahlung associated with a β-group of maximum energy E_β (keV) (thick target approximation) is given by Evans as[11]

$$E_{av} \approx 1.4 \times 10^{-7} \ Z \ E_\beta^2 \ \text{keV per } \beta \tag{3.6-6}$$

In the decay of ^{32}P, where $E_\beta = 1711$ keV, the average energy per disintegration of the external bremsstrahlung is ≈ 0.4 Z keV.

"**Internal**" **bremsstrahlung** — This type of radiation, originating within a decaying atom, is produced by the sudden change of nuclear charge which occurs in β^+-, β^--, or ε-decay. This radiation is not included here because of its low intensity and low average energy. The average energy of the internal bremsstrahlung associated with a β-group of maximum energy E_β (keV) is given for $E_\beta \gg mc^2 (511$ keV) by Schopper.[12]

$$E_{av} \approx 1.5 \times 10^{-3} \ E_\beta \ \log(0.004 E_\beta - 2.2) \ \text{keV per } \beta \tag{3.6-7}$$

In electron capture, the corresponding expression for a transition with energy E_ε (keV) is[13]

$$E_{av} \approx 1.5 \times 10^{-7} \ E_\varepsilon^2 \ \text{keV per capture} \tag{3.6-8}$$

In the decay of ^{32}P, where $E_\beta = 1711$ keV, the average internal bremsstrahlung energy per disintegration is ≈ 1.7 keV.

Table 3.6-1
SUMMARY NUCLIDE DECAY DATA AND DOSE CONSTANT DATA

Index of Nuclides Included in Table

3	H	1	54	Mn	25	99m	Tc	43	169	Yb	70
14	C	6	55	Fe	26	109	Cd	48	198	Au	79
13	N	7	59	Fe	26	113m	In	49	197	Hg	80
15	O	8	58	Co	27	123	I	53	203	Hg	80
18	F	9	60	Co	27	125	I	53	208	Tl	81
22	Na	11	64	Cu	28	129	I	53	212	Pb	82
24	Na	11	65	Zn	29	131	I	53	212	Bi	83
32	P	15	67	Ga	31	132	I	53	212	Po	84
35	S	16	75	Se	34	133	I	53	216	Po	84
40	K	19	85	Kr	36	133	Xe	54	220	Rn	86
42	K	19	85m	Kr	36	133m	Xe	54	224	Ra	88
45	Ca	20	85	Sr	38	134	Cs	55	241	Am	95
47	Ca	20	90	Sr	38	137	Cs	55			
49	Sc	21	90	Y	39	133	Ba	56			
52	Mn	25	99	Mo	42	137m	Ba	56			

From Dillman, L. T. and von der Lage, F. C., *Radionuclide Decay Schemes and Nuclear Parameters for Use in Radiation-dose Estimation*, NM/MIRD Pamphlet No. 10, Society of Nuclear Medicine, New York, September 1975. With permission.

Table 3.6-1 (continued)
SUMMARY NUCLIDE DECAY DATA AND DOSE CONSTANT DATA

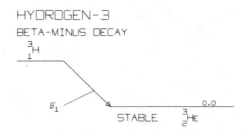

```
HYDROGEN-3
BETA-MINUS DECAY
   3
    H
   1

  β⁻₁                              0.0
          STABLE    3        HE
                    2
```

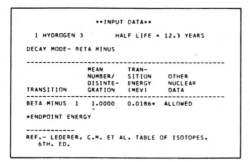

```
            **INPUT DATA**

  1 HYDROGEN 3      HALF LIFE = 12.3 YEARS

DECAY MODE- BETA MINUS

--------------------------------------------
               MEAN     TRAN-
               NUMBER/  SITION   OTHER
               DISINTE- ENERGY   NUCLEAR
TRANSITION     GRATION  (MEV)    DATA
--------------------------------------------
BETA MINUS  1  1.0000   0.0186*  ALLOWED

*ENDPOINT ENERGY

-------------
REF.- LEDERER, C.M. ET AL, TABLE OF ISOTOPES,
      6TH. ED.
```

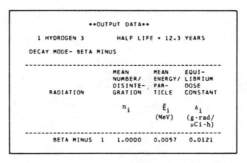

```
            **OUTPUT DATA**

  1 HYDROGEN 3      HALF LIFE = 12.3 YEARS

DECAY MODE- BETA MINUS

--------------------------------------------
               MEAN     MEAN     EQUI-
               NUMBER/  ENERGY/  LIBRIUM
               DISINTE- PAR-     DOSE
RADIATION      GRATION  TICLE    CONSTANT

               nᵢ       Ēᵢ       Δᵢ
                        (MeV)    (g-rad/
                                 µCi-h)
--------------------------------------------
  BETA MINUS  1  1.0000  0.0057  0.0121
```

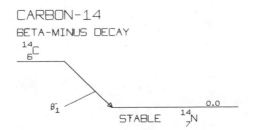

```
CARBON-14
BETA-MINUS DECAY
   14
     C
   6

  β⁻₁                              0.0
          STABLE    14
                     7 N
```

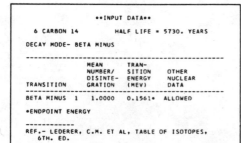

```
            **INPUT DATA**

  6 CARBON 14       HALF LIFE = 5730. YEARS

DECAY MODE- BETA MINUS

--------------------------------------------
               MEAN     TRAN-
               NUMBER/  SITION   OTHER
               DISINTE- ENERGY   NUCLEAR
TRANSITION     GRATION  (MEV)    DATA
--------------------------------------------
BETA MINUS  1  1.0000   0.1561*  ALLOWED

*ENDPOINT ENERGY

-------------
REF.- LEDERER, C.M. ET AL, TABLE OF ISOTOPES,
      6TH. ED.
```

```
            **OUTPUT DATA**

  6 CARBON 14       HALF LIFE = 5730. YEARS

DECAY MODE- BETA MINUS

--------------------------------------------
               MEAN     MEAN     EQUI-
               NUMBER/  ENERGY/  LIBRIUM
               DISINTE- PAR-     DOSE
RADIATION      GRATION  TICLE    CONSTANT

               nᵢ       Ēᵢ       Δᵢ
                        (MeV)    (g-rad/
                                 µCi-h)
--------------------------------------------
  BETA MINUS  1  1.0000  0.0493  0.1050
```

Table 3.6-1 (continued)
SUMMARY NUCLIDE DECAY DATA AND DOSE CONSTANT DATA

NITROGEN-13
BETA-PLUS DECAY

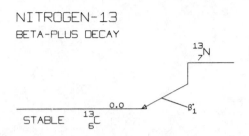

```
            ••INPUT DATA••

  7 NITROGEN 13      HALF LIFE = 10.0 MINUTES

DECAY MODE- BETA PLUS
------------------------------------------------
             MEAN      TRAN-
             NUMBER/   SITION
             DISINTE-  ENERGY   OTHER
             GRATION   (MEV)    NUCLEAR
TRANSITION                      DATA
------------------------------------------------
  BETA PLUS  1  1.0000  1.1900•  ALLOWED

•ENDPOINT ENERGY

------------
REF.- LEDERER, C.M. ET AL, TABLE OF ISOTOPES,
      6TH. ED.
```

```
            ••OUTPUT DATA••

  7 NITROGEN 13      HALF LIFE = 10.0 MINUTES

DECAY MODE- BETA PLUS
------------------------------------------------
                  MEAN     MEAN     EQUI-
                  NUMBER/  ENERGY/  LIBRIUM
                  DISINTE- PAR-     DOSE
   RADIATION      GRATION  TICLE    CONSTANT

                    n_i     Ē_i      Δ_i
                           (MeV)    (g-rad/
                                    μCi-h)
------------------------------------------------
  BETA PLUS  1    1.0000   0.4880   1.0395
ANNIH. RADIATION  2.0000   0.5110   2.1768
```

OXYGEN-15
BETA-PLUS DECAY

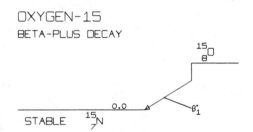

```
            ••INPUT DATA••

  8 OXYGEN 15       HALF LIFE = 124. SECONDS

DECAY MODE- BETA PLUS
------------------------------------------------
             MEAN      TRAN-
             NUMBER/   SITION
             DISINTE-  ENERGY   OTHER
             GRATION   (MEV)    NUCLEAR
TRANSITION                      DATA
------------------------------------------------
  BETA PLUS  1  1.0000  1.7000•  ALLOWED

•ENDPOINT ENERGY

------------
REF.- LEDERER, C.M. ET AL, TABLE OF ISOTOPES,
      6TH. ED.
```

```
            ••OUTPUT DATA••

  8 OXYGEN 15       HALF LIFE = 124. SECONDS

DECAY MODE- BETA PLUS
------------------------------------------------
                  MEAN     MEAN     EQUI-
                  NUMBER/  ENERGY/  LIBRIUM
                  DISINTE- PAR-     DOSE
   RADIATION      GRATION  TICLE    CONSTANT

                    n_i     Ē_i      Δ_i
                           (MeV)    (g-rad/
                                    μCi-h)
------------------------------------------------
  BETA PLUS  1    1.0000   0.7206   1.5349
ANNIH. RADIATION  2.0000   0.5110   2.1768
```

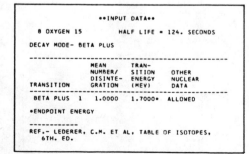

Table 3.6-1 (continued)
SUMMARY NUCLIDE DECAY DATA AND DOSE CONSTANT DATA

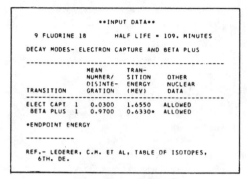

```
                    **INPUT DATA**

   9 FLUORINE 18        HALF LIFE = 109. MINUTES

DECAY MODES- ELECTRON CAPTURE AND BETA PLUS

---------------------------------------------------
                   MEAN       TRAN-
                   NUMBER/    SITION
                   DISINTE-   ENERGY    OTHER
                   GRATION    (MEV)     NUCLEAR
TRANSITION                              DATA
---------------------------------------------------
ELECT CAPT  1      0.0300     1.6550    ALLOWED
BETA PLUS   1      0.9700     0.6330*   ALLOWED

*ENDPOINT ENERGY

-----------
REF.- LEDERER, C.M. ET AL, TABLE OF ISOTOPES,
      6TH. DE.
```

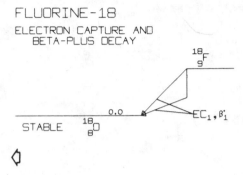

FLUORINE-18
ELECTRON CAPTURE AND
   BETA-PLUS DECAY

```
                    **OUTPUT DATA**

   9 FLUORINE 18        HALF LIFE = 109. MINUTES

DECAY MODES- ELECTRON CAPTURE AND BETA PLUS

---------------------------------------------------
                   MEAN       MEAN      EQUI-
                   NUMBER/    ENERGY/   LIBRIUM
                   DISINTE-   PAR-      DOSE
   RADIATION       GRATION    TICLE     CONSTANT

                     n_i       Ē_i       Δ_i
                               (MeV)    (g-rad/
                                         μCi-h)
---------------------------------------------------
       BETA PLUS 1  0.9700    0.2496    0.5157
  ANNIH. RADIATION  1.9400    0.5110    2.1115
```

```
                    **INPUT DATA**

  11 SODIUM 22         HALF LIFE = 2.60 YEARS

DECAY MODES- ELECTRON CAPTURE AND BETA PLUS

---------------------------------------------------
                   MEAN       TRAN-
                   NUMBER/    SITION
                   DISINTE-   ENERGY    OTHER
                   GRATION    (MEV)     NUCLEAR
TRANSITION                              DATA
---------------------------------------------------
ELECT CAPT  1      0.0940     1.5680    ALLOWED
    GAMMA   1      1.0000     1.2746    E2,  AK(T) =
                                         0.000006
BETA PLUS   1      0.9060     0.5460*   ALLOWED
BETA PLUS   2      0.0006     1.8210*   U 2ND FORBIDDEN

*ENDPOINT ENERGY

-----------
REF.- VATAI, E. AND VARGA, D., NUCL. PHYS.
      A116, 637 (1968).
      LEUTZ, H. AND WENNINGER, H., NUCL. PHYS.
      A99, 55 (1967).
```

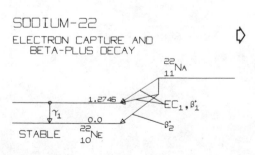

SODIUM-22
ELECTRON CAPTURE AND
   BETA-PLUS DECAY

```
                    **OUTPUT DATA**

  11 SODIUM 22         HALF LIFE = 2.60 YEARS

DECAY MODES- ELECTRON CAPTURE AND BETA PLUS

---------------------------------------------------
                   MEAN       MEAN      EQUI-
                   NUMBER/    ENERGY/   LIBRIUM
                   DISINTE-   PAR-      DOSE
   RADIATION       GRATION    TICLE     CONSTANT

                     n_i       Ē_i       Δ_i
                               (MeV)    (g-rad/
                                         μCi-h)
---------------------------------------------------
          GAMMA   1  0.9999   1.2746    2.7148
 KLL AUGER ELECT     0.0753   0.0008    0.0001
     BETA PLUS   1   0.9060   0.2157    0.4163
     BETA PLUS   2   0.0006   0.8356    0.0010
 ANNIH. RADIATION    1.8132   0.5110    1.9735
```

Table 3.6-1 (continued)
SUMMARY NUCLIDE DECAY DATA AND DOSE CONSTANT DATA

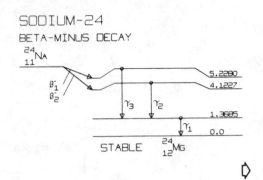

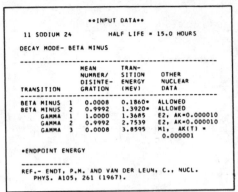

```
                    **INPUT DATA**

   11 SODIUM 24         HALF LIFE = 15.0 HOURS

DECAY MODE- BETA MINUS

-----------------------------------------------------
                 MEAN      TRAN-
                 NUMBER/   SITION    OTHER
                 DISINTE-  ENERGY    NUCLEAR
TRANSITION       GRATION   (MEV)     DATA
-----------------------------------------------------
BETA MINUS  1    0.0008    0.1860*   ALLOWED
BETA MINUS  2    0.9992    1.3920*   ALLOWED
     GAMMA  1    1.0000    1.3685    E2, AK=0.000010
     GAMMA  2    0.9992    2.7539    E2, AK=0.000010
     GAMMA  3    0.0008    3.8595    M1, AK(T) =
                                        0.000001

*ENDPOINT ENERGY
------------
REF.- ENDT, P.M. AND VAN DER LEUN, C., NUCL.
   PHYS. A105, 261 (1967).
```

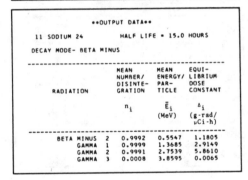

```
                    **OUTPUT DATA**

   11 SODIUM 24         HALF LIFE = 15.0 HOURS

DECAY MODE- BETA MINUS

-----------------------------------------------------
                 MEAN      MEAN      EQUI-
                 NUMBER/   ENERGY/   LIBRIUM
                 DISINTE-  PAR-      DOSE
RADIATION        GRATION   TICLE     CONSTANT

                  n_i       Ē_i       Δ_i
                           (MeV)     (g-rad/
                                     μCi-h)
-----------------------------------------------------
BETA MINUS  2    0.9992    0.5547    1.1805
     GAMMA  1    0.9999    1.3685    2.9149
     GAMMA  2    0.9991    2.7539    5.8610
     GAMMA  3    0.0008    3.8595    0.0065
```

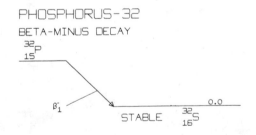

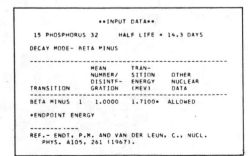

```
          **INPUT DATA**

  15 PHOSPHORUS 32    HALF LIFE = 14.3 DAYS

DECAY MODE- BETA MINUS

-------------------------------------------
                MEAN     TRAN-
                NUMBER/  SITION   OTHER
                DISINTE- ENERGY   NUCLEAR
TRANSITION      GRATION  (MEV)    DATA
-------------------------------------------
BETA MINUS  1   1.0000   1.7100*  ALLOWED

*ENDPOINT ENERGY
---------- ----
REF.- ENDT, P.M. AND VAN DER LEUN, C., NUCL.
   PHYS. A105, 261 (1967).
```

```
          **OUTPUT DATA**

  15 PHOSPHORUS 32    HALF LIFE = 14.3 DAYS

DECAY MODE- BETA MINUS

-------------------------------------------
                MEAN     MEAN     EQUI-
                NUMBER/  ENERGY/  LIBRIUM
                DISINTE- PAR-     DOSE
RADIATION       GRATION  TICLE    CONSTANT

                 n_i      Ē_i      Δ_i
                         (MeV)    (g-rad/
                                  μCi-h)
-------------------------------------------
BETA MINUS  1   1.0000   0.6948   1.4799
```

Table 3.6-1 (continued)
SUMMARY NUCLIDE DECAY DATA AND DOSE CONSTANT DATA

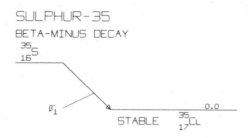

SULPHUR-35

BETA-MINUS DECAY

INPUT DATA

```
      **INPUT DATA**

   16 SULPHUR 35        HALF LIFE = 87.0 DAYS

   DECAY MODE- BETA MINUS

   ---------------------------------------------
                 MEAN      TRAN-
                 NUMBER/   SITION   OTHER
                 DISINTE-  ENERGY   NUCLEAR
   TRANSITION    GRATION   (MEV)    DATA
   ---------------------------------------------
   BETA MINUS 1  1.0000    0.1674*  ALLOWED

   *ENDPOINT ENERGY

   ------------
   REF.- ENDT, P.M. AND VAN DER LEUN, C., NUCL.
         PHYS. A105, 261 (1967).
```

```
      **OUTPUT DATA**

   16 SULPHUR 35        HALF LIFE = 87.0 DAYS

   DECAY MODE- BETA MINUS

   ---------------------------------------------
                 MEAN      MEAN     EQUI-
                 NUMBER/   ENERGY/  LIBRIUM
                 DISINTE-  PAR-     DOSE
   RADIATION     GRATION   TICLE    CONSTANT

                 n_i       Ē_i      Δ_i
                           (MeV)    (g-rad/
                                    µCi-h)
   ---------------------------------------------
   BETA MINUS 1  1.0000    0.0488   0.1039
```

POTASSIUM-40

ELECTRON CAPTURE AND
BETA-MINUS DECAY

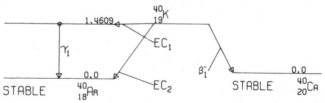

```
      **INPUT DATA**

   19 POTASSIUM 40      HALF LIFE = 1270. MEGAYEAR

   DECAY MODES- BETA MINUS AND ELECTRON CAPTURE

   -----------------------------------------------
                 MEAN      TRAN-
                 NUMBER/   SITION   OTHER
                 DISINTE-  ENERGY   NUCLEAR
   TRANSITION    GRATION   (MEV)    DATA
   -----------------------------------------------
   BETA MINUS  1  0.8951   1.3000*  U 2ND FORBIDDEN
   ELECT CAPT  1  0.1033   0.0410   U 1ST FORBIDDEN
   ELECT CAPT  2  0.0016   1.5010   U 2ND FORBIDDEN
   GAMMA       1  0.1033   1.4609   E2, AK(T) =
                                    0.000027

   *ENDPOINT ENERGY

   ------------
   REF.- MC CANN, M.F. ET AL, NUCL. PHYS. A98,
         577 (1967).
         YOSHIZAWA, Y. ET AL, J. NUCL. SCI. AND
         TECH. (TOKYO) 5, 432 (1968).
         RECHINSALE, R.D. AND GALE, N.H., EARTH AND
         PLANETARY SCI. LETT. 6, 289 (1969).
```

```
      **OUTPUT DATA**

   19 POTASSIUM 40      HALF LIFE = 1270. MEGAYEAR

   DECAY MODES- BETA MINUS AND ELECTRON CAPTURE

   -----------------------------------------------
                    MEAN      MEAN     EQUI-
                    NUMBER/   ENERGY/  LIBRIUM
                    DISINTE-  PAR-     DOSE
   RADIATION        GRATION   TICLE    CONSTANT

                    n_i       Ē_i      Δ_i
                              (MeV)    (g-rad/
                                       µCi-h)
   -----------------------------------------------
   BETA MINUS   1   0.8951    0.5555   1.0591
   GAMMA        1   0.1032    1.4609   0.3214
   KLL AUGER ELECT  0.0606    0.0026   0.0003
```

Table 3.6-1 (continued)
SUMMARY NUCLIDE DECAY DATA AND DOSE CONSTANT DATA

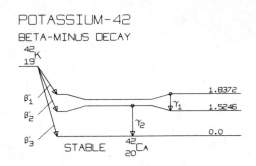

POTASSIUM-42
BETA-MINUS DECAY

```
**INPUT DATA**

 19 POTASSIUM 42     HALF LIFE = 12.4 HOURS

DECAY MODE- BETA MINUS

-----------------------------------------
              MEAN     TRAN-
              NUMBER/  SITION   OTHER
              DISINTE- ENERGY   NUCLEAR
TRANSITION    GRATION  (MEV)    DATA
-----------------------------------------
BETA MINUS 1  0.0018   1.6800*  U 1ST FORBIDDEN
BETA MINUS 2  0.1800   2.0000*  1ST FORBIDDEN
BETA MINUS 3  0.8200   3.5200*  U 1ST FORBIDDEN
   GAMMA   1  0.0018   0.3126   E2, AK(T) =
                                  0.00324
   GAMMA   2  0.1800   1.5246   E2, AK(T) =
                                  0.000034

*ENDPOINT ENERGY

-------------
REF.- KAWADE, K. ET AL, J. PHYS. SOC. JAPAN 29,
  NO. 1, JULY, 1970.
```

```
**OUTPUT DATA**

 19 POTASSIUM 42     HALF LIFE = 12.4 HOURS

DECAY MODE- BETA MINUS

-----------------------------------------
              MEAN     MEAN     EQUI-
              NUMBER/  ENERGY/  LIBRIUM
              DISINTE- PAR-     DOSE
RADIATION     GRATION  TICLE    CONSTANT

                n_i     Ē_i      Δ_i
                       (MEV)    (g-rad/
                                 μCi-h)
-----------------------------------------
BETA MINUS 1  0.0018   0.6994   0.0026
BETA MINUS 2  0.1800   0.8243   0.3160
BETA MINUS 3  0.8200   1.5634   2.7306
   GAMMA   1  0.0017   0.3126   0.0011
   GAMMA   2  0.1799   1.5246   0.5845
```

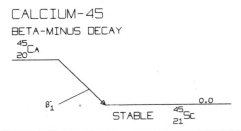

CALCIUM-45
BETA-MINUS DECAY

```
**INPUT DATA**

 20 CALCIUM 45     HALF LIFE = 163. DAYS

DECAY MODE- BETA MINUS

-----------------------------------------
              MEAN     TRAN-
              NUMBER/  SITION   OTHER
              DISINTE- ENERGY   NUCLEAR
TRANSITION    GRATION  (MEV)    DATA
-----------------------------------------
BETA MINUS 1  0.9999   0.2570*  ALLOWED

*ENDPOINT ENERGY

-------------
REF.- NUCLEAR DATA B4, 253 (1970).
```

```
**OUTPUT DATA**

 20 CALCIUM 45     HALF LIFE = 163. DAYS

DECAY MODE- BETA MINUS

-----------------------------------------
              MEAN     MEAN     EQUI-
              NUMBER/  ENERGY/  LIBRIUM
              DISINTE- PAR-     DOSE
RADIATION     GRATION  TICLE    CONSTANT

                n_i     Ē_i      Δ_i
                       (MeV)    (g-rad/
                                 μCi-h)
-----------------------------------------
BETA MINUS 1  0.9999   0.0772   0.1645
```

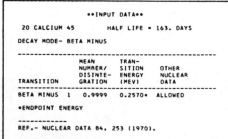

Table 3.6-1 (continued)
SUMMARY NUCLIDE DECAY DATA AND DOSE CONSTANT DATA

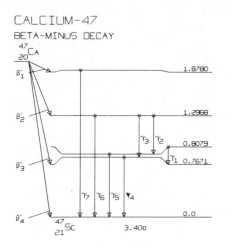

CALCIUM-47
BETA-MINUS DECAY

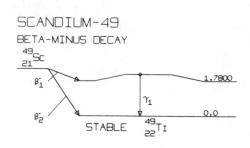

SCANDIUM-49
BETA-MINUS DECAY

```
            **INPUT DATA**

   21 SCANDIUM 49        HALF LIFE = 57.5 MINUTES

DECAY MODE- BETA MINUS

-----------------------------------------------
                MEAN        TRAN-
                NUMBER/     SITION    OTHER
                DISINTE-    ENERGY    NUCLEAR
  TRANSITION    GRATION     (MEV)     DATA
-----------------------------------------------
BETA MINUS  1   0.0003      0.2180*   ALLOWED
BETA MINUS  2   0.9997      2.0080*   ALLOWED
    GAMMA   1   0.0003      1.7800    M1,  AK(T) =
                                      0.000030

*ENDPOINT ENERGY

-------------
REF.- NUCLEAR DATA TABLES A8, 28 (1970).
```

```
            **OUTPUT DATA**

   21 SCANDIUM 49        HALF LIFE = 57.5 MINUTES

DECAY MODE- BETA MINUS

-----------------------------------------------
                MEAN        MEAN      EQUI-
                NUMBER/     ENERGY/   LIBRIUM
                DISINTE-    PAR-      DOSE
  RADIATION     GRATION     TICLE     CONSTANT

                 n_i         Ē_i       Δ_i
                            (MeV)     (g-rad/
                                      μCi-h)
-----------------------------------------------
BETA MINUS  2   0.9997      0.8255    1.7578
    GAMMA   1   0.0003      1.7800    0.0011
```

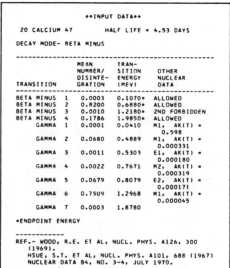

```
            **INPUT DATA**

   20 CALCIUM 47        HALF LIFE = 4.53 DAYS

DECAY MODE- BETA MINUS

-----------------------------------------------
                MEAN        TRAN-
                NUMBER/     SITION    OTHER
                DISINTE-    ENERGY    NUCLEAR
  TRANSITION    GRATION     (MEV)     DATA
-----------------------------------------------
BETA MINUS  1   0.0003      0.1070*   ALLOWED
BETA MINUS  2   0.8200      0.6880*   ALLOWED
BETA MINUS  3   0.0010      1.2180*   2ND FORBIDDEN
BETA MINUS  4   0.1786      1.9850*   ALLOWED
    GAMMA   1   0.0001      0.0410    M1,  AK(T) =
                                      0.598
    GAMMA   2   0.0680      0.4889    M1,  AK(T) =
                                      0.000331
    GAMMA   3   0.0011      0.5303    E1,  AK(T) =
                                      0.000180
    GAMMA   4   0.0022      0.7671    M2,  AK(T) =
                                      0.000319
    GAMMA   5   0.0679      0.8079    E2,  AK(T) =
                                      0.000171
    GAMMA   6   0.7509      1.2968    M1,  AK(T) =
                                      0.000045
    GAMMA   7   0.0003      1.8780

*ENDPOINT ENERGY

-------------
REF.- WOOD, R.E. ET AL, NUCL. PHYS. A126, 300
      (1969).
      HSUE, S.T. ET AL, NUCL. PHYS. A101, 688 (1967)
      NUCLEAR DATA B4, NO. 3-4, JULY 1970.
```

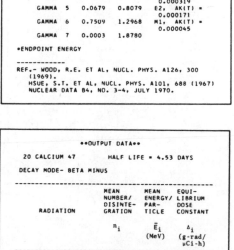

```
            **OUTPUT DATA**

   20 CALCIUM 47        HALF LIFE = 4.53 DAYS

DECAY MODE- BETA MINUS

-----------------------------------------------
                MEAN        MEAN      EQUI-
                NUMBER/     ENERGY/   LIBRIUM
                DISINTE-    PAR-      DOSE
  RADIATION     GRATION     TICLE     CONSTANT

                 n_i         Ē_i       Δ_i
                            (MeV)     (g-rad/
                                      μCi-h)
-----------------------------------------------
BETA MINUS  2   0.8200      0.2398    0.4189
BETA MINUS  3   0.0010      0.4918    0.0010
BETA MINUS  4   0.1786      0.8160    0.3104
    GAMMA   1   0.0000      0.0410    0.0000
    GAMMA   2   0.0679      0.4889    0.0707
    GAMMA   3   0.0011      0.5303    0.0012
    GAMMA   4   0.0022      0.7671    0.0035
    GAMMA   5   0.0678      0.8079    0.1168
    GAMMA   6   0.7508      1.2968    2.0740
    GAMMA   7   0.0003      1.8780    0.0013

-------------
DAUGHTER NUCLIDE, SCANDIUM 47  IS RADIOACTIVE
AND MAY CONTRIBUTE TO THE DOSE.
```

Table 3.6-1 (continued)
SUMMARY NUCLIDE DECAY DATA AND DOSE CONSTANT DATA

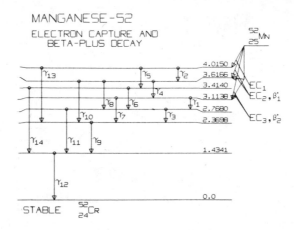

MANGANESE-52

ELECTRON CAPTURE AND
BETA-PLUS DECAY

••OUTPUT DATA••

25 MANGANESE 52   HALF LIFE = 5.67 DAYS

DECAY MODES– ELECTRON CAPTURE AND BETA PLUS

RADIATION		MEAN NUMBER/ DISINTE- GRATION n_i	MEAN ENERGY/ PAR- TICLE $\bar{E}_i$ (MeV)	EQUI- LIBRIUM DOSE CONSTANT Δ_i (g-rad/ μCi-h)
GAMMA	1	0.0097	0.3457	0.0071
GAMMA	2	0.0006	0.3985	0.0005
GAMMA	3	0.0023	0.3985	0.0020
GAMMA	4	0.0101	0.5028	0.0109
GAMMA	5	0.0049	0.6006	0.0063
GAMMA	6	0.0046	0.6470	0.0064
GAMMA	7	0.8999	0.7442	1.4266
K INT CON ELECT		0.0002	0.7382	0.0004
GAMMA	8	0.0342	0.8484	0.0619
GAMMA	9	0.9532	0.9356	1.8997
K INT CON ELECT		0.0001	0.9296	0.0002
GAMMA	10	0.0502	1.2470	0.1335
GAMMA	11	0.0462	1.3337	0.1314
GAMMA	12	0.9994	1.4341	3.0530
K INT CON ELECT		0.0005	1.4282	0.0017
GAMMA	13	0.0004	1.6450	0.0014
GAMMA	14	0.0002	1.9790	0.0011
K ALPHA-1 X-RAY		0.0997	0.0054	0.0011
K ALPHA-2 X-RAY		0.0507	0.0054	0.0005
K BETA-1 X-RAY		0.0168	0.0059	0.0002
KLL AUGER ELECT		0.3784	0.0048	0.0038
KLX AUGER ELECT		0.0870	0.0053	0.0009
LMM AUGER ELECT		1.0575	0.0005	0.0011
MXY AUGER ELECT		2.2239	0.0000	0.0002
BETA PLUS	2	0.2997	0.2414	0.1541
ANNIH. RADIATION		0.5994	0.5110	0.6524

••INPUT DATA••

25 MANGANESE 52   HALF LIFE = 5.67 DAYS

DECAY MODES– ELECTRON CAPTURE AND BETA PLUS

TRANSITION		MEAN NUMBER/ DISINTE- GRATION	TRAN- SITION ENERGY (MEV)	OTHER NUCLEAR DATA
ELECT CAPT	1	0.0059	0.6950	ALLOWED
ELECT CAPT	2	0.0942	1.0940	ALLOWED
ELECT CAPT	3	0.6001	1.5960	ALLOWED
GAMMA	1	0.0097	0.3457	E2, AK= 0.00387
GAMMA	2	0.0006	0.3985	
GAMMA	3	0.0024	0.3985	M1, AK(T) = 0.000806
GAMMA	4	0.0102	0.5028	M1, AK(T) = 0.000481
GAMMA	5	0.0049	0.6006	
GAMMA	6	0.0046	0.6470	M1, AK(T) = 0.000275
GAMMA	7	0.9003	0.7442	E2, AK=0.000324 K/(L+M)= 8.50
GAMMA	8	0.0343	0.8484	M1, AK=0.000186
GAMMA	9	0.9534	0.9356	E2, AK=0.000153 K/(L+M)= 8.90
GAMMA	10	0.0503	1.2470	M1, AK=0.000565
GAMMA	11	0.0463	1.3337	E2, AK=0.000726
GAMMA	12	1.0000	1.4341	E2, AK=0.000567
GAMMA	13	0.0004	1.6450	
GAMMA	14	0.0002	1.9790	
BETA PLUS	1	0.0000	0.0720•	ALLOWED
BETA PLUS	2	0.2997	0.5740•	ALLOWED

•ENDPOINT ENERGY

REF.– NUCLEAR DATA 83, NO. 5-6, APRIL 1970.

Table 3.6-1 (continued)
SUMMARY NUCLIDE DECAY DATA AND DOSE CONSTANT DATA

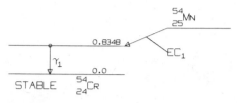

MANGANESE-54
ELECTRON CAPTURE DECAY

```
              ••OUTPUT DATA••

   25 MANGANESE 54       HALF LIFE = 312. DAYS

   DECAY MODE- ELECTRON CAPTURE
```

RADIATION	MEAN NUMBER/ DISINTE- GRATION n_i	MEAN ENERGY/ PAR- TICLE $\bar{E}_i$ (MeV)	EQUI- LIBRIUM DOSE CONSTANT Δ_i (g-rad/ μCi-h)
GAMMA 1	0.9997	0.8348	1.7777
K INT CON ELECT	0.0002	0.8288	0.0004
K ALPHA-1 X-RAY	0.1419	0.0054	0.0016
K ALPHA-2 X-RAY	0.0722	0.0054	0.0008
K BETA-1 X-RAY	0.0239	0.0059	0.0003
KLL AUGER ELECT	0.5388	0.0048	0.0055
KLX AUGER ELECT	0.1239	0.0053	0.0014
LMM AUGER ELECT	1.5071	0.0005	0.0016
MXY AUGER ELECT	3.1695	0.0000	0.0002

```
              ••INPUT DATA••

   25 MANGANESE 54       HALF LIFE = 312. DAYS

   DECAY MODE- ELECTRON CAPTURE
```

TRANSITION	MEAN NUMBER/ DISINTE- GRATION	TRAN- SITION ENERGY (MEV)	OTHER NUCLEAR DATA
ELECT CAPT 1	1.0000	0.5280	ALLOWED
GAMMA 1	1.0000	0.8348	E2, AK(T) = 0.000231

```
   ------------
   REF.- NUCLEAR DATA B3, NO. 5-6, APRIL 1970.
```

IRON-55
ELECTRON CAPTURE DECAY

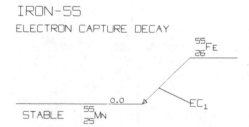

```
              ••OUTPUT DATA••

   26 IRON 55           HALF LIFE = 2.70 YEARS

   DECAY MODE- ELECTRON CAPTURE
```

RADIATION	MEAN NUMBER/ DISINTE- GRATION n_i	MEAN ENERGY/ PAR- TICLE $\bar{E}_i$ (MeV)	EQUI- LIBRIUM DOSE CONSTANT Δ_i (g-rad/ μCi-h)
K ALPHA-1 X-RAY	0.1553	0.0058	0.0019
K ALPHA-2 X-RAY	0.0786	0.0058	0.0009
K BETA-1 X-RAY	0.0257	0.0064	0.0003
KLL AUGER ELECT	0.5151	0.0052	0.0057
KLX AUGER ELECT	0.1220	0.0058	0.0015
LMM AUGER ELECT	1.4807	0.0005	0.0017
MXY AUGER ELECT	3.1174	0.0000	0.0003

```
              ••INPUT DATA••

   26 IRON 55           HALF LIFE = 2.70 YEARS

   DECAY MODE- ELECTRON CAPTURE
```

TRANSITION	MEAN NUMBER/ DISINTE- GRATION	TRAN- SITION ENERGY (MEV)	OTHER NUCLEAR DATA
ELECT CAPT 1	1.0000	0.2250	ALLOWED

```
   ------------
   REF.- NUCLEAR DATA B3, NO. 3-4, JANUARY 1970.
```

Table 3.6-1 (continued)
SUMMARY NUCLIDE DECAY DATA AND DOSE CONSTANT DATA

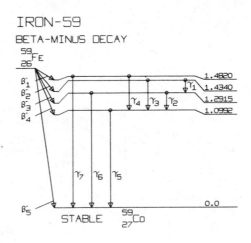

IRON-59

BETA-MINUS DECAY

```
**INPUT DATA**

26 IRON 59          HALF LIFE = 45.0 DAYS

DECAY MODE- BETA MINUS

---------------------------------------------
                MEAN      TRAN-
                NUMBER/   SITION
                DISINTE-  ENERGY   OTHER
                GRATION   (MEV)    NUCLEAR
TRANSITION      GRATION            DATA
---------------------------------------------
BETA MINUS  1   0.0011    0.0910*  ALLOWED
BETA MINUS  2   0.0122    0.1390*  ALLOWED
BETA MINUS  3   0.4609    0.2730*  ALLOWED
BETA MINUS  4   0.5228    0.4670*  ALLOWED
BETA MINUS  5   0.0030    1.5730*  2ND FORBIDDEN
   GAMMA    1   0.0098    0.1420   M1, AK(T) =
                                      0.0143
                                   AL(T) =
                                      0.00116
   GAMMA    2   0.0295    0.1922   M1, AK(T) =
                                      0.00665
                                   AL(T) =
                                      0.000535
   GAMMA    3   0.0024    0.3347   M1, AK(T) =
                                      0.00172
                                   AL(T) =
                                      0.000139
   GAMMA    4   0.0002    0.3810
   GAMMA    5   0.5549    1.0990   E2, AK(T) =
                                      0.000154
                                   AL(T) =
                                      0.000013
   GAMMA    6   0.4412    1.2920   E2, AK(T) =
                                      0.000107
                                   AL(T) =
                                      0.000009
   GAMMA    7   0.0009    1.4810

*ENDPOINT ENERGY

------------
REF.- LEGRAND, J. ET AL, NUCL. PHYS. A142, 63
   (1970).
```

		MEAN NUMBER/ DISINTE- GRATION	MEAN ENERGY/ PAR- TICLE	EQUI- LIBRIUM DOSE CONSTANT
OUTPUT DATA				
26 IRON 59	HALF LIFE = 45.0 DAYS			
DECAY MODE- BETA MINUS				
RADIATION		n_i	$\bar{E}_i$ (MeV)	Δ_i (g-rad/ μCi-h)
BETA MINUS	2	0.0122	0.0381	0.0009
BETA MINUS	3	0.4609	0.0808	0.0793
BETA MINUS	4	0.5228	0.1496	0.1666
BETA MINUS	5	0.0030	0.6396	0.0040
GAMMA	1	0.0096	0.1420	0.0029
GAMMA	2	0.0292	0.1922	0.0119
GAMMA	3	0.0023	0.3347	0.0017
GAMMA	4	0.0002	0.3810	0.0001
GAMMA	5	0.5548	1.0990	1.2987
K INT CON ELECT		0.0000	1.0912	0.0002
GAMMA	6	0.4411	1.2920	1.2140
K INT CON ELECT		0.0000	1.2842	0.0001
GAMMA	7	0.0009	1.4810	0.0028

Table 3.6-1 (continued)
SUMMARY NUCLIDE DECAY DATA AND DOSE CONSTANT DATA

COBALT-58
ELECTRON CAPTURE AND
BETA-PLUS DECAY

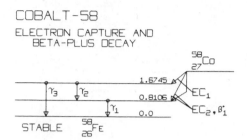

```
                        ••INPUT DATA••

   27 COBALT 58          HALF LIFE = 71.3 DAYS

   DECAY MODES- ELECTRON CAPTURE AND BETA PLUS

   -------------------------------------------------
                    MEAN      TRAN-
                    NUMBER/   SITION   OTHER
                    DISINTE-  ENERGY   NUCLEAR
   TRANSITION       GRATION   (MEV)    DATA
   ELECT CAPT  1    0.0110    0.6320   ALLOWED
   ELECT CAPT  2    0.8340    1.4960   ALLOWED
        GAMMA  1    0.9950    0.8106   E2, AK=0.000300
                                       K/L=11.8
                                       K/(L+M)= 8.90
        GAMMA  2    0.0060    0.8636   M1, AK=0.000260
                                       AL(T) =
                                        0.000015
        GAMMA  3    0.0050    1.6748   E2,  AK(T) =
                                        0.000046
                                       AL(T) =
                                        0.000003
    BETA PLUS  1    0.1550    0.4740•  ALLOWED

   •ENDPOINT ENERGY

   ------------
   REF.- NUCLEAR DATA B3, NO. 3-4, JANUARY 1970.
```

```
                        ••OUTPUT DATA••

   27 COBALT 58          HALF LIFE = 71.3 DAYS

   DECAY MODES- ELECTRON CAPTURE AND BETA PLUS

   -------------------------------------------------
                      MEAN      MEAN     EQUI-
                      NUMBER/   ENERGY/  LIBRIUM
                      DISINTE-  PAR-     DOSE
        RADIATION     GRATION   TICLE    CONSTANT

                        n_i       Ē_i      Δ_i
                                 (MeV)   (g-rad/
                                          µCi-h)
   -------------------------------------------------
           GAMMA  1   0.9946    0.8106   1.7173
   K INT CON ELECT    0.0002    0.8034   0.0005
           GAMMA  2   0.0059    0.8636   0.0110
           GAMMA  3   0.0050    1.6748   0.0178
   K ALPHA-1 X-RAY    0.1441    0.0064   0.0019
   K ALPHA-2 X-RAY    0.0726    0.0063   0.0009
   K BETA-1 X-RAY     0.0233    0.0070   0.0003
   KLL AUGER ELECT    0.4185    0.0056   0.0050
   KLX AUGER ELECT    0.1021    0.0063   0.0013
   LMM AUGER ELECT    1.2330    0.0006   0.0016
   MXY AUGER ELECT    2.5985    0.0000   0.0002
        BETA PLUS  1  0.1550    0.2012   0.0664
   ANNIH. RADIATION   0.3100    0.5110   0.3374
```

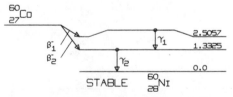

COBALT-60
BETA-MINUS DECAY

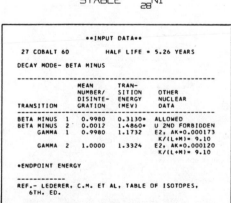

```
                        ••INPUT DATA••

   27 COBALT 60          HALF LIFE = 5.26 YEARS

   DECAY MODE- BETA MINUS

   -------------------------------------------------
                    MEAN      TRAN-
                    NUMBER/   SITION   OTHER
                    DISINTE-  ENERGY   NUCLEAR
   TRANSITION       GRATION   (MEV)    DATA
   BETA MINUS  1    0.9980    0.3130•  ALLOWED
   BETA MINUS  2    0.0012    1.4860•  U 2ND FORBIDDEN
        GAMMA  1    0.9980    1.1732   E2, AK=0.000173
                                       K/(L+M)= 9.10
        GAMMA  2    1.0000    1.3324   E2, AK=0.000120
                                       K/(L+M)= 9.10

   •ENDPOINT ENERGY

   ------------
   REF.- LEDERER, C.M. ET AL, TABLE OF ISOTOPES,
         6TH. ED.
```

```
                        ••OUTPUT DATA••

   27 COBALT 60          HALF LIFE = 5.26 YEARS

   DECAY MODE- BETA MINUS

   -------------------------------------------------
                      MEAN      MEAN     EQUI-
                      NUMBER/   ENERGY/  LIBRIUM
                      DISINTE-  PAR-     DOSE
        RADIATION     GRATION   TICLE    CONSTANT

                        n_i       Ē_i      Δ_i
                                 (MeV)   (g-rad/
                                          µCi-h)
   -------------------------------------------------
         BETA MINUS  1  0.9980  0.0941   0.2000
         BETA MINUS  2  0.0012  0.6243   0.0015
             GAMMA  1   0.9978  1.1732   2.4935
   K INT CON ELECT      0.0001  1.1648   0.0004
             GAMMA  2   0.9998  1.3324   2.8378
   K INT CON ELECT      0.0001  1.3241   0.0003
```

Table 3.6-1 (continued)
SUMMARY NUCLIDE DECAY DATA AND DOSE CONSTANT DATA

COPPER-64

ELECTRON CAPTURE, BETA-PLUS
AND BETA-MINUS DECAY

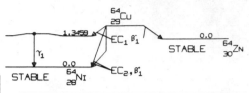

```
                  **INPUT DATA**

   29 COPPER 64         HALF LIFE = 12.8 HOURS

   DECAY MODES- BETA MINUS, BETA PLUS
               AND ELECTRON CAPTURE
   ----------------------------------------------
                   MEAN     TRAN-
                   NUMBER/  SITION
                   DISINTE- ENERGY   OTHER
                   GRATION  (MEV)    NUCLEAR
   TRANSITION      GRATION  (MEV)    DATA
   ----------------------------------------------
   BETA MINUS   1   0.3960   0.5730*  ALLOWED
   ELECT CAPT   1   0.0060   0.3310   ALLOWED
   ELECT CAPT   2   0.4050   1.6770   ALLOWED
       GAMMA    1   0.0060   1.3459   E2,  AK(T) =
                                      0.000109
                                      AL(T) =
                                      0.000009

   BETA PLUS    1   0.1930   0.6550*  ALLOWED

   *ENDPOINT ENERGY

   ------------
   REF.- NUCLEAR DATA 82-3-75 (1967).
       DIXON, W.R. ET AL, CAN. J. PHYS. 48, 483
       (1970).
```

```
                      **OUTPUT DATA**

     29 COPPER 64          HALF LIFE = 12.8 HOURS

   DECAY MODES- BETA MINUS, BETA PLUS
               AND ELECTRON CAPTURE
   ------------------------------------------------
                       MEAN     MEAN     EQUI-
                       NUMBER/  ENERGY/  LIBRIUM
                       DISINTE- PAR-     DOSE
       RADIATION       GRATION  TICLE    CONSTANT

                         n_i      Ē_i     Δ_i
                                 (MeV)   (g-rad/
                                          uCi-h)
   ------------------------------------------------
        BETA MINUS  1   0.3960   0.1882   0.1588
            GAMMA   1   0.0060   1.3459   0.0172
   K ALPHA-1 X-RAY      0.0824   0.0074   0.0013
   K ALPHA-2 X-RAY      0.0412   0.0074   0.0006
   K BETA-1  X-RAY      0.0138   0.0082   0.0002
   KLL AUGER ELECT      0.1829   0.0065   0.0025
   KLX AUGER ELECT      0.0473   0.0073   0.0007
   LMM AUGER ELECT      0.5748   0.0007   0.0009
   MXY AUGER ELECT      1.2171   0.0000   0.0001
          BETA PLUS  1  0.1930   0.2794   0.1148
   ANNIH. RADIATION     0.3860   0.5110   0.4201
```

ZINC-65

ELECTRON CAPTURE AND
BETA-PLUS DECAY

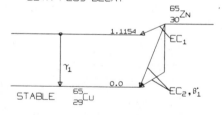

```
                  **INPUT DATA**

   30 ZINC 65          HALF LIFE = 243. DAYS

   DECAY MODES- ELECTRON CAPTURE AND BETA PLUS
   ----------------------------------------------
                   MEAN     TRAN-
                   NUMBER/  SITION
                   DISINTE- ENERGY   OTHER
                   GRATION  (MEV)    NUCLEAR
   TRANSITION      GRATION  (MEV)    DATA
   ----------------------------------------------
   ELECT CAPT   1   0.5060   0.2380   ALLOWED
   ELECT CAPT   2   0.4790   1.3470   ALLOWED
       GAMMA    1   0.5060   1.1154   M1, AK=0.000166
                                      AL(T) =
                                      0.000013
   BETA PLUS    1   0.0150   0.3250*  ALLOWED

   *ENDPOINT ENERGY

   ------------
   REF.- NUCLEAR DATA 82-6-29 (1968).
```

```
                      **OUTPUT DATA**

     30 ZINC 65          HALF LIFE = 243. DAYS

   DECAY MODES- ELECTRON CAPTURE AND BETA PLUS
   ------------------------------------------------
                       MEAN     MEAN     EQUI-
                       NUMBER/  ENERGY/  LIBRIUM
                       DISINTE- PAR-     DOSE
       RADIATION       GRATION  TICLE    CONSTANT

                         n_i      Ē_i     Δ_i
                                 (MeV)   (g-rad/
                                          uCi-h)
   ------------------------------------------------
            GAMMA   1   0.5059   1.1154   1.2019
   K INT CON ELECT      0.0000   1.1064   0.0002
   K ALPHA-1 X-RAY      0.2128   0.0080   0.0036
   K ALPHA-2 X-RAY      0.1063   0.0080   0.0018
   K BETA-1  X-RAY      0.0370   0.0089   0.0007
   KLL AUGER ELECT      0.4111   0.0070   0.0061
   KLX AUGER ELECT      0.1089   0.0079   0.0018
   LMM AUGER ELECT      1.3443   0.0008   0.0023
   MXY AUGER ELECT      2.8531   0.0000   0.0004
          BETA PLUS  1  0.0150   0.1414   0.0045
   ANNIH. RADIATION     0.0300   0.5110   0.0326
```

Table 3.6-1 (continued)
SUMMARY NUCLIDE DECAY DATA AND DOSE CONSTANT DATA

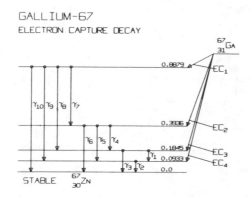

GALLIUM-67
ELECTRON CAPTURE DECAY

```
                                                    67
                                                    31 GA
                                              0.8879 ┐ EC₁
              γ₁₀ γ₉ γ₈ γ₇
                                              0.3936   EC₂
                     γ₆ γ₅ γ₄
                                              0.1845   EC₃
                                          γ₁  0.0933   EC₄
                              γ₃ γ₂              0.0
         STABLE   67
                  30 ZN
```

```
              ••INPUT DATA••

  31 GALLIUM 67        HALF LIFE = 78.1 HOURS

DECAY MODE- ELECTRON CAPTURE

--------------------------------------------
                MEAN      TRAN-
                NUMBER/   SITION
                DISINTE-  ENERGY   OTHER
                GRATION   (MEV)    NUCLEAR
TRANSITION      GRATION   (MEV)    DATA
--------------------------------------------
ELECT CAPT  1   0.0033    0.1170   ALLOWED
ELECT CAPT  2   0.2290    0.6110   ALLOWED
ELECT CAPT  3   0.2520    0.8200   ALLOWED
ELECT CAPT  4   0.5157    0.9120   ALLOWED
    GAMMA   1   0.0351    0.0913   M1, AK= 0.0660
                                   AL= 0.00760
    GAMMA   2   0.7135    0.0933   E2, AK= 0.770
                                   K/M=58.0
                                   AL= 0.100
    GAMMA   3   0.2420    0.1846   M1, AK= 0.0156
                                   K/M=65.0
                                   AL= 0.00170
    GAMMA   4   0.0250    0.2090   M1, AK= 0.00750
                                   K/L=10.6
                                   K/M=61.0
    GAMMA   5   0.1620    0.3002   M1, AK= 0.00337
                                   K/L= 9.50
                                   K/M=75.0
    GAMMA   6   0.0430    0.3936   M1, AK= 0.00192
                                   AL(T) =
                                   0.000144
    GAMMA   7   0.0010    0.4943   M1, AK= 0.00119
                                   AL(T) =
                                   0.000085
    GAMMA   8   0.0001    0.7036   M1, AK(T) =
                                   0.000461
                                   AL(T) =
                                   0.000039
    GAMMA   9   0.0006    0.7947   M1, AK(T) =
                                   0.000354
                                   AL(T) =
                                   0.000030
    GAMMA  10   0.0015    0.8880   M1, AK=0.000337
                                   AL(T) =
                                   0.000024

------------
REF.- LI-SCHOLZ, A. AND BAKHRU, H., PHYS. REV.
   1778 1629 (1969).
   FREEDMAN, M.S. ET AL, PHYS. REV. 151, 886
   (1966).
```

```
              ••OUTPUT DATA••

  31 GALLIUM 67        HALF LIFE = 78.1 HOURS

DECAY MODE- ELECTRON CAPTURE

--------------------------------------------
                  MEAN      MEAN      EQUI-
                  NUMBER/   ENERGY/   LIBRIUM
                  DISINTE-  PAR-      DOSE
   RADIATION      GRATION   TICLE     CONSTANT

                   nᵢ        Ēᵢ        Δᵢ
                            (MEV)     (g-rad/
                                      µCi-h)
--------------------------------------------
        GAMMA   1   0.0326   0.0913   0.0063
K INT CON ELECT     0.0021   0.0816   0.0003
        GAMMA   2   0.3797   0.0933   0.0754
K INT CON ELECT     0.2830   0.0836   0.0504
L INT CON ELECT     0.0379   0.0922   0.0074
M INT CON ELECT     0.0126   0.0932   0.0025
        GAMMA   3   0.2388   0.1846   0.0939
K INT CON ELECT     0.0026   0.1749   0.0009
L INT CON ELECT     0.0004   0.1835   0.0001
        GAMMA   4   0.0247   0.2090   0.0110
        GAMMA   5   0.1613   0.3002   0.1031
K INT CON ELECT     0.0005   0.2905   0.0003
        GAMMA   6   0.0429   0.3936   0.0359
        GAMMA   7   0.0009   0.4943   0.0010
        GAMMA   8   0.0001   0.7036   0.0002
        GAMMA   9   0.0006   0.7947   0.0010
        GAMMA  10   0.0015   0.8880   0.0029
K ALPHA-1 X-RAY     0.3075   0.0086   0.0056
K ALPHA-2 X-RAY     0.1534   0.0086   0.0028
K BETA-1 X-RAY      0.0553   0.0095   0.0011
KLL AUGER ELECT     0.5185   0.0075   0.0083
KLX AUGER ELECT     0.1410   0.0085   0.0025
KXY AUGER ELECT     0.0067   0.0094   0.0001
LMM AUGER ELECT     1.7722   0.0008   0.0033
MXY AUGER ELECT     3.7779   0.0000   0.0006
```

Table 3.6-1 (continued)
SUMMARY NUCLIDE DECAY DATA AND DOSE CONSTANT DATA

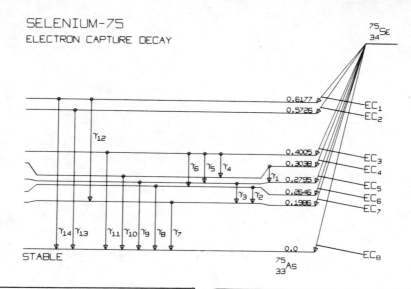

SELENIUM-75
ELECTRON CAPTURE DECAY

••INPUT DATA••

34 SELENIUM 75   HALF LIFE = 120. DAYS

DECAY MODE- ELECTRON CAPTURE

TRANSITION	MEAN NUMBER/ DISINTE- GRATION	TRAN- SITION ENERGY (MEV)	OTHER NUCLEAR DATA
ELECT CAPT 1	0.0001	0.2470	1ST FORBIDDEN
ELECT CAPT 2	0.0003	0.2920	1ST FORBIDDEN
ELECT CAPT 3	0.8957	0.4640	ALLOWED
ELECT CAPT 4	0.0090	0.5610	2ND FORBIDDEN
ELECT CAPT 5	0.0230	0.5860	1ST FORBIDDEN
ELECT CAPT 6	0.0310	0.6000	1ST FORBIDDEN
ELECT CAPT 7	0.0007	0.6660	U 1ST FORBIDDEN
ELECT CAPT 8	0.0400	0.8650	1ST FORBIDDEN
GAMMA 1	0.0547	0.0244	M2, AK=178. K/(L+M)= 3.33
GAMMA 2	0.0127	0.0660	M1, AK= 0.263 K/(L+M)=45.2
GAMMA 3	0.0002	0.0810	E2, AK= 1.65 AL(T) = 0.214
GAMMA 4	0.0589	0.0967	E2, AK= 0.880 K/(L+M)= 6.13
GAMMA 5	0.1635	0.1211	E1, AK= 0.0405 K/(L+M)= 8.16
GAMMA 6	0.5567	0.1360	E1, AK= 0.0286 K/(L+M)= 8.07
GAMMA 7	0.0138	0.1986	M1, AK= 0.0192 K/(L+M)= 9.61
GAMMA 8	0.5750	0.2646	M1, AK= 0.00622 K/(L+M)= 8.55
GAMMA 9	0.2409	0.2795	M1, AK= 0.00778 K/(L+M)= 8.07
GAMMA 10	0.0130	0.3038	E3, AK= 0.0472 K/(L+M)= 6.48
GAMMA 11	0.1166	0.4005	E1, AK= 0.00114 K/(L+M)= 8.56
GAMMA 12	0.0001	0.4193	M1, AK= 0.00179 K/(L+M)= 5.07
GAMMA 13	0.0003	0.5726	M1, AK= 0.00100 AL(T) = 0.000108
GAMMA 14	0.0000	0.6177	M1, AK=0.000700 AL(T) = 0.000091

REF.- PARADELLIS, TH. AND HONTZEAS, S., CAN. J. PHYS. 48, 2256 (1970).

••OUTPUT DATA••

34 SELENIUM 75   HALF LIFE = 120. DAYS

DECAY MODE- ELECTRON CAPTURE

RADIATION	MEAN NUMBER/ DISINTE- GRATION n_i	MEAN ENERGY/ PAR- TICLE $\bar{E}_i$ (MeV)	EQUI- LIBRIUM DOSE CONSTANT Δ_i (g-rad/ µCi-h)
GAMMA 1	0.0002	0.0244	0.0000
K INT CON ELECT	0.0418	0.0125	0.0011
L INT CON ELECT	0.0094	0.0230	0.0004
M INT CON ELECT	0.0031	0.0242	0.0001
GAMMA 2	0.0100	0.0660	0.0014
K INT CON ELECT	0.0026	0.0541	0.0003
GAMMA 3	0.0000	0.0810	0.0000
GAMMA 4	0.0291	0.0967	0.0060
K INT CON ELECT	0.0256	0.0848	0.0046
L INT CON ELECT	0.0031	0.0953	0.0006
M INT CON ELECT	0.0010	0.0965	0.0002
GAMMA 5	0.1563	0.1211	0.0403
K INT CON ELECT	0.0063	0.1092	0.0014
L INT CON ELECT	0.0005	0.1197	0.0001
GAMMA 6	0.5393	0.1360	0.1562
K INT CON ELECT	0.0154	0.1241	0.0040
L INT CON ELECT	0.0014	0.1346	0.0004
M INT CON ELECT	0.0004	0.1358	0.0001
GAMMA 7	0.0135	0.1986	0.0057
K INT CON ELECT	0.0002	0.1867	0.0001
GAMMA 8	0.5710	0.2646	0.3218
K INT CON ELECT	0.0035	0.2527	0.0019
L INT CON ELECT	0.0002	0.2632	0.0001
GAMMA 9	0.2388	0.2795	0.1421
K INT CON ELECT	0.0018	0.2676	0.0010
L INT CON ELECT	0.0001	0.2781	0.0001
GAMMA 10	0.0123	0.3038	0.0080
K INT CON ELECT	0.0005	0.2919	0.0003
GAMMA 11	0.1164	0.4005	0.0993
K INT CON ELECT	0.0001	0.3886	0.0001
GAMMA 12	0.0001	0.4193	0.0001
GAMMA 13	0.0003	0.5726	0.0004
GAMMA 14	0.0000	0.6177	0.0000
K ALPHA-1 X-RAY	0.3175	0.0105	0.0071
K ALPHA-2 X-RAY	0.1581	0.0105	0.0035
K BETA-1 X-RAY	0.0625	0.0117	0.0015
KLL AUGER ELECT	0.3394	0.0091	0.0065
KLX AUGER ELECT	0.0994	0.0103	0.0021
KXY AUGER ELECT	0.0076	0.0115	0.0001
LMM AUGER ELECT	1.3522	0.0010	0.0031
MXY AUGER ELECT	2.9144	0.0001	0.0008

Table 3.6-1 (continued)
SUMMARY NUCLIDE DECAY DATA AND DOSE CONSTANT DATA

KRYPTON-85
BETA-MINUS DECAY

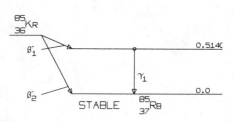

```
        **INPUT DATA**

   36 KRYPTON 85      HALF LIFE = 10.7 YEARS

   DECAY MODE- BETA MINUS

   --------------------------------------------
                  MEAN      TRAN-
                  NUMBER/   SITION    OTHER
                  DISINTE-  ENERGY    NUCLEAR
   TRANSITION     GRATION   (MEV)     DATA
   --------------------------------------------
   BETA MINUS  1  0.0043    0.1500*   ALLOWED
   BETA MINUS  2  0.9957    0.6720*   U 1ST FORBIDDEN
      GAMMA    1  0.0043    0.5140    M2, AK= 0.00700
                                      K/L=12.0

   *ENDPOINT ENERGY (MEV).

   ------------
   REF.- NUCLEAR DATA B5, 142 (1971).
```

```
                    **OUTPUT DATA**

   36 KRYPTON 85      HALF LIFE = 10.7 YEARS

   DECAY MODE- BETA MINUS

   --------------------------------------------
                     MEAN      MEAN      EQUI-
                     NUMBER/   ENERGY/   LIBRIUM
                     DISINTE-  PAR-      DOSE
      RADIATION      GRATION   TICLE     CONSTANT

                     n_i       Ē_i       Δ_i
                               (MeV)     (g-rad/
                                          μCi-h)
   --------------------------------------------
      BETA MINUS  1  0.0043    0.0407    0.0003
      BETA MINUS  2  0.9957    0.2460    0.5218
         GAMMA    1  0.0042    0.5140    0.0046
```

KRYPTON-85M
ISOMERIC LEVEL AND
BETA-MINUS DECAY

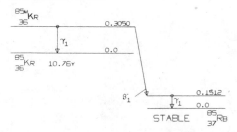

```
                    **OUTPUT DATA**

   36 KRYPTON 85M      HALF LIFE = 4.36 HOURS

   DECAY MODES- BETA MINUS AND ISOMERIC LEVEL

   --------------------------------------------
                        MEAN      MEAN      EQUI-
                        NUMBER/   ENERGY/   LIBRIUM
                        DISINTE-  PAR-      DOSE
      RADIATION         GRATION   TICLE     CONSTANT

                        n_i       Ē_i       Δ_i
                                  (MeV)     (g-rad/
                                             μCi-h)
   --------------------------------------------
        BETA MINUS   1  0.7960    0.2837    0.4811
           GAMMA     1  0.7606    0.1512    0.2450
   K INT CON ELECT      0.0304    0.1360    0.0088
   L INT CON ELECT      0.0037    0.1494    0.0011
   M INT CON ELECT      0.0012    0.1510    0.0003
   K ALPHA-1 X-RAY      0.0118    0.0133    0.0003
   K ALPHA-2 X-RAY      0.0059    0.0133    0.0001
   KLL AUGER ELECT      0.0071    0.0114    0.0001
   LMM AUGER ELECT      0.0368    0.0013    0.0001
           GAMMA     1  0.1356    0.3050    0.0881
   K INT CON ELECT      0.0585    0.2906    0.0362
   L INT CON ELECT      0.0073    0.3032    0.0047
   M INT CON ELECT      0.0024    0.3047    0.0015
   K ALPHA-1 X-RAY      0.0221    0.0126    0.0005
   K ALPHA-2 X-RAY      0.0110    0.0125    0.0002
    K BETA-1 X-RAY      0.0046    0.0141    0.0001
   KLL AUGER ELECT      0.0150    0.0108    0.0003
   KLX AUGER ELECT      0.0047    0.0123    0.0001
   LMM AUGER ELECT      0.0731    0.0013    0.0002

   ------------
   DAUGHTER NUCLIDE, KRYPTON 85  IS RADIOACTIVE
   AND MAY CONTRIBUTE TO THE DOSE.
```

```
        **INPUT DATA**

   36 KRYPTON 85M      HALF LIFE = 4.36 HOURS

   DECAY MODES- BETA MINUS AND ISOMERIC LEVEL

   --------------------------------------------
                  MEAN      TRAN-
                  NUMBER/   SITION    OTHER
                  DISINTE-  ENERGY    NUCLEAR
   TRANSITION     GRATION   (MEV)     DATA
   --------------------------------------------
   BETA MINUS  1  0.7960    0.8240*   ALLOWED
      GAMMA    1  0.7960    0.1512    M1, AK= 0.0400
                                      AL(T) =
                                      0.00488
      GAMMA    1  0.2040    0.3050    M4, AK= 0.432
                                      K/(L+M)= 6.00

   *ENDPOINT ENERGY (MEV).

   ------------
   REF.- NUCLEAR DATA B5, 142 (1971).
```

Table 3.6-1 (continued)
SUMMARY NUCLIDE DECAY DATA AND DOSE CONSTANT DATA

STRONTIUM-85
ELECTRON CAPTURE DECAY

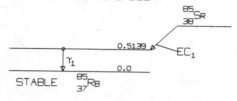

```
              ••INPUT DATA••

   38 STRONTIUM 85      HALF LIFE = 65.1 DAYS

DECAY MODE- ELECTRON CAPTURE

-----------------------------------------
                 MEAN      TRAN-
                 NUMBER/   SITION
                 DISINTE-  ENERGY   OTHER
                 GRATION   (MEV)    NUCLEAR
TRANSITION                          DATA
-----------------------------------------
ELECT CAPT  1    0.9999    0.5560   ALLOWED
     GAMMA  1    0.9999    0.5139   M2, AK= 0.00700
                                    AL(T) =
                                    0.000725
-------------
REF.- NUCLEAR DATA 85, 143 (1971).
```

```
                    ••OUTPUT DATA••

   38 STRONTIUM 85        HALF LIFE = 65.1 DAYS

DECAY MODE- ELECTRON CAPTURE
```

	MEAN NUMBER/ DISINTE- GRATION n_i	MEAN ENERGY/ PAR- TICLE $\bar{E}_i$ (MeV)	EQUI- LIBRIUM DOSE CONSTANT Δ_i (g-rad/ μCi-h)
GAMMA 1	0.9919	0.5139	1.0860
K INT CON ELECT	0.0069	0.4987	0.0073
L INT CON ELECT	0.0007	0.5121	0.0007
M INT CON ELECT	0.0002	0.5137	0.0002
K ALPHA-1 X-RAY	0.3481	0.0133	0.0099
K ALPHA-2 X-RAY	0.1739	0.0133	0.0049
K BETA-1 X-RAY	0.0752	0.0149	0.0023
K BETA-2 X-RAY	0.0085	0.0151	0.0002
L X-RAYS	0.0400	0.0016	0.0001
KLL AUGER ELECT	0.2101	0.0114	0.0051
KLX AUGER ELECT	0.0678	0.0130	0.0018
KXY AUGER ELECT	0.0071	0.0147	0.0002
LMM AUGER ELECT	1.0721	0.0013	0.0031
MXY AUGER ELECT	2.3564	0.0002	0.0011

STRONTIUM-90 [a]
BETA-MINUS DECAY

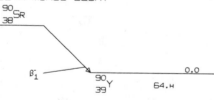

```
              ••INPUT DATA••

   38 STRONTIUM 90      HALF LIFE = 28.1 YEARS

DECAY MODE- BETA MINUS

-----------------------------------------
                 MEAN      TRAN-
                 NUMBER/   SITION
                 DISINTE-  ENERGY   OTHER
                 GRATION   (MEV)    NUCLEAR
TRANSITION                          DATA
-----------------------------------------
BETA MINUS  1    1.0000    0.5460•  U 1ST FORBIDDEN

•ENDPOINT ENERGY (MEV).
-------------
REF.- LEDERER, C.M. ET AL, TABLE OF ISOTOPES,
   6TH. ED.
```

```
                    ••OUTPUT DATA••

   38 STRONTIUM 90        HALF LIFE = 28.1 YEARS

DECAY MODE- BETA MINUS
```

	MEAN NUMBER/ DISINTE- GRATION n_i	MEAN ENERGY/ PAR- TICLE $\bar{E}_i$ (MeV)	EQUI- LIBRIUM DOSE CONSTANT Δ_i (g-rad/ μCi-h)
BETA MINUS 1	1.0000	0.1961	0.4177

```
-------------
DAUGHTER NUCLIDE, YTTRIUM 90  IS RADIOACTIVE
AND MAY CONTRIBUTE TO THE DOSE.
```

[a]   The daughter nuclide Y-90 usually is in secular equilibrium and contributes the larger part of the internal dose.

Table 3.6-1 (continued)
SUMMARY NUCLIDE DECAY DATA AND DOSE CONSTANT DATA

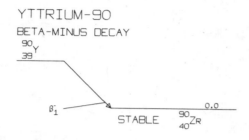

YTTRIUM-90
BETA-MINUS DECAY

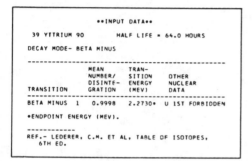

```
            ••INPUT DATA••

  39 YTTRIUM 90        HALF LIFE = 64.0 HOURS

  DECAY MODE- BETA MINUS

  --------------------------------------------
                MEAN      TRAN-
                NUMBER/   SITION    OTHER
                DISINTE-  ENERGY    NUCLEAR
  TRANSITION    GRATION   (MEV)     DATA
  --------------------------------------------
  BETA MINUS 1  0.9998    2.2730•   U 1ST FORBIDDEN

  •ENDPOINT ENERGY (MEV).

  ------------
  REF.- LEDERER, C.M. ET AL, TABLE OF ISOTOPES,
     6TH ED.
```

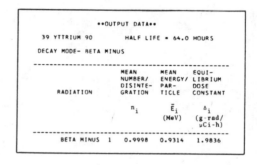

```
            ••OUTPUT DATA••

  39 YTTRIUM 90        HALF LIFE = 64.0 HOURS

  DECAY MODE- BETA MINUS

  --------------------------------------------
                MEAN      MEAN      EQUI-
                NUMBER/   ENERGY/   LIBRIUM
                DISINTE-  PAR-      DOSE
  RADIATION     GRATION   TICLE     CONSTANT
```

$$n_i \qquad \bar{E}_i \qquad \Delta_i$$
$$(\text{MeV}) \quad (\text{g-rad/}\mu Ci\text{-}h)$$

```
  --------------------------------------------
  BETA MINUS 1  0.9998    0.9314    1.9836
```

MOLYBDENUM-99
BETA-MINUS DECAY

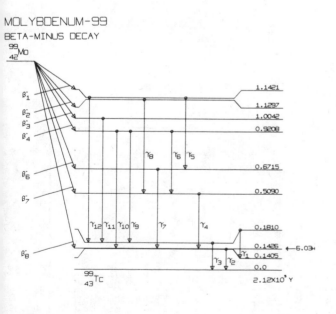

```
            ••INPUT DATA••

  42 MOLYBDENUM 99      HALF LIFE = 66.7 HOURS

  DECAY MODE- BETA MINUS

  --------------------------------------------------
                MEAN      TRAN-
                NUMBER/   SITION    OTHER
                DISINTE-  ENERGY    NUCLEAR
  TRANSITION    GRATION   (MEV)     DATA
  --------------------------------------------------
  BETA MINUS 1  0.0012    0.2340•   1ST FORBIDDEN
  BETA MINUS 2  0.0001    0.2470•   1ST FORBIDDEN
  BETA MINUS 3  0.0014    0.3730•   1ST FORBIDDEN
  BETA MINUS 4  0.1850    0.4560•   ALLOWED
  BETA MINUS 5  0.0001    0.6150•   U 1ST FORBIDDEN
  BETA MINUS 6  0.0004    0.7050•   U 1ST FORBIDDEN
  BETA MINUS 7  0.0143    0.8680•   1ST FORBIDDEN
  BETA MINUS 8  0.7970    1.2340•   1ST FORBIDDEN
  GAMMA      1  0.0630    0.0405    M1, AK(T) =
                                       3.28
                                    AL(T) =
                                       0.40R
  GAMMA      2  0.0633    0.1405    M1, AK= 0.104
                                    K/L= 7.70
  GAMMA      3  0.0760    0.1810    E2, AK= 0.130
                                    AL(T) =
                                       0.0189
  GAMMA      4  0.0145    0.3664    M1, AK(T) =
                                       0.00792
                                    AL(T) =
                                       0.000932
  GAMMA      5  0.0001    0.3807    M1, AK(T) =
                                       0.00721
                                    AL(T) =
                                       0.000847
  GAMMA      6  0.0002    0.4115    E1, AK(T) =
                                       0.00224
                                    AL(T) =
                                       0.000254
```

Table 3.6-1 (continued)
SUMMARY NUCLIDE DECAY DATA AND DOSE CONSTANT DATA

••INPUT DATA••

42 MOLYBDENUM 99 (CONTINUED)

TRANSITION	MEAN NUMBER/ DISINTE- GRATION	TRAN- SITION ENERGY (MEV)	OTHER NUCLEAR DATA
GAMMA 7	0.0005	0.5289	E2, AK(T) = 0.00373 AL(T) = 0.000457
GAMMA 8	0.0002	0.6207	E1, AK(T) = 0.000841 AL(T) = 0.000095
GAMMA 9	0.1370	0.7397	M1, AK= 0.00158 AL(T) = 0.000176
GAMMA 10	0.0480	0.7782	M1, AK(T) = 0.00134 AL(T) = 0.000157
GAMMA 11	0.0014	0.8231	F1, AK(T) = 0.000459 AL(T) = 0.000052
GAMMA 12	0.0011	0.9610	E1, AK(T) = 0.000337 AL(T) = 0.000038

•ENDPOINT ENERGY (MEV).

REF.- VAN EIJK, C.W. ET AL, NUCL. PHYS. A121, 440 (1968).

••OUTPUT DATA••

42 MOLYBDENUM 99   HALF LIFE = 66.7 HOURS

DECAY MODE- BETA MINUS

RADIATION	MEAN NUMBER/ DISINTE- GRATION n_i	MEAN ENERGY/ PAR- TICLE $\bar{E}_i$ (MeV)	EQUI- LIBRIUM DOSE CONSTANT Δ_i (g-rad/ µCi-h)
BETA MINUS 1	0.0012	0.0658	0.0001
BETA MINUS 3	0.0014	0.1112	0.0003
BETA MINUS 4	0.1850	0.1401	0.0552
BETA MINUS 6	0.0004	0.2541	0.0002
BETA MINUS 7	0.0143	0.2981	0.0090
BETA MINUS 8	0.7970	0.4519	0.7673
GAMMA 1	0.0130	0.0405	0.0011
K INT CON ELECT	0.0428	0.0195	0.0017
L INT CON ELECT	0.0053	0.0377	0.0004
M INT CON ELECT	0.0017	0.0401	0.0001
GAMMA 2	0.0564	0.1405	0.0168
K INT CON ELECT	0.0058	0.1194	0.0014
L INT CON ELECT	0.0007	0.1377	0.0002
GAMMA 3	0.0657	0.1810	0.0253
K INT CON ELECT	0.0085	0.1600	0.0029
L INT CON ELECT	0.0012	0.1782	0.0004
M INT CON ELECT	0.0004	0.1806	0.0001
GAMMA 4	0.0143	0.3664	0.0112
GAMMA 5	0.0001	0.3807	0.0000
GAMMA 6	0.0002	0.4115	0.0002
GAMMA 7	0.0005	0.5289	0.0005
GAMMA 8	0.0002	0.6207	0.0003
GAMMA 9	0.1367	0.7397	0.2154
K INT CON ELECT	0.0002	0.7186	0.0003
GAMMA 10	0.0479	0.7782	0.0794
K INT CON ELECT	0.0000	0.7571	0.0001
GAMMA 11	0.0014	0.8231	0.0024
GAMMA 12	0.0011	0.9610	0.0022
K ALPHA-1 X-RAY	0.0253	0.0183	0.0009
K ALPHA-2 X-RAY	0.0127	0.0182	0.0004
K BETA-1 X-RAY	0.0060	0.0206	0.0002
KLL AUGER ELECT	0.0087	0.0154	0.0002
KLX AUGER ELECT	0.0032	0.0178	0.0001
LMM AUGER ELECT	0.0615	0.0019	0.0002
MXY AUGER ELECT	0.1403	0.0004	0.0001

DAUGHTER NUCLIDE, TECHNETIUM 99M IS RADIOACTIVE AND MAY CONTRIBUTE TO THE DOSE. BRANCHING TO 0.1426 MEV, 6.03 HOUR HALF LIFE, ISOMERIC LEVEL IN TECHNETIUM-99 IS 0.860 PER DISINTEGRATION OF MOLYBDENUM-99.

TECHNETIUM-99M
ISOMERIC LEVEL DECAY

••INPUT DATA••

43 TECHNETIUM 99M   HALF LIFE = 6.03 HOURS

DECAY MODE- ISOMERIC LEVEL

TRANSITION	MEAN NUMBER/ DISINTE- GRATION	TRAN- SITION ENERGY (MEV)	OTHER NUCLEAR DATA
GAMMA 1	0.9860	0.0021	E3
GAMMA 2	0.9860	0.1405	M1, AK= 0.104 K/L= 7.70
GAMMA 3	0.0140	0.1426	M4, AK=23.0 AL(T) = 9.21

REF.- LEDERER, C.M. ET AL, TABLE OF ISOTOPES, 6TH ED.
LEGRAND, J. ET AL, INT. J. APPL. RAD. AND ISOTOPES 21, 139 (1970).

••OUTPUT DATA••

43 TECHNETIUM 99M   HALF LIFE = 6.03 HOURS

DECAY MODE- ISOMERIC LEVEL

RADIATION	MEAN NUMBER/ DISINTE- GRATION n_i	MEAN ENERGY/ PAR- TICLE $\bar{E}_i$ (MeV)	EQUI- LIBRIUM DOSE CONSTANT Δ_i (g-rad/ µCi-h)
GAMMA 1	0.0000	0.0021	0.0000
M INT CON ELECT	0.9860	0.0016	0.0035
GAMMA 2	0.8787	0.1405	0.2630
K INT CON ELECT	0.0913	0.1194	0.0232
L INT CON ELECT	0.0118	0.1377	0.0034
M INT CON ELECT	0.0039	0.1400	0.0011
GAMMA 3	0.0003	0.1426	0.0001
K INT CON ELECT	0.0088	0.1215	0.0022
L INT CON ELECT	0.0035	0.1398	0.0010
M INT CON ELECT	0.0011	0.1422	0.0003
K ALPHA-1 X-RAY	0.0441	0.0183	0.0017
K ALPHA-2 X-RAY	0.0221	0.0182	0.0008
K BETA-1 X-RAY	0.0105	0.0206	0.0004
KLL AUGER ELECT	0.0152	0.0154	0.0005
KLX AUGER ELECT	0.0055	0.0178	0.0002
LMM AUGER ELECT	0.1093	0.0019	0.0004
MXY AUGER ELECT	1.2359	0.0004	0.0011

Table 3.6-1 (continued)
SUMMARY NUCLIDE DECAY DATA AND DOSE CONSTANT DATA

CADMIUM-109
ELECTRON CAPTURE DECAY

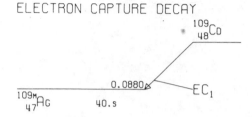

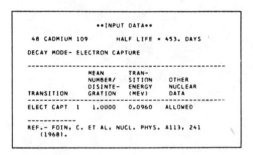

```
              **INPUT DATA**

    48 CADMIUM 109       HALF LIFE = 453. DAYS

DECAY MODE- ELECTRON CAPTURE

-------------------------------------------
              MEAN      TRAN-
              NUMBER/   SITION
              DISINTE-  ENERGY   OTHER
              GRATION   (MEV)    NUCLEAR
TRANSITION                       DATA
-------------------------------------------
ELECT CAPT 1  1.0000    0.0960   ALLOWED

------------
REF.- FOIN, C. ET AL, NUCL. PHYS. A113, 241
    (1968).
```

```
                    **OUTPUT DATA**

     48 CADMIUM 109       HALF LIFE = 453. DAYS

DECAY MODE- ELECTRON CAPTURE

-------------------------------------------------
                   MEAN       MEAN      EQUI-
                   NUMBER/    ENERGY/   LIBRIUM
                   DISINTE-   PAR-      DOSE
                   GRATION    TICLE     CONSTANT
   RADIATION
                    n_i        Ē_i       Δ_i
                              (MeV)     (g-rad/
                                         μCi-h)
-------------------------------------------------
 K ALPHA-1 X-RAY   0.3704    0.0221    0.0174
 K ALPHA-2 X-RAY   0.1874    0.0219    0.0087
  K BETA-1 X-RAY   0.0939    0.0249    0.0049
  K BETA-2 X-RAY   0.0183    0.0255    0.0009
     L X-RAYS      0.0907    0.0029    0.0005
 KLL AUGER ELECT   0.0932    0.0184    0.0036
 KLX AUGER ELECT   0.0365    0.0214    0.0016
 KXY AUGER ELECT   0.0055    0.0243    0.0002
 LMM AUGER ELECT   0.8547    0.0023    0.0043
 MXY AUGER ELECT   1.9717    0.0005    0.0023
------------
DAUGHTER NUCLIDE, SILVER 109M IS RADIOACTIVE
AND MAY CONTRIBUTE TO THE DOSE.
BRANCHING TO 0.0880 MEV, 40.0 SECOND HALF LIFE,
    ISOMERIC LEVEL IN SILVER-109 IS 1.00 PER
    DISINTEGRATION OF CADMIUM-109.
```

INDIUM-113M
ISOMERIC LEVEL DECAY

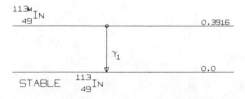

```
              **INPUT DATA**

    49 INDIUM 113M       HALF LIFE = 99.4 MINUTES

DECAY MODE- ISOMERIC LEVEL

-------------------------------------------
              MEAN      TRAN-
              NUMBER/   SITION
              DISINTE-  ENERGY   OTHER
              GRATION   (MEV)    NUCLEAR
TRANSITION                       DATA
-------------------------------------------
   GAMMA  1   1.0000    0.3916   M4, AK= 0.430
                                 K/L= 5.30
                                 K/M= 4.30

------------
REF.- NUCLEAR DATA B5, 195 (1971).
```

```
                    **OUTPUT DATA**

     49 INDIUM 113M       HALF LIFE = 99.4 MINUTES

DECAY MODE- ISOMERIC LEVEL

-------------------------------------------------
                   MEAN       MEAN      EQUI-
                   NUMBER/    ENERGY/   LIBRIUM
                   DISINTE-   PAR-      DOSE
                   GRATION    TICLE     CONSTANT
   RADIATION
                    n_i        Ē_i       Δ_i
                              (MeV)     (g-rad/
                                         μCi-h)
-------------------------------------------------
       GAMMA  1    0.6206    0.3916    0.5178
 K INT CON ELECT   0.2668    0.3637    0.2067
 L INT CON ELECT   0.0503    0.3877    0.0415
 M INT CON ELECT   0.0620    0.3910    0.0516
 K ALPHA-1 X-RAY   0.1243    0.0242    0.0064
 K ALPHA-2 X-RAY   0.0632    0.0240    0.0032
  K BETA-1 X-RAY   0.0324    0.0272    0.0018
  K BETA-2 X-RAY   0.0065    0.0279    0.0003
     L X-RAYS      0.0325    0.0032    0.0002
 KLL AUGER ELECT   0.0274    0.0201    0.0011
 KLX AUGER ELECT   0.0110    0.0233    0.0005
 LMM AUGER ELECT   0.2715    0.0025    0.0014
 MXY AUGER ELECT   0.6846    0.0006    0.0009
```

Table 3.6-1 (continued)
SUMMARY NUCLIDE DECAY DATA AND DOSE CONSTANT DATA

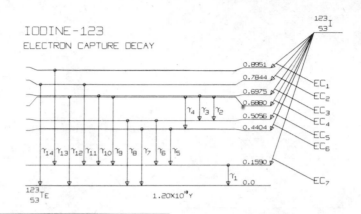

IODINE-123
ELECTRON CAPTURE DECAY

INPUT DATA

53 IODINE 123   HALF LIFE = 13.0 HOURS

DECAY MODE- ELECTRON CAPTURE

TRANSITION		MEAN NUMBER/ DISINTE- GRATION	TRAN- SITION ENERGY (MEV)	OTHER NUCLEAR DATA
ELECT CAPT	1	0.0003	0.3000	ALLOWED
ELECT CAPT	2	0.0011	0.4200	ALLOWED
ELECT CAPT	3	0.0030	0.5000	ALLOWED
ELECT CAPT	4	0.0118	0.5100	ALLOWED
ELECT CAPT	5	0.0031	0.6900	ALLOWED
ELECT CAPT	6	0.0035	0.7600	ALLOWED
ELECT CAPT	7	0.9772	1.0400	ALLOWED
GAMMA	1	0.9932	0.1591	M1, AK(T) = 0.160 AL(T) = 0.0208
GAMMA	2	0.0002	0.1837	M1, AK(T) = 0.108 AL(T) = 0.0140
GAMMA	3	0.0002	0.1927	M1, AK(T) = 0.0954 AL(T) = 0.0123
GAMMA	4	0.0007	0.2483	M1, AK(T) = 0.0485 AL(T) = 0.00625
GAMMA	5	0.0006	0.2810	M1, AK(T) = 0.0351 AL(T) = 0.00451
GAMMA	6	0.0010	0.3466	M1, AK(T) = 0.0204 AL(T) = 0.00262
GAMMA	7	0.0035	0.4404	M1, AK(T) = 0.0111 AL(T) = 0.00142
GAMMA	8	0.0026	0.5056	E2, AK(T) = 0.00674 AL(T) = 0.000954
GAMMA	9	0.0106	0.5290	M1, AK(T) = 0.00710 AL(T) = 0.000908
GAMMA	10	0.0027	0.5385	M1, AK(T) = 0.00680 AL(T) = 0.000870
GAMMA	11	0.0006	0.6249	M1, AK(T) = 0.00474 AL(T) = 0.000607
GAMMA	12	0.0002	0.6877	E2, AK(T) = 0.00299 AL(T) = 0.000402
GAMMA	13	0.0003	0.7361	M1, AK(T) = 0.00322 AL(T) = 0.000412

INPUT DATA

53 IODINE 123 (CONTINUED)

TRANSITION		MEAN NUMBER/ DISINTE- GRATION	TRAN- SITION ENERGY (MEV)	OTHER NUCLEAR DATA
GAMMA	14	0.0004	0.7844	M1, AK(T) = 0.00277 AL(T) = 0.000355

REF.- RAGAINI, R.C. ET AL, NUCL. PHYS. A115, 611 (1968).
SPEJWSKI, E.H. ET AL, NUCL. PHYS. A146, 182 (1970).
HUPF, H.B. ET AL, INT. J. APPL. RAD. AND ISOTOPES 19, 345 (1968).

OUTPUT DATA

53 IODINE 123   HALF LIFE = 13.0 HOURS

DECAY MODE- ELECTRON CAPTURE

RADIATION		MEAN NUMBER/ DISINTE- GRATION n_i	MEAN ENERGY/ PAR- TICLE E_i (MeV)	EQUI- LIBRIUM DOSE CONSTANT Δ_i (g-rad/ μCi-h)
GAMMA	1	0.8356	0.1591	0.2831
K INT CON ELECT		0.1343	0.1272	0.0364
L INT CON ELECT		0.0174	0.1545	0.0057
M INT CON ELECT		0.0058	0.1582	0.0019
GAMMA	2	0.0002	0.1837	0.0000
GAMMA	3	0.0002	0.1927	0.0001
GAMMA	4	0.0006	0.2483	0.0003
GAMMA	5	0.0006	0.2810	0.0003
GAMMA	6	0.0010	0.3466	0.0007
GAMMA	7	0.0034	0.4404	0.0032
GAMMA	8	0.0026	0.5056	0.0028
GAMMA	9	0.0105	0.5290	0.0118
GAMMA	10	0.0026	0.5385	0.0030
GAMMA	11	0.0006	0.6249	0.0009
GAMMA	12	0.0002	0.6877	0.0003
GAMMA	13	0.0002	0.7361	0.0004
GAMMA	14	0.0004	0.7844	0.0007
K ALPHA-1 X-RAY		0.4715	0.0274	0.0275
K ALPHA-2 X-RAY		0.2419	0.0272	0.0140
K BETA-1 X-RAY		0.1273	0.0309	0.0084
K BETA-2 X-RAY		0.0264	0.0318	0.0017
L X-RAYS		0.1332	0.0037	0.0010
KLL AUGER ELECT		0.0877	0.0226	0.0042
KLX AUGER ELECT		0.0370	0.0264	0.0020
KXY AUGER ELECT		0.0059	0.0301	0.0003
LMM AUGER ELECT		0.9242	0.0029	0.0057
MXY AUGER ELECT		2.1864	0.0008	0.0038

Table 3.6-1 (continued)
SUMMARY NUCLIDE DECAY DATA AND DOSE CONSTANT DATA

IODINE-125
ELECTRON CAPTURE DECAY

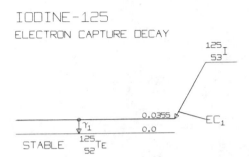

```
**INPUT DATA**

53 IODINE 125        HALF LIFE = 60.2 DAYS

DECAY MODE- ELECTRON CAPTURE

-------------------------------------------
            MEAN      TRAN-
            NUMBER/   SITION
            DISINTE-  ENERGY   OTHER
            GRATION   (MEV)    NUCLEAR
TRANSITION                     DATA
-------------------------------------------
ELECT CAPT 1  1.0000  0.1420   ALLOWED
     GAMMA 1  1.0000  0.0354   M1, AK=12.0
                               K/L= 7.00
                               K/(L+M)= 6.00

-----------
REF.- KARTTUNEN, E. ET AL, NUCL. PHYS. A131, 343
   (1969).
```

```
**OUTPUT DATA**

53 IODINE 125        HALF LIFE = 60.2 DAYS

DECAY MODE- ELECTRON CAPTURE

----------------------------------------------------
                      MEAN      MEAN     EQUI-
                      NUMBER/   ENERGY/  LIBRIUM
                      DISINTE-  PAR-     DOSE
            RADIATION GRATION   TICLE    CONSTANT

                      n_i       Ē_i      Δ_i
                                (MeV)    (g-rad/
                                         μCi-h)
----------------------------------------------------
        GAMMA 1  0.0666  0.0354  0.0050
K INT CON ELECT  0.8000  0.0036  0.0062
L INT CON ELECT  0.1142  0.0309  0.0075
M INT CON ELECT  0.0190  0.0346  0.0014
K ALPHA-1 X-RAY  0.7615  0.0274  0.0445
K ALPHA-2 X-RAY  0.3906  0.0272  0.0226
 K BETA-1 X-RAY  0.2056  0.0309  0.0135
 K BETA-2 X-RAY  0.0426  0.0318  0.0028
      L X-RAYS   0.2226  0.0037  0.0017
 KLL AUGER ELECT 0.1416  0.0226  0.0068
 KLX AUGER ELECT 0.0597  0.0264  0.0033
 KXY AUGER ELECT 0.0096  0.0301  0.0006
 LMM AUGER ELECT 1.5442  0.0029  0.0096
 MXY AUGER ELECT 3.6461  0.0008  0.0063
```

IODINE-129
BETA-MINUS DECAY

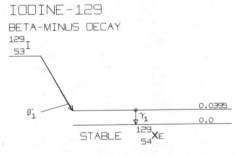

```
**INPUT DATA**

53 IODINE 129        HALF LIFE = 15.7 MEGAYEAR

DECAY MODE- BETA MINUS

-------------------------------------------
            MEAN      TRAN-
            NUMBER/   SITION
            DISINTE-  ENERGY   OTHER
            GRATION   (MEV)    NUCLEAR
TRANSITION                     DATA
-------------------------------------------
BETA MINUS 1  1.0000  0.1500*  U 1ST FORBIDDEN
     GAMMA 1  1.0000  0.0395   M1, AK=10.0
                               K/(L+M)= 6.80

*ENDPOINT ENERGY (MEV).

-----------
REF.- LEDERER, C.M. ET AL, TABLE OF ISOTOPES,
   6TH. ED.
```

```
**OUTPUT DATA**

53 IODINE 129        HALF LIFE = 15.7 MEGAYEAR

DECAY MODE- BETA MINUS

----------------------------------------------------
                      MEAN      MEAN     EQUI-
                      NUMBER/   ENERGY/  LIBRIUM
                      DISINTE-  PAR-     DOSE
            RADIATION GRATION   TICLE    CONSTANT

                      n_i       Ē_i      Δ_i
                                (MeV)    (g-rad/
                                         μCi-h)
----------------------------------------------------
   BETA MINUS 1  1.0000  0.0484  0.1032
        GAMMA 1  0.0801  0.0395  0.0067
K INT CON ELECT  0.8018  0.0050  0.0085
L INT CON ELECT  0.0884  0.0345  0.0065
M INT CON ELECT  0.0294  0.0386  0.0024
K ALPHA-1 X-RAY  0.3814  0.0297  0.0241
K ALPHA-2 X-RAY  0.1968  0.0294  0.0123
 K BETA-1 X-RAY  0.1052  0.0336  0.0075
 K BETA-2 X-RAY  0.0221  0.0345  0.0016
      L X-RAYS   0.1134  0.0041  0.0009
 KLL AUGER ELECT 0.0639  0.0244  0.0033
 KLX AUGER ELECT 0.0277  0.0285  0.0016
 KXY AUGER ELECT 0.0045  0.0326  0.0003
 LMM AUGER ELECT 0.7087  0.0031  0.0048
 MXY AUGER ELECT 1.7026  0.0009  0.0033
```

Table 3.6-1 (continued)
SUMMARY NUCLIDE DECAY DATA AND DOSE CONSTANT DATA

IODINE-131

BETA-MINUS DECAY

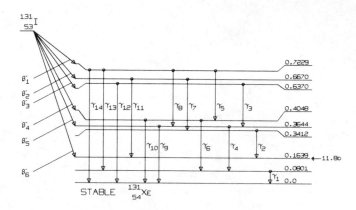

STABLE $^{131}_{54}$Xe

•• INPUT DATA ••

53 IODINE 131   HALF LIFE = 8.06 DAYS

DECAY MODE- BETA MINUS

TRANSITION		MEAN NUMBER/ DISINTE- GRATION	TRAN- SITION ENERGY (MEV)	OTHER NUCLEAR DATA
BETA MINUS	1	0.0200	0.2470*	ALLOWED
BETA MINUS	2	0.0067	0.3030*	1ST FORBIDDEN
BETA MINUS	3	0.0664	0.3330*	ALLOWED
BETA MINUS	4	0.0000	0.5650*	U 1ST FORBIDDEN
BETA MINUS	5	0.8980	0.6060*	ALLOWED
BETA MINUS	6	0.0080	0.8060*	U 1ST FORBIDDEN
GAMMA	1	0.0660	0.0801	M1, AK= 1.33 AL= 0.170
GAMMA	2	0.0035	0.1772	E2, AK= 0.155 K/L= 5.10
GAMMA	3	0.0006	0.2723	M1, AK(T) = 0.0453 AL(T) = 0.00594
GAMMA	4	0.0607	0.2843	E2, AK= 0.0410 K/L= 5.50
GAMMA	5	0.0011	0.3180	E2, AK(T) = 0.0288 AL(T) = 0.00484
GAMMA	6	0.0003	0.3250	M1, AK(T) = 0.0286 AL(T) = 0.00374
GAMMA	7	0.0037	0.3257	M1, AK= 0.0236 K/L= 6.20
GAMMA	8	0.0001	0.3585	M1, AK(T) = 0.0222 AL(T) = 0.00291
GAMMA	9	0.8380	0.3644	E2, AK= 0.0180 K/L= 6.30
GAMMA	10	0.0006	0.4048	M1, AK(T) = 0.0163 AL(T) = 0.00213
GAMMA	11	0.0029	0.5029	E2, AK= 0.00850 AL(T) = 0.00109
GAMMA	12	0.0657	0.6367	E2, AK= 0.00430 AL(T) = 0.000564
GAMMA	13	0.0014	0.6430	E2, AK(T) = 0.00392 AL(T) = 0.000549
GAMMA	14	0.0174	0.7228	M1, AK= 0.00370 AL(T) = 0.000514

*ENDPOINT ENERGY (MEV).

REF.- GRAEFFE, G. AND WALTERS, W.B., PHYS. REV.
   153, 1321 (1967).
   LEDERER, C.M. ET AL, TABLE OF ISOTOPES, 6TH.
   ED.

•• OUTPUT DATA ••

53 IODINE 131   HALF LIFE = 8.06 DAYS

DECAY MODE- BETA MINUS

RADIATION		MEAN NUMBER/ DISINTE- GRATION n_i	MEAN ENERGY/ PAR- TICLE $\bar{E}_i$ (MeV)	EQUI- LIBRIUM DOSE CONSTANT Δ_i (g·rad/ uCi·h)
BETA MINUS	1	0.0200	0.0691	0.0029
BETA MINUS	2	0.0067	0.0867	0.0012
BETA MINUS	3	0.0664	0.0964	0.0136
BETA MINUS	5	0.8980	0.1916	0.3666
BETA MINUS	6	0.0080	0.2839	0.0048
GAMMA	1	0.0258	0.0801	0.0044
K INT CON ELECT		0.0343	0.0456	0.0033
L INT CON ELECT		0.0043	0.0751	0.0007
M INT CON ELECT		0.0014	0.0792	0.0002
GAMMA	2	0.0029	0.1772	0.0011
K INT CON ELECT		0.0004	0.1426	0.0001
GAMMA	3	0.0006	0.2723	0.0003
GAMMA	4	0.0578	0.2843	0.0350
K INT CON ELECT		0.0023	0.2497	0.0012
L INT CON ELECT		0.0004	0.2792	0.0002
GAMMA	5	0.0010	0.3180	0.0007
GAMMA	6	0.0003	0.3250	0.0002
GAMMA	7	0.0036	0.3257	0.0025
GAMMA	8	0.0001	0.3585	0.0001
GAMMA	9	0.8201	0.3644	0.6366
K INT CON ELECT		0.0147	0.3299	0.0103
L INT CON ELECT		0.0023	0.3594	0.0017
M INT CON ELECT		0.0007	0.3635	0.0006
GAMMA	10	0.0006	0.4048	0.0005
GAMMA	11	0.0029	0.5029	0.0031
GAMMA	12	0.0653	0.6367	0.0886
K INT CON ELECT		0.0002	0.6021	0.0003
GAMMA	13	0.0014	0.6430	0.0020
GAMMA	14	0.0173	0.7228	0.0267
K ALPHA-1 X-RAY		0.0249	0.0297	0.0015
K ALPHA-2 X-RAY		0.0128	0.0294	0.0008
K BETA-1 X-RAY		0.0068	0.0336	0.0004
K BETA-2 X-RAY		0.0014	0.0345	0.0001
KLL AUGER ELECT		0.0041	0.0244	0.0002
KLX AUGER ELECT		0.0018	0.0285	0.0001
LMM AUGER ELECT		0.0477	0.0031	0.0003
MXY AUGER ELECT		0.1147	0.0009	0.0002

DAUGHTER NUCLIDE, XENON 131M IS RADIOACTIVE
AND MAY CONTRIBUTE TO THE DOSE.
BRANCHING TO 0.1639 MEV, 11.8 DAY HALF LIFE,
   ISOMERIC LEVEL IN XENON-131 IS 0.0144 PER
   DISINTEGRATION OF IODINE-131.

Table 3.6-1 (continued)
SUMMARY NUCLIDE DECAY DATA AND DOSE CONSTANT DATA

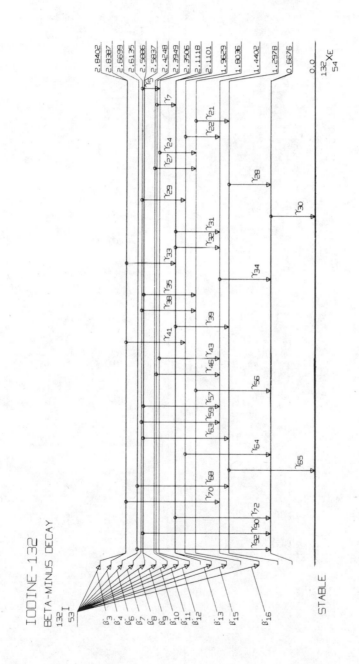

Table 3.6-1 (continued)
SUMMARY NUCLIDE DECAY DATA AND DOSE CONSTANT DATA

```
            ••INPUT DATA••

  53 IODINE 132       HALF LIFE = 2.38 HOURS

DECAY MODE- BETA MINUS

-------------------------------------------------
                 MEAN     TRAN-
                 NUMBER/  SITION
                 DISINTE- ENERGY  OTHER
                 GRATION  (MEV)   NUCLEAR
  TRANSITION                      DATA
-------------------------------------------------
BETA MINUS  1    0.0030   0.5010*  ALLOWED
BETA MINUS  2    0.0080   0.6690*  ALLOWED
BETA MINUS  3    0.0190   0.7200*  ALLOWED
BETA MINUS  4    0.1410   0.7210*  ALLOWED
BETA MINUS  5    0.0050   0.8060*  ALLOWED
BETA MINUS  6    0.0350   0.8900*  ALLOWED
BETA MINUS  7    0.0650   0.9460*  ALLOWED
BETA MINUS  8    0.0300   0.9710*  ALLOWED
BETA MINUS  9    0.0260   0.9760*  ALLOWED
BETA MINUS 10    0.0290   1.1350*  ALLOWED
BETA MINUS 11    0.1840   1.1650*  ALLOWED
BETA MINUS 12    0.0170   1.2090*  ALLOWED
BETA MINUS 13    0.1080   1.4500*  ALLOWED
BETA MINUS 14    0.0010   1.5750*  ALLOWED
BETA MINUS 15    0.1270   1.5970*  ALLOWED
BETA MINUS 16    0.1970   2.1200*  ALLOWED
  GAMMA     1    0.0012   0.1366    , AK= 0.306
  GAMMA     2    0.0028   0.1472   E2, AK= 0.135
                                   AL(T) =
                                   0.0933
  GAMMA     3    0.0018   0.1833    AK= 0.127
  GAMMA     4    0.0021   0.2548    AK= 0.0770
  GAMMA     5    0.0155   0.2627    AK= 0.0440
  GAMMA     6    0.0004   0.2789
  GAMMA     7    0.0080   0.2848   M1, AK= 0.00310
                                   AL(T) =
                                   0.00528
  GAMMA     8    0.0011   0.3066   M1, AK(T) =
                                   0.0332
                                   AL(T) =
                                   0.00435
  GAMMA     9    0.0009   0.3100
  GAMMA    10    0.0016   0.3165
  GAMMA    11    0.0010   0.3436
  GAMMA    12    0.0008   0.3518
  GAMMA    13    0.0050   0.3635   M1, AK(T) =
                                   0.0214
                                   AL(T) =
                                   0.00280
  GAMMA    14    0.0017   0.3878
  GAMMA    15    0.0047   0.4168
  GAMMA    16    0.0046   0.4319   M1, AK(T) =
                                   0.0138
                                   AL(T) =
                                   0.00181
  GAMMA    17    0.0068   0.4460   M1, AK(T) =
                                   0.0128

                                   AL(T) =
                                   0.00167
  GAMMA    18    0.0027   0.4734   M1, AK(T) =
                                   0.0110
                                   AL(T) =
                                   0.00144
  GAMMA    19    0.0010   0.4779   M1, AK(T) =
                                   0.0107
                                   AL(T) =
                                   0.00140
  GAMMA    20    0.0018   0.4875
  GAMMA    21    0.0510   0.5059   M1, AK= 0.00620
                                   L/M= 1.69
                                   AL(T) =
                                   0.00122
  GAMMA    22    0.1630   0.5226   M1, AK= 0.00790
                                   L/M= 0.0625
                                   AL(T) =
                                   0.00112
  GAMMA    23    0.0053   0.5355
  GAMMA    24    0.0127   0.5471
  GAMMA    25    0.0006   0.5909   M1, AK(T) =
                                   0.00640
                                   AL(T) =
                                   0.000835
  GAMMA    26    0.0009   0.6005
  GAMMA    27    0.0200   0.6210   M1, AK= 0.00880
                                   AL(T) =
                                   0.000740
  GAMMA    28    0.1390   0.6302   M1, AK= 0.00420
                                   L/M=28.1
                                   AL(T) =
                                   0.000715
  GAMMA    29    0.0270   0.6506   M1, AK= 0.00650
                                   AL(T) =
                                   0.000662
```

```
            ••INPUT DATA••
  53 IODINE 132 (CONTINUED)

-------------------------------------------------
                 MEAN     TRAN-
                 NUMBER/  SITION
                 DISINTE- ENERGY  OTHER
                 GRATION  (MEV)   NUCLEAR
  TRANSITION                      DATA
-------------------------------------------------
  GAMMA    30    0.9900   0.6676   E2, AK= 0.00350
                                   AL(T) =
                                   0.000496
  GAMMA    31    0.0500   0.6698   M1, AK= 0.00490
                                   AL(T) =
                                   0.000617
  GAMMA    32    0.0530   0.6716   E2, AK= 0.00400
                                   AL(T) =
                                   0.000489
  GAMMA    33    0.0660   0.7271   M1, AK= 0.00270
                                   L/M= 0.102
                                   AL(T) =
                                   0.000507
  GAMMA    34    0.7720   0.7726   E2, AK= 0.00273
                                   AL(T) =
                                   0.000341
  GAMMA    35    0.0125   0.7802   E2, AK(T) =
                                   0.00245
                                   AL(T) =
                                   0.000332
  GAMMA    36    0.0043   0.7845   M1, AK(T) =
                                   0.00325
                                   AL(T) =
                                   0.000424
  GAMMA    37    0.0009   0.7921   M1, AK(T) =
                                   0.00318
                                   AL(T) =
                                   0.000415
  GAMMA    38    0.0290   0.8098   M1, AK= 0.00290
                                   AL(T) =
                                   0.000394
  GAMMA    39    0.0570   0.8122   E2, AK= 0.00220
                                   AL(T) =
                                   0.000301
  GAMMA    40    0.0060   0.8633
  GAMMA    41    0.0109   0.8768   M1, AK(T) =
                                   0.00251
                                   AL(T) =
                                   0.000328
  GAMMA    42    0.0004   0.8890
  GAMMA    43    0.0093   0.9103
  GAMMA    44    0.0045   0.9276   M1, AK(T) =
                                   0.00220
                                   AL(T) =
                                   0.000288
  GAMMA    45    0.0008   0.9486   M1, AK(T) =
                                   0.00209
                                   AL(T) =
                                   0.000273
  GAMMA    46    0.1830   0.9545   M1, AK= 0.00199
                                   L/M= 0.0225
                                   AL(T) =
                                   0.000269
  GAMMA    47    0.0057   0.9845   M1, AK(T) =
                                   0.00192
                                   AL(T) =
                                   0.000251
  GAMMA    48    0.0005   1.0162
  GAMMA    49    0.0058   1.0347   E2, AK= 0.00151
                                   L/M= 0.810
                                   AL(T) =
                                   0.000171
  GAMMA    50    0.0004   1.0499
  GAMMA    51    0.0003   1.0655
  GAMMA    52    0.0007   1.0863   M1, AK(T) =
                                   0.00154
                                   AL(T) =
                                   0.000201
  GAMMA    53    0.0003   1.0968   E2, AK(T) =
                                   0.00115
                                   AL(T) =
                                   0.000150
  GAMMA    54    0.0006   1.1125
  GAMMA    55    0.0005   1.1266
  GAMMA    56    0.0300   1.1360   M1, AK(T) =
                                   0.00139
                                   AL(T) =
                                   0.000181
  GAMMA    57    0.0140   1.1434   M1, AK= 0.00205
                                   L/M= 0.00160
                                   AL(T) =
                                   0.000179
  GAMMA    58    0.0021   1.1482   M1, AK(T) =
                                   0.00136
                                   AL(T) =
                                   0.000177
```

Table 3.6-1 (continued)
SUMMARY NUCLIDE DECAY DATA AND DOSE CONSTANT DATA

INPUT DATA

53 IODINE 132 (CONTINUED)

TRANSITION	MEAN NUMBER/ DISINTE- GRATION	TRAN- SITION ENERGY (MEV)	OTHER NUCLEAR DATA
GAMMA 59	0.0110	1.1732	M1, AK= 0.00124 L/M= 0.160 AL(T) = 0.000168
GAMMA 60	0.0004	1.2541	M1, AK(T) = 0.00111 AL(T) = 0.000145
GAMMA 61	0.0002	1.2637	
GAMMA 62	0.0015	1.2727	
GAMMA 63	0.0115	1.2907	M1,L/M=0.000100 AK(T) = 0.00104 AL(T) = 0.000136
GAMMA 64	0.0200	1.2953	E2, AK(T) = 0.000817 AL(T) = 0.000104
GAMMA 65	0.0090	1.2982	E2, AK(T) = 0.000813 AL(T) = 0.000104
GAMMA 66	0.0006	1.3143	M1, AK(T) = 0.00100 AL(T) = 0.000130
GAMMA 67	0.0012	1.3171	M1, AK(T) = 0.00100 AL(T) = 0.000130
GAMMA 68	0.0250	1.3720	AK=0.000980
GAMMA 69	0.0024	1.3925	
GAMMA 70	0.0720	1.3985	M1, AK=0.000930 L/M= 0.00490 AL(T) = 0.000114
GAMMA 71	0.0006	1.4105	
GAMMA 72	0.0144	1.4425	E2, AK(T) = 0.000659 AL(T) = 0.000083
GAMMA 73	0.0004	1.4565	
GAMMA 74	0.0014	1.4768	
GAMMA 75	0.0005	1.5197	
GAMMA 76	0.0001	1.5422	
GAMMA 77	0.0004	1.5931	
GAMMA 78	0.0003	1.6206	
GAMMA 79	0.0001	1.6378	
GAMMA 80	0.0001	1.6616	
GAMMA 81	0.0005	1.7155	
GAMMA 82	0.0005	1.7208	
GAMMA 83	0.0006	1.7273	E2, AK(T) = 0.000466 AL(T) = 0.000058
GAMMA 84	0.0038	1.7575	
GAMMA 85	0.0006	1.7784	
GAMMA 86	0.0001	1.8144	
GAMMA 87	0.0002	1.8300	
GAMMA 88	0.0001	1.8792	
GAMMA 89	0.0006	1.9143	
GAMMA 90	0.0120	1.9210	M3, AK(T) = 0.00157 AL(T) = 0.000209
GAMMA 91	0.0000	1.9855	E2, AK(T) = 0.000360 AL(T) = 0.000044
GAMMA 92	0.0110	2.0023	
GAMMA 93	0.0025	2.0868	
GAMMA 94	0.0019	2.1726	
GAMMA 95	0.0012	2.2231	
GAMMA 96	0.0003	2.2491	
GAMMA 97	0.0001	2.2904	
GAMMA 98	0.0017	2.3904	
GAMMA 99	0.0001	2.4089	
GAMMA100	0.0003	2.5251	

*ENDPOINT ENERGY (MEV).

REF.- CARTER, H.K. ET AL, PHYS. REV. C1, 649 (1970).
    HAMILTON, J.H. ET AL, PHYS. REV. C1, 666 (1970).

OUTPUT DATA

53 IODINE 132   HALF LIFE = 2.38 HOURS

DECAY MODE- BETA MINUS

RADIATION		MEAN NUMBER/ DISINTE- GRATION n_i	MEAN ENERGY/ PAR- TICLE $\bar{E}_i$ (MeV)	EQUI- LIBRIUM DOSE CONSTANT Δ_i (g-rad/ μCi-h)
BETA MINUS	1	0.0030	0.1536	0.0009
BETA MINUS	2	0.0080	0.2152	0.0036
BETA MINUS	3	0.0190	0.2345	0.0094
BETA MINUS	4	0.1410	0.2349	0.0705
BETA MINUS	5	0.0050	0.2679	0.0028
BETA MINUS	6	0.0350	0.3012	0.0224
BETA MINUS	7	0.0650	0.3237	0.0448
BETA MINUS	8	0.0300	0.3338	0.0213
BETA MINUS	9	0.0260	0.3359	0.0186
BETA MINUS	10	0.0290	0.4016	0.0248
BETA MINUS	11	0.1840	0.4142	0.1623
BETA MINUS	12	0.0170	0.4327	0.0156
BETA MINUS	13	0.1080	0.5363	0.1233
BETA MINUS	14	0.0010	0.5911	0.0012
BETA MINUS	15	0.1270	0.6008	0.1625
BETA MINUS	16	0.1970	0.8360	0.3508
GAMMA	1	0.0009	0.1366	0.0002
GAMMA	2	0.0022	0.1472	0.0006
GAMMA	3	0.0016	0.1833	0.0006
GAMMA	4	0.0019	0.2548	0.0010
GAMMA	5	0.0148	0.2627	0.0083
K INT CON ELECT		0.0006	0.2281	0.0003
GAMMA	6	0.0004	0.2789	0.0002
GAMMA	7	0.0079	0.2848	0.0048
GAMMA	8	0.0010	0.3066	0.0006
GAMMA	9	0.0009	0.3100	0.0005
GAMMA	10	0.0016	0.3165	0.0010
GAMMA	11	0.0010	0.3436	0.0007
GAMMA	12	0.0008	0.3518	0.0006
GAMMA	13	0.0048	0.3635	0.0037
GAMMA	14	0.0017	0.3878	0.0014
GAMMA	15	0.0047	0.4168	0.0041
GAMMA	16	0.0045	0.4319	0.0041
GAMMA	17	0.0067	0.4460	0.0063
GAMMA	18	0.0026	0.4734	0.0026
GAMMA	19	0.0009	0.4779	0.0010
GAMMA	20	0.0018	0.4875	0.0018
GAMMA	21	0.0505	0.5059	0.0545
K INT CON ELECT		0.0003	0.4713	0.0003
GAMMA	22	0.1587	0.5226	0.1766
K INT CON ELECT		0.0012	0.4880	0.0013
L INT CON ELECT		0.0001	0.5175	0.0001
M INT CON ELECT		0.0028	0.5217	0.0031
GAMMA	23	0.0053	0.5355	0.0060
GAMMA	24	0.0127	0.5471	0.0148
GAMMA	25	0.0005	0.5909	0.0007
GAMMA	26	0.0009	0.6005	0.0011
GAMMA	27	0.0198	0.6210	0.0261
K INT CON ELECT		0.0001	0.5864	0.0002
GAMMA	28	0.1383	0.6302	0.1856
K INT CON ELECT		0.0005	0.5956	0.0007
L INT CON ELECT		0.0000	0.6251	0.0001
GAMMA	29	0.0268	0.6506	0.0371
K INT CON ELECT		0.0001	0.6160	0.0002
GAMMA	30	0.9858	0.6676	1.4021
K INT CON ELECT		0.0034	0.6331	0.0046
L INT CON ELECT		0.0004	0.6626	0.0006
M INT CON ELECT		0.0001	0.6667	0.0002
GAMMA	31	0.0497	0.6698	0.0709
K INT CON ELECT		0.0002	0.6352	0.0003
GAMMA	32	0.0527	0.6716	0.0754
K INT CON ELECT		0.0002	0.6370	0.0002
GAMMA	33	0.0654	0.7271	0.1013
K INT CON ELECT		0.0001	0.6925	0.0002
M INT CON ELECT		0.0003	0.7261	0.0005
GAMMA	34	0.7695	0.7726	1.2664
K INT CON ELECT		0.0021	0.7380	0.0033
L INT CON ELECT		0.0002	0.7675	0.0004
M INT CON ELECT		0.0000	0.7716	0.0001
GAMMA	35	0.0124	0.7802	0.0207
GAMMA	36	0.0042	0.7845	0.0071
GAMMA	37	0.0008	0.7921	0.0015
GAMMA	38	0.0289	0.8098	0.0498
K INT CON ELECT		0.0000	0.7752	0.0001
GAMMA	39	0.0568	0.8122	0.0983
K INT CON ELECT		0.0001	0.7776	0.0002
GAMMA	40	0.0060	0.8633	0.0110
GAMMA	41	0.0108	0.8768	0.0202
GAMMA	42	0.0004	0.8890	0.0007

Table 3.6-1 (continued)
SUMMARY NUCLIDE DECAY DATA AND DOSE CONSTANT DATA

```
                    **OUTPUT DATA**

            53 IODINE 132 (CONTINUED)
--------------------------------------------------
                       MEAN      MEAN    EQUI-
                       NUMBER/   ENERGY/ LIBRIUM
                       DISINTE-  PAR-    DOSE
          RADIATION    GRATION   TICLE   CONSTANT

                        n_i       Ē_i     Δ_i
                                 (MeV)   (g-rad/
                                          µCi-h)
--------------------------------------------------
```

RADIATION	n_i	$\bar{E}_i$ (MeV)	Δ_i (g-rad/µCi-h)
GAMMA 43	0.0093	0.9103	0.0180
GAMMA 44	0.0044	0.9276	0.0088
GAMMA 45	0.0007	0.9486	0.0016
GAMMA 46	0.1804	0.9545	0.3668
K INT CON ELECT	0.0003	0.9199	0.0007
M INT CON ELECT	0.0021	0.9536	0.0043
GAMMA 47	0.0056	0.9845	0.0119
GAMMA 48	0.0005	1.0162	0.0010
GAMMA 49	0.0057	1.0347	0.0127
GAMMA 50	0.0004	1.0499	0.0010
GAMMA 51	0.0003	1.0655	0.0007
GAMMA 52	0.0006	1.0863	0.0016
GAMMA 53	0.0003	1.0968	0.0008
GAMMA 54	0.0006	1.1125	0.0014
GAMMA 55	0.0005	1.1266	0.0012
GAMMA 56	0.0299	1.1360	0.0724
GAMMA 57	0.0125	1.1434	0.0306
M INT CON ELECT	0.0014	1.1424	0.0034
GAMMA 58	0.0020	1.1482	0.0051
GAMMA 59	0.0109	1.1732	0.0274
GAMMA 60	0.0004	1.2541	0.0012
GAMMA 61	0.0002	1.2637	0.0006
GAMMA 62	0.0015	1.2727	0.0040
GAMMA 63	0.0114	1.2907	0.0315
GAMMA 64	0.0199	1.2953	0.0551
GAMMA 65	0.0089	1.2982	0.0248
GAMMA 66	0.0006	1.3143	0.0016
GAMMA 67	0.0011	1.3171	0.0033
GAMMA 68	0.0249	1.3720	0.0729
GAMMA 69	0.0024	1.3925	0.0071
GAMMA 70	0.0702	1.3985	0.2093
K INT CON ELECT	0.0000	1.3640	0.0001
M INT CON ELECT	0.0016	1.3976	0.0048
GAMMA 71	0.0006	1.4105	0.0018
GAMMA 72	0.0143	1.4425	0.0442
GAMMA 73	0.0004	1.4565	0.0015
GAMMA 74	0.0014	1.4768	0.0044
GAMMA 75	0.0005	1.5197	0.0016
GAMMA 76	0.0001	1.5422	0.0003
GAMMA 77	0.0004	1.5931	0.0015
GAMMA 78	0.0003	1.6206	0.0010
GAMMA 79	0.0001	1.6378	0.0005
GAMMA 80	0.0001	1.6616	0.0006
GAMMA 81	0.0005	1.7155	0.0019
GAMMA 82	0.0005	1.7208	0.0020
GAMMA 83	0.0006	1.7273	0.0023
GAMMA 84	0.0038	1.7575	0.0142
GAMMA 85	0.0006	1.7784	0.0022
GAMMA 86	0.0001	1.8144	0.0003
GAMMA 87	0.0002	1.8300	0.0011
GAMMA 88	0.0001	1.8792	0.0006
GAMMA 89	0.0006	1.9143	0.0024
GAMMA 90	0.0119	1.9210	0.0490
GAMMA 91	0.0000	1.9855	0.0003
GAMMA 92	0.0110	2.0023	0.0469
GAMMA 93	0.0025	2.0868	0.0111
GAMMA 94	0.0019	2.1726	0.0087
GAMMA 95	0.0012	2.2231	0.0056
GAMMA 96	0.0003	2.2491	0.0014
GAMMA 97	0.0001	2.2904	0.0008
GAMMA 98	0.0017	2.3904	0.0086
GAMMA 99	0.0001	2.4089	0.0005
GAMMA100	0.0003	2.5251	0.0019
K ALPHA-1 X-RAY	0.0054	0.0297	0.0003
K ALPHA-2 X-RAY	0.0028	0.0294	0.0001
K BETA-1 X-RAY	0.0015	0.0336	0.0001

Table 3.6-1 (continued)
SUMMARY NUCLIDE DECAY DATA AND DOSE CONSTANT DATA

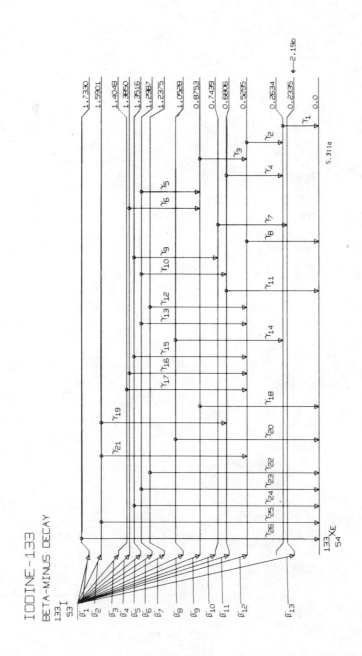

Table 3.6-1 (continued)
SUMMARY NUCLIDE DECAY DATA AND DOSE CONSTANT DATA

```
                    ••INPUT DATA••

    53 IODINE 133          HALF LIFE = 20.8 HOURS

    DECAY MODE- BETA MINUS
    --------------------------------------------------
                    MEAN      TRAN-
                    NUMBER/   SITION     OTHER
                    DISINTE-  ENERGY     NUCLEAR
    TRANSITION      GRATION   (MEV)      DATA
    --------------------------------------------------
    BETA MINUS   1   0.0002   0.0270*   ALLOWED
    BETA MINUS   2   0.0050   0.1700*   ALLOWED
    BETA MINUS   3   0.0050   0.3550*   ALLOWED
    BETA MINUS   4   0.0110   0.3750*   ALLOWED
    BETA MINUS   5   0.0030   0.4080*   ALLOWED
    BETA MINUS   6   0.0290   0.4610*   ALLOWED
    BETA MINUS   7   0.0320   0.5230*   ALLOWED
    BETA MINUS   8   0.0050   0.7072*   1ST FORBIDDEN
    BETA MINUS   9   0.0350   0.8850*   ALLOWED
    BETA MINUS  10   0.0230   1.0160*   1ST FORBIDDEN
    BETA MINUS  11   0.0020   1.0790*   U 1ST FORBIDDEN
    BETA MINUS  12   0.8320   1.2310*   ALLOWED
    BETA MINUS  13   0.0140   1.5270*   U 1ST FORBIDDEN
    GAMMA        1   0.0040   0.2634    M1, AK(T) =
                                        0.0494
                                        AL(T) =
                                        0.00649
    GAMMA        2   0.0013   0.2663    E2, AK(T) =
                                        0.0503
                                        AL(T) =
                                        0.00918
    GAMMA        3   0.0023   0.3440    M1, AK(T) =
                                        0.0247
                                        AL(T) =
                                        0.00323
    GAMMA        4   0.0014   0.4172
    GAMMA        5   0.0023   0.4229    M1, AK(T) =
                                        0.0146
                                        AL(T) =
                                        0.00191
    GAMMA        6   0.0007   0.5104    M1, AK(T) =
                                        0.00916
                                        AL(T) =
                                        0.00119
    GAMMA        7   0.0150   0.5104    M1, AK(T) =
                                        0.00916
                                        AL(T) =
                                        0.00119
    GAMMA        8   0.8710   0.5295    M1, AK= 0.00800
                                        L/M= 2.25
                                        AL(T) =
                                        0.00109
    GAMMA        9   0.0003   0.6080
    GAMMA       10   0.0032   0.6180
    GAMMA       11   0.0075   0.6808    M1, AK(T) =
                                        0.00455
                                        AL(T) =
                                        0.000594
    GAMMA       12   0.0160   0.7074    M1, AK(T) =
                                        0.00415
                                        AL(T) =
                                        0.000542
    GAMMA       13   0.0041   0.7691    M1, AK(T) =
                                        0.00341
                                        AL(T) =
                                        0.000444
    GAMMA       14   0.0004   0.7899
    GAMMA       15   0.0014   0.8209    M1, AK(T) =
                                        0.00293
                                        AL(T) =
                                        0.000382
    GAMMA       16   0.0102   0.8561
    GAMMA       17   0.0050   0.8753
    GAMMA       18   0.0357   0.8753
    GAMMA       19   0.0037   0.9105
    GAMMA       20   0.0015   1.0528    M1, AK(T) =
                                        0.00165
                                        AL(T) =
                                        0.000215
    GAMMA       21   0.0008   1.0611    M1, AK(T) =
                                        0.00162
                                        AL(T) =
                                        0.000212
    GAMMA       22   0.0160   1.2375
    GAMMA       23   0.0193   1.2989
    GAMMA       24   0.0013   1.3516
    GAMMA       25   0.0004   1.5901
    GAMMA       26   0.0002   1.7330

    •ENDPOINT ENERGY (MEV).

    ------------
    REF.- SAXENA, R.N. AND SHAMA, H.D., NUCL. PHYS.
    A171, 593 (1971).
```

```
                    ••INPUT DATA••

        53 IODINE 133 (CONTINUED)
    ------------------------------------------------
                    MEAN      TRAN-
                    NUMBER/   SITION    OTHER
                    DISINTE-  ENERGY    NUCLEAR
    TRANSITION      GRATION   (MEV)     DATA
    ------------------------------------------------

    ------------
    BRANCHING TO 0.2335 MEV, 2.19 DAY HALF LIFE,
      ISOMERIC LEVEL IN XENON-133 IS 0.0147 PER
      DISINTEGRATION OF IODINE-133.
```

```
                        ••OUTPUT DATA••

        53 IODINE 133          HALF LIFE = 20.8 HOURS

    DECAY MODE- BETA MINUS
    ----------------------------------------------------------
                        MEAN      MEAN      EQUI-
                        NUMBER/   ENERGY/   LIBRIUM
                        DISINTE-  PAR-      DOSE
        RADIATION       GRATION   TICLE     CONSTANT

                          n_i      Ē_i       Δ_i
                                  (MeV)    (g-rad/
                                            μCi·h)
    ----------------------------------------------------------
    BETA MINUS    2      0.0050    0.0460    0.0004
    BETA MINUS    3      0.0050    0.1036    0.0011
    BETA MINUS    4      0.0110    0.1102    0.0025
    BETA MINUS    5      0.0030    0.1213    0.0007
    BETA MINUS    6      0.0290    0.1396    0.0086
    BETA MINUS    7      0.0320    0.1615    0.0110
    BETA MINUS    8      0.0050    0.2297    0.0024
    BETA MINUS    9      0.0350    0.2992    0.0223
    BETA MINUS   10      0.0230    0.3522    0.0172
    BETA MINUS   11      0.0020    0.3906    0.0016
    BETA MINUS   12      0.8320    0.4421    0.7834
    BETA MINUS   13      0.0140    0.5748    0.0171
    GAMMA         1      0.0037    0.2634    0.0021
    GAMMA         2      0.0012    0.2663    0.0006
    GAMMA         3      0.0022    0.3440    0.0016
    GAMMA         4      0.0014    0.4172    0.0012
    GAMMA         5      0.0022    0.4229    0.0020
    GAMMA         6      0.0007    0.5104    0.0007
    GAMMA         7      0.0148    0.5104    0.0161
    K INT CON ELECT      0.0001    0.4758    0.0001
    GAMMA         8      0.8627    0.5295    0.9730
    K INT CON ELECT      0.0069    0.4949    0.0072
    L INT CON ELECT      0.0009    0.5244    0.0010
    M INT CON ELECT      0.0004    0.5285    0.0004
    GAMMA         9      0.0003    0.6080    0.0004
    GAMMA        10      0.0032    0.6180    0.0042
    GAMMA        11      0.0074    0.6808    0.0108
    GAMMA        12      0.0159    0.7074    0.0239
    GAMMA        13      0.0041    0.7691    0.0067
    GAMMA        14      0.0004    0.7899    0.0008
    GAMMA        15      0.0013    0.8209    0.0024
    GAMMA        16      0.0102    0.8561    0.0186
    GAMMA        17      0.0050    0.8753    0.0093
    GAMMA        18      0.0357    0.8753    0.0665
    GAMMA        19      0.0037    0.9105    0.0071
    GAMMA        20      0.0014    1.0528    0.0033
    GAMMA        21      0.0008    1.0611    0.0019
    GAMMA        22      0.0160    1.2375    0.0421
    GAMMA        23      0.0193    1.2989    0.0533
    GAMMA        24      0.0013    1.3516    0.0037
    GAMMA        25      0.0004    1.5901    0.0014
    GAMMA        26      0.0002    1.7330    0.0009
    K ALPHA-1 X-RAY      0.0035    0.0297    0.0002
    K ALPHA-2 X-RAY      0.0018    0.0294    0.0001

    ------------
    DAUGHTER NUCLIDE, XENON 133M IS RADIOACTIVE
    AND MAY CONTRIBUTE TO THE DOSE.
```

Table 3.6-1 (continued)
SUMMARY NUCLIDE DECAY DATA AND DOSE CONSTANT DATA

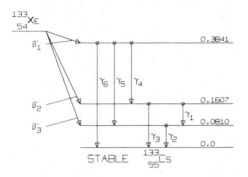

XENON-133
BETA-MINUS DECAY

```
              **INPUT DATA**

   54 XENON 133        HALF LIFE = 5.31 DAYS

DECAY MODE- BETA MINUS

-----------------------------------------------
                    MEAN     TRAN-
                    NUMBER/  SITION   OTHER
                    DISINTE- ENERGY   NUCLEAR
   TRANSITION       GRATION  (MEV)    DATA
-----------------------------------------------
BETA MINUS  1       0.0002   0.0429*  ALLOWED
BETA MINUS  2       0.0163   0.2660*  ALLOWED
BETA MINUS  3       0.9830   0.3460*  ALLOWED
   GAMMA    1       0.0162   0.0796   M1, AK= 1.36
                                      AL(T) =
                                      0.200
   GAMMA    2       0.9997   0.0809   M1, AK= 1.46
                                      K/(L+M)= 4.65
   GAMMA    3       0.0000   0.1606   M1, AK= 0.210
                                      K/(L+M)= 3.75
   GAMMA    4       0.0000   0.2230   M1,  AK(T) =
                                      0.0839
                                      AL(T) =
                                      0.0112
   GAMMA    5       0.0000   0.3028   M1, AK= 0.0390
                                      AL(T) =
                                      0.00494
   GAMMA    6       0.0002   0.3839   E2, AK= 0.0170
                                      AL(T) =
                                      0.00270

*ENDPOINT ENERGY (MEV).

-------------
REF.- ALEXANDER, P. AND LAU, J.P., NUCL. PHYS.
   A121, 612 (1968).
```

```
                   **OUTPUT DATA**

   54 XENON 133          HALF LIFE = 5.31 DAYS

DECAY MODE- BETA MINUS

----------------------------------------------------
                     MEAN      MEAN     EQUI-
                     NUMBER/   ENERGY/  LIBRIUM
                     DISINTE-  PAR-     DOSE
      RADIATION      GRATION   TICLE    CONSTANT

                       n_i      Ē_i      Δ_i
                               (MeV)    (g-rad/
                                        μCi·h)
----------------------------------------------------
       BETA MINUS  2  0.0163   0.0750   0.0026
       BETA MINUS  3  0.9830   0.1006   0.2106
          GAMMA    1  0.0061   0.0796   0.0010
 K INT CON ELECT     0.0084   0.0436   0.0007
 L INT CON ELECT     0.0012   0.0742   0.0001
          GAMMA    2  0.3603   0.0809   0.0621
 K INT CON ELECT     0.5261   0.0450   0.0504
 L INT CON ELECT     0.0848   0.0756   0.0136
 M INT CON ELECT     0.0282   0.0799   0.0048
          GAMMA    3  0.0000   0.1606   0.0000
          GAMMA    4  0.0000   0.2230   0.0000
          GAMMA    5  0.0000   0.3028   0.0000
          GAMMA    6  0.0002   0.3839   0.0001
 K ALPHA-1 X-RAY     0.2552   0.0309   0.0168
 K ALPHA-2 X-RAY     0.1321   0.0306   0.0086
 K BETA-1 X-RAY      0.0712   0.0349   0.0053
 K BETA-2 X-RAY      0.0150   0.0359   0.0011
 L X-RAYS            0.0823   0.0043   0.0007
 KLL AUGER ELECT     0.0402   0.0253   0.0021
 KLX AUGER ELECT     0.0177   0.0296   0.0011
 KXY AUGER ELECT     0.0029   0.0339   0.0002
 LMM AUGER ELECT     0.4894   0.0033   0.0034
 MXY AUGER ELECT     1.1847   0.0009   0.0025
```

XENON-133M
ISOMERIC LEVEL DECAY

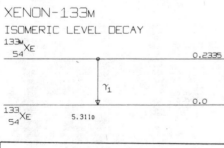

```
              **INPUT DATA**

   54 XENON 133M       HALF LIFE = 2.19 DAYS

DECAY MODE- ISOMERIC LEVEL

-----------------------------------------------
                    MEAN     TRAN-
                    NUMBER/  SITION   OTHER
                    DISINTE- ENERGY   NUCLEAR
   TRANSITION       GRATION  (MEV)    DATA
-----------------------------------------------
   GAMMA    1       1.0000   0.2335   M4, AK= 7.68
                                      K/(L+M)= 2.04

-------------
REF.- ALEXANDER, P. AND LAU, J.P., NUCL. PHYS.
   A121, 612 (1968).
```

```
                   **OUTPUT DATA**

   54 XENON 133M         HALF LIFE = 2.19 DAYS

DECAY MODE- ISOMERIC LEVEL

----------------------------------------------------
                     MEAN      MEAN     EQUI-
                     NUMBER/   ENERGY/  LIBRIUM
                     DISINTE-  PAR-     DOSE
      RADIATION      GRATION   TICLE    CONSTANT

                       n_i      Ē_i      Δ_i
                               (MeV)    (g-rad/
                                        μCi·h)
----------------------------------------------------
          GAMMA    1  0.0803   0.2335   0.0399
 K INT CON ELECT     0.6171   0.1989   0.2615
 L INT CON ELECT     0.2268   0.2284   0.1104
 M INT CON ELECT     0.0756   0.2325   0.0374
 K ALPHA-1 X-RAY     0.2935   0.0297   0.0186
 K ALPHA-2 X-RAY     0.1514   0.0294   0.0095
 K BETA-1 X-RAY      0.0810   0.0336   0.0058
 K BETA-2 X-RAY      0.0170   0.0345   0.0012
 L X-RAYS            0.1092   0.0041   0.0009
 KLL AUGER ELECT     0.0491   0.0244   0.0025
 KLX AUGER ELECT     0.0213   0.0285   0.0012
 KXY AUGER ELECT     0.0035   0.0326   0.0002
 LMM AUGER ELECT     0.6823   0.0031   0.0046
 MXY AUGER ELECT     1.6590   0.0009   0.0033

-------------
DAUGHTER NUCLIDE, XENON 133  IS RADIOACTIVE
AND MAY CONTRIBUTE TO THE DOSE.
```

Table 3.6-1 (continued)
SUMMARY NUCLIDE DECAY DATA AND DOSE CONSTANT DATA

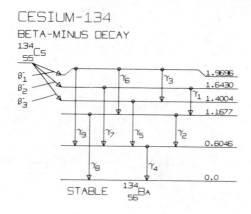

CESIUM-134

BETA-MINUS DECAY

INPUT DATA

55 CESIUM 134   HALF LIFE = 2.05 YEARS

DECAY MODE- BETA MINUS

TRANSITION		MEAN NUMBER/ DISINTE- GRATION	TRAN- SITION ENERGY (MEV)	OTHER NUCLEAR DATA
BETA MINUS	1	0.2470	0.0920*	ALLOWED
BETA MINUS	2	0.0160	0.4190*	ALLOWED
BETA MINUS	3	0.7370	0.6620*	ALLOWED
GAMMA	1	0.0160	0.4755	E2, AK= 0.00900 AL(T) = 0.00147
GAMMA	2	0.0860	0.5630	F2, AK= 0.00600 AL(T) = 0.000901
GAMMA	3	0.1330	0.5693	M1, AK= 0.00800 AL(T) = 0.00109
GAMMA	4	0.9750	0.6048	E2, AK= 0.00480 AL(T) = 0.000737
GAMMA	5	0.8700	0.7960	E2, AK= 0.00250 AL(T) = 0.000359
GAMMA	6	0.0790	0.8019	E2, AK= 0.00260 AL(T) = 0.000353
GAMMA	7	0.0100	1.0381	M1, AK= 0.00160 DEL= 2.40 AL(T) = 0.000214
GAMMA	8	0.0200	1.1684	E2, AK= 0.00100 AL(T) = 0.000149
GAMMA	9	0.0350	1.3654	E2, AK=0.000700 AL(T) = 0.000106

*ENDPOINT ENERGY (MEV).

REF.- NAGPAL, T.S., CAN. J. PHYS. 46, 2579 (1968).
LEDERER, C.M. ET AL, TABLE OF ISOTOPES, 6TH. ED.

OUTPUT DATA

55 CESIUM 134   HALF LIFE = 2.05 YEARS

DECAY MODE- BETA MINUS

RADIATION		MEAN NUMBER/ DISINTE- GRATION n_i	MEAN ENERGY/ PAR- TICLE $\bar{E}_i$ (MeV)	EQUI- LIBRIUM DOSE CONSTANT Δ_i (g-rad/ μCi-h)
BETA MINUS	1	0.2470	0.0240	0.0126
BETA MINUS	2	0.0160	0.1248	0.0042
BETA MINUS	3	0.7370	0.2119	0.3327
GAMMA	1	0.0158	0.4755	0.0160
K INT CON ELECT		0.0001	0.4381	0.0001
GAMMA	2	0.0853	0.5630	0.1024
K INT CON ELECT		0.0005	0.5256	0.0005
GAMMA	3	0.1317	0.5693	0.1597
K INT CON ELECT		0.0010	0.5318	0.0011
L INT CON ELECT		0.0001	0.5637	0.0001
GAMMA	4	0.9693	0.6048	1.2488
K INT CON ELECT		0.0046	0.5673	0.0056
L INT CON ELECT		0.0007	0.5992	0.0009
M INT CON ELECT		0.0002	0.6037	0.0003
GAMMA	5	0.8674	0.7960	1.4707
K INT CON ELECT		0.0021	0.7585	0.0035
L INT CON ELECT		0.0003	0.7904	0.0005
M INT CON ELECT		0.0001	0.7949	0.0001
GAMMA	6	0.0787	0.8019	0.1345
K INT CON ELECT		0.0002	0.7644	0.0003
GAMMA	7	0.0099	1.0381	0.0220
GAMMA	8	0.0199	1.1684	0.0497
GAMMA	9	0.0349	1.3654	0.1017
K ALPHA-1 X-RAY		0.0042	0.0321	0.0002
K ALPHA-2 X-RAY		0.0021	0.0318	0.0001

Table 3.6-1 (continued)
SUMMARY NUCLIDE DECAY DATA AND DOSE CONSTANT DATA

CESIUM-137 [b]
BETA-MINUS DECAY

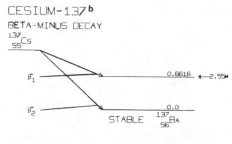

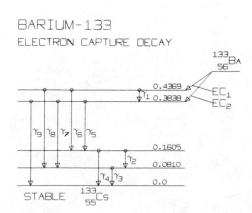

```
              **INPUT DATA**

    55 CESIUM 137       HALF LIFE = 30.0 YEARS

  DECAY MODE- BETA MINUS

  ---------------------------------------------
                 MEAN      TRAN-
                 NUMBER/   SITION
                 DISINTE-  ENERGY   OTHER
                 GRATION   (MEV)    NUCLEAR
  TRANSITION                        DATA
  ---------------------------------------------
  BETA MINUS  1  0.9460    0.5140*  U 1ST FORBIDDEN
  BETA MINUS  2  0.0540    1.1760*  2ND FORBIDDEN

  *ENDPOINT ENERGY (MEV).

  ------------
  REF.- HANSEN, H.H. ET AL, Z. PHYSIK 218, 25
        (1969).
        KARTASHOV, V.M. ET AL, SOV. J. NUCL. PHYS. 6,
        656 (1968).
```

```
                  **OUTPUT DATA**

    55 CESIUM 137        HALF LIFE = 30.0 YEARS

  DECAY MODE- BETA MINUS

  --------------------------------------------------
                   MEAN      MEAN     EQUI-
                   NUMBER/   ENERGY/  LIBRIUM
                   DISINTE-  PAR-     DOSE
    RADIATION      GRATION   TICLE    CONSTANT

                     n_i       Ē_i      Δ_i
                             (MEV)    (g·rad/
                                      μCi·h)
  --------------------------------------------------
    BETA MINUS  1  0.9460    0.1747   0.3520
    BETA MINUS  2  0.0540    0.4269   0.0491

  ------------
  DAUGHTER NUCLIDE, BARIUM 137M IS RADIOACTIVE
  AND MAY CONTRIBUTE TO THE DOSE.
  BRANCHING TO 0.6616 MEV, 2.55 MINUTE HALF LIFE,
    ISOMERIC LEVEL IN BARIUM-137 IS 0.946 PER
    DISINTEGRATION OF CESIUM-137.
```

[b]   Include gamma emissions from
       Ba-137m in dose calculations.

BARIUM-133
ELECTRON CAPTURE DECAY

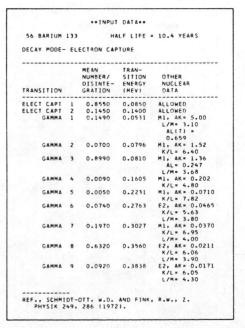

```
                  **INPUT DATA**

    56 BARIUM 133        HALF LIFE = 10.4 YEARS

  DECAY MODE- ELECTRON CAPTURE

  --------------------------------------------------
                 MEAN      TRAN-
                 NUMBER/   SITION
                 DISINTE-  ENERGY   OTHER
                 GRATION   (MEV)    NUCLEAR
  TRANSITION                        DATA
  --------------------------------------------------
  ELECT CAPT  1  0.8550    0.0850   ALLOWED
  ELECT CAPT  2  0.1450    0.1400   ALLOWED
     GAMMA    1  0.1490    0.0531   M1, AK= 5.00
                                    L/M= 3.10
                                    AL(T) =
                                      0.659
     GAMMA    2  0.0700    0.0796   M1, AK= 1.52
                                    K/L= 6.40
     GAMMA    3  0.8990    0.0810   M1, AK= 1.36
                                    AL= 0.247
                                    L/M= 3.68
     GAMMA    4  0.0090    0.1605   M1, AK= 0.202
                                    K/L= 4.80
     GAMMA    5  0.0050    0.2231   M1, AK= 0.0710
                                    K/L= 7.82
     GAMMA    6  0.0740    0.2763   E2, AK= 0.0465
                                    K/L= 5.63
                                    L/M= 3.80
     GAMMA    7  0.1970    0.3027   M1, AK= 0.0370
                                    K/L= 6.95
                                    L/M= 4.00
     GAMMA    8  0.6320    0.3560   E2, AK= 0.0211
                                    K/L= 6.06
                                    L/M= 3.90
     GAMMA    9  0.0920    0.3838   E2, AK= 0.0171
                                    K/L= 6.05
                                    L/M= 4.30

  ------------
  REF., SCHMIDT-OTT, W.D. AND FINK, R.W., Z.
       PHYSIK 249, 286 (1972).
```

Table 3.6-1 (continued)
SUMMARY NUCLIDE DECAY DATA AND DOSE CONSTANT DATA

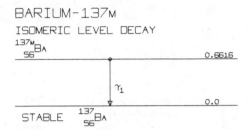

BARIUM-137m

ISOMERIC LEVEL DECAY

$^{137m}_{56}$Ba

0.6616

γ_1

0.0

STABLE $^{137}_{56}$Ba

```
          ••INPUT DATA••

  56 BARIUM 137M       HALF LIFE = 2.55 MINUTES

  DECAY MODE- ISOMERIC LEVEL

  ---------------------------------------------
                 MEAN      TRAN-
                 NUMBER/   SITION
                 DISINTE-  ENERGY   OTHER
                 GRATION   (MEV)    NUCLEAR
  TRANSITION                        DATA
  ---------------------------------------------
     GAMMA  1    1.0000    0.6616   M4, AK= 0.0914
                                    K/L= 5.55
  ------------
  REF.- NUCLEAR DATA TABLES A8, 109 (1970).
```

```
                   ••OUTPUT DATA••

      56 BARIUM 137M        HALF LIFE = 2.55 MINUTES

      DECAY MODE- ISOMERIC LEVEL
```

RADIATION	MEAN NUMBER/ DISINTE- GRATION n_i	MEAN ENERGY/ PAR- TICLE $\bar{E}_i$ (MeV)	EQUI- LIBRIUM DOSE CONSTANT Δ_i (g-rad/ μCi-h)
GAMMA 1	0.8981	0.6616	1.2658
K INT CON ELECT	0.0820	0.6241	0.1091
L INT CON ELECT	0.0147	0.6560	0.0206
M INT CON ELECT	0.0049	0.6605	0.0069
K ALPHA-1 X-RAY	0.0392	0.0321	0.0026
K ALPHA-2 X-RAY	0.0203	0.0318	0.0013
K BETA-1 X-RAY	0.0110	0.0363	0.0008
K BETA-2 X-RAY	0.0023	0.0374	0.0001
L X-RAYS	0.0133	0.0044	0.0001
KLL AUGER ELECT	0.0059	0.0263	0.0003
KLX AUGER ELECT	0.0026	0.0308	0.0001
LMM AUGER ELECT	0.0756	0.0034	0.0005
MXY AUGER ELECT	0.1841	0.0010	0.0004

```
                   ••OUTPUT DATA••

      56 BARIUM 133        HALF LIFE = 10.4 YEARS

      DECAY MODE- ELECTRON CAPTURE
```

RADIATION	MEAN NUMBER/ DISINTE- GRATION n_i	MEAN ENERGY/ PAR- TICLE $\bar{E}_i$ (MeV)	EQUI- LIBRIUM DOSE CONSTANT Δ_i (g-rad/ μCi-h)
GAMMA 1	0.0216	0.0531	0.0024
K INT CON ELECT	0.1084	0.0171	0.0039
L INT CON ELECT	0.0142	0.0478	0.0014
M INT CON ELECT	0.0046	0.0521	0.0005
GAMMA 2	0.0246	0.0796	0.0041
K INT CON ELECT	0.0375	0.0436	0.0034
L INT CON ELECT	0.0058	0.0743	0.0009
M INT CON ELECT	0.0019	0.0786	0.0003
GAMMA 3	0.3361	0.0810	0.0580
K INT CON ELECT	0.4572	0.0450	0.0438
L INT CON ELECT	0.0830	0.0757	0.0133
M INT CON ELECT	0.0225	0.0800	0.0038
GAMMA 4	0.0071	0.1605	0.0024
K INT CON ELECT	0.0014	0.1245	0.0003
GAMMA 5	0.0046	0.2231	0.0021
K INT CON ELECT	0.0003	0.1871	0.0001
GAMMA 6	0.0700	0.2763	0.0412
K INT CON ELECT	0.0032	0.2403	0.0016
L INT CON ELECT	0.0005	0.2710	0.0003
GAMMA 7	0.1887	0.3027	0.1217
K INT CON ELECT	0.0069	0.2667	0.0039
L INT CON ELECT	0.0010	0.2974	0.0006
M INT CON ELECT	0.0002	0.3017	0.0001
GAMMA 8	0.6163	0.3560	0.4673
K INT CON ELECT	0.0130	0.3200	0.0088
L INT CON ELECT	0.0021	0.3507	0.0016
M INT CON ELECT	0.0005	0.3550	0.0004
GAMMA 9	0.0901	0.3838	0.0736
K INT CON ELECT	0.0015	0.3478	0.0011
L INT CON ELECT	0.0002	0.3785	0.0002
K ALPHA-1 X-RAY	0.6459	0.0309	0.0426
K ALPHA-2 X-RAY	0.3342	0.0306	0.0218
K BETA-1 X-RAY	0.1802	0.0349	0.0134
K BETA-2 X-RAY	0.0381	0.0359	0.0029
L X-RAYS	0.2254	0.0043	0.0020
KLL AUGER ELECT	0.1019	0.0253	0.0055
KLX AUGER ELECT	0.0448	0.0296	0.0028
KXY AUGER ELECT	0.0074	0.0339	0.0005
LMM AUGER ELECT	1.3403	0.0033	0.0094
MXY AUGER ELECT	3.2239	0.0009	0.0068

Table 3.6-1 (continued)
SUMMARY NUCLIDE DECAY DATA AND DOSE CONSTANT DATA

YTTERBIUM-169
ELECTRON CAPTURE DECAY

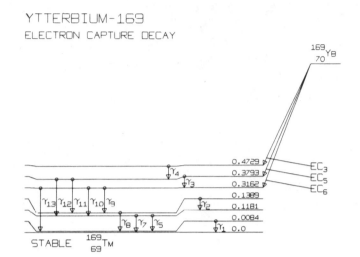

INPUT DATA

70 YTTERBIUM 169   HALF LIFE = 32.0 DAYS

DECAY MODE- ELECTRON CAPTURE

TRANSITION	MEAN NUMBER/ DISINTE- GRATION	TRAN- SITION ENERGY (MEV)	OTHER NUCLEAR DATA
ELECT CAPT 1	0.0000	0.4820	ALLOWED
ELECT CAPT 2	0.0001	0.5670	ALLOWED
ELECT CAPT 3	0.1300	0.7270	1ST FORBIDDEN
ELECT CAPT 4	0.0012	0.7670	ALLOWED
ELECT CAPT 5	0.8400	0.8210	1ST FORBIDDEN
ELECT CAPT 6	0.0297	0.8840	ALLOWED
GAMMA 1	0.9540	0.0084	M1
GAMMA 2	0.1250	0.0207	M1, AL(T) = 42.2
GAMMA 3	0.9560	0.0631	E1, AL= 0.160 AK(T) = 0.903
GAMMA 4	0.1310	0.0936	M1, AK= 3.07 L/M= 0.0400 AL(T) = 0.486
GAMMA 5	0.6000	0.1097	M1, AK= 2.06 L/M= 0.0250 AL(T) = 0.307
GAMMA 6	0.0012	0.1172	M1, AK(T) = 1.65 AL(T) = 0.254
GAMMA 7	0.0510	0.1181	E2, AK= 0.694 AL(T) = 0.733
GAMMA 8	0.2490	0.1305	E2, AK= 0.550 AL(T) = 0.472
GAMMA 9	0.3500	0.1771	M1, AK= 0.480 L/M= 0.170 AL(T) = 0.0786
GAMMA 10	0.5190	0.1979	M1, AK= 0.350 L/M= 0.100 AL(T) = 0.0575
GAMMA 11	0.0013	0.2404	E1, AK= 0.0350 AL(T) = 0.00426
GAMMA 12	0.0180	0.2610	E1, AK= 0.0270 AL(T) = 0.00343
GAMMA 13	0.1180	0.3076	E2, AK= 0.0490 AL(T) = 0.0143

REF.- NUCLEAR DATA GROUP, ORNL, PRIVATE COMMUNICATION.
AGNIHOTRY, A.P. ET AL, PHYS. REV. C6, 321 (1972).

OUTPUT DATA

70 YTTERBIUM 169   HALF LIFE = 32.0 DAYS

DECAY MODE- ELECTRON CAPTURE

RADIATION	MEAN NUMBER/ DISINTE- GRATION n_i	MEAN ENERGY/ PAR- TICLE $\bar{E}_i$ (MeV)	EQUI- LIBRIUM DOSE CONSTANT Δ_i (g-rad/ µCi-h)
GAMMA 1	0.0000	0.0084	0.0000
M INT CON ELECT	0.9540	0.0065	0.0132
GAMMA 2	0.0021	0.0207	0.0000
L INT CON ELECT	0.0921	0.0113	0.0022
M INT CON ELECT	0.0307	0.0188	0.0012
GAMMA 3	0.4516	0.0631	0.0607
K INT CON ELECT	0.4080	0.0037	0.0032
L INT CON ELECT	0.0722	0.0537	0.0082
M INT CON ELECT	0.0240	0.0612	0.0031
GAMMA 4	0.0078	0.0936	0.0015
K INT CON ELECT	0.0240	0.0342	0.0017
L INT CON ELECT	0.0038	0.0842	0.0006
M INT CON ELECT	0.0953	0.0917	0.0186
GAMMA 5	0.0382	0.1097	0.0089
K INT CON ELECT	0.0788	0.0503	0.0084
L INT CON ELECT	0.0117	0.1003	0.0025
M INT CON ELECT	0.4711	0.1078	0.1082
GAMMA 6	0.0004	0.1172	0.0001
GAMMA 7	0.0190	0.1181	0.0048
K INT CON ELECT	0.0132	0.0587	0.0016
L INT CON ELECT	0.0140	0.1087	0.0032
M INT CON ELECT	0.0046	0.1162	0.0011
GAMMA 8	0.1142	0.1305	0.0317
K INT CON ELECT	0.0628	0.0711	0.0095
L INT CON ELECT	0.0539	0.1211	0.0139
M INT CON ELECT	0.0179	0.1286	0.0049
GAMMA 9	0.1731	0.1771	0.0653
K INT CON ELECT	0.0831	0.1177	0.0208
L INT CON ELECT	0.0136	0.1677	0.0048
M INT CON ELECT	0.0801	0.1752	0.0299
GAMMA 10	0.2616	0.1979	0.1103
K INT CON ELECT	0.0915	0.1385	0.0270
L INT CON ELECT	0.0150	0.1885	0.0060
M INT CON ELECT	0.1506	0.1960	0.0629
GAMMA 11	0.0012	0.2404	0.0006
GAMMA 12	0.0174	0.2610	0.0097
K INT CON ELECT	0.0004	0.2016	0.0002
GAMMA 13	0.1104	0.3076	0.0724
K INT CON ELECT	0.0054	0.2482	0.0028
L INT CON ELECT	0.0015	0.2982	0.0010
M INT CON ELECT	0.0005	0.3057	0.0003
K ALPHA-1 X-RAY	0.7750	0.0507	0.0837
K ALPHA-2 X-RAY	0.4154	0.0497	0.0440
K BETA-1 X-RAY	0.2472	0.0575	0.0302
K BETA-2 X-RAY	0.0569	0.0593	0.0072
L X-RAYS	0.4027	0.0075	0.0064
KLL AUGER ELECT	0.0626	0.0405	0.0054
KLX AUGER ELECT	0.0322	0.0481	0.0033
KXY AUGER ELECT	0.0055	0.0556	0.0006
LMM AUGER ELECT	1.3635	0.0056	0.0163
MXY AUGER ELECT	5.2840	0.0018	0.0212

Table 3.6-1 (continued)
SUMMARY NUCLIDE DECAY DATA AND DOSE CONSTANT DATA

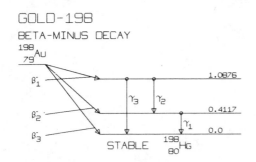

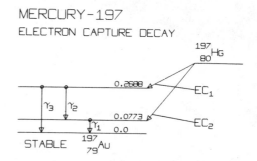

GOLD-198
BETA-MINUS DECAY

MERCURY-197
ELECTRON CAPTURE DECAY

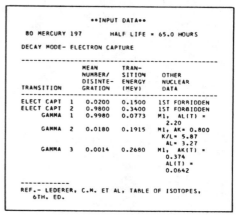

```
                    ••INPUT DATA••

    79 GOLD 198          HALF LIFE = 2.69 DAYS

  DECAY MODE- BETA MINUS

  -------------------------------------------------
                 MEAN      TRAN-
                 NUMBER/   SITION
                 DISINTE-  ENERGY    OTHER
                 GRATION   (MEV)     NUCLEAR
  TRANSITION     GRATION             DATA
  -------------------------------------------------
  BETA MINUS  1  0.0130    0.2900•   1ST FORBIDDEN
  BETA MINUS  2  0.9860    0.9612•   1ST FORBIDDEN
  BETA MINUS  3  0.0002    1.3710•   U 1ST FORBIDDEN
  GAMMA       1  0.9970    0.4117    E2, AK= 0.0301
                                     AL= 0.0100
  GAMMA       2  0.0110    0.6758    M1, AK= 0.0224
                                     K/L= 5.70
  GAMMA       3  0.0023    1.0876    E2, AK= 0.00450
                                     K/L= 6.30

  •ENDPOINT ENERGY (MEV).

  ------------
  REF.- NUCLEAR DATA 86, 328 (1971).
```

```
                    ••INPUT DATA••

    80 MERCURY 197        HALF LIFE = 65.0 HOURS

  DECAY MODE- ELECTRON CAPTURE

  -------------------------------------------------
                 MEAN      TRAN-
                 NUMBER/   SITION
                 DISINTE-  ENERGY    OTHER
                 GRATION   (MEV)     NUCLEAR
  TRANSITION     GRATION             DATA
  -------------------------------------------------
  ELECT CAPT  1  0.0200    0.1500    1ST FORBIDDEN
  ELECT CAPT  2  0.9800    0.3400    1ST FORBIDDEN
  GAMMA       1  0.9980    0.0773    M1,  AL(T) =
                                     2.20
  GAMMA       2  0.0180    0.1915    M1, AK= 0.800
                                     K/L= 5.87
                                     AL= 3.27
  GAMMA       3  0.0014    0.2680    M1,  AK(T) =
                                     0.374
                                     AL(T) =
                                     0.0642

  ------------
  REF.- LEDERER, C.M. ET AL, TABLE OF ISOTOPES,
  6TH. ED.
```

```
                    ••OUTPUT DATA••

    79 GOLD 198          HALF LIFE = 2.69 DAYS

  DECAY MODE- BETA MINUS

  -------------------------------------------------
                      MEAN     MEAN     EQUI-
                      NUMBER/  ENERGY/  LIBRIUM
                      DISINTE- PAR-     DOSE
      RADIATION       GRATION  TICLE    CONSTANT

                        nᵢ       Ēᵢ       Δᵢ
                                (MeV)    (g-rad/
                                         µCi-h)
  -------------------------------------------------
        BETA MINUS  1  0.0130   0.0811   0.0022
        BETA MINUS  2  0.9860   0.3163   0.6643
        BETA MINUS  3  0.0002   0.4648   0.0002
        GAMMA       1  0.9555   0.4117   0.8380
  K INT CON ELECT      0.0287   0.3286   0.0201
  L INT CON ELECT      0.0095   0.3979   0.0081
  M INT CON ELECT      0.0031   0.4089   0.0027
        GAMMA       2  0.0107   0.6758   0.0154
  K INT CON ELECT      0.0002   0.5927   0.0003
        GAMMA       3  0.0022   1.0876   0.0053
  K ALPHA-1 X-RAY      0.0139   0.0708   0.0021
  K ALPHA-2 X-RAY      0.0076   0.0688   0.0011
  K BETA-1  X-RAY      0.0048   0.0802   0.0008
  K BETA-2  X-RAY      0.0013   0.0831   0.0002
  L ALPHA X-RAYS       0.0060   0.0099   0.0001
  L BETA X-RAYS        0.0056   0.0118   0.0001
  LMM AUGER ELECT      0.0206   0.0081   0.0003
  MXY AUGER ELECT      0.0622   0.0028   0.0003
```

```
                    ••OUTPUT DATA••

    80 MERCURY 197        HALF LIFE = 65.0 HOURS

  DECAY MODE- ELECTRON CAPTURE

  -------------------------------------------------
                      MEAN     MEAN     EQUI-
                      NUMBER/  ENERGY/  LIBRIUM
                      DISINTE- PAR-     DOSE
      RADIATION       GRATION  TICLE    CONSTANT

                        nᵢ       Ēᵢ       Δᵢ
                                (MeV)    (g-rad/
                                         µCi-h)
  -------------------------------------------------
        GAMMA       1  0.2535   0.0773   0.0417
  L INT CON ELECT      0.5583   0.0639   0.0760
  M INT CON ELECT      0.1861   0.0745   0.0295
        GAMMA       2  0.0029   0.1915   0.0011
  K INT CON ELECT      0.0023   0.1107   0.0005
  L INT CON ELECT      0.0095   0.1781   0.0036
  M INT CON ELECT      0.0031   0.1887   0.0012
        GAMMA       3  0.0009   0.2680   0.0005
  K INT CON ELECT      0.0003   0.1872   0.0001
  K ALPHA-1 X-RAY      0.3609   0.0688   0.0529
  K ALPHA-2 X-RAY      0.1980   0.0669   0.0282
  K BETA-1  X-RAY      0.1250   0.0779   0.0207
  K BETA-2  X-RAY      0.0335   0.0807   0.0057
  L ALPHA X-RAYS       0.2434   0.0097   0.0050
  L BETA X-RAYS        0.2280   0.0114   0.0055
  L GAMMA X-RAYS       0.0307   0.0133   0.0008
  KLL AUGER ELECT      0.0189   0.0540   0.0021
  KLX AUGER ELECT      0.0107   0.0646   0.0014
  KXY AUGER ELECT      0.0017   0.0752   0.0002
  LMN AUGER ELECT      0.8736   0.0078   0.0146
  MXY AUGER ELECT      2.6315   0.0027   0.0153
```

Table 3.6-1 (continued)
SUMMARY NUCLIDE DECAY DATA AND DOSE CONSTANT DATA

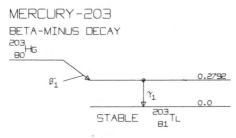

MERCURY-203
BETA-MINUS DECAY

$^{203}_{80}Hg$

β^-_1 0.2792
γ_1 0.0

STABLE $^{203}_{81}Tl$

•INPUT DATA••

80 MERCURY 203   HALF LIFE = 46.5 DAYS

DECAY MODE- BETA MINUS

TRANSITION	MEAN NUMBER/ DISINTE- GRATION	TRAN- SITION ENERGY (MEV)	OTHER NUCLEAR DATA
BETA MINUS 1	1.0000	0.2120*	ALLOWED
GAMMA 1	1.0000	0.2792	E2, ΔK= 0.163
			K/L = 3.36
			K/M=13.7

*ENDPOINT ENERGY (MEV).

REF.- NUCLEAR DATA B5, 53A (1971).

••OUTPUT DATA••

80 MERCURY 203   HALF LIFE = 46.5 DAYS

DECAY MODE- BETA MINUS

RADIATION	MEAN NUMBER/ DISINTE- GRATION n_i	MEAN ENERGY/ PAR- TICLE $\bar{E}_i$ (MeV)	EQUI- LIBRIUM DOSE CONSTANT Δ_i (g-rad/ μCi-h)
BETA MINUS 1	1.0000	0.0577	0.1229
GAMMA 1	0.8173	0.2792	0.4860
K INT CON ELECT	0.1332	0.1936	0.0549
L INT CON ELECT	0.0396	0.2649	0.0223
M INT CON ELECT	0.0097	0.2762	0.0057
K ALPHA-1 X-RAY	0.0638	0.0728	0.0099
K ALPHA-2 X-RAY	0.0352	0.0708	0.0053
K BETA-1 X-RAY	0.0224	0.0825	0.0039
K BETA-2 X-RAY	0.0063	0.0855	0.0011
L ALPHA X-RAYS	0.0275	0.0102	0.0006
L BETA X-RAYS	0.0258	0.0122	0.0006
L GAMMA X-RAYS	0.0034	0.0142	0.0001
KLL AUGER ELECT	0.0031	0.0569	0.0003
KLX AUGER ELECT	0.0018	0.0683	0.0002
LMM AUGER ELECT	0.0900	0.0083	0.0016
MXY AUGER ELECT	0.2717	0.0029	0.0017

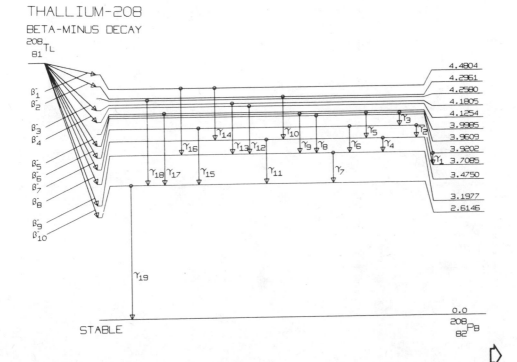

THALLIUM-208
BETA-MINUS DECAY

$^{208}_{81}Tl$

Table 3.6-1 (continued)
SUMMARY NUCLIDE DECAY DATA AND DOSE CONSTANT DATA

```
                    **INPUT DATA**

    81 THALLIUM 208      HALF LIFE = 3.07 MINUTES

    DECAY MODE- BETA MINUS

    -------------------------------------------------
                    MEAN      TRAN-
                    NUMBER/   SITION
                    DISINTE-  ENERGY    OTHER
                    GRATION   (MEV)     NUCLEAR
    TRANSITION                          DATA
    -------------------------------------------------
    BETA MINUS  1   0.0006    0.5130*   1ST FORBIDDEN
    BETA MINUS  2   0.0009    0.6970*   1ST FORBIDDEN
    BETA MINUS  3   0.0020    0.8130*   1ST FORBIDDEN
    BETA MINUS  4   0.0013    0.8680*   1ST FORBIDDEN
    BETA MINUS  5   0.0340    1.0320*   1ST FORBIDDEN
    BETA MINUS  6   0.0070    1.0730*   1ST FORBIDDEN
    BETA MINUS  7   0.2400    1.2850*   1ST FORBIDDEN
    BETA MINUS  8   0.2200    1.5180*   1ST FORBIDDEN
    BETA MINUS  9   0.5000    1.7950*   1ST FORBIDDEN
    BETA MINUS 10   0.0003    2.3780*   U 1ST FORBIDDEN
       GAMMA    1   0.0040    0.2115    M1, AK= 0.890
                                        AL(T) =
                                        0.163
       GAMMA    2   0.0058    0.2335    M1, AK= 0.470
                                        AL(T) =
                                        0.124
       GAMMA    3   0.0140    0.2526    M1, AK= 0.530
                                        AL(T) =
                                        0.100
       GAMMA    4   0.1060    0.2773    M1, AK= 0.390
                                        K/L= 4.70
       GAMMA    5   0.0007    0.4862    M1, AK= 0.180
                                        AL(T) =
                                        0.0169
       GAMMA    6   0.2500    0.5108    M1, AK= 0.0820
                                        K/L= 5.65
```

```
                    **INPUT DATA**

       81 THALLIUM 208 (CONTINUED)

    -------------------------------------------------
                    MEAN      TRAN-
                    NUMBER/   SITION
                    DISINTE-  ENERGY    OTHER
                    GRATION   (MEV)     NUCLEAR
    TRANSITION                          DATA
    -------------------------------------------------
       GAMMA    7   0.8800    0.5831    E2, AK= 0.0151
                                        AL(T) =
                                        0.00433
       GAMMA    8   0.0030    0.7223
       GAMMA    9   0.0190    0.7631    M1, AK= 0.0280
                                        AL(T) =
                                        0.00522
       GAMMA   10   0.0009    0.8211
       GAMMA   11   0.1260    0.8603    M1, AK= 0.0240
                                        AL(T) =
                                        0.00383
       GAMMA   12   0.0013    0.9277
       GAMMA   13   0.0020    0.9828
       GAMMA   14   0.0001    1.0040
       GAMMA   15   0.0040    1.0931    E2,  AK(T) =
                                        0.00451
                                        AL(T) =
                                        0.000891
       GAMMA   16   0.0005    1.2827
       GAMMA   17   0.0002    1.3840
       GAMMA   18   0.0001    1.6430
       GAMMA   19   1.0000    2.6146    E3, AK= 0.00180
                                        AL(T) =
                                        0.000295

    *ENDPOINT ENERGY (MEV).

    ------------
    REF.- NUCLEAR DATA B5, 251 (1971).
```

```
                         **OUTPUT DATA**

    81 THALLIUM 208     HALF LIFE = 3.07 MINUTES

    DECAY MODE- BETA MINUS

    ------------------------------------------------------
                        MEAN      MEAN      EQUI-
                        NUMBER/   ENERGY/   LIBRIUM
                        DISINTE-  PAR-      DOSE
         RADIATION      GRATION   TICLE     CONSTANT

                        n_i       E_i       Δ_i
                                  (MEV)     (g-rad/
                                            μCi-h)
    ------------------------------------------------------
         BETA MINUS  1   0.0006    0.1530    0.0001
         BETA MINUS  2   0.0009    0.2172    0.0004
         BETA MINUS  3   0.0020    0.2594    0.0011
         BETA MINUS  4   0.0013    0.2799    0.0007
         BETA MINUS  5   0.0340    0.3424    0.0248
         BETA MINUS  6   0.0070    0.3583    0.0053
         BETA MINUS  7   0.2400    0.4423    0.2261
         BETA MINUS  8   0.2200    0.5373    0.2518
         BETA MINUS  9   0.5000    0.6531    0.6955
         BETA MINUS 10   0.0003    0.8578    0.0005
            GAMMA    1   0.0018    0.2115    0.0008
    K INT CON ELECT      0.0016    0.1234    0.0004
    L INT CON ELECT      0.0003    0.1967    0.0001
            GAMMA    2   0.0035    0.2335    0.0017
    K INT CON ELECT      0.0016    0.1454    0.0005
    L INT CON ELECT      0.0004    0.2187    0.0002
            GAMMA    3   0.0084    0.2526    0.0045
    K INT CON ELECT      0.0044    0.1645    0.0015
    L INT CON ELECT      0.0008    0.2378    0.0004
    M INT CON ELECT      0.0002    0.2495    0.0001
            GAMMA    4   0.0706    0.2773    0.0417
    K INT CON ELECT      0.0275    0.1893    0.0111
    L INT CON ELECT      0.0058    0.2626    0.0032
    M INT CON ELECT      0.0019    0.2742    0.0011
            GAMMA    5   0.0005    0.4862    0.0006
            GAMMA    6   0.2269    0.5108    0.2469
    K INT CON ELECT      0.0186    0.4227    0.0167
    L INT CON ELECT      0.0032    0.4960    0.0034
    M INT CON ELECT      0.0010    0.5077    0.0011
            GAMMA    7   0.8620    0.5831    1.0706
    K INT CON ELECT      0.0130    0.4951    0.0137
    L INT CON ELECT      0.0037    0.5683    0.0045
    M INT CON ELECT      0.0012    0.5800    0.0015
            GAMMA    8   0.0030    0.7223    0.0046
            GAMMA    9   0.0183    0.7631    0.0298
    K INT CON ELECT      0.0005    0.6751    0.0007
    L INT CON ELECT      0.0000    0.7483    0.0001
            GAMMA   10   0.0009    0.8211    0.0015
            GAMMA   11   0.1224    0.8603    0.2243
    K INT CON ELECT      0.0029    0.7723    0.0048
    L INT CON ELECT      0.0004    0.8456    0.0008
    M INT CON ELECT      0.0001    0.8573    0.0002
            GAMMA   12   0.0013    0.9277    0.0025
            GAMMA   13   0.0020    0.9828    0.0041
            GAMMA   14   0.0001    1.0040    0.0002
            GAMMA   15   0.0039    1.0931    0.0092
            GAMMA   16   0.0005    1.2827    0.0013
            GAMMA   17   0.0002    1.3840    0.0005
            GAMMA   18   0.0001    1.6430    0.0003
            GAMMA   19   0.9978    2.6146    5.5569
    K INT CON *ELECT     0.0017    2.5265    0.0096
    L INT CON ELECT      0.0002    2.5998    0.0016
    M INT CON ELECT      0.0000    2.6115    0.0005
    K ALPHA-1 X-RAY      0.0345    0.0749    0.0055
    K ALPHA-2 X-RAY      0.0191    0.0728    0.0029
    K BETA-1 X-RAY       0.0122    0.0849    0.0022
    K BETA-2 X-RAY       0.0035    0.0880    0.0006
    L ALPHA X-RAYS       0.0141    0.0105    0.0003
    L BETA X-RAYS        0.0132    0.0126    0.0003
    KLL AUGER ELECT      0.0016    0.0585    0.0002
    KLX AUGER ELECT      0.0009    0.0701    0.0001
    LMM AUGER ELECT      0.0442    0.0086    0.0008
    MXY AUGER ELECT      0.1364    0.0030    0.0008
```

Table 3.6-1 (continued)
SUMMARY NUCLIDE DECAY DATA AND DOSE CONSTANT DATA

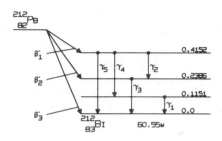

LEAD-212

BETA-MINUS DECAY

```
            ••INPUT DATA••

   82 LEAD 212        HALF LIFE = 10.6 HOURS

 DECAY MODE- BETA MINUS

 -----------------------------------------------
                MEAN      TRAN-
                NUMBER/   SITION
                DISINTE-  ENERGY   OTHER
                GRATION   (MEV)    NUCLEAR
 TRANSITION               DATA

 BETA MINUS  1  0.0510   0.1550•  1ST FORBIDDEN
 BETA MINUS  2  0.8300   0.3320•  1ST FORBIDDEN
 BETA MINUS  3  0.1300   0.5710•  1ST FORBIDDEN
 GAMMA       1  0.0470   0.1151   M1, AK= 5.80
                                  K/L= 5.70
 GAMMA       2  0.0016   0.1766   M1, AK= 1.70
                                  K/L= 5.70
 GAMMA       3  0.8230   0.2386   M1, AK= 0.740
                                  K/L= 5.75
 GAMMA       4  0.0490   0.3000   M1, AK= 0.390
                                  DEL= 3.50
                                  AL(T) =
                                  0.0483
 GAMMA       5  0.0010   0.4152   M1, AK= 0.200
                                  AL(T) =
                                  0.0283

 •ENDPOINT ENERGY (MEV).

 -------------
 REF.- NUCLEAR DATA B8, 168 (1972).
```

```
                  ••OUTPUT DATA••

   82 LEAD 212          HALF LIFE = 10.6 HOURS

 DECAY MODE- BETA MINUS

 -----------------------------------------------
                     MEAN      MEAN    EQUI-
                     NUMBER/   ENERGY/ LIBRIUM
                     DISINTE-  PAR-    DOSE
     RADIATION       GRATION   TICLE   CONSTANT

                       n         E        Δ
                        1         1        1
                               (MEV)   (g-rad/
                                       μC1-h)

 -----------------------------------------------
         BETA MINUS 1  0.0510  0.0413  0.0044
         BETA MINUS 2  0.8300  0.0939  0.1661
         BETA MINUS 3  0.1300  0.1726  0.0477
              GAMMA  1  0.0057  0.1151  0.0014
 K INT CON ELECT       0.0334  0.0246  0.0017
 L INT CON ELECT       0.0058  0.0999  0.0012
 M INT CON ELECT       0.0019  0.1119  0.0004
              GAMMA  2  0.0005  0.1766  0.0001
 K INT CON ELECT       0.0008  0.0861  0.0001
              GAMMA  3  0.4305  0.2386  0.2188
 K INT CON ELECT       0.3185  0.1481  0.1005
 L INT CON ELECT       0.0554  0.2233  0.0263
 M INT CON ELECT       0.0184  0.2354  0.0092
              GAMMA  4  0.0336  0.3000  0.0215
 K INT CON ELECT       0.0131  0.2095  0.0058
 L INT CON ELECT       0.0016  0.2848  0.0009
 M INT CON ELECT       0.0005  0.2969  0.0003
              GAMMA  5  0.0008  0.4152  0.0007
 K INT CON ELECT       0.0001  0.3246  0.0001
 K ALPHA-1 X-RAY       0.1746  0.0771  0.0286
 K ALPHA-2 X-RAY       0.0967  0.0748  0.0154
 K BETA-1 X-RAY        0.0624  0.0873  0.0116
 K BETA-2 X-RAY        0.0183  0.0905  0.0035
 L ALPHA X-RAYS        0.0698  0.0108  0.0016
 L BETA X-RAYS         0.0654  0.0129  0.0018
 L GAMMA X-RAYS        0.0088  0.0152  0.0002
 KLL AUGER ELECT       0.0082  0.0600  0.0010
 KLX AUGER ELECT       0.0048  0.0721  0.0007
 KXY AUGER ELECT       0.0007  0.0841  0.0001
 LMM AUGER ELECT       0.2118  0.0088  0.0040
 MXY AUGER ELECT       0.6576  0.0031  0.0044

 -------------
 DAUGHTER NUCLIDE, BISMUTH 212  IS RADIOACTIVE
 AND MAY CONTRIBUTE TO THE DOSE.
```

Table 3.6-1 (continued)
SUMMARY NUCLIDE DECAY DATA AND DOSE CONSTANT DATA

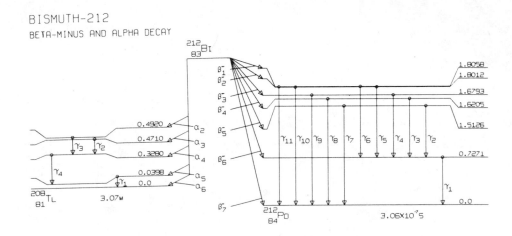

BISMUTH-212
BETA-MINUS AND ALPHA DECAY

••INPUT DATA••

83 BISMUTH 212   HALF LIFE = 60.5 MINUTES

DECAY MODES— BETA MINUS AND ALPHA

TRANSITION		MEAN NUMBER/ DISINTE- GRATION	TRAN- SITION ENERGY (MEV)	OTHER NUCLEAR DATA
BETA MINUS	1	0.0066	0.4450*	1ST FORBIDDEN
BETA MINUS	2	0.0002	0.4470*	1ST FORBIDDEN
BETA MINUS	3	0.0025	0.5720*	1ST FORBIDDEN
BETA MINUS	4	0.0194	0.6300*	1ST FORBIDDEN
BETA MINUS	5	0.0149	0.7380*	1ST FORBIDDEN
BETA MINUS	6	0.0445	1.5240*	1ST FORBIDDEN
BETA MINUS	7	0.5530	2.2510*	1ST FORBIDDEN
GAMMA	1	0.0675	0.7271	E2, AK= 0.0140 K/L= 4.40
GAMMA	2	0.0117	0.7854	M1, AK= 0.0350 K/L= 7.80
GAMMA	3	0.0038	0.8933	M1, AK= 0.0360 K/L= 4.00
GAMMA	4	0.0018	0.9521	
GAMMA	5	0.0001	1.0740	
GAMMA	6	0.0055	1.0786	M1, AK= 0.0190 AL(T) = 0.00254
GAMMA	7	0.0032	1.5127	
GAMMA	8	0.0156	1.6205	M1, AK= 0.00650 AL(T) = 0.000887
GAMMA	9	0.0007	1.6798	
GAMMA	10	0.0001	1.8000	E0
GAMMA	11	0.0011	1.8060	, AK= 0.0240
ALPHA	1	0.0000	5.5875	
ALPHA	2	0.0040	5.7109	
ALPHA	3	0.0005	5.7302	
ALPHA	4	0.0060	5.8749	
ALPHA	5	0.2520	6.1628	
ALPHA	6	0.0963	6.2028	
GAMMA	1	0.2626	0.0398	M1, AL(T) = 18.6
GAMMA	2	0.0006	0.1449	
GAMMA	3	0.0040	0.1640	E2, AK= 0.260 AL(T) = 0.435
GAMMA	4	0.0106	0.2880	M1, AK= 0.390 K/L= 5.50 DEL= 4.50

*ENDPOINT ENERGY (MEV).

REF.— NUCLEAR DATA B8, 171 (1972).

••OUTPUT DATA••

83 BISMUTH 212   HALF LIFE = 60.5 MINUTES

DECAY MODES— BETA MINUS AND ALPHA

RADIATION		MEAN NUMBER/ DISINTE- GRATION n_i	MEAN ENERGY/ PAR- TICLE $\bar{E}_i$ (MeV)	EQUI- LIBRIUM DOSE CONSTANT Δ_i (g-rad/ μCi-h)
BETA MINUS	1	0.0066	0.1300	0.0018
BETA MINUS	3	0.0025	0.1727	0.0009
BETA MINUS	4	0.0194	0.1928	0.0079
BETA MINUS	5	0.0149	0.2313	0.0073
BETA MINUS	6	0.0445	0.5375	0.0509
BETA MINUS	7	0.5530	0.8445	0.9948
GAMMA	1	0.0662	0.7271	0.1026
K INT CON ELECT		0.0009	0.6340	0.0012
L INT CON ELECT		0.0002	0.7114	0.0003
M INT CON ELECT		0.0000	0.7238	0.0001
GAMMA	2	0.0112	0.7854	0.0188
K INT CON ELECT		0.0003	0.6923	0.0005
GAMMA	3	0.0036	0.8933	0.0069
K INT CON ELECT		0.0001	0.8002	0.0002
GAMMA	4	0.0018	0.9521	0.0036
GAMMA	5	0.0001	1.0740	0.0003
GAMMA	6	0.0053	1.0786	0.0123
K INT CON ELECT		0.0001	0.9855	0.0002
GAMMA	7	0.0032	1.5127	0.0103
GAMMA	8	0.0154	1.6205	0.0534
K INT CON ELECT		0.0001	1.5274	0.0003
GAMMA	9	0.0007	1.6798	0.0025
GAMMA	10	0.0000	1.8000	0.0000
K INT CON ELECT		0.0001	1.7068	0.0004
GAMMA	11	0.0010	1.8060	0.0041
K ALPHA-1 X-RAY		0.0000	0.0792	0.0001
ALPHA	1	0.0000	5.4860	0.0006
ALPHA	2	0.0040	5.6071	0.0480
RECOIL ATOM		0.0040	0.1078	0.0009
ALPHA	3	0.0005	5.6261	0.0068
RECOIL ATOM		0.0005	0.1081	0.0001
ALPHA	4	0.0060	5.7681	0.0737
RECOIL ATOM		0.0060	0.1109	0.0014
ALPHA	5	0.2520	6.0507	3.2478
RECOIL ATOM		0.2520	0.1163	0.0624
ALPHA	6	0.0963	6.0900	1.2491
RECOIL ATOM		0.0963	0.1171	0.0240
GAMMA	1	0.0101	0.0398	0.0008
L INT CON ELECT		0.1893	0.0255	0.0103
M INT CON ELECT		0.0631	0.0369	0.0049
GAMMA	2	0.0006	0.1449	0.0001
GAMMA	3	0.0021	0.1640	0.0007
L INT CON ELECT		0.0009	0.1497	0.0003
M INT CON ELECT		0.0003	0.1610	0.0001
GAMMA	4	0.0071	0.2880	0.0043
K INT CON ELECT		0.0027	0.2025	0.0012

Table 3.6-1 (continued)
SUMMARY NUCLIDE DECAY DATA AND DOSE CONSTANT DATA

```
                  ••OUTPUT DATA••

        83 BISMUTH 212 (CONTINUED)

-------------------------------------------
                   MEAN      MEAN    EQUI-
                   NUMBER/   ENERGY/ LIBRIUM
                   DISINTE-  PAR-    DOSE
      RADIATION    GRATION   TICLE   CONSTANT

                    n_i      Ē_i     Δ_i
                            (MeV)    (g-rad/
                                      μCi-h)
-------------------------------------------
   L INT CON ELECT  0.0005   0.2738  0.0002
   M INT CON ELECT  0.0001   0.2851  0.0001
   K ALPHA-1 X-RAY  0.0016   0.0728  0.0002
   K ALPHA-2 X-RAY  0.0008   0.0708  0.0001
   L ALPHA X-RAYS   0.0362   0.0107  0.0007
   L BETA X-RAYS    0.0340   0.0122  0.0008
   L GAMMA X-RAYS   0.0045   0.0142  0.0001
   LMM AUGER ELECT  0.1186   0.0083  0.0021
   MXY AUGER ELECT  0.3763   0.0029  0.0023

   ------------
   DAUGHTER NUCLIDE, THALLIUM 212 IS
   RADIOACTIVE AND MAY CONTRIBUTE TO THE DOSE.
```

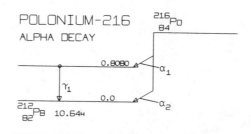

POLONIUM-216 $^{216}_{84}$Po

ALPHA DECAY

```
                  ••INPUT DATA••

     84 POLONIUM 216      HALF LIFE =0.150 SECONDS

   DECAY MODE- ALPHA

   -------------------------------------------
                   MEAN     TRAN-
                   NUMBER/  SITION   OTHER
                   DISINTE- ENERGY   NUCLEAR
   TRANSITION      GRATION  (MEV)    DATA
   -------------------------------------------
      ALPHA 1      0.0000   6.0927
      ALPHA 2      0.9999   6.9002
      GAMMA 1      0.0000   0.8080   E2, AK(T) =
                                     0.00796
                                     AL(T) =
                                     0.00179

   ------------
   REF.- LEDERER, C.M. ET AL, TABLE OF ISOTOPES,
     6TH. ED.
```

```
                  ••OUTPUT DATA••

     84 POLONIUM 216      HALF LIFE =0.150 SECONDS

   DECAY MODE- ALPHA

   -------------------------------------------
                   MEAN     MEAN    EQUI-
                   NUMBER/  ENERGY/ LIBRIUM
                   DISINTE- PAR-    DOSE
      RADIATION    GRATION  TICLE   CONSTANT

                    n_i     Ē_i     Δ_i
                           (MeV)    (g-rad/
                                     μCi-h)
   -------------------------------------------
      ALPHA   1    0.0000   5.9840   0.0002
      ALPHA   2    0.9999   6.7770  14.4347
   RECOIL ATOM     0.9999   0.1278   0.2723
      GAMMA   1    0.0000   0.8080   0.0000

   ------------
   DAUGHTER NUCLIDE, LEAD 212 IS RADIOACTIVE
   AND MAY CONTRIBUTE TO THE DOSE.
```

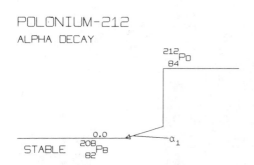

POLONIUM-212

ALPHA DECAY

```
                  ••INPUT DATA••

     84 POLONIUM 212      HALF LIFE =0.306 MICROSEC

   DECAY MODE- ALPHA

   -------------------------------------------
                   MEAN     TRAN-
                   NUMBER/  SITION   OTHER
                   DISINTE- ENERGY   NUCLEAR
   TRANSITION      GRATION  (MEV)    DATA
   -------------------------------------------
      ALPHA 1      1.0000   8.9474

   ------------
   REF.- NUCLEAR DATA B8, 180 (1972).
```

```
                  ••OUTPUT DATA••

     84 POLONIUM 212      HALF LIFE =0.306 MICROSEC

   DECAY MODE- ALPHA

   -------------------------------------------
                   MEAN     MEAN    EQUI-
                   NUMBER/  ENERGY/ LIBRIUM
                   DISINTE- PAR-    DOSE
      RADIATION    GRATION  TICLE   CONSTANT

                    n_i     Ē_i     Δ_i
                           (MeV)    (g-rad/
                                     μCi-h)
   -------------------------------------------
      ALPHA   1    1.0000   8.7848  18.7116
   RECOIL ATOM     1.0000   0.1689   0.3598
```

Table 3.6-1 (continued)
SUMMARY NUCLIDE DECAY DATA AND DOSE CONSTANT DATA

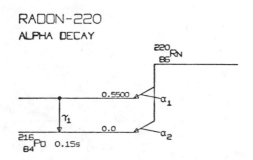

RADON-220
ALPHA DECAY

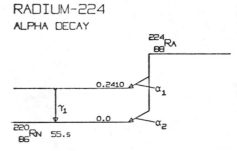

RADIUM-224
ALPHA DECAY

RADON-220 INPUT DATA

```
           ••INPUT DATA••

   86 RADON 220       HALF LIFE = 55.0 SECONDS

DECAY MODE- ALPHA

---------------------------------------------
                 MEAN      TRAN-
                 NUMBER/   SITION
                 DISINTE-  ENERGY   OTHER
                 GRATION   (MEV)    NUCLEAR
 TRANSITION                         DATA

     ALPHA  1    0.0007    5.8496
     ALPHA  2    0.9993    6.3992
     GAMMA  1    0.0007    0.5500   E2,  AK(T)
                                    0.0183
                                    AL(T) =
                                    0.00597

------------
REF.- LEDERER, C.M. ET AL, TABLE OF ISOTOPES,
   6TH. ED.
```

RADIUM-224 INPUT DATA

```
           ••INPUT DATA••

   88 RADIUM 224      HALF LIFE = 3.64 DAYS

DECAY MODE- ALPHA

---------------------------------------------
                 MEAN      TRAN-
                 NUMBER/   SITION
                 DISINTE-  ENERGY   OTHER
                 GRATION   (MEV)    NUCLEAR
 TRANSITION                         DATA

     ALPHA  1    0.0520    5.5427
     ALPHA  2    0.9480    5.7837
     GAMMA  1    0.0520    0.2410   E2,K/L= 0.790
                                    AK(T) =
                                    0.110

------------
REF.- LEDERER, C.M. ET AL, TABLE OF ISOTOPES,
   6TH. ED.
```

RADON-220 OUTPUT DATA

```
           ••OUTPUT DATA••

   86 RADON 220       HALF LIFE = 55.0 SECONDS

DECAY MODE- ALPHA

---------------------------------------------
                 MEAN      MEAN      EQUI-
                 NUMBER/   ENERGY/   LIBRIUM
                 DISINTE-  PAR-      DOSE
   RADIATION     GRATION   TICLE     CONSTANT

                  n_i       Ē_i       Δ_i
                           (MEV)     (g-rad/
                                     μCi-h)
---------------------------------------------
     ALPHA   1   0.0007    5.7470    0.0085
RECOIL ATOM      0.0007    0.1064    0.0001
     ALPHA   2   0.9993    6.2870   13.3819
RECOIL ATOM      0.9993    0.1164    0.2478
     GAMMA   1   0.0006    0.5500    0.0008

------------
DAUGHTER NUCLIDE, POLONIUM 216 IS
RADIOACTIVE AND MAY CONTRIBUTE TO THE DOSE.
```

RADIUM-224 OUTPUT DATA

```
           ••OUTPUT DATA••

   88 RADIUM 224      HALF LIFE = 3.64 DAYS

DECAY MODE- ALPHA

---------------------------------------------
                   MEAN      MEAN      EQUI-
                   NUMBER/   ENERGY/   LIBRIUM
                   DISINTE-  PAR-      DOSE
   RADIATION       GRATION   TICLE     CONSTANT

                    n_i       Ē_i       Δ_i
                             (MEV)     (g-rad/
                                       μCi-h)
---------------------------------------------
         ALPHA  1  0.0520    5.4472    0.6033
    RECOIL ATOM    0.0520    0.0990    0.0109
         ALPHA  2  0.9480    5.6840   11.4773
    RECOIL ATOM    0.9480    0.1033    0.2086
         GAMMA  1  0.0400    0.2410    0.0205
K INT CON ELECT    0.0044    0.1425    0.0013
L INT CON ELECT    0.0056    0.2242    0.0026
M INT CON ELECT    0.0018    0.2374    0.0009
K ALPHA-1 X-RAY    0.0020    0.0837    0.0003
K ALPHA-2 X-RAY    0.0011    0.0810    0.0002
  K BETA-1 X-RAY   0.0007    0.0948    0.0001
LMM AUGER ELECT    0.0052    0.0097    0.0001
MXY AUGER ELECT    0.0170    0.0035    0.0001

------------
DAUGHTER NUCLIDE, RADON 220 IS RADIOACTIVE
AND MAY CONTRIBUTE TO THE DOSE.
```

Table 3.6-1 (continued)
SUMMARY NUCLIDE DECAY DATA AND DOSE CONSTANT DATA

AMERICIUM-241

ALPHA DECAY

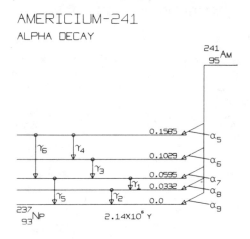

```
                    ••INPUT DATA••

   95 AMERICIUM 241     HALF LIFE = 433. YEARS

   DECAY MODE- ALPHA

   -----------------------------------------------
                    MEAN      TRAN-
                    NUMBER/   SITION
                    DISINTE-  ENERGY   OTHER
                    GRATION   (MEV)    NUCLEAR
   TRANSITION                          DATA
   -----------------------------------------------
      ALPHA   1     0.0000    5.3062
      ALPHA   2     0.0000    5.3265
      ALPHA   3     0.0001    5.4078
      ALPHA   4     0.0159    5.4749
      ALPHA   5     0.0001    5.5034
      ALPHA   6     0.1257    5.5306
      ALPHA   7     0.8527    5.5738
      ALPHA   8     0.0020    5.6009
      ALPHA   9     0.0034    5.6334
      GAMMA   1     0.2010    0.0263   E1, AL= 3.82
                                       L/M= 2.42
      GAMMA   2     0.2030    0.0332   M1, AL=133.
                                       L/M= 4.20
      GAMMA   3     0.1388    0.0434   M1, AL=125.
                                       L/M= 5.50
      GAMMA   4     0.0126    0.0555   M1, AL=50.6
                                       L/M= 2.80
      GAMMA   5     0.7938    0.0595   E1, AL= 0.840
                                       L/M= 3.68
      GAMMA   6     0.0033    0.0989   E2, AL=12.1
                                       L/M= 3.24

   ------------
   REF.- NUCLEAR DATA B6, 635 (1971).
      LEDERER, C.M. ET AL. NUCL. PHYS. 84, 481
      (1966)
```

```
                    ••OUTPUT DATA••

   95 AMERICIUM 241     HALF LIFE = 433. YEARS

   DECAY MODE- ALPHA

   -----------------------------------------------
                    MEAN      MEAN     EQUI-
                    NUMBER/   ENERGY/  LIBRIUM
                    DISINTE-  PAR-     DOSE
   RADIATION        GRATION   TICLE    CONSTANT

                      n_i       Ē_i      Δ_i
                              (MEV)    (g-rad/
                                       μCi-h)
   -----------------------------------------------
        ALPHA   1    0.0000    5.2210   0.0001
        ALPHA   2    0.0000    5.2410   0.0002
        ALPHA   3    0.0001    5.3210   0.0017
        ALPHA   4    0.0159    5.3870   0.1826
   RECOIL ATOM       0.0159    0.0909   0.0030
        ALPHA   5    0.0001    5.4150   0.0011
        ALPHA   6    0.1257    5.4418   1.4570
   RECOIL ATOM       0.1257    0.0918   0.0245
        ALPHA   7    0.8527    5.4843   9.9608
   RECOIL ATOM       0.8527    0.0925   0.1681
        ALPHA   8    0.0020    5.5110   0.0234
   RECOIL ATOM       0.0020    0.0930   0.0003
        ALPHA   9    0.0034    5.5430   0.0401
   RECOIL ATOM       0.0034    0.0935   0.0006
        GAMMA   1    0.0314    0.0263   0.0017
   L INT CON ELECT   0.1200    0.0054   0.0013
   M INT CON ELECT   0.0495    0.0219   0.0023
        GAMMA   2    0.0012    0.0332   0.0000
   L INT CON ELECT   0.1629    0.0122   0.0042
   M INT CON ELECT   0.0388    0.0287   0.0023
        GAMMA   3    0.0009    0.0434   0.0000
   L INT CON ELECT   0.1166    0.0225   0.0055
   M INT CON ELECT   0.0212    0.0389   0.0017
        GAMMA   4    0.0001    0.0555   0.0000
   L INT CON ELECT   0.0091    0.0346   0.0006
   M INT CON ELECT   0.0032    0.0511   0.0003
        GAMMA   5    0.3838    0.0595   0.0486
   L INT CON ELECT   0.3223    0.0386   0.0265
   M INT CON ELECT   0.0876    0.0551   0.0102
        GAMMA   6    0.0001    0.0989   0.0000
   L INT CON ELECT   0.0023    0.0780   0.0003
   M INT CON ELECT   0.0007    0.0945   0.0001
   L ALPHA X-RAYS    0.1603    0.0139   0.0047
    L BETA X-RAYS    0.1502    0.0174   0.0055
   L GAMMA X-RAYS    0.0202    0.0207   0.0008
   LMM AUGER ELECT   0.4027    0.0120   0.0103
   MXY AUGER ELECT   1.3374    0.0044   0.0126
```

Table 3.6-2
NUCLEAR DECAY SCHEME DATA AND DOSE CONSTANTS FOR SELECTED NUCLIDES (1977)

Index of Nuclides Included in Table

From Martin, M. J., Nuclear Decay Data for Selected Radionuclides, ORNL-5114 (data update from the Evaluated Nuclear Structure Data File (ENSDF) of the Nuclear Data Project as of June 1977), Oak Ridge National Laboratory, operated by the Union Carbide Corporation for the Department of Energy, Oak Ridge, Tenn., March 1976. With permission.

3H B- DECAY (12.35 Y 1)   I(MIN) = 0.10%

Radiation Type	Energy (keV)	Intensity (%)	Δ (g-rad/ μCi-h)
β⁻ 1 max	18.600 20		
avg	5.680 10	100	0.0121

16N B- DECAY (7.12 S 2)   I(MIN) = 0.10%

Radiation Type	Energy (keV)	Intensity (%)	Δ (g-rad/ μCi-h)
β⁻ 1 max	1546.9 24		
avg	630.5 11	1.00 20	0.0134
β⁻ 2 max	3299.9 24		
avg	1460.6 12	4.9 4	0.152
β⁻ 3 max	4287.9 23		
avg	1941.0 12	68.0 20	2.81
β⁻ 4 max	10418.6 23		
avg	4979.2 12	26.0 20	2.76
total β⁻			
avg	2695.0 15	100 3	5.74

1 weak β's omitted (ΣIβ = 0.01%)

γ	3	1753.0 6	0.13 4	0.0049
γ	5	2741.0 6	0.76 15	0.0444
γ	8	6129.39 18	69.0 20	9.01
γ	10	7117.0 4	5.0 4	0.758

7 weak γ's omitted (ΣIγ = 0.16%)

Table 3.6-2 (continued)
NUCLEAR DECAY SCHEME DATA AND DOSE CONSTANTS FOR SELECTED NUCLIDES (1977)

18F B+ DECAY (109.74 M 4)   I(MIN) = 0.10%

Radiation Type	Energy (keV)	Intensity (%)	Δ (g-rad/ μCi-h)
Auger-K	0.52	2.88 20	≈0
β+ 1 max	633.5 6		
avg	249.4 3	96.90 17	0.515

Maximum γ±-intensity =193.80%

22NA B+ DECAY (2.602 Y 2)   I(MIN) = 0.10%

Radiation Type	Energy (keV)	Intensity (%)	Δ (g-rad/ μCi-h)
Auger-K	0.82	9.20 5	0.0002
β+ 1 max	545.5 5		
avg	215.54 21	89.84 11	0.412

1 weak β's omitted (ΣIβ = 0.06%)

K X-ray	0.84	0.12 4	≈0
γ 1	1274.540 20	99.940 20	2.71

Maximum γ±-intensity =179.80%

24NA B- DECAY (15.00 H 4)   I(MIN) = 0.10%

Radiation Type	Energy (keV)	Intensity (%)	Δ (g-rad/ μCi-h)
β- 1 max	1390.2 7		
avg	553.9 4	99.935 4	1.18

3 weak β's omitted (ΣIβ = 0.07%)

γ 2	1368.53 5	100	2.91
γ 3	2754.09 5	99.863 5	5.86

4 weak γ's omitted (ΣIγ = 0.06%)

32P B- DECAY (14.29 D 3)   I(MIN) = 0.10%

Radiation Type	Energy (keV)	Intensity (%)	Δ (g-rad/ μCi-h)
β- 1 max	1710.4 6		
avg	695.0 3	100	1.48

35S B- DECAY (87.44 D 7)   I(MIN) = 0.10%

Radiation Type	Energy (keV)	Intensity (%)	Δ (g-rad/ μCi-h)
β- 1 max	167.47 19		
avg	48.80 10	100	0.104

45CA B- DECAY (164 D 1)   I(MIN) = 0.10%

Radiation Type	Energy (keV)	Intensity (%)	Δ (g-rad/ μCi-h)
β- 1 max	256.9 10		
avg	77.3 5	100	0.165

47CA B- DECAY (4.536 D 2)   I(MIN) = 0.10%

Radiation Type	Energy (keV)	Intensity (%)	Δ (g-rad/ μCi-h)
β- 1 max	691 3		
avg	241.2 10	82.0 20	0.421
β- 2 max	1988.3 25		
avg	817.2 12	18.0 20	0.313
total β- avg	344.9 13	100 3	0.735

2 weak β's omitted (ΣIβ = 0.11%)

γ 2	489.23 10	6.7 3	0.0702
γ 3	530.4 3	0.105 16	0.0012
γ 4	767.0 3	0.195 16	0.0032
γ 5	807.86 10	6.9 3	0.119
γ 6	1297.09 10	74.9 19	2.07

2 weak γ's omitted (ΣIγ = 0.03%)

47SC B- DECAY (3.351 D 2)   I(MIN) = 0.10%

Radiation Type	Energy (keV)	Intensity (%)	Δ (g-rad/ μCi-h)
Auger-L	0.42	0.347 15	≈0
Auger-K	4	0.175 8	≈0
ce-K- 1	154.42 5	0.224 8	0.0007
β- 1 max	440.6 20		
avg	142.5 8	68.0 20	0.206
β- 2 max	600.0 20		
avg	203.8 8	32.0 20	0.139
total β- avg	162.1 9	100 3	0.345
γ 1	159.39 5	68.0 20	0.231

Table 3.6-2 (continued)
NUCLEAR DECAY SCHEME DATA AND DOSE CONSTANTS FOR
SELECTED NUCLIDES (1977)

CR EC DECAY (27.704 D 2)   I(MIN) = 0.10%

Radiation Type	Energy (keV)	Intensity (%)	Δ (g-rad/μCi-h)
Auger-L	0.47	144.7 12	0.0014
Auger-K	4.38	66.9 7	0.0062
X-ray L	0.5	0.33 12	≈0
X-ray Kα₂	4.94464	6.59 21	0.0007
X-ray Kα₁	4.95220	13.1 4	0.0014
X-ray Kβ	5.43	2.62 9	0.0003
γ 1	320.078 8	9.80 10	0.0668

52MN IT DECAY (21.1 M 2)   I(MIN) = 0.10%
%IT=1.68 4
SEE ALSO 52MN B+ DECAY (21.1 M)
FEEDING OF 21.1-M 52MN IN 52FE B+ DECAY=100%

Radiation Type	Energy (keV)	Intensity (%)	Δ (g-rad/μCi-h)
γ 1	377.738 11	1.68 4	0.0135

52FE B+ DECAY (8.275 H 8)   I(MIN) = 0.10%
SEE ALSO 52MN IT DECAY (21.1 M)
SEE ALSO 52MN B+ DECAY (21.1 M)

Radiation Type	Energy (keV)	Intensity (%)	Δ (g-rad/μCi-h)
Auger-L	0.6	61.6 23	0.0008
Auger-K	5.19	26.9 12	0.0030
β+ 1 max	804 12		
avg	340 6	56.0 13	0.406
X-ray L	0.64	0.19 7	≈0
X-ray Kα₂	5.88765	3.6 3	0.0005
X-ray Kα₁	5.89875	7.2 6	0.0009
X-ray Kβ	6.49	1.45 12	0.0002
γ 1	168.684 11	100	0.359

Maximum γ±-intensity =112.00%

52MN B+ DECAY (21.1 M 2)   I(MIN) = 0.10%
%(EC+B+)=98.32 4
SEE ALSO 52MN IT DECAY (21.1 M)
FEEDING OF 21.1-M 52MN IN 52FE B+ DECAY=100%

Radiation Type	Energy (keV)	Intensity (%)	Δ (g-rad/μCi-h)
Auger-L	0.54	2.37 5	≈0
Auger-K	4.78	1.068 24	0.0001
β+ 1 max	905.2 23		
avg	383.0 10	0.164 8	0.0013
β+ 2 max	2632.8 23		
avg	1173.8 11	96.5 20	2.41
total β+			
avg	1172.1 11	96.7 20	2.41

3 weak β's omitted (ΣIβ = 0.05%)

X-ray Kα₂	5.40551	0.124 5	≈0
X-ray Kα₁	5.41472	0.246 8	≈0
γ 4	1434.060 10	98.3 20	3.00
γ 6	1727.53 7	0.216 10	0.0080

12 weak γ's omitted (ΣIγ = 0.17%)
Maximum γ±-intensity =193.43%

52MN EC DECAY (5.60 D 1)   I(MIN) = 0.10%

Radiation Type	Energy (keV)	Intensity (%)	Δ (g-rad/μCi-h)
Auger-L	0.54	99.4 20	0.0011
Auger-K	4.78	44.7 10	0.0046
β+ 1 max	575.3 23		
avg	241.6 10	29.4 7	0.151
X-ray L	0.57	0.26 9	≈0
X-ray Kα₂	5.40551	5.20 18	0.0006
X-ray Kα₁	5.41472	10.3 4	0.0012
X-ray Kβ	6	2.06 8	0.0003
γ 2	346.03 3	0.980 20	0.0072
γ 4	399.56 5	0.183 8	0.0016
γ 5	502.05 5	0.210 20	0.0022
γ 6	600.18 4	0.390 20	0.0050
γ 7	647.50 4	0.400 20	0.0055
γ 8	744.214 11	90.0 19	1.43
γ 9	848.13 3	3.32 8	0.0600
γ 11	935.520 20	94.5 20	1.88
γ 13	1246.246 12	4.21 10	0.112
γ 14	1247.85 9	0.38 4	0.0101
γ 15	1333.615 16	5.07 11	0.144
γ 16	1434.056 16	100	3.05

9 weak γ's omitted (ΣIγ = 0.37%)
Maximum γ±-intensity = 58.80%

55FE EC DECAY (2.7 Y 1)   I(MIN) = 0.10%

Radiation Type	Energy (keV)	Intensity (%)	Δ (g-rad/μCi-h)
Auger-L	0.6	139 4	0.0018
Auger-K	5.19	60.7 22	0.0067
X-ray L	0.64	0.42 14	≈0
X-ray Kα₂	5.88765	8.2 7	0.0010
X-ray Kα₁	5.89875	16.3 12	0.0020
X-ray Kβ	6.49	3.3 3	0.0005

Table 3.6-2 (continued)
NUCLEAR DECAY SCHEME DATA AND DOSE CONSTANTS FOR SELECTED NUCLIDES (1977)

57CO EC DECAY (270.9 D 6)   I(MIN) = 0.10%

Radiation Type	Energy (keV)	Intensity (%)	Δ (g-rad/ μCi-h)
Auger-L	0.67	249 3	0.0036
Auger-K	5.62	105.5 14	0.0126
e-K- 1	7.3007 10	69.6 4	0.0108
e-L- 1	13.5666 6	7.79 22	0.0023
e-MNO- 1	14.3198 10	1.15 7	0.0004
e-K- 2	114.951 4	1.87 10	0.0046
e-L- 2	121.217 3	0.183 10	0.0005
e-K- 3	129.364 4	1.40 12	0.0039
e-L- 3	135.630 3	0.140 12	0.0004
X-ray L	0.7	0.8 3	≈0
X-ray Kα₂	6.39084	16.6 5	0.0023
X-ray Kα₁	6.40384	32.8 8	0.0045
X-ray Kβ	7	6.63 21	0.0010
γ 1	14.4127 4	9.54 13	0.0029
γ 2	122.063 3	85.59 19	0.223
γ 3	136.476 3	10.61 18	0.0309
γ 9	692.00 3	0.160 5	0.0024

6 weak γ's omitted (ΣIγ = 0.03%)

63NI B- DECAY (96 Y 4)   I(MIN) = 0.10%

Radiation Type	Energy (keV)	Intensity (%)	Δ (g-rad/ μCi-h)
β⁻ 1 max	65.87 20		
avg	17.13 5	100	0.0365

65ZN EC DECAY (243.9 D 1)   I(MIN) = 0.10%

Radiation Type	Energy (keV)	Intensity (%)	Δ (g-rad/ μCi-h)
Auger-L	0.92	126.7 18	0.0025
Auger-K	7	48.3 8	0.0072
β⁺ 1 max	329.9 11		
avg	143.0 5	1.460 20	0.0044
X-ray L	0.93	0.57 20	≈0
X-ray Kα₂	8.027830	11.5 3	0.0020
X-ray Kα₁	8.047780	22.6 5	0.0039
X-ray Kβ	9	4.61 13	0.0009
γ 3	1115.52 3	50.75 10	1.21

Maximum γ±-intensity = 2.92%

58CO B+ DECAY (70.78 D 10)   I(MIN) = 0.10%

Radiation Type	Energy (keV)	Intensity (%)	Δ (g-rad/ μCi-h)
Auger-L	0.67	116.5 13	0.0017
Auger-K	5.62	49.4 6	0.0059
β⁺ 1 max	474.6 14		
avg	201.2 6	15.00 5	0.0643
X-ray L	0.7	0.36 12	≈0
X-ray Kα₂	6.39084	7.78 21	0.0011
X-ray Kα₁	6.40384	15.4 4	0.0021
X-ray Kβ	7	3.10 10	0.0005
γ 1	810.757 18	99.4 3	1.72
γ 2	863.935 18	0.676 10	0.0124
γ 3	1674.68 4	0.517 10	0.0184

Maximum γ±-intensity = 30.00%

60CO B- DECAY (5.271 Y 1)   I(MIN) = 0.10%

Radiation Type	Energy (keV)	Intensity (%)	Δ (g-rad/ μCi-h)
β⁻ 1 max	317.87 11		
avg	95.80 10	99.920 20	0.204

2 weak β's omitted (ΣIβ = 0.09%)

| γ   3 | 1173.210 10 | 99.900 20 | 2.50 |
| γ   4 | 1332.470 10 | 99.9827 | 2.84 |

4 weak γ's omitted (ΣIγ = 0.02%)

85KR B- DECAY (10.72 Y 1)   I(MIN) = 0.10%

Radiation Type	Energy (keV)	Intensity (%)	Δ (g-rad/ μCi-h)
β⁻ 1 max	173.0 20		
avg	47.5 6	0.430 10	0.0004
β⁻ 2 max	687.0 20		
avg	251.4 8	99.570 10	0.533
total β⁻ avg	250.5 8	100.000 15	0.534
γ 1	513.990 10	0.430 10	0.0047

Table 3.6-2 (continued)
NUCLEAR DECAY SCHEME DATA AND DOSE CONSTANTS FOR
SELECTED NUCLIDES (1977)

85KR B- DECAY (4.48 H 1)   I(MIN) = 0.10%
          %B-=78.9 6
          SEE ALSO 85KR IT DECAY (4.48 H)

Radiation Type	Energy (keV)	Intensity (%)	Δ (g-rad/μCi-h)
Auger-L	1.68	3.69 10	0.0001
Auger-K	11.4	1.00 4	0.0002
ce-K- 2	135.980 10	3.02 7	0.0087
ce-L- 2	149.115 10	0.340 16	0.0011
β⁻ 1 max	710.8 20		
avg	238.3 7	0.290 10	0.0015
β⁻ 2 max	840.7 20		
avg	290.4 7	78.6 6	0.486
total β⁻			
avg	290.2 7	78.9 6	0.488

          1 weak β's omitted (ΣIβ = 0.02%)

Radiation Type	Energy (keV)	Intensity (%)	Δ (g-rad/μCi-h)
X-ray Kα₂	13.33580 2	0.587 18	0.0002
X-ray Kα₁	13.39530 2	1.14 3	0.0003
X-ray Kβ	15	0.298 10	≈0
γ 1	129.850 20	0.300 11	0.0008
γ 2	151.180 10	75.5 9	0.243

          2 weak γ's omitted (ΣIγ = 0.02%)

85KR IT DECAY (4.48 H 1)   I(MIN) = 0.10%
          %IT=21.1 6
          SEE ALSO 85KR B- DECAY (4.48 H)

Radiation Type	Energy (keV)	Intensity (%)	Δ (g-rad/μCi-h)
Auger-L	1.5	7.8 5	0.0002
Auger-K	10.8	2.14 21	0.0005
ce-K- 1	290.544 20	6.0 3	0.0374
ce-L- 1	302.949 20	0.90 4	0.0058
ce-MNO- 1	304.582 20	0.182 8	0.0012
X-ray L	1.59	0.10 4	≈0
X-ray Kα₂	12.59880 20	1.14 8	0.0003
X-ray Kα₁	12.6490 20	2.21 15	0.0006
X-ray Kβ	14	0.56 4	0.0002
γ 1	304.870 20	14.0 6	0.0909

85SR EC DECAY (64.84 D 3)   I(MIN) = 0.10%

Radiation Type	Energy (keV)	Intensity (%)	Δ (g-rad/μCi-h)
Auger-L	1.68	108.2 23	0.0039
Auger-K	11.4	29.1 9	0.0071
ce-K- 1	498.790 10	0.617 13	0.0066
X-ray L	1.69	1.6 6	≈0
X-ray Kα₂	13.33580 2	17.1 4	0.0048
X-ray Kα₁	13.39530 2	33.0 7	0.0094
X-ray Kβ	15	8.66 24	0.0028
γ 1	513.990 10	98.0 10	1.07

          1 weak γ's omitted (ΣIγ = 0.01%)

89SR B-DECAY (50.5 D 1)   I(MIN) = 0.10%

Radiation Type	Energy (keV)	Intensity (%)	Δ (g-rad/μCi-h)
β⁻ 1 max	1492 4		
avg	583.1 13	99.985 5	1.24

          1 weak β's omitted (ΣIβ = 0.02%)

          1 weak γ's omitted (ΣIγ = 0.02%)

90SR B- DECAY (29.12 Y 24)   I(MIN) = 0.10%

Radiation Type	Energy (keV)	Intensity (%)	Δ (g-rad/μCi-h)
β⁻ 1 max	546.0 20		
avg	195.8 8	100	0.417

90Y B- DECAY (64.0 H 1)   I(MIN) = 0.10%

Radiation Type	Energy (keV)	Intensity (%)	Δ (g-rad/μCi-h)
β⁻ 1 max	2284 4		
avg	934.8 18	99.984 3	1.99

          1 weak β's omitted (ΣIβ = 0.02%)

Table 3.6-2 (continued)
NUCLEAR DECAY SCHEME DATA AND DOSE CONSTANTS FOR SELECTED NUCLIDES (1977)

99MO B- DECAY (66.0 H 2)   I(MIN) = 0.10%
SEE ALSO 99TC IT DECAY (6.02 H)

Radiation Type	Energy (keV)	Intensity (%)	Δ (g-rad/ μCi-h)
Auger-L	2.17	5.4 3	0.0002
Auger-K	15.5	1.11 17	0.0004
ce-K- 2	19.5400 22	3.77 8	0.0016
ce-L- 2	37.54 15 21	0.456 16	0.0004
ce-MNO- 2	40.0400 23	0.110 6	≈0
ce-K- 4	119.435 23	0.490 24	0.0012
ce-K- 8	160.023 11	0.763 17	0.0026
ce-L- 8	178.024 11	0.1138 2	0.0004
β- 1 max	214.9 10		
avg	59.9 3	0.111 3	0.0001
β- 2 max	352.7 10		
avg	104.3 4	0.134 4	0.0003
β- 3 max	436.2 10		
avg	133.0 4	16.53 7	0.0468
β- 4 max	847.6 10		
avg	289.6 4	1.17 3	0.0072
β- 5 max	1214.0 10		
avg	442.7 5	82.0 10	0.773
total β- avg	388.8 6	100.0 10	0.828

2 weak β's omitted (ΣIβ = 0.01%)

X-ray L	2.42	0.25 9	≈0
X-ray Kα₂	18.2508 8	1.12 6	0.0004
X-ray Kα₁	18.3671 8	2.14 10	0.0008
X-ray Kβ	20.6	0.64 3	0.0003
γ 2	40.5840 20	1.14 4	0.0010
γ 4	140.479 23	4.95 9	0.0148
γ 8	181.067 11	6.05 8	0.0233
γ 11	366.440 11	1.191 24	0.0093
γ 25	739.477 14	12.180 20	0.192
γ 27	777.892 16	4.31 7	0.0714
γ 28	822.957 13	0.133 4	0.0023

25 weak γ's omitted (ΣIγ = 0.27%)

103RU B- DECAY (39.28 D 1)   I(MIN) = 0.10%
SEE ALSO 103RH IT DECAY (56.12 M)

Radiation Type	Energy (keV)	Intensity (%)	Δ (g-rad/ μCi-h)
Auger-L	2.39	0.98 9	≈0
Auger-K	17	0.20 4	≈0
ce-K- 3	30.065 5	0.66 7	0.0004
ce-K- 13	473.860 10	0.397 13	0.0040
β- 1 max	112 4		
avg	29.5 10	6.0 3	0.0038
β- 2 max	225 4		
avg	62.8 11	87.1 15	0.117
β- 3 max	467 4		
avg	143.4 13	0.231 10	0.0007
β- 4 max	669 4		
avg	218.1 14	0.10	0.0005
β- 5 max	722 4		
avg	238.7 14	6 3	0.0325
total β- avg	72.4 13	100 4	0.154

3 weak β's omitted (ΣIβ = 0.10%)

X-ray Kα₂	20.07370 2	0.244 19	0.0001
X-ray Kα₁	20.21610 2	0.46 4	0.0002
X-ray Kβ	22.7	0.145 11	≈0
γ 3	53.285 5	0.36 4	0.0004
γ 9	294.980 10	0.242 11	0.0015
γ 12	443.800 10	0.311 11	0.0029
γ 13	497.080 10	86.4 24	0.915
γ 15	557.040 20	0.76 3	0.0090
γ 17	610.330 10	5.3 3	0.0685

13 weak γ's omitted (ΣIγ = 0.12%)

99TC IT DECAY (6.02 H 3)   I(MIN) = 0.10%
FEEDING OF 6.02-H 99TC IN 99MO DECAY=87.6% 10

Radiation Type	Energy (keV)	Intensity (%)	Δ (g-rad/ μCi-h)
ce-M- 1	1.630 5	86.84 16	0.0030
ce-NOP- 1	2.106 5	12.18 20	0.0005
Auger-L	2.17	10.4 7	0.0005
Auger-K	15.5	2.1 4	0.0007
ce-K- 2	119.435 23	8.8 4	0.0224
ce-K- 3	121.63 3	0.69 4	0.0018
ce-L- 2	137.436 23	1.06 4	0.0031
ce-L- 3	139.63 3	0.215 14	0.0006
ce-M- 2	139.935 23	0.198 10	0.0006
X-ray L	2.42	0.49 17	≈0
X-ray Kα₂	18.2508 8	2.12 13	0.0008
X-ray Kα₁	18.3671 8	4.06 24	0.0016
X-ray Kβ	20.6	1.22 8	0.0005
γ 2	140.479 23	88.97 24	0.266

2 weak γ's omitted (ΣIγ = 0.02%)

103RH IT DECAY (56.12 M 1)   I(MIN) = 0.10%
FEEDING OF 56.12-M 103RH IN 103RU DECAY=99.75% 1
FEEDING OF 56.12-M 103RH IN 103PD DECAY=99.975% 1

Radiation Type	Energy (keV)	Intensity (%)	Δ (g-rad/ μCi-h)
Auger-L	2.39	76.6 15	0.0039
ce-K- 1	16.530 7	9.5 3	0.0033
Auger-K	17	1.8 3	0.0007
ce-L- 1	36.338 7	71.290 10	0.0552
ce-M- 1	39.123 7	14.4 4	0.0120
ce-NOP- 1	39.669 7	4.70 20	0.0040
X-ray L	2.7	4.0 13	0.0002
X-ray Kα₂	20.07370 2	2.19 12	0.0009
X-ray Kα₁	20.21610 2	4.17 21	0.0018
X-ray Kβ	22.7	1.30 7	0.0006

1 weak γ's omitted (ΣIγ = 0.07%)

Table 3.6-2 (continued)
NUCLEAR DECAY SCHEME DATA AND DOSE CONSTANTS FOR SELECTED NUCLIDES (1977)

110AG EC DECAY   (24.6 S 2)   I(MIN) = 0.10%
%EC=0.30 6
SEE ALSO 110AG B- DECAY (24.6 S)

Radiation Type	Energy (keV)	Intensity (%)	Δ (g-rad/ μCi-h)
Auger-L	2.5	0.23 5	≈0
X-ray Kα₁	21.17710 2	0.115 24	≈0

110AG B- DECAY   (24.6 S 2)   I(MIN) = 0.10%
%B-=99.70 6
SEE ALSO 110AG EC DECAY (24.6 S)

Radiation Type	Energy (keV)	Intensity (%)	Δ (g-rad/ μCi-h)
β- 1 max	2235.0 19		
avg	894.1 9	4.4 3	0.0838
β- 2 max	2892.8 19		
avg	1199.3 9	95.2 3	2.43
total β-			
avg	1185.1 9	99.7 5	2.52

9 weak β's omitted (ΣIβ = 0.09%)

| γ   2 | 657.749 10 | 4.50 23 | 0.0630 |

12 weak γ's omitted (ΣIγ = 0.10%)

110AG B- DECAY   (249.9 D 1)   I(MIN) = 0.10%
%B-=98.67 10
SEE ALSO 110AG IT DECAY (249.9 D)

Radiation Type	Energy (keV)	Intensity (%)	Δ (g-rad/ μCi-h)
β- 1 max	83.9 19		
avg	21.8 6	67.5 4	0.0313
β- 2 max	133.8 19		
avg	35.7 6	0.408 12	0.0003
β- 3 max	530.7 19		
avg	165.6 7	30.6 4	0.108
total β-			
avg	66.6 14	98.7 6	0.140

4 weak β's omitted (ΣIβ = 0.18%)

γ 12	365.441 15	0.106 9	0.0008
γ 17	446.797 8	3.66 4	0.0348
γ 23	620.346 11	2.78 3	0.0367
γ 24	626.246 10	0.235 7	0.0031
γ 26	657.749 10	94.7 10	1.33
γ 27	676.60 10	0.142 19	0.0020
γ 28	677.606 11	10.72 11	0.155
γ 29	686.988 11	6.49 7	0.0950
γ 30	706.670 13	16.74 12	0.252
γ 31	708.115 20	0.28 10	0.0043
γ 32	744.260 13	4.66 5	0.0739
γ 33	763.928 13	22.36 23	0.364
γ 35	818.016 12	7.32 8	0.128
γ 36	884.667 13	72.9 8	1.37
γ 37	937.478 13	34.3 4	0.685
γ 39	997.233 18	0.125 5	0.0027
γ 49	1334.304 17	0.133 10	0.0038
γ 50	1384.270 13	24.35 25	0.718
γ 52	1475.759 22	3.99 4	0.125
γ 53	1505.001 21	13.11 14	0.420
γ 54	1562.266 22	1.184 13	0.0394

40 weak γ's omitted (ΣIγ = 0.92%)

110AG IT DECAY   (249.9 D 1)   I(MIN) = 0.10%
%IT=1.33 10
SEE ALSO 110AG B- DECAY (249.9 D)

Radiation Type	Energy (keV)	Intensity (%)	Δ (g-rad/ μCi-h)
ce-M- 1	0.56 10	1.00 8	≈0
ce-NOP- 1	1.18 10	0.332 25	≈0
Auger-L	2.6	1.11 8	≈0
Auger-K	18.5	0.141 24	≈0
ce-K- 2	90.97 5	0.83 7	0.0016
ce-L- 2	112.67 5	0.39 3	0.0009
X-ray Kα₂	21.9903 3	0.196 17	≈0
X-ray Kα₁	22.16290 1	0.37 3	0.0002
X-ray Kβ	24.9	0.120 10	≈0

125I EC DECAY (60.14 D 11)   I(MIN) = 0.10%

Radiation Type	Energy (keV)	Intensity (%)	Δ (g-rad/ μCi-h)
Auger-L	3.19	156 9	0.0106
ce-K- 1	3.65 3	80.0 6	0.0062
Auger-K	22.7	20 5	0.0097
ce-L- 1	30.52 3	10.5 3	0.0068
ce-M- 1	34.45 3	2.20 20	0.0016
ce-NOP- 1	35.29 3	0.70 20	0.0005
X-ray L	3.77	15 6	0.0012
X-ray Kα₂	27.20170 2	39.8 14	0.0230
X-ray Kα₁	27.47230 2	74.2 25	0.0434
X-ray Kβ	31	25.8 10	0.0170
γ 1	35.46 3	6.67 22	0.0050

129I B- DECAY (1.57E7 Y 4)   I(MIN) = 0.10%

Radiation Type	Energy (keV)	Intensity (%)	Δ (g-rad/ μCi-h)
Auger-L	3.43	74 4	0.0054
ce-K- 1	5.02 3	79.10 20	0.0085
Auger-K	24.6	8.8 16	0.0046
ce-L- 1	34.13 3	10.6 3	0.0077
ce-M- 1	38.44 3	2.10 10	0.0017
ce-NOP- 1	39.37 3	0.70 10	0.0006
β- 1 max	150 5		
avg	40 5	100	0.0852
X-ray L	4.1	8.2 25	0.0007
X-ray Kα₂	29.4580 10	20.0 6	0.0126
X-ray Kα₁	29.7790 10	37.1 9	0.0235
X-ray Kβ	33.6	13.2 4	0.0094
γ 1	39.58 3	7.50 20	0.0063

Table 3.6-2 (continued)
NUCLEAR DECAY SCHEME DATA AND DOSE CONSTANTS FOR
SELECTED NUCLIDES (1977)

131I B- DECAY (8.04 D 1)   I(MIN) = 0.10%
          SEE ALSO 131XE IT DECAY

Radiation Type	Energy (keV)	Intensity (%)	Δ (g-rad/μCi-h)
Auger-L	3.43	5.1 3	0.0004
Auger-K	24.6	0.60 11	0.0003
ce-K- 1	45.622 10	3.54 9	0.0034
ce-L- 1	74.730 10	0.472 12	0.0008
ce-MNO- 1	79.041 10	0.1310 2	0.0002
ce-K- 7	249.737 11	0.248 6	0.0013
ce-K- 14	329.919 11	1.54 4	0.0108
ce-L- 14	359.027 11	0.244 6	0.0019
β- 1 max	247.9 6		
avg	69.40 20	2.13 3	0.0031
β- 2 max	303.9 6		
avg	87.00 20	0.620 20	0.0011
β- 3 max	333.8 6		
avg	96.60 20	7.36 10	0.0151
β- 4 max	606.3 6		
avg	191.6 3	89.4 10	0.365
β- 5 max	806.9 6		
avg	283.2 3	0.420 20	0.0025
total β-			
avg	181.7 3	100.0 10	0.387

1 weak β's omitted (ΣIβ = 0.06%)

x-ray L	4.1	0.56 17	≈0
x-ray Kα₂	29.4580 10	1.37 5	0.0009
x-ray Kα₁	29.7790 10	2.54 8	0.0016
x-ray Kβ	33.6	0.90 3	0.0006
γ 1	80.183 10	2.62 5	0.0045
γ 4	177.210 10	0.265 4	0.0010
γ 7	284.298 11	6.06 9	0.0367
γ 12	325.781 11	0.251 6	0.0017
γ 14	364.480 11	81.2 12	0.631
γ 16	502.991 11	0.361 7	0.0039
γ 17	636.973 10	7.27 11	0.0986
γ 18	642.703 11	0.220 4	0.0030
γ 19	722.893 10	1.80 3	0.0278

10 weak γ's omitted (ΣIγ = 0.23%)

131XE IT DECAY (11.9 D 1)   I(MIN) = 0.10%
          FEEDING OF 11.9-D 131XE IN 131I
                    DECAY=1.086% 13

Radiation Type	Energy (keV)	Intensity (%)	Δ (g-rad/μCi-h)
Auger-L	3.43	75 4	0.0055
Auger-K	24.6	6.8 13	0.0036
ce-K- 1	129.369 8	61.2 7	0.169
ce-L- 1	158.477 8	28.6 6	0.0965
ce-M- 1	162.788 8	6.50 20	0.0225
ce-NOP- 1	163.722 8	1.78 6	0.0062
x-ray L	4.1	8 3	0.0007
x-ray Kα₂	29.4580 10	15.5 5	0.0097
x-ray Kα₁	29.7790 10	28.7 8	0.0182
x-ray Kβ	33.6	10.2 4	0.0073
γ 1	163.930 8	1.96 6	0.0068

132I B- DECAY (2.30 H 3)   I(MIN) = 0.10%

Radiation Type	Energy (keV)	Intensity (%)	Δ (g-rad/μCi-h)
Auger-L	3.43	0.89 7	≈0
Auger-K	24.6	0.118 22	≈0
ce-K- 24	488.09 9	0.127 14	0.0013
ce-K- 34	633.13 8	0.35 3	0.0047
ce-K- 41	738.04 8	0.206 24	0.0032
β- 1 max	226 20		
avg	63 7	0.12 6	0.0002
β- 2 max	320 20		
avg	92 7	0.261 3	0.0005
β- 3 max	353 20		
avg	103 7	0.12 5	0.0003
β- 4 max	366 20		
avg	107 7	0.19	0.0004
β- 5 max	425 20		
avg	127 7	0.21 4	0.0006
β- 6 max	458 20		
avg	138 7	0.38	0.0011
β- 7 max	504 20		
avg	154 8	0.24 6	0.0008
β- 8 max	522 20		
avg	161 8	0.34 7	0.0012
β- 9 max	689 20		
avg	223 8	0.88	0.0042
β-10 max	740 20		
avg	242 8	1.81 10	0.0093
β-11 max	741 20		
avg	242 8	12.8 8	0.0660
β-12 max	826 20		
avg	275 8	0.37 6	0.0022
β-13 max	910 20		
avg	309 8	3.60 20	0.0237
β-14 max	967 20		
avg	331 9	8.1 4	0.0571
β-15 max	996 20		
avg	344 9	3.79 16	0.0278
β-16 max	1155 20		
avg	409 9	2.11 7	0.0184
β-17 max	1185 20		
avg	422 9	19.0 7	0.171
β-18 max	1276 20		
avg	440 9	0.9	0.0085
β-19 max	1413 20		
avg	519 9	1.7 6	0.0188
β-20 max	1468 20		
avg	543 9	1.9 8	0.0220
β-21 max	1470 20		
avg	543 9	10.2 10	0.118
β-22 max	1617 20		
avg	608 9	12.7 7	0.164
β-23 max	2140 20		
avg	841 9	17.6 22	0.315
total β-			
avg	486 11	100 3	1.03

Table 3.6-2 (continued)
NUCLEAR DECAY SCHEME DATA AND DOSE CONSTANTS FOR SELECTED NUCLIDES (1977)

11 weak β's omitted (ΣIβ = 0.53%)

X-ray	Kα$_2$	29.4580 10	0.269 15	0.0002
X-ray	Kα$_1$	29.7790 10	0.50 3	0.0003
X-ray	Kβ	33.6	0.177 10	0.0001
γ	2	147.20 10	0.237 20	0.0007
γ	3	183.3	0.1579	0.0006
γ	4	254.80 20	0.19 3	0.0010
γ	5	262.70 10	1.44 9	0.0081
γ	7	284.80 10	0.79 7	0.0048
γ	9	306.6 4	0.10857	0.0007
γ	10	306.6 4	0.11 4	0.0007
γ	12	316.5 4	0.16 4	0.0011
γ	15	363.5 4	0.49 10	0.0038
γ	16	387.8 4	0.17 3	0.0014
γ	17	416.8 4	0.46 9	0.0041
γ	18	431.9 4	0.45 9	0.0042
γ	19	446.0 4	0.67 8	0.0064
γ	20	473.4 7	0.27 5	0.0027
γ	22	487.5 7	0.18 5	0.0018
γ	23	505.90 15	5.03 20	0.0542
γ	24	522.65 9	16.1 6	0.179
γ	25	535.5 4	0.52 8	0.0060
γ	26	547.10 20	1.25 9	0.0146
γ	30	620.8 10	0.3948	0.0052
γ	31	621.2 10	1.579 4	0.0209
γ	32	630.22 9	13.7 6	0.184
γ	33	650.60 20	2.66 20	0.0369
γ	34	667.69 8	98.70 20	1.40
γ	35	669.8 3	4.9 8	0.0704
γ	36	671.6 3	5.2 4	0.0748
γ	37	727	2.2 6	0.0336
γ	38	727.2 10	3.2 6	0.0489
γ	39	729.5 4	1.1 3	0.0169
γ	41	772.60 8	76.2 18	1.25
γ	42	780.2 3	1.23 6	0.0205
γ	43	784.5 4	0.42 5	0.0071
γ	45	809.80 20	2.9 3	0.0494
γ	46	812.20 20	5.6 5	0.0973
γ	47	863.30 20	0.59 5	0.0109
γ	48	876.80 20	1.08 5	0.0201
γ	50	910.30 20	0.92 5	0.0178
γ	51	927.6 3	0.44 8	0.0088
γ	53	954.55 9	18.1 6	0.367
γ	54	984.50 20	0.56 6	0.0118
γ	58	1034.70 20	0.57 5	0.0126
γ	64	1136.03 12	2.96 20	0.0716
γ	65	1138	0.2961	0.0072
γ	66	1143.40 20	1.38 10	0.0337
γ	67	1148.2 7	0.21 5	0.0051
γ	68	1173.20 20	1.09 10	0.0271
γ	71	1272.7 4	0.15 3	0.0040
γ	72	1290.7 3	1.14 6	0.0312
γ	73	1295.3 3	1.97 10	0.0545
γ	74	1298.2 5	0.89 10	0.0246
γ	76	1317.1 7	0.118 20	0.0033
γ	77	1372.07 13	2.47 10	0.0721
γ	78	1398.57 10	7.1 3	0.212
γ	80	1442.56 10	1.42 6	0.0437
γ	82	1476.80 20	0.138 20	0.0043
γ	94	1757.50 20	0.38 3	0.0140
γ	101	1921.08 12	1.18 9	0.0485
γ	103	2002.30 12	1.09 10	0.0463
γ	104	2086.82 15	0.25 4	0.0110
γ	105	2172.68 15	0.19 3	0.0087
γ	107	2223.17 15	0.118 20	0.0056
γ	110	2390.48 15	0.168 20	0.0085

58 weak γ's omitted (ΣIγ = 2.23%)

132CS B+ DECAY (6.475 D 10)   I(MIN) = 0.10%
%(EC+B+)=98.0 1
SEE ALSO 132CS B- DECAY

Radiation Type	Energy (keV)	Intensity (%)	Δ(g-rad/μCi-h)
Auger-L	3.43	78 4	0.0057
Auger-K	24.6	9.2 17	0.0048
β+ 1 max	403 23		
avg	190 10	0.30 8	0.0012
X-ray L	4.1	9 3	0.0008
X-ray Kα$_2$	29.4580 10	21.0 6	0.0132
X-ray Kα$_1$	29.7790 10	39.0 10	0.0248
X-ray Kβ	33.6	13.9 4	0.0099
γ 2	505.90 15	0.80 10	0.0086
γ 3	630.22 9	1.01 8	0.0136
γ 4	667.67 6	97.47 11	1.39
γ 6	1136.03 12	0.51 4	0.0123
γ 8	1317.80 20	0.58 5	0.0164

4 weak γ's omitted (ΣIγ = 0.27%)
Maximum γ±-intensity = 0.60%

132CS B- DECAY (6.475 D 10)   I(MIN) = 0.10%
%B-=2.0 1
SEE ALSO 132CS B+ DECAY

Radiation Type	Energy (keV)	Intensity (%)	Δ(g-rad/μCi-h)
β- 1 max	240 30		
avg	67 8	0.37 3	0.0005
β- 2 max	810 30		
avg	267 11	1.57 12	0.0089
total β-			
avg	223 13	2.00 13	0.0095

1 weak β's omitted (ΣIβ = 0.06%)

γ	1	464.55 6	1.87 14	0.0185
γ	2	567.14 3	0.24 4	0.0029
γ	4	1031.70 3	0.122 12	0.0027

1 weak γ's omitted (ΣIγ = 0.06%)

Table 3.6-2 (continued)
NUCLEAR DECAY SCHEME DATA AND DOSE CONSTANTS FOR
SELECTED NUCLIDES (1977)

133I B- DECAY (20.8 H 1)   I(MIN) = 0.10%
SEE ALSO 133XE IT DECAY (2.25 D)

Radiation Type	Energy (keV)	Intensity (%)	Δ (g-rad/μCi-h)
β- 1 max	170 30		
avg	46 9	0.414 14	0.0004
β- 2 max	370 30		
avg	110 10	1.24 4	0.0029
β- 3 max	410 30		
avg	122 11	0.396 10	0.0010
β- 4 max	460 30		
avg	140 11	3.74 6	0.0112
β- 5 max	520 30		
avg	162 11	3.12 6	0.0108
β- 6 max	710 30		
avg	230 12	0.541 19	0.0027
β- 7 max	880 30		
avg	299 12	4.15 11	0.0264
β- 8 max	1020 30		
avg	352 13	1.81 5	0.0136
β- 9 max	1230 30		
avg	441 13	83.5 18	0.784
β-10 max	1530 30		
avg	573 13	1.07 5	0.0131
total β-			
avg	407 15	100.0 18	0.866
γ 6	262.702 12	0.356 10	0.0020
γ 7	267.173 22	0.117 6	0.0007
γ 8	345.43 5	0.104 18	0.0008
γ 9	361.09 6	0.11 4	0.0009
γ 13	417.556	0.153 11	0.0014
γ 14	422.910 16	0.309 9	0.0028
γ 16	510.530 11	1.81 5	0.0197
γ 19	529.872 11	86.3 18	0.974
γ 24	617.974 17	0.539 13	0.0071
γ 28	680.247 15	0.645 16	0.0093
γ 29	706.578 13	1.49 4	0.0225
γ 30	768.382 18	0.457 13	0.0075
γ 32	820.506 24	0.154 6	0.0027
γ 33	856.278 12	1.23 4	0.0225
γ 34	875.329 11	4.47 10	0.0834
γ 35	909.67 3	0.212 8	0.0041
γ 39	1052.296 21	0.552 13	0.0124
γ 40	1060.07 6	0.137 6	0.0031
γ 42	1236.411 12	1.49 4	0.0393
γ 43	1298.223 11	2.33 6	0.0644
γ 45	1350.38 3	0.148 5	0.0043

26 weak γ's omitted (ΣIγ = 0.71%)

133XE B- DECAY   (5.245 D 6)   I(MIN) = 0.10%

Radiation Type	Energy (keV)	Intensity (%)	Δ (g-rad/μCi-h)
Auger-L	3.55	49.0 20	0.0037
Auger-K	25.5	5.5 7	0.0030
ce-K- 1	43.636 11	0.30 6	0.0003
ce-K- 2	45.0124 4	52.0 3	0.0498
ce-L- 2	75.2827 4	8.49 20	0.0136
ce-MNO- 2	79.7799 4	2.3 3	0.0039
β- 1 max	266 3		
avg	75.0 10	0.66 10	0.0011
β- 2 max	346 3		
avg	100.5 10	99.34 10	0.213
total β-			
avg	100.3 10	100.01 15	0.214
X-ray L	4.29	6.1 17	0.0006
X-ray Kα₂	30.6251 3	13.3 3	0.0087
X-ray Kα₁	30.9728 3	24.6 5	0.0163
X-ray Kβ	35	8.84 20	0.0066
γ 1	79.621 11	0.22 6	0.0004
γ 2	81	37.1 4	0.0640

4 weak γ's omitted (ΣIγ = 0.07%)

133XE IT DECAY   (2.19 D 1)   I(MIN) = 0.10%
FEEDING OF 2.188-D 133XE IN 133I
DECAY=2.883% 23

Radiation Type	Energy (keV)	Intensity (%)	Δ (g-rad/μCi-h)
Auger-L	3.43	70 4	0.0051
Auger-K	24.6	7.1 13	0.0037
ce-K- 1	198.62 4	63.7 8	0.269
ce-L- 1	227.73 4	20.8 6	0.101
ce-M- 1	232.04 4	4.16 13	0.0206
ce-NOP- 1	232.97 4	1.04 17	0.0052
X-ray L	4.1	7.8 24	0.0007
X-ray Kα₂	29.4580 10	16.1 5	0.0101
X-ray Kα₁	29.7790 10	29.9 8	0.0190
X-ray Kβ	33.6	10.6 4	0.0076
γ 1	233.18 4	10.3 3	0.0512

Table 3.6-2 (continued)
NUCLEAR DECAY SCHEME DATA AND DOSE CONSTANTS FOR SELECTED NUCLIDES (1977)

135XE B- DECAY   (9.09 H 1)   I(MIN) = 0.10%

Radiation Type	Energy (keV)	Intensity (%)	Δ (g-rad/ μCi-h)
Auger-L	3.55	5.12 22	0.0004
Auger-K	25.5	0.59 7	0.0003
ce-K- 3	213.809 15	5.63 10	0.0257
ce-L- 3	244.080 15	0.739 13	0.0038
β- 1 max	96 9		
avg	25.0 25	0.123 4	≈0
β- 2 max	550 9		
avg	171 4	3.12 10	0.0114
β- 3 max	750 9		
avg	245 4	0.68 3	0.0035
β- 4 max	908 9		
avg	307 4	95.98 3	0.628
total β-			
avg	302 4	99.98 11	0.643

1 weak β's omitted (ΣIβ = 0.08%)

X-ray L	4.29	0.63 18	≈0
X-ray Kα₂	30.6251 3	1.43 4	0.0009
X-ray Kα₁	30.9728 3	2.66 7	0.0018
X-ray Kβ	35	0.95 3	0.0007
γ 1	158.197 18	0.289 10	0.0010
γ 3	249.794 15	90.13 8	0.480
γ 4	358.39 4	0.221 9	0.0017
γ 6	407.990 20	0.359 13	0.0031
γ 9	608.185 16	2.90 9	0.0376

8 weak γ's omitted (ΣIγ = 0.21%)

135XE IT DECAY   (15.29 M 3)   I(MIN) = 0.10%
FEEDING OF 15.29-M 135XE IN 135I
DECAY=15.8% 5

Radiation Type	Energy (keV)	Intensity (%)	Δ (g-rad/ μCi-h)
Auger-L	3.43	14.9 8	0.0011
Auger-K	24.6	1.7 3	0.0009
ce-K- 1	492.010 17	15.2 3	0.159
ce-L- 1	521.118 17	2.88 9	0.0320
ce-M- 1	525.429 17	0.610 20	0.0068
ce-NOP- 1	526.363 17	0.200 10	0.0022
X-ray L	4.1	1.7 5	0.0001
X-ray Kα₂	29.4580 10	3.84 13	0.0024
X-ray Kα₁	29.7790 10	7.13 23	0.0045
X-ray Kβ	33.6	2.54 9	0.0018
γ 1	526.571 17	81.2 5	0.911

137CS B- DECAY (30.0 Y 2)   I(MIN) = 0.10%
SEE ALSO 137BA IT DECAY

Radiation Type	Energy (keV)	Intensity (%)	Δ (g-rad/ μCi-h)
β- 1 max	511.553 9		
avg	173.5 3	94.6 3	0.350
β- 2 max	1173.2		
avg	415.4 3	5.4 3	0.0478
total β-			
avg	186.6 4	100.0 5	0.397

137BA IT DECAY (2.552 M 2)   I(MIN) = 0.10%
FEEDING OF 2.552-M 137BA IN 137CS
DECAY=94.6% 3

Radiation Type	Energy (keV)	Intensity (%)	Δ (g-rad/ μCi-h)
Auger-L	3.67	7.8 5	0.0006
Auger-K	26.4	0.82 22	0.0005
ce-K- 1	624.204 9	8.24 5	0.110
ce-L- 1	655.656 9	1.50 3	0.0209
ce-MNO- 1	660.352 9	0.500 20	0.0070
X-ray L	4.47	1.1 3	0.0001
X-ray Kα₂	31.8171 3	2.11 7	0.0014
X-ray Kα₁	32.1936 3	3.90 12	0.0027
X-ray Kβ	36.4	1.42 5	0.0011
γ 1	661.645 9	89.9 4	1.27

141CE B- DECAY (32.51 D 2)   I(MIN) = 0.10%

Radiation Type	Energy (keV)	Intensity (%)	Δ (g-rad/ μCi-h)
Auger-L	4	16.1 10	0.0014
Auger-K	29.4	1.6 5	0.0010
ce-K- 1	103.449 10	18.6 8	0.0410
ce-L- 1	138.605 10	2.54 12	0.0075
ce-MNO- 1	143.929 10	0.72 6	0.0022
β- 1 max	435.4 15		
avg	130.8 6	70 3	0.195
β- 2 max	580.8 15		
avg	181.0 6	30 3	0.116
total β-			
avg	145.9 7	100 5	0.311
X-ray L	5	2.6 4	0.0003
X-ray Kα₂	35.55020 2	4.8 3	0.0037
X-ray Kα₁	36.02630 2	8.9 5	0.0068
X-ray Kβ	40.7	3.33 18	0.0029
γ 1	145.440 10	48.0 20	0.149

144PR B- DECAY   (17.28 M 3)   I(MIN) = 0.10%

Radiation Type	Energy (keV)	Intensity (%)	Δ (g-rad/ μCi-h)
β- 1 max	811 3		
avg	267.0 12	1.08 5	0.0061
β- 2 max	2301 3		
avg	894.8 14	1.17 5	0.0223
β- 3 max	2997 3		
avg	1221.8 14	97.75 10	2.54
total β-			
avg	1207.6 15	100.01 13	2.57
γ 3	696.490 20	1.48 6	0.0220
γ 7	1489.15 5	0.300 13	0.0095
γ 9	2185.70 6	0.77 4	0.0360

7 weak γ's omitted (ΣIγ = 0.02%)

Table 3.6-2 (continued)
NUCLEAR DECAY SCHEME DATA AND DOSE CONSTANTS FOR
SELECTED NUCLIDES (1977)

147ND B- DECAY (11.06 D 3)   I(MIN) = 0.10%

Radiation Type	Energy (keV)	Intensity (%)	Δ (g-rad/μCi-h)
Auger-L	4.38	41.1 21	0.0038
Auger-K	31.5	3.7 12	0.0025
ce-K- 3	45.9210 22	48.3 10	0.0473
ce-K- 4	75.30 5	0.31 4	0.0005
ce-L- 3	83.6771 22	7.10 14	0.0126
ce-MNO- 3	89.456 6	2.375 18	0.0045
β- 1 max	209.9 9		
avg	57.6 3	2.10 10	0.0026
β- 2 max	364.8 9		
avg	106.1 3	15.0 10	0.0339
β- 3 max	406.5 9		
avg	119.9 3	0.80 10	0.0020
β- 4 max	485.3 9		
avg	146.7 4	0.50 20	0.0016
β- 5 max	804.7 9		
avg	264.0 4	81.0 10	0.455
β- 6 max	895.8 9		
avg	299.5 4	0.2	0.0013
total β- avg	234.1 5	99.7 15	0.497

1 weak β's omitted (ΣIβ = 0.07%)

X-ray L	5.43	7.8 11	0.0009
X-ray Kα₂	38.1712 5	12.8 5	0.0104
X-ray Kα₁	38.7247 5	23.2 8	0.0192
X-ray Kβ	43.8	8.9 4	0.0083
γ 3	91.1050 20	27.94 21	0.0542
γ 4	120.48 5	0.40 5	0.0010
γ 10	196.64 4	0.204 17	0.0009
γ 15	275.374 15	0.80 5	0.0047
γ 17	319.411 18	1.96 12	0.0133
γ 21	398.155 20	0.87 6	0.0074
γ 23	410.48 3	0.140 9	0.0012
γ 24	439.895 22	1.20 9	0.0113
γ 27	489.24 3	0.154 9	0.0016
γ 28	531.016 22	13.1 8	0.148
γ 30	594.803 3	0.265 17	0.0034
γ 33	685.90 4	0.81 5	0.0119

21 weak γ's omitted (ΣIγ = 0.07%)

198AU B- DECAY (2.696 D 2)   I(MIN) = 0.10%

Radiation Type	Energy (keV)	Intensity (%)	Δ (g-rad/μCi-h)
Auger-L	7.6	2.10 16	0.0003
ce-K- 1	328.693 9	2.87 3	0.0201
ce-L- 1	396.956 9	1.031 20	0.0087
ce-M- 1	408.233 9	0.267 19	0.0023
ce-NOP- 1	410.995 9	0.101 7	0.0009
β- 1 max	285.3 6		
avg	79.60 20	1.3 3	0.0022
β- 2 max	961.2 6		
avg	314.80 20	98.7 3	0.662
total β- avg	311.78 21	100.0 5	0.664

1 weak β's omitted (ΣIβ = 0.03%)

X-ray L	10	1.29 15	0.0003
X-ray Kα₂	68.8950 20	0.812 22	0.0012
X-ray Kα₁	70.8190 20	1.38 4	0.0021
X-ray Kβ	80.3	0.607 18	0.0010
γ 1	411.795 9	95.47 8	0.837
γ 2	675.878 18	1.06 23	0.0153
γ 3	1087.69 3	0.23 6	0.0053

239NP B- DECAY (2.355 D 4)   I(MIN) = 0.10%

Radiation Type	Energy (keV)	Intensity (%)	Δ (g-rad/μCi-h)
ce-M- 1	1.9171 14	6.6 6	0.0003
ce-NOP- 1	6.2914 8	2.19 19	0.0003
Auger-L	10.3	40 7	0.0088
ce-L- 3	21.553 20	7.0 18	0.0032
ce-L- 4	26.313 20	9.5 21	0.0053
ce-L- 5	34.163 20	25 4	0.0180
ce-L- 7	38.383 5	0.33 5	0.0003
ce-M- 3	38.717 20	1.6 4	0.0013
ce-NOP- 3	43.091 20	0.43 11	0.0004
ce-M- 4	43.477 20	2.6 6	0.0024
ce-L- 8	44.78 3	6.6 17	0.0063
ce-NOP- 4	47.851 20	0.80 18	0.0008
ce-M- 5	51.327 20	6.9 10	0.0076
ce-MNO- 7	55.547 5	0.105 16	0.0001
ce-NOP- 5	55.701 20	2.6 4	0.0030
ce-K- 14	59.89 8	0.33 7	0.0004
ce-M- 8	61.95 3	1.8 5	0.0024
ce-NOP- 8	66.32 3	0.63 16	0.0009
Auger-K	76	0.9 10	0.0015
ce-L- 10	83.033 11	4.3 8	0.0076
ce-L- 11	83.37 4	0.42 8	0.0007
ce-K- 15	87.93 5	7.8 9	0.0146
ce-M- 10	100.197 10	1.14 20	0.0024
ce-MNO-11	100.54 4	0.139 25	0.0003
ce-NOP-10	104.571 10	0.39 8	0.0009
ce-K- 16	104.60 10	0.6 4	0.0013
ce-K- 17	106.37 5	20.9 20	0.0473
ce-K- 18	132.59 10	0.18 5	0.0005
ce-K- 19	151.02 9	0.116 22	0.0004
ce-K- 20	155.78 6	17.2 6	0.0571
ce-L- 15	186.653 11	1.62 18	0.0064
ce-L- 16	203.32 8	0.139 19	0.0006
ce-L- 17	205.093 11	4.4 5	0.0192
ce-MNO-17	222.257 10	1.50 9	0.0071
ce-L- 20	254.50 3	3.52 12	0.0191
ce-MNO-20	271.67 3	1.13 4	0.0065
β- 1 max	211 3		
avg	57.0 10	1.80 20	0.0022
β- 2 max	332 3		
avg	93.0 10	33.0 20	0.0654
β- 3 max	393 3		
avg	112.0 10	7 3	0.0167
β- 4 max	438 3		
avg	126.0 10	53 5	0.142
β- 5 max	666 3		
avg	201.0 10	2.0 10	0.0086
β- 6 max	715 3		
avg	219.0 10	4.0 20	0.0187
total β- avg	118.1 11	101 7	0.254

4 weak β's omitted (ΣIβ = 0.03%)

X-ray L	14.3	60 7	0.0183
γ 4	49.410 20	0.100 22	0.0001
γ 5	57.260 20	0.151 21	0.0002
γ 7	61.480 4	0.96 15	0.0013
X-ray Kα₂	99.55 5	13.8 8	0.0292
X-ray Kα₁	103.76 5	22.1 12	0.0489
γ 10	106.130 10	22.7 13	0.0513
X-ray Kβ	117	10.4 6	0.0259
γ 14	181.71 6	0.111 15	0.0004
γ 15	209.750 10	3.24 25	0.0145
γ 16	226.42 8	0.34 5	0.0016
γ 17	228.190 10	10.7 7	0.0521
γ 18	254.41 8	0.100 18	0.0005
γ 20	277.60 3	14.1 4	0.0834
γ 21	285.41 3	0.78 8	0.0047
γ 22	315.88 4	1.59 11	0.0107
γ 23	334.30 5	2.03 18	0.0145

20 weak γ's omitted (ΣIγ = 0.37%)

REFERENCES

1. Lederer, C. M., Hollander, J. M., and Perlman, I., *Table of Isotopes,* 6th ed., John Wiley & Sons, New York, 1967.
2. *Radiological Health Handbook,* U. S. Department of Health, Education, and Welfare, Bureau of Radiological Health, Rockville, Md., January 1970.
3. *Nucl. Data Sheets,* Nuclear Data Group of Oak Ridge National Laboratory, Eds., serial publication by John Wiley & Sons, New York.
4. Dillman, L. T. and von der Lage, F. C., Radionuclide Decay Schemes and Nuclear Parameters for Use in Radiation-dose Estimation, *NM/MIRD Pamphlet No. 10, Society of Nuclear Medicine,* New York, September 1975.
5. Martin, M. J., Nuclear Decay Data for Selected Radionuclides, ORNL-5114, Oak Ridge National Laboratory, Oak Ridge, Tenn., March 1976.
6. Bambynek, W., Crasemann, B., Fink, R. W., Freund, H. V., Mark, H., Swift, C. D., Price, R. E., and Rao, P. V., X-Ray fluorescence yields, Auger, and Coster-Kronig transition probabilities, *Rev. Mod. Phys.,* 44, 716, 1972.
7. Bearden, J. A. and Burr, A. F., Reevaluation of X-ray atomic energy levels, *Rev. Mod. Phys.,* 39, 125, 1967.
8. Rao, P. V., Chen, M. S., and Crasemann, B., Atomic vacancy distribution produced by inner-shell ionization, *Phys. Rev. A,* 5, 977, 1972.
9. Wapstra, A. H., Nijgh, G. J., and van Lieshout, R., *Nuclear Spectroscopy Tables,* Interscience, New York, 1959.
10. Bergstrom, I., Nordling, C., Snell, A. H., Wilson, R., and Pettersson, B. C., Some 'internal' effects in nuclear decay, in *Alpha-, Beta-, and Gamma-ray Spectroscopy,* Vol. 2, Siegbahn, K., Ed., North-Holland, Amsterdam, 1965, chap. 25.
11. Evans, R. D., *The Atomic Nucleus,* McGraw-Hill, New York, 1955, 619.
12. Schopper, H. F., *Weak Interactions and Nuclear Beta Decay,* North-Holland, Amsterdam, 1966, 76—84.

3.7. SELECTED PARTICLE AND PHOTON SPECTRAL DATA

Allen Brodsky

This section contains spectral data and related information useful in measuring spectra of interest in radiation protection. Further sources of data may be found in References 1 to 15,* which include some references giving fundamental descriptions of the origins and interactions of the various types of ionizing radiation. The reader should note that for more recent and accurate data on dosimetry calculations, Section 3.6 by L. T. Dillman should be used. Some older spectral data and decay schemes are presented in this section for purposes of comparing certain data and illustrating decay modes of the important radionuclides. Dillman has presented more complete and updated decay data for the purpose of internal dosimetry.

3.7.1. Spectral Output by Radionuclide

The table of nuclides in Section 3.2 provides summary data on the energies of radiations emitted by radionuclides. Section 3.6 contains more detailed decay schemes and energies for radionuclides most frequently encountered as well as some information on energies and frequencies of internal conversion, Auger electrons, and X-rays as calculated by L. T. Dillman for obtaining accurate estimates of internal dose from internal emitters. Dillman[16] has recently refined methods of calculating energies and frequencies of radiations emitted in decay of certain nuclides of interest and is extending these calculations to nuclides encountered in industry as well as medicine.

For purposes of visualizing decay schemes and beta spectral data, part of an earlier compendium by Slack and Way[14] of nuclear decay data for a limited number of nuclides of interest is presented in Table 3.7-1.** In this table, the average energy per disintegration for beta particles from isotopes having only one group is given by:[14]

$$\bar{E}_\beta = \frac{\int_0^{E_0} E N(E) dE}{\int_0^{E_0} N(E) dE} \qquad (3.7\text{-}1)$$

where $N(E)dE$ is the number of particles having a kinetic energy in the energy range between E and $E + dE$, and E_0 is the maximum energy of the β-particle spectrum.

If more than one β-particle group is present, for the decay one has

$$\bar{E}_\beta = \sum_i \bar{E}_{\beta_i} w_i \qquad (3.7\text{-}2)$$

where $\bar{E}_{\beta_i}$ is the average energy of the ith group computed using Equation 3.7-1 and w_i is the fraction of the disintegrations that take place via the ith beta group.

Figures 3.7-1 and 3.7-2*** present Loevinger's ratio of average beta energy to actual beta energy as presented by Slack and Way. Dillman[16] has extended the calculations of decay scheme data since the earlier report of Slack and Way. Examples of recent results from his calculations are given in Tables 3.7-2 and 3.7-3 and Figures 3.7-3 through 3.7-5.

---

* Reference 1 is from the journal *Nuclear Data Sheets,* edited by the Nuclear Data Group, W. B. Eubank, Director, Oak Ridge National Laboratory, P.O. Box X, Oak Ridge, Tenn. 37830; it contains a reaction index, with each reaction heading arranged according to increasing mass and charge of the target(s). The Nuclear Data Group also provides specialized bibliographies from their computer-based reference system.

** All tables appear at end of the text.

*** All figures appear at the end of the text on page.

3.7.2. Alpha Spectra

Since alpha particles are emitted in discrete energies, alpha spectra can be found from the detailed decay scheme data presented in Section 3.6, Reference 18, or the summary information in the table of nuclides of Section 3.2. Table 3.7-4 presents selected alpha particle calibration energies.[49] Although alpha spectra of individual nuclides contain discrete energies, these spectra can sometimes contain complex groups of alpha energies that are not resolvable under given experimental conditions. Examples of complex alpha decay schemes, taken from Evans,[7] that also indicate the competing beta and alpha decay modes of ThC, are shown in Figures 3.7-6 and 3.7-7. Table 3.7-5 presents an example of some of the highest energy alpha particles, emitted in the decay of [212]Po. Evans[7] should be consulted for an excellent discussion of the characteristics of alpha spectra.

Considerable advances have been made in recent years in the measurement of alpha spectra using solid-state detectors.[17] An example of the resolution obtainable with a silicon surface-barrier detector is shown in Figure 3.7-8. Some alpha spectra measurements are still carried out, however, with ionization-type detectors (see Section 13.2). These measurements may be compared with published nuclear data[1] or summaries such as those in Section 3.2 and 3.6 to identify alpha-emitting radionuclides detected in environmental surveys or bioassays.

3.7.3. Beta and Electron Spectra

Figure 3.7-9 presents the beta spectra of most of the radionuclides of interest in radiation protection with average and median energies, half-lives, maximum endpoint energies, and frequencies of emission of betas from each endpoint energy, as calculated by Hogan, Zigman, and Mackin.[42] The reader is cautioned to use the multiplying factors shown on the bottom energy scale. These beta spectra are in good agreement (usually within 5%) with those of Slack and Way[41] presented with the decay scheme data for certain nuclides in Table 3.7-1. The beta spectra are identified by chemical symbol, mass number, and the isomer designation where appropriate. Nonunique forbidden transitions were treated as allowed (F = O); unique forbidden transitions were treated as unique first-forbidden. Original plots were prepared using a plotter precise to 1/100 in.,[42] but they were reduced to one half the original size in Reference 42. These spectral plots have been further reduced for this handbook since it is not expected that these plots will be used for more than qualitative reference; for most present dosimetry calculations, the newer summary data such as those provided in Section 3.6 are now available.

As described in Reference 42, the probability distribution function for a beta particle is a function of the endpoint energy of the transition and the atomic number of the daughter nuclide. For the kth beta transition of a nuclide of atomic number Z_j, this function was taken as:[42]

$$P_{jk}(W) = \frac{F(Z_j + 1, W) \cdot (W^2 - 1)^{1/2} \cdot (W_{ok} - W)^2 \cdot W \cdot C}{\int_1^{W_{ok}} F(Z_j + 1, W) \cdot (W^2 - 1)^{1/2} \cdot (W_{ok} - W)^2 \cdot W \cdot C \cdot dW} \qquad (3.7-3)$$

where W_{ok} = the endpoint energy of the kth beta transition, in units of mc^2, $F(Z_j+1, W)$ = the relativistic nuclear Coulomb factor, and C = the unique shape factor. For allowed and nonunique forbidden transitions, this factor was taken to be one, while for unique first-forbidden transitions it has the form: $(W^2 - 1) + (W_{ok} - W)^2$. For unique second- and third-forbidden transitions, the unique shape factors are more complicated. However, for practical purposes, these may be approximated with the shape factor for the unique first-forbidden transition, as the error so introduced is small relative to the error due to the experimental uncertainty in W_{ok}.

The beta spectrum, $B_j(W)$, of a nuclide having more than one beta transition is obtained[42] by summing the values obtained from Equation 3.7-3 over all of the k transitions:

$$B_j(W) = \Sigma P_{jk}(W) \cdot \Delta_{jk} \qquad (3.7\text{-}4)$$

where Δ_{jk} = the probability that the kth transition will occur. A FORTRAN program[42] performed the computations at up to 400 uniformly spaced energy points. The spectra produced were normalized so that the integral from E = O to E = E_{max} (the maximum endpoint energy) equals the total probability of beta emission. The probability that a beta particle will have an energy within any energy range may be obtained by integrating from the lower limit to the upper limit of the range.

Average, and median, energies were computed using Equations 3.7-5 and 3.7-6. The average, or expected, energy is given by:

$$\overline{E} = \frac{\int_0^{E_{max}} E \cdot B(E) \cdot dE}{\int_0^{E_{max}} B(E) \cdot dE} \qquad (3.7\text{-}5)$$

The median energy is that value of the energy that bisects the area under the spectrum. Its value appears as the upper limit, $E_{1/2}$, of the first integral in the defining equation:

$$\int_0^{E_{1/2}} B(E) \cdot dE = \frac{1}{2} \int_0^{E_{max}} B(E) \cdot dE \qquad (3.7\text{-}6)$$

(In the spectral plots, there is a systematic error in the median energy such that it is too high by $(1/200)E_{max}$.[42])

The parent nuclides in Figure 3.7-9 are arranged in order of increasing mass number. Where two or more nuclides have the same mass number, they are ordered by increasing atomic number. Isomers are arranged in order of increasing energy level. Nuclides in the ground state are identified by their chemical symbol, followed by the mass number. The first excited state is designated by an "A" following the mass number. The second excited state is designated by a "B," and so on.

Hogan et al.[42] have also presented some curves of the ratio $\overline{E}/E_{max}$ vs. atomic number and E_{max}. More recent curves of $\overline{E}/E_{max}$ by Dillman are presented in Figures 3.7-10 through 3.7-15 for allowed and forbidden negatron and positron decays.

The average number and energy of the total beta decay chains per thermal neutron fission for ^{235}U and ^{233}U are given in Table 3.7-6. An unlimited variety of electron spectra are encountered in radiation measurements. For single interactions of photons with atomic electrons by well-known phenomena such as the Compton effect, the resulting primary electron spectra can be calculated theoretically. Examples of such calculations are given in Section 3.7.8. Spectra of secondary and slowed-down electrons in various media have been measured and calculated for many conditions of interest. Figure 3.7-16 shows calculations of the electron spectra produced in water by initial electrons of various energies normalized so that an integral over the numbers of electrons per square centimeter per keV would, for a given initial electron energy, represent the number of secondary electrons per square centimeter that would be equivalent to a dose of 1 rad.[31] It should be noted that all curves of Figure 3.7-16 come closer together at the low-energy end. This is true since the spectra produced by either high-energy or low-energy electrons contain about the same number of very low-energy electrons per square centimeter.

Figure 3.7-17 shows an example of low-energy electron spectra within five materials with respect to the Fermi level. It is seen that these spectra are almost independent of the type of the material as well as the initial primary photon energy generating these electrons.

Spectra of secondary electrons emitted from the walls of vacuum chambers used for radiation measurement[29] are shown in Figure 3.7-18. These electrons also have very low average energies since they are generated by slowing down of electrons generated within the wall by primary radiation. Electron fluence distributions as a function of kinetic energy produced by several types of primary photon spectra in water[19] are given in Figure 3.7-19.

Figure 3.7-20 shows electron spectra produced at various depths in water for 20-MeV incident electrons. These curves show the increasing spread in electron energies as a high-energy beam of electrons penetrates more deeply into tissue. The high-energy electrons used more recently in therapy are obtained from betatrons or linear accelerators and generally deliver energy to targets at depths greater than 2 cm in tissues (electrons greater than 5 MeV). In Figure 3.7-20, digital computer techniques were used to remove distortion by the nonzero energy resolution of the spectrometer. The spectra of Figure 3.7-20 represent the fluence of all electrons 2 MeV or greater passing through a small area at a given depth including electrons passing through this area at any angle.[32] Figure 3.7-21 shows the slowing-down spectra for [198]Au in an infinite gold medium. Further discussion of theoretical and experimental electron slowing-down spectra may be found in sources cited in References 8 to 11 and 29.

3.7.4. Gamma Spectra

Tables 3.7-7 through 3.7-9 present well-known gamma-ray energies from sources that can be used for calibration.[49] An extensive list of gamma-ray energies of radionuclides in order of increasing energy is given in Reference 44, which is useful for identifying nuclides from their gamma radiation spectrum. Only the lower energy and higher energy range of the tabulations in Reference 44 are presented in Table 3.7-10 to show the extensive information in this reference and to provide suggestions for useful calibration energies at the extreme ends of the gamma-ray energy range. Figure 3.7-22 presents the gamma spectrum from radium in equilibrium in daughter products. Measurements as well as calculations of dose have been made for radium sources contained in various thicknesses of platinum.

Energies and numbers of prompt gamma ray per fission are given in Table 3.7-11. Moteff[45] has provided an extensive set of graphs showing decay gamma energy spectra for various reactor-operating and decay times. A selected sample of his figures for 1000 hr of operation is presented in Figures 3.7-23 through 3.7-28. Major contributions to the gamma ray spectra for longer cooling times are obtained from [140]La, the daughter of [140]Ba, as shown also in Table 3.7-12 from Reference 26.

Sources of information on spectra scattered from large ^{60}Co sources may be found in Reference 21. Figure 3.7-29 shows the spectrum of scattered radiation from a source of cobalt pellets with given characteristics at a source to detector distance of 80 cm. Reference 21 gives additional information on the relative dose contributions of scattered and primary radiation from such sources. Figure 3.7-30 presents the total spectrum of photons reaching a detector after scatter from a lead collimator placed near a ^{60}Co source. This is a sample spectrum only, since as discussed in Reference 21, the spectra depend on the collimator design and geometrical dimensions such as those in Figure 3.7-31. However, Figure 3.7-30 serves to indicate that an appreciable fraction of the dose at a point outside of the collimator may be contributed by scattered photons of somewhat lower energies than the primary 1.17 and 1.33 MeV photons of [60]Co.

Figure 3.7-32 presents energy spectra of photons reflected by backscatter.[27] The upper set of curves illustrates dependence of the backscattered spectrum on type of material for 1-MeV photons normally incident on semi-infinite slabs; the second row illustrates dependence of backscatter spectra on incident photon energy for photons normally incident on water; and the third row indicates spectral dependence of backscattered radiation on angle of incidence of 1-MeV photons on a plane of water. Further details on backscatter radiation are presented in Reference 27.

Calculations of spectra and dose of scattered gamma rays often depend heavily on use of the Compton equations for energies of scattered radiation vs. angle as well as the Klein-Nishina differential cross-sections for the possibility of scattering at various angles.[50] A plot of the Klein-Nishina cross-section for several energies both for photon number scattering and energy scattering is presented in Section 3.5. Figure 3.7-33 from Reference 50 shows the geometrical relationship between the incident and scattered photons and the Compton recoil electrons. As shown in Figure 3.7-33, the electron recoils at an angle ϕ with kinetic energy T, and the Compton scattered photon is emitted at an angle θ with a remaining energy

$$h\nu' = h\nu'_0 - T.$$

The following information giving useful equations relevant to Compton scattering is quoted from Evans:[50]

The *Compton shift* is the difference between the wavelength λ_0 (or the energy $h\nu_0$) of the incident photon and the wavelength λ' (or energy $h\nu'$ of the Compton scattered photon and is

$$\frac{1}{h\nu'} - \frac{1}{h\nu_0} = \frac{1}{m_0 c^2} (1 - \cos \vartheta) \qquad (3.7\text{-}7)$$

or

$$\lambda' - \lambda_0 = \frac{h}{m_0 c} (1 - \cos \vartheta) \qquad (3.7\text{-}8)$$

Note that the Compton shift in wavelength $(\lambda' - \lambda_0)$ in any particular direction is independent of the energy of the incident photon but that the Compton shift in energy $(h\nu_0 - h\nu')$ increases very strongly as $h\nu_0$ increases. The length $h/m_0 c = \lambda_0 = 2.426 \times 10^{-10}$ cm is called the Compton wavelength for an electron. It is equal to the wavelength of a photon whose energy is just equal to the rest energy of the electron $m_0 c^2 = 0.5110$ MeV.

Writing the incident energy in terms of the dimensionless quantity

$$\alpha \equiv \frac{h\nu_0}{m_0 c^2} \qquad (3.7\text{-}9)$$

the conservation laws give, for the energy of the Compton scattered photon,

$$\frac{\nu'}{\nu_0} = \frac{1}{1 + \alpha(1 - \cos \vartheta)} \qquad (3.7\text{-}10)$$

$$(h\nu')_{min} = m_0 c^2 \frac{\alpha}{1 + 2\alpha} = h\nu_0 \frac{1}{1 + 2\alpha} \text{ for } \vartheta = 180 \text{ deg} \qquad (3.7\text{-}11)$$

Figure 3.7-34* gives $h\nu'$ vs. $h\nu_0$ for 10 values of ϑ. The curve for $\vartheta = 180$ deg gives $(h\nu')_{min}$ and thus evaluates the backscatter peak and the energy separation between the Compton edge and the total energy peak in γ-ray scintillation spectroscopy.

The energy of the backscattered photon $(h\nu'_{min})$ approaches its maximum value of $m_0 c^2 / 2 = 0.25$ MeV for high-energy incident photons, $\alpha \gg 1$.

The angle φ and the kinetic energy T of the Compton recoil electron are related to the photon scattering angle ϑ by

$$\cot \varphi = (1 + \alpha) \tan \frac{\varphi}{2} \qquad (3.7\text{-}12)$$

$$T = h\nu_0 - h\nu' = h\nu_0 \frac{\alpha(1 - \cos \vartheta)}{1 + \alpha(1 - \cos \vartheta)} \qquad (3.7\text{-}13)$$

* The figure number has been changed from that in the quoted Reference 50; a figure from Reference 60 has been used for Figure 3.7-34.

$$T_{max} = h\nu_0 - (h\nu')_{min} = h\nu_0 \frac{2\alpha}{1 + 2\alpha} \text{ for } \vartheta = 180 \text{ deg} \qquad (3.7\text{-}14)$$

Klein-Vishina Collision Differential Cross Section. The *collision* differential cross section $d(_e\sigma)$, where the subscript connotes per electron in the attenuator, refers to the *number* of collisions of a particular type. The *number* of photons which are scattered in a particular direction, as a fraction of the *number* of incident photons, is $d(_e\sigma)$ and has the dimensions

$$d(_e\sigma) = \frac{\text{number scattered } [\text{number/(sec} \cdot \text{electron)}]}{\text{number incident } [\text{number/(cm}^2 \cdot \text{sec)}]} = \frac{\text{cm}^2}{\text{electron}} \qquad (3.7\text{-}15)$$

As first shown by Klein and Nishina, the collision differential cross section for unpolarized photons striking unbound, randomly oriented electrons is

$$d(_e\sigma) = \frac{r_0^2}{2} d\Omega \left(\frac{\nu'}{\nu_0}\right)^2 \left(\frac{\nu_0}{\nu'} + \frac{\nu'}{\nu_0} - \sin^2 \vartheta\right) = \frac{\text{cm}^2}{\text{electron}} \qquad (3.7\text{-}16)$$

Sample characteristics of recoil electron spectra from Compton interactions are given in Section 3.7.8. An example of the shape of a pulse-height spectrum in NaI(Tl) resulting from combined Compton electron, backscatter photon, photoelectron, and spectral spreading phenomena is shown in Figure 3.7-35, and relative fractions of pulses in the photo-electric peaks for various crystal sizes are shown in Figure 3.7-36 from Evans.[50]

3.7.5. X-Ray Spectra

3.7.5.1. X-Ray Critical Absorption and Emission Energies

Table 3.7-13 summarizes, for all elements, the compilation of K and L absorption and emission energies by Fine and Hendee.[47] The K-adsorption edge is plotted in Figure 3.7-37 as a function of atomic number Z. Where wavelength data were converted to energies the following equation was used:

$$E(keV) = (12.39644 \pm 0.00017)/\lambda(\text{Å})$$
$$= 12.39644/1.002020 \; \lambda(\text{kX unit}) \qquad (3.7\text{-}17)$$

Values in the table have been listed uniformly to 1 eV, but chemical form may shift adsorption edges as much as 10 to 20 eV. Moseley's law was used to check computational errors; where a few irregularities appeared, values fitting Moseley's law more smoothly are footnoted.

3.7.5.2. Detailed Tabulation of X-Ray Wavelengths for the Elements

The following portion of the section on the X-ray wavelengths is reproduced from Bearden in the *Handbook of Chemistry and Physics:*[46]

X-RAY WAVELENGTHS

Table 13a was originally published as the final report to the U.S. Atomic Energy Commission as Report NYO-10586 in partial fulfillment of Contract AT(30-1)-2543. The tables were later reproduced in *Review of Modern Physics*. The data may also be obtained from the Superintendent of Documents, U.S. Government Printing Office, Washington, D.C. 20402 in the publication NSRDS-NBS 14. Persons seeking discussion of the experimental work, conventions, secondary standards, etc. will find these in *Review of Modern Physics*, Vol. 39, No. 1, 78–124, January 1967.

THE W $K\alpha_1$ WAVELENGTH STANDARD

A wavelength standard should possess characteristics which permit its ready redetermination in other laboratories by different techniques. Considering all of the factors involved in the selection of a wavelength standard, the W $K\alpha_1$ line is superior to any other X-ray or γray wavelength. Its advantages as the X-ray wavelength standard are discussed in *Review of Modern Physics*, Vol. 39, page 82 (1967).

$$\lambda W \, K_{\alpha_1} = 0.2090100 \text{ Å } \pm 5 \text{ ppm.}$$

This numerical value of the wavelength of the W $K\alpha_1$ line is used to define the *X-ray wavelength standard* by the relation

$$\lambda(W K_{\alpha_1}) = 0.2090100 \text{ Å} *$$

This is a new unit of length which may differ from the angstrom by ±5 ppm (probable error), *but as a wavelength standard it has no error.* In order to clearly indicate that this unit is not exactly an angstrom, it has been designated Å*.

Wavelengths tabulated normally refer to the pure element in its solid form. However, there are many instances in which such data are not available. For example, rare gases are of necessity almost always used in the gaseous form, while the rare-earth elements are customarily used in the form of salts.

In the study of the X-ray literature, the wavelengths of a number of lines were noted which appeared inconsistent with the remaining data. A Moseley-type diagram was constructed, and if the value was clearly outside estimated probable error, it was assumed that an experimental or typographical error had occurred and the interpolated value was listed in the table. Such cases are marked with a dagger (†) as a superscript to the wavelength. For elements of atomic number 85 through 89 and 91, there are no measured lines of the K series and very few of other series except for 88 radium and 91 protactinium. Likewise, there are very few measurements for 43 technetium and 54 xenon. In these cases, interpolated values are listed for the more prominent lines and marked with a dagger (†).

In high precision work there is some ambiguity as to exactly what feature of a line profile should be taken to be the "true wavelength." In double-crystal work the line peak is usually employed. In crystallography the centroid is widely used; in photographic work with visual observation of the plates, there is involved some subjective criterion of the observer which it is difficult to define precisely. In this survey the peak of the line profile has been adopted as the standard criterion.

3.7.5.3. Bremsstrahlung Spectra

Kramers (1923) derived the following expression for the X-ray energy production resulting from the bombardment of a thick target by electrons:[30]

$$F(E) = CZ(E_0 - E) \qquad\qquad (3.7\text{-}18)$$

where $F(E)$ is the X-ray energy produced per unit energy interval at energy E by electrons of energy E_0 falling on a target of atomic number Z. C is a constant which may be deduced from the measurements of Dyson (1959)[30] as 2.76×10^{-6} (keV interval-electron)$^{-1}$ if energy is expressed in keV. The distribution of $F(E)$ with respect to energy rises linearly from zero at $E = E_0$ to a maximum at $E = 0$. However, in practice, X-rays produced are attenuated within the target of the X-ray tube and by the tube window and any external filter. The attenuation in the tube window and the external filter is readily calculated using tabulated attenuation data, but attenuation in the target is more difficult to assess.

References 30 and 31 discuss various approximations for calculating X-ray spectra produced by thick targets, since it is not possible to determine exactly the depth in the target at which each photon is produced.

For a "thin" target and low-energy X-rays (less than 100 keV), Heitler (1954) and Evans (1955)[31] show that the spectral distribution of Bremsstrahlung intensity is constant from zero to $El = T_0$, and is given by:

$$
\begin{aligned}
I(E)dE &= CJdE \quad \text{for } E < E_1 \\
&= 0 \qquad\quad \text{for } E > E_1
\end{aligned}
\qquad (3.7\text{-}19)
$$

where $I(E)\, dE$ is the intensity of X-rays occurring between quantum energies E and $E + dE$, C is a constant of proportionality, J is the bombarding electron current, and E_1 is the maximum quantum energy generated, which is equal to the incident electron energy T_0. However, in practical X-ray tubes, the targets are not thin and bombarding electrons slow down. This produces a spectrum of broad energy distribution. When the bombarding electron energy rises above the characteristic K edge for the target (for example, tungsten), characteristic K X-ray spectra become superimposed on the Bremsstrahlung spectrum and the K lines under certain conditions may contribute up to 20% of the intensity. In a practical tube, the L X-ray lines are so low in energy that they contribute practically nothing to the emitted X-ray intensity. In order to obtain the maximum amount of characteristic radiation relative to Bremsstrahlung, the tube potential should be about 1.5 times the energy of the absorption edge. Table 3.7-14 presents the emission lines for tungsten.

For high-energy electrons (> 10 MeV) bombarding a thin target, the Bremsstrahlung spectrum in the forward direction has been calculated by Schiff (1951) to be:[31]

$$I(E)dE = C[2(1 - \eta)(\ln \alpha - 1) + \eta^2 (\ln \alpha - \tfrac{1}{2})] dE$$

where $\alpha^2 = \alpha_1{}^2 \alpha_2{}^2 / (\alpha_1{}^2 + \alpha_2{}^2)$, in which

$$\alpha_1 = 2T_0 (1 - \eta)/\epsilon\eta \text{ and } \alpha_2 = 111 Z^{-1/3} \qquad (3.7-20)$$

In this expression, E is the rest energy of the electron, T'_0 is the total energy $(T_0 + E)$ of the bombarding electron, Z is the atomic number of the target, $n = E/T'_0$, and I(E)dE is the intensity radiated between E and E + dE. Spectral distributions calculated from this equation are summarized in some of the ICRU references given at the end of this section. The angular distribution of X-ray intensity produced by electron bombardment of a thin foil is shown in Figure 3.7-37 for electron energies of 34 keV on aluminum and 10 and 20 MeV on Tungsten. For thick targets where the electrons are deviated widely from the original direction, no detailed theoretical analysis of angular distribution is available.[31] Target absorption and angular distributions depend strongly on target geometry. Up to 400-kV tubes are designed to emit maximum intensity at about 90° to the electron beam direction. For potentials of 1 MeV or more, the intensity in the forward direction tends to be more intense, and the X-ray beam transmitted through the target is used. Under these circumstances, the target is usually held at ground potential, is water cooled, and acts as a filter for the X-ray beam. For energies of the order of 20 MeV, as in betatrons, effectively "thin" targets are used and angular distributions are similar to the theoretical thin-target spectrum. For energies more than 4 MeV, angular variations may be large in the forward direction, and for medical applications, "beam-flattening" filters are often needed to give a uniform field over the area required.[31]

Figures 3.7-38 to 3.7-40 show several ways of plotting the continuous spectra emitted by X-ray tubes. Figure 3.7-41 shows calculated spectral distributions of continuous X-rays generated at 10, 15, and 20 kV for electrons incident at a 20° angle on a Tungsten target and filtered by the target and 1-mm Be. The discontinuities are due to the effects on target self-absorption resulting from absorption edges; characteristic radiation must be added to obtain the complete spectra.[30] Similar spectra for 25, 30, 40, and 50 kV are shown in Figures 3.7-42 and 3.7-43.[30] Other examples of spectral X-ray phenomena are shown in Figures 3.7-44 through 3.7-64.

3.7.5.4. X-Ray Spectra in Absorbing Media

Figures 3.7-65 through 3.7-67 and Table 3.7-15 show X-ray spectral changes within absorbing media as taken from References 20 and 24.

3.7.5.5. Half-value Layer as a Crude Measure of X-ray Spectral Quality

The thickness of materials required to reduce radiation intensity has been used as an indicator of radiation X-ray quality. Figure 3.7-68 shows the variation of half-value layer with total filtration of aluminum for X-ray tubes of different constant and sine wave voltages. Table 3.7-16 shows differences obtained in estimating X-ray quality from HVL measurements as opposed to measurements of exposure reduction in typical component materials as described in Reference 24. Table 3.7-17 gives characteristic half-value layers for diagnostic X-ray tubes with full-wave rectified, sine-wave, and constant potential voltage forms. Figures 3.7-69 through 3.7-71 show the variation of HVL measurements with geometrical characteristics of the measurement.

3.7.5.6. Bremsstrahlung Spectra from Beta-ray and Electron Absorption

The energy radiated as X-radiation per beta-ray absorbed is:[48]

$$B = 1.23 \times 10^{-4} \, (\overline{Z} + 3) \, E^2, \text{Mev/beta} \qquad (3.7-21)$$

where E is the maximum beta energy in MeV and $\overline{Z}$ is the effective atomic number given by

$$\overline{Z} = \frac{\Sigma f_a Z_a^2}{\Sigma f_a Z_a} \qquad (3.7-22)$$

where f_a is the fraction of the number of atoms of atomic number Z_a. The Bremsstrahlung spectrum is given in Table 3.7-18.

A simpler rule of thumb, useful as an approximation to the fractional amount, F, of monoenergetic electron energy that is converted to Bremsstrahlung (X-rays) is

$$F = ZE/800 \qquad (3.7-23)$$

where Z is the atomic number of the target material and E is the (monoenergetic) electron energy in megaelectronvolts. This fraction, integrated over an allowed beta energy spectrum, would give a result similar to that of Equation 3.7-21). However, in accident situations with electron accelerators, Equation 3.7-23 has been found useful in estimating average exposure rates in the vicinity of the target.

3.7.5.7. LET Spectra

Figure 3.7-72 shows the linear energy transfer of electrons of different energies when delta rays of either 100 or 500 eV or greater are not considered part of the track influencing the region of biological interest.[31] The "mother" track in radiation biology usually includes all delta rays with an energy of 100 eV or less, which means the track effectively is confined to a diameter of about 30 Å.

The results of Figure 3.7-72 were used to obtain the LET distribution due to electrons of all energies resulting from X-rays with an HVL of 1.25 mm at the surface and at 10 cm depth in water (Figure 3.7-73). The radiations at the two depths are quite different, the primary corresponding to an HVL of 1.25 mm Cu and the second to an HVL of 0.46 mm Cu. Although these are large differences in X-ray spectra, the LET spectra are seen to be almost identical.

Two types of averages are in common use for averaging LET: the number average $\overline{L}_N$ and the energy average $\overline{L}_T$. These are defined as follows:[31]

$$\overline{L}_N = \frac{\int_0^\infty LN(L)dL}{\int_0^\infty N(L)dL} \qquad (3.7-24)$$

$$\overline{L}_T = \frac{\int_0^\infty L^2 N(L)dL}{\int_0^\infty LN(L)dL} \qquad (3.7-25)$$

The number average is the value of L weighted by the number of tracks N(L) in the interval L to L + dL and is the average LET of electron tracks passing through an area.

Johns[31] indicates that it is more meaningful for dosimetry to weight the tracks by the energy transfer per micron to give the energy average $\overline{L}_T$. Average LET values from both averaging methods are presented in Table 3.7-19 for all of the radiations studied by Bruce et al. (1963).[31] In this table, it is seen that the differences between primary and primary-plus-scattered radiation are very small for either method of averaging, and we see further that the energy average is not greatly dependent on the type of radiation. $\overline{L}_T$ changes less than a factor of 1.5 from the soft scattered radiation produced with low-energy X-rays up to gamma rays from ^{60}Co. Thus, as Johns points out,[31] LET is not very useful as a parameter in the radiobiology of X-rays and electrons but becomes more important for heavy particles in terms of determining the relative biological effectiveness (RBE) for various biological effects. However, more recent literature should be examined in regard to variations in radiobiological effect between types of radiation of low LET, examined both theoretically according to LET distribution and for microdosimetric considerations of the variations in specific energy deposited in target volumes of sizes on the order of 1 micron (see References 53 to 57).

For all ionizing radiations, heavy particles as well as electrons, photons, and their secondary and tertiary particles, the detailed spatial distribution of energy distributions for a given absorbed dose depends on the velocity and changes of all ionizing particles traversing a given point in the medium. Rather similar distribution patterns may in fact be produced by electrons and by photons of certain higher energies. It is usually impossible to obtain complete theoretical information on the spatial and angular distribution of particles of all energies. The wide variation of LET along the tracks of individual secondary particles and between tracks of different initial energy, even for a homogeneous primary radiation, can therefore best be represented by the distribution of LET throughout the irradiated object.[25] For all chemical and many biological effects, the proportion of the total dose deposited in different LET intervals is the main quantity of interest and has been found to be related empirically and semiempirically to a quality factor or RBE for various biological endpoints. Figure 3.7-74 shows LET distributions for recoil protons produced in water in first collisions of monoenergetic neutrons. Here, a rapid variation of energy-averaged LET with energy of the incident neutrons is apparent. As neutron energies, or incident photon energies, tend toward 20 MeV or higher, the quality factors for some biological endpoints do in fact approach those of x- or gamma radiation. Figure 3.7-75 shows experimentally determined LET distributions in a test sphere designed to simulate a cell of diameter of 0.75 μ in soft tissue. Although these curves were determined for nearly monoenergetic neutrons, it can be seen that the peaks have been broadened as a result of the statistical fluctuations in energy deposition within a small volume, which are not taken into account in the deterministic theories of stopping power. Also, this is partly because recoiling heavier nuclei are not included in the theoretical treatment.[25] Some additional LET spectra are presented in Figures 3.7-76 through 3.7-80 to illustrate the wide variation in shapes of spectra of interest as well as the average LET of the associated particles.

Figure 3.7-81 shows the augmented Bragg peaks of π^- and π^+ mesons as they penetrate into water or tissue.[33] The augmentation for the π^- meson is much greater as a result of its capture by a constituent nucleus of the medium, which then explodes into a "star" of short-range and heavily ionizing fragments of very high LET capable of delivering a large, localized radiation dose. The π^+ meson comes to rest and decays into a $\mu+$ meson which, in turn, decays into a positron. In some larger accelerator centers, radiations having augmented Bragg peaks are used to treat tumors deep in the body. Since the doses delivered are not only higher locally in the tumor relative to doses to nearby healthy tissue but also the tumor doses are delivered by radiations of very high LET, the tumoricidal effect is enhanced also by the fact that the less oxygenated tumor tissue is no longer protected as it would be for lower LET radiation.[33,58]

3.7.7. Neutron Energy Spectra

In this section, examples of various types of neutron energy spectra are presented. Such spectra are subject to wide variation in any given experimental situation, and in some cases such as the nuclear reactor environment, they are very difficult to measure with precision. Figures 3.7-82 through 3.7-84 show neutron energy spectra from several types of nuclear reactors. For the case of enriched fuel reactors, the primary neutron spectra for ^{235}U fission has been measured by Cranberg (1956) and has the form:[34]

$$\Phi'_c(E) = 0.454 \exp(-E/0.965) \sinh(2.29E)^{1/2} \qquad (3.7-26)$$

where $\varphi'_c(E)$ is given in neutrons cm^{-2} sec^{-1} MeV^{-1}, and E is in MeV. The function has been normalized so that $\int_0^{14} \varphi'_c(E)\, dE$ = unity. Moteff[34] also presents another form for the primary spectrum, which is attributed to Watt (1948):

$$\Phi'_w(E) = 0.483 \exp(-E) \sinh(2E)^{1/2} \qquad (3.7-27)$$

Table 3.7-20, also taken from Moteff,[34] shows that the two functions representing the primary neutron spectrum are not very different from each other. Figure 3.7-85[34] presents a graphic illustration of the Watt spectrum. Figure 3.7-86 shows that the actual spectra in a reactor or critical assembly differ markedly from the primary neutron spectrum. Figure 3.7-87 shows a neutron spectrum in a fast critical assembly. Figure 3.7-88 compares neutron spectrum within a fast and thermal reactor. Tables 3.7-21 and 3.7-22, from Loveland,[51] present the average number of prompt neutrons per fission for various nuclides and the average neutron energies for some of the fissionable nuclides. Although there is some difference in the average number emitted per fission, the average energies of the primary spectra are nearly the same.

As Moteff points out,[34] the spectrum within a light water reactor is similar to the primary spectrum above about 1 MeV, but the lower part of the spectrum depends on the effects of neutron scattering and slowing down within the reactor environment. The peak at lower energies represents neutrons in thermal equilibrium within moderator. This equilibrium spectrum has a Maxwellian distribution in media with low neutron absorption cross sections. This neutron energy distribution, in terms of the differential neutron density, n(E) (in neutrons cm^{-3} eV^{-1}), depends on temperature as follows:

$$n(E) = n_0 \frac{2\pi}{(\pi kT)^{3/2}} E^{1/2} \exp(-\frac{E}{kT}) \qquad (3.7-28)$$

where n_0 is the total number of neutrons per cubic centimeter, T is the temperature in degrees Kelvin, E is the neutron energy in electron volts, and k the Boltzmann constant, 8.617×10^{-5} eV/°K. Figure 3.7-89 illustrates this differential neutron density for T = 298°K, or 25°C. However, the differential neutron flux density $\varphi'(E)$ is related to the differential neutron density through the equation:

$$\Phi'(E)\ [neutrons\ cm^{-2}\ sec^{-1}\ eV^{-1}] =$$

$$n(E)\ [neutrons\ cm^{-3}\ eV^{-1}]\ \nu(E)\ [cm\ sec^{-1}] \qquad (3.7-29)$$

$\varphi'(E)$ also has a Maxwellian energy distribution, illustrated in Figure 3.7-90, that shows a somewhat broader spectrum than that of Figure 3.7-89. Figures 3.7-91 through 3.7-113 show spectra of individual isotopic neutron sources used in various applications. While the Be(α, n) reactions provide the highest emission rates, their complex spectra limit their value for some applications.[36] The use of target materials other than Be provides spectra with narrower energy bands (as shown in Figures 3.7-101, 3.7-105 through 3.7-109, 3.7-112, and 3.7-113). Figure 3.7-97 shows that high-energy deuterons incident on Be can also produce rather simple spectra. Figure 3.7-113 shows a mock fission neutron spectrum produced by ^{210}Po alphas bombarding a mixture of 90.3% LiF, 9.4% B, and 0.3% Be. Figures 3.7-114 and 3.7-115 show that considerable variability may be obtained in different measurements of the same spectrum.

Photoneutron sources may be made by bombarding Be and D with gamma rays from radioisotopes. The energy of the emitted neutron will be the difference between the photon energy and the neutron binding energy, which for D is 2.226 MeV. Since gamma rays from useable isotopes are usually less than 3 MeV, neutrons produced will be of low energy.[36] It is also difficult to obtain monoenergetic neutrons from (γ, n) sources since reaction cross sections near threshold are on the order of millibarns for both Be and D. Thus, large amounts of target material are required causing a spread in the gamma photon energy and, thus, in the neutron energy.[36] Ra-Be (γ, n) sources have a long half-life and an energy spectrum extending up to 0.7 MeV with a mean energy of 0.2 MeV (DePangher, 1960).[36] The National Bureau of Standards (NBS) maintains a 1-Ci Ra-Be (γ,n) source as a primary standard of neutron emission rate. (The NBS primary standard has a source strength of 1.257×10^6 neutrons/sec, with a quoted error of 1% (Noyce et al., 1963).[36] Of course, these sources have a very high gamma-ray background restricting their use to detectors highly insensitive to gamma rays.

Antimony-124, with a half-life of 60 days and a gamma-ray energy of 1.7 MeV, may be used with Be to obtain 0.024 MeV, which is a lower energy than generally obtainable with accelerators. The neutron yield, when surrounded with 1 cm of Be, is about 3×10^6 n/Ci-sec, while the effective gamma-ray exposure rate is about 1.5 R/hr at 1 m. With intense neutron irradiation, Sb-Be (γ, n) sources have been produced with an initial neutron output of 1.3×10^{10} n/sec.[36]

Particle accelerators can produce a variety of neutron spectra including monoenergetic spectra. Figures 3.7-116 and 3.7-117 give brief summary information on some of the accelerator neutron spectra available.[35,36] Figure 3.7-118 compares the neutron spectrum from 12-MeV protons accelerated by cyclotron and impinging on Be with three fission spectra. Information on other charged particle reactions available with accelerators may be found in the references of this and other sections.

3.7.8. Data Useful in Spectral Determinations

In this section, some supplementary data useful in measuring or determining various radiation spectra will be presented. Range and cross section vs. energy data are often useful in determining or estimating energies of various radiations, and some fundamental information has been given in Sections 3.3 and 3.4 for obtaining range and cross-section data. Additional data of interest are presented in Tables 3.7-23 through 3.7-36 and Figures 3.7-119 through 3.7-131.

Threshold detectors have been widely used for reactor fast neutron dosimetry connected with various radiation effects studies.[34] Moteff[34] has summarized the concepts of "effective threshold" and "effective cross section" used in threshold detector studies. A cross section of an ideal threshold detector is illustrated in Figure 3.7-132A. In this case, induced number of interactions per cm^3 of material would be given by the equation:

$$A_s = N_t \, \sigma \, (E_i) \int_{E_i}^{\infty} \Phi'(E)dE \tag{3.7-30}$$

where A_s is the number of interactions producing the activated atoms or the recoil particles of interest per cm^3 of detector, N_t is the number of threshold detector atoms per cm^3, σ (E_i) is the cross section above the threshold energy E_i, and φ' (E) is the differential neutron flux density in neutrons per square centimeter per unit energy interval. If one defines ϕ (E_i) as the flux density above energy E_i then

$$\Phi(E_i) = \int_{E_i}^{\infty} \Phi'(E)dE = \frac{A_s}{N_t \sigma(E_i)} \qquad (3.7\text{-}31)$$

Effective energies (E_{eff}) have been defined with reference to neutron spectral distributions similar to the expected spectrum to be measured. Denoting the differential neutron flux density of a reference spectrum by φ_r' (E), the effective threshold cross section for the assumed spectrum is defined as:

$$\sigma(E_{eff}) = \frac{\int_0^{\infty} \sigma(E) \; \Phi_r'(E)dE}{\int_{E_{eff}}^{\infty} \Phi_r'(E)dE} \qquad (3.7\text{-}32)$$

The choice of E_{eff} is somewhat arbitrary, but it is usually chosen in the steeply rising portion of the excitation function. One method of selecting E_{eff} is illustrated in Figure 3.7-132B.[34] Once the effective cross section is determined for a given spectrum then the flux density above the effective threshold may be calculated from the measured saturated activity using Equation 3.7-30 with E_{eff} substituted for E_i. Further discussion of methodology and literature on the use of threshold detectors for neutron spectral measurements may be found in Moteff.[34]

Stamatelatos and England[26] have recently calculated fission product gamma-ray and photoneutron spectra for gamma rays of energies greater than the photoneutron threshold in ^{9}Be (1.67 MeV). The induced photoneutron spectra in ^{9}Be and ^{2}H for thermal and fast fission of uranium, plutonium, and thorium isotopes were calculated at cooling times between 1 and 5000 hr following fission at constant neutron irradiation periods of 1 hr, 1 day, 1 month, and 1 year prior to shutdown for ^{235}U, and for 1 month irradiation for the other fuels. The authors' report and unpublished information should be consulted for the detailed spectral calculations and results. In their calculations they have made use of the evaluated nuclear data files for 825 fission product nuclides. Table 3.7-37 presents the fission product chains and Table 3.7-38 presents a summary of fission product nuclides released in various types of situations. Other and old fission product data may also be presented elsewhere in this handbook where they are used for specific calculations. Such calculations may be updated, where necessary, by reference to Tables 3.7-37 and 3.7-38. Figure 3.7-133 presents an example[26] of a spectrum of fission product gamma radiation for gamma rays greater than 1.67 MeV and a particular irradiation and decay time. It should be noted that many gamma rays of lower energy are not shown since they are not of interest in the calculation of photoneutron spectra. However, they are of interest in the determination of exposure rates from fission products, and some sample data of the complete gamma ray spectrum vs. operating and decay time have been given in Section 3.7.4.

Another tabulation of thermal neutron fission yields is presented in Table 3.7-39. A diagram of the fission product yield curve is shown in Figure 3.7-134.[51] A much more detailed tabulation of fission product yields has recently been published by Voigt.[59] This tabulation is too extensive for reproduction in this handbook but includes yields, single particle model states, observed spins and parities, decay modes and half-lives for products of thermal neutron fission of ^{235}U. The tabulation includes calculated absolute cumulative yields, taking into account the latest data on fission product chains and other nuclear data. Figures 3.7-135 through 3.7-137 provide data useful for determining Compton-recoil electron spectra.

Table 3.7-1
SUMMARY OF DECAY SCHEME DATA BY NUCLIDES (1959)*

Hydrogen-3 ($_1$H^3)

Decay Scheme

Half-life: 12.262 ± 0.004 years

Usually produced by ^{6}Li (n, α) for which the thermal neutron cross section is 945 barns the natural isotopic abundance of ^{6}LI is ~ 7.5%; it has been found to vary slightly with the source

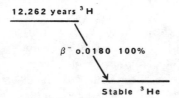

12.262 years ^{3}H

β^- 0.0180 100%

Stable ^{3}He

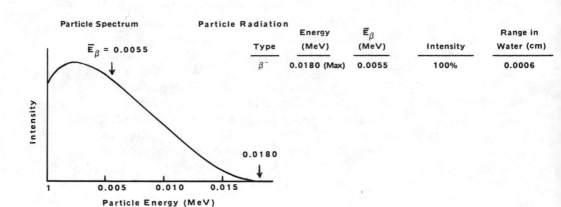

Particle Spectrum	Particle Radiation				
	Type	Energy (MeV)	$\overline{E}_\beta$ (MeV)	Intensity	Range in Water (cm)
	β^-	0.0180 (Max)	0.0055	100%	0.0006

$\overline{E}_\beta = 0.0055$

Table 3.7-1 (continued)
SUMMARY OF DECAY SCHEME DATA BY NUCLIDES (1959)[e]

Carbon-14 ($_6C^{14}$)

Decay Scheme

Half-life: 12,262 ± 0.004 years

Usually produced by ^{6}Li (n, α) for which the thermal
neutron cross section is 945 barns the natural
isotopic abundance of ^{6}Li is ~ 7.5%; it has been
found to vary slightly with the source

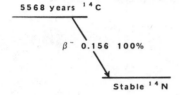

5568 years ^{14}C

β^- 0.156 100%

Stable ^{14}N

Particle Radiation

	Type	Energy (MeV)	$\bar{E}_\beta$ (MeV)	Intensity	Range in water (cm)
Particle Spectrum	β^-	0.156 (Max)	0.050	100%	0.029

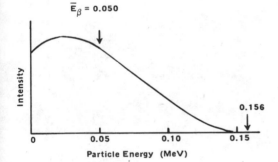

$\bar{E}_\beta = 0.050$

0.156

Intensity

0   0.05   0.10   0.15

Particle Energy (MeV)

Table 3.7-1 (continued)
SUMMARY OF DECAY SCHEME DATA BY NUCLIDES (1959)ᵉ

Sodium-22 ($_{11}$Na^{22})

Decay Scheme

Half-life: 2.58 ± 0.03 years

Usually produced by ^{24}**Mg d, α)**

Note: Two 0.511-MeV γ rays are produced when a positron emitted in the nuclear disintegration is annihilated in the medium surrounding the source; the dose rate from this annihilation radiation has been calculated under the assumption that the radiation can be considered as coming from the source itself

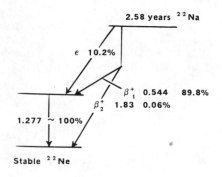

2.58 years ^{22}Na

ε 10.2%

β_1^+ 0.544 89.8%
β_2^+ 1.83 0.06%

1.277 ~ 100%

Stable ^{22}Ne

Particle Spectrum

$\overline{E}_\beta = 0.214$

0.544

Intensity

0   0.1   0.2   0.3   0.4   0.5   0.6

Particle Energy (MeV)

Particle Radiation

Type	Energy (MeV)	$\overline{E}_\beta$ (MeV)	Intensity	Range in water (cm)
^eA_K	~ 0.001		10.2%	< 0.0001
β_1^+	0.544 (Max)	0.214	89.8%	0.18
ce	1.28		< 0.1%	0.56
β_2^+	1.83 (Max)		0.06%	0.88

Electromagnetic Radiation

Dose Rate

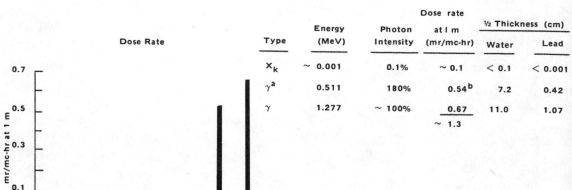

0.7
0.5
0.3
0.1
0

mr/mc-hr at 1 m

0.001   0.01   0.1   0.2   0.5   1.0   2.0

Energy of Radiation (MeV)

Type	Energy (MeV)	Photon Intensity	Dose rate at l m (mr/mc-hr)	½ Thickness (cm) Water	½ Thickness (cm) Lead
X_k	~ 0.001	0.1%	~ 0.1	< 0.1	< 0.001
γ^a	0.511	180%	0.54^b	7.2	0.42
γ	1.277	~ 100%	0.67	11.0	1.07
			~ 1.3		

Table 3.7-1 (continued)
SUMMARY OF DECAY SCHEME DATA BY NUCLIDES (1959)^q

Sodium-24 ($_{11}Na^{24}$)

Decay Scheme

Half-life: 14.97 ± 0.02 hr

Usually produced by ^{23}Na (n, γ) for which the thermal neutron cross section is 0.54 barns; the natural isotopic abundance of ^{23}Na is 100%

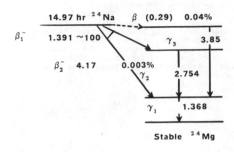

Particle Radiation

Particle Spectrum

$\overline{E}_\beta = 0.56$

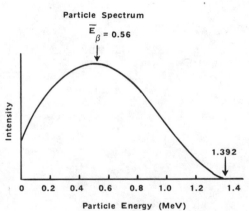

Type	Energy (MeV)	$\overline{E}_\beta$ (MeV)	Intensity	Range in water (cm)
β_1^-	1.392 (Max)	0.56	~ 100%	0.62
β_2^-	4.17 (Max)		0.003%	2.12

Electromagnetic Radiation

Dose Rate

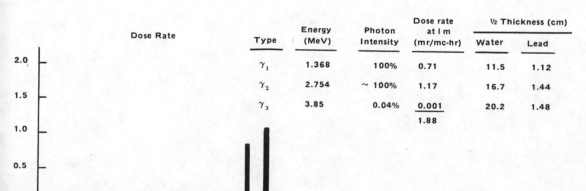

Type	Energy (MeV)	Photon Intensity	Dose rate at l m (mr/mc-hr)	½ Thickness (cm) Water	Lead
γ_1	1.368	100%	0.71	11.5	1.12
γ_2	2.754	~ 100%	1.17	16.7	1.44
γ_3	3.85	0.04%	0.001	20.2	1.48
			1.88		

Table 3.7-1 (continued)
SUMMARY OF DECAY SCHEME DATA BY NUCLIDES (1959)ᵉ

Phosphorus-32 ($_{15}P^{32}$)

Decay Scheme

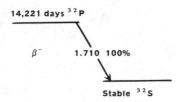

Half-life: 14,221 ± 0.005 days

Usually produced by ^{31}P (n, γ) for which the thermal
neutron cross section is 0.19 barns; the natural
isotopic abundance of ^{31}P is 100%

Note: Inner bremsstrahlung with a continuous energy
distribution is the only electromagnetic
radiation coming from the radioactive atom;
although the energy spectrum extends to the
maximum energy of the betas, the average
energy of the bremsstrahlung amounts to only
1.5 keV per beta; external bremsstrahlung
produced in neighboring atoms is present as
usual

Particle Radiation

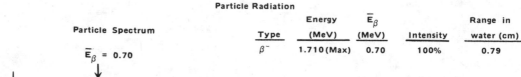

Type	Energy (MeV)	$\overline{E}_\beta$ (MeV)	Intensity	Range in water (cm)
β^-	1.710 (Max)	0.70	100%	0.79

Particle Spectrum

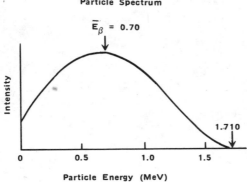

Particle Energy (MeV)

Table 3.7-1 (continued)
SUMMARY OF DECAY SCHEME DATA BY NUCLIDES (1959)·

Decay Scheme

Half-life: 87.2 ± 0.1 days

Usually produced by ^{35}Cl (n,p) for which the thermal neutron cross section is 0.40 barns; the natural isotopic abundance of ^{35}Cl is 75.53%

Note: Inner bremsstrahlung with a continous energy distribution is the only electromagnetic radiation coming from the radioactive atom; although the energy spectrum extends to the maximum energy of the betas, the average energy of the bremsstrahlung amounts to only ~ 11 eV per disintegration; external bremsstrahlung, though soft because of the low energy of the beta particles, is present as usual

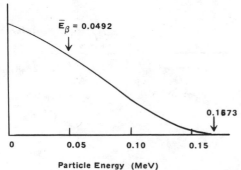

87,2 days S^{35}

β^- 0.1673 100%

Stable ^{35}Cl

Particle Radiation

Particle Spectrum	Type	Energy (MeV)	$\bar{E}_\beta$ (MeV)	Intensity	Range in water (cm)
	β^-	0.1673 (Max)	0.0492	100%	0.032

$\bar{E}_\beta = 0.0492$

0.1673

| 0 | 0.05 | 0.10 | 0.15 |

Particle Energy (MeV)

Table 3.7-1 (continued)
SUMMARY OF DECAY SCHEME DATA BY NUCLIDES (1959)ᵉ

Chlorine-36 ($_{17}CL^{36}$)

Decay Scheme

Half-life: **3.03 ± 0.03 x 10⁵ years**

Usually produced by ^{35}Cl (n, γ) for which the thermal neutron cross section is 43.5 barns; the natural isotopic abundance of ^{35}Cl is 75.53%

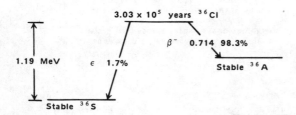

Particle Radiation

Particle Spectrum

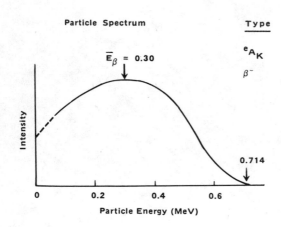

Type	Energy (MeV)	$\bar{E}_\beta$ (MeV)	Intensity	Range in water (cm)
$^e A_K$	~ 0.002		1.6%	< 0.0001
β^-	0.714 (Max)	0.30	98.3%	0.27

Dose Rate

Electromagnetic Radiation

Type	Energy (MeV)	Photon Intensity	Dose rate at l m (mr/mc-hr)	½ Thickness (cm) Water	½ Thickness (cm) Lead
X_k	~ 0.002	~ 0.1%	~ 0.025	< 0.1	< 0.001

Table 3.7-1 (continued)
SUMMARY OF DECAY SCHEME DATA BY NUCLIDES (1959)[e]

Potassium-42 ($_{19}K^{42}$)

Decay Scheme

Half-life: 12.42 ± 0.03 hr

Usually produced by ^{41}K (n, γ) for which the thermal neutron cross section is 1.2 barns; the natural isotopic abundance of ^{41}K is 6.9%

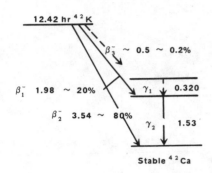

Particle Radiation

Particle Spectrum

$\overline{E}_\beta = 1.42$

Type	Energy (MeV)	$\overline{E}_\beta$ (MeV)	Intensity	Range in water (cm)
β_1^-	1.98 (Max)	0.82	~ 20%	0.95
β_2^-	3.54 (Max)	1.57	~ 80%	1.8
		1.42 per disintegration		
β_3^-	~ 0.5 (Max)		~ 0.2%	0.16

3.54

Electromagnetic Radiation

Dose Rate

Type	Energy (MeV)	Photon Intensity	Dose rate at 1 m (mr/mc-hr)	½ Thickness (cm) Water	½ Thickness (cm) Lead
γ_1	0.320	~ 0.2%	< 0.001	6.0	6.20
γ_2	1.53	~ 20%	0.154	12.1	1.21
			0.154		

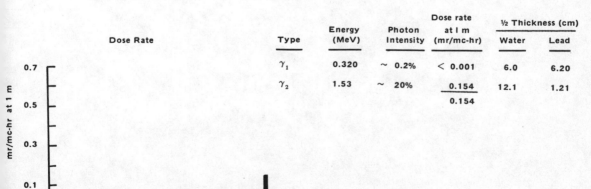

Table 3.7-1 (continued)
SUMMARY OF DECAY SCHEME DATA BY NUCLIDES (1959)ᵉ

Calcium-45 ($_{20}Ca^{45}$)

Decay Scheme

Half-life: 164 ± 4 days

Usually produced by ^{44}Ca (n, γ) for which the thermal neutron cross section is 0.82 barns; the natural isotopic abundance of ^{44}Ca is 2.08%

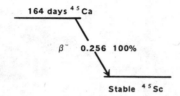

164 days ^{45}Ca

β^- 0.256 100%

Stable ^{45}Sc

Particle Radiation

Particle Spectrum

Type	Energy (MeV)	$\overline{E}_\beta$ (MeV)	Intensity	Range in water (cm)
β^-	0.256 (Max)	0.077	100%	0.061

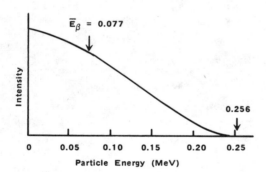

Table 3.7-1 (continued)
SUMMARY OF DECAY SCHEME DATA BY NUCLIDES (1959)[q]

Chromium-51 ($_{24}Cr^{51}$)

Decay Scheme

Half-life: 27.8 ± 0.1 days

Usually produced by ^{60}Cr (n, γ) for which the thermal neutron cross section is 17 barns; the natural isotopic abundance of ^{50}Cr is 4.3%

Note: No positrons are emitted in the decay; the disintegration energy of 0.756 MeV is carried off chiefly by neutrinos

27.8 days ^{51}Cr

ε 10%

ε 90%

0.321   10%

0.756 MeV

Stable ^{51}V

Particle Radiation

Type	Energy (MeV)	E_β (MeV)	Intensity	Range in water (cm)
e_{A_L}	~0.0005		151%	< 0.0001
e_{A_K}	~0.005		72%	< 0.0001
ce	0.316		< 0.1%	0.083

Electromagnetic Radiation

Type	Energy (MeV)	Photon Intensity	Dose rate at I m (mr/mc-hr)	½ Thickness (cm) Water	½ Thickness (cm) Lead
x_K	~0.005	20%	0.71	< 0.1	< 0.001
γ	0.321	10%	0.018	0.60	0.20
			0.73		

Dose Rate

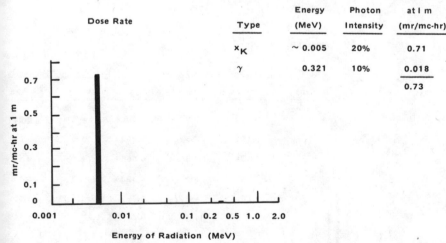

mr/mc-hr at 1 m

0.7
0.5
0.3
0.1
0

0.001   0.01   0.1 0.2 0.5 1.0 2.0

Energy of Radiation (MeV)

Table 3.7-1 (continued)
SUMMARY OF DECAY SCHEME DATA BY NUCLIDES (1959)*

Iron-59 ($_{26}$Fe^{59})

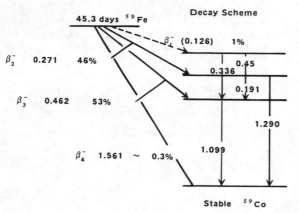

Decay Scheme

Half-life: 45.3 ± 0.2 days

Usually produced by ^{58}Fe (n, γ) for which the thermal neutron cross section is 1.1 barns; the natural isotopic abundance of ^{58}Fe is 0.31%, but enriched targets are ordinarily used

Particle Radiation

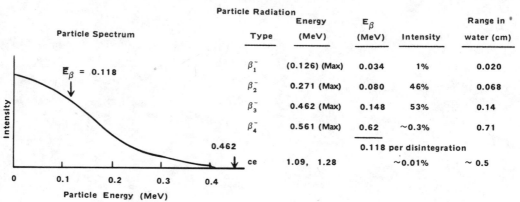

Particle Spectrum

$E_\beta = 0.118$

Type	Energy (MeV)	E_β (MeV)	Intensity	Range in * water (cm)
β_1^-	(0.126) (Max)	0.034	1%	0.020
β_2^-	0.271 (Max)	0.080	46%	0.068
β_3^-	0.462 (Max)	0.148	53%	0.14
β_4^-	0.561 (Max)	0.62	~0.3%	0.71
			0.118 per disintegration	
ce	1.09, 1.28		~0.01%	~ 0.5

Electromagnetic Radiation

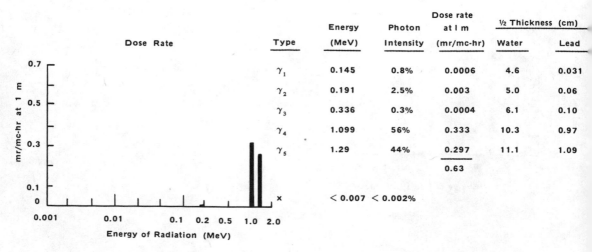

Dose Rate

Type	Energy (MeV)	Photon Intensity	Dose rate at I m (mr/mc-hr)	½ Thickness (cm) Water	Lead
γ_1	0.145	0.8%	0.0006	4.6	0.031
γ_2	0.191	2.5%	0.003	5.0	0.06
γ_3	0.336	0.3%	0.0004	6.1	0.10
γ_4	1.099	56%	0.333	10.3	0.97
γ_5	1.29	44%	0.297	11.1	1.09
			0.63		
x		< 0.007 < 0.002%			

Table 3.7-1 (continued)
SUMMARY OF DECAY SCHEME DATA BY NUCLIDES (1959)ᵉ

Cobalt-60 ($_{27}Co^{60}$)

Decay Scheme

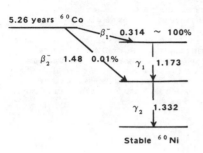

Half-life: 5.26 ± 0.02 years

Usually produced by ^{59}CO (n, γ) for which the thermal neutron cross section is 36.7 barns; the natural isotopic abundance of ^{69}CO is 100%

Note: Thermal neutron bombardment of ^{59}Co also produces the 10.5-min isomeric state of ^{60}Co (not shown); > 99% of the decays of this isomeric state lead to the 5.27-year ground state of ^{60}Co; the effective cross section for production of the 5.26-year activity is therefore the sum of the cross sections for the direct production of the 10.5-min and 5.26-year states.

Particle Radiation

Particle Spectrum

$\overline{E}_\beta = 0.093$

0.314

Type	Energy (MeV)	$\overline{E}_\beta$ (MeV)	Intensity	Range in water (cm)
β_1^-	0.314 (Max)	0.093	~ 100%	0.082
β_2^-	1.48 (Max)		0.01%	0.67

Intensity

| 0 | 0.05 | 0.10 | 0.15 | 0.20 | 0.25 | 0.30 |

Particle Energy (MeV)

Electromagnetic Radiation

Dose Rate

Type	Energy (MeV)	Photon Intensity	Dose rate at 1 m (mr/mc-hr)	½ Thickness (cm) Water	½ Thickness (cm) Lead
γ_1	1.173	~ 100%	0.63	10.7	1.02
γ_2	1.332	100%	0.70	11.3	1.11
			1.33		

0.7

0.5

0.3

0.1

0

| 0.001 | 0.01 | 0.1 | 0.2 | 0.5 | 1.0 | 2.0 |

Energy of Radiation (MeV)

Table 3.7-1 (continued)
SUMMARY OF DECAY SCHEME DATA BY NUCLIDES (1959)ᵍ

Zinc-65 ($_{30}Zn^{65}$)

Decay Scheme

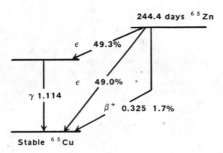

Half-life: 244.4 ± 0.5 days

Usually produced by ^{64}Zn (n, γ) for which the thermal neutron cross section is 0.46 barn; the natural isotopic abundance of ^{64}Zn is 48.9%

Note: Two 0.511-MeV γ rays are produced when a positron, emitted in a nuclear disintegration, is annihilated in the medium surrounding the source; the dose rate from the annihilation radiation has been calculated under the assumption that it can be considered as coming from the source itself.

Particle Radiation

Type	Energy (MeV)	$\overline{E}_\beta$ (MeV)	Intensity	Range in water (cm)
^eA_L	~ 0.001		130%	< 0.0001
^eA_k	~ 0.007		53%	< 0.0001
β^+	0.325 (Max)	0.143	1.7%	0.088
ce	~ 1.10		~ 0.01%	0.44

Particle Spectrum

$\overline{E}_\beta = 0.143$

Electromagnetic Radiation

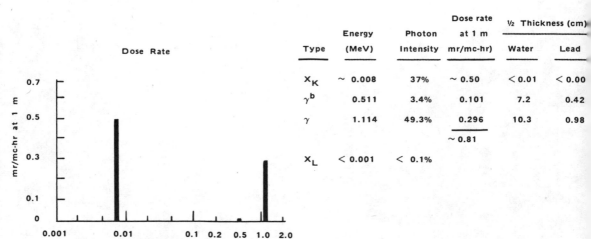

Type	Energy (MeV)	Photon Intensity	Dose rate at 1 m mr/mc-hr	½ Thickness (cm) Water	½ Thickness (cm) Lead
X_K	~ 0.008	37%	~ 0.50	< 0.01	< 0.00
γ^b	0.511	3.4%	0.101	7.2	0.42
γ	1.114	49.3%	0.296	10.3	0.98
			~ 0.81		
X_L	< 0.001	< 0.1%			

Table 3.7-1 (continued)
SUMMARY OF DECAY SCHEME DATA BY NUCLIDES (1959)*

Rubidium-86 ($_{37}Rb^{86}$)

Decay Scheme

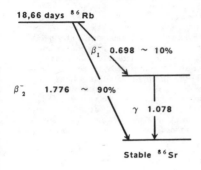

Half-life: 18.66 ± 0.02 days

Usually produced by ^{85}Rb (n, γ) for which the thermal neutron cross section is 0.95 barns; the natural isotopic abundance of ^{85}Rb is 72. 15%

Note: Thermal neutron bombardment of ^{85}Rb also produces a 1.02-min isomeric state of ^{86}Rb, this state (not shown on the decay scheme) decays to the ground state of ^{86}Rb by means of a 0.53-MeV γ ray; all neutron capture in ^{85}Rb thus results in the production of the 18.7-day activity

Particle Radiation

Particle Spectrum

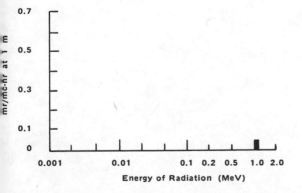

Type	Energy (MeV)	$\overline{E}_\beta$ (MeV)	Intensity	Range in water (cm)
β_1^-	0.698 (Max)	0.234	~ 10%	0.26
β_2^-	1.776 (Max)	0.71	~ 90%	0.84
		0.66	per disintegration	

Electromagnetic Radiation

Dose Rate

Type	Energy (MeV)	Photon Intensity	Dose rate at l m (mr/me-hr)	½ Thickness (cm) Water	Lead
γ	1.078	~ 10%	0.058	10.2	0.96

Table 3.7-1 (continued)
SUMMARY OF DECAY SCHEME DATA BY NUCLIDES (1959)ᵉ

Strontium-90 ($_{38}Sr^{90}$) — Yttrium-90 ($_{39}Y^{90}$)

Decay Scheme

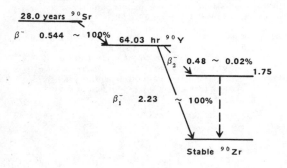

Half-life: **28.0** ± **0.3 years** ^{90}Sr

Obtained by separation from fission products; the yield in fission of ^{235}U by slow neutrons is 5.8%

Note: ^{90}Y, the daughter of ^{90}Sr, has a half-life of 64.03 hr, a value much less than that of its ^{90}Sr parent; ^{90}Y is thus constantly being created and destroyed in a ^{90}Sr source; most sources consist of mixtures in which the number of ^{90}Y nuclei decaying per second is equal to the number created, that is, "equilibrium mixtures;" the number of disintegrations per second in such a source is equal to twice the number of ^{90}Sr disintegrations; the 1.75-MeV level in ^{90}Zr decays by emission of conversion electrons and electron-positron pairs only

Particle Radiation

Particle Spectrum

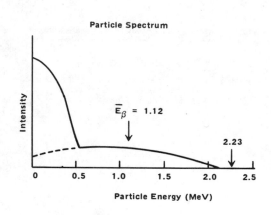

Particle Energy (MeV)

Type	Energy (MeV)	$\overline{E}_\beta$ (MeV)	Intensity	Range in Water (cm)
β (Sr)	0.544 (Max)	0.194	100%	0.18
β_1^- (Y)	2.25 (Max)	0.93	~ 100%	1.09
			1.12 per ^{90}Sr & ^{90}Y disintegration	
β_2^- (Y)	0.48 (Max)		~ 0.02%	0.16
ce⁻	1.75		~ 0.02%	0.81

Table 3.7-1 (continued)
SUMMARY OF DECAY SCHEME DATA BY NUCLIDES (1959)[e]

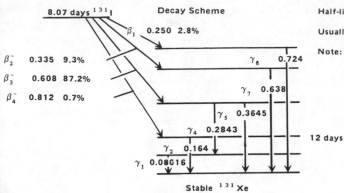

Decay Scheme

8.07 days ^{131}I

β_1 0.250 2.8%

β_2^- 0.335   9.3%

β_3^- 0.608   87.2%

β_4^- 0.812   0.7%

γ_8 0.724

γ_7 0.638

γ_5 0.3645

γ_4 0.2843

γ_2 0.164

γ_1 0.08016

12 days ^{131}Xe

Stable ^{131}Xe

Half-life: 8.07 ± 0.02 days

Usually obtained by separation from fission products

Note: E_β per disintegration includes contributions from conversion electrons; conversion electrons of energies 0.128, 0.249, 0.278, 0.359, 0.603, 0.633, 0.687, and 0.717 all have intensities < 0.5%

Particle Radiation[a]

Particle Spectrum

$\bar{E}_\beta = 0.186$

0.608

Intensity

Type	Energy (MeV)	$\bar{E}_\beta$ (MeV)	Intensity	Range in water (cm)
e_{A_L}	~ 0.003		5.1%	< 0.0001
e_{A_k}	~ 0.025		0.7%	0.001
ce_k	0.045		3.6%	0.003
ce_{LMN}	~ 0.075		0.7%	0.008
ce_K	0.329		1.4%	0.089
β_1^-	0.250 (Max)	0.071	2.8%	0.060
β_2^-	0.335 (Max)	0.098	9.3%	0.090
β_3^-	0.608 (Max)	0.193	87.2 %	0.21
β_4^-	0.812 (Max)	0.272	0.7%	0.31

0.188 per disintegration

0   0.1   0.2   0.3   0.4   0.5   0.6

Particle Energy (MeV)

Electromagnetic Radiation

Dose Rate

Type	Energy (MeV)	Photon intensity	Dose rate at 1 m (mr/mc-hr)	½ Thickness (cm) Water	Lead
x_L	~ 0.005	0.6%	0.021	< 0.1	< 0.001
x_K	~ 0.030	5%	0.004	2.1	~ 0.003
γ_1	0.08016	2%	0.001	4.0	0.044
γ_2	0.164	~ 0.7%	~ 0.001	4.8	0.042
γ_3	0.1771				
γ_4	0.2843	6%	0.010	5.8	0.15
γ_5	0.3645	79%	0.166	6.3	0.25
γ_6	0.51				
γ_7	0.638	9%	0.032	7.9	0.58
γ_8	0.724	3%	0.012	8.4	0.67

0.247

0.20

0.15

0.10

0.05

0

mr/mc-hr at 1 m

0.001   0.01   0.1 0.2 0.5 1.0 2.0

Energy of Radiation (MeV)

Table 3.7-1 (continued)
SUMMARY OF DECAY SCHEME DATA BY NUCLIDES (1959)[c]

Gold-198 ($_{79}Au^{198}$)

Decay Scheme

Half-life: 2.696 ± 0.002 days

Usually produced by ^{197}Au (n, γ) for which the thermal neutron cross section is 98.4 barns; the natural isotopic abundance of ^{197}Au is 100%

Note: The intensities of the conversion electrons, due to γ_2 and γ_β, which have energies of approximately 0.60 MeV and 1.00 MeV respectively, are < 0.1 per 100 disintegrations; E_β per disintegration includes contributions from conversion electrons

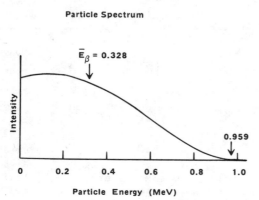

2.696 days ^{198}Au

β_1^- 0.282 0.98%

γ_2 0.676

β_2^- 0.959 99%

γ_3 1.089

β_3^- 1.371 0.025%

γ_1 0.4118

Stable ^{198}Hg

Particle Radiation[a]

Particle Spectrum

$\bar{E}_\beta$ = 0.328

0.959

Type	Energy (MeV)	$\bar{E}_\beta$ (MeV)	Intensity	Range in water (cm)
e_{AL}	~ 0.007		2.0%	< 0.0001
e_{AK}	~ 0.056		~ 0.2%	0.005
ce_K	0.328		3%	0.089
ce_{LMN}	~ 0.40		1.6%	0.118
β_1^-	0.282 (Max)		9.98%	0.071
β_2^-	0.959 (Max)		99%	0.38
β_3^-	1.371 (Max)		0.025%	0.61

Electromagnetic Radiation

Dose Rate

Type	Energy (MeV)	Photon intensity	Dose rate at 1 m (mr/mc-hr)	½ Thickness (cm) Water	½ Thickness (cm) Lead
X_L	~ 0.009	1.3%	~ 0.013	~ 0.1	< 0.001
X_K	~ 0.070	2.7%	~ 0.001	3.8	0.3
γ_1	0.4118	95%	0.228	6.6	0.31
γ_2	0.676	0.82%	0.004	8.1	0.62
γ_3	1.089	0.16%	0.001	10.2	0.97
			2.25		

Note: Beta spectra from parent-daughter particles, such as $^{90}Sr-\ ^{90}Y$, present the probability distribution of beta emission from both nuclides in secular equilibrium.

[a] As stated in text, the reader is cautioned that much of the data in this table has been superseded. Reference should be made to Section 3.6 for current decay scheme data for use in internal dose calculations.

From Slack, L. and Way, K., *Radiations from Radioactive Atoms in Frequent Use*, U. S. Atomic Energy Commission, Washington, D. C., February 1959. With permission.

TABLE 3.7-2
SPECIFIC GAMMA-RAY CONSTANT DETERMINATION FOR ^{125}I (HALF-LIFE, 60.14 DAYS)

Radiation type	Mean number per decay	Mean energy (MeV)	Energy absorption coefficient (cm²/g)	Specific gamma-ray constant (R cm² mCi⁻¹hr⁻¹)
Gamma 1	6.68E−2	3.546E−2	9.194E−2	4.25E−2
K alpha 1 X-ray	7.41E−1	2.747E−2	1.963E−1	7.81E−1
K alpha 2 X-ray	3.98E−1	2.720E−2	2.032E−1	4.30E−1
K alpha 3 X-ray	3.16E−5	2.687E−2	2.117E−1	3.52E−5
K beta 1 X-ray	1.40E−1	3.099E−2	1.311E−1	1.11E−1
K beta 2 X-ray	4.40E−1	3.171E−2	1.235E−1	3.29E−2
K beta 3 X-ray	7.20E−2	3.094E−2	1.317E−1	5.74E−2
K beta 5 X-ray	1.44E−3	3.124E−2	1.285E−1	1.13E−3

Note: Total specific gamma-ray constant, 4.25E−2; total specific X-ray constant, 1.41EO; total exposure-rate constant, 1.45EO.

From Dillman, L. T., in *Health Physics Division Annual Progress Report, Period Ending June 30, 1976,* ORNL-5171, Oak Ridge National Laboratory, Oak Ridge, Tenn., October 1976. With permission.

TABLE 3.7-3
SPONTANEOUS FISSION DECAY DATA FOR ^{252}Cf (HALF-LIFE, 2.58 YEARS)

Radiation type	Mean number per decay	Mean energy (MeV)
Spontaneous fission neutrons	1.16E−1	2.164EO
Spontaneous fission fragments	6.20E−2	9.514E1
Spontaneous fission prompt gammas	2.68E−1	8.847E−1
Spontaneous fission delayed gammas	2.52E−1	9.578E−1
Spontaneous fission betas	1.93E−1	1.280EO

Note: Decay mode, spontaneous fission and alpha; only spontaneous fission data shown here.

From Dillman, L. T., in *Health Physics Division Annual Progress Report, Period Ending June 30, 1976,* ORNL-5171, Oak Ridge National Laboratory, Oak Ridge, Tenn., October 1976. With permission.

TABLE 3.7-4
ALPHA-PARTICLE CALIBRATION ENERGIES

Source	Bϱ (G-cm)	Energy (keV)
^{210}Po	331,722 ± 15	5304.5 ± 0.5
^{211}Bi	370,720 ± 40	6,621.9 ± 1.4
^{211}Po	393,190 ± 50	7,448.1 ± 1.9
^{212}Bi (ThC αo)[a]	354,326 ± 20	6,049.6 ± 0.7
^{212}Bi (ThC α1)[a]	355,475 ± 20	6,088.9 ± 0.7
^{212}Po (ThC′)	427,060 ± 20	8,785.0 ± 0.8
^{214}Bi	338,170 ± 70	5,510.9 ± 2.3
^{214}Po	399,488 ± 16	7,688.4 ± 0.6
^{215}Po	391,490 ± 40	7,383.9 ± 1.5
^{216}Po	375,050 ± 40	6,777.3 ± 1.5
^{218}Po	352,870 ± 70	6,000.1 ± 2.4
^{219}Rn	376,160 ± 40	6,817.5 ± 1.5
^{220}Rn	361,260 ± 60	6,288.5 ± 2.1
^{222}Rn	337,410 ± 70	5,486.2 ± 2.3
^{223}Ra	349,010 ± 50	5,869.6 ± 1.7
^{224}Ra	343,450 ± 40	5,684.2 ± 1.3
^{226}Ra	314,990 ± 80	4,781.8 ± 2.4
^{227}Th	354,070 ± 60	6,040.9 ± 2.0
^{228}Th	335,570 ± 60	5,426.6 ± 2.0
^{230}Th	311,960 ± 160	4,690.3 ± 4.8

From Marion, J. B., in *American Institute of Physics Handbook,* 3rd ed., Gray, D. E., Ed., McGraw-Hill, New York, 1972, 8-3. With permission.

TABLE 3.7-5

THE LONG-RANGE α-RAY SPECTRUM OF ThC'$\rightarrow$ ThD($_{84}$Po^{212} $\rightarrow_{82}$Pb^{208},^{26} + He^4) ACCORDING TO RYTZ (R54)

Group	Relative abundance	α-Ray energy E_α(MeV)	Disintegration energy E_0 (MeV)	Excitation energy in parent ThC', (MeV)
Normal: α_0	10^5	8.776	8.946	0
Long-range: α_2	35	9.489	9.671	0.725
Long-range: α_3	20	10.417	10.617	1.671
Long-range: α_1	170	10.536	10.739	1.793

From Loveland, W. D., in *American Institute of Physics Handbook*, 3rd ed., Gray, D. E., Ed., McGraw-Hill, New York, 1972. With permission.

Table 3.7-6

AVERAGE NUMBER AND ENERGY OF BETA DECAYS PER FISSION FOR ^{235}U and ^{233}U THERMAL NEUTRON FISSION

^{235}U				^{233}U	
N_β	Ref.	E_β	Ref.	N_β	Ref.
6.6 $\pm$ 0.9	1				
6.9 $\pm$ 0.4	2	8.1 $\pm$ 0.4	2		
6.6 $\pm$ 0.2	3				
5.93 $\pm$ 0.2	4			5.25 $\pm$ 0.2	4
6.10^a				5.27	
		7.6 $\pm$ 0.5	5		

^{a} Calculated value.

From Gindler, J. and Hulzenga, J. R., *Nuclear Chemistry*, Vol. 2, Yaffe, L., Ed., Academic Press, New York, 1968, as given in Loveland, W. D., in *American Inshitute of Physics Handbook*, 3rd ed., Gray, D. E., Ed., McGraw-Hill, New York, 1972. With permission.

REFERENCES

1. Alzmann, G. *Nukleonik*, 3, 295, 1961.
2. Armbruster, P. and Meister, H., *Z. Phys.*, 170, 274, 1962.
3. Armbruster, P., Hovestadt, D., Meister, H., and Specht, H. J., *Nucl. Phys.*, 54, 586, 1964.
4. Specht, H. J. and Seyfarth, H., *Physics and Chemistry of Fission*, Vol. 2, International Atomic Energy Agency, Vienna, 1965, 253.
5. Perkins, J. F. and King, R. W., *Nucl. Sci. Eng.*, 3, 726, 1958.

TABLE 3.7-7
GAMMA RAYS FROM
RADIOACTIVE SOURCES

Source	Energy (keV)	Half-life
$m_0 o^2$	511.006 ± 0.002	
^{7B}e	477.57 ± 0.05	53 days
^{22}Na	1274.55 ± 0.04	2.60 years
^{24}Na	1368.526 ± 0.044	
	2753.92 ± 0.12	15.0 hr
^{51}Cr	320.080 ± 0.013	27.8 days
^{54}Md	834.81 ± 0.03	314 days
	1173.23 ± 0.04	
^{60}Co	1332.49 ± 0.04	5.26 years
^{65}Zn	1115.40 ± 0.12	246 days
	898.04 ± 0.04	
^{88}Y	1836.13 ± 0.04	106.6 days
^{137}Cs	661.635 ± 0.076	30 years
^{198}Au	411.795 ± 0.009	2.70 days
	569.62 ± 0.06	
^{207}Bi	1063.44 ± 0.09	
	1769.71 ± 0.13	30 years
	510.723 ± 0.020	
^{208}Ti	583.139 ± 0.023	
(ThC″)	2614.47 ± 0.10	(1.91 years)
	26.348 ± 0.010	
^{241}Am	59.543 ± 0.015	433 years

From Marion, J. B., in *American Institute of Physics Handbook,* 3rd ed., Gray, D. E., Ed., McGraw-Hill, New York, 1972. With permission.

TABLE 3.7-8
GAMMA RAYS FROM ^{56}Co

Energy, keV	Relative intensity	Energy (keV)	Relative intensity
733.79 ± 0.19	0.1 ± 0.05	2015.49 ± 0.20	2.93 ± 0.16
787.92 ± 0.15	0.40 ± 0.11	2035.03 ± 0.12	7.33 ± 0.30
846.76 ± 0.05	100	2113.00 ± 0.10	0.37 ± 0.08
977.47 ± 0.13	1.52 ± 0.16	2598.80 ± 0.12	16.77 ± 0.57
1037.97 ± 0.07	13.02 ± 0.35	3009.99 ± 0.24	0.84 ± 0.16
1175.26 ± 0.13	1.86 ± 0.23	3202.25 ± 0.19	3.15 ± 0.16
1238.34 ± 0.09	69.35 ± 1.47	3253.82 ± 0.15	7.70 ± 0.34
1360.35 ± 0.09	4.38 ± 0.16	3273.38 ± 0.18	1.55 ± 0.11
1771.57 ± 0.10	15.30 ± 0.53	3452.18 ± 0.22	0.88 ± 0.10
1964.88 ± 0.45	0.72 ± 0.08	3548.11 ± 0.25	0.18 ± 0.10

From Marion, J. B., in *American Institute of Physics Handbook,* 3rd ed., Gray, D. E., Ed., McGraw-Hill, New York, 1972. With permission.

TABLE 3.7-9
GAMMA RAYS FROM NUCLEAR REACTIONS

Nucleus	γ-ray energy (keV)	Nucleus	γ-ray energy (keV)
^{17}F	495.33 ± 0.10	^{13}C	4945.46 ± 0.17[c]
^{18}F	658.75 ± 0.7[a]	^{14}N	5104.87 ± 0.18
^{17}O	870.81 ± 0.22	^{15}O	5240.53 ± 0.52
^{12}B	953.10 ± 0.60	^{15}N	5268.9 ± 0.2
^{12}B	1673.52 ± 0.60	^{15}N	5297.9 ± 0.2[a]
^{14}N	2312.68 ± 0.10[b]	^{16}O	6129.3 ± 0.4
^{10}Xe	2589.9 ± 0.25[c]	^{10}Be	6809.4 ± 0.4[e]
^{14}N	2792.68 ± 0.15[d]	^{16}O	7117.02 ± 0.49
^{10}Be	3367.4 ± 0.2[e]	^{209}Pb	7367.5 ± 1[e]
^{12}C	4439.0 ± 0.2[f]	^{14}N	9173 ± 1[h]
		^{15}N	10829.2 ± 0.4[c]

[a]   From 1.70- to 1.04-MeV decay.

[b]   Doppler shifted unless formed in O^{14} (β^+) N^{14}.

[c]   From 5.96- to 3.37-MeV decay (thermal neutron capture).

[d]   From 5.10- to 2.31-MeV decay.

[e]   From thermal neutron capture.

[f]   Doppler shifted unless formed in ^{12}B (β^-) C^{12}.

[g]   Doppler shifted unless formed in ^{15}O (β^-) ^{15}N or by thermal neutron capture.

[h]   Calculated from ^{13}C (P,γ) ^{14}N resonance energy (1747.6 ± 0.9 keV) and 1964 masses: value given for observation at 0° to beam direction.

From Marion, J. B., in *American Institute of Physics Handbook,* 3rd ed., Gray, D. E., Ed., McGraw-Hill, New York, 1972. With permission.

TABLE 3.7-10

LOW- and HIGH-ENERGY RANGE OF TABLES FROM SLATER[a] SHOWING EXTENSIVE NUMBERS OF GAMMA RAYS OF RADIONUCLIDES LISTED IN ORDER OF INCREASING PHOTON ENERGY

γ-Energy (MeV)	Nuclide	Half-life	Some modes of formation	Percent abundance of relevant stable isotope	Thermal neutron activation cross section (barns) or fission yield	Percent abundance of γ-radiation	Radioactive parent or daughter of
0.000075	^{235m}U	26.5 min	^{238}Pu (n,γ)	b	403 ± 10		D^{239}Pu
0.003(?)	^{239}Pu	24360 years	^{238}U (α,3n)	99.276^c	10 ± 3		D^{239}Np P^{235m}U
0.0039	^{181}Hf	44.6 days	^{180}Hf (n,γ) ^{181}Ta (n,p)	35.22 99.9877			
0.0051	^{171}Er	7.8 hr	^{170}Er (n,γ)	14.88	9 ± 2		P^{171}Tm
0.0075	^{85m}Sr	70 min	^{84}Sr (n,γ) ^{85}Rb (p,n) ^{85}Rb (d,2n)	0.56 72.15 72.15	< 1		
0.0084	^{169}Yb	31.8 days	^{168}Yb (n,γ) ^{169}Tm (d,2n)	0.140 100	11,000 ± 3,000^a		D^{169}Lu
~0.0084	^{169}Er	9.4 days	^{168}Er (n,γ) ^{170}Er (γ,n)	27.07 14.88	2.0 ± 0.4		
0.0093	^{83m}Kr	114 min	^{82}Kr (n,γ) ^{82}Kr (d,p)	11.56 11.56	45 ± 15		D^{83}Br D^{83}Rb
0.0093	^{83}Rb	83 days	^{80}Se (α,n) ^{81}Br (α,2n)	49.82 49.48			P^{83m}Kr D^{83}Sr
0.010	^{137}Ce	8.7 hr	^{136}Ce (n,γ) ^{139}La (d,4n) ^{139}La (p,3n)	0.193 99.911 99.911	6.3 ± 1.5		D^{137}Pr
0.0105	^{134m}Cs	3.2 hr	^{133}Cs (n,γ) ^{133}Cs (d,p) ^{134}Xe (d,2n) ^{134}Xe (p,n)	100 100 10.44 10.44	17 ± 4mb		
0.012	^{124m_1}Sb	1.3 min	^{123}Sb (n,γ)	42.75	30 ± 15mb		
0.012	^{81}Kr	2.1 × 10^5 years					
0.0122(?)	^{243}Pu	4.98 hr	^{242}Pu (n,γ)	b	19 ± 1		P^{243}Am
0.0124	^{171}Er	7.8 hr	^{170}Er (n,γ)	14.88	9 ± 2		P^{171}Tm
0.0125	^{239}Pu	24360 years	^{238}Pu (n,γ) ^{238}U (α,3n)	b 99.276^c	403 ± 10		P^{235m}U D^{239}Np
0.0127	^{193m}Pt	4.3 days	^{192}Pt (n,γ) ^{192}Pt (d,p)	0.78 0.78	90 ± 40		D^{193}Au D^{193m}Au (0.03%)

Table 3.7-10 (continued)

LOW- AND HIGH-ENERGY RANGE OF TABLES FROM SLATER[44] SHOWING EXTENSIVE NUMBERS OF GAMMA RAYS OF RADIONUCLIDES LISTED IN ORDER OF INCREASING PHOTON ENERGY

γ-Energy (MeV)	Nuclide	Half-life	Some modes of formation	Percent abundance of relevant stable isotope	Thermal neutron activation cross section (barns) or fission yield	Percent abundance of γ-radiation	Radioactive parent or daughter of
0.020(?)	^{237}Np	2.20 × 10^6 years	^{182}W (p,n) ^{182}W (d,2n)	26.4 26.4			D ^{237}U P ^{233}Pa
0.0207	^{169}Yb	31.8 days	^{168}Yb (n,γ) ^{169}Tm (d,2n)	0.140 100	11,000 ± 3,000[a]		D ^{169}Lu
0.021	^{151}Sm	~93 years	^{235}U (n,f)		0.445%		
0.0210	^{155}Tb	5.6 days					
0.0216	^{151}Gd	150 days	^{151}Eu (d,2n)	47.77			
0.023(?)	^{77}Br	57 hr	^{76}Se (d,n)	9.02			P ^{77m}Se
0.0230	^{71}As	62 hr	^{75}As (α,2n) ^{74}Se (α,p) ^{70}Ge (d,n) ^{69}Ga (α,2n)	0.87 100 20.55 60.2			P ^{71}Ge
0.02383	^{119}Sb	38.0 hr	^{118}Sn (d,n) ^{119}Sn (p,n)	24.01 8.58			D ^{119}Te
0.0242	^{77}Kr	1.1 hr	^{74}Se (α,n)	0.87			
0.0242	^{119m}Sn	250 days	^{118}Sn (n,γ)	24.01	10 ± 6mb		
0.025	^{75}Se	121 days	^{74}Se (n,γ) ^{75}As (d,2n) ^{75}As (p,n)	0.87 100 100	26 ± 6		
0.025	^{58m}Co	9.2 hr	^{58}Fe (p,n) ^{59}Co (γ,n) ^{58}Ni (n,p)	0.31 100 67.76			
0.0255	^{105}Cd	55 min	^{102}Pd (α,n) ^{107}Ag (p,3n) ^{106}Cd (n,2n)	0.96 51.35 1.22			
0.0255	^{187}Ir	13 hr	^{185}Re (α,2n) ^{188}Os (d,3n)	37.07 13.3			
0.0256	^{231}Th (UY)	25.64 hr.	^{230}Th (n,γ)	c	35 × 10[a]	13	D ^{235}U (AcU) P ^{231}Pa

TABLE A.xx

LOW- and HIGH-ENERGY RANGE OF TABLES FROM SLATER[44] SHOWING EXTENSIVE NUMBERS OF GAMMA RAYS OF RADIONUCLIDES LISTED IN ORDER OF INCREASING PHOTON ENERGY

γ-Energy (MeV)	Nuclide	Half-life	Some modes of formation	Percent abundance of relevant stable isotope	Thermal neutron activation cross section (barns) or fission yield	Percent abundance of γ-radiation	Radioactive parent or daughter of
0.0127	^{193}Au	15.8 hr	^{194}Pt (n,2n)	32.9			D^{193}Hg
			^{193}Ir (d,2n)	61.5			D^{193m}Hg
			^{191}Ir (α,2n)	38.5			P^{193}Pt
			^{192}Pt (d,n)	0.78			
			^{194}Pt (d,3n)	32.9			
0.0135	^{73}As	76 days	^{70}Ge (α,p)	20.55			P^{73m}Ge
0.0135	^{73}Ga	5.0 hr	^{76}Se (p,α)	9.02			
			^{73}Ge (n,p)	7.67			
0.0135	^{73m}Ge	0.53 sec	^{74}Ge (γ,p)	36.74			
			^{235}U (n,f)		1.0×10^{-4}%		
0.01437	^{57}Co	270 days	^{56}Fe (d,n)	91.68			D^{73}As
			^{56}Fe (p,γ)	91.68			D^{57}Ni
0.016	^{233}Pa	27.0 days	^{232}Pa (n,γ)	b	760 ± 100[a]		D^{233}Th
			^{237}Th (d,n)	100[c]			P^{233}U
			^{232}Th (α,p2n)	100[c]			D^{237}Np
0.0172	^{231}Th(UY)	25.64 hr	^{230}Th (n,γ)	c	35 ± 10[c]		D^{235}U(AcU)
							P^{231}Pa
0.0181	^{231}Th(UY)	25.64 hr	^{230}Th (n,γ)	c	35 ± 10[a]		D^{235}U(AcU)
							P^{231}Pa
0.0181	^{231}U	4.3 days	^{231}Pa (d,2n)	c			P^{227}Th
			^{231}Pa (α,p3n)	c			P^{231}Pa
			^{232}Th (α,5n)	100[c]			
0.0185	^{112}Pd	21 hr	^{235}U (n,f)		8.3×10^{-3}%		P^{112}Ag
0.0185	^{124m}Sb	21 min	^{123}Sb (n,γ)	42.75	30 ± 15 mb		
0.0188	^{155}Tb	5.6 days					
0.0189(?)	^{231}Pa	3.43×10^4 years	^{232}Th (d,3n)	100[c]			D^{231}Th(UY)
0.0189	^{155}Eu	1.7 years	^{155}Eu (n,γ)	b	1500 ± 400[a]		P^{227}Ac
			^{154}Sm (d,n)	22.53			D^{231}U
			^{235}U (n,f)		~0.03%		
0.0199	^{182}Re	64.0 hr	^{181}Ta (α,3n)	99.9877			D^{153}Sm

Table 3.7-10 (continued)
LOW- AND HIGH-ENERGY RANGE OF TABLES FROM SLATER[44] SHOWING EXTENSIVE NUMBERS OF GAMMA RAYS OF RADIONUCLIDES LISTED IN ORDER OF INCREASING PHOTON ENERGY

γ-Energy (MeV)	Nuclide	Half-life	Some modes of formation	Percent abundance of relevant stable isotope	Thermal neutron activation cross section (barns) or fission yield	Percent abundance of γ-radiation	Radioactive parent or daughter of
0.0256	²³¹U	4.3 days	²³¹Pa (d,2n)	c			P²²⁷Th
			²³¹Pa (α,p3n)	c			P²³¹Pa
			²³²Th (α,5n)	100ᶜ			
0.0256	¹⁶¹Tb	6.88 days	¹⁶⁰Tb (n,γ)	b	525 ± 100ᵃ		D¹⁶¹Gd
			¹⁶⁰Gd (d,n)	21.9			
			¹⁶⁰Dy (γ,p) (?)	25.53			
0.0256	¹⁶¹Ho	2.5 hr	¹⁵⁹Tb (α,2n)	100			D¹⁶¹Er
			¹⁶¹Dy (p,n)	18.88			
			¹⁶⁰Dy (d,n)	2.294			
0.026	²³⁵Np	410 days	²³⁵U (d,2n)	0.720ᶜ			
			²³³U (α,pn)	b			
0.0264	²⁴¹Am	458 years	²³⁶U (n,γ)	b			D²⁴¹Pu
0.0264	²³⁷U	6.75 days	²³⁸U (n,2n)	99.276ᶜ	6 ± 1		P²³⁷Np
							D²⁴¹Pu
0.0265	¹⁵⁵Eu	1.7 years	¹⁵⁴Eu (n,γ)	b	1500 ± 400ᵃ		D¹⁵⁵Sm
			¹⁵⁴Sm (d,n)	22.53			
			²³⁵U (n,f)		~0.03%		
0.0268	¹²⁹Te	33.5 days	¹²⁸Te (n,γ)	31.79	15 ± 5mb		P¹²⁹Te
			¹²⁸Te (d,p)	31.79			
			¹³⁰Te (n,2n)	34.49			
²³⁵U (n,f)		0.19%					
0.0268	¹²⁹ᵐTe	72 min	¹²⁸Te (n,γ)	31.79	0.13 ± 0.03		D¹²⁹ᵐTe
			¹²⁸Te (d,p)	31.79			D¹²⁹Sb
			¹³⁰Te (n,2n)	34.49			
			¹³⁰Te (γ,n)	34.49			
0.027	²⁴⁶Pu	10.85 days	²⁴⁵Pu (n,γ)	b	260 ± 150ᵃ		P²⁴⁶Am
0.0275	²²⁷Ra	41.2 min	²²⁶Ra (n,γ)	c	20 ± 3ᵃ		P²²⁷Ac
							D²³¹Th
0.0275	²³¹Pa	3.43 × 10⁴ years	²³²Th (d,3n)	100ᶜ			(UY)
							P²²⁷Ac
							D²³¹U
0.0276 (?)	¹⁸²Os	21.9 hr	¹⁸⁵Re (p,4n)	37.07			P(13 hr) ¹⁸²Re

Table 5. A16 (continued)

LOW- AND HIGH-ENERGY RANGE OF TABLES FROM SLATER[44] SHOWING EXTENSIVE NUMBERS OF GAMMA RAYS OF RADIONUCLIDES LISTED IN ORDER OF INCREASING PHOTON ENERGY

γ-Energy (MeV)	Nuclide	Half-life	Some modes of formation	Percent abundance of relevant stable isotope	Thermal neutron activation cross section (barns) or fission yield	Percent abundance of γ-radiation	Radioactive parent or daughter of
0.0277	^{105}Cd	55 min	^{102}Pd (α,n)	0.96			
			^{107}Ag (p,3n)	51.35			
			^{106}Cd (n,2n)	1.22			
0.0277	^{161}Tb	6.88 days	^{160}Tb (n,γ)	b	525 ± 100^a		D^{161}Gd
			^{160}Gd (d,n)	21.9			
			^{162}Dy (γ,p)(?)	25.53			
0.028	^{76}Kr	9.7 hr					P^{76}Br
0.028	^{88}Kr	2.77 hr					P^{88}Rb
			^{87}Kr (n,γ)		< 600		D^{88}Br
0.028	^{233}Pa	27.0 days	^{232}Pa (n,γ)	b	760 ± 100^a		D^{233}Th
			^{232}Th (d,n)	100^c			P^{233}U
			^{232}Th (α,p2n)	100^c			D^{237}Np
0.0286	^{189}Hg	23 min					P^{189}Au
0.0290	^{234}Th (UX)	24.10 days	^{233}Th (n,γ)	b	1,400 ± 200^a		D^{238}U
							P^{234m}Pa (UX$_2$)
							P^{234}Pa (UZ) (0.63%)
0.0292	^{233}Th	22.12 min	^{232}Th (n,γ)	100^c	7.33 ± 0.12	2.1	P^{233}Pa
			^{232}Th (d,p)	100^c			
0.0296	^{140}Ba	12.80 days	^{139}Ba (n,γ)	b	4 ± 1		P^{140}La
			^{235}U (n,f)		6.17%		Descendant ^{140}Xe
0.0297	^{237}Np	2.20 × 10^6 years				14	D^{237}U
							P^{233}Pa
0.0297(?)	^{243}Pu	4.98 hr	^{242}Pu (n,γ)	b	19 ± 1		P^{243}Am
0.0299	^{231}Pa	3.43 × 10^4 years	^{232}Th (d,3n)	100^b			D^{231}Th (UY)
							P^{227}Ac
							D^{231}U
0.030	^{179}W	30 min	^{181}Ta (p,3n)	99.9877	3.7 ± 1.2		
0.030	^{149}Nd	2.0 hr	^{148}Nd (n,γ)	5.72			P^{149}Pm
			^{148}Nd (d,p)	5.72			
			^{150}Nd (n,2n)	5.60			

Table 3.7-10 (continued)

LOW- AND HIGH-ENERGY RANGE OF TABLES FROM SLATER[44] SHOWING EXTENSIVE NUMBERS OF
GAMMA RAYS OF RADIONUCLIDES LISTED IN ORDER OF INCREASING PHOTON ENERGY

γ-Energy (MeV)	Nuclide	Half-life	Some modes of formation	Percent abundance of relevant stable isotope	Thermal neutron activation cross section (barns) or fission yield	Percent abundance of γ-radiation	Radioactive parent or daughter of
0.0300	^{227}Th (Rd Ac)	18.17 days					D^{227}Ac
							P^{223}Ra (AcX)
							D^{227}Pa
							D^{231}U
0.0304	^{93}Zr	1.1×10^6 years					P^{93m}Nb
0.0304	^{93m}Nb	12 years					D^{93}Zr
3.09	^{37}S	5.04 min	^{36}S (n,γ)	0.017	0.14 ± 0.04		
			^{37}Cl (n,p)	24.47			
3.10	^{49}Ca	8.8 min	^{48}Ca (n,γ)	0.185	1.1 ± 0.1		
			^{48}Ca (d,p)	0.185			
3.13	^{60}Cu	23.4 min	^{60}Ni (p,n)	26.16		3.7	D^{60}Zn
			^{60}Ni (d,2n)	26.16			
			^{58}Ni (α,pn)	67.76			
3.18	^{136}I	86 sec	^{235}U (n,f)		3.1%		
3.22	^{34m}Cl	32.40 min	^{31}P (α,n)	100			P^{34}Cl
			^{34}S (p,n)	4.215			
			^{33}S (d,n)	0.750			
3.24	^{88}Rb	17.8 min	^{87}Rb (n,γ)	27.85	0.12 ± 0.03		D^{88}Kr
			^{87}Rb (d,p)	27.85			
			^{88}Sr (n,p)	82.56			
3.24	^{66}Ga	9.45 hr	^{63}Cu (α,n)	69.1			D^{66}Ge
			^{66}Zn (d,2n)	27.81			
			^{66}Zn (p,n)	27.81			
3.25	^{64}Ga	2.6 min	^{64}Zn (p,n)	48.89			
			^{64}Zn (d,2n)	48.89			
3.25	^{56}Co	77.3 days	^{56}Fe (d,2n)	91.68			D^{56}Ni
			^{58}Ni (γ,pn)	67.76			
			^{54}Fe (α,np)	5.84			
3.27	^{94}Tc	53 min	^{94}Mo (p,n)	9.12			D^{94}Ru
			^{94}Mo (d,2n)	9.12			

Table 3.7.18 (continued)

LOW- AND HIGH-ENERGY RANGE OF TABLES FROM SLATER[44] SHOWING EXTENSIVE NUMBERS OF GAMMA RAYS OF RADIONUCLIDES LISTED IN ORDER OF INCREASING PHOTON ENERGY

γ-Energy (MeV)	Nuclide	Half-life	Some modes of formation	Percent abundance of relevant stable isotope	Thermal neutron activation cross section (barns) or fission yield	Percent abundance of γ-radiation	Radioactive parent or daughter of
3.28	^{84}Br	31.8 min	^{87}Rb (n,α)	27.85			
3.34	^{138}Cs	32.2 min	^{138}Ba (n,p)	71.66		0.5	D^{138}Xe
			^{137}Cs (n,γ)	b	<2		Descendant ^{138}I
3.350	^{72}Ga	14.3 hr	^{71}Ga (n,γ)	39.8	4.0 ± 0.7		D^{72}Zn
			^{71}Ga (d,p)	39.8			
			^{72}Ge (n,p)	27.37			
^{72}Zn β⁻							
3.4	^{140}La	77m	^{63}Cu (α,n)	69.1			D^{142}Ba
3.41	^{66}Ga	9.45 hr	^{66}Zn (d,2n)	27.81			D^{66}Ge
			^{66}Zn (p,n)	27.81			
3.47	^{56}Co	77.3 days	^{55}Fe (d,2n)	91.68			D^{56}Ni
			^{58}Ni (γ,pn)	67.76			
			^{54}Fe (γ,np)	5.84			
3.5	^{38}Ca	0.66 sec		b			
3.52	^{89}Rb	15.4 min	^{88}Rb (n,γ)	b	1.0 ± 0.2		D^{89}Kr P^{89}Sr
3.52	^{88}Rb	17.8 min	^{87}Rb (n,γ)	27.85	0.12 ± 0.03		D^{88}Kr
			^{87}Rb (d,p)	27.85			
			^{88}Sr (n,p)	82.56			
3.52	^{60}Cu	23.4 min	^{60}Ni (p,n)	26.16		2.0	D^{60}Zn
			^{60}Ni (d,2n)	26.16			
			^{58}Ni (α,pn)	67.76			
3.6(?)	^{44}K	22.0 min	^{44}Ca (n,p)	2.06			
3.68	^{88}Rb	17.8 min	^{87}Rb (n,γ)	27.85	0.12 ± 0.03		D^{88}Kr
			^{87}Rb (d,p)	27.85			
			^{88}Sr (n,p)	82.56			
3.74	^{72}As	26 hr	^{69}Ga (α,n)	60.2			D^{72}Se
			^{72}Ge (p,n)	27.37			
			^{72}Ge (d,2n)	27.37			
			^{74}Se (d,α)	0.87			
3.75	^{40}Sc	0.22 sec					
3.78	^{66}Ga	9.45 hr	^{63}Cu (α,n)	69.1			D^{66}Ge

Table 3.7-10 (continued)
LOW- and HIGH-ENERGY RANGE OF TABLES FROM SLATER[a] SHOWING EXTENSIVE NUMBERS OF GAMMA RAYS OF RADIONUCLIDES LISTED IN ORDER OF INCREASING PHOTON ENERGY

γ-Energy (MeV)	Nuclide	Half-life	Some modes of formation	Percent abundance of relevant stable isotope	Thermal neutron activation cross section (barns) or fission yield	Percent abundance of γ-radiation	Radioactive parent or daughter of
3.79(?)	^{32}Cl	0.306 sec	^{66}Zn (d,2n)	27.81		~10	
			^{66}Zn (p,n)	27.81			
3.93	^{84}Br	31.8 min	^{87}Rb (n,α)	27.85			
4.0	^{34}P	12.4 sec	^{34}S (n,p)	4.215		0.2	
			^{37}Cl (n,α)	24.47			
4.0	^{60}Cu	23.4 min	^{60}Ni (p,n)	26.16		1.1	D^{60}Zn
			^{60}Ni (d,2n)	26.16			
			^{58}Ni (α,pn)	67.76			
4.05	^{40}Ca	8.8 min	^{48}Ca (n,γ)	0.185	1.1 ± 0.1		
			^{48}CA (d,p)	0.185			
4.12	^{66}Ga	9.45 hr	^{63}Cu (α,n)	69.1			D^{66}Ge
			^{66}Zn (d,2n)	27.81			
			^{66}Zn (p,n)	27.81			
4.2	^{24}Al	2.1 sec	^{24}Mg (p,n)	78.6			
4.27	^{32}Cl	0.306 sec				7	
4.33	^{66}Ga	9.45 hr	^{63}Cu (α,n)	69.1			D^{66}Ge
			^{66}Zn (d,2n)	27.81			
			^{66}Zn (p,n)	27.81			
4.44	^{28}P	0.28 sec				10	
~4.5	^{12}B	0.019s	^{14}C (d,α)	b			
			^{15}N (n,α)	0.365			
4.68	^{49}Ca	8.8 min	^{48}Ca (n,γ)	0.185	1.1 ± 0.1		
			^{48}Ca (d,p)	0.185			
4.77	^{32}Cl	0.306 sec					
4.83	^{66}Ga	9.45 hr	^{63}Cu (α,n)	69.1		14	D^{66}Ge
			^{66}Zn (d,2n)	27.81			
			^{66}Zn (p,n)	27.81			
4.87	^{88}Rb	17.8 min	^{87}Rb (n,γ)	27.85	0.12 ± 0.03		D^{88}Kr
			^{87}Rb (d,p)	27.85			
			^{88}Sr (n,p)	82.56			
4.93(?)	^{28}P	0.28 sec		b			
5.3	^{15}C	2.25 sec	^{14}C (d,p)				
5.4	^{24}Al	2.1 sec	^{24}Mg (p,n)	78.6			

LOW- AND HIGH-ENERGY RANGE OF TABLES FROM SLATER[a] SHOWING EXTENSIVE NUMBERS OF GAMMA RAYS OF RADIONUCLIDES LISTED IN ORDER OF INCREASING PHOTON ENERGY

γ-Energy (MeV)	Nuclide	Half-life	Some modes of formation	Percent abundance of relevant stable isotope	Thermal neutron activation cross section (barns) or fission yield	Percent abundance of γ-radiation	Radioactive parent or daughter of
6.0	^{40}Cl	1.4 min	^{76}Se (d,n)	9.02			$P^{86}Kr$ (2% of disintegrations)
6.1	^{28}P	0.28 sec					
6.13	^{16}N	7.35 sec	^{15}N (n,γ)	0.365	$24 \pm 8\mu b$		
			^{15}N (d,p)	0.365			
			^{16}O (n,p)	99.759			
6.7	^{28}P	0.28 sec					
7.0	^{28}P	0.28 sec					
7.10	^{16}N	7.35 sec	^{15}N (n,γ)	0.365	$24 \pm 8\ \mu b$		
			^{15}N (d,p)	0.365			
			^{16}O (n,p)	99.759			
7.1	^{24}Al	2.1 sec	^{24}Mg (p,n)	78.6		0.5	
7.6	^{28}P	0.28 sec					
8.88	^{16}N	7.35 sec	^{15}N (n,γ)	0.365	$24 \pm 8\mu b$		
			^{15}N (d,p)	0.365			
			^{16}O (n,p)	99.759			

[a] Pile neutrons.
[b] Artificially active.
[c] Naturally active.

From Slater, D. N., *Gamma-rays of Radionuclides in Order of Increasing Energy*, Butterworths, Washington, 1962. With permission.

TABLE 3.7-11
CHARACTERISTICS OF PROMPT GAMMA RADIATIONS EMITTED IN FISSION

Fissioning nuclide	Average number of photons per fission	Average photon energy released per fission	Ref.
^{235}U + nth	7.5	7.46	1
^{235}U + nth	7.2 ± 0.8	7.4 ± 0.8	2
^{235}U + nth	7.93 ± 0.48	9.51 ± 0.23	3
^{252}Cf	10	9	4
^{252}Cf	10.3	8.2	5

From Gindler, J. and Huizenga, J. R., *Nuclear Chemistry,* Vol. 2, Yaffe, L., Ed., Academic Press, New York, 1968, as given in Loveland, W. P., in *American Institute of Physics Handbook,* 3rd ed., Gray, D. E., Ed., McGraw-Hill, New York, 1972. With permission.

REFERENCES

1. **Francis, J. and Gamble, R.,** Oak Ridge National Laboratory Report, ORNL-1879, unpublished.
2. **Maienshein, F. C., Peele, R. W., Zohel, W., and Love, T. A.,** *Proc. U.N. Int. Conf. Peaceful Uses of Atomic Energy, Geneva,* 0.3 ⩽ Er ⩽ 10 MeV, 15, 366, 1958.
3. **Rau, F. E. W.,** *Ann. Phys.,* 10, 252, 1963.
4. **Bowman, H. R. and Thompson, S. G.,** *Proc. U.N. Int. Conf. Peaceful Uses of Atomic Energy, Geneva,* 15, 212, 1958.
5. **Smith, A., Fields, P., Friedman, A., Cox, S., and Sjoblom, R.,** *Proc. U.N. Int. Conf. Peaceful Uses of Atomic Energy, Geneva,* 15, 392, 1958.

TABLE 3.7-12
MAJOR CONTRIBUTORS TO PEAKS IN GAMMA SPECTRA (^{235}U THERMAL FISSION)

Energy bin (MeV)	Nuclide(s)
1-Hr Cooling	
1.67—1.75	^{135}I, ^{98}Nb, ^{134}I
1.75—1.80	^{135}I, ^{142}La, ^{97}Zr
1.80—1.85	^{88}Rb, ^{134}I
1.90—1.95	^{142}La, ^{93}Y, ^{132}I, ^{135}I
2.00—2.05	^{88}Kr, ^{142}La
2.15—2.20	^{88}Kr, ^{142}La, ^{89}Rb
2.20—2.25	^{138}Cs, ^{88}Kr
2.35—2.40	^{88}Kr, ^{142}La
2.50—2.55	^{142}La, ^{140}La
10-Hr Cooling	
1.67—1.75	^{135}I
1.75—1.80	^{135}I, ^{97}Zr
1.80—1.85	^{88}Rb, ^{135}I
1.90—1.95	^{132}I, ^{93}Y
2.00—2.05	^{132}I, ^{88}Kr, ^{135}I, ^{131m}Te
2.15—2.20	^{88}Kr, ^{93}Y, ^{132}I
2.35—2.40	^{88}Kr, ^{142}La
2.50—2.55	^{140}La

Energy bin (MeV)	Nuclide(s)
100-Hr Cooling	
1.90—1.95	^{132}I
2.00—2.05	^{132}I
2.30—2.35	^{140}La
2.50—2.55	^{140}La
1000-Hr Cooling	
2.15—2.20	^{144}Pt
2.30—2.35	^{140}La
2.50—2.55	^{140}La

From Stamatelatos and England, T. R., Fission-product Gamma-ray and Photoneutron Spectra and Energy-integrated Sources Informal Report NUREG-0155, U.S. Nuclear Regulatory Commission, Los Alamos Scientific Laboratory, University of California, Los Alamos, N.M., December, 1976.

Table 3.7.13

X-RAY CRITICAL-ABSORPTION AND EMISSION ENERGIES IN keV

Atomic number	Element	K series K_{ab}	K_{β_2}	K_{β_1}	K_{α_1}	K_{α_2}	L series L_{Iab}	L_{IIab}	L_{IIIab}	L_{γ_1}	L_{β_2}	L_{β_1}	L_{α_1}	L_{α_2}
1	Hydrogen	0.0136[c]												
2	Helium	0.0246[c]												
3	Lithium	0.055			0.052									
4	Beryllium	0.116[a]			0.110									
5	Boron	0.192[b]			0.185									
6	Carbon	0.283			0.282									
7	Nitrogen	0.399			0.392									
8	Oxygen	0.531			0.523									
9	Fluorine	0.687[b]			0.677									
10	Neon	0.874[a]			0.851[a]		0.048[b]	0.022[b]	0.022[b]					
11	Sodium	1.08[a]		1.067	1.041		0.055[c]	0.034[a]	0.034[a]					
12	Magnesium	1.303		1.297	1.254		0.063	0.050	0.049					
13	Aluminum	1.559		1.553	1.487	1.486	0.087	0.073[f]	0.072[f]					
14	Silicon	1.838		1.832	1.740	1.739	0.118[a]	0.099[f]	0.098[f]					
15	Phosphorus	2.142		2.136	2.015[a]	2.014[f]	0.153[a]	0.129[a]	0.128[a]					
16	Sulphur	2.470		2.464	2.308	2.306	0.193[a]	0.164[f]	0.163[f]					
17	Chlorine	2.819[d]		2.815	2.622	2.621	0.238[a]	0.203[a]	0.202[a]					
18	Argon	3.203		3.192[a]	2.957	2.955	0.287[a]	0.247[f]	0.245[f]					
19	Potassium	3.607		3.589	3.313	3.310	0.341[a]	0.297[f]	0.294[f]					
20	Calcium	4.038		4.012	3.691	3.688	0.399[a]	0.352	0.349			0.344	0.341	
21	Scandium	4.496		4.460	4.090	4.085	0.462[a]	0.411[f]	0.406[f]			0.399	0.395	
22	Titanium	4.964		−4.931	4.510	4.504	0.530[a]	0.460[f]	0.454[f]			0.458	0.452	
23	Vanadium	5.463		−5.427	4.952	4.944	0.604[a]	0.519[f]	0.512[f]			0.519	0.510	
24	Chromium	5.988		−5.946	5.414	5.405	0.679[a]	0.583[f]	0.574[f]			0.581	0.571	
25	Manganese	6.537		6.490	5.898	5.887	0.762[a]	0.650[f]	0.639[f]			0.647	0.636	
26	Iron	7.111		7.057	6.403	6.390	0.849[a]	0.721[f]	0.708[f]			0.717	0.704	
27	Cobalt	7.709		7.649	6.930	6.915	0.929[a]	0.794[f]	0.779[f]			0.790	0.775	
28	Nickel	8.331	8.328	8.264	7.477	7.460	1.015[a]	0.871[f]	0.853[f]			0.866	0.849	
29	Copper	8.980	8.976	8.904	8.047	8.027	1.100[a]	0.953	0.933			0.948	0.928	
30	Zinc	9.660	9.657	9.571	8.638	8.615	1.200[a]	1.045	1.022			1.032	1.009	

Table 3.7-13 (continued)
X-RAY CRITICAL-ABSORPTION AND EMISSION ENERGIES IN keV

Atomic number	Element	K series					L series							
		K_{ab}	$K\beta_2$	$K\beta_1$	$K\alpha_1$	$K\alpha_2$	L_{Iab}	L_{IIab}	L_{IIIab}	$L\gamma_1$	$L\beta_2$	$L\beta_1$	$L\alpha_1$	$L\alpha_2$
31	Gallium	10.368	10.365	10.263	9.251	9.234	1.30ᵃ	1.134ᶠ	1.117ᶠ			1.122	1.096	
32	Germanium	11.103	11.100	10.981	9.885	9.854	1.42ᵃ	1.248ᶠ	1.217ᶠ			1.216	1.186	
33	Arsenic	11.863	11.863	11.725	10.543	10.507	1.529	1.359	1.323			1.317	1.282	
34	Selenium	12.652	12.651	12.495	11.221	11.181	1.652	1.473	1.434			1.419	1.379	
35	Bromine	13.475	13.465	13.290	11.923	11.877	1.794ᵃ	1.599ᶠ	1.552ᶠ			1.526	1.480	
36	Krypton	14.323	14.313	14.112	12.648	12.597	1.931ᵃ	1.727ᶠ	1.675ᶠ			1.638ᶠ	1.587ᶠ	
37	Rubidium	15.201	15.184	14.960	13.394	13.335	2.067	1.866	1.806			1.752	1.694	1.692
38	Strontium	16.106	16.083	15.834	14.164	14.097	2.221	2.008	1.941			1.872	1.806	1.805
39	Yttrium	17.037	17.011	16.736	14.957	14.882	2.369	2.154	2.079			1.996	1.922	1.920
40	Zirconium	17.998	17.969	17.666	15.774	15.690	2.547	2.305	2.220	2.302	2.219	2.124	2.042	2.040
41	Niobium	18.987	18.951	18.621	16.614	16.520	2.706	2.467ᶠ	2.374	2.462	2.367	2.257	2.166	2.163
42	Molybdenum	20.002	19.964	19.607	17.478	17.373	2.884	2.627	2.523	2.623	2.518	2.395	2.293	2.290
43	Technetium	21.054ᵃ	21.012ᵃ	20.585ᵈ	18.410ᵈ	18.328ᵈ	3.054ᵃ	2.795ᵃ	2.677ᵃ	2.792ᵃ	2.674ᵃ	2.538ᵃ	2.424ᵃ	2.420ᵃ
44	Ruthenium	22.118	22.072	21.655	19.278	19.149	3.236ᵃ	2.966	2.837	2.964	2.836	2.683	2.558	2.554
45	Rhodium	23.224	23.169	22.721	20.214	20.072	3.419	3.145	3.002	3.144	3.001	2.834	2.696	2.692
46	Palladium	24.347	24.297	23.816	21.175	21.018	3.617	3.329	3.172	3.328	3.172	2.990	2.838	2.833
47	Silver	25.517	25.454	24.942	22.162	21.988	3.810	3.528	3.352	3.519	3.348	3.151	2.984	2.978
48	Cadmium	26.712	26.641	26.093	23.172	22.982	4.019	3.727	3.538	3.716	3.528	3.316	3.133	3.127
49	Indium	27.928	27.859	27.274	24.207	24.000	4.237	3.939	3.729	3.920	3.713	3.487	3.287	3.279
50	Tin	29.190	29.106	28.483	25.270	25.042	4.464	4.157	3.928	4.131	3.904	3.662	3.444	3.435
51	Antimony	30.486	30.387	29.723	26.357	26.109	4.697	4.381	4.132	4.347	4.100	3.843	3.605	3.595
52	Tellurium	31.809	31.698	30.993	27.471	27.200	4.938	4.613	4.341	4.570	4.301	4.029	3.769	3.758
53	Iodine	33.164	33.016	32.292	28.610	28.315	5.190	4.856	4.559	4.800	4.507	4.220	3.937	3.926
54	Xenon	34.579	34.446ᵈ	33.644	29.802ᵃ	29.485ᵈ	5.452	5.104	4.782	5.036ᵃ	4.720ᵃ	4.422ᵃ	4.111ᵃ	4.098ᵃ
55	Cesium	35.959	35.819	34.984	30.970	30.623	5.720	5.358	5.011	5.280	4.936	4.620	4.286	4.272
56	Barium	37.410	37.255	36.376	32.191	31.815	5.995	5.623	5.247	5.531	5.156	4.828	4.467	4.451
57	Lanthanum	38.931	38.728	37.799	33.440	33.033	6.283	53894	53489	5.789	5.384	5.043	4.651	4.635
58	Cerium	40.449	40.231	39.255	34.717	34.276	6.561	6.165ᵇ	5.729	6.052	5.613	5.262	4.840	4.823
59	Praseodymium	41.998	41.772	40.746	36.023	35.548	6.846	6.443	5.968	6.322	5.850	5.489	5.034	5.014
60	Neodymium	43.571	43.298ᵈ	42.269	37.359	36.845	7.144	6.727	6.215	6.602	6.090	5.722	5.230	5.208

X-RAY CRITICAL-ABSORPTION AND EMISSION ENERGIES IN keV

Atomic number	Element	K series					L series							
		K_{ab}	$K\beta_2$	$K\beta_1$	$K\alpha_1$	$K\alpha_2$	L_{Iab}	L_{IIab}	L_{IIIab}	$L\gamma_1$	$L\beta_2$	$L\beta_1$	$L\alpha_1$	$L\alpha_2$
61	Promethium	45.207ᵈ	44.955ᵉ	43.945ᵈ	38.649ᵈ	38.160ᵉ	7.448ᵈ	7.018ᵉ	6.460ᵉ	6.891ᵉ	6.336ᵉ	5.956	5.431	5.408ᵉ
62	Samarium	46.846	46.553ᵈ	45.400	40.124	39.523	7.754	7.281ᵉ	6.721	7.180	6.587	6.206	5.636	5.609
63	Europium	48.515	48.241	47.027	41.529	40.877	8.069	7.624	6.983	7.478	6.842	6.456	5.846	5.816
64	Gadolinium	50.229	49.961	48.718	42.983	42.280	8.393	7.940	7.252	7.788	7.102	6.714	6.059	6.027
65	Terbium	51.998	51.737	50.391	44.470	43.737	8.724	8.258	7.519	8.104	7.368	63979	6.275	6.241
66	Dysprosiun	53.789	53.491	52.178	45.985	45.193	9.083	8.621ᵈ	7.850ᵈ	8.418	7.638	7.249	6.495	6.457
67	Holmium	55.615	55.292ᶠ	53.934ᵉ	47.528	46.686	9.411	8.920	8.074	8.748	7.912	7.528	6.720	6.680
68	Erbium	57.483	57.088	55.690	49.099	48.205	9.776	9.263	8.364	9.089	8.188	7.810	6.918	6.004
69	Thulium	59.005ᵈ	58.909ᶠ	57.576ᵈ	50.730	49.762	10.144	9.628	8.652	9.424	8.472	8.103	7.181	7.135
70	Ytterbium	61.303	60.959	59.352	52.360	51.326	10.486	9.977	8.943	9.779	8.758	8.401	7.414	7.367
71	Lutecium	63.304	62.946	61.282	54.063	52.959	10.867	10.345	9.241	10.142	9.048	8.708	7.654	7.604
72	Hafnium	65.313	64.936	63.209	55.757	54.579	11.264	10.734	9.556	10.514	9.346	9.021	7.898	7.843
73	Tantalum	67.400	66.999	65.210	57.524	56.270	11.676	11.130	9.876	10.892	9.649	9.341	8.145	8.087
74	Tungsten	69.508	69.090	67.233	59.310	57.973	12.090	11.535	10.198	11.283	9.959	9.670	8.396	8.333
75	Rhenium	71.622	71.220	69.298	61.131	59.707	12.522	11.955	10.531	11.684	10.273	10.008	8.651	8.584
76	Osmium	73.860	73.393	71.404	62.991	61.477	12.965	12.383	10.869	12.094	10.596	10.354	8.910	8.840
77	Iridium	76.097	75.605	73.549	64.886	63.278	13.413	12.819	11.211	12.509	10.918	10.706	9.173	9.098
78	Platinum	78.379	77.866	75.736	66.820	65.111	13.873	13.268	11.559	12.939	11.249	11.069	9.441	9.360
79	Gold	80.713	80.165	77.968	68.794	66.980	14.353	13.733	11.919	13.379	11.582	11.439	9.711	9.625
80	Mercury	83.106	82.526	80.258	70.821	68.894	14.841	14.212	12.285	13.828	11.923	11.823	9.987	9.896
81	Thallium	85.517	84.904	82.558	72.860	70.820	15.346	14.697	12.657	14.288	12.268	12.210	10.266	10.170
82	Lead	8 8.001	87.343	84.922	74.957	72.794	15.870	15.207	13.044	14.762	12.620	12.611	10.549	10.448
83	Bismuth	90.521	89.833	87.335	77.097	74.805	16.393	15.716	13.424	15.244	12.977	13.021	10.836	10.729
84	Polonium	93.112	92.386	89.809	79.296	76.868	16.935	16.244	13.817	15.740	13.338	13.441	11.128	11.014
85	Astatine	95.740	94.976	92.319	81.525	78.956	17.490	16.784	14.215	16.248	13.705	13.873	11.424	11.304
86	Radon	98.418	97.616	94.877	83.800	81.080	18.058	17.387	14.618	16.768	14.077	14.316	11.724	11.597
87	Francium	101.147	100.305	97.483	86.119	83.243	18.638	17.904	15.028	17.301	14.459	14.770	12.029	11.894
88	Radium	103.927	103.048	100.136	88.485	85.446	19.233	18.481	15.442	17.845	14.839	15.233	12.338	12.194
89	Actinium	106.759	105.838	102.846	90.894	87.681	19.842	19.078	15.865	18.405	15.227	15.712	12.650	12.499
90	Thorium	109.630	108.671	105.592	93.334	89.942	20.460	19.688	16.296	18.977	15.620	16.200	12.966	12.808

Table 3.7-13 (continued)
X-RAY CRITICAL-ABSORPTION AND EMISSION ENERGIES IN keV

Atomic number	Element	K series					L series							
		K_{ab}	$K\beta_2$	$K\beta_1$	$K\alpha_1$	$K\alpha_2$	L_{Iab}	L_{IIab}	L_{IIIab}	$L\gamma_1$	$L\beta_2$	$L\beta_1$	$L\alpha_1$	$L\alpha_2$
91	Protactinium	112.581	111.575	108.408	95.851	92.271	21.102	20.311	16.731	19.559	16.022	16.700	13.291	13.120
92	Uranium	115.591	114.549	111.289	98.428	94.648	21.753	20.943	17.163	20.163	16.425	17.218	13.613	13.438
93	Neptunium	118.619	117.533	114.181	101.005	97.023	22.417	21.596	17.614	20.774	16.837	17.740	13.945	13.758
94	Plutonium	121.720	120.592	117.146	103.653	99.457	23.097	22.262	18.066	21.401	17.254	18.278	14.279	14.082
95	Americium	124.876	123.706	120.163	106.351	101.932	23.793	22.944	18.525	22.042	17.677	18.829	14.618	14.411
96	Curium	128.088	126.875	123.235	109.098	104.448	24.503	23.640	18.990	22.699	18.106	19.393	14.961	14.743
97	Berkelium	131.357	130.101	126.362	111.896	107.023	25.230	24.352	19.461	23.370	18.540	19.971	15.309	15.079
98	Californium	134.683	133.383	129.544	114.745	109.603	25.971	25.080	19.938	24.056	18.980	20.562	15.661	15.420
99		138.067	136.724	132.781	117.646	112.244	26.729	25.824	20.422	24.758	19.426	21.166	16.018	15.764
100		141.510	140.122	136.075	120.598	114.926	27.503	26.584	20.912	25.475	19.879	21.785	16.379	16.133

Note: For $\leqslant 69$, values without symbols are derived from Cauchois and Hulubei; values prefixed with a $-$ sign are $K\beta_{1,+8}$; for $2 \geqslant 70$, absorption-edge values are from Cauchois, in the case of $Z = 70, 80, 83, 88, 90,$ and 92; remaining absorption edges to $Z = 100$ are obtained from these by least-squares quadratic fitting. All emission values for $Z \geqslant 70$ are derived from the preceding absorption edges and others based on Cauchois using the transition relations $K\alpha_1 = K_{ab} - L_{III}$, $K\alpha_2 = K_{ab} - L_{II}$, $K\beta_1 = K_{ab} - M_{III}$, etc.

a Obtained from Hill, Church, and Mihelich.[5]

b Derived from Compton and Allison.[2]

c Derived from Moore.[3]

d Values derived from Cauchois and Hulubei[1] which deviate from the Moseley law. Better-fitting values are: $Z = 17$, $K_{ab} = 2.826$, $Z = 43$, $K\alpha_1 = 18.370$, $K\alpha_2 = 18.250$, $K\beta_1 = 20.612$; $Z = 54$, $K\alpha_1 = 29.779$, $K\alpha_2 = 29.463$, $K\beta_1 = 34.398$; $Z = 60$, $K\beta_1 = 43.349$; $Z = 61$ $K\alpha_1 = 43.811$, $K\beta_2 = 46.581$, $L_{II} = 7.312$; $Z = 66$, $L_{III} = 8.591$, $L_{III} = 7.790$; $Z = 69$, $K_{ab} = 59.382$, $K\beta_1 = 57.487$.

e Calculated by method of least squares.

f Calculated by transition relations.

From Fine, S. and Hendee, C. F., *Nucleonics*, 13, 37, 1955. With permission.

REFERENCES

1. Cauchois, Y. and Hulubei, H., *Tables de Constantes et Donnees Numeriques I. Longueurs D'Onde des Emissions X et des Discontinuites D'Absorption X*, Hermann et Cie, Paris, 1947.
2. Compton, A. H. and Allison, S. K., *X-Rays in Theory and Experiment*, Van Nostrand, New York, 1951.
3. Moore, C. E., Atomic Energy Levels, NBS467, National Bureau of Standards, U. S. Department of Commerce, Washington, D.C. 1949.
4. Couchois, Y., *J. Phys. Radium*, 13, 113, 1952.
5. Hill, R. D., Church, E. L., and Mihelich, J. W., *Rev. Sci. Instrum.*, 23, 523, 1952.

Table 3.7-13 (continued)
X-RAY CRITICAL-ABSORPTION AND EMISSION ENERGIES IN keV

X-RAY WAVELENGTHS

J. A. Bearden

These tables were originally published as the final report to the U.S. Atomic Energy Commission as Report NYO-10586 in partial fulfillment of Contract AT(30-1)-2543. The tables were later reproduced in *Review of Modern Physics*. The data may also be obtained from the Superintendent of Documents, U.S. Government Printing Office, Washington, D. C. 20402 in the publication NSRDS-NBS 14. Persons seeking discussion of the experimental work, conventions, secondary standards, etc. will find these in *Review of Modern Physics*, Vol. 39, No. 1, 78-124, January 1967.

THE W $K\alpha_1$ WAVELENGTH STANDARD

A wavelength standard should possess characteristics which permit its ready redetermination in other laboratories by different techniques. Considering all of the factors involved in the selection of a wavelength standard, the W $K\alpha_1$ line is superior to any other x-ray or γ-ray wavelength. Its advantages as the x-ray wavelength standard are discussed in *Review of Modern Physics* Vol. 39, page 82 (1967).

$$\lambda \text{W } K\alpha_1 = 0.2090100 \text{ Å} \pm 5 \text{ ppm.}$$

This numerical value of the wavelength of the W $K\alpha_1$ line is used to define the *x-ray wavelength standard* by the relation

$$\lambda(\text{W } K\alpha_1) = 0.2090100 \text{ Å}^*.$$

This is a new unit of length which may differ from the angstrom by ± 5 ppm (probable error), *but as a wavelength standard it has no error*. In order to clearly indicate that this unit is not exactly an angstrom, it has been designated Å*.

Wavelengths tabulated normally refer to the pure element in its solid form. However, there are many instances in which such data are not available. For example, rare gases are of necessity almost always used in the gaseous form, while the rare-earth elements were customarily used in the form of salts.

In high precision work there is some ambiguity as to exactly what feature of a line profile should be taken to be the "true wavelength." In double-crystal work the line peak is usually employed. In crystallography the centroid is widely used; in photographic work with visual observation of the plates, there is involved some subjective criterion of the observer which it is difficult to define precisely. In this survey the peak of the line profile has been adopted as the standard criterion.

In the study of the X-ray literature, the wavelengths of a number of lines were noted which appeared inconsistent with the remaining data. A Moseley-type diagram was constructed, and if the value was clearly outside estimated probable error, it was assumed that an experimental or typographical error had occurred, and the interpolated value was listed in the table. Such cases are marked with a dagger (†) as a superscript to the wavelength. For elements of atomic number 85 through 89 and 91, there are no measured lines of the K series and very few of other series except for 88 radium and 91 protactinium. Likewise there are very few measurements for 43 technetium and 54 xenon. In these cases, interpolated values are listed for the more prominent lines and marked with a dagger (†).

Table 3.7-13 (continued)
X-RAY CRITICAL-ABSORPTION AND EMISSION ENERGIES IN keV

X-RAY WAVELENGTHS

X-ray wavelengths in Å* units and in keV. The probable error (p.e.) is the error in the last digit of wavelength. Designation indicates both conventional Siegbahn notation (if applicable) and transition, e.g., $\beta_1\,L_{II}M_{IV}$ denotes a transition between the L_{II} and M_{IV} levels, which is the $L\beta_1$ line in Siegbahn notation.

Designation	Å*	p.e.	keV	Å*	p.e.	keV
	3 Lithium			**4 Beryllium**		
$\alpha\,KL$	228.	1	0.0543	114.	1	0.1085
	5 Boron			**6 Carbon**		
$\alpha\,KL$	67.6	3	0.1833	44.7	3	0.277
	7 Nitrogen			**8 Oxygen**		
$\alpha\,KL$	31.6	4	0.3924	23.62	3	0.5249
	9 Fluorine			**10 Neon**		
$\alpha_{1,2}\,KL_{II,III}$	18.32	2	0.6768	14.610	3	0.8486
$\beta\,KM$				14.452	5	0.8579
	11 Sodium			**12 Magnesium**		
$\alpha_{1,2}\,KL_{II,III}$	11.9101	9	1.0410	9.8900	2	1.25360
$\beta\,KM$	11.575	2	1.0711	9.521	2	1.3022
$L_{II,III}M$	407.1	5	0.03045	251.5	5	0.0493
$L_I L_{II,III}$	376	1	0.0330	317	1	0.0392
	13 Aluminum			**14 Silicon**		
$\alpha_2\,KL_{II}$	8.34173	9	1.48627	7.12791	9	1.73938
$\alpha_1\,KL_{III}$	8.33934	9	1.48670	7.12542	9	1.73998
$\beta\,KM$	7.960	2	1.5574	6.753	1	1.8359
$L_{II,III}$	171.4	5	0.0724	135.5	4	0.0915
$L_I L_{II,III}$	290.	1	0.0428			
	15 Phosphorus			**16 Sulfur**		
$\alpha_2\,KL_{II}$	6.160†	1	2.0127	5.37496	8	2.30664
$\alpha_1\,KL_{III}$	6.157†	1	2.0137	5.37216	7	2.30784
$\beta\,KM$	5.796	2	2.1390			
$\beta_1\,KM$				5.0316	2	2.4640
$\beta_2\,KM$				5.0233	3	2.4681
$L_{II,III}M$	103.8	4	0.1194			
$l,\eta\,L_{II,III}M_I$				83.4	3	0.1487
	17 Chlorine			**18 Argon**		
$\alpha_2\,KL_{II}$	4.7307	1	2.62078	4.19474	5	2.95563
$\alpha_1\,KL_{III}$	4.7278	1	2.62239	4.19180	5	2.95770
$\beta\,KM$	4.4034	3	2.8156			
$\beta_{1,3}\,KM_{II,III}$				3.8860	2	3.1905
$\eta\,L_{II}M_I$	67.33	9	0.1841	55.9†	1	0.2217
$l\,L_{III}M_I$	67.90	9	0.1826	56.3†	1	0.2201
	19 Potassium			**20 Calcium**		
$\alpha_2\,KL_{II}$	3.7445	2	3.3111	3.36166	3	3.68809
$\alpha_1\,KL_{III}$	3.7414	2	3.3138	3.35839	3	3.69168
$\beta_{1,3}\,KM_{II,III}$	3.4539	2	3.5896	3.0897	2	4.0127
$\beta_5\,KM_{IV,V}$	3.4413	4	3.6027	3.0746	3	4.0325

Designation	Å*	p.e.	keV	Å*	p.e.	keV
	19 Potassium (*Cont.*)			**20 Calcium** (*Cont.*)		
$\eta\,L_{II}M_I$	47.24	2	0.2625	40.46	2	0.3064
β_1				35.94	2	0.3449
$l\,L_{III}M_I$	47.74	1	0.25971	40.96	2	0.3027
$\alpha_{1,2}\,L_{III}M_{IV,V}$				36.33	2	0.3413
$M_{II,III}N_I$	692	9	0.0179	525.	9	0.0236
	21 Scandium			**22 Titanium**		
$\alpha_2\,KL_{II}$	3.0342	1	4.0861	2.75216	2	4.5048
$\alpha_1\,KL_{III}$	3.0309†	1	4.0906	2.74851	2	4.5108
$\beta_{1,3}\,KM_{II,III}$	2.7796	2	4.4605	2.51391	2	4.9318
$\beta_5\,KM_{IV,V}$	2.7634	3	4.4865	2.4985	2	4.9623
$\eta\,L_{II}M_I$	35.13	2	0.3529	30.89	3	0.4013
$\beta_1\,L_{II}M_{IV}$	31.02	2	0.3996	27.05	2	0.4584
$l\,L_{III}M_I$	35.59	3	0.3483	31.36	2	0.3953
$\alpha_{1,2}\,L_{III}M_{IV,V}$	31.35	3	0.3954	27.42	2	0.4522
	23 Vanadium			**24 Chromium**		
$\alpha_2\,KL_{II}$	2.50738	2	4.94464	2.293606	3	5.4055
$\alpha_1\,KL_{III}$	2.50356	2	4.95220	2.28970	2	5.4147
$\beta_{1,3}\,KM_{II,III}$	2.28440	2	5.42729	2.08487	2	5.9467
$\beta_5\,KM_{IV,V}$	2.26951	6	5.4629	2.07087	6	5.9869
$\beta_{2,4}\,L_I M_{II,III}$	21.19†	9	0.585	18.96	2	0.654
$\eta\,L_{II}M_I$	27.34	3	0.4535	24.30	3	0.5102
$\beta_1\,L_{II}M_{IV}$	23.88	4	0.5192	21.27	1	0.5828
$l\,L_{III}M_I$	27.77	1	0.4465	24.78	1	0.5003
$\alpha_{1,2}\,L_{III}M_{IV,V}$	24.25	3	0.5113	21.64	3	0.5728
$M_{II,III}M_{IV,V}$	337.	9	0.037	309.	9	0.040
	25 Manganese			**26 Iron**		
$\alpha_2\,KL_{II}$	2.10578	2	5.88765	1.939980	9	6.3908
$\alpha_1\,KL_{III}$	2.101820	9	5.89875	1.936042	9	6.4038
$\beta_{1,3}\,KM_{II,III}$	1.91021	2	6.49045	1.75661	2	7.0579
$\beta_5\,KM_{IV,V}$	1.8971	1	6.5352	1.7442	1	7.1081
$\beta_{2,4}\,L_I M_{II,III}$	17.19	2	0.721	15.65	2	0.792
$\eta\,L_{II}M_I$	21.85	2	0.5675	19.75	4	0.628
$\beta_1\,L_{II}M_{IV}$	19.11	2	0.6488	17.26	1	0.7185
$l\,L_{III}M_I$	22.29	1	0.5563	20.15	1	0.6152
$\alpha_{1,2}\,L_{III}M_{IV,V}$	19.45	1	0.6374	17.59	2	0.7050
$M_{II,III}M_{IV,V}$	273.	6	0.045	243.	5	0.051
	27 Cobalt			**28 Nickel**		
$\alpha_2\,KL_{II}$	1.792850	9	6.91530	1.661747	8	7.4608
$\alpha_1\,KL_{III}$	1.788965	9	6.93032	1.657910	8	7.4781
$\beta_{1,3}\,KM_{II,III}$	1.62079	2	7.64943	1.500135	8	8.2646
$\beta_5\,KM_{IV,V}$	1.60891	3	7.7059	1.48862	4	8.3286
$\beta_{2,4}\,L_I M_{II,III}$	14.31	3	0.870	13.18	1	0.941
$\eta\,L_{II}M_I$	17.87	3	0.694	16.27	3	0.762
$\beta_1\,L_{II}M_{IV}$	15.666	8	0.7914	14.271	6	0.8688
$l\,L_{III}M_I$	18.292	8	0.6778	16.693	9	0.7427
$\alpha_{1,2}\,L_{III}M_{IV,V}$	15.972	6	0.7762	14.561	3	0.8515
$M_{II,III}M_{IV,V}$	214.	6	0.058	190.	2	0.0651

Table 3.7-13 (continued)
X-RAY CRITICAL-ABSORPTION AND EMISSION ENERGIES IN keV

X-Ray Wavelengths

Left half

Desig-nation	Å*	p.e.	keV	Å*	p.e.	keV
29 Copper				**30 Zinc**		
KL_{II}	1.544390	2	8.02783	1.439000	8	8.61578
KL_{III}	1.540562	2	8.04778	1.435155	7	8.63886
KM_{II}	1.3926	1	8.9029			
$KM_{II,III}$	1.392218	9	8.90529	1.29525	2	9.5720
$KN_{II,III}$				1.28372	2	9.6580
$KM_{IV,V}$	1.38109	3	8.9770	1.2848	1	9.6501
$L_IM_{II,III}$	12.122	8	1.0228	11.200	7	1.1070
$L_{II}M_I$	14.90	2	0.832	13.68	2	0.906
$L_{II}M_{IV}$	13.053	3	0.9498	11.983	3	1.0347
$L_{III}M_I$	15.286	9	0.8111	14.02	2	0.884
$L_{III}M_{IV,V}$	13.336	3	0.9297	12.254	3	1.0117
$_{II,III}M_{V,V}$	173.	3	0.072	157.	3	0.079
31 Gallium				**32 Germanium**		
KL_{II}	1.34399	1	9.22482	1.258011	9	9.85532
KL_{III}	1.340083	9	9.25174	1.254054	9	9.88642
KM_{II}	1.20835	5	10.2603	1.12936	9	10.9780
KM_{III}	1.20789	2	10.2642	1.12894	2	10.9821
$KN_{II,III}$	1.19600	2	10.3663	1.11686	2	11.1008
$KM_{IV,V}$	1.1981	2	10.348	1.1195	1	11.0745
L_IM_{II}				9.640	2	1.2861
L_IM_{III}				9.581	2	1.2941
$L_IM_{II,III}$	10.359†	8	1.197			
$L_{II}M_I$	12.597	2	0.9842	11.609	2	1.0680
$L_{II}M_{IV}$	11.023	2	1.1248	10.175	1	1.2185
$_{III}M_I$	12.953	2	0.9572	11.965	4	1.0362
$L_{III}M_{IV,V}$	11.292	1	1.09792	10.4361	8	1.18800
33 Arsenic				**34 Selenium**		
KL_{II}	1.17987	1	10.50799	1.10882	2	11.1814
KL_{III}	1.17588	1	10.54372	1.10477	2	11.2224
KM_{II}	1.05783	5	11.7203	0.99268	5	12.4896
KM_{III}	1.05730	2	11.7262	0.99218	3	12.4959
$KN_{II,III}$	1.04500	3	11.8642	0.97992	2	12.6522
$KM_{IV,V}$	1.0488	1	11.822	0.9843	1	12.595
$L_IM_{II,III}$	8.929	1	1.3884	8.321†	9	1.490
$_{II}M_I$	10.734	1	1.1550	9.962	1	1.2446
$L_{II}M_{IV}$	9.4141	8	1.3170	8.7358	5	1.41923
$_{III}M_I$	11.072	1	1.1198	10.294	1	1.2044
$L_{III}M_{IV,V}$	9.6709	8	1.2820	8.9900	5	1.37910
N_{III}				230.	2	0.0538
35 Bromine				**36 Krypton**		
KL_{II}	1.04382	2	11.8776	0.9841	1	12.598
KL_{III}	1.03974	2	11.9242	0.9801	1	12.649
KM_{II}	0.93327	5	13.2845	0.8790	1	14.104
KM_{III}	0.93279	2	13.2914	0.8785	1	14.112
$KN_{II,III}$	0.92046	2	13.4695	0.8661	1	14.315
$KM_{IV,V}$	0.9255	1	13.396	0.8708	2	14.238
$KN_{IV,V}$				0.8653	2	14.328
$_1M_{II}$				7.304	5	1.697
$_1M_{III}$				7.264	5	1.707

Right half

Desig-nation	Å*	p.e.	keV	Å*	p.e.	keV
35 Bromine (*Cont.*)				**36 Krypton** (*Cont.*)		
$\beta_{3,4}\,L_IM_{II,III}$	7.767†	9	1.596			
$\eta\,L_{II}M_I$	9.255	1	1.3396			
$\beta_1\,L_{II}M_{IV}$	8.1251	5	1.52590	7.576†	3	1.6366
γ_5				7.279	5	1.703
$l\,L_{III}M_I$	9.585	1	1.2935			
$\alpha_{1,2}\,L_{III}M_{IV,V}$	8.3746	5	1.48043	7.817†	3	1.5860
β_6				7.510	4	1.6510
$L_{III}N_{III}$				7.250	5	1.710
M_IM_{II}	184.6	3	0.0672			
M_IM_{III}	164.7	3	0.0753			
$M_{II}M_{IV}$	109.4	3	0.1133			
$M_{II}N_I$	76.9	2	0.1613			
$M_{III}M_{IV,V}$	113.8	3	0.1089			
	79.8	3	0.1554			
$\zeta_2\,M_{IV}N_{II}$	191.1	2	0.06488			
$M_{IV}N_{III}$	189.5	3	0.0654			
$\zeta_1\,M_VN_{III}$	192.6	2	0.06437			
37 Rubidium				**38 Strontium**		
$\alpha_2\,KL_{II}$	0.92969	1	13.3358	0.87943	1	14.0979
$\alpha_1\,KL_{III}$	0.925553	9	13.3953	0.87526	1	14.1650
$\beta_3\,KM_{II}$	0.82921	3	14.9517	0.78345	3	15.8249
$\beta_1\,KM_{III}$	0.82868	2	14.9613	0.78292	2	15.8357
$\beta_2\,KN_{II,III}$	0.81645	3	15.1854	0.77081	3	16.0846
$\beta_5\,KM_{IV,V}$	0.8219	1	15.085	0.7764	1	15.969
$\beta_4\,KN_{IV,V}$	0.8154	2	15.205	0.76989	5	16.104
$\beta_4\,L_IM_{II}$	6.8207	3	1.81771	6.4026	3	1.93643
$\beta_3\,L_IM_{III}$	6.7876	3	1.82659	6.3672	3	1.94719
$\gamma_{2,3}\,L_IN_{II,III}$	6.0458	3	2.0507	5.6445	3	2.1965
$\eta\,L_{II}M_I$	8.0415	4	1.54177	7.5171	3	1.64933
$\beta_1\,L_{II}M_{IV}$	7.0759	3	1.75217	6.6239	3	1.87172
$\gamma_5\,L_{II}N_{IV}$	6.7553	3	1.83532	6.2961	3	1.96916
$l\,L_{III}M_I$	8.3636	4	1.48238	7.8362	3	1.58215
$\alpha_2\,L_{III}M_{IV}$	7.3251	3	1.69256	6.8697	3	1.80474
$\alpha_1\,L_{III}M_V$	7.3183	2	1.69413	6.8628	2	1.80656
$\beta_6\,L_{III}N_I$	6.9842	3	1.77517	6.5191	3	1.90181
M_IM_{II}	144.4	3	0.0859			
$M_{II}M_{IV}$	91.5	3	0.1355	85.7	2	0.1447
$M_{II}N_I$	57.0	2	0.2174	51.3	1	0.2416
$M_{III}M_{IV,V}$	96.7	2	0.1282	91.4	2	0.1357
$M_{III}N_I$	59.5	2	0.2083	53.6	1	0.2313
$\zeta_2\,M_{IV}N_{II}$	127.8	2	0.0970			
$M_{IV}N_{II}$	126.8	2	0.0978			
$\zeta_2\,M_{IV}N_{II,III}$				108.0	2	0.1148
$\zeta_1\,M_VN_{III}$	128.7	2	0.0964	108.7	1	0.1140
39 Yttrium				**40 Zirconium**		
$\alpha_2\,KL_{II}$	0.83305	1	14.8829	0.79015	1	15.6909
$\alpha_1\,KL_{III}$	0.82884	1	14.9584	0.78593	1	15.7751
$\beta_3\,KM_{II}$	0.74126	3	16.7258	0.70228	4	17.654
$\beta_1\,KM_{III}$	0.74072	2	16.7378	0.70173	3	17.6678
$\beta_2\,KN_{II,III}$	0.72864	4	17.0154	0.68993	4	17.970
$\beta_5\,KM_{IV,V}$	0.7345	1	16.879	0.6959	1	17.815

Table 3.7-13 (continued)
X-RAY CRITICAL-ABSORPTION AND EMISSION ENERGIES IN keV

X-Ray Wavelengths

39 Yttrium (Cont.) / 40 Zirconium (Cont.)

Designation	Å*	p.e.	keV	Å*	p.e.	keV
$\beta_4\ KN_{IV,V}$	0.72776	5	17.036	0.68901	5	17.994
$\beta_4\ L_IM_{II}$	6.0186	3	2.0600	5.6681	3	2.1873
$\beta_3\ L_IM_{III}$	5.9832	3	2.0722	5.6330	3	2.2010
$\gamma_{2,3}\ L_IN_{II,III}$	5.2830	3	2.3468	4.9536	3	2.5029
$\eta\ L_IM_I$	7.0406	3	1.76095	6.6069	3	1.87654
$\beta_1\ L_{III}M_{IV}$	6.2120	3	1.99584	5.8360	3	2.1244
$\gamma_5\ L_{II}N_I$	5.8754	3	2.1102	5.4977	3	2.2551
$\gamma_1\ L_{II}N_{IV}$				5.3843	3	2.3027
$l\ L_{III}M_I$	7.3563	3	1.68536	6.9185	3	1.79201
$\alpha_2\ L_{III}M_{IV}$	6.4558	3	1.92047	6.0778	3	2.0399
$\alpha_1\ L_{III}M_V$	6.4488	2	1.92256	6.0705	2	2.04236
$\beta_6\ L_{III}N_I$	6.0942	3	2.0344	5.7101	3	2.1712
$\beta_{2,15}$				5.5863	3	2.2194
$M_{II}M_{IV}$	81.5	2	0.1522	76.7	2	0.1617
$M_{II}N_I$	46.48	9	0.267			
$M_{III}M_V$				80.9	3	0.1533
$M_{III}N_I$	48.5	2	0.256			
$M_{III}M_{IV,V}$	86.5	2	0.1434			
$\zeta\ M_{IV,V}N_{II,III}$	93.4	2	0.1328	82.1	2	0.1511
$M_{IV,V}O_{II,III}$				70.0	4	0.177

41 Niobium / 42 Molybdenum

Designation	Å*	p.e.	keV	Å*	p.e.	keV
$\alpha_2\ KL_{II}$	0.75044	1	16.5210	0.713590	6	17.3743
$\alpha_1\ KL_{III}$	0.74620	1	16.6151	0.709300	1	17.47934
$\beta_3\ KM_{II}$	0.66634	3	18.6063	0.632872	9	19.5903
$\beta_1\ KM_{III}$	0.66576	2	18.6225	0.632288	9	19.6083
β_2^{II}				0.62107	5	19.963
$\beta_2\ KN_{II,III}$	0.65416	4	18.953	0.62099	2	19.9652
$\beta_4\ KN_{IV,V}$	0.65318	5	18.981			
$\beta_5^{II}\ KM_{IV}$				0.62708	5	19.771
$\beta_5^{I}\ KM_V$				0.62692	5	19.776
$\beta_4\ KN_{IV,V}$				0.62001	9	19.996
$\beta_4\ L_IM_{II}$	5.3455	3	2.3194	5.0488	3	2.4557
$\beta_3\ L_IM_{III}$	5.3102	3	2.3348	5.0133	3	2.4730
$\gamma_{2,3}\ L_IN_{II,III}$	4.6542	2	2.6638	4.3800	2	2.8306
$\eta\ L_{II}M_I$	6.2109	3	1.99620	5.8475	3	2.1202
$\beta_1\ L_{II}M_{IV}$	5.4923	3	2.2574	5.17708	8	2.39481
$\gamma_5\ L_{II}N_I$	5.1517	3	2.4066	4.8369	2	2.5632
$\gamma_1\ L_{II}N_{IV}$	5.0361	3	2.4618	4.7258	2	2.6235
$l\ L_{III}M_I$	6.5176	3	1.90225	6.1508	3	2.01568
$\alpha_2\ L_{III}M_{IV}$	5.7319	3	2.1630	5.41437	8	2.28985
$\alpha_1\ L_{III}M_V$	5.7243	3	2.16589	5.40655	8	2.29316
$\beta_6\ L_{III}N_I$	5.3613	3	2.3125	5.0488	5	2.4557
$\beta_{2,15}\ L_{III}N_{IV,V}$	5.2379	3	2.3670	4.9232	2	2.5183
$M_{II}M_{IV}$	72.1	3	0.1718	68.9	2	0.1798
$M_{II}N_I$	38.4	3	0.323	35.3	3	0.351
$M_{II}N_{IV}$	33.1	2	0.375			
$M_{III}M_V$	78.4	2	0.1582	74.9	1	0.1656
$M_{III}N_I$	40.7	2	0.305	37.5	2	0.331
$\gamma\ M_{III}N_{IV,V}$	34.9	2	0.356			
$\zeta\ M_{IV,V}N_{II,III}$	72.19	9	0.1717	64.38	7	0.1926
$M_{IV,V}O_{II,III}$	61.9	2	0.2002	54.8	2	0.2262

43 Technetium / 44 Ruthenium

Designation	Å*	p.e.	keV	Å*	p.e.	keV
$\alpha_2\ KL_{II}$	0.67932†	3	18.2508	0.647408	5	19.1504
$\alpha_1\ KL_{III}$	0.67502†	3	18.3671	0.643083	4	19.2792
$\beta_3\ KM_{II}$	0.60188†	4	20.599	0.573067	4	21.6346
$\beta_1\ KM_{III}$	0.60130†	4	20.619	0.572482	4	21.6568
$\beta_2\ KN_{II,III}$	0.59024†	5	21.005	0.56166	3	22.074
$\beta_5^{II}\ KM_{IV}$				0.5680	2	21.829
$\beta_5^{I}\ KM_V$				0.56785	9	21.834
β_4				0.56089	9	22.104
$\beta_4\ L_{II}M_{IV}$				4.5230	2	2.7411
$\beta_3\ L_IM_{III}$				4.4866	3	2.7634
$\gamma_{2,3}\ L_IN_{II,III}$				3.8977	2	3.1809
$\eta\ L_{II}M_I$				5.2050	2	2.3819
$\beta_1\ L_{II}M_{IV}$	4.8873†	8	2.5368	4.62058	3	2.6832
$\gamma_5\ L_{II}N_I$				4.2873	2	2.8918
$\gamma_1\ L_{II}N_{IV}$				4.1822	2	2.9645
$l\ L_{III}M_I$				5.5035	3	2.2528
$\alpha_2\ L_{III}M_{IV}$				4.85381	7	2.5543
$\alpha_1\ L_{III}M_V$	5.1148†	3	2.4240	4.84575	5	2.5585
$\beta_6\ L_{III}N_I$				4.4866	3	2.7634
$\beta_{2,15}\ L_{III}N_{IV,V}$				4.3718	2	2.8360
$M_{II}M_{IV}$				62.2	1	0.1992
$M_{II}N_I$				32.3	2	0.384
$M_{II}N_{IV}$				25.50	9	0.486
$M_{III}M_V$				68.3	1	0.1814
$\gamma\ M_{III}N_{IV,V}$				26.9	1	0.462
$\zeta\ M_{IV,V}N_{II,III}$				52.34	7	0.2369
$M_{IV,V}O_{II,III}$				44.8	1	0.2763

45 Rhodium / 46 Palladium

Designation	Å*	p.e.	keV	Å*	p.e.	keV
$\alpha_2\ KL_{II}$	0.617630	4	20.0737	0.589821	3	21.0203
$\alpha_1\ KL_{III}$	0.613279	4	20.2161	0.585448	3	21.1771
$\beta_3\ KM_{II}$	0.546200	4	22.6989	0.521123	4	23.791
$\beta_1\ KM_{III}$	0.545605	4	22.7236	0.520520	4	23.818
$\beta_2^{II}\ KN_{II}$	0.53513	5	23.168			
$\beta_2\ KN_{II,III}$	0.53503	2	23.1728	0.510228	4	24.299
$\beta_5^{II}\ KM_{IV}$	0.54118	9	22.909			
$\beta_5^{I}\ KM_V$	0.54101	9	22.917			
$\beta_4\ KN_{IV,V}$	0.53401	9	23.217	0.51670	9	23.995
$\beta_6\ KM_{IV,V}$				0.5093	2	24.346
$\beta_4\ L_IM_{II}$	4.2888	2	2.8908	4.0711	2	3.045
$\beta_3\ L_IM_{III}$	4.2522	2	2.9157	4.0346	2	3.073
$\gamma_{2,3}\ L_IN_{II,III}$	3.6855	2	3.3640	3.4892	2	3.553
$\eta\ L_{II}M_I$	4.9217	2	2.5191	4.6605	2	2.660
$\beta_1\ L_{II}M_{IV}$	4.37414	4	2.83441	4.14622	5	2.990
$\gamma_5\ L_{II}N_I$	4.0451	2	3.0650	3.8222	2	3.243
$\gamma_1\ L_{II}N_{IV}$	3.9437	2	3.1438	3.7246	2	3.328
$l\ L_{III}M_I$	5.2169	2	2.3765	4.9525	3	2.503
$\alpha_2\ L_{III}M_{IV}$	4.60545	9	2.69205	4.37588	7	2.833
$\alpha_1\ L_{III}M_V$	4.59743	9	2.69674	4.36767	5	2.838
$\beta_6\ L_{III}N_I$	4.2417	2	2.9229	4.0162	2	3.087
$\beta_{2,15}\ L_{III}N_{IV,V}$	4.1310	2	3.0013	3.90887	4	3.171
$\beta_{10}\ L_IM_{IV}$				3.7988	2	3.263

Table 3.7-13 (continued)
X-RAY CRITICAL-ABSORPTION AND EMISSION ENERGIES IN keV

X-Ray Wavelengths

Left group — 45 Rhodium / 46 Palladium / 47 Silver / 48 Cadmium / 49 Indium / 50 Tin

Designation	Å*	p.e.	keV	Å*	p.e.	keV
45 Rhodium (Cont.)				**46 Palladium (Cont.)**		
$\beta_9\,L_IM_V$				3.7920	2	3.2696
$\beta_1 N_{II,III}$				20.1	2	0.616
$\beta_{II} M_{IV}$	59.3	1	0.2090	56.5	1	0.2194
$\beta_{II} N_I$	28.1	2	0.442	26.2	2	0.474
$\beta_{II} N_{IV}$				22.1	1	0.560
$\beta_{II}\,M_V$	65.5	1	0.1892	62.9	1	0.1970
$\beta_{II}{}^{T} N_I$	29.8	1	0.417	27.9	1	0.445
$M_{III} N_{IV,V}$	25.01	9	0.496	23.3†	1	0.531
$M_{IV,V} N_{II,III}$	47.67	9	0.2601	43.6	1	0.2844
$M_{IV,V} O_{II,III}$	40.9	2	0.303	37.4	2	0.332
47 Silver				**48 Cadmium**		
$\alpha_2\,KL_{II}$	0.563798	4	21.9903	0.539422	3	22.9841
$\alpha_1\,KL_{III}$	0.5594075	6	22.16292	0.535010	3	23.1736
$\beta_1\,KM_{II}$	0.497685	4	24.9115	0.475730	5	26.0612
KM_{III}	0.497069	4	24.9424	0.475105	6	26.0955
$\beta\,KN_{II,III}$	0.487032	4	25.4564	0.465328	7	26.6438
$KM_{IV,V}$	0.49306	2	25.145			
$KN_{IV,V}$	0.48598	3	25.512			
L_IM_{II}	3.87023	5	3.20346	3.68203	9	3.36719
L_IM_{III}	3.83313	9	3.23446	3.64495	9	3.40145
L_IN_{II}	3.31216	9	3.7432	3.1377	2	3.9513
L_IN_{III}	3.30635	9	3.7498			
$L_{II}M_I$	4.4183	2	2.8061	4.19315	9	2.95675
$L_{II}M_{IV}$	3.93473	3	3.15094	3.73823	4	3.31657
$L_{II}N_I$	3.61638	9	3.42832	3.42551	9	3.61935
$L_{III}N_{IV}$	3.52260	4	3.51959	3.33564	6	3.71686
$L_{III}M_I$	4.7076	2	2.6337	4.48014	9	2.76735
$L_{III}M_{IV}$	4.16294	5	2.97821	3.96496	6	3.12691
$L_{III}M_V$	4.15443	3	2.98431	3.95635	4	3.13373
$L_{III}N_I$	3.80774	9	3.25603	3.61467	9	3.42994
$L_{III}N_{IV,V}$	3.70335	3	3.34781	3.51408	4	3.52812
L_IM_{IV}	3.61158	9	3.43287	3.4367	2	3.6075
L_IM_V	3.60497	9	3.43917	3.43015	9	3.61445
$N_{II,III}$	18.8	2	0.658			
M_{IV}	54.0	1	0.2295	52.0	2	0.2384
N_I				22.9	2	0.540
N_{IV}	20.66	7	0.600	19.40	7	0.639
M_V	60.5	1	0.2048	58.7	2	0.2111
N_I	26.0	1	0.478	24.5	1	0.507
$M_{III}N_{IV,V}$	21.82	7	0.568	20.47	7	0.606
$O_{II,III}$				30.4	1	0.408
$M_{IV,V}N_{II,III}$	39.77	7	0.3117	36.8	1	0.3371
N_I	24.4	2	0.509			
O_{III}				30.8	1	0.403
$M_{IV,V}O_{II,III}$	33.5	3	0.370			
49 Indium				**50 Tin**		
KL_{II}	0.516544	3	24.0020	0.495053	3	25.0440
KL_{III}	0.512113	3	24.2097	0.490599	3	25.2713
KM_{II}	0.455181	4	27.2377	0.435877	5	28.4440

Right group — 49 Indium / 50 Tin / 51 Antimony / 52 Tellurium

Designation	Å*	p.e.	keV	Å*	p.e.	keV
49 Indium (Cont.)				**50 Tin (Cont.)**		
$\beta_1\,KM_{III}$	0.454545	4	27.2759	0.435236	5	28.4860
$\beta_2\,KN_{II,III}$	0.44500	1	27.8608	0.425915	8	29.1093
$KO_{II,III}$	0.44374	3	27.940	0.42467	3	29.195
$\beta_5{}^{II}\,KM_{IV}$	0.45098	2	27.491	0.43184	3	28.710
$\beta_5{}^{I}\,KM_V$	0.45086	2	27.499	0.43175	3	28.716
$\beta_4\,KN_{IV,V}$	0.44393	4	27.928	0.42495	3	29.175
$\beta_3\,L_IM_{II}$	3.50697	9	3.5353	3.34335	9	3.7083
$\beta_4\,L_IM_{III}$	3.46984	9	3.5731	3.30585	3	3.7500
$\gamma_{2,3}\,L_IN_{II,III}$	2.9800	2	4.1605	2.8327	2	4.3768
$\gamma_4\,L_IO_{II,III}$	2.9264	2	4.2367	2.7775	2	4.4638
$\eta\,L_{II}M_I$	3.98327	9	3.11254	3.78876	9	3.27234
$\beta_1\,L_{II}M_{IV}$	3.55531	4	3.48721	3.38487	3	3.66280
$\gamma_5\,L_{II}N_I$	3.24907	9	3.8159	3.08475	9	4.0192
$\gamma_1\,L_{II}N_{IV}$	3.16213	4	3.92081	3.00115	3	4.13112
$l\,L_{III}M_I$	4.26873	9	2.90440	4.07165	9	3.04499
$\alpha_2\,L_{III}M_{IV}$	3.78073	6	3.27929	3.60891	4	3.43542
$\alpha_1\,L_{III}M_V$	3.77192	4	3.28694	3.59994	3	3.44398
$\beta_6\,L_{III}N_I$	3.43606	9	3.60823	3.26901	9	3.7926
$\beta_{2,15}\,L_{III}N_{IV,V}$	3.33838	3	3.71381	3.17505	3	3.90486
$\beta_7\,L_{III}O_I$	3.324	4	3.730	3.1564	3	3.9279
$\beta_{10}\,L_IM_{IV}$	3.27404	9	3.7868	3.12170	9	3.9716
$\beta_9\,L_IM_V$	3.26763	9	3.7942	3.11513	9	3.9800
$M_{II}M_{IV}$				47.3	1	0.2621
$M_{II}N_I$				20.0	1	0.619
$M_{II}N_{IV}$				16.93	5	0.733
$M_{III}M_V$				54.2	1	0.2287
$M_{III}N_I$				21.5	1	0.575
$\gamma\,M_{III}N_{IV,V}$				17.94	5	0.691
$M_{IV}O_{II,III}$				25.3	1	0.491
$\zeta\,M_{IV,V}N_{II,III}$				31.24	9	0.397
M_VO_{III}				25.7	1	0.483
51 Antimony				**52 Tellurium**		
$\alpha_2\,KL_{II}$	0.474827	3	26.1108	0.455784	3	27.2017
$\alpha_1\,KL_{III}$	0.470354	3	26.3591	0.451295	3	27.4723
$\beta_3\,KM_{II}$	0.417737	4	29.6792	0.400659	4	30.9443
$\beta_1\,KM_{III}$	0.417085	3	29.7256	0.399995	5	30.9957
$\beta_2\,KN_{II,III}$	0.407973	5	30.3895	0.391102	6	31.7004
$KO_{II,III}$	0.40666	1	30.4875	0.38974	1	31.8114
$\beta_5{}^{II}\,KM_{IV}$	0.41388	1	29.9560			
$\beta_5{}^{I}\,KM_V$	0.41378	1	29.9632			
$\beta_4\,KN_{IV,V}$	0.40702	1	30.4604			
$\beta_3\,L_IM_{II}$	3.19014	9	3.8864	3.04661	9	4.0695
$\beta_4\,L_IM_{III}$	3.15258	9	3.9327	3.00893	9	4.1204
$\gamma_{2,3}\,L_IN_{II,III}$	2.6953	2	4.5999	2.5674	2	4.8290
$\gamma_4\,L_IO_{II,III}$	2.6398	2	4.6967	2.5113	2	4.9369
$\eta\,L_{II}M_I$	3.60765	9	3.43661	3.43832	9	3.60586
$\beta_1\,L_{II}M_{IV}$	3.22567	4	3.84357	3.07677	6	4.02958
$\gamma_5\,L_{II}N_I$	2.93187	9	4.2287	2.79007	9	4.4437
$\gamma_1\,L_{II}N_{IV}$	2.85159	3	4.34779	2.71241	6	4.5709
$l\,L_{III}M_I$	3.88826	9	3.18860	3.71696	9	3.33555
$\alpha_2\,L_{III}M_{IV}$	3.44840	6	3.59532	3.29846	9	3.7588

Table 3.7-13 (continued)
X-RAY CRITICAL-ABSORPTION AND EMISSION ENERGIES IN keV

X-Ray Wavelengths

Left half

Designation	Å*	p.e.	keV	Å*	p.e.	keV
51 Antimony (*Cont.*)				**52 Tellurium (*Cont.*)**		
$\alpha_1\ L_{III}M_V$	3.43941	4	3.60472	3.28920	6	3.76933
$\beta_6\ L_{III}N_I$	3.11513	9	3.9800	2.97088	9	4.1732
$\beta_{2,15}\ L_{III}N_{IV,V}$	3.02335	3	4.10078	2.88217	8	4.3017
$\beta_7\ L_{III}O_I$	3.0052	3	4.1255	2.8634	3	4.3298
$\beta_{10}\ L_I M_{IV}$	2.97917	9	4.1616	2.84679	9	4.3551
$\beta_9\ L_I M_V$	2.97261	9	4.1708	2.83897	9	4.3671
$M_{II}M_{IV}$	45.2	1	0.2743			
$M_{II}N_I$	18.8	1	0.658	17.6	1	0.703
$M_{II}N_{IV}$	15.98	5	0.776			
$M_{III}M_V$	52.2	1	0.2375	50.3	1	0.2465
$M_{III}N_I$	20.2	1	0.612	19.1	1	0.648
$\gamma\ M_{III}N_{IV,V}$	16.92	4	0.733	15.93	4	0.778
$M_{IV}O_{II,III}$				21.34	5	0.581
$\zeta\ M_{IV,V}N_{II,III}$	28.88	8	0.429	26.72	9	0.464
$M_V O_{III}$				21.78	5	0.569
53 Iodine				**54 Xenon**		
$\alpha_2\ K L_{II}$	0.437829	7	28.3172	0.42087†	2	29.458
$\alpha_1\ K L_{III}$	0.433318	5	28.6120	0.41634†	2	29.779
$\beta_3\ K M_{II}$	0.384564	4	32.2394	0.36941†	2	33.562
$\beta_1\ K M_{III}$	0.383905	4	32.2947	0.36872†	2	33.624
$\beta_2\ K N_{II,III}$	0.37523†	2	33.042	0.36026†	3	34.415
$\beta_4\ L_I M_{II}$	2.91207	9	4.2575			
$\beta_3\ L_I M_{III}$	2.87429	9	4.3134			
$\gamma_{2,3}\ L_I N_{II,III}$	2.4475	2	5.0657			
$\gamma_4\ L_I O_{II,III}$	2.3913	2	5.1848			
$\eta\ L_{II}M_I$	3.27979	9	3.7801			
$\beta_1\ L_{II}M_{IV}$	2.93744	6	4.22072			
$\gamma_5\ L_{II}N_I$	2.65710	9	4.6660			
$\gamma_1\ L_{II}N_{IV}$	2.58244	8	4.8009			
$l\ L_{III}M_I$	3.55754	9	3.48502			
$\alpha_2\ L_{III}M_{IV}$	3.15791	6	3.92604			
$\alpha_1\ L_{III}M_V$	3.14860	6	3.93765	3.0166†	2	4.1099
$\beta_6\ L_{III}N_I$	2.83672	9	4.3706			
$\beta_{2,15}\ L_{III}N_{IV,V}$	2.75053	8	4.5075			
$\beta_7\ L_{III}O_I$	2.7288	3	4.5435			
$\beta_{10}\ L_I M_{IV}$	2.72104	9	4.5564			
$\beta_9\ L_I M_V$	2.71352	9	4.5690			
55 Cesium				**56 Barium**		
$\alpha_2\ K L_{II}$	0.404835	4	30.6251	0.389668	5	31.8171
$\alpha_1\ K L_{III}$	0.400290	4	30.9728	0.385111	4	32.1936
$\beta_3\ K M_{II}$	0.355050	4	34.9194	0.341507	4	36.3040
$\beta_1\ K M_{III}$	0.354364	7	34.9869	0.340811	3	36.3782
$\beta_2\ K N_{II,III}$	0.34611	2	35.822	0.33277	1	37.257
$K O_{II,III}$				0.33127	2	37.446
$\beta_5^{II}\ K M_{IV}$				0.33835	2	36.643
$\beta_5^{I}\ K M_V$				0.33814	2	36.666
$\beta_4\ K N_{IV,V}$				0.33229	2	37.311
$\beta_4\ L_I M_{II}$	2.6666	2	4.6494	2.5553	2	4.8519
$\beta_3\ L_I M_{III}$	2.6285	2	4.7167	2.5164	2	4.9269
$\gamma_2\ L_I N_{II}$	2.2371	2	5.5420	2.1387	2	5.7969
$\gamma_3\ L_I N_{III}$	2.2328	2	5.5527	2.1342	2	5.8092

Right half

Designation	Å*	p.e.	keV	Å*	p.e.	keV
55 Cesium (*Cont.*)				**56 Barium (*Cont.*)**		
$\gamma_4\ L_I O_{II,III}$	2.1741	2	5.7026	2.0756	3	5.9733
$\eta\ L_{II}M_I$	2.9932	2	4.1421	2.8627	3	4.3309
$\beta_1\ L_{II}M_{IV}$	2.6837	2	4.6198	2.56821	5	4.8275
$\gamma_5\ L_{II}N_I$	2.4174	2	5.1287	2.3085	3	5.3707
$\gamma_1\ L_{II}N_{IV}$	2.3480	2	5.2804	2.2415	2	5.5311
$l\ L_{III}M_I$	3.2670	2	3.7950	3.1355	2	3.9541
$\alpha_2\ L_{III}M_V$	2.9020	2	4.2722	2.78553	5	4.4509
$\alpha_1\ L_{III}M_V$	2.8924	2	4.2865	2.77595	5	4.4662
$\beta_6\ L_{III}N_I$	2.5932	2	4.7811	2.4826	2	4.9939
$\beta_{2,15}\ L_{III}N_{IV,V}$	2.5118	2	4.9359	2.40435	6	5.1565
$\beta_7\ L_{III}O_I$	2.4849	2	4.9893	2.3806	2	5.2079
$\beta_{10}\ L_I M_{IV}$	2.4920	2	4.9752	2.3869	2	5.1941
$\beta_9\ L_I M_V$	2.4783	2	5.0026	2.3764	2	5.2171
$\gamma\ M_{III}N_{IV,V}$				12.75	3	0.973
$M_{IV}O_{II}$				15.91	5	0.779
$M_{IV}O_{III}$				15.72	9	0.789
$\zeta\ M_V N_{III}$				20.64	4	0.601
$M_V O_{III}$				16.20	5	0.765
$N_{IV}O_{II}$	188.6	1	0.06574	163.3	2	0.0759
$N_{IV}O_{III}$	183.8	1	0.06746	159.0	2	0.0779
$N_V O_{III}$	190.3	1	0.06515	164.6	2	0.0753
57 Lanthanum				**58 Cerium**		
$\alpha_2\ K L_{II}$	0.375313	2	33.0341	0.361683	2	34.2786
$\alpha_1\ K L_{III}$	0.370737	2	33.4418	0.357092	2	34.719
$\beta_3\ K M_{II}$	0.328686	4	37.7202	0.316520	4	39.1706
$\beta_1\ K M_{III}$	0.327983	3	37.8010	0.315816	2	39.257
$\beta_2\ K N_{II,III}$	0.320117	7	38.7299	0.30816	1	40.233
$K O_{II,III}$	0.31864	2	38.909	0.30668	2	40.447
$\beta_5^{II}\ K M_{IV}$	0.32563	2	38.074	0.31357	2	39.539
$\beta_5^{I}\ K M_V$	0.32546	2	38.094	0.31342	2	39.558
$\beta_4\ K N_{IV,V}$	0.31931	2	38.828	0.30737	2	40.337
$\beta_4\ L_I M_{II}$	2.4493	3	5.0620	2.3497	4	5.276
$\beta_3\ L_I M_{III}$	2.4105	3	5.1434	2.3109	3	5.365
$\gamma_2\ L_I N_{II}$	2.0460	4	6.060	1.9602	3	6.325
$\gamma_3\ L_I N_{III}$	2.0410	4	6.074	1.9553	3	6.340
$\gamma_4\ L_I O_{II,III}$	1.9830	4	6.252	1.8991	4	6.528
$\eta\ L_{II}M_I$	2.740	3	4.525	2.6203	4	4.731
$\beta_1\ L_{II}M_{IV}$	2.45891	5	5.0421	2.3561	3	5.262
$\gamma_5\ L_{II}N_I$	2.2056	4	5.621	2.1103	3	5.875
$\gamma_1\ L_{II}N_{IV}$	2.1418	3	5.7885	2.0487	4	6.052
$\gamma_8\ L_{II}O_I$				2.0237	4	6.120
$l\ L_{III}M_I$	3.006	3	4.124	2.8917	4	4.28
$\alpha_2\ L_{III}M_{IV}$	2.67533	5	4.63423	2.5706	3	4.82
$\alpha_1\ L_{III}M_V$	2.66570	5	4.65097	2.5615	2	4.84
$\beta_6\ L_{III}N_I$	2.3790	4	5.2114	2.2818	3	5.43
$\beta_{2,15}\ L_{III}N_{IV,V}$	2.3030	3	5.3835	2.2087	2	5.61
$\beta_7\ L_{III}O_I$	2.275	3	5.450	2.1701	2	5.71
$\beta_{10}\ L_I M_{IV}$	2.290	3	5.415	2.1958	5	5.64
$\beta_9\ L_I M_V$	2.282	3	5.434	2.1885	3	5.66
$\gamma\ M_{III}N_{IV,V}$	12.08	4	1.027	11.53	1	1.07
$\beta\ M_{IV}N_{VI}$	14.51	5	0.854	13.75	2	0.90
$\zeta\ M_V N_{III}$	19.44	5	0.638	18.35	4	0.67
$\alpha\ M_V N_{VI,VII}$	14.88	5	0.833	14.04	2	0.88

Table 3.7-13 (continued)
X-RAY CRITICAL-ABSORPTION AND EMISSION ENERGIES IN keV

X-Ray Wavelengths

Designation	Å*	p.e.	keV	Å*	p.e.	keV
	57 Lanthanum (*Cont.*)			**58 Cerium** (*Cont.*)		
$M_V O_{II,III}$				14.39	5	0.862
$N_{IV,V} O_{II,III}$	152.6	6	0.0812	144.4	6	0.0859
	59 Praseodymium			**60 Neodymium**		
$\alpha_2\ KL_{II}$	0.348749	2	35.5502	0.336472	2	36.8474
$\alpha_1\ KL_{III}$	0.344140	2	36.0263	0.331846	2	37.3610
$\beta_3\ KM_{II}$	0.304975	5	40.6529	0.294027	3	42.1665
$\beta_1\ KM_{III}$	0.304261	4	40.7482	0.293299	2	42.2713
$\beta_2\ KN_{II,III}$	0.29679	2	41.773	0.2861†	1	43.33
$\beta_4\ L_I M_{II}$	2.2550	4	5.4981	2.1669	3	5.7216
$\beta_3\ L_I M_{III}$	2.2172	3	5.5918	2.1268	2	5.8294
$\gamma_2\ L_I N_{II}$	1.8791	4	6.598	1.8013	4	6.883
$\gamma_3\ L_I N_{III}$	1.8740	4	6.616	1.7964	4	6.902
$\gamma_4\ L_I O_{II,III}$	1.8193	4	6.815	1.7445	4	7.107
$\eta\ L_{II} M_I$	2.512	3	4.935	2.4094	4	5.1457
$\beta_1\ L_{II} M_{IV}$	2.2588	3	5.4889	2.1669	3	5.7216
$\gamma_5\ L_{II} N_I$	2.0205	4	6.136	1.9355	4	6.406
$\gamma_1\ L_{II} N_{IV}$	1.9611	3	6.3221	1.8779	2	6.6021
$\gamma_8\ L_{II} O_I$	1.9362	4	6.403	1.8552	5	6.683
$l\ L_{III} M_I$	2.7841	4	4.4532	2.6760	4	4.6330
$\alpha_2\ L_{III} M_{IV}$	2.4729	3	5.0135	2.3807	3	5.2077
$\alpha_1\ L_{III} M_V$	2.4630	2	5.0337	2.3704	2	5.2304
$\beta_6\ L_{III} N_I$	2.1906	4	5.660	2.1039	3	5.8930
$\beta_{2,15}\ L_{III} N_{IV,V}$	2.1194	4	5.850	2.0360	3	6.0894
$\beta_7\ L_{III} O_I$	2.0919	4	5.927	2.0092	3	6.1708
$\beta_{10}\ L_I M_{IV}$	2.1071	4	5.884	2.0237	3	6.1265
$\beta_9\ L_I M_V$	2.1004	4	5.903	2.0165	3	6.1484
$\gamma\ M_{III} N_{IV,V}$	10.998	9	1.1273	10.505	9	1.180
$\beta\ M_{IV} N_{VI}$	13.06	2	0.950	12.44	2	0.997
$\zeta\ M_V N_{III}$	17.38	4	0.714	16.46	4	0.753
$\alpha\ M_V N_{VI,VII}$	13.343	5	0.9292	12.68	2	0.978
$N_{IV,V} N_{VI,VII}$	113.	1	0.1095	107.	1	0.116
$N_{IV,V} O_{II,III}$	136.5	4	0.0908	128.9	7	0.0962
	61 Promethium			**62 Samarium**		
$\alpha_2\ KL_{II}$	0.324803	4	38.1712	0.313698	2	39.5224
$\alpha_1\ KL_{III}$	0.320160	4	38.7247	0.309040	2	40.1181
$\beta_3\ KM_{II}$	0.28363†	4	43.713	0.27376	2	45.289
$\beta_1\ KM_{III}$	0.28290†	3	43.826	0.27301	2	45.413
$\beta_2\ KN_{II,III}$	0.2759†	1	44.94	0.2662	1	46.58
$KO_{II,III}$				0.26491	3	46.801
$\beta_5\ KM_{IV,V}$				0.27111	3	45.731
$\beta_4\ L_I M_{II}$				2.00095	6	6.1963
$\beta_3\ L_I M_{III}$	2.0421	4	6.071	1.96241	3	6.3180
$\gamma_2\ L_I N_{II}$				1.66044	6	7.4668
$\gamma_3\ L_I N_{III}$				1.65601	3	7.4867
$\gamma_4\ L_I O_{II,III}$				1.60728	3	7.7137
$\eta\ L_{II} M_I$				2.21824	3	5.5892
$\beta_1\ L_{II} M_{IV}$	2.0797	4	5.961	1.99806	3	6.2051
$\gamma_5\ L_{II} N_I$				1.77934	3	6.9678
$\gamma_1\ L_{II} N_{IV}$	1.7989	9	6.892	1.72724	3	7.1780
$\gamma_6\ L_{II} O_{IV}$				1.6966	9	7.3076
$l\ L_{III} M_I$				2.4823	4	4.9945

Designation	Å*	p.e.	keV	Å*	p.e.	keV
	61 Promethium (*Cont.*)			**62 Samarium** (*Cont.*)		
$\alpha_2\ L_{III} M_{IV}$	2.2926	4	5.4078	2.21062	3	5.6084
$\alpha_1\ L_{III} M_V$	2.2822	3	5.4325	2.1998	2	5.6361
$\beta_6\ L_{III} N_I$				1.94643	3	6.3697
$\beta_{2,15}\ L_{III} N_{IV,V}$	1.9559	6	6.339	1.88221	3	6.5870
$\beta_7\ L_{III} O_I$				1.85626	3	6.6791
$\beta_5 L_{III} O_{IV,V}$				1.84700	9	6.7126
$\beta_{10}\ L_I M_{IV}$				1.86990	3	6.6304
$\beta_9\ L_I M_V$				1.86166	3	6.6597
$\gamma\ M_{III} N_{IV,V}$				9.600	9	1.291
$\beta\ M_{IV} N_{VI}$				11.27	1	1.0998
$\zeta\ M_V N_{III}$				14.91	4	0.831
$\alpha\ M_V N_{VI,VII}$				11.47	3	1.081
$N_{IV,V} N_{VI,VII}$				98.	1	0.126
$N_{IV,V} O_{II,III}$				117.4	4	0.1056
	63 Europium			**64 Gadolinium**		
$\alpha_2\ KL_{II}$	0.303118	2	40.9019	0.293038	2	42.3089
$\alpha_1\ KL_{III}$	0.298446	2	41.5422	0.288353	2	42.9962
$\beta_3\ KM_{II}$	0.264332	5	46.9036	0.25534	2	48.555
$\beta_1\ KM_{III}$	0.263577	5	47.0379	0.25460	2	48.697
$\beta_2\ KN_{II,III}$	0.256923	8	48.256	0.24816	3	49.959
$KO_{II,III}$	0.255645	7	48.497	0.24687	3	50.221
$\beta_5\ KM_{IV,V}$				0.25275	3	49.052
$\beta_4\ L_I M_{II}$	1.9255	2	6.4389	1.8540	2	6.6871
$\beta_3\ L_I M_{III}$	1.8867	2	6.5713	1.8150	2	6.8311
$\gamma_2\ L_I N_{II}$	1.5961	2	7.7677	1.5331	2	8.087
$\gamma_3\ L_I N_{III}$	1.5903	2	7.7961	1.5297	2	8.105
$\gamma_4\ L_I O_{II,III}$	1.5439	1	8.0304	1.4839	2	8.355
$\eta\ L_{II} M_I$	2.1315	2	5.8166	2.0494	1	6.0495
$\beta_1\ L_{II} M_{IV}$	1.9203	2	6.4564	1.8468	2	6.7132
$\gamma_5\ L_{II} N_I$	1.7085	2	7.2566	1.6412	2	7.5543
$\gamma_1\ L_{II} N_{IV}$	1.6574	2	7.4803	1.5924	2	7.7858
$\gamma_8\ L_{II} O_I$	1.6346	2	7.5849	1.5707	2	7.894
$\gamma_6\ L_{II} O_{IV}$	1.6282	2	7.6147	1.5644	2	7.925
$l\ L_{III} M_I$	2.3948	2	5.1772	2.3122	2	5.3621
$\alpha_2\ L_{III} M_{IV}$	2.1315	2	5.8166	2.0578	2	6.0250
$\alpha_1\ L_{III} M_V$	2.1209	2	5.8457	2.0468	2	6.0572
$\beta_6\ L_{III} N_I$	1.8737	2	6.6170	1.8054	2	6.8671
$\beta_{2,15}\ L_{III} N_{IV,V}$	1.8118	2	6.8432	1.7455	2	7.1028
$\beta_7\ L_{III} O_I$	1.7851	2	6.9453	1.7203	2	7.2071
$\beta_5\ L_{III} O_{IV,V}$	1.7772	2	6.9763	1.7130	2	7.2374
$\beta_{10}\ L_I M_{IV}$	1.7993	3	6.890	1.7315	3	7.160
$\beta_9\ L_I M_V$	1.7916	3	6.920	1.7240	3	7.192
$L_I O_{IV,V}$				1.4807	3	8.373
$\gamma\ M_{III} N_{IV,V}$	9.211	9	1.346	8.844	9	1.402
$\beta\ M_{IV} N_{VI}$	10.750	7	1.1533	10.254	6	1.2091
$\zeta\ M_V N_{III}$	14.22	2	0.872	13.57	2	0.914
$\alpha\ M_V N_{VI,VII}$	10.96	3	1.131	10.46	3	1.185
$N_{IV,V} O_{II,III}$	112.0	6	0.1107			
	65 Terbium			**66 Dysprosium**		
$\alpha_2\ KL_{II}$	0.283423	2	43.7441	0.274247	2	45.2078
$\alpha_1\ KL_{III}$	0.278724	2	44.4816	0.269533	2	45.9984
$\beta\ KM_{II}$	0.24683	2	50.229	0.23862	2	51.957

Table 3.7-13 (continued)
X-RAY CRITICAL-ABSORPTION AND EMISSION ENERGIES IN keV

X-Ray Wavelengths

Designation	65 Terbium (Cont.)			66 Dysprosium (Cont.)		
	Å*	p.e.	keV	Å*	p.e.	keV
$\beta_1\ KM_{III}$	0.24608	2	50.382	0.23788	2	52.119
$\beta_2\ KN_{II,III}$	0.2397†	2	51.72	0.2317†	2	53.51
$KO_{II,III}$	0.23858	3	51.965	0.23056	3	53.774
$\beta_5\ KM_{IV,V}$				0.23618	3	52.494
$\beta_4\ L_IM_{II}$	1.7864	2	6.9403	1.72103	7	7.2039
$\beta_3\ L_IM_{III}$	1.7472	2	7.0959	1.6822	2	7.3702
$\gamma_2\ L_IN_{II}$	1.4764	2	8.398	1.42278	7	8.7140
$\gamma_3\ L_IN_{III}$	1.4718	2	8.423	1.41640	7	8.7532
$\gamma_4\ L_IO_{II,III}$	1.4276	2	8.685	1.37459	7	9.0195
$\eta\ L_{II}M_I$	1.9730	2	6.2839	1.89743	7	6.5342
$\beta_1\ L_{II}M_{IV}$	1.7768	3	6.978	1.71062	7	7.2477
$\gamma_5\ L_{II}N_I$	1.5787	2	7.8535	1.51824	7	8.1661
$\gamma_1\ L_{II}N_{IV}$	1.5303	2	8.102	1.47266	7	8.4188
$\gamma_8\ L_{II}O_I$	1.5097	2	8.212			
$\gamma_6\ L_{II}O_{IV}$	1.5035	2	8.246	1.44579	7	8.5753
$l\ L_{III}M_I$	2.2352	2	5.5467	2.15877	7	5.7431
$\alpha_2\ L_{III}M_{IV}$	1.9875	2	6.2380	1.91991	3	6.4577
$\alpha_1\ L_{II}M_V$	1.9765	2	6.2728	1.90881	3	6.4952
$\beta_6\ L_{III}N_I$	1.7422	2	7.1163	1.68213	7	7.3705
$\beta_{2,15}\ L_{III}N_{IV,V}$	1.6830	2	7.3667	1.62369	7	7.6357
$\beta_7\ L_{III}O_I$	1.6585	2	7.4753	1.60447	7	7.7272
$\beta_5\ L_{III}O_{IV,V}$	1.6510	2	7.5094	1.58837	7	7.8055
$\beta_{10}\ L_IM_{IV}$	1.6673	3	7.436	1.60743	9	7.7130
$\beta_9\ L_IM_V$				1.59973	9	7.7501
$L_IO_{IV,V}$	1.4228	3	8.714			
$\gamma\ M_{III}N_{IV,V}$	8.486	9	1.461	8.144	9	1.522
$\beta\ M_{IV}N_{VI}$	9.792	6	1.2661	9.357	6	1.3250
$\zeta\ M_VN_{III}$	12.98	2	0.955	12.43	2	0.998
$\alpha\ M_VN_{VI,VII}$	10.00	2	1.240	9.59	2	1.293
$N_{IV,V}N_{VI,VII}$	86.	1	0.144	83.	1	0.149
$N_{IV,V}O_{II,III}$	102.2	4	0.1213	97.2	8	0.128

Designation	67 Holmium			68 Erbium		
	Å*	p.e.	keV	Å*	p.e.	keV
$\alpha_2\ KL_{II}$	0.265486	2	46.6997	0.257110	2	48.2211
$\alpha_1\ KL_{III}$	0.260756	2	47.5467	0.252365	2	49.1277
$\beta_3\ KM_{II}$	0.23083	2	53.711	0.22341	2	55.494
$\beta_1\ KM_{III}$	0.23012	2	53.877	0.22266	2	55.681
$\beta_2\ KN_{II,III}$	0.2241†	2	55.32	0.2167†	2	57.21
$KO_{II,III}$	0.22305	3	55.584	0.21581	3	57.450
$\beta_5\ KM_{IV,V}$	0.22855	3	54.246	0.22124	3	56.040
$\beta_4\ L_IM_{II}$	1.6595	2	7.4708	1.6007	1	7.7453
$\beta_3\ L_IM_{III}$	1.6203	2	7.6519	1.5616	1	7.9392
$\gamma_2\ L_IN_{II}$	1.3698	2	9.051	1.3210	2	9.385
$\gamma_3\ L_IN_{III}$	1.3643	2	9.087	1.3146	1	9.4309
$\gamma_4\ L_IO_{II,III}$	1.3225	2	9.374	1.2752	2	9.722
$\eta\ L_{II}M_I$	1.8264	2	6.7883	1.7566	1	7.0579
$\beta_1\ L_{II}M_{IV}$	1.6475	2	7.5253	1.5873	1	7.8109
$\gamma_5\ L_{II}N_I$	1.4618	2	8.481	1.4067	3	8.814
$\gamma_1\ L_{II}N_{IV}$	1.4174	2	8.747	1.3641	2	9.089
$\gamma_8\ L_{II}O_I$	1.3983	2	8.867			
$\gamma_6\ L_{II}O_{IV}$	1.3923	2	8.905	1.3397	3	9.255
$l\ L_{III}M_I$	2.0860	2	5.9434	2.015	1	6.152
$\alpha_2\ L_{III}M_{IV}$	1.8561	2	6.6795	1.7955	6	6.9050
$\alpha_1\ L_{III}M_V$	1.8450	2	6.7198	1.78425	9	6.9487
$\beta_6\ L_{III}N_I$	1.6237	2	7.6359	1.5675	2	7.909
$\beta_{2,15}\ L_{III}N_{IV,V}$	1.5671	2	7.911	1.51399	9	8.1890
$\beta_7\ L_{III}O_I$				1.4941	3	8.298
$\beta_5\ L_{III}O_{IV,V}$	1.5378	2	8.062	1.4848	3	8.350

Designation	67 Holmium (Cont.)			68 Erbium (Cont.)		
	Å*	p.e.	keV	Å*	p.e.	keV
$\beta_{10}\ L_IM_{IV}$	1.5486	3	8.006	1.4941	3	8.298
$L_IO_{IV,V}$	1.3208	3	9.387			
$\beta_9\ L_IM_V$				1.4855	5	8.346
$M_{II}N_{IV}$				7.60	1	1.632
$\gamma\ M_{III}N_{IV,V}$	7.865	9	1.576			
$\gamma\ M_{III}$ v				7.546	8	1.643
$\beta\ M_{IV}N_{VI}$	8.965	4	1.3830	8.592	3	1.4430
$\zeta\ M_VN_{III}$	11.86	1	1.0450	11.37	1	1.0901
$\alpha\ M_VN_{VI,VII}$	9.20	2	1.348	8.82	1	1.406
$N_{IV}N_{VI}$				72.7	9	0.171
$N_VN_{VI,VII}$				76.3	7	0.163

Designation	69 Thulium			70 Ytterbium		
	Å*	p.e.	keV	Å*	p.e.	keV
$\alpha_2\ KL_{II}$	0.249095	2	49.7726	0.241424	2	51.3540
$\alpha_1\ KL_{III}$	0.244338	2	50.7416	0.236655	2	52.3889
$\beta_3\ KM_{II}$	0.21636	2	57.304	0.2096†	1	59.14
$\beta_1\ KM_{III}$	0.21556	2	57.517	0.20884	8	59.37
$\beta_2\ KN_{II,III}$	0.2098†	2	59.09	0.2033†	2	60.98
$KO_{II,III}$	0.20891	2	59.346	0.20226	2	61.298
$\beta_5\ KM_{IV,V}$	0.21404	2	57.923	0.20739	2	59.782
$\beta_4\ L_IM_{II}$	1.5448	2	8.026	1.49138	3	8.3132
$\beta_3\ L_IM_{III}$	1.5063	2	8.231	1.45233	5	8.5367
$\gamma_2\ L_IN_{II}$	1.2742	2	9.730	1.22879	7	10.0897
$\gamma_3\ L_IN_{III}$	1.2678	2	9.779	1.22232	5	10.1431
$\gamma_4\ L_IO_{II,III}$	1.2294	2	10.084	1.1853	1	10.4603
$\eta\ L_{II}M_I$	1.6963	2	7.3088	1.63560	5	7.5802
$\beta_1\ L_{II}M_{IV}$	1.5304	2	8.101	1.47565	5	8.4018
$\gamma_5\ L_{II}N_I$	1.3558	2	9.144	1.3063	1	9.4910
$\gamma_1\ L_{II}N_{IV}$	1.3153	2	9.426	1.26769	5	9.8701
$\gamma_8\ L_{II}O_I$				1.24923	5	9.9246
$\gamma_6\ L_{II}O_{IV}$	1.2905	2	9.607	1.24271	3	9.9766
$l\ L_{III}M_I$	1.9550	2	6.3419	1.89415	5	6.5455
$\alpha_2\ L_{III}M_{IV}$	1.7381	2	7.1331	1.68285	5	7.3673
$\alpha_1\ L_{III}M_V$	1.7268†	2	7.1799	1.67189	4	7.4156
$\beta_6\ L_{III}N_I$	1.5162	2	8.177	1.4661	1	8.4563
$\beta_{2,15}\ L_{III}N_{IV,V}$	1.4640	2	8.468	1.41550	5	8.7588
$\beta_7\ L_{III}O_I$				1.3948	1	8.8889
$\beta_5\ L_{III}O_{IV,V}$	1.4349	2	8.641	1.38696	7	8.9390
$\beta_{10}\ L_IM_{IV}$	1.4410	3	8.604	1.3915	1	8.9100
$\beta_9\ L_IM_V$	1.4336	3	8.648	1.3838	1	8.9597
L_IO_I				1.1886	1	10.4312
$L_IO_{IV,V}$	1.2263	3	10.110	1.1827	1	10.4833
$L_{II}M_{II}$				1.58844	9	7.8052
$L_{II}O_{II,III}$				1.2453	1	9.9561
$t\ L_{III}M_{II}$				1.83091	9	6.7715
$L_{III}O_{II,III}$				1.3898	1	9.8209
$M_{III}N_I$				8.470	9	1.464
$\gamma\ M_{III}N_V$				7.024	8	1.765
$\beta\ M_{IV}N_{VI}$	8.249	7	1.503	7.909	2	1.5675
$\zeta\ M_VN_{III}$				10.48	1	1.183
$\alpha\ M_VN_{VI,VII}$	8.48	1	1.462	8.149	5	1.5214
$N_{IV}N_{VI}$				65.1	7	0.190
$N_VN_{VI,VII}$				69.3	5	0.179

Designation	71 Lutetium			72 Hafnium		
	Å*	p.e.	keV	Å*	p.e.	keV
$\alpha_2\ KL_{II}$	0.234081	2	52.9650	0.227024	3	54.6114
$\alpha_1\ KL_{III}$	0.229298	2	54.0698	0.222227	3	55.7902
$\beta_3\ KM_{II}$	0.20309†	4	61.05	0.19686†	4	62.98

Table 3.7-13 (continued)
X-RAY CRITICAL-ABSORPTION AND EMISSION ENERGIES IN keV

X-Ray Wavelengths

Designation	Å*	p.e.	keV	Å*	p.e.	keV	Designation	Å*	p.e.	keV	Å*	p.e.	keV
	71 Lutetium (*Cont.*)			**72 Hafnium** (*Cont.*)				**73 Tantalum** (*Cont.*)			**74 Tungsten** (*Cont.*)		
$\beta_1 KM_{III}$	0.20231†	3	61.283	0.19607†	3	63.234	$KO_{II,III}$	0.184031	7	67.370	0.178444	5	69.479
$\beta_2 KN_{II,III}$	0.1969†	2	62.97	0.1908†	2	64.98	KL_I				0.21592	4	57.42
$KO_{II,III}$	0.19589	2	63.293				$\beta_3^{II} KM_{IV}$	0.188920	6	65.626	0.183264	5	67.652
$\beta_5 KM_{IV,V}$	0.20084	2	61.732				$\beta_4^{I} KM_V$	0.188757	6	65.683	0.183092	7	67.715
$\beta_4 L_I M_{II}$	1.44056	5	8.6064	1.39220	5	8.9054	$\beta_4 KN_{IV,V}$	0.18451	1	67.194	0.17892	2	69.294
$\beta_3 L_I M_{III}$	1.40140	5	8.8469	1.35300	5	9.1634	$\beta_4 L_I M_{II}$	1.34581	3	9.2124	1.30162	5	9.5252
$\gamma_2 L_I N_{II}$	1.1853	2	10.460	1.14442	5	10.8335	$\beta_3 L_I M_{III}$	1.30678	3	9.4875	1.26269	5	9.8188
$\gamma_3 L_I N_{III}$	1.17953	4	10.5110	1.13841	5	10.8907	$\gamma_2 L_I N_{II}$	1.1053	1	11.217	1.06806	3	11.6080
$\gamma'_4 L_I O_{II}$				1.10376	5	11.2326	$\gamma_3 L_I N_{III}$	1.09936	4	11.2776	1.06200	6	11.6743
$\gamma_4 L_I O_{II,III}$	1.1435	1	10.8425	1.10303	5	11.2401	$\gamma'_4 L_I O_{II}$	1.06544	3	11.6366	1.02863	3	12.0530
$\eta L_{II} M_I$	1.5779	1	7.8575	1.52325	5	8.1393	$\gamma_4 L_I O_{III}$	1.06467	3	11.6451	1.02775	3	12.0634
$\beta_1 L_{II} M_{IV}$	1.42359	3	8.7090	1.37410	5	9.0227	$\eta L_{II} M_I$	1.47106	5	8.4280	1.42110	3	8.7243
$\gamma_5 L_{II} N_I$	1.2596	1	9.8428	1.21537	5	10.2011	$\beta_1 L_{II} M_{IV}$	1.32698	3	9.3431	1.281809	9	9.67235
$\gamma_1 L_{II} N_{IV}$	1.22228	4	10.1434	1.17900	5	10.5158	$\gamma_5 L_{II} N_I$	1.1729	1	10.5702	1.13235	3	10.9490
$\gamma_8 L_{II} O_I$	1.2047	1	10.2915	1.16138	5	10.6754	$\gamma_1 L_{II} N_{IV}$	1.13794	3	10.8952	1.09855	3	11.2895
$\gamma_6 L_{II} O_{IV}$	1.1987	1	10.3431	1.15519	5	10.7325	$\gamma_8 L_{II} O_I$	1.1205	1	11.0646	1.08113	4	11.4677
$L_{III} M_I$	1.8360	1	6.7528	1.78145	5	6.9596	$\gamma_6 L_{II} O_{IV}$	1.11388	3	11.1306	1.07448	5	11.5387
$\alpha_2 L_{III} M_{IV}$	1.63029	5	7.6049	1.58046	5	7.8446	$l L_{III} M_I$	1.72841	5	7.1731	1.6782	1	7.3878
$\alpha_1 L_{III} M_{IV}$	1.61951	3	7.6555	1.56958	5	7.8990	$\alpha_2 L_{III} M_{IV}$	1.53293	2	8.0879	1.48743	2	8.3352
$\beta_6 L_{III} N_I$	1.4189	1	8.7376	1.37410	5	9.0227	$\alpha_1 L_{III} M_V$	1.52197	2	8.1461	1.47639	2	8.3976
$\beta_{15} L_{III} N_{IV}$	1.3715	1	9.0395	1.32783	5	9.3371	$\beta_6 L_{III} N_I$	1.33094	8	9.3153	1.28989	7	9.6117
$\beta_2 L_{III} N_V$	1.37012	3	9.0489	1.32639	5	9.3473	$\beta_{15} L_{III} N_{IV}$	1.28619	5	9.6394	1.24631	3	9.9478
$\beta_7 L_{III} O_I$	1.34949	5	9.1873	1.30564	5	9.4958	$\beta_2 L_{III} N_V$	1.28454	2	9.6518	1.24460	3	9.9615
$\beta_5 L_{III} O_{IV,V}$	1.34183	7	9.2397	1.29761	5	9.5546	$\beta_7 L_{III} O_I$	1.26385	5	9.8098	1.22400	4	10.1292
$L_I M_I$				1.43025	9	8.6685	$\beta_5 L_{III} O_{IV,V}$	1.2555	1	9.8750	1.21545	3	10.2004
$\beta_{10} L_I M_{IV}$	1.3430	2	9.232	1.29819	9	9.5503	$L_I M_I$				1.3365	3	9.277
$\beta_9 L_I M_V$	1.3358	1	9.2816	1.29025	9	9.6090	$\beta_{10} L_I M_{IV}$	1.2537	2	9.889	1.21218	3	10.2279
$L_I N_{IV}$	1.16227	9	10.6672	1.12250	9	11.0451	$\beta_9 L_I M_V$	1.2466	2	9.946	1.20479	7	10.2907
$\gamma_{11} L_I N_V$	1.16107	9	10.6782	1.12146	9	11.0553	$L_I N_I$	1.11521	9	11.1173			
$L_I O_I$				1.10664	9	11.2034	$L_I N_{IV}$	1.08377	7	11.4398	1.0468	2	11.844
$L_I O_{IV}$				1.10086	9	11.2622	$\gamma_{11} L_I N_V$	1.08205	7	11.4580	1.0458	1	11.856
$L_{II} M_{II}$	1.53333	9	8.0858	1.48064	9	8.3735	$L_I N_{VI,VII}$	1.06357	9	11.6570			
$\beta_{17} L_{II} M_{III}$				1.43643	9	8.6312	$L_I O_I$	1.06771	9	11.6118	1.0317	3	12.017
$L_{II} N_V$				1.17788	9	10.5258	$L_I O_{IV,V}$	1.06192	9	11.6752	1.0250	2	12.095
$L_{II} N_{VI}$				1.15830	9	10.7037	$L_{II} M_{II}$	1.43048	9	8.6671			
$L_{II} O_{II,III}$	1.2014	1	10.3198				$\beta_{17} L_{II} M_{III}$	1.3864	1	8.9428	1.3387	2	9.261
$L_{III} M_{II}$	1.7760	1	6.9810	1.72305	9	7.1954	$L_{II} M_V$	1.31897	9	9.3998	1.2728	2	9.741
$L_{III} M_{III}$				1.66346	9	7.4532	$L_{II} N_I$	1.1600	2	10.688	1.1218	3	11.052
$L_{III} N_{II}$				1.35887	9	9.1239	$L_{II} N_{III}$	1.1553	1	10.7316	1.1149	2	11.120
$L_{III} N_{III}$				1.35053	9	9.1802	$L_{II} N_V$	1.13687	9	10.9055			
$L_{III} N_{VI,VII}$				1.30165	9	9.5249	$\nu L_{II} N_{VI}$	1.1158	1	11.1113	1.0771	1	11.510
$L_{III} O_{II,III}$	1.34524	9	9.2163				$L_{II} O_{II}$	1.11789	9	11.0907			
$M_{II} N_I$				7.887	9	1.572	$L_{II} O_{III}$	1.11693	9	11.1001	1.0792	2	11.488
$M_{III} N_V$	6.768	6	1.832	6.544	4	1.895	$t L_{III} M_{II}$	1.67265	9	7.4123	1.6244	3	7.632
				9.686	7	1.2800	$s L_{III} M_{III}$	1.61264	9	7.6881	1.5642	3	7.926
$M_{IV} N_{VI}$	7.601	2	1.6312	7.303	1	1.6976	$L_{III} N_{II}$	1.3167	1	9.4158	1.2765	2	9.712
				9.686	7	1.2800	$L_{III} N_{III}$	1.3086	1	9.4742	1.2672	2	9.784
$M_V N_{VI,VII}$	7.840	2	1.5813	7.539	1	1.6446	$u L_{III} N_{VI,VII}$	1.25778	4	9.8572	1.21868	5	10.1733
$N_{IV} N_{VI}$	63.0	5	0.197				$L_{III} O_{II,III}$	1.2601	3	9.839	1.2211	2	10.153
$\nu N_{VI,VII}$	65.7	2	0.1886				$M_I N_{III}$	5.40	2	2.295	5.172	9	2.397
	73 Tantalum			**74 Tungsten**			$M_I O_{II,III}$				4.44	2	2.79
KL_{II}	0.220305	8	56.277	0.213828	2	57.9817	$M_{II} N_I$				6.28	2	1.973
KL_{III}	0.215497	4	57.532	0.2090100	Std	59.31824	$M_{II} N_{IV}$	5.570	4	2.226	5.357	2	2.314
KM_{II}	0.190890	2	64.9488	0.185181	2	66.9514	$M_{III} N_I$	7.612	9	1.629	7.360	8	1.684
KM_{III}	0.190089	4	65.223	0.184374	2	67.2443	$M_{III} N_{IV}$	6.353	5	1.951	6.134	4	2.021
KN_{II}	0.185188	9	66.949	0.17960	1	69.031	$\gamma M_{III} N_V$	6.312	4	1.964	6.092	3	2.035
KN_{III}	0.185011	8	67.013	0.179421	7	69.101	$M_{III} O_I$	5.83	2	2.126	5.628	8	2.203
							$M_{III} O_{IV,V}$	5.67	3	2.19			

Table 3.7-13 (continued)
X-RAY CRITICAL-ABSORPTION AND EMISSION ENERGIES IN keV

X-Ray Wavelengths

73 Tantalum (Cont.) / 74 Tungsten (Cont.)

Designation	Å*	p.e.	keV	Å*	p.e.	keV
$\zeta_2\ M_{IV}N_{II}$	9.330	5	1.3288	8.993	5	1.3787
$M_{IV}N_{III}$	8.90	2	1.393	8.573	8	1.446
$\beta\ M_{IV}N_{VI}$	7.023	1	1.7655	6.757	1	1.8349
$M_{IV}O_{II}$	7.09	2	1.748	6.806	9	1.822
$\zeta_1\ M_V N_{III}$	9.316	4	1.3308	8.962	4	1.3835
$\alpha\ M_V N_{VI,VII}$	7.252	1	1.7096			
$\alpha_2\ M_V N_{VI}$				6.992	2	1.7731
$\alpha_1\ M_V N_{VII}$				6.983	1	1.7754
$M_V O_{III}$	7.30	2	1.700	7.005	9	1.770
$N_{II}N_{IV}$				54.0	2	0.2295
$N_{IV}N_{VI}$	58.2	1	0.2130	55.8	1	0.2221
$N_V N_{VI,VII}$	61.1	2	0.2028			
$N_V N_{VI}$				59.5	3	0.208
$N_V N_{VII}$				58.4	1	0.2122

75 Rhenium / 76 Osmium

Designation	Å*	p.e.	keV	Å*	p.e.	keV
$\alpha_2\ K L_{II}$	0.207611	1	59.7179	0.201639	2	61.4867
$\alpha_1\ K L_{III}$	0.202781	2	61.1403	0.196794	2	63.0005
$\beta_3\ K M_{II}$	0.179697	3	68.994	0.174431	3	71.077
$\beta_1\ K M_{III}$	0.178880	3	69.310	0.173611	3	71.413
$\beta_2^{II}\ K N_{II}$	0.17425	1	71.151	0.16910	1	73.318
$\beta_2^{I}\ K N_{III}$	0.174054	6	71.232	0.168906	6	73.402
$K O_{II,III}$	0.17308	1	71.633	0.16798	1	73.808
$\beta_5^{II}\ K M_{IV}$	0.17783	1	69.719	0.17262	1	71.824
$\beta_5^{I}\ K M_V$	0.17766	1	69.786	0.17245	1	71.895
$\beta_4\ K N_{IV,V}$	0.17362	2	71.410	0.16842	2	73.615
$\beta_4\ L_I M_{II}$	1.25917	5	9.8463	1.21844	5	10.1754
$\beta_3\ L_I M_{III}$	1.22031	5	10.1598	1.17955	7	10.5108
$\gamma_2\ L_I N_{II}$	1.03233	5	12.0098	0.99805	5	12.4224
$\gamma_3\ L_I N_{III}$	1.02613	7	12.0824	0.99186	5	12.4998
$\gamma'_4\ L_I O_{II}$	0.99334	5	12.4813	0.96033	8	12.910
$\gamma_4\ L_I O_{III}$	0.99249	5	12.4920	0.95938	8	12.923
$\eta\ L_{II}M_I$	1.37342	5	9.0272	1.32785	7	9.3370
$\beta_1\ L_{II}M_{IV}$	1.23858	2	10.0100	1.19727	7	10.3553
$\gamma_5\ L_{II}N_I$	1.09388	5	11.3341	1.05693	5	11.7303
$\gamma_1\ L_{II}N_{IV}$	1.06099	5	11.6854	1.02503	5	12.0953
$\gamma_8\ L_{II}O_I$	1.04398	5	11.8758	1.00788	5	12.3012
$\gamma_6\ L_{II}O_{IV}$	1.03699	9	11.956	1.00107	5	12.3848
$\iota\ L_{III}M_I$	1.63056	5	7.6036	1.58498	7	7.8222
$\alpha_2\ L_{III}M_{IV}$	1.44396	5	8.5862	1.40234	5	8.8410
$\alpha_1\ L_{III}M_V$	1.43290	4	8.6525	1.39121	5	8.9117
$\beta_6\ L_{III}N_I$	1.25100	5	9.9105	1.21349	5	10.2169
$\beta_{15}\ L_{III}N_{IV}$	1.20819	5	10.2617	1.17167	5	10.5816
$\beta_2\ L_{III}N_V$	1.20660	4	10.2752	1.16979	8	10.5985
$\beta_7\ L_{III}O_I$	1.18610	5	10.4529	1.14933	8	10.7872
$\beta_5\ L_{III}O_{IV,V}$	1.17721	5	10.5318	1.1405	1	10.8711
$\beta_{10}\ L_I M_{IV}$	1.17218	5	10.5770	1.13353	5	10.9376
$\beta_9\ L_I M_V$	1.16487	4	10.6433	1.12637	6	11.0071
$L_I N_I$	1.0420	1	11.899			
$L_I N_{IV}$	1.0119	1	12.252	0.9772	3	12.687
$\gamma_{11}\ L_I N_V$	1.0108	1	12.266	0.9765	3	12.696
$L_I O_I$	0.9965	1	12.442	0.96318	7	12.8721
$L_I O_{IV,V}$	0.9900	1	12.524	0.95603	5	12.9683
$L_{II}M_{II}$	1.3366	1	9.2761	1.2934	2	9.586
$\beta_{17}\ L_{II}M_{III}$	1.2927	1	9.5910	1.2480	2	9.934

75 Rhenium (Cont.) / 76 Osmium (Cont.)

Designation	Å*	p.e.	keV	Å*	p.e.	keV
$L_{II}M_V$	1.2305	1	10.0753	1.18977	7	10.4205
$L_{II}N_{II}$	1.0839	1	11.438			
$L_{II}N_{III}$	1.0767	1	11.515	1.03973	5	11.9243
$v\ L_{II}N_{VI}$	1.0404	1	11.917	1.0050	2	12.337
$L_{II}O_{III}$	1.0397	1	11.925	1.0047	2	12.340
$t\ L_{III}M_{II}$	1.5789	1	7.8525	1.5347	2	8.079
$s\ L_{III}M_{III}$	1.5178	1	8.1682	1.4735	2	8.414
$L_{III}N_I$				1.20086	7	10.3244
$L_{III}N_{III}$	1.2283	1	10.0933			
$u\ L_{III}N_{VI,VII}$	1.1815	1	10.4931	1.14537	7	10.8245
$M_I N_{III}$				4.79	2	2.59
$M_{II}N_I$				5.81	2	2.133
$M_{II}N_{IV}$				4.955	4	2.502
$M_{III}N_I$				6.89	2	1.798
$M_{III}N_{IV}$	5.931	5	2.090	5.724	5	2.166
$\gamma\ M_{III}N_V$	5.885	2	2.1067	5.682	4	2.182
$\zeta_2\ M_{IV}N_{II}$	8.664	5	1.4310	8.359	5	1.4831
$M_{IV}N_{III}$	8.239	8	1.505			
$\beta\ M_{IV}N_{VI}$	6.504	1	1.9061	6.267	1	1.9783
$\zeta_1\ M_V N_{III}$	8.629	4	1.4368	8.310	4	1.4919
$\alpha\ M_V N_{VI,VII}$	6.729	1	1.8425	6.490	1	1.9102
$N_{IV}N_{VI}$				51.9	1	0.2388
$N_V N_{VI,VII}$				54.7	2	0.2266

77 Iridium / 78 Platinum

Designation	Å*	p.e.	keV	Å*	p.e.	keV
$\alpha_2\ K L_{II}$	0.195904	2	63.2867	0.190381	4	65.122
$\alpha_1\ K L_{III}$	0.191047	2	64.8956	0.185511	4	66.832
$\beta_3\ K M_{II}$	0.169367	2	73.2027	0.164501	3	75.368
$\beta_1\ K M_{III}$	0.168542	2	73.5608	0.163675	3	75.748
$\beta_2^{II}\ K N_{II}$	0.16415	1	75.529	0.15939	1	77.785
$\beta_2^{I}\ K N_{III}$	0.163956	7	75.619	0.15920	1	77.878
$K O_{II,III}$	0.163019	6	76.053	0.15826	1	78.341
$\beta_5^{II}\ K M_{IV}$	0.16759	2	73.980	0.16271	2	76.199
$\beta_5^{I}\ K M_V$	0.167373	9	74.075	0.16255	3	76.27
$\beta_4\ K N_{IV,V}$	0.16352	2	75.821	0.15881	2	78.069
$\beta_4\ L_I M_{II}$	1.17958	3	10.5106	1.14223	5	10.8543
$\beta_3\ L_I M_{III}$	1.14085	3	10.8674	1.10394	5	11.2308
$\gamma_2\ L_I N_{II}$	0.96545	3	12.8418	0.93427	5	13.270
$\gamma_3\ L_I N_{III}$	0.95931	5	12.9240	0.92791	5	13.361
$\gamma'_4\ L_I O_{II}$	0.92831	3	13.3555	0.89747	4	13.814
$\gamma_4\ L_I O_{III}$	0.92744	3	13.3681	0.89659	4	13.828
$\eta\ L_{II}M_I$	1.28448	3	9.6522	1.2429	2	9.975
$\beta_1\ L_{II}M_{IV}$	1.15781	3	10.7083	1.11990	2	11.070
$\gamma_5\ L_{II}N_I$	1.02175	5	12.1342	0.9877	2	12.552
$\gamma_1\ L_{II}N_{IV}$	0.99085	3	12.5126	0.95797	3	12.942
$\gamma_8\ L_{II}O_I$	0.97409	3	12.7279	0.9411	1	13.173
$\gamma_6\ L_{II}O_{IV}$	0.96708	4	12.8201	0.9342	2	13.271
$\iota\ L_{III}M_I$	1.54094	3	8.0458	1.4995	2	8.268
$\alpha_2\ L_{III}M_{IV}$	1.36250	5	9.0995	1.32432	2	9.361
$\alpha_1\ L_{III}M_V$	1.35128	3	9.1751	1.31304	3	9.442
$\beta_6\ L_{III}N_I$	1.17796	3	10.5251	1.14355	5	10.841
$\beta_{15}\ L_{III}N_{IV}$	1.13707	3	10.9036			
$\beta_2\ L_{III}N_V$	1.13532	3	10.9203	1.10200	3	11.250
$\beta_7\ L_{III}O_I$	1.11489	3	11.1205	1.08168	3	11.461

Table 3.7-13 (continued)
X-RAY CRITICAL-ABSORPTION AND EMISSION ENERGIES IN keV

X-Ray Wavelengths

Left section

Designation	Å*	p.e.	keV	Å*	p.e.	keV
77 Iridium (*Cont.*)				**78 Platinum** (*Cont.*)		
$\beta_5\ L_{III}O_{IV,V}$	1.10585	3	11.2114	1.0724	2	11.561
$L_I M_I$	1.2102	2	10.245	1.16962	9	10.6001
$\beta_{10}\ L_I M_{IV}$	1.09702	4	11.3016	1.06183	7	11.6762
$\beta_9\ L_I M_V$	1.08975	5	11.3770	1.05446	5	11.7577
$L_I N_I$	0.9766	2	12.695	0.9455	2	13.113
$L_I N_{IV}$	0.9459	2	13.108			
$\gamma_{11}\ L_I N_V$	0.9446	2	13.126	0.9143	2	13.560
$L_I O_{IV,V}$	0.9243	3	13.413			
$L_I O_I$				0.8995	2	13.784
$L_I O_{IV}$				0.8943	1	13.864
$L_I O_V$				0.8934	1	13.878
$L_{II} M_{II}$	1.2502	3	9.917	1.213	1	10.225
$\beta_{17}\ L_{II} M_{III}$	1.2069	2	10.273	1.1667	1	10.6265
$L_{II} M_V$	1.1489	2	10.791	1.1129	2	11.140
$L_{II} N_{II}$	1.0120	2	12.251	0.9792	2	12.661
$L_{II} N_{III}$	1.0054	3	12.332	0.97173	4	12.7588
$v\ L_{II} N_{VI}$	0.97161	6	12.7603	0.93931	5	13.1992
$L_{II} O_{III}$	0.96979	5	12.7843			
$t\ L_{III} M_{II}$	1.4930	3	8.304	1.4530	2	8.533
$s\ L_{III} M_{III}$	1.4318	2	8.659	1.3895	2	8.923
$L_{III} N_{II}$	1.16545	5	10.6380	1.1310	2	10.962
$L_{III} N_{III}$	1.1560	3	10.725	1.1226	2	11.044
$u\ L_{III} N_{VI,VII}$	1.11145	4	11.1549	1.07896	5	11.4908
$L_{III} O_{II,III}$	1.10923	6	11.1772	1.0761	3	11.521
$M_I N_{III}$	4.631†	9	2.677	4.460	2	2.780
$M_{II} N_{IV}$	4.780	4	2.594	4.601	4	2.695
$M_{III} N_I$	6.669	9	1.859	6.455	9	1.921
$M_{III} N_{IV}$	5.540	5	2.238	5.357	5	2.314
$\gamma\ M_{II} N_V$	5.500	4	2.254	5.319	4	2.331
$M_{III} O_I$				4.876	9	2.543
$M_{III} O_{IV,V}$	4.869	9	2.546	4.694	8	2.641
$\zeta_2\ M_{IV} N_{II}$	8.065	5	1.5373	7.790	5	1.592
$M_{IV} N_{III}$	7.645	8	1.622	7.371	8	1.682
$\beta\ M_{IV} N_{VI}$	6.038	1	2.0535	5.828	1	2.1273
$\zeta_1\ M_V N_{III}$	8.021	4	1.5458	7.738	4	1.6022
$\alpha_2\ M_V N_{VI}$	6.275	3	1.9758	6.058	3	2.047
$\alpha_1\ M_V N_{VII}$	6.262	1	1.9799	6.047	1	2.0505
$M_V O_{III}$				5.987	9	2.071
$N_{IV} N_{VI}$	50.2	1	0.2470	48.1	2	0.258
$N_V N_{VI,VII}$	52.8	1	0.2348	50.9	2	0.2436
79 Gold				**80 Mercury**		
$\alpha_2\ K L_{II}$	0.185075	2	66.9895	0.179958	3	68.895
$\alpha_1\ K L_{III}$	0.180195	2	68.8037	0.175068	3	70.819
$\beta_2\ K M_{II}$	0.159810	2	77.580	0.155321	3	79.822
$\beta_1\ K M_{III}$	0.158982	3	77.984	0.154487	3	80.253
$\beta_2^{II}\ K N_{II}$	0.15483	2	80.08	0.15040	2	82.43
$\beta_2^{I}\ K N_{III}$	0.154618	9	80.185	0.15020	2	82.54
$K O_{II,III}$	0.153694	7	80.667	0.14931	2	83.04
$K L_I$	0.18672	4	66.40			
$\beta_5^{II}\ K M_{IV}$	0.158062	7	78.438			
$\beta_5^{I}\ K M_V$	0.157880	5	78.529			
$\beta_5\ K M_{IV,V}$				0.15353	2	80.75
$\beta_4\ K N_{IV,V}$	0.154224	5	80.391	0.14978	2	82.78
$\beta_4\ L_I M_{II}$	1.10651	3	11.2047	1.07222	7	11.5630
$\beta_3\ L_I M_{III}$	1.06785	9	11.6103	1.03358	7	11.9953

Right section

Designation	Å*	p.e.	keV	Å*	p.e.	keV
79 Gold (*Cont.*)				**80 Mercury** (*Cont.*)		
$\gamma_2\ L_I N_{II}$	0.90434	3	13.7095	0.87544	7	14.162
$\gamma_3\ L_I N_{III}$	0.89783	5	13.8090	0.86915	7	14.265
$\gamma'_4\ L_I O_{II}$	0.86816	4	14.2809	0.84013	7	14.757
$\gamma_4\ L_I O_{III}$	0.86703	4	14.2996	0.83894	7	14.778
$\eta\ L_{II} M_I$	1.20273	3	10.3083	1.1640	1	10.6512
$\beta_1\ L_{II} M_{IV}$	1.08353	3	11.4423	1.04868	5	11.8226
$\gamma_5\ L_{II} N_I$	0.95559	3	12.9743	0.92453	7	13.410
$\gamma_1\ L_{II} N_{IV}$	0.92650	3	13.3817	0.89646	5	13.8301
$\gamma_8\ L_{II} O_I$	0.90989	5	13.6260	0.87995	7	14.090
$\gamma_6\ L_{II} O_{IV}$	0.90297	3	13.7304	0.87319	7	14.199
$l\ L_{III} M_I$	1.45964	9	8.4939	1.4216	1	8.7210
$\alpha_2\ L_{III} M_{IV}$	1.28772	3	9.6280	1.25264	7	9.8976
$\alpha_1\ L_{III} M_V$	1.27640	3	9.7133	1.24120	5	9.9888
$\beta_6\ L_{III} N_I$	1.11092	3	11.1602	1.07975	7	11.4824
$\beta_{15}\ L_{III} N_{IV}$	1.07188	5	11.5667	1.04151	7	11.9040
$\beta_2\ L_{III} N_V$	1.07022	3	11.5847	1.03975	7	11.9241
$\beta_7\ L_{III} O_I$	1.04974	8	11.8106	1.01937	7	12.1625
$\beta_5\ L_{III} O_{IV,V}$	1.04044	3	11.9163	1.00987	7	12.2769
$L_I M_I$	1.13525	5	10.9210	1.0999	2	11.272
$\beta_{10}\ L_I M_{IV}$	1.02789	7	12.0617	0.9962	2	12.446
$\beta_9\ L_I M_V$	1.02063	7	12.1474	0.9871	2	12.560
$L_I N_I$	0.9131	1	13.578	0.8827	2	14.045
$L_I N_{IV}$	0.88563	7	13.999			
$\gamma_{11}\ L_I N_V$	0.88433	7	14.020	0.85657	7	14.474
$L_I O_I$	0.87074	5	14.2385	0.8452	2	14.670
$L_I O_{IV,V}$	0.86400	5	14.3497	0.8350	2	14.847
$L_{II} M_{II}$	1.1708	1	10.5892	1.1387	5	10.888
$\beta_{17}\ L_{II} M_{III}$	1.12798	5	10.9915	1.0916	5	11.358
$L_{II} M_V$	1.0756	2	11.526			
$L_{II} N_{III}$	0.9402	2	13.186	0.90894	7	13.640
$v\ L_{II} N_{VI}$	0.90837	5	13.6487	0.87885	7	14.107
$L_{II} O_{II}$	0.90746	7	13.662	0.8784	1	14.114
$L_{II} O_{III}$	0.90638	7	13.679	0.8758	1	14.156
$t\ L_{III} M_{II}$	1.41366	7	8.7702	1.3746	2	9.019
$s\ L_{III} M_{III}$	1.35131	7	9.1749	1.3112	2	9.455
$L_{III} N_{II}$	1.09968	7	11.2743	1.0649	2	11.642
$L_{III} N_{III}$	1.09026	7	11.3717	1.0585	1	11.713
$u\ L_{III} N_{VI,VII}$	1.04752	5	11.8357			
$u'\ L_{III} N_{VI}$				1.01769	7	12.1826
$u\ L_{III} N_{VII}$				1.01674	7	12.1940
$L_{III} O_{II,III}$	1.0450	2	11.865			
$L_{III} O_{II}$				1.01558	7	12.2079
$L_{III} O_{III}$				1.01404	7	12.2264
$L_{III} P_{II,III}$	1.03876	7	11.9355			
$M_I N_{III}$	4.300	9	2.883			
$M_{II} N_{IV}$	4.432	4	2.797			
$M_{III} N_I$	6.259	9	1.981	6.09	2	2.036
$M_{III} N_{IV}$	5.186	5	2.391			
$\gamma\ M_{III} N_V$	5.145	4	2.410	4.984†	2	2.4875
$M_{III} O_I$	4.703	9	2.636			
$M_{III} O_{IV,V}$	4.522	6	2.742			
$\zeta_2\ M_{IV} N_{II}$	7.523	5	1.648			
$M_{IV} N_{IV}$	7.101	8	1.746	6.87	2	1.805
$\beta\ M_{IV} N_{VI}$	5.624	1	2.2046	5.4318†	9	2.2825
$\zeta_1\ M_V N_{III}$	7.466	4	1.6605			
$\alpha_2\ M_V N_{VI}$	5.854	3	2.118			

Table 3.7-13 (continued)
X-RAY CRITICAL-ABSORPTION AND EMISSION ENERGIES IN keV

X-Ray Wavelengths

Desig-nation	Å*	p.e.	keV	Å*	p.e.	keV	Desig-nation	Å*	p.e.	keV	Å*	p.e.	keV
								81 Thallium (*Cont.*)			**82 Lead** (*Cont.*)		
	79 Gold (*Cont.*)			**80 Mercury** (*Cont.*)			$L_{II}N_{III}$	0.87996	5	14.0893	0.85192	7	14.553
							$L_{II}N_V$				0.8382	2	14.791
$\alpha_1\,M_VN_{VII}$	5.840	1	2.1229	5.6476†	9	2.1953	$v\,L_{II}N_{VI}$	0.85048	5	14.5777	0.82327	7	15.060
M_VO_{III}	5.767	9	2.150				$L_{II}O_{II}$	0.8490	1	14.604			
$N_{IV}N_{VI}$	46.8	2	0.265	45.2†	3	0.274	$L_{II}O_{III}$				0.8200	1	15.120
$N_VN_{VI,VII}$	49.4	1	0.2510	47.9†	3	0.259	$t\,L_{III}M_{II}$	1.34154	5	9.2417	1.30767	7	9.4811
							$s\,L_{III}M_{III}$	1.27807	5	9.7007	1.24385	7	9.9675
	81 Thallium			**82 Lead**			$L_{III}N_{II}$	1.01040	7	12.2705			
							$L_{III}N_{III}$	1.0286	1	12.053	1.0005	1	12.392
$\alpha_2\,KL_{II}$	0.175036	2	70.8319	0.170294	2	72.8042	$u\,L_{III}N_{VI,VII}$	0.9888	1	12.538	0.96133	7	12.8968
$\alpha_1\,KL_{III}$	0.170136	2	72.8715	0.165376	2	74.9694	$L_{III}O_{II}$	0.98738	5	12.5566	0.9586	1	12.934
$\beta_3\,KM_{II}$	0.150980	6	82.118	0.146810	4	84.450	$L_{III}O_{III}$	0.98538	5	12.5820	0.9578	1	12.945
$\beta_1\,KM_{III}$	0.150142	5	82.576	0.145970	6	84.936	$L_{III}P_{II,III}$	0.97926	5	12.6607	0.95118	7	13.0344
$\beta_2^{II}\,KN_{II}$	0.14614	1	84.836	0.14212	2	87.23	M_IN_{III}	4.013	9	3.089	3.872	9	3.202
$\beta_2^{I}\,KN_{III}$	0.14595	1	84.946	0.14191	1	87.364	$M_{II}N_I$				4.655	8	2.664
$KO_{II,III}$	0.14509	1	85.451	0.141012	8	87.922	$M_{II}N_{IV}$	4.116	4	3.013	3.968	5	3.124
KP				0.1408	1	88.06	$M_{III}N_I$	5.884	8	2.107	5.704	8	2.174
$\beta_5\,KM_{IV,V}$	0.14917	1	83.114				$M_{III}N_{IV}$	4.865	8	2.548	4.715	3	2.630
$\beta_5^{II}\,KM_{IV}$				0.14512	2	85.43	$\gamma\,M_{III}N_V$	4.823	4	2.571	4.674	1	2.6527
$\beta_5^{I}\,KM_V$				0.14495	3	85.53	$M_{III}O_I$				4.244	9	2.921
$\beta_4\,KN_{IV,V}$	0.14553	2	85.19	0.14155	3	87.59	$M_{II}O_{IV,V}$	4.216	6	2.941	4.069	6	3.047
$\beta_4\,L_IM_{II}$	1.03918	3	11.9306	1.0075	1	12.306	$\zeta_2\,M_{IV}N_{II}$	7.032	5	1.763	6.802	5	1.823
$\beta_3\,L_IM_{III}$	1.00062	3	12.3904	0.96911	7	12.7933	$M_{IV}N_{III}$				6.384	7	1.942
$\gamma_2\,L_IN_{II}$	0.84773	5	14.6251	0.8210	2	15.101	$\beta\,M_{IV}N_{VI}$	5.249	1	2.3621	5.076	1	2.4427
$\gamma_3\,L_IN_{III}$	0.84130	4	14.7368	0.8147	1	15.218	$M_{IV}O_{II}$	5.196	9	2.386	5.004	9	2.477
$\gamma'_4\,L_IO_{II}$	0.81308	5	15.2482	0.78706	7	15.752	$\zeta_1\,M_VN_{III}$	6.974	4	1.778	6.740	3	1.8395
$\gamma_4\,L_IO_{III}$	0.81184	5	15.2716	0.7858	1	15.777	$\alpha_2\,M_VN_{VI}$	5.472	2	2.2656	5.299	2	2.3397
$\eta\,L_{II}M_I$	1.12769	3	10.9943	1.09241	7	11.3493	$\alpha_1\,M_VN_{VII}$	5.460	1	2.2706	5.286	1	2.3455
$\beta_1\,L_{II}M_{IV}$	1.01513	4	12.2133	0.98291	3	12.6137	M_VO_{III}				5.168	9	2.399
$\gamma_5\,L_{II}N_I$	0.89500	4	13.8526	0.86655	5	14.3075	$N_{IV}N_{VI}$				42.3	2	0.293
$\gamma_1\,L_{II}N_{IV}$	0.86752	3	14.2915	0.83973	3	14.7644	$N_VN_{VI,VII}$	46.5	2	0.267	45.0	1	0.2756
$\gamma_8\,L_{II}O_I$	0.8513	2	14.564	0.82365	5	15.0527	$N_{VI}O_{IV}$	115.3	2	0.1075	102.4	1	0.1211
$\gamma_6\,L_{II}O_{IV}$	0.8442	2	14.685	0.81683	5	15.1783	$N_{VI}O_V$	113.0	1	0.10968	100.2	2	0.1237
$L_{II}P_I$				0.81583	5	15.1969	$N_{VII}O_V$	117.7	1	0.10530	104.3	1	0.1189
$l\,L_{III}M_I$	1.38477	3	8.9532	1.34990	7	9.1845							
$\alpha_2\,L_{III}M_{IV}$	1.21875	3	10.1728	1.18648	5	10.4495		**83 Bismuth**			**84 Polonium**		
$\alpha_1\,L_{III}M_V$	1.20739	4	10.2685	1.17501	2	10.5515	$\alpha_2\,KL_{II}$	0.165717	2	74.8148	0.16130†	1	76.862
$\beta_6\,L_{III}N_I$	1.04963	5	11.8118	1.0210	1	12.143	$\alpha_1\,KL_{III}$	0.160789	2	77.1079	0.15636†	1	79.290
$\beta_{15}\,L_{III}N_{IV}$	1.01201	3	12.2510	0.98389	7	12.6011	$\beta_3\,KM_{II}$	0.142779	7	86.834	0.13892†	2	89.25
$\beta_2\,L_{III}N_V$	1.01031	3	12.2715	0.98221	7	12.6226	$\beta_1\,KM_{III}$	0.141948	3	87.343	0.13807†	2	89.80
$\beta_7\,L_{III}O_I$	0.99017	5	12.5212	0.9620	1	12.888	$\beta_2^{II}\,KN_{II}$	0.13817	1	89.733	0.13438†	2	92.26
$\beta_5\,L_{III}O_{IV,V}$	0.98058	3	12.6436	0.9526	1	13.015	$\beta_2^{I}\,KN_{III}$	0.13797	1	89.864	0.13418†	2	92.40
L_IM_I	1.0644	2	11.648	1.0323	2	12.010	$KO_{II,III}$	0.13709	1	90.435			
$\beta_{10}\,L_IM_{IV}$	0.96389	7	12.8626	0.9339	2	13.275	$\beta_5\,KM_{IV,V}$	0.14111	1	87.860			
$\beta_9\,L_IM_V$	0.95675	7	12.9585	0.9268	1	13.377	$\beta_4\,KN_{IV,V}$	0.13759	2	90.11			
L_IN_I	0.8549	1	14.503	0.82859	7	14.963	$\beta_4\,L_IM_{II}$	0.97690	4	12.6912	0.9475	3	13.086
L_IN_{IV}	0.83001	7	14.937	0.80364	7	15.427	$\beta_3\,L_IM_{III}$	0.93855	3	13.2098	0.9091	3	13.638
$\gamma_{11}\,L_IN_V$	0.82879	5	14.9593	0.80233	9	15.453	$\gamma_2\,L_IN_{II}$	0.79565	3	15.5824	0.772	1	16.07
$L_IN_{VI,VII}$				0.7884	1	15.725	$\gamma_3\,L_IN_{III}$	0.78917	5	15.7102			
L_IO_I	0.8158	1	15.198	0.7897	1	15.699	$\gamma'_4\,L_IO_{II}$	0.76198	3	16.2709			
$L_IO_{IV,V}$	0.80861	5	15.3327	0.78257	7	15.843	$\gamma_4\,L_IO_{III}$	0.76087	3	16.2947			
$L_{II}M_{II}$	1.0997	1	11.274	1.0644	2	11.648	$\gamma_{13}\,L_IP_{II,III}$	0.75690	3	16.3802			
$\beta_{17}\,L_{II}M_{III}$	1.05609	7	11.7397	1.0223	1	12.127	$\eta\,L_{II}M_I$	1.05856	3	11.7122			
$L_{II}M_V$	1.00722	5	12.3093	0.9747	1	12.720	$\beta_1\,L_{II}M_{IV}$	0.951978	9	13.0235	0.9220	2	13.447
$L_{II}N_{II}$	0.882	2	14.057	0.8585	3	14.442	$\gamma_5\,L_{II}N_I$	0.83923	5	14.7732			

Table 3.7-13 (continued)
X-RAY CRITICAL-ABSORPTION AND EMISSION ENERGIES IN keV

X-Ray Wavelengths

Designation	83 Bismuth (Cont.) Å*	p.e.	keV	84 Polonium (Cont.) Å*	p.e.	keV
$\gamma_1\,L_{II}N_{IV}$	0.81311	2	15.2477	0.78748	9	15.744
$\gamma_8\,L_{II}O_I$	0.7973	1	15.551			
$\gamma_6\,L_{II}O_{IV}$	0.79043	3	15.6853	0.7645	2	16.218
$l\,L_{III}M_I$	1.31610	7	9.4204	1.2829	5	9.664
$\alpha_2\,L_{III}M_{IV}$	1.15536	1	10.73091	1.12548†	5	11.0158
$\alpha_1\,L_{III}M_V$	1.14386	2	10.8388	1.11386	4	11.1308
$\beta_6\,L_{III}N_I$	0.99331	3	12.4816	0.9672	2	12.819
$\beta_{15}\,L_{III}N_{IV}$	0.95702	5	12.9549	0.9312	2	13.314
$\beta_2\,L_{III}N_V$	0.95518	4	12.9799	0.92937	5	13.3404
$\beta_7\,L_{III}O_I$	0.93505	5	13.2593			
$\beta_5\,L_{III}O_{IV,V}$	0.92556	3	13.3953	0.8996	2	13.782
L_1M_I	1.0005	9	12.39			
$\beta_{10}\,L_1M_{IV}$	0.90495	4	13.7002			
$\beta_9\,L_1M_V$	0.89791	3	13.8077			
L_1N_I	0.8022	1	15.456			
L_1N_{IV}	0.7795	5	15.904			
$\gamma_{11}\,L_1N_V$	0.77728	5	15.951			
$L_1N_{VI,VII}$	0.7641	5	16.23			
$L_1O_{IV,V}$	0.75791	5	16.358			
$L_{II}M_{II}$	1.0346	9	11.98			
$\beta_{17}\,L_{II}M_{III}$	0.98913	5	12.5344			
$L_{II}M_V$	0.94419	5	13.1310			
$L_{II}N_{II}$	0.8344	9	14.86			
$L_{II}N_{III}$	0.8248	1	15.031			
$v\,L_{II}N_{VI}$	0.79721	9	15.552			
$L_{II}O_{III}$	0.79384	5	15.6178			
$t\,L_{III}M_{II}$	1.2748	1	9.7252			
$s\,L_{III}M_{III}$	1.2105	1	10.2421			
$L_{III}N_{II}$	0.98280	5	12.6151			
$L_{III}N_{III}$	0.97321	5	12.7394			
$u\,L_{III}N_{VI,VII}$	0.93505	5	13.2593			
$L_{III}O_{II}$	0.9323	2	13.298			
$L_{III}O_{III}$	0.9302	2	13.328			
$L_{III}P_{II,III}$	0.92413	3	13.4159			
M_1N_{II}	3.892	9	3.185			
M_1N_{III}	3.740	9	3.315			
$M_{II}N_{IV}$	3.834	4	3.234			
$M_{III}N_I$	5.537	8	2.239			
$M_{III}N_{IV}$	4.571	5	2.712			
$\gamma\,M_{III}N_V$	4.532	2	2.735			
$M_{III}O_I$	4.105	9	3.021			
$M_{III}O_{IV,V}$	3.932	6	3.153			
$\zeta_2\,M_{IV}N_{II}$	6.585	5	1.883			
$M_{IV}N_{III}$	6.162	8	2.012			
$\beta\,M_{IV}N_{VI}$	4.909	1	2.5255			
$M_{IV}O_{II}$	4.823	3	2.571			
$M_{IV}P_{II,III}$	4.59	2	2.70			
$\zeta_1\,M_VN_{III}$	6.521	4	1.901			
$\alpha_2\,M_VN_{VI}$	5.130	2	2.4170			
$\alpha_1\,M_VN_{VII}$	5.118	1	2.4226			
$N_1P_{II,III}$	13.30	6	0.932			
$N_{VI}O_{IV}$	91.6	1	0.1354			
$N_{VII}O_V$	93.2	1	0.1330			

Designation	85 Astatine Å*	p.e.	keV	86 Radon Å*	p.e.	keV
$\alpha_2\,KL_{II}$	0.15705†	2	78.95	0.15294†	3	81.07
$\alpha_1\,KL_{III}$	0.15210†	2	81.52	0.14798†	3	83.78
$\beta_3\,KM_{II}$	0.13517†	4	91.72	0.13155†	5	94.24
$\beta_1\,KM_{III}$	0.13432†	4	92.30	0.13069†	5	94.87
$\beta_2{}^{II}\,KN_{II}$	0.13072†	4	94.84	0.12719†	5	97.47
$\beta_2{}^{I}\,KN_{III}$	0.13052†	4	94.99	0.12698†	5	97.64
$\beta_3\,L_1M_{III}$	0.88135†	9	14.067	0.85436†	9	14.512
$\beta_4\,L_{II}M_{IV}$	0.89349†	9	13.876	0.86605†	9	14.316
$\gamma_1\,L_{II}N_{IV}$	0.76289†	9	16.251	0.73928†	9	16.770
$\alpha_2\,L_{III}M_{IV}$	1.09671†	5	11.3048	1.06899†	5	11.5979
$\alpha_1\,L_{III}M_V$	1.08500†	5	11.4268	1.05723†	5	11.7270

Designation	87 Francium Å*	p.e.	keV	88 Radium Å*	p.e.	keV
$\alpha_2\,KL_{II}$	0.14896†	3	83.23	0.14512†	2	85.43
$\alpha_1\,KL_{III}$	0.14399†	3	86.10	0.14014†	2	88.47
$\beta_3\,KM_{II}$	0.12807†	5	96.81	0.12469†	3	99.43
$\beta_1\,KM_{III}$	0.12719†	5	97.47	0.12382†	3	100.13
$\beta_2{}^{II}\,KN_{II}$	0.12379†	5	100.16	0.12050†	3	102.89
$\beta_2{}^{I}\,KN_{III}$	0.12358†	5	100.33	0.12029†	3	103.07
$\beta_4\,L_1M_{II}$				0.84071	5	14.7472
$\beta_3\,L_1M_{III}$	0.82789†	9	14.976	0.80273	5	15.4449
$\gamma_2\,L_1N_{II}$				0.68199	5	18.179
$\gamma_3\,L_1N_{III}$				0.67538	5	18.357
$\gamma'_4\,L_1O_{II}$				0.65131	5	19.036
$\gamma_4\,L_1O_{III}$				0.64965	5	19.084
$\gamma_{13}\,L_1P_{II,III}$				0.64513	5	19.218
$\eta\,L_{II}M_I$				0.90742	5	13.6630
$\beta_1\,L_{II}M_{IV}$	0.83940†	9	14.770	0.81375	5	15.2358
$\gamma_5\,L_{II}N_I$				0.71774	5	17.274
$\gamma_1\,L_{II}N_{IV}$	0.71652†	9	17.303	0.69463	5	17.849
γ_8				0.6801	1	18.230
$\gamma_6\,L_{II}O_{IV}$				0.67328	5	18.414
$L_{II}P_I$				0.6724	1	18.439
$l\,L_{III}M_I$				1.16719	5	10.6222
$\alpha_2\,L_{III}M_{IV}$	1.04230	5	11.8950	1.01656	5	12.1962
$\alpha_1\,L_{III}M_V$	1.03049	5	12.0313	1.00473	5	12.3397
$\beta_6\,L_{III}N_I$				0.87088	5	14.2362
$\beta_{15}\,L_{III}N_{IV}$				0.83722	5	14.8086
$\beta_2\,L_{III}N_V$	0.858	2	14.45	0.83537	5	14.8414
$\beta_7\,L_{III}O_I$				0.8162	1	15.190
$\beta_5\,L_{III}O_{IV,V}$				0.80627	5	15.3771
$L_{III}P_I$				0.8050	1	15.402
$\beta_{10}\,L_1M_{IV}$				0.77546	5	15.988
$\beta_9\,L_1M_V$				0.76857	5	16.131
L_1N_I				0.6874	1	18.036
L_1N_{IV}				0.6666	1	18.600
$\gamma_{11}\,L_1N_V$				0.6654	1	18.633
$L_1O_{IV,V}$				0.6468	1	19.167
$\beta_{17}\,L_{II}M_{III}$				0.8438	1	14.692
$L_{II}N_{III}$				0.7043	1	17.604
$L_{II}N_V$				0.6932	1	17.884
$L_{II}O_{II}$				0.6780	1	18.286

Table 3.7-13 (continued)
X-RAY CRITICAL-ABSORPTION AND EMISSION ENERGIES IN keV

X-Ray Wavelengths

Designation	Å*	p.e.	keV	Å*	p.e.	keV
	87 Francium (*Cont.*)			**88 Radium** (*Cont.*)		
$L_{II}O_{III}$				0.6764	1	18.330
$L_{II}P_{II,III}$				0.6714	1	18.466
$L_{III}N_{II}$				0.8618	1	14.387
$L_{III}N_{III}$				0.8512	1	14.566
$u\ L_{III}N_{VI,VII}$				0.8186	1	15.146
$L_{III}P_{II,III}$				0.8038	1	15.425
	89 Actinium			**90 Thorium**		
$\alpha_2\ KL_{II}$	0.14141†	2	87.67	0.137829	2	89.953
$\alpha_1\ KL_{III}$	0.136417†	8	90.884	0.132813	2	93.350
$\beta_3\ KM_{II}$	0.12143†	2	102.10	0.118268	3	104.831
$\beta_1\ KM_{III}$	0.12055†	2	102.85	0.117396	9	105.609
$\beta_2^{II}\ KN_{II}$	0.11732†	2	105.67	0.11426	1	108.511
$\beta_2^{I}\ KN_{III}$	0.11711†	2	105.86	0.114040	8	108.717
$KO_{II,III}$				0.11322	1	109.500
$\beta_5\ KM_{IV,V}$				0.116667	9	106.269
$\beta_4\ KN_{IV,V}$				0.11366	2	109.08
$\beta_4\ L_I M_{II}$				0.79257	4	15.6429
$\beta_3\ L_I M_{III}$	0.77822†	9	15.931	0.75479	3	16.4258
$\gamma_2\ L_{II}N_{II}$				0.64221	4	19.305
$\gamma_3\ L_I N_{III}$				0.63559	4	19.507
$\gamma'_4\ L_I O_{II}$				0.61251	4	20.242
$\gamma_4\ L_I O_{III}$				0.61098	4	20.292
$\gamma_{13}\ L_I P_{II,III}$				0.60705	8	20.424
$\eta\ L_{II}M_I$				0.85446	4	14.5099
$\beta_1\ L_{II}M_{IV}$	0.78903†	9	15.713	0.765210	9	16.2022
$\gamma_5\ L_{II}N_I$				0.67491	4	18.370
$\gamma_1\ L_{II}N_{IV}$	0.67351†	9	18.408	0.65313	3	18.9825
$\gamma_8\ L_{II}O_I$				0.63898	5	19.403
$\gamma_6\ L_{II}O_{IV}$				0.63258	4	19.599
$L_{II}P_I$				0.6316	1	19.629
$L_{II}P_{IV}$				0.62991	9	19.682
$l\ L_{III}M_I$				1.11508	4	11.1186
$\alpha_2\ L_{III}M_{IV}$	0.99178†	5	12.5008	0.96788	2	12.8096
$\alpha_1\ L_{III}M_V$	0.97993†	5	12.6520	0.95600	3	12.9687
$\beta_6\ L_{III}N_I$				0.82790	8	14.975
$\beta_{15}\ L_{III}N_{IV}$				0.79539	5	15.5875
$\beta_2\ L_{III}N_V$				0.79354	3	15.6237
$\beta_7\ L_{III}O_I$				0.77437	4	16.0105
$\beta_5\ L_{III}O_{IV,V}$				0.76468	5	16.213
$L_{III}P_I$				0.76338	5	16.241
$L_{III}P_{IV,V}$				0.76087	9	16.295
$\beta_{10}\ L_I M_{IV}$				0.7301	1	16.981
$\beta_9\ L_I M_V$				0.7234	1	17.139
$L_I N_I$				0.64755	5	19.146
$L_I N_{IV}$				0.6276	1	19.755
$\gamma_{11}\ L_I N_V$				0.62636	9	19.794
$L_I N_{VI,VII}$				0.6160	1	20.128
$L_I O_I$				0.6146	1	20.174
$L_I O_{IV,V}$				0.6083	1	20.383
$L_{II}M_{II}$				0.8338	1	14.869
$\beta_{17}\ L_{II}M_{III}$				0.79257	4	15.6429
$L_{II}M_V$				0.7579	1	16.359
$L_{II}N_{III}$				0.6620	1	18.729
$L_{II}N_V$				0.6521	1	19.014

Designation	Å*	p.e.	keV	Å*	p.e.	keV
	89 Actinium (*Cont.*)			**90 Thorium** (*Cont.*)		
$v\ L_{II}N_{VI}$				0.64064	9	19.353
$L_{II}O_{II}$				0.6369	1	19.466
$L_{II}O_{III}$				0.6356	1	19.506
$L_{II}P_{II,III}$				0.6312	1	19.642
$t\ L_{III}M_I$				1.08009	9	11.4788
$s\ L_{III}M_{II}$				1.0112	1	12.261
$L_{III}N_{II}$				0.8190	2	15.138
$L_{III}N_{III}$				0.8082	1	15.341
$u\ L_{III}N_{VI,VII}$				0.77661	5	15.964
$L_{III}O_{II}$				0.7713	1	16.074
$L_{III}O_{III}$				0.7690	1	16.123
$L_{III}P_{II,III}$				0.7625	2	16.260
$M_I N_{III}$				2.934	8	4.23
$M_I O_{III}$				2.442	9	5.08
$M_{II}N_I$				3.537	9	3.505
$M_{II}N_{IV}$				3.011	2	4.117
$M_{II}O_{IV}$				2.618	5	4.735
$M_{III}N_I$				4.568	5	2.714
$M_{III}N_{IV}$				3.718	3	3.335
$\gamma\ M_{II}N_V$				3.679	2	3.370
$M_{III}O_I$				3.283	9	3.78
$M_{III}O_{IV,V}$				3.131	3	3.959
$\zeta_2\ M_V N_{II}$				5.340	5	2.322
$M_{IV}N_{III}$				4.911	5	2.524
$\beta\ M_{IV}N_{VI}$				3.941	1	3.1458
$M_{IV}O_{II}$				3.808	4	3.256
$\zeta_1\ M_V N_{III}$				5.245	5	2.364
$\alpha_2\ M_V N_{VI}$				4.151	2	2.987
$\alpha_1\ M_V N_{VII}$				4.1381	9	2.9961
$M_V P_{III}$				3.760	9	3.298
$N\ P_{II}$				9.44	7	1.313
$N\ P_{III}$				9.40	7	1.1319
$N_{II}O_{IV}$				11.56	5	1.072
$N_I P_I$				11.07	7	1.120
$N_{III}O_V$				13.8	1	0.897
$N_{IV}N_{VI}$				33.57	9	0.3693
$N_V N_{VI,VII}$				36.32	9	0.3414
$N_{VI}O_{IV}$				49.5	1	0.2505
$N_{VI}O_V$				48.2	1	0.2572
$N_{VII}O_V$				50.0	1	0.2479
$O_{III}P_{IV,V}$				68.2	3	0.1817
$O_{IV,V}Q_{II,III}$				181.	5	0.068
	91 Protactinium			**92 Uranium**		
$\alpha_2\ KL_{II}$	0.134343†	9	92.287	0.130968	4	94.665
$\alpha_1\ KL_{III}$	0.129325†	3	95.868	0.125947	3	98.439
$\beta_3\ KM_{II}$	0.11523†	2	107.60	0.112296	4	110.406
$\beta_1\ KM_{III}$	0.114345†	8	108.427	0.111394	5	111.300
$\beta_2^{II}\ KN_{II}$	0.11129†	2	111.40	0.10837	1	114.40
$\beta_2^{I}\ KN_{III}$	0.11107†	2	111.62	0.10818	1	114.60
$KO_{II,III}$				0.10744	1	115.39
$\beta_5\ KM_{IV,V}$				0.11069	1	112.01
$\beta_4\ KN_{IV,V}$				0.10780	2	115.01
$\beta_4\ L_I M_I$	0.7699	1	16.104	0.747985	9	16.5753
$\beta_3\ L_I M_{III}$	0.73230	5	16.930	0.71029	2	17.4550

Table 3.7-13 (continued)
X-RAY CRITICAL-ABSORPTION AND EMISSION ENERGIES IN keV

X-Ray Wavelengths

Designation	91 Protactinium (Cont.) Å*	p.e.	keV	92 Uranium (Cont.) Å*	p.e.	keV
$\gamma_2\,L_IN_{II}$	0.6239	1	19.872	0.605237	9	20.4847
$\gamma_3\,L_IN_{III}$	0.6169	1	20.098	0.598574	9	20.7127
$\gamma'_4\,L_IO_{II}$				0.576700	9	21.4984
$\gamma_4\,L_IO_{II,III}$	0.5937	1	20.882	0.57499	9	21.562
γ_8				0.5706	1	21.729
$\eta\,L_{II}M_I$	0.8295	1	14.946	0.80509	2	15.3997
$\beta_1\,L_{II}M_{IV}$	0.74232	5	16.702	0.719984	8	17.2200
$\gamma_5\,L_{II}N_I$	0.6550	1	18.930	0.63557	2	19.5072
$\gamma_1\,L_{II}N_{IV}$	0.63358†	9	19.568	0.614770	9	20.1671
$\gamma_8\,L_{II}O_I$				0.60125	5	20.621
$\gamma_6\,L_{II}O_{IV}$	0.6133	1	20.216	0.594845	9	20.8426
$L_{II}P_{IV}$				0.59203	5	20.942
$l\,L_{III}M_I$	1.0908	1	11.366	1.06712	2	11.6183
$\alpha_2\,L_{III}M_{IV}$	0.94482†	5	13.1222	0.922558	9	13.4388
$\alpha_1\,L_{III}M_V$	0.93284	5	13.2907	0.910639	9	13.6147
$\beta_6\,L_{III}N_I$	0.8079	1	15.347	0.78838	2	15.7260
$\beta_{15}\,L_{III}N_{IV}$				0.756642	9	16.3857
$\beta_2\,L_{III}N_V$	0.7737	1	16.024	0.754681	9	16.4283
$\beta_7\,L_{III}O_I$	0.7546	2	16.431	0.73602	6	16.845
$\beta_5\,L_{III}O_{IV,V}$	0.7452	2	16.636	0.726305	9	17.0701
$L_{III}P_I$				0.72521	5	17.096
$L_{III}P_{IV,V}$				0.72240	5	17.162
$\beta_{10}\,L_IM_V$	0.7088	2	17.492	0.68760	5	18.031
$\beta_9\,L_IM_V$	0.7018	1	17.667	0.681014	8	18.2054
L_IN_{IV}				0.59096	5	20.979
$\gamma_{11}\,L_IN_V$				0.58986	5	21.019
$L_IO_{IV,V}$				0.5725	1	21.657
$\beta_{17}\,L_{II}M_{III}$				0.74503	5	16.641
$L_{II}N_{III}$				0.6228	1	19.907
$v\,L_{II}N_{VI}$				0.6031	1	20.556
$L_{III}O_{III}$				0.59728	5	20.758
$L_{II}P_{II,III}$				0.5930	2	20.906
$t\,L_{III}M_{II}$				1.0347	1	11.982
$s\,L_{III}M_{III}$				0.9636	1	12.866
$L_{III}N_{II}$				0.78078	9	15.892
$L_{III}N_{III}$				0.7691	1	16.120
$u\,L_{III}N_{VI,VII}$				0.738603	9	16.7859
$L_{III}O_{II}$				0.7333	1	16.907
$L_{III}O_{III}$				0.7309	1	16.962
$L_{III}P_{II,III}$				0.72426	5	17.118
M_IN_{II}				2.92	2	4.25
M_IN_{III}				2.753	8	4.50
M_IO_{III}				2.304	7	5.38
M_IP_{III}				2.253	6	5.50
$M_{II}N_I$	3.441	5	3.603	3.329	4	3.724
$M_{II}N_{IV}$	2.910	2	4.260	2.817	2	4.401
$M_{II}O_{IV}$	2.527	4	4.906	2.443	4	5.075
$M_{III}N_I$	4.450	4	2.786	4.330	2	2.863
$M_{III}N_{IV}$	3.614	2	3.430	3.521	2	3.521
$\gamma\,M_{III}N_V$	3.577	1	3.4657	3.479	1	3.563
$M_{III}O_I$	3.245	9	3.82	3.115	7	3.980
$M_{III}O_{IV,V}$	3.038	2	4.081	2.948	2	4.205
$\zeta_2\,M_{IV}N_{II}$	5.193	2	2.3876	5.050	2	2.4548
$M_{IV}N_{III}$				4.625	5	2.681
$\beta\,M_{IV}N_{VI}$	3.827	1	3.2397	3.716	1	3.3367

Designation	91 Protactinium (Cont.) Å*	p.e.	keV	92 Uranium (Cont.) Å*	p.e.	keV
$M_{IV}O_{II}$	3.691	2	3.359	3.576	1	3.4666
$\zeta_1\,M_VN_{III}$	5.092	2	2.4350	4.946	2	2.507
$\alpha_2\,M_VN_{VI}$	4.035	3	3.072	3.924	1	3.1595
$\alpha_1\,M_VN_{VII}$	4.022	1	3.0823	3.910	1	3.1708
N_IO_{III}				10.09	7	1.229
N_IP_{II}				8.81	7	1.41
N_IP_{III}				8.76	7	1.42
$N_{II}P_I$				10.40	7	1.192
$N_{III}O_V$				12.90	9	0.961
$N_{IV}N_{VI}$				31.8	1	0.390
$N_VN_{VI,VII}$				34.8	1	0.357
$N_{IV}O_{IV}$				43.3	2	0.286
$N_{VI}O_V$				42.1	2	0.295
$N_IP_{IV,V}$				8.60	7	1.44

Designation	93 Neptunium Å*	p.e.	keV	94 Plutonium Å*	p.e.	keV
$\beta_4\,L_IM_{II}$	0.72671	2	17.0607	0.70620	2	17.5560
$\beta_3\,L_IM_{III}$	0.68920†	9	17.989	0.66871	2	18.5405
$\gamma_2\,L_IN_{II}$	0.5873	5	21.11	0.57068	2	21.7251
$\gamma_3\,L_IN_{III}$	0.5810	5	21.34	0.564001	9	21.9824
$\gamma'_4\,L_IO_{II}$				0.5432	1	22.823
$\gamma_4\,L_IO_{II,III}$	0.5585	5	22.20	0.5416	1	22.891
$\eta\,L_{II}M_I$	0.7809	2	15.876	0.7591	1	16.333
$\beta_1\,L_{II}M_{IV}$	0.698478	9	17.7502	0.67772	2	18.2937
$\gamma_5\,L_{II}N_I$	0.616	1	20.12	0.5988	1	20.704
$\gamma_1\,L_{II}N_{IV}$	0.596498	9	20.7848	0.578882	9	21.4173
γ_8				0.5658	1	21.914
$\gamma_6\,L_{II}O_{IV}$	0.57699	5	21.488	0.55973	2	22.1502
$l\,L_{III}M_I$	1.0428	6	11.890	1.0226	1	12.124
$\alpha_2\,L_{III}M_{IV}$	0.901045	9	13.7597	0.88028	2	14.0842
$\alpha_1\,L_{III}M_V$	0.889128	9	13.9441	0.86830	2	14.2786
$\beta_6\,L_{III}N_I$	0.769	1	16.13	0.75148	2	16.4983
$\beta_{15}\,L_{III}N_{IV}$				0.7205	1	17.208
$\beta_2\,L_{III}N_V$	0.736230	9	16.8400	0.71851	2	17.2553
$\beta_7\,L_{III}O_I$				0.7003	1	17.705
$\beta_5\,L_{III}O_{IV,V}$	0.70814	2	17.5081	0.69068	2	17.9506
$\beta_{10}\,L_IM_{IV}$				0.6482	1	19.126
$\beta_9\,L_IM_V$				0.6416	1	19.323
$u\,L_{III}N_{VI,VII}$				0.7031	1	17.635

Designation	95 Americium Å*	p.e.	keV
$\beta_4\,L_IM_{II}$	0.68639	2	18.0627
$\beta_3\,L_IM_{III}$	0.64891	2	19.1059
$\gamma_2\,L_IN_{II}$	0.5544	2	22.361
$\beta_1\,L_{II}M_{IV}$	0.657655	9	18.8520
$\gamma_1\,L_{II}N_{IV}$	0.561886	9	22.0652
$\gamma_6\,L_{II}O_{IV}$	0.54311	2	22.8282
$l\,L_{III}M_I$	1.0012	6	12.384
$\alpha_2\,L_{III}M_{IV}$	0.860266	9	14.4119
$\alpha_1\,L_{III}M_V$	0.848187	9	14.6172
$\beta_6\,L_{III}N_I$	0.73418	2	16.8870
$\beta_{15}\,L_{III}N_{IV}$	0.70341	2	17.6258
$\beta_2\,L_{III}N_V$	0.701390	9	17.6765
$\beta_5\,L_{III}O_{IV,V}$	0.67383	2	18.3996

Table 3.7-14[a]

PRINCIPAL CHARACTERISTIC EMISSION LINES FOR TUNGSTEN

Transition	Symbol	Energy (keV)	λ (A)	Intensity	Transition	Symbol	Energy (keV)	λ (A)
$N_{II}N_{III}$-K	$K_{\beta 2}$	69.068	0.17950	15	N_{IV}-L_{II}	$L_{\gamma 1}$	11.286	1.0985
M_{III}-K	$K_{\beta 1}$	67.243	0.18438	35	N_V-L_{III}	$L_{\beta 2}$	9.962	1.2445
					M_{IV}-L_{II}	$L_{\beta 1}$	9.671	1.2820
L_{III}-K	$K_{\alpha 1}$	59.320	0.20900	100	M_V-L_{III}	$L_{\alpha 1}$	8.395	1.4768
L_{II}-K	$K_{\alpha 2}$	57.982	0.21382	50	M_{IV}-L_{III}	$L_{\alpha 2}$	8.333	1.4878

[a]   Calculated from Table I. Relative Intensities of Compton and Allison, *X-Rays in Theory and Experiment,* Van Nostrand, Princeton, N.J., 1935.

From Johns, H. E., X-Rays and teleisotope Y-rays, in *Radiation Dosimetry,* Vol. 3, 2nd ed., Attix, F. H. and Tochilin, E., Eds., Academic Press, New York, 1969. With permission.

TABLE 3.7-15

HVLS OF RADIATION WITHIN A WATER PHANTOM IRRADIATED
BY X-RAYS PRODUCED AT 90 kV WITH HVL OF 2.9 mm Al

HVL (mm Al)

	2[a]			6[a]			10[a]		
	50[b]	100[b]	500[b]	50[b]	100[b]	500[b]	50[b]	100[b]	500[b]
Primary radiation	3.6	3.6	3.6	4.8	4.8	4.8	5.6	5.6	5.6
Scattered radiation	3.1	3.1	3.1	3.8	3.7	3.7	4.2	4.1	3.9
Total radiation	3.4	3.4	3.4	4.1	4.0	3.9	4.5	4.4	4.2

[a]   Depth in water (cm).
[b]   Beam area (cm²).

From *Radiation Dosimetry: X-Rays Generated at Potentials of 5 to 150 kV,* ICRU Report 17, International Commission on Radiation Units and Measurements, Washington, D. C., June 15, 1970. With permission.

TABLE 3.7-16
EQUIVALENT INHERENT
FILTRATION OF A DIAGNOSTIC
X-RAY TUBE

	Equivalent inherent filtration	
Tube voltage, constant potential (kV)	From measurements of reduction in exposure dose (mm Al)	From HVL measurements (mm Al)
50	1.7	1.4
100	1.9	1.8
150	2.1	1.8

From Report of the International Commission on Radiological Units and Measurements, *Nat. Bur. Stand. Handb.,* 78, 1959. With permission.

TABLE 3.7-17
RADIATION QUALITY OF A
DIAGNOSTIC X-RAY TUBE
(WITHOUT EXTRA
FILTRATION)

	Half-value layer (mm Al)	
Tube voltage (Kv) (Wave-form)	Constant Potential	Full wave rectified
50	1.25	1.16
100	2.75	1.9
125	3.2	2.3

From Report of the International Commission on Radiological Units and Measurements, *Nat. Bur. Stand. Handb.,* 78, 1959. With permission.

TABLE 3.7-18
BREMSSTRAHLUNG SPECTRUM FROM BETA ABSORPTION

Photon energy intervals in fractions of the maximum beta energy	Percent of total intensity[a] contributed by photons in energy intervals
0—0.1	43.5
0.1—0.2	25.8
0.2—0.3	15.2
0.3—0.4	8.3
0.4—0.5	4.3
0.5—0.6	2.0
0.6—0.7	0.7
0.7—0.8	0.2
0.8—0.9	0.03
0.9—1.0	<0.01

[a] As an example of the total intensities emitted, when beta particles from a 1-Ci source of ^{90}Sr—^{90}Y are absorbed in aluminum, the bremsstrahlung intensity is approximately equal to the gamma intensity from 12 mg of radium. The average bremsstrahlung energy is about 300 keV (Haybittle, *Phys. Med. Biol.*, 1(3), 270, 1956). The bremsstrahlung from a 1-Ci ^{32}P aqueous solution in a glass bottle is about 3 mr/hr at 1 m.

From Wyard, S. J., *Nucleonics*, 13(7), July 1955. With permission.

TABLE 3.7-19
AVERAGE LET RATES FOR X- AND γ-RAY SOURCES FOR A VARIETY OF RADIATIONS

Radiation source	L_D(keV/μ)[a] δ = 100 eV	δ = 500 eV	L_T(keV/μ)[b] δ = 100 eV	δ = 500 eV
X-Rays				
1.25 mm Cu HVL				
P[c]	1.3	1.3	10.4	21.3
P + S[d]	1.4	1.5	10.7	21.9
3.2 mm Cu HVL				
P	1.2	1.2	10.6	21.0
P + S	1.3	1.4	11.0	21.8
γ-Rays				
^{137}Cs				
P	0.32	0.30	8.4	14.8
P + S	0.40	0.40	9.4	17.2
^{60}Co				
P	0.30	0.28	7.9	14.4
P + S	0.32	0.30	8.3	15.0

Note: Data derived by Bruce et al. (1963).

[a] LET averaged over number of tracks in energy interval.

[b] LET averaged over energy deposited in track in energy interval.

[c] P signifies primary only.

[d] P + S signifies primary and scattered at a depth of 10 cm in water for a 400-cm² field.

From Johns, H. E., in *Radiation Dosimetry*, Vol. 3, 2nd ed., Attix, F. H. and Tochilin, E., Eds., Academic Press, New York, 1969. With permission.

TABLE 3.7-20
WATT[a] AND CRANBERG[b] FISSION SPECTRA, WITH INTEGRALS NORMALIZED TO UNITY[c]

Neutron energy, E (MeV)	Differential fission spectra. ϕ' (E) (neutrons cm^{-2}sec^{-1}MeV^{-1})		Integral fission spectra, $\phi(E)$ (neutrons cm^{-2}sec^{-1} above E	
	Watt	Cranberg	Watt	Cranberg
0.1	0.202	0.203	0.9860	0.989
0.5	0.345	0.348	0.8679	0.870
0.7	0.355	0.360	0.7973	0.799
0.9	0.351	0.351	0.7271	0.728
1.0	0.344	0.346	0.6918	0.693
1.6	0.284	0.285	0.5017	0.501
2.0	0.237	0.239	0.3972	0.396
2.5	0.184	—	0.2927	—
3.0	0.138	0.138	0.2124	0.209
4.0	0.0748	0.0742	0.1086	0.106
4.5	0.0538	—	0.0808	—
5.0	0.0384	0.0375	0.0537	0.0515
6.0	0.0191	0.0184	0.0261	0.0245
7.0	9.31×10^{-3}	8.80×10^{-3}	0.0124	0.0114
8.0	4.42×10^{-3}	4.11×10^{-3}	0.0057	5.21×10^{-3}
9.0	2.07×10^{-3}	1.89×10^{-3}	0.0027	2.33×10^{-3}
10.0	9.53×10^{-4}	8.30×10^{-4}	1.21×10^{-3}	1.02×10^{-3}
12.0	1.99×10^{-4}	1.76×10^{-4}	2.47×10^{-4}	1.625×10^{-4}
14.0	4.00×10^{-5}	3.32×10^{-5}	4.87×10^{-5}	

[a] Watt (1948).
[b] Cranberg (1956).
[c] See Grundl (1968) for discussion on fission spectra.

From Moteff, J., in *Radiation Dosimetry,* Vol. 3, 2nd ed., Attix, F. H. and Tochilin, E., Eds., Academic Press, New York, 1969. With permission.

Table 3.7-21
AVERAGE NUMBER OF PROMPT NEUTRONS EMITTED PER FISSION FOR VARIOUS NUCLIDES

Fissioning nucleus	Bondarenko (1958)[1]	Leachman (1958)[2]	Recent values	Ref.
		Spontaneous Fission		
²³⁸Pu	2.30 ± 0.20		1.97 ± 0.07	3
²³⁶Pu	2.17 ± 0.20			
²³⁸Pu	2.28 ± 0.10			
²⁴⁰Pu	2.23 ± 0.05	2.26 ± 0.05	2.154 ± 0.028	3
		2.22 ± 0.11	2.189 ± 0.026	4
²⁴²Pu	2.28 ± 0.13	2.18 ± 0.09		
²⁴²Cm	2.59 ± 0.11			
²⁴⁴Cm	2.82 ± 0.09			
²⁴⁹Bk	3.72 ± 0.16			
²⁴⁶Cf	2.92 ± 0.19			
²⁵²Cf	3.84 ± 0.12		3.771 ± 0.031	4
			3.799 ± 0.034	5
			3.704 ± 0.015	6
²⁵⁴Cf	3.90 ± 0.14			
²⁵⁴Fm	4.05 ± 0.19	**Thermal Neutron Fission**		
²³⁰Th	21.3 ± 0.03			
²³⁴V	2.52 ± 0.03	2.54 ± 0.04	2.473 ± 0.026	4
		2.55 ± 0.05		
²³⁶U	2.47 ± 0.03	2.47 ± 0.05	2.425 ± 0.020	4
		2.46 ± 0.03	2.369 ± 0.015	6
			2.417 ± 0.015	7
²⁴⁰Pu		2.88 ± 0.04	2.813 ± 0.028	4
		2.95 ± 0.06		
²⁴²Pu		3.03 ± 0.06	3.14 ± 0.06	8
			2.96 ± 0.08	9
²⁴²Am	3.14 ± 0.04			

From Gindler, J. and Huizenga, J. R., *Nuclear Chemistry,* Vol. 2, Yaffe, L., Ed., Academic Press, New York, 1968, as given in Loveland, W. D., in *American Institute of Physics Handbook,* 3rd ed., Gray, D. E., Ed., McGraw-Hill, New York, 1972. With permission.

REFERENCES

1. Bondarenko, I. I., Kuzminov, B. D., Kutsayeva, L. S., Prokhorova, L. I., and Smirenkin, G. N., *Proc. U. N. Int. Conf. Peaceful Uses of Atomic Energy, Geneva,* 15, 353, 1958.
2. Leachman, R. B., *Proc. U. N. Int. Conf. Peaceful Uses of Atomic Energy, Geneva,* 15, 229, 1958.
3. Asplund-Nilsson, I., Condé, H., and Starfelt, N., *Nucl. Sci. Eng.,* 15, 213, 1963.
4. Hopkins, J. C. and Diven, B. C., *Nucl. Phys.,* 48, 433, 1963.
5. Asplund-Nilsson, I., Condé, H., and Starfelt, N., *Nucl. Sci. Eng.,* 16, 124, 1963.
6. Colvin, D. W. and Sowerby, M. G., *Physics and Chemistry of Fission,* Vol. 2, International Atomic Energy Agency, Vienna, 1965, 25.
7. Condé, H. and Holmberg, M., *Physics and Chemistry of Fission,* Vol. 2, International Atomic Energy Agency, Vienna, 1965, 57.
8. de Saussure, G. and Silver, E. G., *Nucl. Sci. Eng.,* 5, 49, 1959.
9. Jaffey, A. H., Hibdon, C. T., and Sjoblom, R., *J. Nucl. Energy, Part A,* 11, 21, 1959.

Table 3.7-22
CHARACTERISTICS OF FISSION NEUTRON SPECTRA

Fissile nuclide	Average energy F (MeV)	Ref.	Maxwellian temperature, $T = 2 \bar{E}/3$ (MeV)
$^{233}U + n$th	1.98 ± 0.05	1	1.32 ± 0.03
$^{235}U + n$th	1.95 ± 0.05	1	1.30 ± 0.03
$^{239}Pu + n$th	2.03 ± 0.05	1	1.35 ± 0.03
$^{241}Pu + n$th	2.002 ± 0.051	2	1.335 ± 0.034
^{252}Cf	2.15 ± 0.08	1	1.43 ± 0.05

From Gindler, J. and Huizenga, J. R., *Nuclear Chemistry,* Vol. 2, Yaffe, L., Ed., Academic Press, New York, 1968, as given in Loveland, W. D., in *American Institute of Physics Handbook,* 3rd ed., Gray, D. E., Ed., McGraw-Hill, New York, 1972. With permission.

REFERENCES

1. Terrell, J., *Phys. Rev.,* 127, 880, 1967.
2. Smith, A. B., Sjoblom, R., and Roberts, J. H., *Phys. Rev.,* 123, 2140, 1961.

TABLE 3.7-23
α-PARTICLE RANGES (R) IN VARIOUS MATERIALS (g/cm²)[a]

α Energy (MeV) $A^{1/2}=$	3.0	H₂O 3.0	C 3.47	Al 5.18	Cu 7.97	Ag 10.4	Pb 14.4
3.97	0.0029	0.0039	0.0039	0.0042	0.0061	0.0079	0.0115
5.56	0.0049	0.0055	0.0059	0.0068	0.0097	0.0126	0.0184
7.15	0.0074	0.0075	0.0082	0.0098	0.0140	0.0182	0.0263
8.74	0.0103	0.0100	0.0110	0.0134	0.0190	0.0244	0.0352

[a]   Derived from Table 9, Chapter 4 (Volume I) of Mayneord and Hill[38] on the assumption that an α particle of energy 3.97 E has a range equal to 0.993 times the range of a proton of energy E. The range in air is practically the same (in g/cm²) as that given for carbon.

From Mayneord, W. V. and Hill, C. R., in *Radiation Dosimetry,* Vol. 3, 2nd ed., Attix, F. H. and Tochilin, E., Eds., Academic Press, New York, 1969. With permission.

TABLE 5.1-24

VALUES OF RANGE AND MASS STOPPING POWER FOR ELECTRONS IN WATER[a,b]

E/MeV	$R\rho$/(g/cm²)[c]	Mass stopping power; $\frac{1}{\rho}\left(\frac{dE}{dl}\right)$ /MeV·cm²/g					
		$\frac{1}{\rho}\left(\frac{dE}{dl}\right)_{tot}$	$\frac{1}{\rho}\left(\frac{dE}{dl}\right)_{rad}$	$\frac{1}{\rho}\left(\frac{dE}{dl}\right)_{tot}$	$\frac{1}{\rho}\left(\frac{dE}{dl}\right)_{100\,eV}$	$\frac{1}{\rho}\left(\frac{dE}{dl}\right)_{1000\,eV}$	$\frac{1}{\rho}\left(\frac{dE}{dl}\right)_{10,000\,eV}$
1×10^{-5}	$\sim4\times10^{-8}$			$\sim3{-}30$			
2	8.2×10^{-8}	133		133	133	133	133
3	1.55×10^{-7}	139		139	139	139	139
5	2.85	232		232	232	232	232
1×10^{-4}	4.48	303		303	303	303	303
2	8.00	220		220	220	220	220
3	1.23×10^{-6}	215		215	211	215	215
5	2.20	195		195	183	195	195
1×10^{-3}	5.34	130		130	112	130	130
2	1.64×10^{-5}	77.5		77.5	60	77.5	77.5
3	3.22	57.8		57.8	42.2	56.6	57.8
5	7.70	39.2		39.2	27.1	36.9	39.2
1×10^{-2}	2.50×10^{-4}	23.2		23.2	15.1	20.2	23.2
2	8.33	13.5		13.5	8.5	11.1	13.5
3	1.71×10^{-3}	9.88		9.88	6.12	7.9	9.7
5	4.22	6.75		6.75	4.12	5.26	6.35
1×10^{-1}	1.40×10^{-2}	4.20		4.20	2.52	3.15	3.78
2	4.40	2.84_4	0.006	2.85	1.67	2.08	2.44
3	8.26	2.39_4	0.007_5	2.40	1.39	1.72	2.01
5	1.74×10^{-1}	2.06	0.01	2.07	1.17	1.44	1.69
1×10^{0}	4.30	1.87_6	0.01_7	1.89	1.05	1.28	1.48
2	9.61	1.86_4	0.03_2	1.89	1.02	1.23	1.41
3	1.49×10^{0}	1.88	0.04_8	1.93	1.01	1.22	1.41
5	2.50	1.93	0.08	2.01	1.00	1.23	1.41
1×10^{1}	4.88	2.00	0.18	2.18	1.00	1.24	1.42
2	9.18	2.06	0.41	2.47	1.02	1.25	1.42
3	1.30×10^{1}	2.10	0.64	2.74	1.03	1.25	1.42
5	1.97	2.14	1.13	3.27	1.04	1.26	1.43
1×10^{2}	3.25	2.20	2.40	4.61	1.06	1.27	1.44
2	4.96	2.26	5.01	7.27	1.07	1.28	1.46
3	6.13	2.30	7.65	9.95	1.08	1.29	1.47
5	7.74	2.34	12.9_6	15.30	1.09	1.30	1.49
1×10^{3}	1.01×10^{2}	2.40	26.3	28.7	1.12	1.31	1.50
2	2.50	2.50	49.0	51.5	1.125	1.32	1.51

TABLE 3.7-24 (continued)
VALUES OF RANGE AND MASS STOPPING POWER FOR ELECTRONS IN WATER[a,b]

E/MeV	R_Q/(g/cm²)[c]	Mass stopping power; $\frac{1}{\rho}\left(\frac{dE}{dl}\right)$ /MeV·cm²/g					
		$\frac{1}{\rho}\left(\frac{dE}{dl}\right)_{col}$	$\frac{1}{\rho}\left(\frac{dE}{dl}\right)_{rad}$	$\frac{1}{\rho}\left(\frac{dE}{dl}\right)_{tot}$	$\frac{1}{\rho}\left(\frac{dE}{dl}\right)_{100\,eV}$	$\frac{1}{\rho}\left(\frac{dE}{dl}\right)_{1000\,eV}$	$\frac{1}{\rho}\left(\frac{dE}{dl}\right)_{10,000\,eV}$
3		2.55	72.0	74.5	1.13	1.34	1.53
5		2.61	117.4	121	1.14	1.34	1.54
1×10^4		2.70	230	233	1.17	1.36	1.57

a This table appears in three sections. The top section is based on experimental determinations of dE/dR only (Appendix 3.5 of Reference 19), the middle section is based both on experimental data and theoretical calculations, and the bottom section on calculations only. Data for electron energies below 10^{-4} MeV are to be considered tentative at present.

b Multiply values of mass stopping power in MeV·cm²/g by 0.1 in order to obtain values of LET in keV/μm in unit density material i.e., a mass stopping power of 10 MeV·cm²/g corresponds to an LET of 1 keV/μm in unit density material.

c R represents an approximation to R_{csda} and is obtained from various theoretical and experimental determinations as described in Appendix 2 of Reference 19.

From *Neutron Fluence, Neutron Spectra, and Kerma*, ICRU Report 13, International Commission on Radiation Units and Measurements, Washington, D.C., September 15, 1969. With permission.

Table 5.1-23

VALUES OF MASS ATTENUATION COEFFICIENTS

Mass attenuation coefficient $\mu/\varrho/m^2kg^{-1}$

Photon energy (keV)	H	C	N	O	Na	Mg	Al	P	S	Ar	K	Ca
1	0.860	210	329	463	—	—	—	—	—	—	—	—
1.5	0.287	67.0	106	153	341	441	—	—	—	—	—	—
2	0.143	29.5	46.8	68.0	152	197	239	—	—	—	—	—
3	0.0717	8.91	14.3	20.9	48.5	63.4	76.8	114	139	—	—	—
4	0.0514	3.77	5.98	8.88	21.5	28.2	34.6	51.5	63.5	79.0	98.4	—
5	0.0445	1.91	3.06	4.59	11.4	15.1	18.5	28.0	34.4	43.0	53.6	63.3
6	0.0420	1.07	1.75	2.65	6.79	9.07	11.1	16.9	20.8	26.1	32.6	38.5
8	0.0395	0.439	0.722	1.12	2.96	4.01	5.0	7.62	9.45	11.9	14.9	17.6
10	0.0387	0.225	0.369	0.576	1.54	2.10	2.63	4.06	5.05	6.42	8.01	9.48
15	0.0376	0.0767	0.117	0.176	0.459	0.627	0.783	1.23	1.54	1.99	2.49	2.98
20	0.0369	0.0424	0.0589	0.0824	0.199	0.268	0.335	0.523	0.657	0.851	1.07	1.29
30	0.0357	0.0250	0.0297	0.0365	0.0694	0.0897	0.108	0.163	0.203	0.262	0.333	0.40
40	0.0346	0.0206	0.0225	0.0254	0.0388	0.0473	0.0549	0.0776	0.0947	0.118	0.149	0.179
50	0.0335	0.0186	0.0196	0.0211	0.0275	0.0322	0.0359	0.0474	0.0565	0.0677	0.0842	0.0998
60	0.0326	0.0175	0.0181	0.0190	0.0225	0.0253	0.0272	0.0342	0.0394	0.0454	0.0553	0.0643
80	0.0309	0.0161	0.0164	0.0167	0.0179	0.0194	0.0201	0.0231	0.0255	0.0273	0.0322	0.0362
100	0.0295	0.0151	0.0153	0.0155	0.0159	0.0169	0.0171	0.0186	0.0201	0.0187	0.0233	0.0256
150	0.0265	0.0134	0.0135	0.0136	0.0134	0.0140	0.0138	0.0144	0.0150	0.0143	0.0159	0.0168

a Multiply by 10 if cm² g⁻¹ required.

From *Radiation Dosimetry: X-Rays Generated at Potentials of 5 to 150 kV*, ICRU Report 17, International Commission on Radiation Units and Measurements, Washington, D. C., June 15, 1970.

Table 3.7-26a
VALUES OF MASS ATTENUATION COEFFICIENTS

Mass attenuation coefficient $(\mu/\varrho/m^2kg^{-1})^a$

Photon energy (keV)	Polystyrene $(C_8H_8)_n$	Perspex, plexiglass, lucite $(C_5H_8O_2)_n$	Polyethylene $(CH_2)_n$	Bakelite $(C_{43}H_{38}O_7)_n$	Water	Air[b]	Compact bone[b]	Muscle[b]	Fricke dosimeter solution $(0.4$ mol/1 $H_2SO_4)$
1	193	277	180	242	412	—	—	—	—
1.5	61.6	89.1	57.1	77.6	135	—	—	—	—
2	27.2	39.4	25.2	34.2	60.1	—	—	—	—
3	8.21	12.0	7.63	10.4	18.5	—	—	17.9	19.9
4	3.48	5.10	3.24	4.42	7.89	7.60	—	7.90	8.63
5	1.77	2.62	1.64	2.25	4.08	3.93	13.9	4.10	4.49
6	0.990	1.49	0.922	1.28	2.36	2.28	8.34	2.38	2.61
8	0.408	0.624	0.382	0.531	0.999	0.960	3.75	1.01	1.11
10	0.211	0.322	0.198	0.273	0.516	0.496	2.00	0.524	0.576
15	0.0737	0.105	0.0711	0.0911	0.161	0.155	0.628	0.164	0.179
20	0.0420	0.0547	0.0416	0.0488	0.0773	0.0747	0.278	0.0790	0.0850
30	0.0258	0.0295	0.0265	0.0276	0.0364	0.0343	0.0958	0.0368	0.0386
40	0.0217	0.0233	0.0226	0.0222	0.0264	0.0244	0.0510	0.0265	0.0273
50	0.0198	0.0206	0.0207	0.0199	0.0225	0.0206	0.0347	0.0224	0.0229
60	0.0187	0.0192	0.0197	0.0186	0.0205	0.0187	0.0272	0.0204	0.0207
80	0.0173	0.0175	0.0182	0.0171	0.0183	0.0166	0.0208	0.0182	0.0184
100	0.0162	0.0164	0.0172	0.0160	0.0171	0.0154	0.0180	0.0169	0.0171
150	0.0144	0.0145	0.0153	0.0142	0.0150	0.0135	0.0149	0.0149	0.0150

[a] Multiply by 10 if cm² g⁻¹ required.
[b] See Table 3.7-26b.

From *Radiation Dosimetry: X-Rays Generated at Potentials of 5 to 150 kV*, ICRU Report 17, International Commission on Radiation Units and Measurements, Washington, D.C., June 15, 1970. With permission.

TABLE 3.7-26b
ASSUMED ELEMENTARY COMPOSITION (PERCENT BY WEIGHT)

Element	Muscle	Compact bone	Air
H	10.2	6.4	—
C	12.3	27.8	—
N	3.5	2.7	75.5
O	72.9	41.0	23.2
Na	0.08	—	—
Mg	0.02	0.2	—
P	0.2	7.0	—
S	0.5	0.2	—
Ar	—	—	1.3
K	0.3	—	—
Ca	0.007	14.7	—

From *Radiation Dosimetry: X-Rays Generated at Potentials of 5 to 150 kV*, ICRU Report 17, International Commission on Radiation Units and Measurements, Washington, D. C., June 15, 1970. With permission.

Table 3.1-21

VALUES OF MASS ENERGY ABSORPTION COEFFICIENTS

Mass energy absorption coefficient $(\mu_m/\varrho/\text{m}^2\text{kg}^{-1})^a$

Photon energy (keV)	H	C	N	O	Na	Mg	Al	P	S	Ar	K	Ca
1	0.819	209	328	459	—	—	—	—	—	—	—	—
1.5	0.250	66.9	106	152	335	430	—	—	—	—	—	—
2	0.106	29.3	46.7	67.7	150	193	233	—	—	—	—	—
3	0.0312	8.85	14.2	20.8	47.8	62.4	75.2	110	132	—	—	—
4	0.0122	3.71	5.88	8.79	21.2	27.7	33.9	50.0	60.9	73.3	88.9	—
5	0.00586	1.85	3.00	4.51	11.2	14.9	18.2	27.2	33.1	40.4	49.4	56.8
6	0.00326	1.03	1.69	2.58	6.66	8.91	10.9	16.5	20.1	24.7	30.4	35.1
8	0.00150	0.404	0.683	1.07	2.87	3.92	4.88	7.43	9.18	11.4	14.1	16.4
10	0.00109	0.194	0.338	0.534	1.48	2.02	2.55	3.94	4.89	6.16	7.61	8.91
15	0.00107	0.0515	0.0903	0.146	0.419	0.581	0.737	1.17	1.47	1.89	2.37	2.81
20	0.00131	0.0202	0.0354	0.0574	0.168	0.234	0.299	0.482	0.610	0.796	1.01	1.21
30	0.00184	0.00578	0.00979	0.0157	0.0463	0.0649	0.0830	0.135	0.173	0.231	0.296	0.358
40	0.00230	0.00296	0.00449	0.00681	0.0191	0.0263	0.0338	0.0549	0.0704	0.0945	0.122	0.149
50	0.00270	0.00217	0.00294	0.00404	0.00988	0.0136	0.0173	0.0278	0.0357	0.0477	0.0620	0.0761
60	0.00303	0.00199	0.00238	0.00301	0.00623	0.00820	0.0103	0.0165	0.0207	0.0277	0.0358	0.0439
80	0.00362	0.00199	0.00213	0.00238	0.00363	0.00451	0.00532	0.00788	0.00958	0.0124	0.0159	0.0193
100	0.00406	0.00212	0.00219	0.00232	0.00288	0.00335	0.00372	0.00496	0.00589	0.00720	0.00909	0.0108
150	0.00482	0.00241	0.00245	0.00248	0.00258	0.00276	0.00281	0.00318	0.00344	0.00370	0.00437	0.00490

^a Multiply by 10 if cm² g⁻¹ required.

a Multiply by 10 if cm² g⁻¹ required.

From *Radiation Dosimetry: X-Rays Generated at Potentials of 5 to 150 kV*, ICRU Report 17, International Commission of Radiation Units and Measurements, Washington, D.C., June 15, 1970. With permission.

Table 3.7-28
VALUES OF THE MASS ENERGY ABSORPTION COEFFICIENTS AND OF THE FACTOR f

Photon energy (keV)	Mass energy absorption coefficient ($\mu_{en}/\rho/m^2\,kg^{-1}$)[a]									$f = 0.869$ $\dfrac{rad/R}{}$ = $\dfrac{(\mu_{en}/\rho)_{med.}[b]}{(\mu_{en}/\rho)_{air}}$ Medium			
	Poly-styrene $(C_8H_8)_n$	Perspex, plexiglass, lucite $(C_5H_8O_2)_n$	Poly-ethylene $(CH_2)_n$	Bakelite $(C_{43}H_{38}O_7)_n$	Water	Air[b]	Compact bone[b]	Muscle[b]	Fricke dosimeter solution (0.4 mol/l H_2SO_4)	Water	Compact bone	Muscle	Fricke solution
1	192	275	179	241	409	—	—	—	—	—	—	—	—
1.5	61.5	88.7	57.1	77.3	135	—	—	—	—	—	—	—	—
2	27.0	39.2	25.0	34.0	59.9	—	—	—	—	—	—	—	—
3	8.15	11.9	7.58	10.3	18.4	—	—	17.8	19.7	—	—	—	—
4	3.42	5.04	3.18	4.35	7.81	7.43	—	7.77	8.51	0.913	—	0.909	0.995
5	1.71	2.55	1.59	2.19	4.01	3.84	12.8	4.01	4.39	0.907	2.90	0.907	0.995
6	0.951	1.44	0.883	1.23	2.29	2.20	7.76	2.30	2.53	0.907	3.07	0.911	1.002
8	0.373	0.584	0.346	0.493	0.950	0.912	3.53	0.961	1.06	0.905	3.36	0.916	1.009
10	0.179	0.287	0.166	0.240	0.474	0.459	1.88	0.483	0.533	0.898	3.56	0.913	1.008
15	0.0476	0.0776	0.0443	0.0645	0.130	0.127	0.576	0.134	0.148	0.891	3.95	0.917	1.012
20	0.0187	0.0306	0.0175	0.0254	0.0511	0.0504	0.244	0.0530	0.0585	0.881	4.20	0.914	1.010
30	0.00548	0.00863	0.00521	0.00722	0.0142	0.0140	0.0710	0.0148	0.0163	0.876	4.39	0.915	1.005
40	0.00291	0.00414	0.00287	0.00357	0.00631	0.00620	0.0298	0.00658	0.00715	0.884	4.18	0.923	1.003
50	0.00221	0.00281	0.00225	0.00252	0.00389	0.00378	0.0157	0.00403	0.00431	0.895	3.62	0.926	0.991
60	0.00207	0.00240	0.00214	0.00222	0.00301	0.00286	0.00971	0.00309	0.00324	0.917	2.96	0.939	0.987
80	0.00212	0.00225	0.00222	0.00215	0.00252	0.00232	0.00524	0.00254	0.00261	0.943	1.96	0.951	0.977
100	0.00227	0.00234	0.00240	0.00227	0.00252	0.00229	0.00381	0.00251	0.00256	0.957	1.45	0.956	0.972
150	0.00260	0.00263	0.00276	0.00256	0.00274	0.00247	0.00302	0.00272	0.00275	0.964	1.06	0.956	0.965

[a] Multiply by 10 if cm² g⁻¹ required.
[b] See Table 3.7-26b

From *Radiation Dosimetry: X-Rays Generated at Potentials of 5 to 150 kV*, ICRU Report 17, International Commission on Radiation Units and Measurements, Washington, D. C., June 15, 1970. With permission.

Table 3.7-29
VALUES OF RANGE AND MASS STOPPING POWER FOR PROTONS IN WATER[a,b,c]

E/MeV	Rρ/(g/cm²)	Mass stopping power; $\frac{1}{\rho}\left(\frac{dE}{dl}\right)$ /MeV·cm²/g					
		$\frac{1}{\rho}\left(\frac{dE}{dl}\right)_{nucl}$	$\frac{1}{\rho}\left(\frac{dE}{dl}\right)_{col}$	$\frac{1}{\rho}\left(\frac{dE}{dl}\right)_{tot}$	$\frac{1}{\rho}\left(\frac{dE}{dl}\right)_{100\,eV}$	$\frac{1}{\rho}\left(\frac{dE}{dl}\right)_{1000\,eV}$	$\frac{1}{\rho}\left(\frac{dE}{dl}\right)_{10,000\,eV}$
$\times 10^{-5}$	9×10^{-8}	96	18	114	114	114	114
2	1.68×10^{-7}	100	26	126	126	126	126
5	3.85	99	41	140	140	140	140
$\times 10^{-4}$	7.2	90	58	148	148	148	148
2	1.37×10^{-6}	78	82	160	160	160	160
5	3.14	58	130	188	188	188	188
$\times 10^{-3}$	5.58	43	180	223	223	223	223
2	9.43	29	260	289	289	289	289
5	1.80×10^{-5}	14.7	410	425	425	425	425
$\times 10^{-2}$	2.79	7.5	580	587	587	587	587
2	4.22	~4.5	730	734	734	734	734
5	7.3		910	910	910	910	910
$\times 10^{-1}$	1.25×10^{-4}		915	915	715	915	915
2	2.42		750	750	488	750	750
5	8.0		437	437	254	433	437
$\times 10^{0}$	2.25×10^{-3}		268	268	147	240	268
2	7.13		170	170	91.8	141	170
5	2.30×10^{-2}		100	100	54	78	100
	4.71		71.4	71.4	38.5	54.2	69.2
$\times 10^{1}$	1.18×10^{-1}		46.8	46.8	25.3	34.2	43.0
	8.64		19.2	19.2	10.3	13.4	16.5
	2.18×10^{0}		12.7	12.7	6.88	8.76	10.8
$\times 10^{2}$	7.57		7.42	7.42	4.00	5.04	6.08
.5	1.55×10^{1}		5.54	5.54	3.00	3.77	4.55
	5.06		3.58	3.58	1.93	2.46	2.90
	1.15×10^{2}		2.79	2.79	1.50	1.90	2.25
$\times 10^{3}$	3.21		2.24	2.24	1.23	1.52	1.81
	1.29×10^{3}		2.03	2.03	1.13	1.39	1.63
	2.26		2.07	2.07	1.18	1.42	1.67
$\times 10^{4}$	4.62		2.17	2.17	1.29	1.54	1.78
	1.335×10^{4}		2.37	2.37	1.41	1.68	1.94
	2.162		2.46	2.46	1.46	1.74	2.02
$\times 10^{5}$	4.142		2.58	2.58	1.54	1.83	2.12

[a] This table appears in three sections. The top section and the bottom section depend upon theoretical calculations only, the middle section is based both on experimental data and theoretical calculations.

[b] Multiply values of mass stopping power in MeV·cm²/g by 0.1 in order to obtain values of LET in keV/μm in unit density material, i.e., a mass stopping power of 10 MeV·cm²/g corresponds to an LET of 1 keV/μm in unit density material.

[c] Above certain limits,[19] this table may also be used for particles other than protons if the first column is treated as a listing of values of ε_M (MeV/amu) and the second column is considered to be a listing of values of normalized range $(z^2/M_i) \cdot R\rho$, and the stopping powers considered to be a listing of the normalized form, $(1/\rho)(dE/dl) \cdot (1/z^2)$.

From *Neutron Fluence, Neutron Spectra, and Kerma,* ICRU Report 13, International Commission on Radiation Units and Measurements, Washington, D.C., September 15, 1969. With permission.

TABLE 3.7-30
RANGE OF APPLICABILITY OF DATA IN TABLE 3.7-29 FOR HEAVY PARTICLES OTHER THAN PROTONS

Muons	$1 \leqslant \varepsilon_M \leqslant 10^5$
Pions	$1 \leqslant \varepsilon_M \leqslant 10^5$
Kaons	$1 \leqslant \varepsilon_M \leqslant 10^5$
Protons	$1 \leqslant \varepsilon_M \leqslant 10^5$
Hyperons	$1 \leqslant \varepsilon_M \leqslant 10^5$
Deuterons	$1 \leqslant \varepsilon_M \leqslant 10^5$
Tritons	$1 \leqslant \varepsilon_M \leqslant 10^5$
α-Particles	$1 \leqslant \varepsilon_M \leqslant 10^5$
Nuclei with $z \leqslant 15$	$10 \leqslant \varepsilon_M \leqslant 10^5$

[a]   In this range, charge exchange and nuclear and radiation losses are negligible.

From *Neutron Fluence, Neutron Spectra, and Kerma,* ICRU Report 13, International Commission on Radiation Units and Measurements, Washington, D.C., September 15, 1969. With permission.

TABLE 3.7-31
USEFUL THERMAL NEUTRON DETECTORS[a]

Nuclide or element	Reaction	Half-life of nuclide produced	$\sigma_0 \times 10^{24}$ for production (cm²)	Westcott g(20°C)	Westcott s(20°C)	Application
^{3}He	(n,p)^{3}H	12.3 years	5,327	1.0		PC, SD
^{6}Li	(n,t)^{4}He	Stable	945	1.0		SC, SD
^{10}B	(n,α)^{7}Li	Stable	3,837	1.0	Negligible	PC, PT, SC, IC, SD
^{23}Na(IS)	(n,γ)^{24}Na	15.0 hr	0.534	1.0	0.15[b]	BC
^{45}Sc(IS)	(n,γ)^{46}Sc	85 days	22.3			FC
^{51}V	(n,γ)^{52}V	3.8 min	4.9	1.0	0.083[c]	FC
^{55}Mn(IS)	(n,γ)^{56}Mn	2.58 hr	13.3	1.0	0.666	BC, FC
^{59}Co(IS)	(n,γ)^{60}Co	5.24 years	36.6	1.0	1.736	FC
^{63}Cu	(n,γ)^{64}Cu	12.8 hr	4.5	1.0	0.77[b]	FC
^{115}In	(n,γ)^{116}In^m	54 min	157	1.019	19.8	FC
^{157}Gd	(n,γ)^{158}Gd	Stable	242,000	0.854	−0.85	SD
^{197}Au(IS)	(n,γ)^{198}Au	2.70 days	98.8	1.005	17.3	FC
^{235}U	Fission	Many	577	0.976	−0.04	PC, IC, PT, SD

Note: (IS) denotes natural monoisotopic element. Abbreviations for main applications:

    BC: bath counting, incorporated into salt solution
    FC: foil counting
    PT: particle track production in insulating materials
    IC: use in ionization chamber
    PC: proportional counter
    SC: incorporated into scintillator
    SD: converter for semiconductor detector

[a]   Data from BNL 325 and Westcott (1960) except where noted otherwise.
[b]   Calculated from excess resonance integrals of Dahlberg et al. (1961).
[c]   Geiger and Van der Zwan (1967).

From *Neutron Fluence, Neutron Spectra, and Kerma,* ICRU Report 13, International Commission on Radiation Units and Measurements, Washington, D.C., September 15, 1969. With permission.

TABLE 3.7-32
THRESHOLD DETECTORS

	E_{min} (MeV)	E_{opr} (MeV)	E_{oeff} (MeV)	σ_{eff} (mb)	$\bar{\sigma}_{fission}$ (mb)	$\bar{\sigma}_{fission}$ for most recent fission spectrum (mb)
$^{103}Rh(n,n')^{103}Rh^{ma}$	0.04					
$^{115}In(n,n')^{115}In^{mb}$	0.34	0.6	1.12	282	178	212
$^{31}P(n,p)^{31}Si$	0.7	2.0	2.70	126	31.1	40.7
$^{32}S(n,p)^{32}P$	1.0	2.2	2.70	234	57.3	76.1
$^{58}Ni(n,p)^{58}Co + {}^{58}Co^{m}$	0.0	1.9	2.79	478	97 (R)	120(F)
$^{54}Fe(n,p)^{54}Mn$	0.0	2.4			66 (R)	89(F)
$^{27}Al(n,p)^{27}Mg$	1.9	3.8			3.66	4.67
$^{56}Fe(n,p)^{56}Mn$	3.0	5.7	6.34	54	1.08	1.26
$^{46}Ti(n,p)^{46}Sc$	1.6				11 (R)	13.0(F)
$^{24}Mg(n,p)^{24}Na$	4.9	6.4	7.0	120	1.45 (R)	1.62(F)
$^{27}Al(n,\alpha)^{24}Na$			$\begin{cases}7.49\\7.42\end{cases}$	$\begin{cases}83(B)\\62(Z)\end{cases}$		
	3.2	6.6			0.683	0.765
$^{63}Cu(n,2n)^{62}Cu$	11.0	12.0	$\begin{cases}12.7\\12.84\end{cases}$	$\begin{cases}830(B)\\650(Z)\end{cases}$	0.116	0.125
$^{237}Np(n,f)$	0.0	0.6	0.63	1640	1269	1368
$^{238}U(n,f)$	0.0	1.3	1.62	582	277	335

Note: E_{min} indicates the minimum neutron energy required for the reaction calculated on the basis of nuclear masses. E_{opr} indicates the "practical" threshold energy. As such, the energy at which the cross section has one tenth of the value it has at the beginning of the plateau, has been chosen. As the latter point is not clearly defined, the value of E_{opr} is somewhat vague. It does, however, give an idea of the energy range covered by the detector reaction. The two columns headed $\bar{\sigma}_{fission}$ indicate the average cross-section for neutrons produced in the fission of ^{235}U by thermal neutrons. The best values for these cross-sections have changed appreciably in recent years, since it has become known that the energy spectrum of fission neutrons does not agree well with equation III.2 of Reference 19 (Grench and Menlove, 1968; McElroy, 1969). Most of the values listed have been obtained by McElroy, mainly on the basis of measurements of Grundl (1968) and of Fabry (1967). Column 6 contains $\bar{\sigma}$-values for a spectrum corresponding to equation III.2 of Reference 19 with $E_n = 1.29$ MeV, column 7 $\bar{\sigma}$-values which would fit the most accurate spectrum available for neutrons from the thermal fission of ^{235}U. For this spectrum, some data have been included taken from Fabry (1968) for detectors not included in McElroy's list. These data are marked (F). Corresponding figures for a spectrum of equation III.2 of Reference marked (R), have been recalculated on the basis of relative values from earlier work (Senaux, 1965; Zijp, 1965). The quantities σ_{eff} and E_{oeff} are defined by the relation:

$$\sigma_{\text{fission}} \int_0^\infty \frac{d\phi(E)}{dE}\, dE = \sigma_{\text{eff}} \int_{E_{0\,\text{eff}}}^\infty \frac{d\phi(E)}{dE}\, dE$$

where $d\phi(E)/dE$ represents the energy spectrum of fission neutrons, produced by the action of thermal neutrons on ^{235}U. In principle an infinite number of combinations of σ_{eff} and E_{oeff} can be made to fulfill this condition, but the combination selected is the one which does not cause a change in the calculated activity if a slight variation occurs in the exponential term in equation III.2 of Reference 19, i.e., in E_n (Grundl and Usner, 1960). The values used for $d\phi(E)/dE$ are once more equal to those given by equation III.2 of Reference 19 for $dN(E)/dE$. However, the present values of σ_{eff} have been recalculated from the earlier determinations (Béauge, 1963; Zijp, 1965) to correspond to the σ-values in column 6 of Table III.II of Reference 19. In cases where the two earlier values lead to very different values of σ_{eff}, both have been listed and identified by (B) and (Z).

[a] $^{103}Rh^m$ is difficult to measure absolutely. A method for this purpose has been described recently (Nagel and Aten, 1966).

[b] $^{115}In^m$ is measured by means of the 0.335-MeV γ-ray line, which occurs in 50 ± 2% of the transitions (Heertje et al., 1964; Grench and Menlove, 1968).

From *Neutron Fluence, Neutron Spectra, and Kerma*, ICRU Report 13, International Commission on Radiation Units and Measurements, Washington, D.C., September 15, 1969. With permission.

TABLE 3.7-33
RESONANCE DETECTORS USEFUL FOR THE MEASUREMENT OF A NEUTRON SLOWING DOWN SPECTRUM

Detector	Activity measured	Half-life	Resonance energy (eV)	Westcott parameter g
^{115}In	^{116}In^m	54.1 min	1.46	1.02
^{197}Au	^{198}Au	2.70 days	4.9	1.005
^{186}W	^{187}W	24.0 hr	18.8	1.00
^{139}La	^{140}La	40.2 hr	73.5	1.00
^{198}Pt	^{199}Pt	30 min	95	1.00
^{55}Mn	^{56}Mn	2.58 hr	337	1.00
^{100}Mo	^{101}Mo	15 min	367	1.00
	^{101}Tc			
^{23}Na	^{24}Na	15.0 hr	2850	1.00

Note: Only the energy for the main resonance is given.

From *Neutron Fluence, Neutron Spectra, and Kerma,* ICRU Report 13, International Commission on Radiation Units and Measurements, Washington, D.C., September 15, 1969. With permission.

TABLE 3.7-34
DETECTORS BASED ON REACTIONS IN ^{3}He, ^{6}Li, AND ^{10}B AND ON FISSION IN ^{235}U or ^{239}Pu[a]

^{3}He (n,p)T	(Q = 0.77 MeV),	used in a proportional counter or in association with a surface barrier counter
^{6}Li (n,α)T	(Q = 4.8 MeV),	used in a scintillator or in association with a surface barrier counter
^{10}B (n,α)^{7}Li	(Q = 2.8 MeV),	used in a proportional
^{10}B (n,α)^{7}Li[a]	(Q = 2.3 MeV)	counter, ion chamber, or scintillation counter
Fission in ^{235}U or ^{239}Pu	(Q $\sim$ 200 MeV),	used in a fission chamber, as a foil in which the activity is determined after irradiation, or for fission track registration

[a] These reactions are characterized by the relatively large amount of energy produced in the reaction products. This is an advantage in discriminating against γ-ray effects in the counters.

From *Neutron Fluence, Neutron Spectra, and Kerma,* ICRU Report 13, International Commission on Radiation Units and Measurements, Washington, D. C., September 15, 1969. With permission.

INTERMEDIATE-ENERGY NEUTRON ACTIVATION DETECTORS

Reaction	Isotopic composition (%)	Neutron resonance energy[a] (eV) Uthe (1957)	Neutron resonance energy[a] (eV) Hughes and Schwartz (1958)	Half-life	γ-Ray energy[a] (MeV)	Resonance parameters[b] α_1	Resonance parameters[b] α_2	Resonance integrals $\int\infty E_c \sigma^n(E)\,dE/E$ (b)[c]	Resonance integrals $\int\infty E_c \sigma_{ra}(E)\,dE/E$ (b)[d]
^{197}Au(n,γ)^{198}Au	98.8	—	4.6	2.7 days	0.41	—	—	49.4	1490
^{186}W(n,γ)^{187}W	28.4	—	18.8	24 hr	0.134	—	—	17	562
^{59}Co(n,γ)^{60}Co	100	120	132	5.26 years	1.17 1.33	2.0	—	18.5	61.1
^{55}Mn(n,γ)^{56}Mn	100	260	337	2.6 hr	0.85 1.81 2.12	1.0	1.3	6.6	14.6
^{63}Cu(n,γ)^{64}Cu	69.1	570	580	12.8 hr	1.34	1.3	0.93	2.2	5.42
^{23}Na(n,γ)^{24}Na	100	1710	2950	15.0 hr	2.75 1.37	~0	0.24	0.27	0.32
^{27}Al(n,γ)^{28}Al	100	9100	5800	2.3 min	1.78	0.68	0.16	0.11	0.17

[a] Since the neutron resonance structure and the radioactivity decay schemes are complex for many of the isotopes considered, only the prominent values are listed.

[b] From Uthe (1957). Based on the equation, $\phi = k_\phi / (CR - 1)(1 + \alpha_1)$, where $k_i = 2.27$ is used with α_1 for the case of 0.02-in. cadmium thickness and $k_i = 1.43$ is used with α_2 for the case of 0.01-in. cadmium thickness.

[c] The $1/v$ part of the resonance integral was calculated by assuming $\phi(E) = \phi/E$ and $E_c = Q.4$ eV, i.e., $\int\infty E_c \alpha 1'/v(E) \pm dE/E$ equal to $0.5\sigma_0$, which is obtained by using Equation 20 in Reference 34, with $\sigma_i'/v(E) = \sigma_0$ for $E = 0.025$ eV for determining k_0.

[d] The resonance absorption integrals (without the $1/v$ contribution) were obtained from the detailed review (91 references) by Drake (1966). Uncertainties still exist in these values owing to the various indirect methods employed in correlating and (or) normalizing the experimental data.

From Moteff, J., in *Radiation Dosimetry*, Vol. 3, 2nd ed., Attix, F. H. and Tochilin, E., Eds., Academic Press, New York, 1969. With permission.

TABLE 3.7-36
FAST-NEUTRON ACTIVATION DETECTORS

Reaction	Half-life[1]	γ-Ray energy (MeV)	Effective energy,[a] E_{eff} (MeV)	Fission average cross section,[b] σ_f (mb)	Effective cross section,[a] $\sigma(E_{eff})$ (mb)
$^{58}Ni(n,p),^{58}Co$	71 days	0.81	—	91.5 ± 3.2	—
$^{58}Ni(n,p)^{58m}Co$	9.0 hr	0.025	—	29.5 ± 1.0	—
			2.9	121	530
$^{32}S(n,p)^{32}P$	14.3 days	1.71 (β)	3.0	65	310
$^{64}zn(n,p)^{64}Cu$	12.8 days	1.35	4.0	27	250
$^{54}Fe(n,p)^{54}MN$	291 days	0.835	4.2	65.0 ± 2.3	680
$^{46}Ti(n,p)^{46}So$	84.0 days	1.12	5.5	10.2 ± 0.4	270
$^{24}Mg(n,p)^{24}Na$	15.0 hr	2.75	6.3	1.26 ± 0.04	60
$^{63}Cu(n,\alpha)^{60}Co$	5.26 years	1.17, 1.33	6.7	0.40	25
$^{56}Fe(n,p)^{56}MN$	2.58 hr	0.85	7.5	0.93 ± 0.032	110
$^{27}Al(n,\alpha)^{24}Na$	15.0 hr	2.75	8.1	0.61	110

[a]   Effective energy and cross section are approximate values based on a Watt (1948) fission spectrum and the tabulated values for the fission average cross section. See Equations 27 and 28 of Reference 34.

[b]   Cross sections which include the limits of uncertainty are from Bresesti et al. (1967). The other cross sections represent averages of those reported in the literature. All data normalized to the $Al^{27}(n,\alpha)Na^{24}$ reaction cross section.

From Moteff, J., in *Radiation Dosimetry,* Vol. 3, 2nd ed., Attix, F. H. and Tochilin, E., Eds., Academic Press, New York, 1969. With permission.

REFERENCE

1.   **Weast, R.C., Ed.,** *Handbook of Chemistry and Physics,* 48th ed., CRC Press, Cleveland, Ohio, 1967—1968.

Table 3.7-37
FISSION-PRODUCT CHAINS

CHAIN

1. 0.9504   0.1825

    SE0840 ——————→ BR0840
    3.4 min   31.8 min

2. 2.54

    KR0870
    76 min

3. 3.594   0.0322

    KR0880 ———→ RB0880
    2.8 hr   17.7 min

4. 5.91   0.0002222

    SR0900 ——————→ Y 0900
    29 years   64 hr

5. 5.952   0.0078

    SR0920 ——————→ Y 0920
    2.69 hr   3.53 hr

6. 6.371

    Y 0930
    10.2 hr

7. 5.948

    ZR0970
    16.8 hr

8. 0.3912   3.462 x 10⁻⁷

    RU1060 ——————→ RH1060
    369 days   29.9 sec

9. 0.1266   7.604 x 10⁻⁷

    PD1120 ———→ AG1120
    20.1h   3.12h

10. 2.531   0.191

    ⌐ SB1310 ¬ α = 0.07 TE1311
    23 min   30 hr

11. 4.231   0.017

    TE1320 ——————→ I 1320
    78 hr   2.285 h

12. 6.776   0.4286

    TE1340 ——————→ I 1340
    42 m   52.6 m

13. 6.349

    I 1350
    6.585 hr

14. 6.252   0.3122

    XE1380 ——————→ CS1380
    14.2 min   32.3 min

15. 6.32   0.005671

    BA1400 ——————→ LA1400
    12.79 days   40.23 hr

16. 5.829   .1013

    BA1420 ——————→ LA1420
    10.7 min   92.4 min

17. 5.457   6.854

    CE1440 ——————→ PR1440
    284.4 day   17.3 min

18. 4.678   0.1655

    KR0890 ——————→ RB0890
    3.16 min   15.2 min

19. 6.058   0.3626

    SR0940 ——————→ Y 0940
    1.29 min   20.3 min

20. 5.484   0.951

    SR0950 ——————→ Y 0950
    0.26 sec   10.5 min

21. 5.705   0.04783

    ZR0980 ——————→ NB0980
    31 sec   51 min

22. 1.77   0.05372

    MO1040 ——————→ TC1040
    1.6 min   18 min

Table 3.7-37(continued)
FISSION-PRODUCT CHAINS

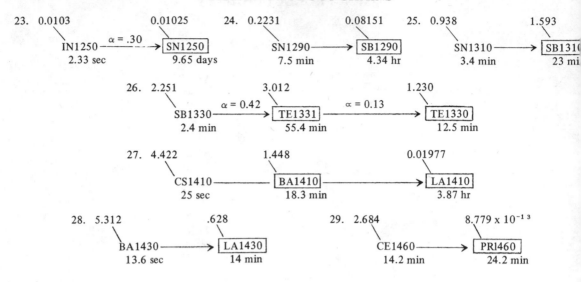

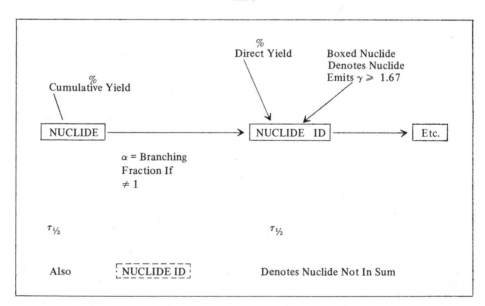

From Stamatelatos, M. G., and England, T. R., Fission-product Gamma-ray and Photoneutron Spectra and Energy-integrated Sources Informal Report, NUREG-0155, U. S. Nuclear Regulatory Commission, Los Alamos Scientific Laboratory, University of California, Los Alamos, N. M., December 1976.

TABLE 3.7-38
FISSION YIELDS (%)

Nuclide	^{235}U thermal	^{235}U fast	^{235}U high energy	^{238}U fast	^{238}U high energy
^{84}SE	0.9504	0.9915	0.8722	0.7476	0.8920
^{84}BR	0.1825	0.1680E−01	0.1196E−01	0.2704E−02	0.3548E−02
^{87}KR	2.544	2.412	2.411	1.523	1.616
^{88}KR	3.594	3.555	3.297	2.290	2.145
^{88}RB	0.3222E−01	0.3063E−01	0.1703E−01	0.2533E−02	0.1869E−02
^{90}SR	5.910	5.506	4.597	3.354	3.111
^{90}Y	0.2220E−03	0.2880E−03	0.4927E−05	0.2580E−06	0.8726E−07
^{92}SR	5.952	5.700	5.113	4.203	3.900
^{92}Y	0.7800E−02	0.2416E−01	0.7186E−02	0.8201E−02	0.1689E−03
^{93}Y	6.371	6.096	5.210	4.843	4.386
^{97}ZR	5.948	5.881	5.903	5.518	5.270
^{106}RU	0.3912	5.753	1.529	2.752	2.429
^{106}RH	0.3462E−06	0.5481E−06	0.8294E−05	0.5609E−05	0.2139E−04
^{112}PD	0.1266E−01	0.3799E−01	0.9786	0.5131E−01	0.9507
^{112}AG	0.7604E−06	0.5831E−05	0.5330E−01	0.1060E−06	0.7317E−04
^{131}SB	2.531	2.899	2.250	3.185	3.620
^{131m}TE	0.1910	0.1665	0.8384	0.1370E−01	0.3312
^{132}TE	4.231	4.622	4.026	5.027	4.794
^{132}I	0.1700E−01	0.3150E−01	0.5954	0.1225E−02	0.1585
^{134}TE	6.776	6.343	2.635	6.905	5.628
^{134}I	0.4286	0.6165	2.120	0.9616E−01	0.5056
^{135}I	6.349	5.955	3.555	6.548	5.345
^{138}XE	6.252	6.051	3.100	5.888	4.310
^{138}CS	0.3122	0.1966	1.409	0.6070E−01	0.4284
^{140}BA	6.320	6.012	4.397	5.889	4.611
^{140}LA	0.5671E−02	0.8009E−02	0.3247E−01	0.6069E−05	0.4879E−02
^{142}BA	5.829	5.375	4.508	4.928	4.371
^{142}LA	0.1013	0.8453E−01	0.3633	0.2466E−02	0.5362E−01
^{144}CE	5.459	5.267	3.228	4.751	3.441
^{144}PR	0.6854E−04	0.1750E−04	0.3238E−03	0.5100E−05	0.6937E−05
^{89}KR	4.678	4.359	3.715	2.899	2.748
^{89}RB	0.1655	0.2017	0.1788	0.7576E−01	0.2277E−01
^{94}SR	6.058	5.905	4.985	4.478	3.965
^{94}Y	0.3626	0.2791	0.2050	0.4846E−01	0.5766E−01
^{95}SR	5.484	5.423	4.289	5.089	4.972
^{95}Y	0.9510	0.9317	0.7331	0.2322	0.2259
^{98}ZR	5.705	5.820	4.835	5.551	5.228
^{98}NB	0.4783E−01	0.2435E−01	0.2037E−01	0.1422E−02	0.2193E−02
^{104}MO	1.770	2.230	2.100	5.399	4.470
^{104}TC	0.5372E−01	0.8149E−01	0.1732	0.3712E−01	0.4087E−01
^{125}IN	0.1030E−01	0.4413E−01	0.4859	0.4559E−01	0.6037
^{125}SN	0.1025E−01	0.6131E−02	0.2762	0.4980E−03	0.2168
^{129}SN	0.2231	0.3698	0.6058	0.6549	1.258
^{129}SB	0.8151E−01	0.3246	0.7829	0.8092E−01	0.1550
^{131}SN	0.9380	1.324	0.3638	2.534	1.825
^{131}SB	1.593	1.575	1.886	0.6501	1.795
^{133}SB	2.251	2.297	0.5093	4.820	2.461
^{133m}TE	3.012	1.954	1.787	0.7571	1.776
^{133}TE	1.230	1.952	1.552	0.7572	1.789

TABLE 3.7-38 (continued)
FISSION YIELDS (%)

Nuclide	^{235}U thermal	^{235}U fast	^{235}U high energy	^{238}U fast	^{238}U high energy
^{141}CS	4.422	4.477	2.258	5.046	3.641
^{141}BA	1.448	1.503	2.225	0.9633E−01	0.7760
^{141}LA	0.1977E−01	0.1515E−01	0.8690E−01	0.3240E−03	0.6007E−02
^{143}BA	5.312	5.285	3.106	4.806	3.620
^{143}LA	0.6280	0.4072	0.7571	0.2809E−01	0.1969
^{146}CE	2.684	2.889	2.343	3.610	2.732
^{146}PR	0.8779E−02	0.6345E−02	0.2857E−01	0.1490E−03	0.2339E−02

From Stamatelatos, M. G. and England, T. R., Fission-product Gamma-ray and Photoneutron Spectra and Energy-integrated Sources Informal Report, NUREG-0155, U.S. Nuclear Regulatory Commission, Los Alamos Scientific Laboratory, University of California, Los Alamos, N.M., December 1976.

TABLE 3.7-39
THERMAL-NEUTRON-FISSION YIELDS (PERCENT) FROM ^{233}U, ^{235}U, and ^{239}Pu

Fission product	^{233}U	^{235}U	^{239}Pu	Fission product	^{233}U	^{235}U	^{239}Pu
105-day ^{127m}Te		0.035		5×10^{15}-year^{144}Nd	4.61	5.62	3.93
57-min ^{128}Sn		0.37		Stable ^{145}Nd	3.47	3.98	3.13
25.0-min^{128}I		3 × 10^{-5}		Stable ^{146}Nd	2.63	3.07	2.60
37-day ^{129m}Te		0.35		11.1-day ^{147}Nd		~2.7	2.2
1.7 × 10^7-year ^{129}I		0.8		2.6-year ^{147}Pm	1.9		1.94
2.6-min ^{130}Sn		2.0		1.3×10^{11}-year^{147}Sm	1.98	2.36	2.07
12.6-hr ^{130}I		5 × 10^{-4}		Stable ^{148}Nd	1.34	1.71	1.73
30-hr ^{131m}Te		0.44		53.1-hr ^{149}Pm			1.4
8.05-day ^{131}I	2.9	~ 3.1	3.77	Stable ^{149}Sm	0.76	1.13	1.32
Stable ^{131}Xe	3.39	2.93	3.78	Stable ^{150}Nd	0.56	0.67	1.01
77-hr ^{132}Te	4.4	~4.7	5.1	80-year ^{151}Sm	0.335	0.44	0.80
Stable ^{132}Xe	4.64	4.38	5.26	Stable ^{152}Sm	0.220	0.281	0.62
20.8-hr ^{133}I		~6.9	5.2	47-hr ^{153}Sm	0.11	0.15	0.37
5.27-day ^{133}Xe		6.62	6.91	Stable ^{153}Eu	0.13	0.169	
Stable ^{133}Cs	5.78	6.59	6.91	Stable ^{154}Sm	0.045	0.077	0.29
52.5-min ^{134}I		7.8		24-min ^{155}Sm		0.033	0.23
Stable ^{134}Xe	5.95	8.06	7.47	4-year ^{156}Eu		0.033	
6.7-hr^{135}I	5.5	6.1	5.7	15.4-day ^{156}Eu	0.011	0.014	0.11
9.2-hr ^{135}Xe		6.3		15.4-hr ^{157}Eu		0.0078	
2.6 × 10^6 hr ^{135}Cs	6.03	6.41	7.17	60-min ^{158}Eu		0.002	
86-sec ^{136}I	1.8	3.1	2.1	18.0-hr ^{159}Go		0.00107	0.021
Stable ^{136}Xe	6.63	6.46	6.63	6.9-day ^{161}Tb		7.6 × 10^{-5}	0.0039
13-day ^{136}Cs	0.12	0.0068	0.11	82-hr ^{166}Dy			6.8 × 10$^-$
30-year ^{137}Cs	6.58	6.15	6.63	47-hr ^{72}Zn		1.6 × 10^{-5}	1.2 × 10$^-$
Stable ^{138}Ba		5.74	6.31	4.9-hr ^{73}Ga		1.1 × 10^{-4}	
83-min ^{139}Ba	6.45	6.55	5.87	7.8-min ^{74}Ga		3.5 × 10^{-4}	
12.8-day ^{140}Ba	5.4	6.35	5.4	11.3-hr ^{77}Ge	0.011		0.0031
Stable ^{140}Ce	6.47	6.44	5.60	38.7-hr ^{77}As	0.021		0.0083
3.8-hr ^{141}La	7.1	6.4	5.7	2.1-hr ^{78}Ge			0.020
33-day ^{141}Ce		~6.0	5.1	91-min ^{78}As			0.020
Stable ^{141}Pr	6.4		(4.5)	9.0-min ^{79}As			0.056
Stable ^{142}Ce	6.83	6.01	5.01	Total ^{80}Br	3.9 × 10^{-4}	1.0 × 10^{-5}	
33-hr ^{143}Ce		5.7	5.3	57-min ^{81m}Se			0.0084
Stable ^{143}Nd	5.99	6.03	4.57	18.4-min ^{81}Se			0.14
280-day ^{144}Ce	4.5	~6.0	3.79				

TABLE 3.7-39
THERMAL-NEUTRON-FISSION YIELDS (PERCENT) FROM ^{233}U, ^{235}U, and ^{239}Pu

Fission product	^{233}U	^{235}U	^{239}Pu	Fission product	^{233}U	^{235}U	^{239}Pu
35.9-hr ^{82}Br	1.1×10^{-3}	4×10^{-5}		23-hr ^{96}Nb	6.5×10^{-3}	6.1×10^{-3}	3.6×10^{-3}
25-min ^{83}Se		0.22		17.0-hr ^{97}Zr		5.9	5.5
2.4-hr ^{83}Br	0.87	0.51	0.084	Stable ^{97}Mo	5.37	6.09	5.65
Stable ^{83}Kr	1.17	0.544	0.29	52-min ^{98}Nb	0.20	0.064	0.20
6.0-min ^{84}Br		0.019		Stable ^{98}Mo	5.15	5.78	5.89
31.8-min ^{84}Br		0.92		66.5-hr ^{99}Mo	4.80	6.06	6.10
Stable ^{84}Kr	1.95	1.00	0.47	Stable ^{100}Mo	4.41	6.30	7.10
39-sec ^{85}Se		~1.1		Stable ^{101}Ru	2.91	5.0	5.91
10.6-year ^{85}Kr	0.58	0.293	0.127	Stable ^{102}Ru	2.22	4.1	5.99
Stable ^{85}Rb	2.51	1.30	0.539	39.7-day ^{103}Ru	1.8	3.0	5.67
Stable ^{86}Kr	3.27	2.02	0.76	Stable ^{104}Ru	0.94	1.8	5.93
18.6-day ^{86}Rb	2.3×10^{-4}	2.9×10^{-5}	2.3×10^{-5}	4.45-hr ^{105}Ru		0.9	
16-sec ^{87}Se		~2		36-hr ^{105}Rh			3.9
5×10^{10}-year ^{87}Rb	4.56	2.49	0.92	1.01-year ^{106}Ru	0.24	0.38	4.57
Stable ^{88}Sr	5.37	3.57	1.42	22-min ^{107}Rh		0.19	
50.5-day ^{89}Sr	5.86	4.79	1.71	13.4-hr ^{109}Pd	0.044	0.030	1.40
28-year ^{90}Sr	6.43	5.77	2.25	7.6-day ^{111}Ag	0.024	0.019	0.23
9.7-hr ^{91}Sr	5.57	5.81	2.43	21.0-hr ^{112}Pd	0.016	0.010	0.12
58-day ^{91}Y	5.1	~5.4	2.9	43-day ^{115m}Cd	0.0011	0.0007	0.0031
Stable ^{91}Zr	6.43	5.84	2.61	53-hr ^{115}Cd	0.020	0.0097	0.0038
2.7-hr ^{92}Sr		5.3		Total 115	0.021	0.0104	0.041
Stable ^{92}Zr	6.64	6.03	3.14	3.0-hr ^{117m}Cd		0.011	
10.3-hr ^{93}Y		6.1		27.5-hr ^{121}Sn	0.018	0.015	0.043
1.1×10^{6}-year ^{93}Zr	6.98	6.45	3.97	136-day ^{123}Sn		0.0013	
Stable ^{94}Zr	6.68	6.40	4.48	9.6-day ^{125}Sn	0.052	0.013	0.071
65-day ^{95}Zr	6.1	6.2	5.8	2.0-year ^{125}Gb		0.021	
Stable ^{95}Mo	6.11	6.27	5.03	91-hr ^{127}Sb	0.060	0.13	0.39
Stable ^{96}Zr	5.58	6.33	5.17				

Reprinted from Katcoff, S., *Nucleonics*, 18(11), 203, 1960, as given in Loveland, W. D., in *American Institute of Physics Handbook*, 3rd ed., Gray, D. E., Ed., McGraw-Hill, New York, 1972. With permission.

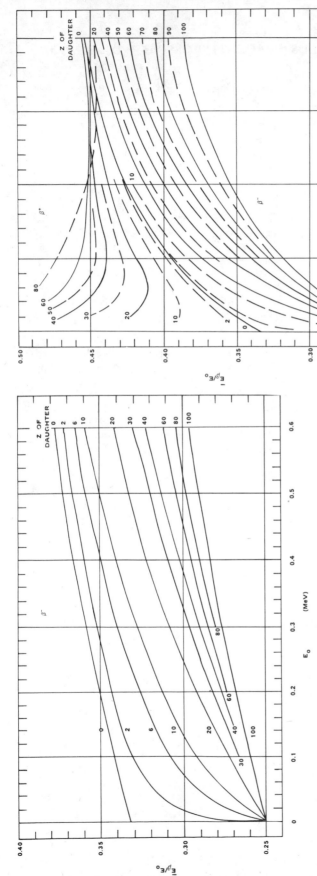

FIGURE 3.7-1. The ratio of the average beta energy, E_β, to the maximum beta energy, E_0, electrons for different values of Z. (These earlier data have been superseded by later calculations of Dillman.[16] See also Section 3.6 and Figures 3.7-10 through 3.7-15.) (From Loevinger, *Phys. Med. Biol.*, 1, 330, 1957, as given in Slack, L. and Way, K., *Radiations From Radioactive Atoms in Frequent Use*, U.S. Atomic Energy Commission, Washington D.C., February 1959. With permission.)

FIGURE 3.7-2. The ratio of the average beta energy, E_β, to the maximum beta energy, E_0 for different Z values for positrons (upper region) and electrons (lower region). (These data have been superseded by more recent calculations of Dillman.[16] See also Section 3.6 and Figures 3.7-10 through 3.7-15.) (From Loevinger, *Phys. Med. Biol.*, 1, 330, 1957, as given in Slack, L. and Way, K., *Radiations From Radioactive Atoms in Frequent Use*, U.S. Atomic Energy Commission, Washington, D.C., February 1959. With permission.)

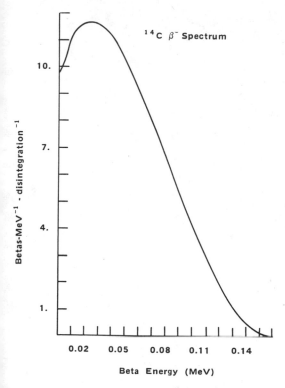

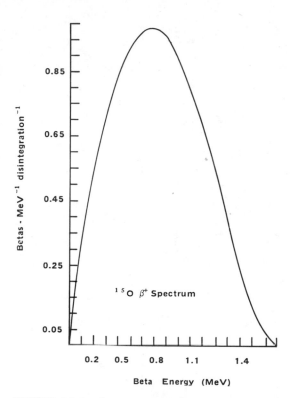

FIGURE 3.7-3.   Carbon-14 beta spectrum. (From Dillman, L. T., in *Health Physics Division Annual Progress Report, Period Ending June 30, 1976,* ORNL-5171, Oak Ridge National Laboratory, Oak Ridge, Tenn., October 1976. With permission.)

FIGURE 3.7-4.   Oxygen-15 beta spectrum. (From Dillman, L. T., in *Health Physics Division Annual Progress Report, Ending June 30, 1976,* ORNL-5171, Oak Ridge National Laboratory, Oak Ridge, Tenn., October 1976. With permission.)

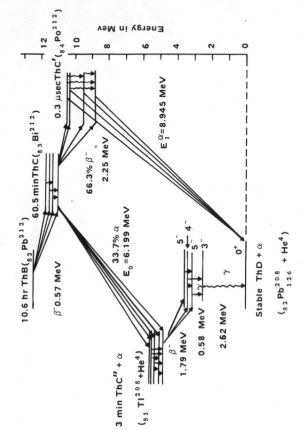

FIGURE 3.7-6. Decay schemes for the principal transitions in the "thorium-active deposit" ThB[*β*]-ThC[*α,β*] → .... Notice that the short-range *α*-rays of ThC are *α* transitions *to* excited levels, while the long-range *α*-rays of ThC' are *α* transitions *from* excited levels. Notice the origin of the very important and useful 2.62-MeV *γ*-ray, which is in cascade with a preceding 0.58-MeV *γ*-ray and a *β* transition. When all *α*-, *β*-, and *γ*-ray energies are summed, the total disintegration energy is the same (11.19 MeV) in the two competing branches ThC[*β*] → ThC'[*α*] → ThD and ThC[*α*] → ThC''[*β*] → ThD. The angular momentum and parity assignments in ThD are as determined by Elliott and coworkers. (From Evans, R. D., *The Atomic Nucleus*, McGraw-Hill, New York, 1955. With permission.)

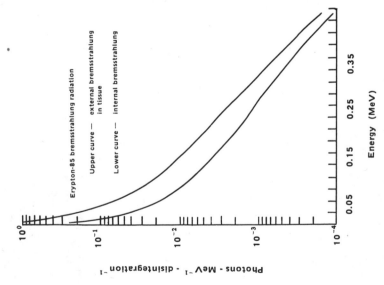

FIGURE 3.7-5.   Krypton-85 bremsstrahlung radiation. (From Dillman, L. T., in *Health Physics Division Annual Progress Report, Period Ending June 30, 1976*, ORNL-5171, Oak Ridge National Laboratory, Oak Ridge, Tenn., October 1976. With permission.)

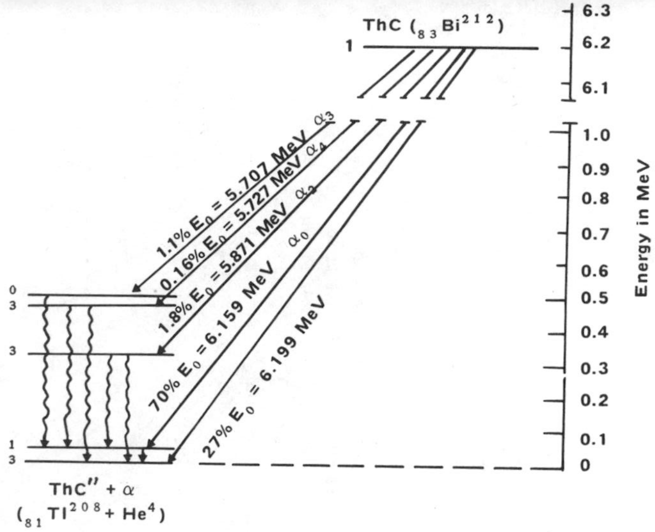

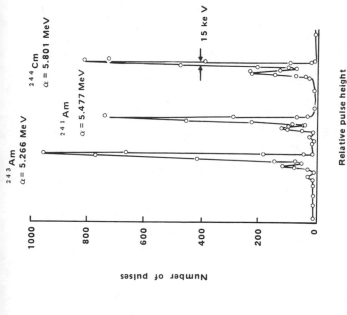

FIGURE 3.7-8. Pulse-height spectra of ^{243}Am, ^{241}Am, and ^{244}Cm taken with a silicon surface-barrier detector of 25 mm^2 sensitive area (Blankenship et al., 1962). Temperature was 23°C. The width of the ^{244}Cm peak could be reduced to 13.5 keV at −195°C and bias 100 V. (From Fowler, J. F., *Radiation Dosimetry*, Vol. 2, 2nd ed., Attix, F. H. and Roesch, W. C., Eds., Academic Press, New York, 1966. With permission.)

FIGURE 3.7-7. Energy-level diagram for the transition ThC″ → ThC′. Tentative assignments of angular momentum for each level are shown, according to the analysis by F. Oppenheimer. Note the ~5-MeV discontinuity in the energy scale. (From Evans, R. D., *The Atomic Nucleus*, McGraw-Hill, New York, 1955. With permission.)

N 1 (Neutron)

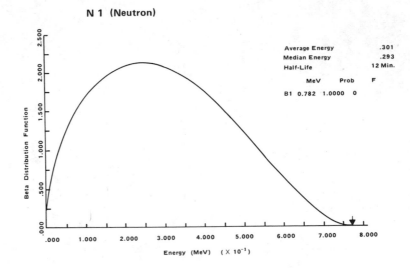

H 3

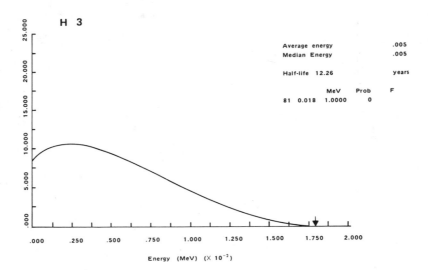

FIGURE 3.7-9.   Beta spectra and half-lives of selected nuclides from Hogan, Zigman, and Mackin.[142] Average and median energies are included. Daughter spectra
are not included, as in Slack and Way,[14] but are plotted separately. Note the multiplying factors on some of the energy scales. (From Hogan, O. H., Zigman, P. E.,
and Mackin, J. L., Beta Spectra II. Spectra of Individual Negatron Emitters,
USNRDL-TR-802, U.S. Naval Radiological Defense Laboratory, San Francisco,
Calif., December, 16, 1964. With permission.)

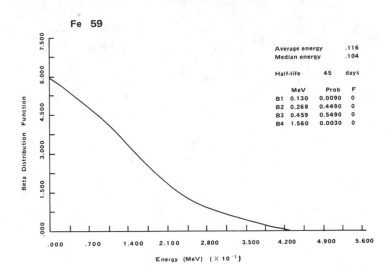

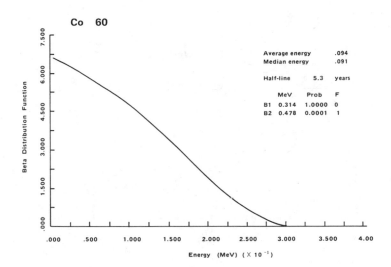

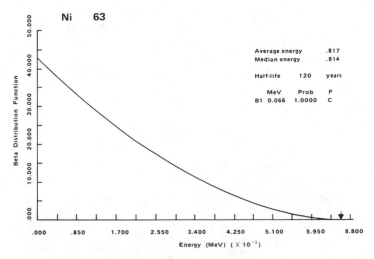

FIGURE 3.7-9.   (Continued)

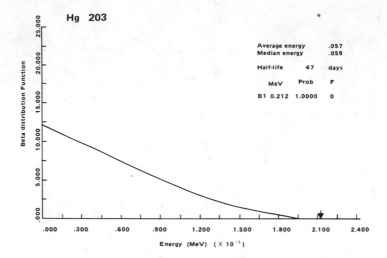

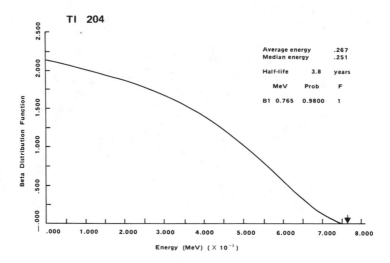

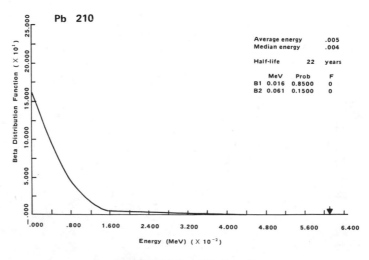

FIGURE 3.7-9.   (Continued)

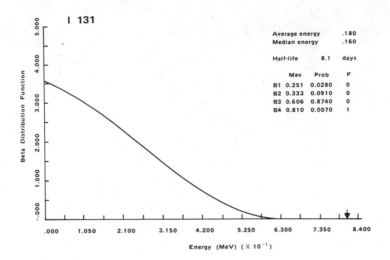

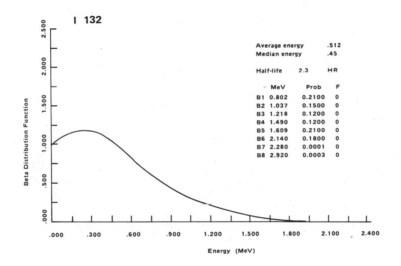

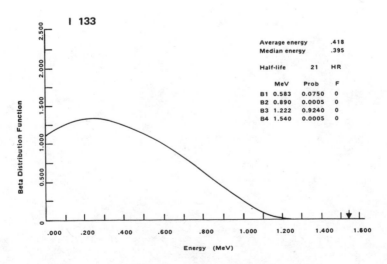

FIGURE 3.7-9.   (Continued)

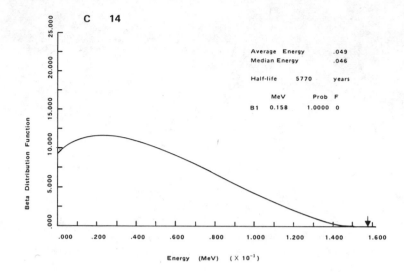

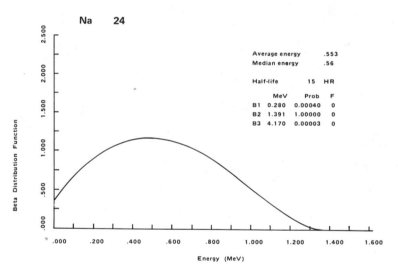

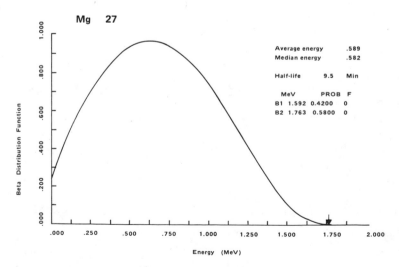

FIGURE 3.7-9.   (Continued)

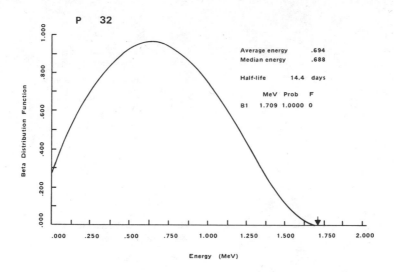

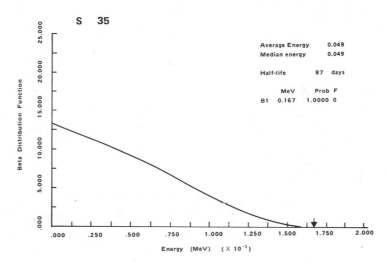

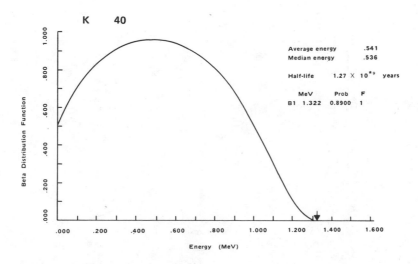

FIGURE 3.7-9.   (Continued)

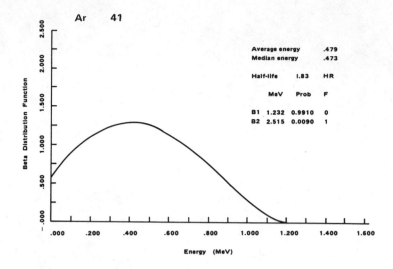

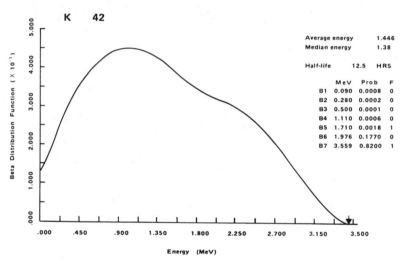

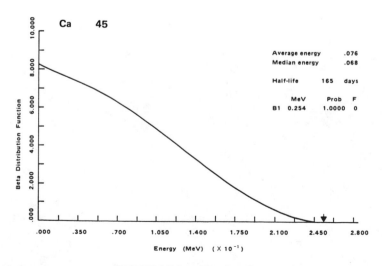

FIGURE 3.7-9.   (Continued)

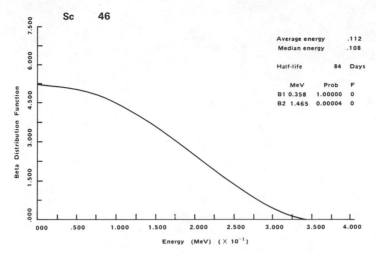

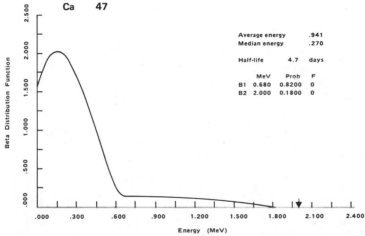

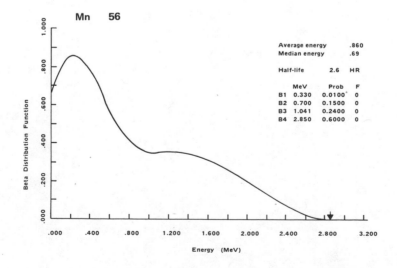

FIGURE 3.7-9.   (Continued)

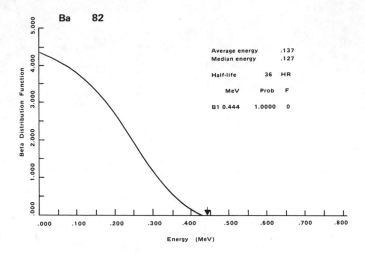

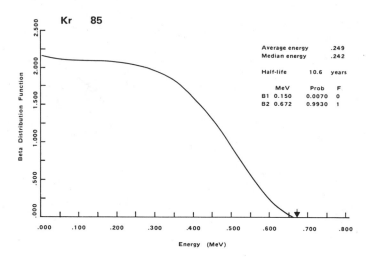

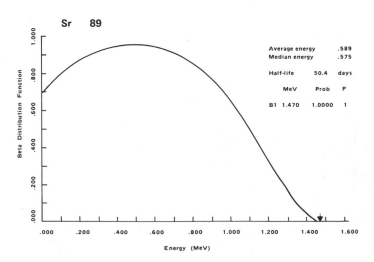

FIGURE 3.7-9.   (Continued)

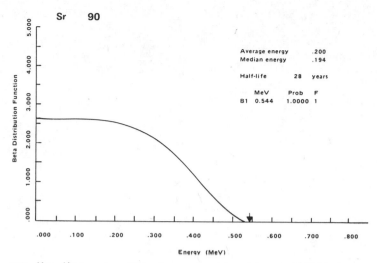

Note: ^{90}Sr — ^{90}Y in secular equilibrium have an average beta spectral energy of 0.57 MeV (See ICRU Handbook 78, 1959, p. 72); however, to calculate dose per disintegration of ^{90}SR, the ^{90}Y average beta contribution should be added to that of ^{90}SR when they are in secular equilibrium.

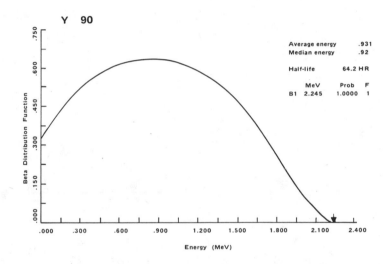

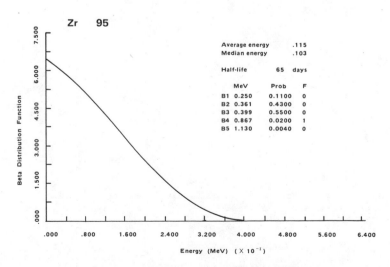

FIGURE 3.7-9.   (Continued)

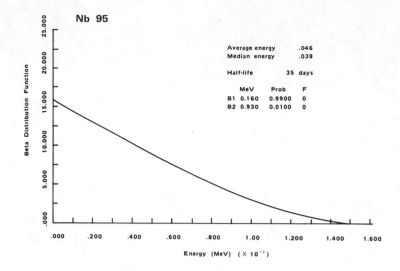

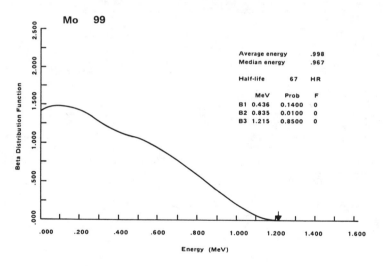

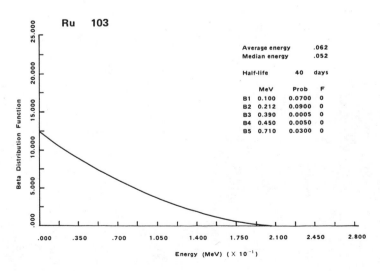

FIGURE 3.7-9.   (Continued)

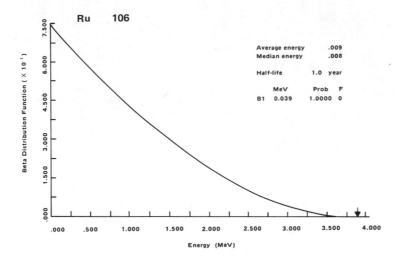

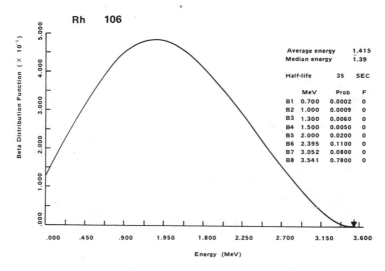

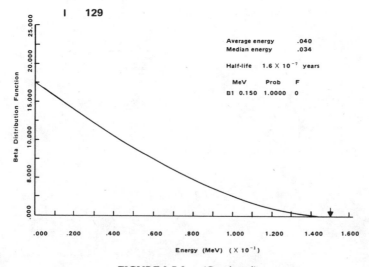

FIGURE 3.7-9.   (Continued)

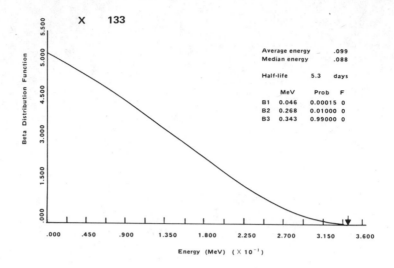

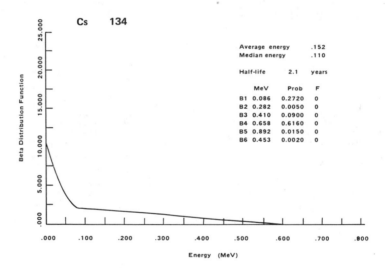

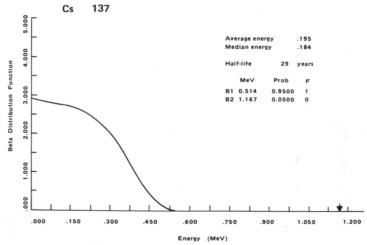

Note: ^{137}Cs includes ^{137m}Ba daughter, but not Auger and conversion electrons (See ICRU Handbook 78, 1959, p. 72).

FIGURE 3.7-9.   (Continued)

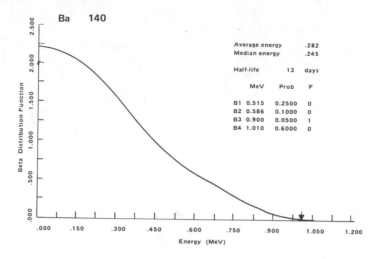

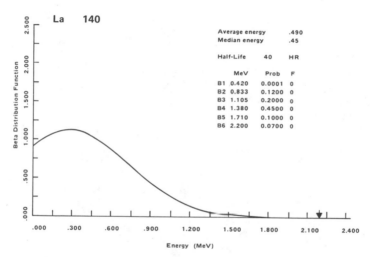

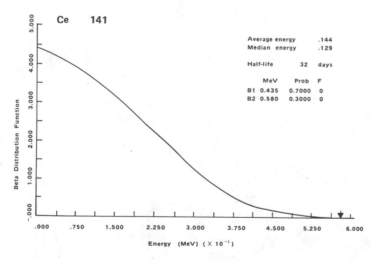

FIGURE 3.7-9.   (Continued)

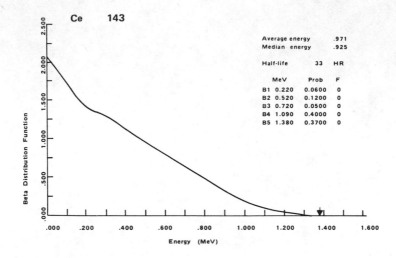

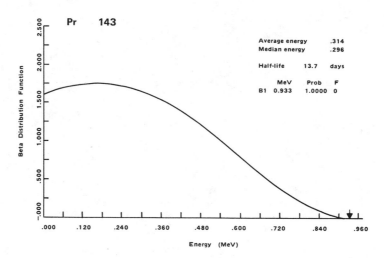

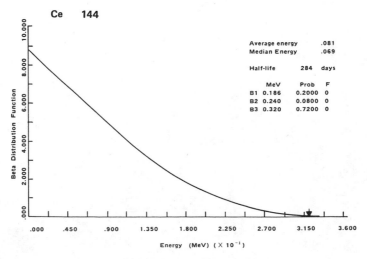

FIGURE 3.7-9.   (Continued)

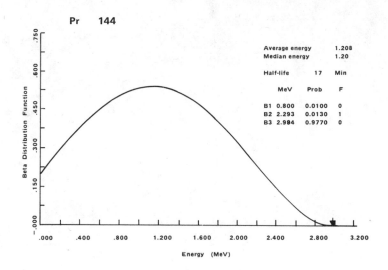

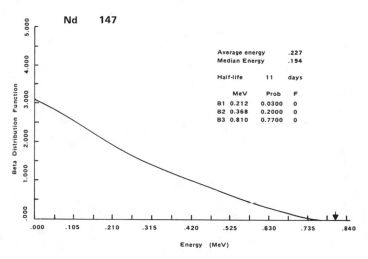

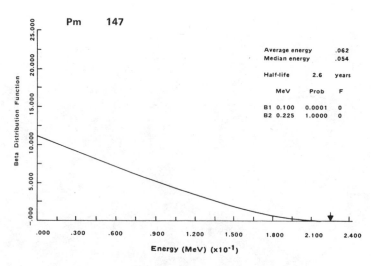

FIGURE 3.7-9.   (Continued)

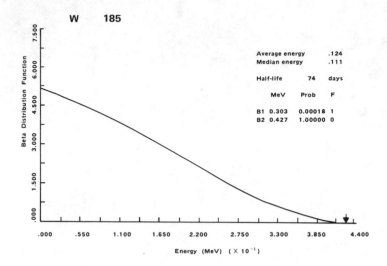

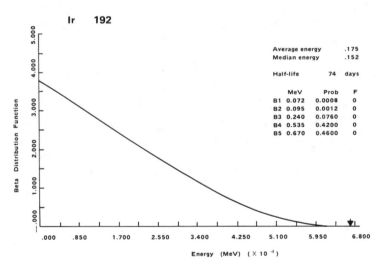

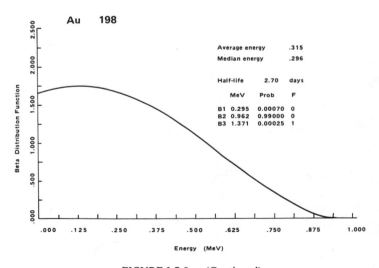

FIGURE 3.7-9.   (Continued)

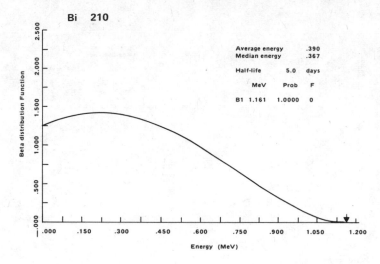

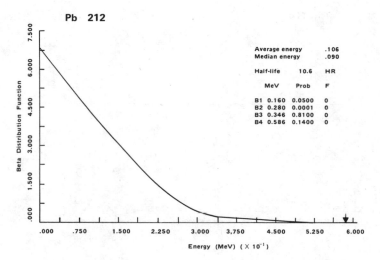

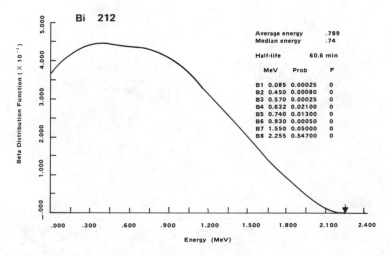

FIGURE 3.7-9.   (Continued)

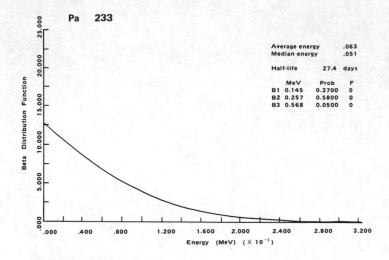

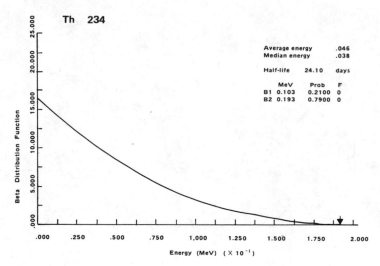

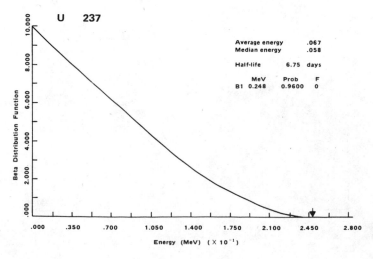

FIGURE 3.7-9.   (Continued)

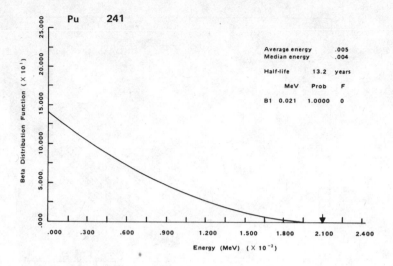

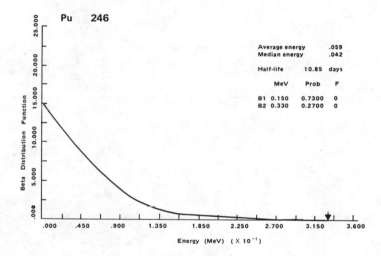

FIGURE 3.7-9.   (Continued)

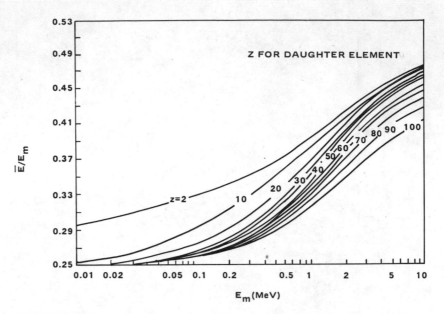

FIGURE 3.7-10.   Allowed β^- transitions. This graph also gives approximate values for the first forbidden nonunique transitions. (From Dillman, L. T., Oak Ridge National Laboratory Report ORNL-4168, Oak Ridge, Tenn., 1967, 233ff, as given in Morgan, K. Z. and Turner, J. E., in *American Institute of Physics Handbook,* 3rd ed., Gray, D. E., Ed., McGraw-Hill, New York, 1972. With permission.)

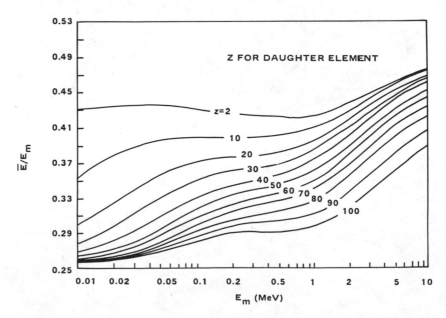

FIGURE 3.7-11.   First forbidden unique β^- transitions. This graph also gives approximate values for the second forbidden nonunique transitions. (From Dillman, L. T., Oak Ridge National Laboratory Report ORNL-4168, Oak Ridge, Tenn., 1967, 233ff, as given in Morgan, K. Z. and Turner, J. E., in *American Institute of Physics Handbook,* 3rd ed., Gray, D. E., Ed., McGraw-Hill, New York, 1972. With permission.)

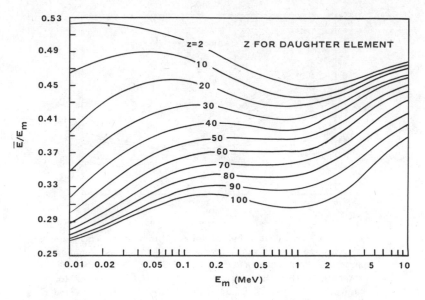

FIGURE 3.7-12.   Second forbidden unique β^- transitions. (From Dillman, L. T., Oak Ridge National Laboratory Report ORNL-4168, Oak Ridge, Tenn., 1967, 233ff, as given in Morgan, K. Z. and Turner, J. E., in *American Institute of Physics Handbook,* 3rd. ed., Gray, D. E., Ed., McGraw-Hill, New York, 1972. With permission.)

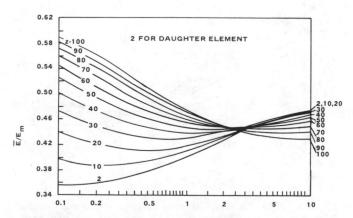

FIGURE 3.7-13.   Allowed β^+ transitions. This graph also gives approximate values for the first forbidden nonunique transitions. (From Dillman, L. T., Oak Ridge National Laboratory Report ORNL-4168, Oak Ridge, Tenn., 1967, 233ff, as given in Morgan, K. Z. and Turner, J. E., in *American Institute of Physics Handbook,* 3rd. ed., Gray, D. E. Ed., McGraw-Hill, New York 1972. With permission.)

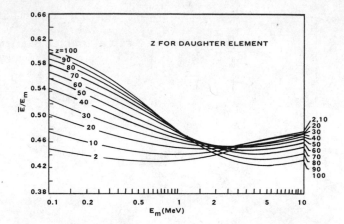

FIGURE 3.7-14. First forbidden unique β^+ transitions. This graph also gives approximate values for the second forbidden nonunique transitions. (From Dillman, L. T., Oak Ridge National Laboratory Report ORNL-4168, Oak Ridge, Tenn., 1967, 233ff, as given in Morgan, K. Z. and Turner, J. E., in *American Institue of Physics Handbook,* 3rd ed., Gray, D. E., Ed., McGraw-Hill, New York, 1972. With permission.)

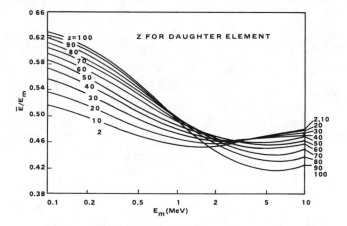

FIGURE 3.7-15. Second forbidden unique β^+ transitions. (From Dillman, L. T., Oak Ridge National Laboratory Report ORNL-4168, Oak Ridge, Tenn., 1967, 233ff, as given in Morgan, K. Z. and Turner, J. E., in *American Institute of Physics Handbook,* 3rd ed., Gray, D. E., Ed., McGraw-Hill, New York, 1972. With permission.)

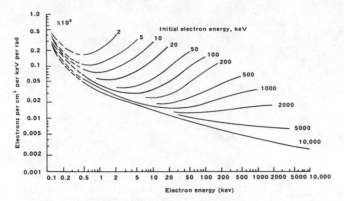

FIGURE 3.7-16. The energy spectra of slowed-down and secondary electrons, in terms of electrons per square centimeter per kiloelectron volt interval, from initial electrons with energy from 2 to 10,000 keV, in an infinite water medium. All spectra are adjusted to give a dose of 1 rad. The ordinate is a function of the initial electron energy T_o and the energy T is represented by $\gamma(T_o, T)$. The solid curve is based on data of McGinnies (1959). The dashed curves are an extension of her data to lower energies (Bruce et al., 1963). (From Johns, H. E., in *Radiation Dosimetry*, 2nd Ed., Vol. 3, Attix, F. H. and Tochilin, E., Eds., Academic Press, New York, 1969. With permission.)

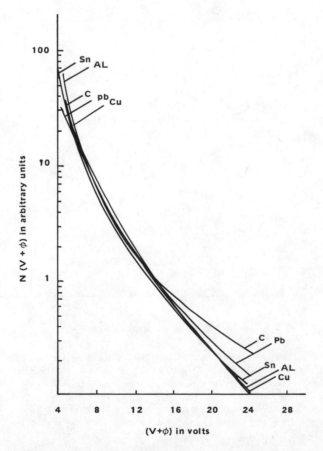

FIGURE 3.7-17. The low-energy electron spectrum with respect to the Fermi level established within five materials. (From Burlin, T. E., in *Topics in Radiation Dosimetry*, Suppl. 1, Attix, F. H., Ed., Academic Press, New York, 1972. With permission.)

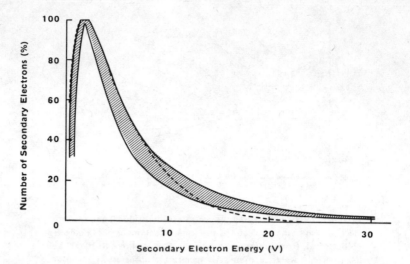

FIGURE 3.7-18. Range of secondary electron energy spectra emitted from ten different metals is shown by the shaded region (Kollath, 1947). The dashed line represents Maxwellian distribution. (From Burlin, T. E., in *Topics in Radiation Dosimetry,* Suppl. 1, Attix, F. H., Ed., Academic Press, New York, 1972. With permission.)

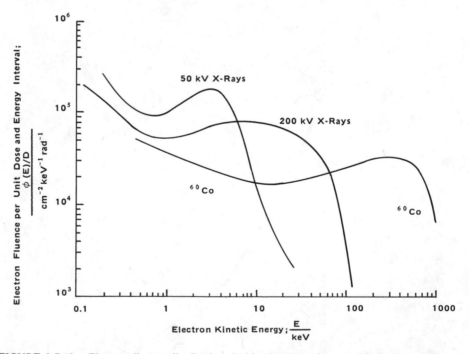

FIGURE 3.7-19.   Electron fluence distributions in kinetic energy for water for various radiations. The electron fluence distribution for 50-kV X-rays is based on the X-ray spectrum given by Burke and Pettit (1960). The electron fluence distribution for 200-kV X-rays was reduced from data of Burch (1957b). The fluence distributions include the delta rays or knock-on electrons. (From *Neutron Fluence, Neutron Spectra, and Kerma,* ICRU Report 13, International Commission on Radiation Units and Measurements, Washington, D.C., September 15, 1969. With permission.)

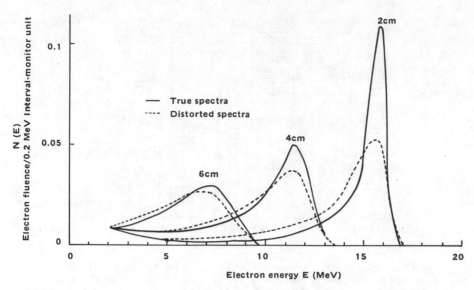

FIGURE 3.7-20.   Electron energy spectra produced at various depths in water 20-MeV incident electrions (Epp et al., 1965). (From Laughlin, J. S., in *Radiation Dosimetry,* 2nd ed., Vol. 3, Attix, F. H. and Tochilin, E., Eds., Academic Press, New York, 1969. With permission.)

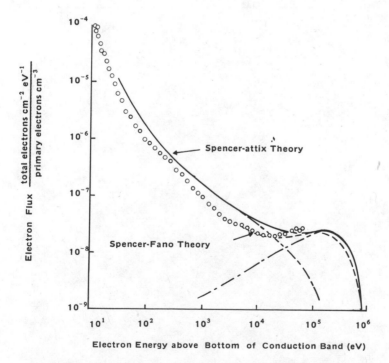

FIGURE 3.7-21.   Slowing-down spectrum for ^{198}Au β-rays in an infinite isotropic gold medium. The experimental points are compared with the Spencer-Attix theory, shown broken down into the primary and secondary contributions (dot-dash lines) and also with the Spencer-Fano theory (McConnell et al., 1966). (From Burlin, T. E., in *Topics in Radiation Dosimetry,* Suppl. 1, Attix, F. H., Ed., Academic Press, New York, 1972. With permission.)

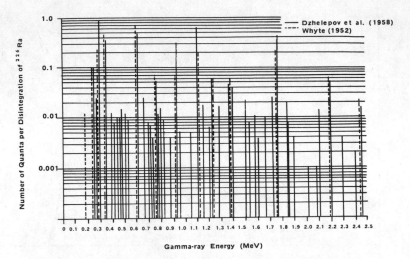

FIGURE 3.7-22.   γ-Ray spectrum of ^{226}Ra in equilibrium with its daughter products, filtered by 0.5 mm of platinum. (From Shalek, R. J. and Stovall, M., in *Radiation Dosimetry*, 2nd ed., Vol. 3, Attix, F. H. and Tochilin, E., Eds., Academic Press, New York, 1969. With permission.)

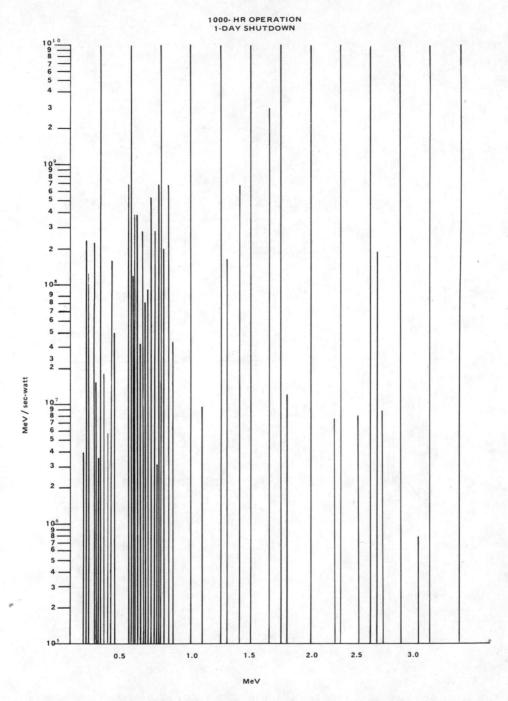

FIGURE 3.7-23.   Decay gamma energy spectrum 1000-hr operation 1-day shutdown. (From Moteff, J., Fission Product Decay Gamma Energy Spectrum, APEX-134, Aircraft Nuclear Propulsion Project, General Electric, June 1953. With permission.)

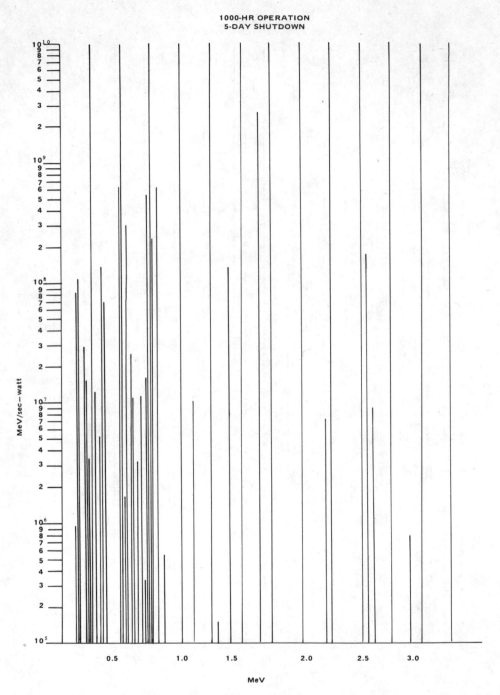

FIGURE 3.7-24.   Decay gamma energy spectrum 1000-hr operation 5-day shutdown. (From Moteff, J., Fission Product Decay Gamma Energy Spectrum, APEX- 134, Aircraft Nuclear Propulsion Project, General Electric, June 1953. With permission.)

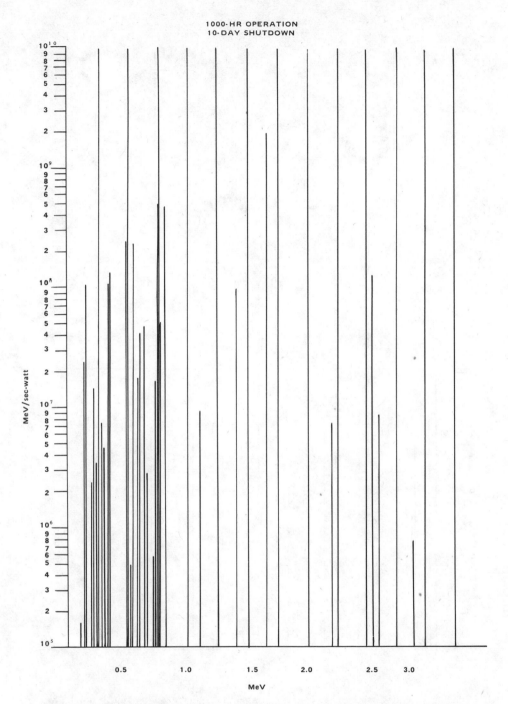

**1000-HR OPERATION
10-DAY SHUTDOWN**

FIGURE 3.7-25.   Decay gamma energy spectrum 1000-hr operation 10-day shutdown. (From Moteff, J., Fission Product Decay Gamma Energy Spectrum, APEX-134, Aircraft Nuclear Propulsion Project, General Electric, June 1953. With permission.)

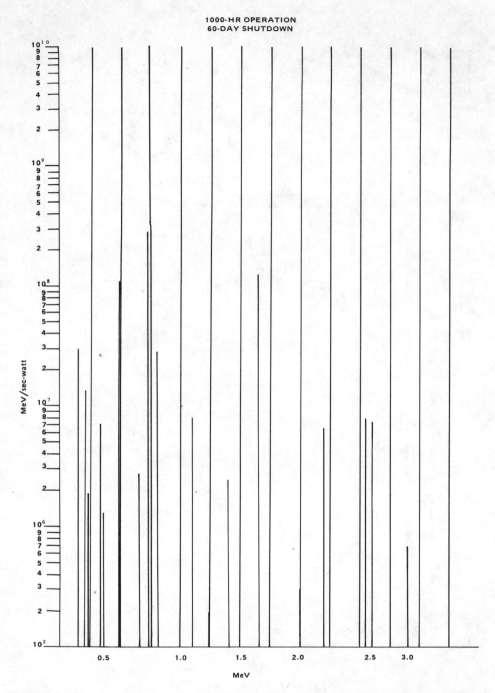

FIGURE 3.7-26.   Decay gamma energy spectrum 1000-hr operation 60-day shutdown. (From Moteff, J., Fission Product Decay Gamma Energy, Spectrum, APEX-134, Aircraft Nuclear Propulsion Project, General Electric, June 1953. With permission.)

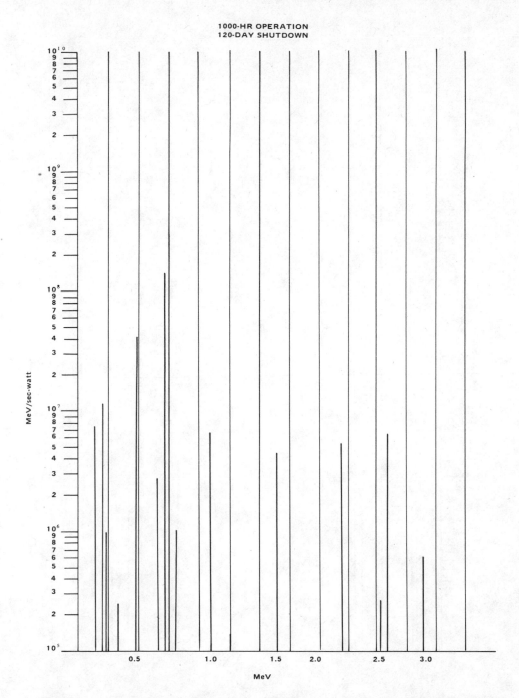

FIGURE 3.7-27.   Decay gamma energy spectrum 1000-hr operation 120-day shutdown. (From Moteff, J. Fission Product Decay Gamma Energy Spectrum, APEX-134, Aircraft Nuclear Propulsion Project, General Electric, June 1953. With permission.)

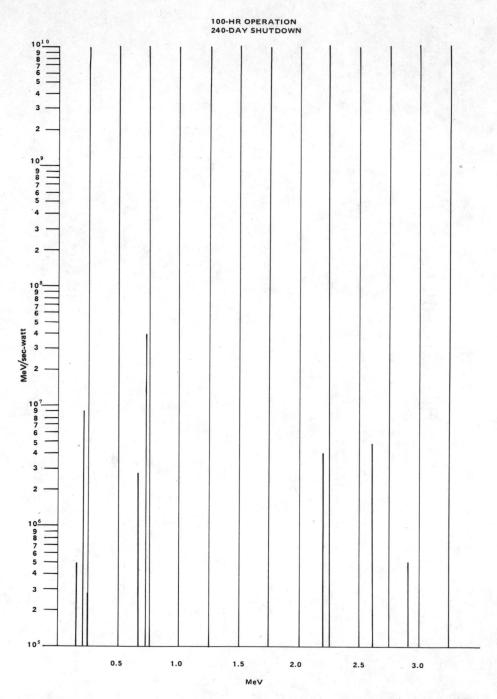

FIGURE 3.7-28.   Decay gamma energy spectrum 1000-hr operation 240-day shutdown. (From Moteff, J., Fission Product Decay Gamma Energy Spectrum, APEX-134, Aircraft Nuclear Propulsion Project, General Electric, June 1953. With permission.)

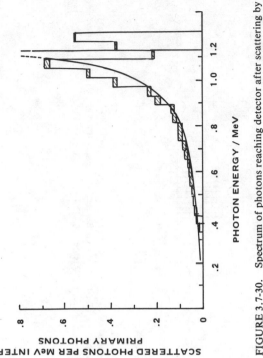

FIGURE 3.7-30. Spectrum of photons reaching detector after scattering by a conical lead collimator placed near a point-isotropic cobalt-60 source. Curve is from a calculation by Cormack and Johns (1958) taking into account single scatter only. Histogram is a Monte Carlo result. Contribution of photons scattered more than once in collimator is indicated by shaded area. Collimator parameters were as follows: $r_1 = 3$ cm; $r_2 = 10$ cm; $t_1 = 60$ cm; $t_1 + t_2 = 100$ cm. (See Figure 7.3-31.) (From *Specification of High Activity Gamma-ray Sources*, ICRU Report 18, International Commission on Radiation Units and Measurements, Washington, D.C., October 15, 1970. With permission.)

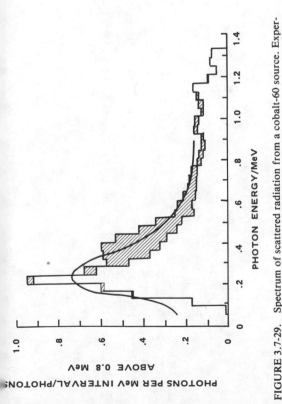

FIGURE 3.7-29. Spectrum of scattered radiation from a cobalt-60 source. Experimental curve was obtained by Scrimger and Cormack (1963) with source capsule and head shown in Reference 21. Calculated histogram is for scattered radiation from encapsulated source only. Parameters for the Monte Carlo calculation were: Steel front plate, $z_F = 0.1$ cm; radioactive core (pellets with effective density of 5.88 g/cm³), $z = 0.74$ cm and $r = 1.27$ cm, steel backing, $z_B = 2.8$ cm; tungsten sleeve, $s = 1.23$ cm. Source-detector distance $t = 80$ cm. Shaded part of histogram represents scattered photons reaching the detector through the front surface of the tungsten sleeve. In the experiment, such photons are to a large extent prevented from reaching the detector because of the presence of the collimator. (From *Specification of High Activity Gamma-ray Sources*, ICRU Report 18, International Commission on Radiation Units and Measurements, Washington, D.C., October 15, 1970. With permission.)

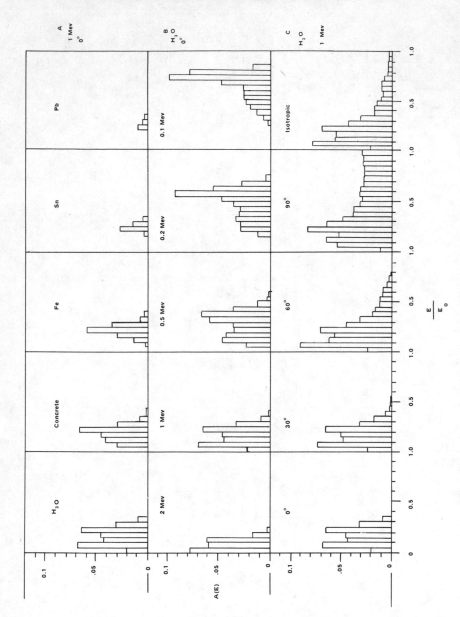

FIGURE 3.7-32.   Energy spectra of reflected radiation. Dependence on A. material, B. source energy, and C. source obliquity. Contents of histogram boxes add up to number albedo A. (From Berger, M. J. and Raso, D. J., *Backscattering of Gamma Rays*, Report NBS 5982, National Bureau of Standards, Washington, D.C., July 25, 1968.)

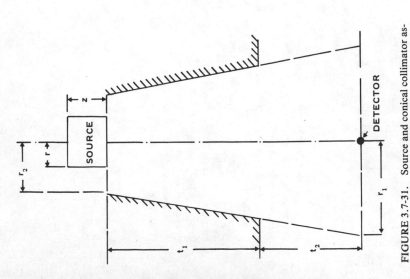

FIGURE 3.7-31.   Source and conical collimator assumed in Monte Carlo calculation of radiation scattered by collimators. (From *Specification of High Activity Gamma-ray Sources*, ICRU Report 18, International Commission on Radiation Units and Measurements, Washington, D.C., October 15, 1970. With permission.)

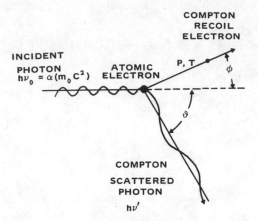

FIGURE 3.7-33. Trajectories in the scattering plane for the incident photon $h\nu_o$,-,2,3 the scattered photon $h\nu'$, and the scattering electron which acquires momentum p and kinetic energy T. (From Evans, R. D., in *American Institute of Physics Handbook*, 3rd ed., Gray, D. E., Ed., McGraw-Hill, New York, 1972. With permission.)

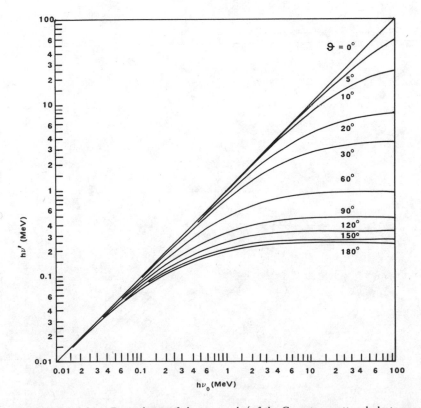

FIGURE 3.7-34. Dependence of the energy $h\nu'$ of the Compton-scattered photon on $h\nu_o$ and the photon scattering angle ϑ. (From Evans, R. D., in *Radiation Dosimetry*, Vol. 1, 2nd ed., Attix, F. H. and Roesch, W. C., Eds., Academic Press, New York, 1968. With permission.)

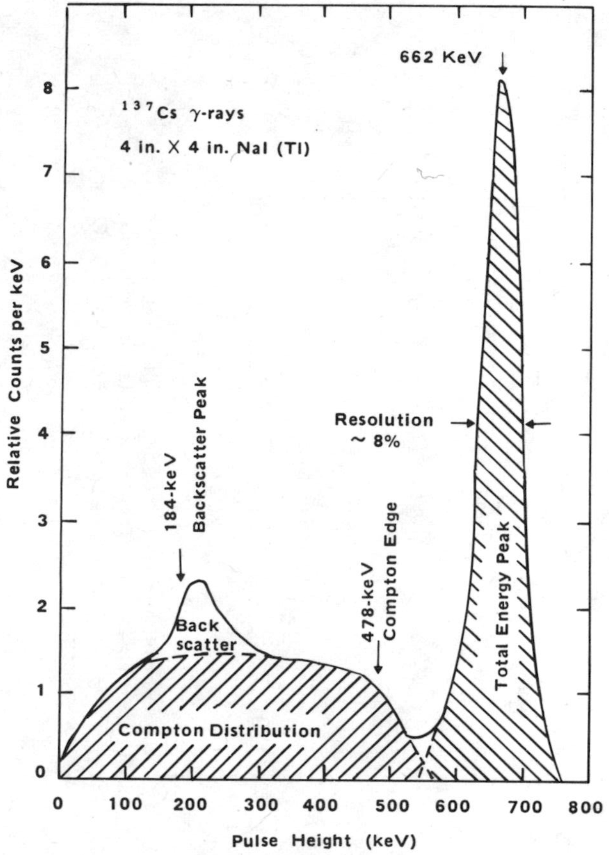

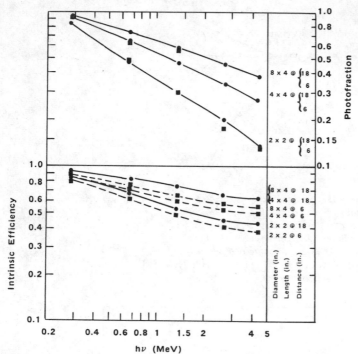

FIGURE 37.-36.   Illustrative calculated values of the *photofraction* (pulses in total-energy peak/pulses in entire distribution) and the *intrinsic efficiency* (pulses per photon striking the crystal face) for point sources of γ-rays located on the axis of NaI(Tl) crystal at 18 in. (circles) or 6 in. (squares) from the crystal face, for crystals whose diameters and lengths in inches are shown opposite the curves. (From Evans, R. D., in *American Institute of Physics Handbook*, 3rd ed., Gray, D. E., Ed., McGraw-Hill, New York, 1972. With permission.)

FIGURE 3.7-35.   Typical pulse-height distribution in a 4- by 4-in. NaI(Tl) scintillator irradiated by 0.662-MeV γ-rays from the decay of ^{137}Cs. The resolution is about 8% of the energy of the total-energy peak. For $hv_o = 662$ KeV, $(hv)_{min} = 184$ keV, and $T_{max} = 478$ keV as marked. (From Evans, R. D., in *American Institute of Physics Handbook*, 3rd ed., Gray, D. E., Ed., McGraw-Hill, New York, 1972. With permission.)

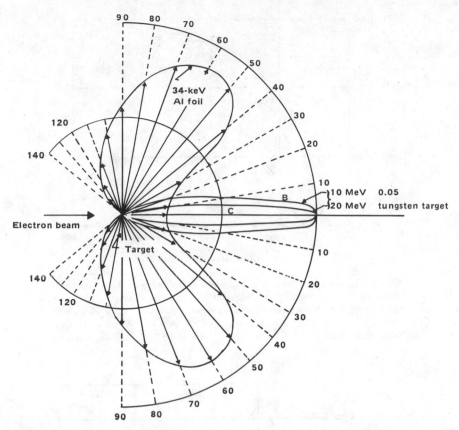

FIGURE 3.7-37.   Diagram showing the variation with angle of the intensity of X-rays produced by the electron bombardment of a thin foil. Curve A is obtained by bombarding a thin foil of aluminium with 34-keV electrons, based on data by Honerjäger (1940). Curves B and C were plotted from data by Schiff (1951). (From Johns, H. E., in *Radiation Dosimetry,* Vol. 3, 2nd ed., Attix, F. H. and Tochilin, E., Eds., Academic Press, New York, 1969. With permission.)

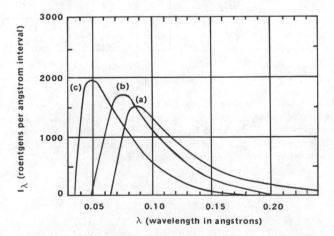

FIGURE 3.7-38.   Spectral distributions for three X-ray beams, showing the relative exposure per angstrom interval, X_x, plotted against the wavelength in angstroms. The curves are normalized so that the area under them equals 100 R. (a) 200 kV, HVL 1.8 mm Cu; (b) 250 kV, HVL 2.6 mm Cu; (c) 400 kV, HVL 4 mm Cu (Johns, 1953). (From Johns, H. E., in *Radiation Dosimetry,* Vol. 3, 2nd ed., Attix, F. H. and Tochilin, E., Eds., Academic Press, New York, 1969. With permission.)

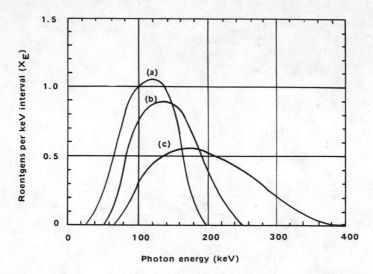

FIGURE 3.7-39.   The same distributions as shown in Figure 3.7-38 but plotted to give the exposure in roentgens per kiloelectronvolt interval, X_E, as a function of the photon energy in kiloelectronvolts (Johns, 1953). (From Johns, H. E., in *Radiation Dosimetry,* Vol. 3, 2nd ed., Attix, F. H. and Tochilin, E., Eds., Academic Press, New York, 1969. With permission.)

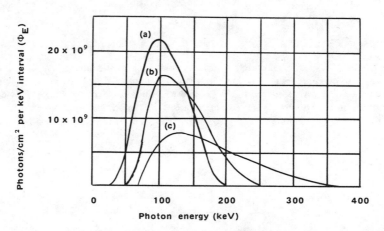

FIGURE 3.7-40.   The same distribution as shown in Figure 3.7-39 but plotted to give the number of photons per kiloelectronvolt interval, Φ_E, as a function of the photon energy in kiloelectronvolts. (From Johns, H. E., in *Radiation Dosimetry,* Vol. 3, 2nd ed., Attix, F. H. and Tochilin, E., Eds., Academic Press, New York, 1969. With permission.)

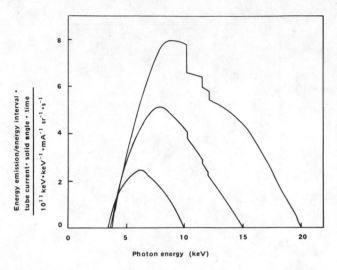

FIGURE 3.7-41.   Calculated spectral distributions of "continuous" X-rays generated at constant potentials of 10, 15, and 20 kV, allowing for filtration by 20° tungsten target and 1 mm Be. Characteristic radiation must be added to obtain complete spectra. (From Unsworth, M. H. and Greening, J. R., *Phys. Med. Biol.,* 15, 621, 1970, as given in Greening, J. R., in *Topics in Radiation Dosimetry,* Suppl. 1, Attix, F. H., Ed., Academic Press, New York, 1972. With permission from the Institute of Physics.)

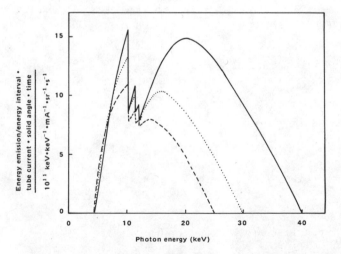

FIGURE 3.7-42.   Calculated spectral distributions of "continuous" X-rays generated at constant potentials of 25, 30, and 40 kV, allowing for filtration by 20° tungsten target and 1 mm Be. Characteristic radiation must be added to obtain complete spectra. (From Unsworth, M. H. and Greening, J. R., *Phys. Med. Biol.,* 15, 621, 1970, as given in Greening, J. R., in *Topics in Radiation Dosimetry,* Suppl. 1, Attix, F. H., Ed., Academic Press, New York, 1972. With permission from the Institute of Physics.)

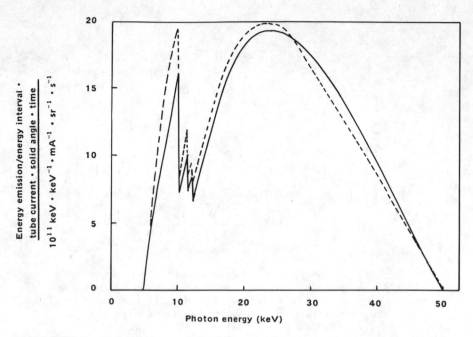

FIGURE 3.7-43.   Calculated spectral distributions of "continuous" X-rays generated at a constant potential of 50 kV, allowing for filtration by 20° tungsten target and 1 mm Be. Full line: multiple production depths; Broken line: single production depth of one quarter electron range. Characteristic radiation must be added to obtain complete spectra. (From Unsworth, M. H. and Greening, J. R., *Phys. Med. Biol.,* 15, 621, 1970, as given in Greening, J. R., in *Topics in Radiation Dosimetry,* Suppl. 1, Attix, F. H., Ed., Academic Press, New York, 1972. With permission from the Institute of Physics.)

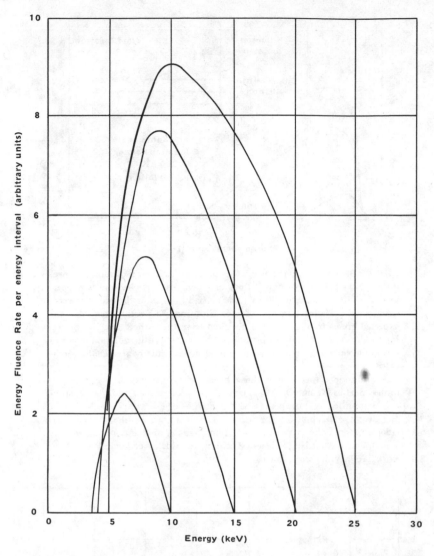

FIGURE 3.7-44.   Continuous X-ray spectra of radiations generated at 10, 15, 20, and 25 kV. Filter 1 mm Be. (To these spectra tungsten L-characteristic radiation should be added. The data were derived from Unsworth and Greening, 1970. (From *Radiation Dosimetry: X-Rays Generated at Potentials of 5 to 150 kV,* ICRU Report 17, International Commission on Radiation Units and Measurements, Washington, D.C., June 15, 1970. With permission.)

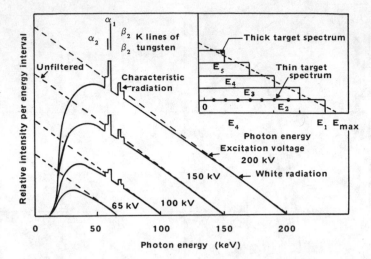

FIGURE 3.7-45.   Graph showing the spectral distribution of radiation produced by the bombardment of a thick tungsten target by electrons with energies of 65, 100, 150, and 200 kV. The dashed lines show the expected spectrum when no correction is made for filtration. The solid lines show the
expected bremsstrahlung spectra obtained from the dashed lines by correcting for the effect of 1 mm Al filtration. The characteristic radiation is shown
at the correct energy but the relative intensity is actually greater than indicated.

The thin-target spectrum produced by electrons of energy $T_o = E_1$ is shown
in the insert as the line $O_1A_1E_1$. The superposition of many thin-target spectra, exemplified by those generated by electrons of energies $T_o = E_2$, E_3, E_4,
E_5, gives the thick-target spectrum which is shown by the dashed line.
Adapted from Johns (1966). (From Johns, H. E., in *Radiation Dosimetry*,
Vol. 3, 2nd ed., Attix, F. H. and Tochilin, E., Eds., Academic Press, New
York, 1969. With permission.)

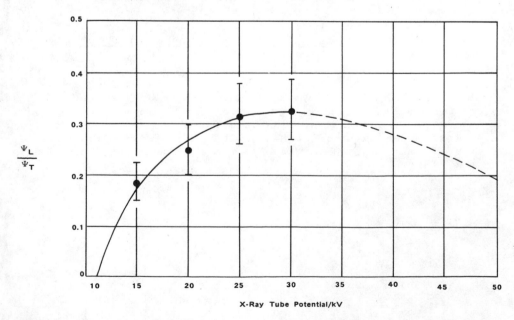

FIGURE 3.7-46.   Ratio of energy fluence, Ψ_B, due to tungsten L radiation to energy fluence, Ψ_T, due
to total radiation for X-rays filtered by 1 mm Be. The extrapolation from 30 to 50 kV is based on theory.
The data were derived from Unsworth and Greening, 1970. (From *Radiation Dosimetry: X-Rays Generated at Potentials of 5 to 150 kV*, ICRU Report 17, International Commission on Radiation Units and
Measurements, Washington, D.C., June 15, 1970. With permission.)

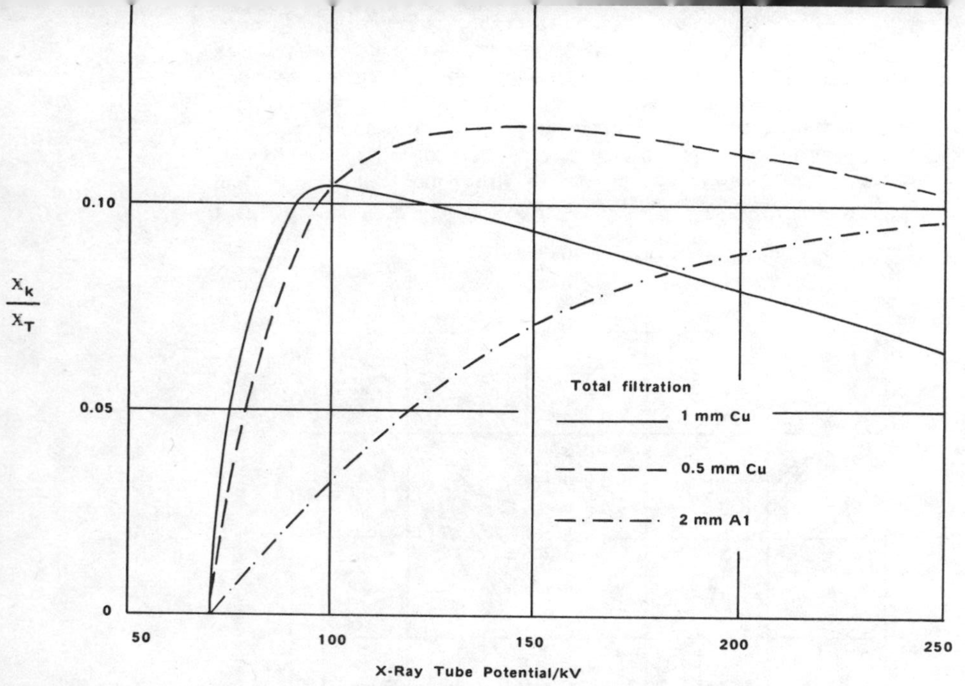

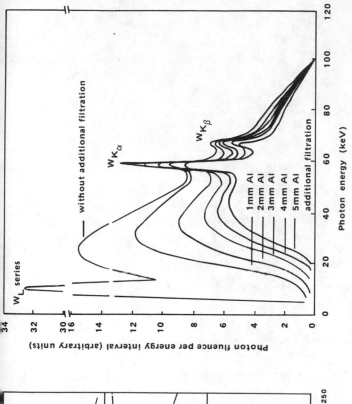

FIGURE 3.7-48.   Experimental photon flux spectra of X-rays generated at a constant potential of 100 kV in a tungsten target tube. Filtration: 1 mm Be, 80 cm of air, and aluminum as indicated. (From Gossrau, M. and Drexler, G., in *Advances in Physical and Biological Radiation Detectors*, International Atomic Energy Commission, Vienna, 1971, 205, as given in Greening, J. R., in *Topics in Radiation Dosimetry*, Suppl. 1, Attix, F. H., Ed., Academic Press, New York, 1972. With permission.)

FIGURE 3.7-47.   Ratio of exposure, X_k, due to tungsten K radiation to exposure, X_r, due to total radiation. (Derived from Tothill, P., *Br. J. Appl. Phys.*, 1, 1093, 1968, as given in *Radiation Dosimetry: X-Rays Generated at Potentials of 5 to 150 kV*, ICRU Report 17, International Commission on Radiation Units and Measurements, Washington, D.C., June 15, 1970. With permission from the Institute of Physics.)

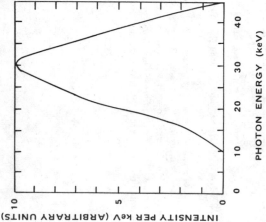
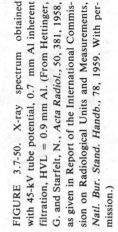

FIGURE 3.7-50. X-ray spectrum obtained with 45-kV tube potential, 0.7 mm Al inherent filtration, HVL = 0.9 mm Al. (From Hettinger, G. and Starfelt, N., *Acta Radiol.*, 50, 381, 1958, as given in Report of the International Commission on Radiological Units and Measurements, *Natl. Bur. Stand. Handb.*, 78, 1959. With permission.)

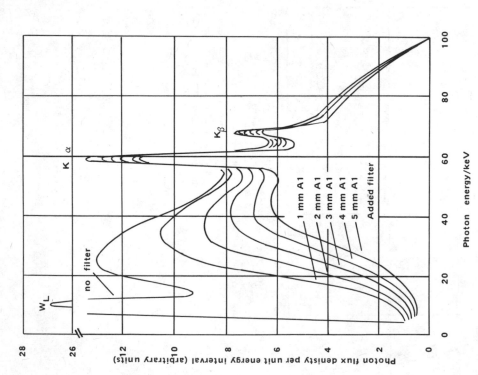

FIGURE 3.7-49.   Spectral distributions of X-rays from a tube with 1 mmBe inherent filtration operated at 100 kV and with various filters (measured after passage through 2 m of air). The data were deived from Drexler, G. and Perzl, F., Methoden zur Spectrometrie neiderenergetischen Bremsstrahlung, in 1st European Congress of Radioprotection, Menton, France, 1968. (From *Radiation Dosimetry: X-Rays Generated at Potentials of 5 to 150 kV*, ICRU Report 17, International Commission on Radiation Units and Measurements, Washington, D.C., June 15, 1970. With permission.)

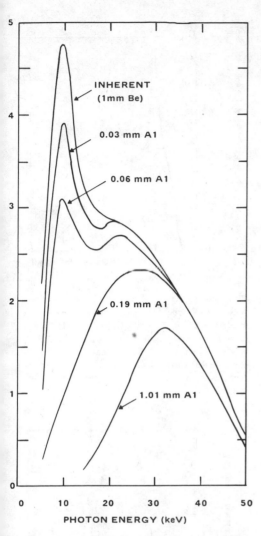

IGURE 3.7-51.   X-Ray spectra obtained with 50-kV
ube potential, an inherent filtration of 1 mm Be and
e indicated added filters of aluminum. (From Ehrlich,
1., *J. Opt. Soc. Am.,* 46, 797, 1956, as given in Report
f the International Commission on Radiological Units
nd Measurements, *Natl. Bur. Stand. Handb.,* 78,
959. With permission.)

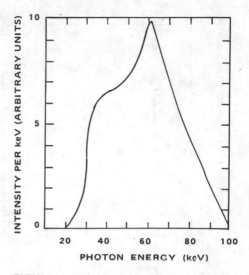

FIGURE 3.7-52.   X-Ray spectrum obtained
with 100-kV tube potential, 4 mm Al inherent filtration, HVL = 0.18 mm Cu. (From Hettinger,
G. and Starfelt, N., *Acta Radiol.,* 50, 381, 1958,
as given in Report of the International Commission on Radiological Units and Measurements,
Natl. Bur. Stand. Handb., 78, 1959. With permission.)

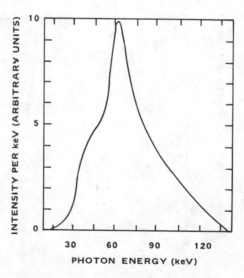

FIGURE 3.7-53.   X-Ray spectrum obtained
wtih 140-kV tube potential, 4 mm Al inherent filtration, HVL = 0.30 mm Cu. (From Hettinger,
G. and Starfelt, N., *Acta Radiol.,* 50, 381, 1958,
as given in Report of the International Commission on Radiological Units and Measurements,
Natl. Bur. Stand. Handb., 78, 1959. With permission.)

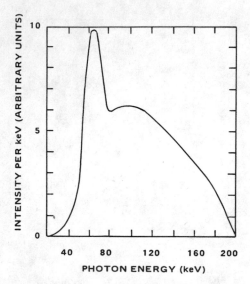

FIGURE 3.7-54. X-Ray spectrum obtained with 200-kV tube potential, 2.5 mm Al inherent filtration, added 1.0 mm Al and 0.5 mm Cu. HVL = 1.2 mm Cu. (From Helle, P., personal communication, as given in Report of the International Commission on Radiological Units and Measurements, *Natl. Bur. Stand. Handb.*, 78, 1959. With permission.)

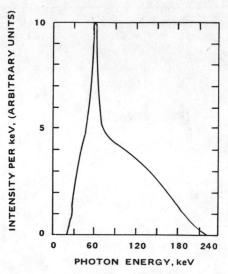

FIGURE 3.7-55. X-Ray spectrum obtained with 220-kV tube potential, inherent filtration 4 mm Al, HVL = 0.55 mm Cu. (From Hettinger, G. and Starfelt, N. *Acta Radiol.*, 50, 381, 1958, as given in Report of the International Commission on Radiological Units and Measurements, *Natl. Bur. Stand. Handb.*, 78, 1959. With permission.)

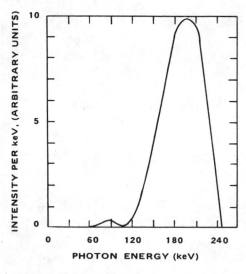

FIGURE 3.7-56. X-Ray spectrum obtained with 250-kV tube potential, 4 mm Al inherent filtration, added 1.32 mm Al, 2.07 mm Cu, 1.10 mm Cd, 1.34 mm Pb, HVL = 4.8 mm Cu. (From Hettinger, G. and Starfelt, N., *Acta Radiol.*, 50, 381, 1998, as given in Report of the International Commission on Radiological Units and Measurements, *Natl. Bur. Stand. Handb.*, 78, 1959. With permission.)

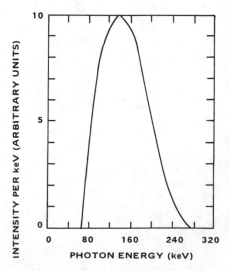

FIGURE 3.7-57. X-Ray spectrum obtained with 280-kvp tube potential, with filtration to give HVL of 2.5 mm Cu. (From Cormack, D. V., Till, J. E., Whitmore, G. F., and Johns, H. E., *Br. J. Radiol.*, 28, 605, 1955, as given in Report of the International Commission on Radiological Units and Measurements, *Natl. Bur. Stand. Handb.*, 78, 1959. With permission.)

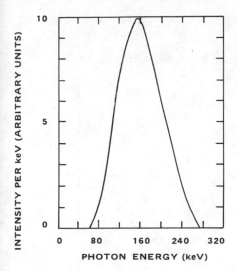

FIGURE 3.7-58. X-Ray spectrum obtained with 280-kvp tube potential, with filtration to give HVL of 3.1 mm Cu. (From Cormack, D. V., Till, J. E., Whitmore, G. F., and Johns, H. E., *Br. J. Radiol.,* 28, 605, 1995, as given in Report of the International Commission on Radiological Units and Measurements, *Natl. Bur. Stand. Hand.,* 78, 1959. With permission.)

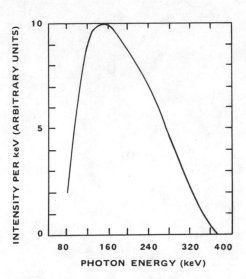

FIGURE 3.7-59. X-Ray spectrum obtained with 400-kV tube potential, added filtration of 1 mm Sn, 0.5 mm Cu, and 1 mm Al; HVL = 4.0 mm Cu. (From Johns, H. E., Fedoruk, S. O., Kornelsen, P. O., Epp, E. R., and Darby, E. K., *Br. J. Radiol.,* 25, 542, 1952, as given in Report of the International Commission on Radiological Units and Measurements, *Natl. Bur. Stand. Handb.,* 78, 1959. With permission.)

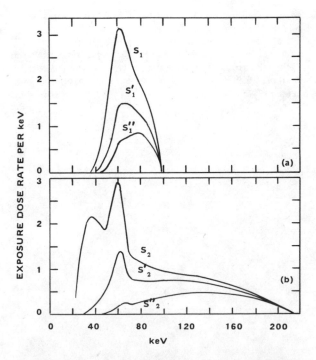

FIGURE 3.7-60. Spectral distribution of X-rays produced by 100-kV accelerating potential (curve S_1) and 220-kV (curve S_2) Curve S' and S'' are the calculated spectral distributions after the radiation has passed through 1 and 2 HVL of copper, respectively. (From Hettinger, G. and Starfelt, N., *Acta Radiol.,* 50, 381, 1958, as given in Report of the International Commission on Radiological Units and Measurements, *Natl. Bur. Stand. Handb.,* 78, 1959. With permission.)

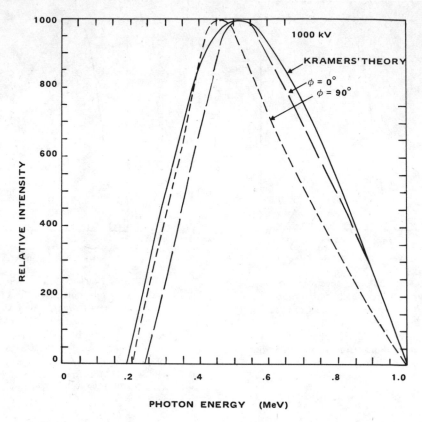

FIGURE 3.7-61.   X-Ray spectra obtained at 1000-kV tube potential. Filtration: 2.8-mm W + 2.8-mm Cu + 18.7-mm water + 2.1-mm brass. Dashed curves: spectra measured by Compton spectrometer (the ordinate is photon energy multiplied by number of photons of that energy) (Miller et al., 1954). Angles refer to direction of X-rays relative to initial electron direction. Solid curve: spectrum calculated by Kramers' method (1923). (From Report of the International Commission on Radiological Units and Measurements, *Natl. Bur. Stand. Handb.*, 78, 1959. With permission.)

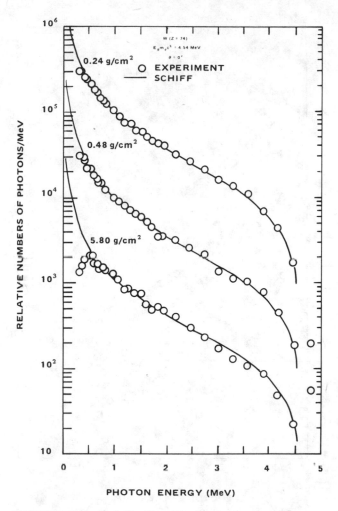

FIGURE 3.7-62.   Relative X-ray spectra (in photon MeV interval) produced by 4.54-MeV electrons from a betatron (Starfelt and Koch, 1956). Curves are given for three tungsten target thicknesses: 0.24 g/cm², 0.48 g/cm², and 5.80 g/cm². Only the curve shapes, not their relative positions, are significant. The points were obtained with a scintillation spectrometer. The solid curves were calculated from the thin targer formula of Schiff (1951) and normalized to the experimental data. (From Starfelt, N. and Koch, H. W., *Phys. Rev.*, 102, 1958, 1956, as given in Report of the International Commission on Radiological Units and Measurements, *Natl. Bur. Stand. Handb.*, 78, 1959. With permission.)

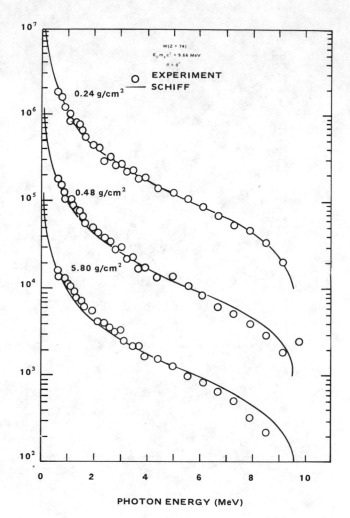

FIGURE 3.7-63.   Relative X-ray spectra (in photon MeV interval) produced by 4.54-MeV electrons from a betatron (Starfelt and Koch, 1956). Curves are given for three tungsten target thicknesses: 0.24 g/cm², 0.48 g/cm², and 5.80 g/cm². Only the curve shapes, not their relative positions, are significant. The points were obtained with a scintillation spectrometer. The solid curves were calculated from the thin target formula of Schiff (1991) and normalized to the experimental data. (From Starfelt, N. and Koch, H. W., *Phys. Rev.,* 102, 1958, 1956, as given in Report of the International Commission on Radiological Units and Measurements, *Natl. Bur. Stand. Handb.,* 78, 1959. With permission.)

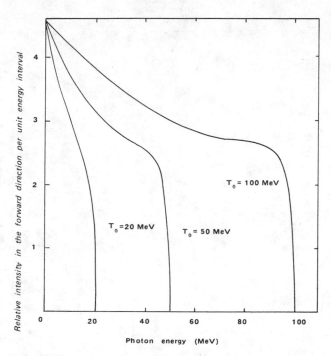

FIGURE 3.7-64.   Relative intensity of the bremsstrahlung in the forward direction produced by bombardment of a thin target of tungsten by high-energy electrons with kinetic energies of 20, 50, and 100 MeV. (From *Natl. Bur. Stand. Handb.*, 55, 1954, as given in Johns, H. E. in *Radiation Dosimetry*, Vol. 3, 2nd ed., Attix, F. H. and Tochilin, E., Eds., Academic Press, New York, 1969. With permission.)

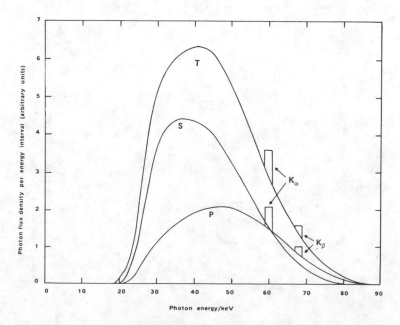

FIGURE 3.7-65.   Spectral distributions of primary (P), scattered (S), and total (T) radiations in water irradiated with X-rays generated at 90 kV with HVL 2.9 mm Al (depth 6 cm; beam area 100 cm²). (Derived from Epp., E. R. and Weiss, H., *Radiat. Res.*, 30, 129—139, 1967, as given in *Radiation Dosimetry: X-Rays Generated at Potentials of 5 to 150 kV,* ICRU Report 17, International Commission on Radiation Units and Measurements, Washington, D.C., June 15, 1970. With permission.)

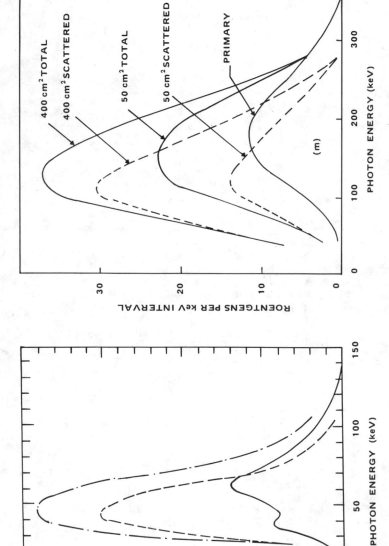

FIGURE 3.7-66. X-ray spectra obtained with 140-kV tube potential at a depth of 5 cm in a water phantom with a field of 300 cm². The solid curve gives the primary spectrum, the dashed curve gives the spectrum of the scattered radiation, and the dot-dash curve gives the sum of the two. The incident beam had a HVL of 1.45 mm Al. (From Report of the International Commission on Radiological Units and Measurements, *Natl. Bur. Stand. Handb.*, 78, 1959. With permission.)

FIGURE 3.7-67... Spectral distribution of the exposure dose at a depth of 10 cm in a water phantom (Cormack et al., 1957), X-rays generated at 400 kvp (HVL = 3.8 mm Cu in absence of phantom). The dashed curves apply to the scattered radiation only for field areas of 50 and 400 cm². The solid curves show the primary distribution and the total distribution combining both primary and secondary radiation. All curves are normalized to a primary surface exposure dose of 1000 R. (From Cormack, D. V., Griffith, T. J., and Johns, H. E., *Br. J. Radiol.*, 30, 129, 157, as given in Report of the International Commission on Radiological Units and Measurements, *Natl. Bur. Stand. Handb.*, 78, 1959. With permission.)

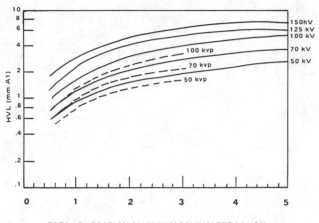

FIGURE 3.7-68.   Variation of HVL with total filtration. Solid curve
— constant potential (Reinsma, 1950). Dotted curve — sine wave potential (Trout et al., 1956). (From Report of the International Commission on Radiological Units and Measurements, *Natl. Bur. Stand.
Handb.*, 78, 1959. With permission.)

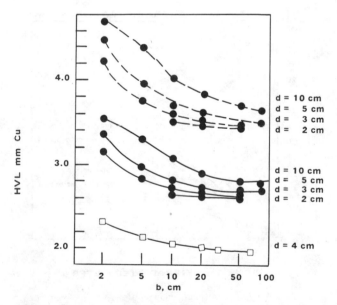

FIGURE 3.7-69.   Variation of half-value layers with parameter d
(diameter of the field at the attenuator) and b (chamber-attenuator
distance). The focus-attenuator distance is F. •—•—• First HVL 250
kV, 2mm Cu filtration, F = 20 cm (Somerwil, 1957). •——• Second
HVL 250 kV, 2 mm Cu filtration, F = 20 cm. △—△—△ First HVL
300 Kv, 1.3 mm Cu inherent filtration, F = 21.5 cm (Farr, 1955).
(From Report of the International Commission on Radiological Units
and Measurements, *Natl. Bur. Stand. Handb.*, 78, 1959. With permission.)

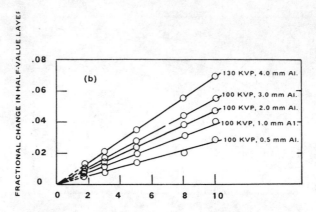

FIGURE 3.7-70.   Fractional change of half-value-layer for the indicated potentials and filtrations with F = 25 cm and b = 25 cm. (See Figure 3.7-69 for definition of F and b.) (From Trout, E. D., Kelley, J. P., and Lucas, A. C., *Am. J. Roentgenol.*, 84, 729, 1960, courtesy of Charles C. Thomas, as given in Report of the International Commission on Radiological Units and Measurements, *Natl. Bur. Stand. Handb.*, 78, 1959.)

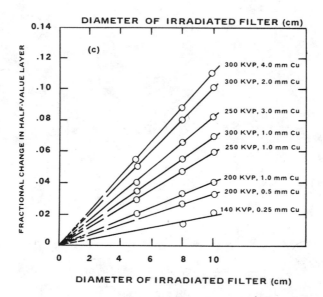

FIGURE 3.7-71.   Fractional change of half-value-layer for the indicated potentials and filtrations with F = 25 cm and b = 25 cm. (See Figure 3.7-69 for definition of F and b.) (From Trout, E. D., Kelley, J. P., and Lucas, A. C., *Am. J. Roentgenol.*, 84, 729, 1960, courtesy of Charles C. Thomas, as given in Report of the International Commission on Radiological Units and Measurements, *Natl. Bur. Stand. Handb.*, 78, 1959.)

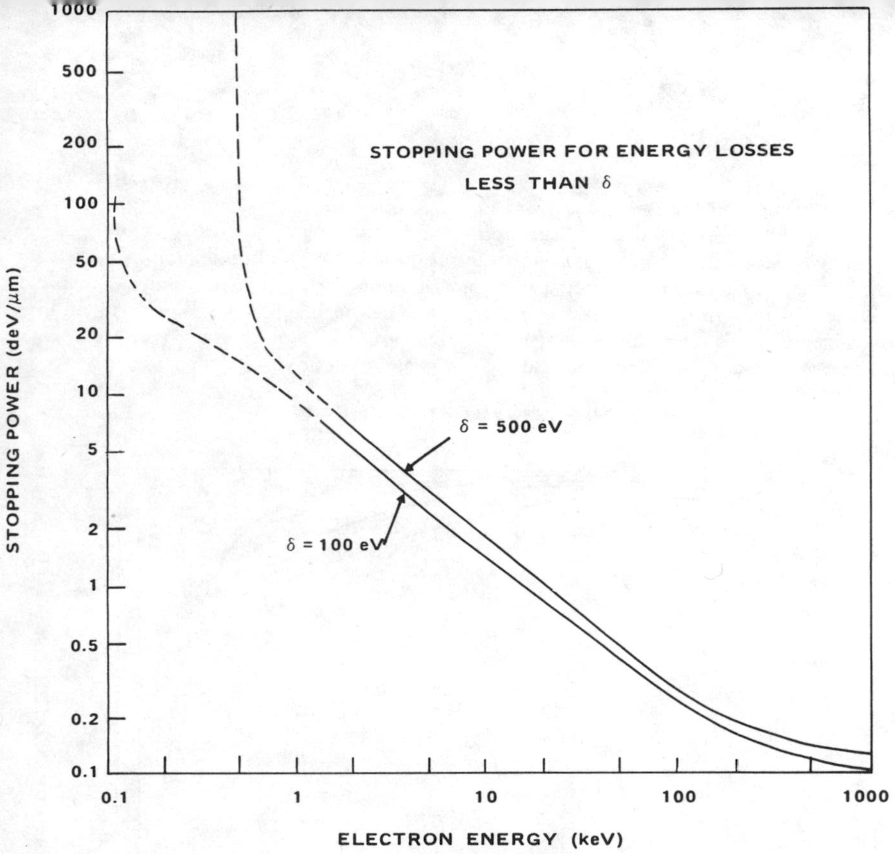

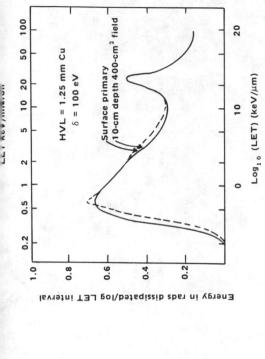

FIGURE 3.7-73.   The linear energy transfer distribution that results from X-rays with an HVL of 1.25 mm at the surface (——) amd at a depth (——) in a water phantom. Energy losses less than $\delta = 100$ eV are assumed to occur "on" the mother track. The total energy dissipated is 1 rad. (From Bruce, W. R., Pearson, M. L., and Freedhoff, H. S., *Radiat. Res.*, 19, 606, 1963, as given in Johns, H. E., in *Radiation Dosimetry*, Vol. 3, 2nd ed. Attix, F. H. and Tochilin, E., Eds., Academic Press, New York, 1969. With permission.)

FIGURE 3.7-72.   The linear energy transfer of electrons in kiloelectronvolts per micron for electrons of energy from 0.1 to 1000 keV, based on the assumption that only energy losses by tertiary delta rays less than $\delta = 100$ eV and $\delta = 500$ eV occur "on" the track (as calculated from the theory of Spencer and Attix, 1955). (From Bruce, W. R., Pearson, M. L., and Freedhoff, H. S., *Radiat. Res.*, 19, 606, 1963, as given in Johns, H. E., in *Radiation Dosimetry*, Vol. 3, 2nd ed., Attix, F. H. and Tochilin, E., Eds., Academic Press, New York, 1969. With permission.)

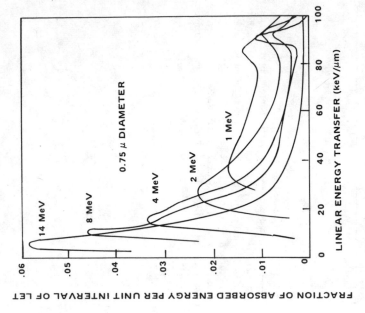

FIGURE 3.7-75. Experimentally determined LET distributions. These were determined for nearly monoenergetic neutrons of the indicated energies and for a test sphere corresponding to a diameter of 0.75 μ in soft tissue. (From Rosenzweig, W. and Rossi, H. H., *Radiat. Res.*, 10, 532, 1959, and private communication, as given in Physical Aspects of Irradiation, Recommendations of the International Commission on Radiological Units and Measurements, Report 106, 1962, *Natl. Bur. Stand. Handb.*, 85, 1964. With permission.)

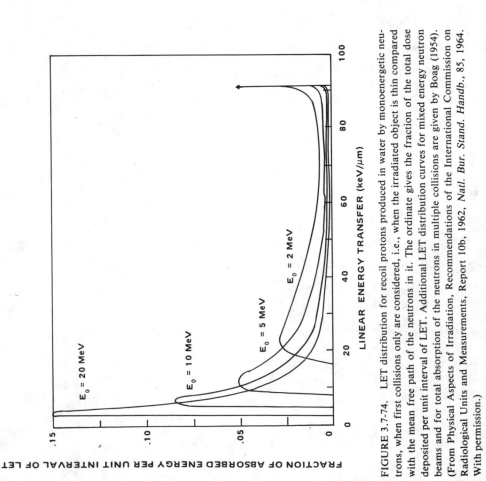

FIGURE 3.7-74. LET distribution for recoil protons produced in water by monoenergetic neutrons, when first collisions only are considered, i.e., when the irradiated object is thin compared with the mean free path of the neutrons in it. The ordinate gives the fraction of the total dose deposited per unit interval of LET. Additional LET distribution curves for mixed energy neutron beams and for total absorption of the neutrons in multiple collisions are given by Boag (1954). (From Physical Aspects of Irradiation, Recommendations of the International Commission on Radiological Units and Measurements, Report 10b, 1962, *Natl. Bur. Stand. Handb.*, 85, 1964. With permission.)

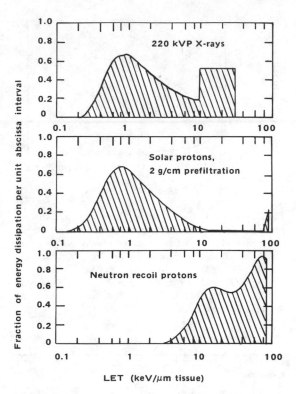

FIGURE 3.7-76.   Calculated distributions in LET of 220-kVp X-Rays, typical solar-flare protons, and fission neutron recoil protons in tissue. (From Schaefer, H. J., in *Symp. Biological Effects Neutron Proton Irradiation,* Upton, N.Y., Vol. 1, IAEA, Vienna, 1963, 297, as given in Sondhaus, C. A. and Evans, R. D., in *Radiation Dosimetry,* Vol. 3, 2nd ed., Attix, F. H. and Tochilin, E., Eds., Academic Press, New York, 1969. With permission.)

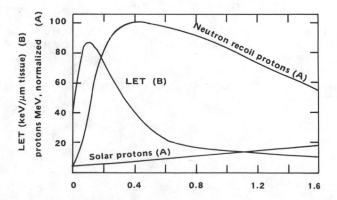

FIGURE 3.7-77.   Calculated spectra (A) of fission neutron recoil protons and of solar-flare protons. Average local LET values for the same energy region are also shown (B). (From Schaefer, H. J., in *Symp. Biological Effects Neutron Proton Irradiation,* Upton, N.Y., Vol. 1, IAEA, Vienna, 1963, 297, as given in Sondhaus, C. A. and Evans, R. D., in *Radiation Dosimetry,* Vol. 3, 2nd ed., Attix, F. H. and Tochilin, E., Eds., Academic Press, New York, 1969. With permission.)

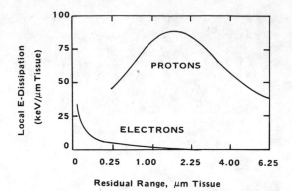

FIGURE 3.7-78. Calculated local average LET for protons and electrons near the ends of their tracks in tissue. (From Schaefer, H. J., in *Proc. Symp. Protection against Radiation Hazards in Space,* Gatlinburg, Tenn., Book 1, USAEC-TID-7652, UASEC, 1962, 393, as given in Sondhaus, C. A. and Evans, R. D., in *Radiation Dosimetry,* Vol. 3, 2nd ed., Attix, F. H. and Tochilin, E., Eds., Academic Press, New York, 1969. With permission.)

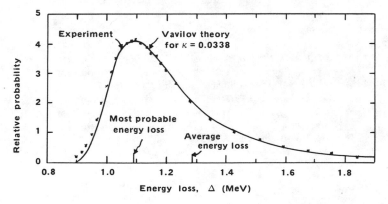

FIGURE 3.7-79. Energy-loss probability, 730-MeV protons in 0.66-g/cm² silicon; x = 0.0338. (From Maccabee, H. D., Raju, M. R., and Tobias, C. A., *IEEE Trans. Nucl. Sci.,* 13(3), 176, 1966, as given in Raju, M. R., Lyman, J. T., Brustad, T., and Tobias, C. A., in *Radiation Dosimetry,* Vol. 3, 2nd ed., Attix. F. H. and Tochilin, E., Eds., Academic Press, New York, 1969. With permission.)

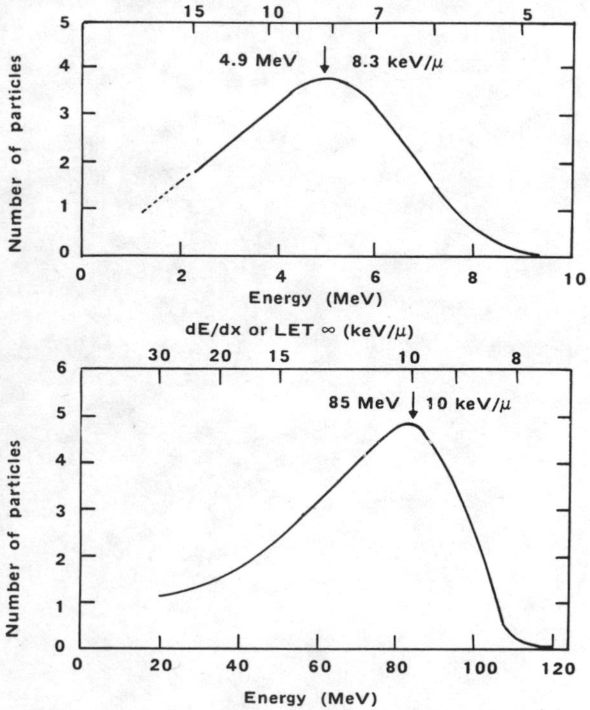

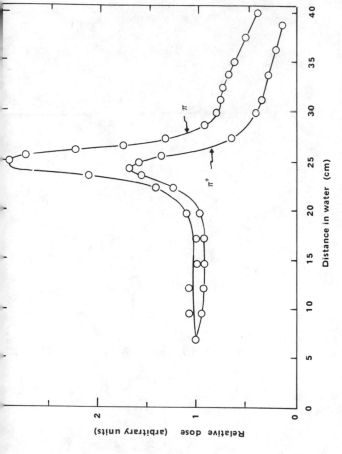

FIGURE 3.7-81.   Depth-dose distribution of 96 MeV π^+ and π^- meson beams, measured with a lithium-drifted silicon detector at various depths in water. (From Raju, M. R., Lyman, J. T., Brustad, T., and Tobias, C. A., in *Radiation Dosimetry*, Vol. 3, 2nd ed., Attix. F. H. and Tochilin, E., Eds., Academic Press, New York, 1969. With permission.)

FIGURE 3.7-80.   Energy distribution of charged particles at the Bragg peak, as measured with a lithium-drifted silicon detector. The upper curve is for a 50-MeV proton beam in aluminum; lower curve is for 910-MeV helium ions in copper. (From Raju, M. R., *Radiat. Res. Suppl.*, 7, 43, 1967, as given in Raju, M. R., Lyman, J. T., Brustad, T., and Tobias, C. A., in *Radiation Dosimetry*, Vol. 3, 2nd ed., Attix, F. H. and Tochilin, E., Eds., Academic Press, New York, 1969. With permission.)

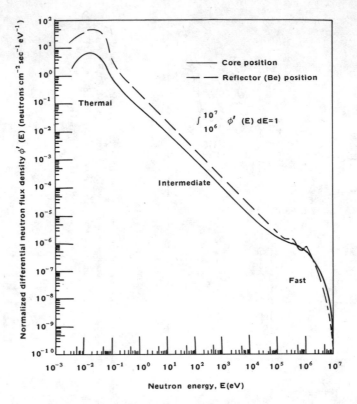

FIGURE 3.7-82.   Differential neutron flux density in a homogeneous water-moderated reactor core, and in the beryllium reflector. (From Moteff, J., in *Radiation Dosimetry,* Vol. 3, 2nd ed., Attix, F. H. and Tochilin, E., Eds., Academic Press, New York, 1969. With permission.)

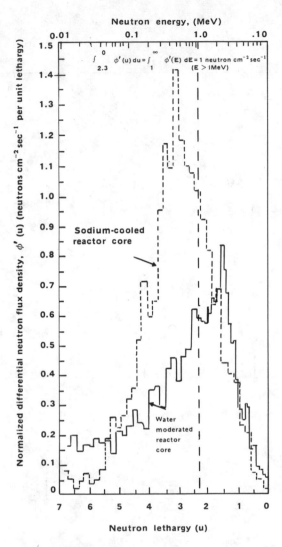

FIGURE 3.7-83.   Monte Carlo analysis of the fast-neutron spectrum in a heterogeneous light water-moderated and a sodium-cooled reactor core. The integral of the flux density greater than 1 MeV is set equal to unity. (From Moteff, J., in *Radiation Dosimetry,* Vol. 3, 2nd ed., Attix, F. H. and Tochilin, E., Eds., Academic Press, New York, 1969. With permission.)

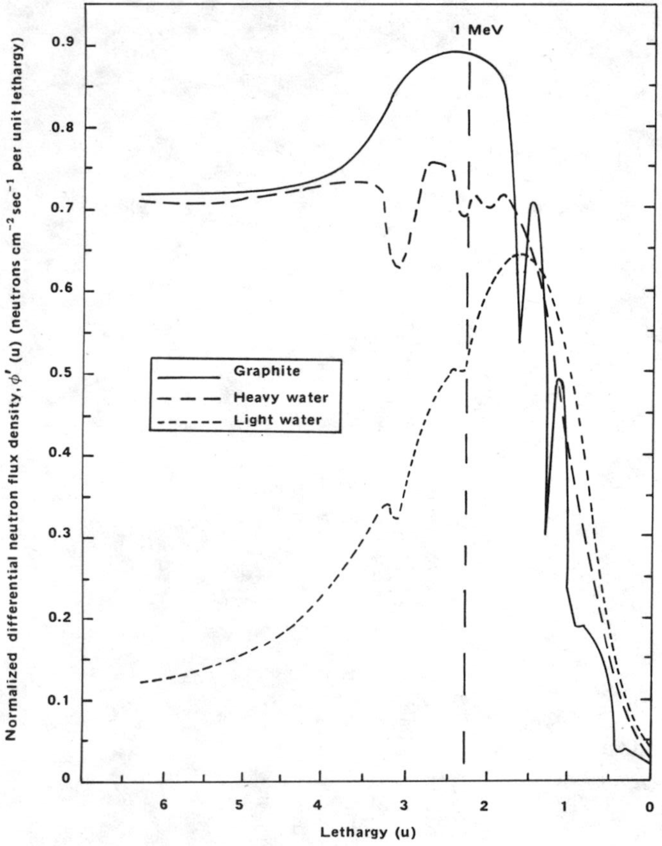

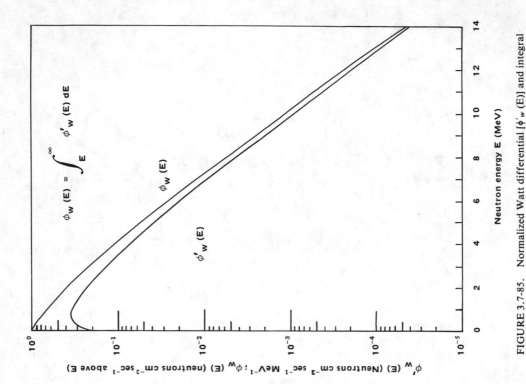

FIGURE 3.7-85.   Normalized Watt differential [ϕ'_w (E)] and integral [ϕ_w (E)] as a function of neutron energy. (From Moteff, J., in *Radiation Dosimetry*, Vol. 3, 2nd ed., Attix, F. H. and Tochilin, E., Eds., Academic Press, New York, 1969. With permission.)

FIGURE 3.7-84.   Fast-neutron spectra typical of a graphite-, heavy water-, and light water-moderated homogeneous reactor core. The integral of the flux density at energies greater than 1 MeV is equal to unity. (From Moteff, J., in *Radiation Dosimetry*, Vol. 3, 2nd ed., Attix, F. H. and Tochilin, E., Eds., Academic Press, New York, 1969. With permission.)

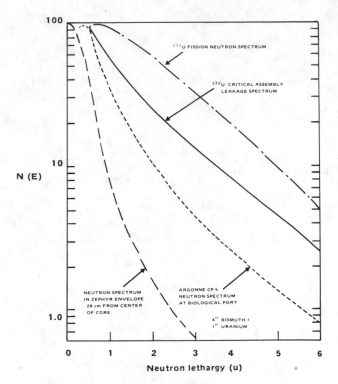

FIGURE 3.7-86.   Neutron fission spectra from reactors. (Watt, 1952; Rosen, 1956; USNRDL measurement on CP-5 spectrum (unpublished); Codd, Sheppard, and Tait, 1956.) (From Report of the International Commission of Radiological Units and Measurements, *Natl. Bur. Stand. Handb.,* 78, 1959. With permission.)

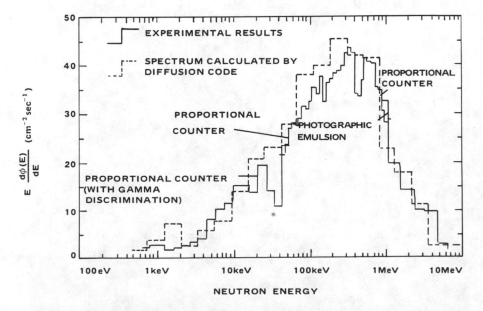

FIGURE 3.7-87.   Neutron spectrum at the centre of a fast critical assembly measured by proton recoil techniques. (From Weale, J. W., Benjamin, P. W., Kemshall, C. D., Paterson, W. I., and Redfearn, J., Proc. Int. Conf. Radiation Measurements in Nuclear Power, Berkeley Nuclear Laboratory, September 1966, 231. With permission from the Institute of Physics.)

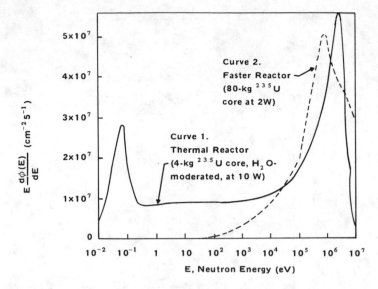

FIGURE 3.7-88.   Comparison of neutron spectra in fast and thermal reactors. (From *Neutron Fluence, Neutron Spectra, and Kerma,* ICRU Report 13, International Commission on Radiation Units and Measurements, Washington, D.C., September 15, 1969. With permission.)

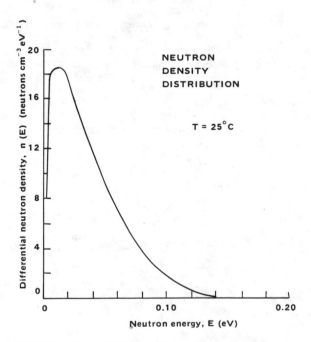

FIGURE 3.7-89.   Differential thermal-neutron density as a function of energy, assuming a Maxwellian distribution, at equilibrium in a medium at 25°C. (From Moteff, J., in *Radiation Dosimetry,* Vol. 3, 2nd ed., Attix, F. H. and Tochilin, E., Eds., Academic Press, New York, 1969. With permission.)

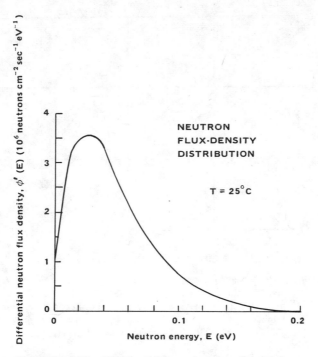

FIGURE 3.7-90.   Differential thermal-neutron flux density as a function of energy, assuming a Maxwellian distributtion, at equilibrium in a medium at 25°C. (From Moteff, J., in *Radiation Dosimetry*, 2nd ed., Vol. 3, Attix, F. H. and Tochilin, E., Eds., Academic Press, New York, 1969. With permission.)

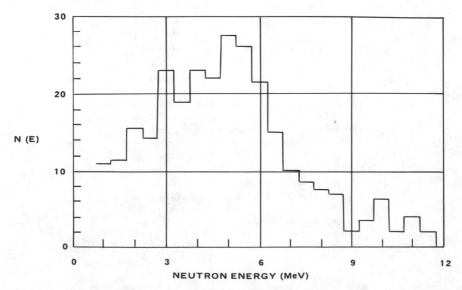

FIGURE 3.7-91.   Neutron energy spectrum of Ac-Be(α,n). Nuclear emulsions were used for the measurement. (From Dixon, W. R., Bielesch, A., and Geiger, K. W., *Can. J. Phys.*, 35, 699, 1957, as given in Physical Aspects of Irradiation, Recommendations of the International Commission on Radiological Units and Measurements, Report 10b, 1962, *Natl. Bur. Stand. Handb.*, 85, 1964. With permission of the National Research Council of Canada.)

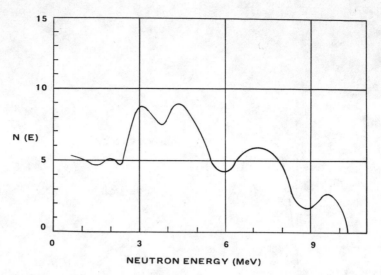

FIGURE 3.7-92.   Neutron energy spectrum of Pu-Be(α,n). Nuclear emulsions were used for the measurement. (From Stewart, L., *Phys. Rev.*, 98, 740, 1955, as given in Physical Aspects of Irradiation, Recommendations of the International Commission on Radiological Units and Measurements, Report 10b, 1962, *Natl. Bur. Stand. Handb.*, 85, 1964. With permission.)

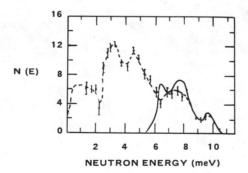

FIGURE 3.7-93.   Neutron energy spectrum of a Pu-Be(α,n) source. An organic scintillation crystal was used for the measurement. The measurement is shown dotted, and the solid curve is a calculated part of the spectrum due to ^{9}Be (α,n) leaving ^{12}Cu in its ground state. (From Broek, H. W. and Anderson, C. E., *Rev. Sci. Instrum.*, 31, 1063, 1960, as given in Physical Aspects of Irradiation, Recommendations of the International Commission on Radiological Units and Measurements, Report 10b, 1962, *Natl. Bur. Stand. Handb.*, 85, 1964.

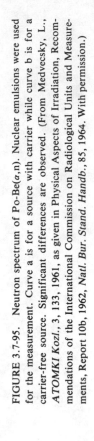

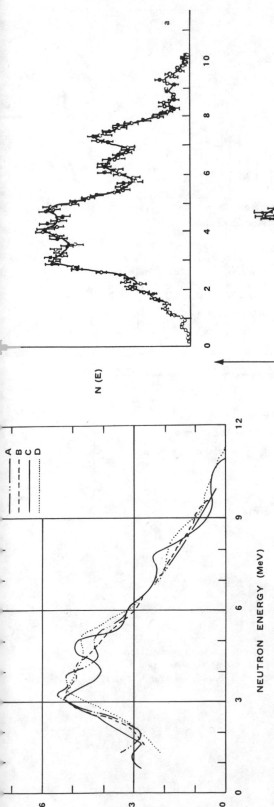

FIGURE 3.7-94.   Neutron energy spectra of Po-Be(α,n). Curve A was measured with a proton recoil proportional counter telescope (Cochran and Henry, 1955); curve B was determined by a similar method (Breen, Hertz, and Wright, 1996); curve C with nuclear emulsions (private communication from author, 1960, indicates that the all quoted energies should be increased by 3 percent, Whitmore and Baker, 1950); curve D with ^{6}LiI scintillation crystal spectrometer (Murray, 1958). (From Physical Aspects of Irradiation, Recommendations of the International Commission on Radiological Units and Measurements, Report 10b, 1962, *Natl. Bur. Stand. Handb.*, 85, 1964. With permission.)

FIGURE 3.7-95.   Neutron spectrum of Po-Be(α,n). Nuclear emulsions were used for the measurement. Curve a is for a source with carrier while curve b is for a carrier-free source. Significant differences are observed. (From Medvecsky, L., *ATOMKI Kozl.*, 3, 133, 1961, as given in Physical Aspects of Irradiation, Recommendations of the International Commission on Radiological Units and Measurements, Report 10b, 1962, *Natl. Bur. Stand. Handb.*, 85, 1964. With permission.)

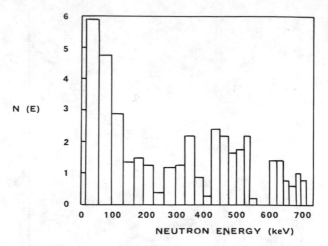

FIGURE 3.7-96.   Neutron energy spectrum of a Ra-Be(γ,n) source. A cloud chamber was used for the measurement. (From Eggler, C. and Hughes, D. J., The Neutron Spectrum of a Radium-Beryllium Photo Source, Report ANL 4476, USAEC, 1950, as given in Physical Aspects of Irradiation, Recommendations of the International Commission on Radiological Units and Measurements, Report 10b, 1962, *Natl. Bur. Stand. Handb.*, 85, 1964. With permission.)

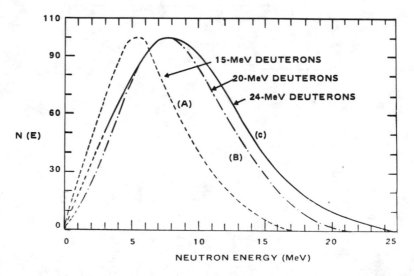

FIGURE 3.7-97.   Neutron energy spectra for 15-, 20-, and 24-MeV deuterons on thick Be target. Measurement of 15 MeV was by Cohen and Falk, 1951; for 20 and 24 MeV by Tochilin and Kohler, 1958. (From Report of the International Commission of Radiological Units and Measurements, *Natl. Bur. Stand. Handb.*, 78, 1959. With permission.)

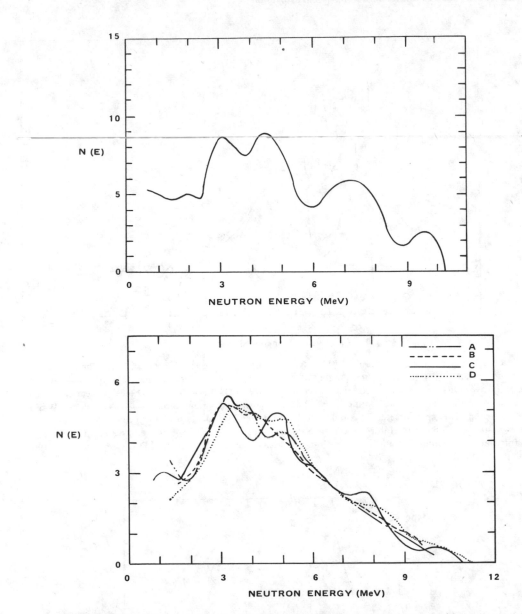

FIGURE 3.7-98.   Neutron energy spectra of Po-Be (α,n). Curve A was measured with a proton recoil proportional counter telescope (Cochran and Henry, 1955); curve B was determined by a similar method (Breen, Hertz, and Wright, 1956); curve C with nuclear emulsions (private communication from author, 1960, indicates that the all quoted energies should be increased by 3 percent, Whitmore and Baker, 1950); curve D with ⁶LiI scintillation crystal spectrometer Murray, 1958). (From Report of the International Commission on Radiological Units and Measurements, *Natl. Bur. Stand. Handb.*, 78, 1959. With permission.)

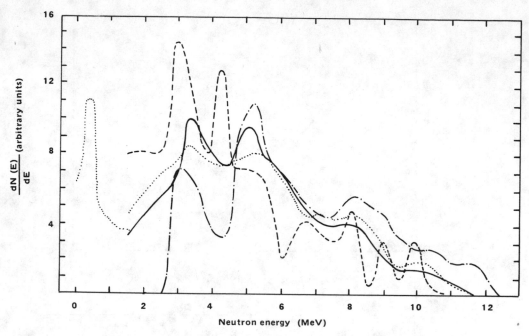

FIGURE 3.7-99.   Energy distribution of ²⁴¹Am-Be neutrons according to different authors. The curves have been normalized to give the same total number of neutrons between energies 2.5 and 10.5 MeV:——Geiger and Hargrove (1964), proton-recoil counter telescope and time of flight;———— Greiss (1968), nuclear emulsions; ...... Thompson and Taylor (1965), stilbene crystal with pulse-shape discrimination;——·——·——Salgir and Walker (1967), proton-recoil spectrometer (silicon detector). The dotted low energy peak has been drawn according to the suggestion of Geiger and Hargrove. (From *Neutron Fluence, Neutron Spectra, and Kerma,* ICRU Report 13, International Commission on Radiation Units and Measurements, Washington, D.C., September 15, 1969. With permission.)

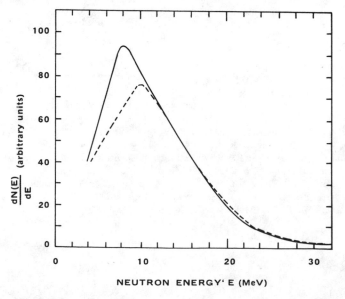

FIGURE 3.7-100.   Neutron spectra of two 1-Kg PuF₄ sources. (From Tochilin, E., Kaerntner Ring II, A-1011, International Atomic Energy Agency, Vienna, 1963, as given in *Neutron Fluence, Neutron Spectra, and Kerma,* ICRU Report 13, International Commission on Radiation Units and Measurements, Washington, D.C., September 15, 1969. With permission.)

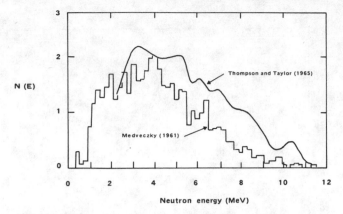

FIGURE 3.7-101. Ra-Be neutron spectrum (Medveczky, 1961; Thompson and Taylor, 1965). (From DePangher, J. and Tochilin, E., in *Radiation Dosimetry,* Vol. 3, 2nd ed., Attix, F. H. and Tochilin, E., Eds., Academic Press, New York, 1969. With permission.)

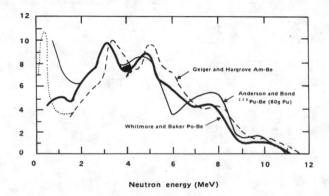

FIGURE 3.7-102. Am-Be, Po-Be, and ^{239}Pu-Be neutron spectra (Geiger and Hargrove, 1964; Whitmore and Baker, 1950; Anderson and Bond, 1963). (From DePangher, J. and Tochilin, E., in *Radiation Dosimetry,* Vol. 3, 2nd ed., Attix, F. H. and Tochilin, E., Eds., Academic Press, New York, 1969. With permission.)

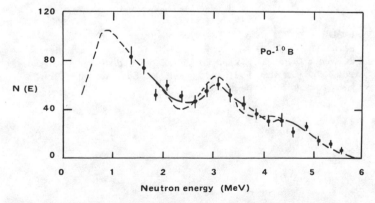

FIGURE 3.7-103. Po-^{10}B netron spectrum. (From Geiger, K. W. and Jarvis, C. J. D., *Can. J. Phys.,* 40, 33, 1962, as given in DePangher, J. and Tochilin, E., in *Radiation Dosimetry,* Vol. 3, 2nd ed., Attix, F. H. and Tochilin, E., Eds., Academic Press, New York, 1969. With permission of the National Research Council of Canada.)

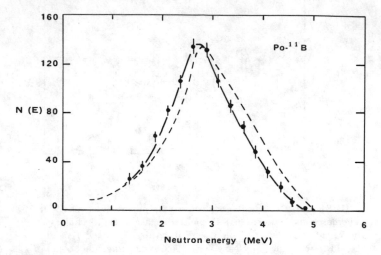

FIGURE 3.7-104.   Po-^{11}B neutron spectrum. (From Geiger, K. W. and Jarvis, C. J. D., *Can. J. Phys.*, 40, 33, 1962, as given in DePangher, J. and Tochilin, E., in *Radiation Dosimetry,* Vol. 3, 2nd ed., Attix, F. H. and Tochilin, E., Eds., Academic Press, New York, 1969. With permission of the National Research Council of Canada.)

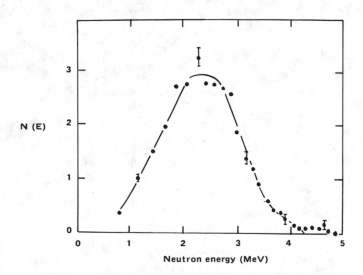

FIGURE 3.7-105.   Po-^{18}O neutron spectrum. (From Khabakhpashev, A. G., *Sov. J. At. Energy,* 1, 591, 1960 (English translation), as given in DePangher, J. and Tochilin, E., in *Radiation Dosimetry*, Vol. 3, 2nd ed., Attix, F. H. and Tochilin, E., Eds., Academic Press, New York, 1969. With permission.)

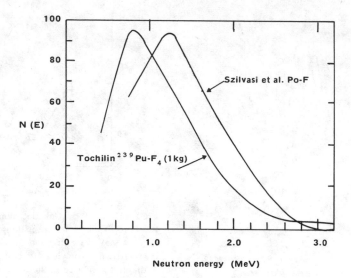

FIGURE 3.7-106.   Po-F and ^{329}Pu-F neutron spectra (Szilvasi et al., 1960; Tochilin, 1963). (From DePangher, J. and Tochilin, E., *Radiation Dosimetry,* Vol. 3, 2nd ed., Attix, F. H. and Tochilin, E., Eds., Academic Press, New York, 1969. With permission.)

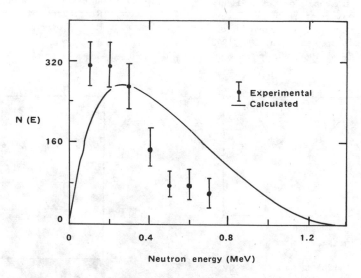

FIGURE 3.7-107.   Po-Li neutron spectrum (Barton, 1953; Hess, 1959). (From DePangher, J. and Tochilin, E., *Radiation Dosimetry,* Vol. 3, 2nd ed., Attix, F. H. and Tochilin, E., Eds., Academic Press, New York, 1969. With permission.)

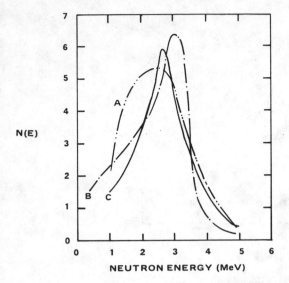

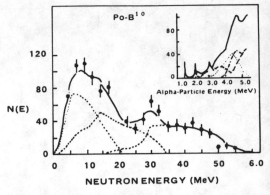

FIGURE 3.7-108.   Neutron spectrum of Po-B(α,n). Curve A is the result of a measurement with nuclear emulsions (Periman, Richards, and Speck, 1946); curve B was also by nuclear emulsions (Staub, 1947); curve C by proton recoil counter telescope (Cochran and Henry, 1955). (From Physical Aspects of Irradiation, Recommendations of the International Commission of Radiological Units and Measurements, Report 10b, 1962, *Natl. Bur. Stand. Handb.*, 85, 1964. With permission.)

FIGURE 3.7-109.   Calculated neutron spectra for a Po-^{10}B(α,n) source for transitions into the ground state (a), 2.4-MeV state (b), and 3.54-MeV state (c), where the relative neutron yields (curves a, b, and c), as shown in the inset, are used in the calculation. Also shown are the sum of these three calculated spectra (solid line) and the experimental points from emulsion data, corrected for a small contribution from the ^{11}B impurity in the actual source (Geiger and Jarvis, 1961). (From Physical Aspects of Irradiation, Recommendations of the International Commission on Radiological Units and Measurements, Report 10b, 1962, *Natl. Bur. Stand. Handb.*, 85, 1964. With permission.)

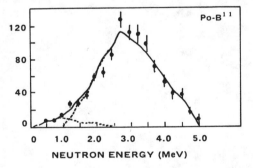

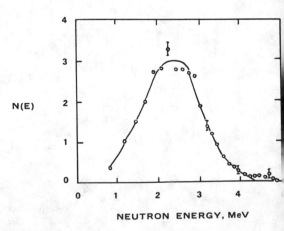

FIGURE 3.7-110.   Calculated neutron spectra for a Po-^{11}B(α,n) source for transitions into the ground state (a) and the 2.3-MeV state (b). The sum of the two curves and the experimental points from emulsion data are also shown (Geiger and Jarvis, 1962). (From Physical Aspects of Irradiation, Recommendations of the International Commission on Radiological Units and Measurements, Report 10b, 1962, *Natl. Bur. Stand. Handb.*, 85, 1964. With permission.)

FIGURE 3.7-111.   Neutron spectrum of ^{210}Po-^{18}O(α,n). (From Khabakhpashev, A. G., *Sov. J. At. Energy*, 1, 591, 1960 (English translation), as given in Physical Aspects of Irradiation, Recommendations of the Interntional Commission on Radiological Units and Measurements, Report 10b, 1962, *Natl. Bur. Stand. Handb.*, 85, 1964. With permission.)

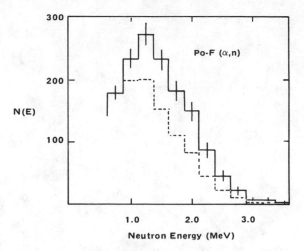

FIGURE 3.7-112.   Neutron spectrum of ^{210}Po-^{19}F(α,n).
Dashed histogram shows the neutron spectrum obtained
from these results after correcting for the neutron-proton
scattering cross section and the escape of proton tracks
from the nuclear emulsion (Szilvasi, Geiger, and Dixon,
1960). (From Physical Aspects of Irradiation, Recommendations of the International Commission on Radiological
Units and Measurements, Report 10b, 1962, *Natl. Bur.
Stand. Handb.*, 85, 1964. With permission.)

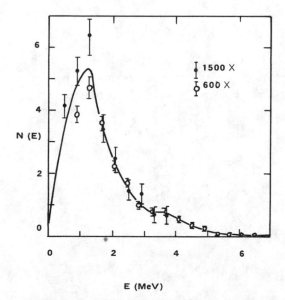

FIGURE 3.7-113.   Mock fission neutron spectrum,
produced by ^{210}Po alphas on a mixture of 90.3% LiF,
9.4% B, and 0.3% Be. (From Tochilin, E. and Alves,
R. V., *Nucleonics,* 16(11), 145, 1968, as given in De-
Pangher, J. and Tochilin, E., in *Radiation Dosimetry,*
Vol. 3, 2nd ed., Attix, F. H. and Tochilin, E., Eds.,
Academic Press, New York, 1969. With permission.)

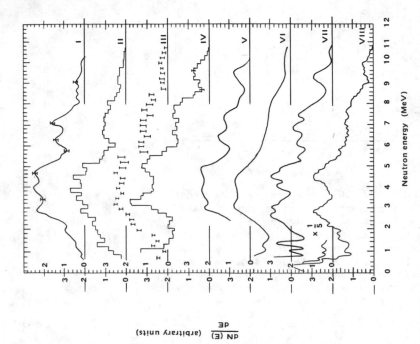

FIGURE 3.7-115.   Energy distribution of Po-Be neutrons according to different authors. (From *Neutron Fluence, Neutron Spectra, and Kerma*, ICRU Report 13, International Commission on Radiation Units and Measurements, Washington, D.C., September 15, 1969. With permission.)

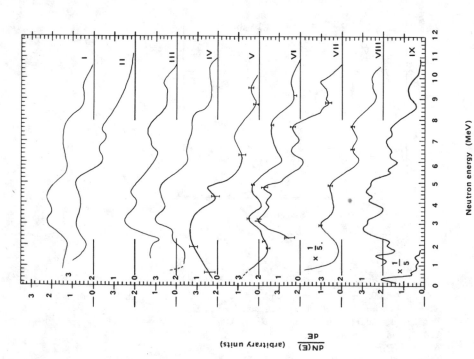

FIGURE 3.7-114.   Energy distribution of Pu-Be neutrons, according to different authors. (From *Neutron Fluence, Neutron Spectra, and Kerma*, ICRU Report 13, International Commission on Radiation Units and Measurements, Washington, D.C., September 15, 1969. With permission.)

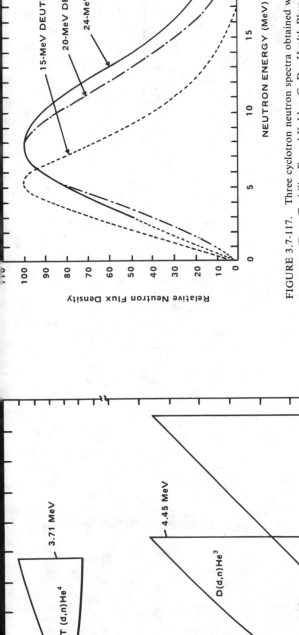

FIGURE 3.7-117.   Three cyclotron neutron spectra obtained with 15-, 20-, and 24-MeV deuterons. (From Tochilin, E. and Kohler, G. D., *Health Phys.*, 1, 332, 1958, as given in Tochilin, E. and Shumway, B. W., in *Radiation Dosimetry*, Vol. 3, 2nd ed., Attix, F. H. and Tochilin, E., Eds., Academic Press, New York, 1969. With permission.)

FIGURE 3.7-116.   Ranges of neutron energies available from the five most commonly used monoenergetic neutron reactions. The neutron energy is maximum at 0° and minimum at 180°. (From De-Pangher, J. and Tochilin, E., in *Radiation Dosimetry*, Vol. 3, 2nd ed., Attix, F. H. and Tochilin, E., Eds., Academic Press, New York, 1969. With permission.)

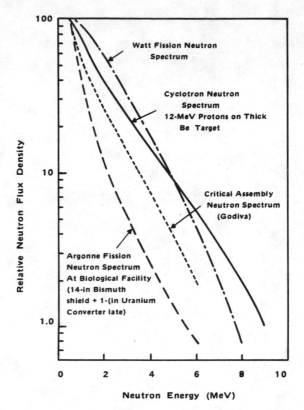

FIGURE 3.7-118.   Comparison of neutron spectrum from 12-MeV protons on Be with three fission spectra. (From Tochilin, E. and Kohler, G. D., *Health Phys.*, 1, 332, 1958, as given in Tochilin, E. and Shumway, B. W., in *Radiation Dosimetry,* Vol. 3, 2nd ed., Attix. F. H. and Tochilin, E., Eds., Academic Press, New York, 1969. With permission.)

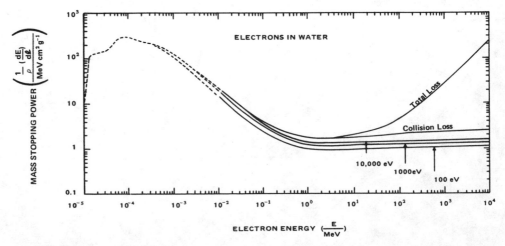

FIGURE 3.7-119.   Mass stopping power for water for electrons vs. electron energy. Below 10 keV theoretical formulae are inapplicable. Limited experimental data only are available; consequently this region is shown dashed. In the region between 10 keV and 100 keV, both measured and calculated values are used. Restricted stopping power of Δ values of 100, 1000, and 10,000 eV are shown. (From *Neutron Fluence, Neutron Spectra, and Kerma,* ICRU Report 13, International Commission on Radiation Units and Measurements, Washington, D.C., September 15, 1969. With permission.)

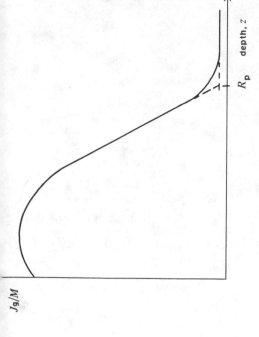

FIGURE 3.7-121.   Definition of the practical range of electrons. Drawn is the measured value of J_g, relative to an arbitrary monitor value, M, as a function of depth, z, in the absorber. (This definition corresponds to that of the "extrapolated projected range" as defined in ICRU Report No. 16 (ICRU, 1970) except that the extrapolation has to be made here to the bremsstrahlung background.) (From *Radiation Dosimetry: Electrons with Initial Energies Between 1 and 50 MeV*, ICRU Report 21, International Commission on Radiation Units and Measurements, Washington, D.C., May 15, 1972. With permission.)

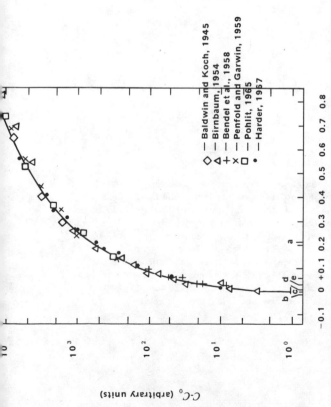

FIGURE 3.7-120.   Average activation curve for the determination of the threshold of nuclear reaction in copper (Harder, 1967). The ordinate is the difference, C − C_o, of the measured count rate and the background count rate, in relative units. The abscissa is the increment, ΔE_o, of electron energy above the threshold energy. The ΔE_o-scale is based on the most probable indication (C). (From *Radiation Dosimetry: Electrons with Initial Energies Between 1 and 50 MeV*, ICRU Report 21, International Commission on Radiation Units and Measurements, Washington, D.C., May 15, 1972. With permission.)

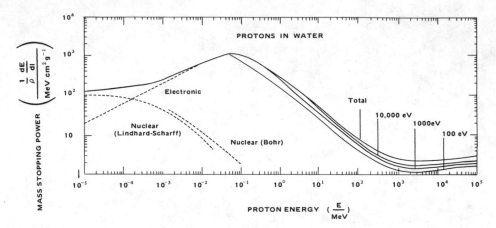

FIGURE 3.7-122.   Mass stopping power for water for protons vs. proton energy. Restricted stopping powers for 100, 1000, and 10,000 eV are also shown. (From *Neutron Fluence, Neutron Spectra, and Kerma,* ICRU Report 13, International Commission on Radiation Units and Measurements, Washington, D.C., September 15, 1969. With permission.)

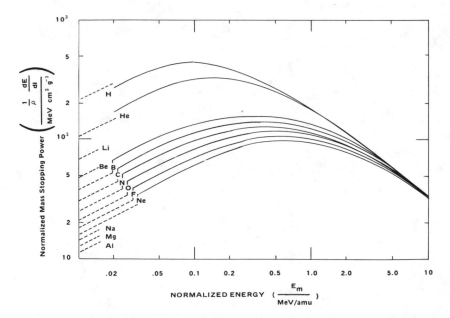

FIGURE 3.7-123.   Smoothed data stopping-power curves for various ions for aluminum. Dashed lines, theoretical result of Lindhard and Scharff, 1960; solid lines, experimental data, Northcliffe, 1964. (From Lindhard, J. and Scharff, M., in *Penetration of Charged Particles in Matter,* Publication 752, National Academy of Sciences-National Research Council, Washington, D.C., 1960, 49; Northcliff, L. C., *Annu. Rev. Nucl. Sci.,* 13, 67, 1964; and Northcliffe, L.C., in *Studies in Penetration of Charged Particles in Matter,* Publication 1133, National Academy of Sciences-National Research Council, Washington, D.C., 1964, 173, as given in *Linear Energy Transfer,* ICRU Report 16, International Commission on Radiation Units and Measurements, Washington, D.C., June 15, 1970. With permission. Copyright © 1964 by Annual Reviews, Inc. All rights reserved.)

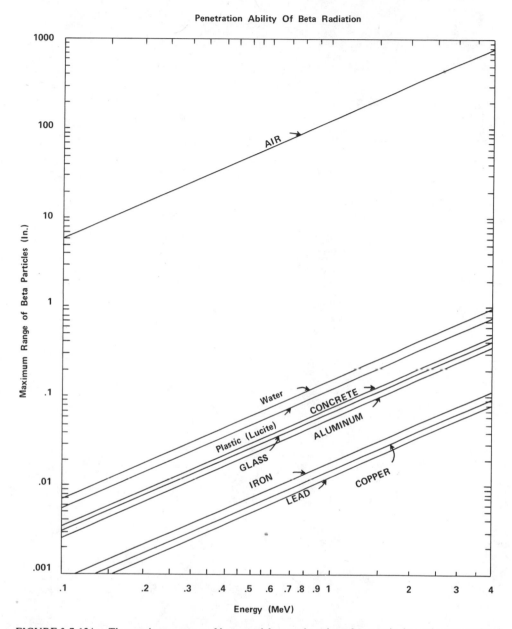

FIGURE 3.7-124.   The maximum range of beta particles as a function of energy in the various materials indicated. (From SRI Report No. 361, "The Industrial Uses of Radioactive Fission Products". With permission of the Stanford Research Institute and the U. S. Atomic Energy Commission, as given in *Radiological Health Handbook,* Bureau of Radiological Health, Training Institute Environmental Control Administration, Public Health Service, Rockville, Md., January 1970.)

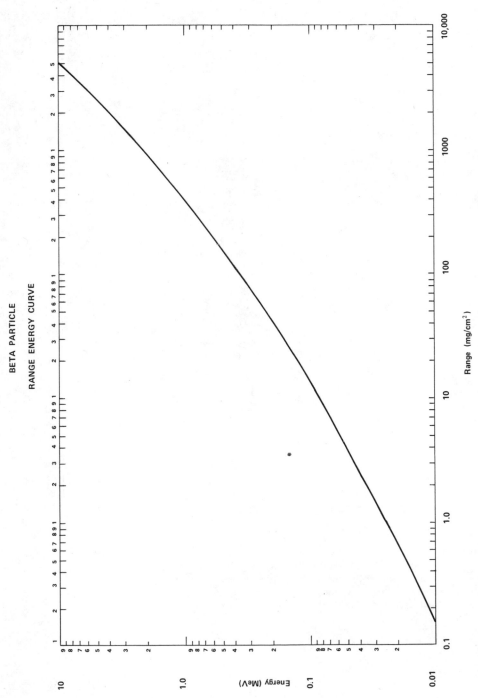

FIGURE 3.7-125. Beta particle range energy curve. (From *Radiological Health Handbook*, Bureau of Radiological Health, Training Institute Environmental Control Administration, Public Health Service, Rockville, Md., January 1970. With permission.)

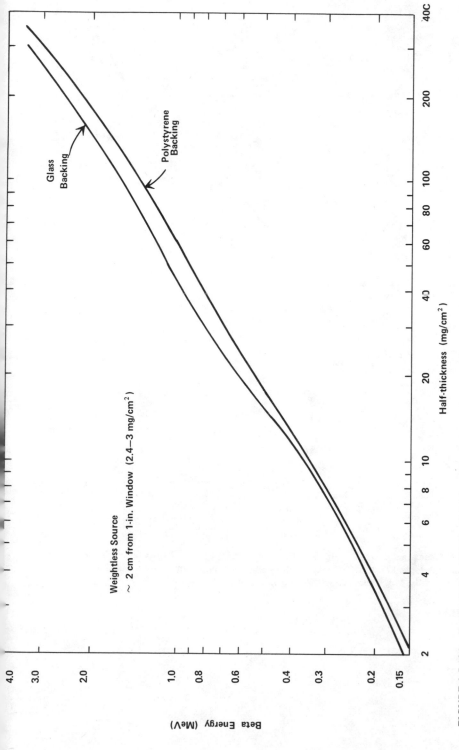

FIGURE 3.7-126.   Beta radiation initial half-thickness in aluminum vs. maximum energy. (From *Radiological Health Handbook*, Bureau of Radiological Health, Training Institute Environmental Control Administration, Public Health Service, Rockville, Md., January 1970. With permission.)

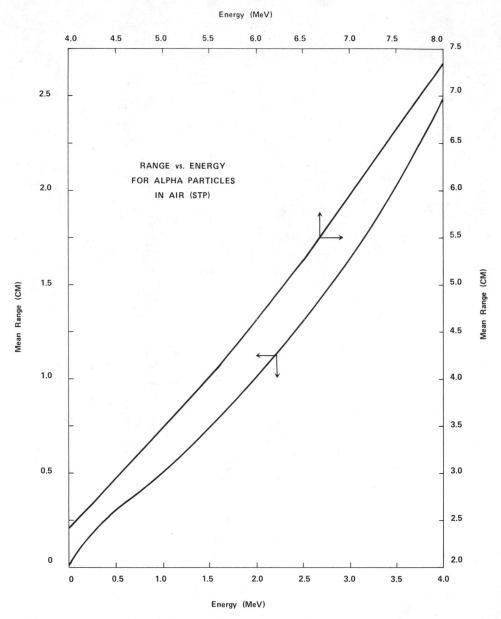

FIGURE 3.7-127.   Range vs. energy for alpha particles in air (STP). (From *Radiological Health Handbook,* Bureau of Radiological Health, Training Institute Environmental Control Administration, Public Health Services, Rockville, Md., Jaunary 1970. With permission.)

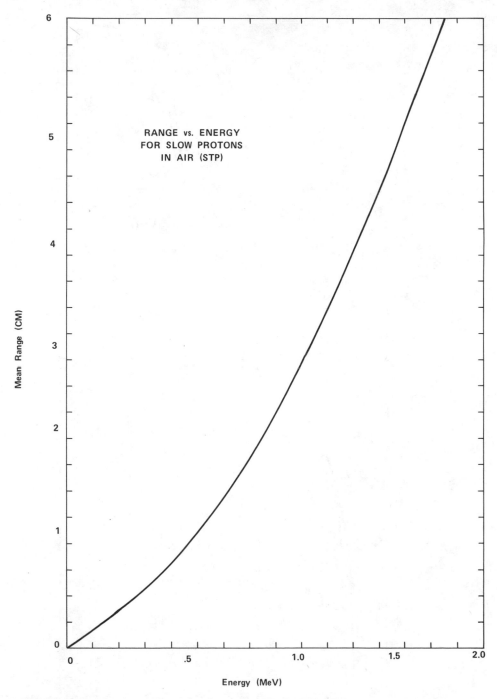

FIGURE 3.7-128.   Range vs. energy for slow protons in air (STP). (From *Radiological Health Handbook,* Bureau of Radiological Health, Training Institute Environmental Control Administration, Public Health Service, Rockville, Md., January 1970. With permission.)

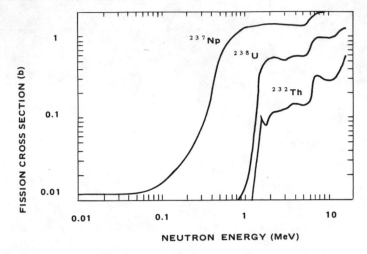

FIGURE 3.7-129. Fission cross sections of different fast-neutron threshold detectors as a function of neutron energy. (From Becker, K., in *Topics in Radiation Dosimetry,* Suppl. 1, Attix, F. H., Ed., Academic Press, New York, 1972. With permission.)

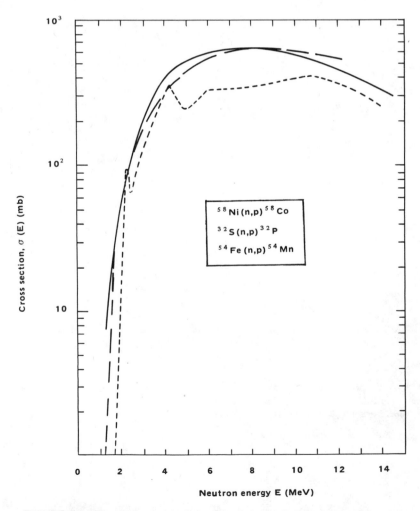

FIGURE 3.7-130. Activation cross sections for several commonly used (n, p) reactions. (From Moteff, J., in *Radiation Dosimetry,* Vol. 3, 2nd ed., Attix, F. H. and Tochilin, Ed., Eds., Academic Press, New York, 1969. With permission.)

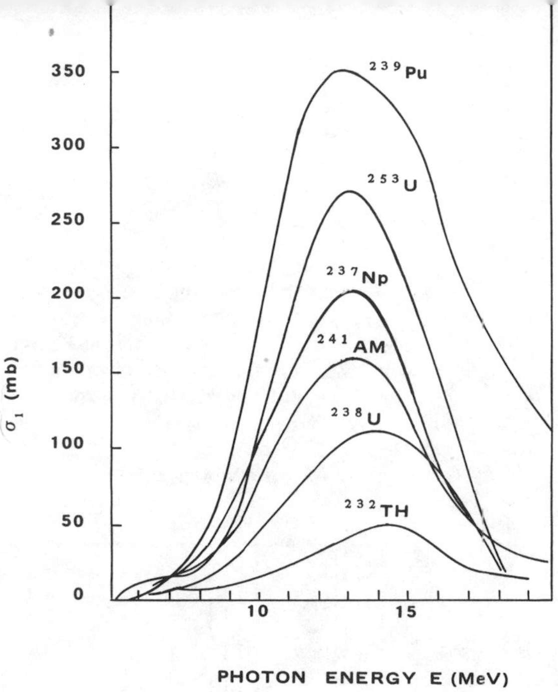

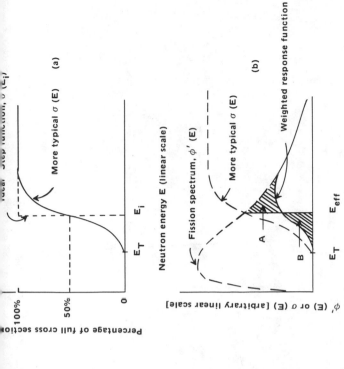

FIGURE 3.7-131. Energy variation of the fission cross section for photon-induced fission of various targets. (From Katz, L., Baerg, A. P., and Brown, F., *Proc. 2nd U.N. Conf. on Peaceful Uses of Atomic Energy*, 15, 200, 1958, as given in Loveland, W. D., in *American Institute of Physics Handbook*, 3rd ed., Gray, D. E., Ed., McGraw-Hill, New York, 1972. With permission.)

FIGURE 3.7-132. A. Comparison of an ideal "step-function" activation cross section with that of a more typical threshold detector. B. The weighted response function of the latter for a fission neutron spectrum. Setting area A = area B is the criterion usually employed to establish the value of the effective threshold E_{eff}. (From Moteff, J., in *Radiation Dosimetry*, Vol. 3, 2nd ed., Attix, F. H. and Tochilin, E., Eds., Academic Press, New York, 1969. With permission.)

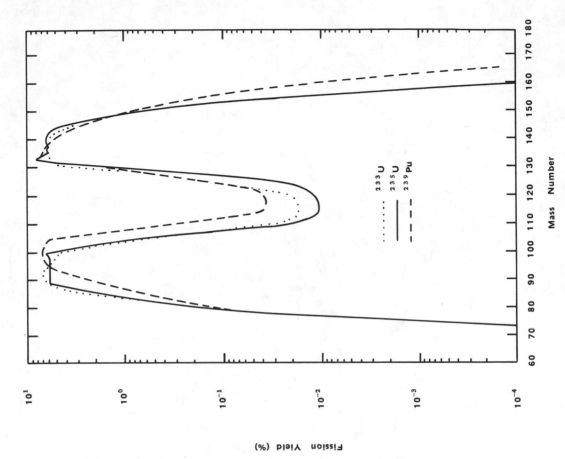

FIGURE 3.7-134.   Fission-fragment mass distributions for the thermal-neutron-fission of [233]U, [235]U and [239]Pu. (From Loveland, W. D., in *American Institute of Physics Handbook,*

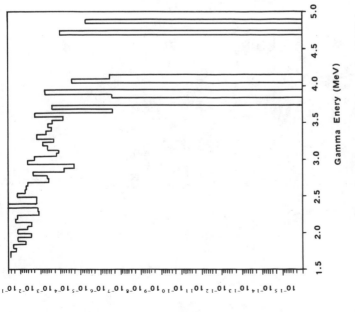

FIGURE 3.7-133.   Gamma-ray spectrum at 721-hr total and 720-hr irradiation normalization 4.798E −02 ([235]U thermal). (From Stamatelatos, M. G. and England, T. R., Fisson-produce Gamma-ray and Photoneutron Spectra and Energy-integrated Sources Informal Report, NUREG-0155, U.S. Nuclear Regulatory Commission, Los Alamos Scientific Laboratory, University of California, Los Alamos, N.M., December, 1976. With permission.)

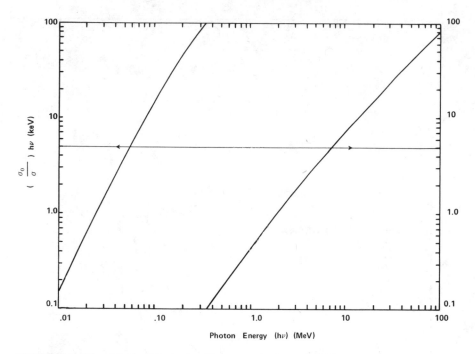

FIGURE 3.7-135.   Graph of $(\sigma_a/\sigma)\,h\nu$, the mean initial energy of the Compton recoil electrons produced by monochromatic γ-rays of quantum energy $h\nu$. (From, Report of the International Commission on Radiological Units and Measurements, *Natl. Bur. Stand. Handb.*, 78, 1959. With permission.)

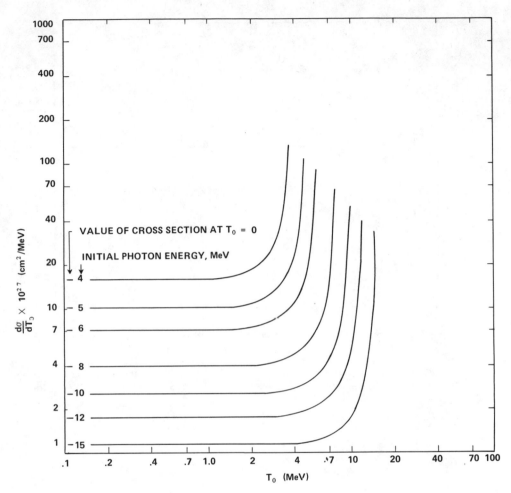

FIGURE 3.7-136.   Starting-energy distribution for Compton-recoil electrons produced by photons with initial energies of 4 to 15 MeV. The ordinate is the absolute differential cross section for giving a free electron a recoil energy in the interval from T_o to $T_o + dT_o$. (From Nelms, A. T., *Natl. Bur. Stand. Circ.*, 512, 1953, as given in Report of the International Commission of Radiological Units and Measurements, *Natl. Bur. Stand. Handb.*, 78, 1959. With permission.)

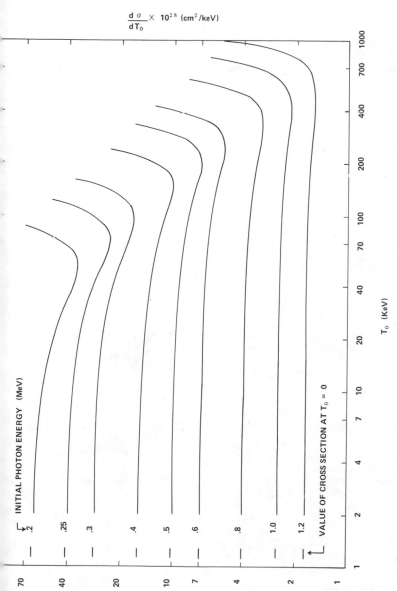

FIGURE 3.7-137.   Starting-energy distributions for Compton-recoil electrons produced by photons with initial energies of 0.2 to 1.2 MeV. The ordinate is the absolute differential cross section for giving free electron a recoil energy in the interval from T_0 to $T_0 + dT_0$. (From Nelms, A. T., *Natl. Bur. Stand. Circ.*, 512, 1953, as given in Report of the International Commission on Radiological Units and Measurements, *Natl. Bur. Stand. Handb.*, 78, 1959. With permission.)

REFERENCES

1. *Nucl. Data Sheets,* 20(1), 1977.
2. **Gray, D. E., Ed.,** *American Institute of Physics Handbook,* 3rd ed., McGraw-Hill, New York, 1972.
3. **Ewbank, W. B., Haese, R. L., Hurley, F. W., and McGinnis, M. R.,** *Nuclear Structure References, 1969–1974,* Academic Press, New York, 1975.
4. **Serpan, C. Z., Jr. and Menke, B. H.,** *Nuclear Reactor Neutron Energy Spectra,* ASTM Publication DS 52, American Society for Testing and Materials, Philadelphia, 1976.
5. **Condon, E. U. and Odishaw, H., Eds.,** *Handbook of Physics,* 2nd ed., McGraw-Hill, New York, 1967.
6. **Etherington, H., Ed.,** *Nuclear Engineering Handbook,* McGraw-Hill, New York, 1958.
7. **Evans, R. D.,** *The Atomic Nucleus,* McGraw-Hill, New York, 1955.
8. **Attix, F. H. and Roesch, W. C., Eds.,** *Radiation Dosimetry,* Vol. 1, 2nd ed., Academic Press, New York, 1968.
9. **Attix, F. H. and Roesch, W. C., Eds.,** *Radiation Dosimetry,* Vol. 2, 2nd ed., Academic Press, New York, 1966.
10. **Attix, F. H. and Tochilin, E., Eds.,** *Radiation Dosimetry,* Vol. 3, 2nd ed., Academic Press, New York, 1969.
11. **Attix, F. H., Ed.,** *Topics in Radiation Dosimetry,* Suppl. 1, Academic Press, New York, 1972.
12. National Neutron Cross Section Center, *Reports to the ERDA Nuclear Data Committee,* BNL-NCS-21501, Brookhaven National Laboratory, Associated Universities, Upton, New York, May 1976.
13. **Garber, D. I. and Kinsey, R. R., Eds.,** *Neutron Cross Sections,* Vol. 2, 3rd. ed., BNL-325, Information Analysis Center Report, National Neutron Cross Section Center, Brookhaven National Laboratory, Upton, New York, January 1976.
14. **Slack, L. and Way, K.,** *Radiations from Radioactive Atoms in Frequent Use,* U.S. Atomic Energy Commission, Washington, D.C., February, 1959.
15. *Neutron Standard Reference Data,* Panel Proceedings Series, International Atomic Energy Agency, Vienna, Austria, 1974.
16. **Dillman, L. T.,** Improved radioactive decay scheme data, in *Health Physics Division Annual Progress Report, Period Ending June 30, 1976,* ORNL-5171, Oak Ridge National Laboratory, Oak Ridge, Tenn., October 1976.
17. **Fowler, J. F.,** Solid state electrical conductivity dosimeters, in *Radiation Dosimetry,* Vol. 2, 2nd ed., Attix, F. H. and Roesch, W. C., Eds., Academic Press, New York, 1966.
18. *Radiological Health Handbook,* Bureau of Radiological Health, Training Institute Environmental Control Administration, Public Health Service, Rockville, Md., January 1970.
19. *Neutron Fluence, Neutron Spectra, and Kerma,* ICRU Report 13, International Commission on Radiation Units and Measurements, Washington, D.C., September 15, 1969.
20. *Radiation Dosimetry: X-Rays Generated at Potentials of 5 to 150 kV,* ICRU Report 17, International Commission on Radiation Units and Measurements, Washington, D.C., June 15, 1970.
21. *Specification of High Activity Gamma-ray Sources,* ICRU Report 18, International Commission on Radiation Units and Measurements, Washington, D.C., October 15, 1970.
22. *Radiation Quantities and Units,* ICRU Report 19, International Commission on Radiation Units and Measurements, Washington, D.C., July 1, 1971.
23. *Radiation Dosimetry: Electrons with Initial Energies Between 1 and 50 MeV,* ICRU Report 21, International Commission on Radiation Units and Measurements, Washington, D.C., May 15, 1972.
24. Report of the International Commission on Radiological Units and Measurements, *Nat. Bur. Stand. Handb.,* 78, 1959.
25. Physical Aspects of Irradiation, Recommendations of the International Commission on Radiological Units and Measurements, Report 106, 1962, *Nat. Bur. Stand. Handb.,* 85, 1964.
26. **Stamatelatos, M. G. and England, T. R.,** Fission-product Gamma-ray and Photoneutron Spectra and Energy-integrated Sources Informal Report, NUREG-0155, U.S. Nuclear Regulatory Commission, Los Alamos Scientific Laboratory, University of California, Los Alamos, N.M., December 1976.
27. **Berger, M. J. and Raso, D. J.,** Backscattering of Gamma Rays, Report NBS 5982, National Bureau of Standards, Washington, D.C., July 25, 1958.
28. **Becker, K.,** Dosimetric applications of track etching, in *Topics in Radiation Dosimetry,* Suppl. 1, Attix, F. H., Ed., Academic Press, New York, 1972.
29. **Burlin, T. E.,** Vacuum chambers for radiation measurement, in *Topics in Radiation Dosimetry,* Suppl. 1, Attix, F. H., Ed., Academic Press, New York, 1972.

30. **Greening, J. R.,** Low-energy X-ray dosimetry, in *Topics in Radiation Dosimetry,* Suppl. 1, Attix, F. H., Ed., Academic Press, New York, 1972.

31. **Johns, H. E.,** X-Rays and teleisotope gamma rays, in *Radiation Dosimetry,* Vol. 3, 2nd ed., Attix, F. H. and Tochilin, E., Eds., Academic Press, New York, 1969.

32. **Laughlin, J. S.,** Electron beams, in *Radiation Dosimetry,* Vol. 3, 2nd ed., Attix, F. H. and Tochilin, E., Eds., Academic Press, New York, 1969.

33. **Raju, M. R., Lyman, J. T., Brustad, T., and Tobias, C. A.,** Heavy charged-particle beams, in *Radiation Dosimetry,* Vol. 3, 2nd ed., Attix, F. H. and Tochilin, E., Eds., Academic Press, New York, 1969.

34. **Moteff, J.,** Neutron dosimetry in irradiation of materials, in *Radiation Dosimetry,* Vol. 3, 2nd ed., Attix, F. H. and Tochilin, E., Eds., Academic Press, New York, 1969.

35. **Tochilin, E. and Shumway, B. W.,** Dosimetry of neutrons and mixed n + γ fields, in *Radiation Dosimetry,* Vol. 3, 2nd ed., Attix, F. H. and Tochilin, E., Eds., Academic Press, New York, 1969.

36. **DePangher, J. and Tochilin, E.,** Neutrons from accelerators, in *Radiation Dosimetry,* Vol. 3, 2nd ed., Attix, F. H. and Tochilin, E., Eds., Academic Press, New York, 1969.

37. **Cook, C. S.,** Radiations from nuclear weapons, in *Radiation Dosimetry,* Vol. 3, 2nd ed., Attix, F. H. and Tochilin, E., Eds., Academic Press, New York, 1969.

38. **Mayneord, W. V. and Hill, C. R.,** Natural and man-made background radiation, in *Radiation Dosimetry,* Vol. 3, 2nd ed., Attix, F. H. and Tochilin, E., Eds., Academic Press, New York, 1969.

39. **Sondhaus, C. A. and Evans, R. D.,** Dosimetry of radiation in space flight, in *Radiation Dosimetry,* Vol. 3, 2nd ed., Attix, F. H. and Tochilin, E., Eds., Academic Press, New York, 1969.

40. **Sinclair, W. K.,** Radiobiological dosimetry, in *Radiation Dosimetry,* Vol. 3, 2nd ed., Attix, F. H. and Tochilin, E., Eds., Academic Press, New York, 1969.

41. **Shalek, R. J. and Stovall, M.,** Dosimetry in implant therapy, in *Radiation Dosimetry,* Vol. 3, 2nd ed., Attix, F. H. and Tochilin, E., Eds., Academic Press, New York, 1969.

42. **Hogan, O. H., Zigman, P. E., and Mackin, J. L.,** Beta Spectra II. Spectra of Individual Negatron Emitters, USNRDL-TR-802, U.S. Naval Radiological Defense Laboratory, San Francisco, Calif., December 16, 1964.

43. **Morgan, K. Z. and Turner, J. E.,** Health Physics, in *American Institute of Physics Handbook,* 3rd ed., Gray, D. E., Ed., McGraw-Hill, New York, 1972.

44. **Slater, D. N.,** *Gamma-rays of Radionuclides in Order of Increasing Energy,* Butterworths, Washington, 1962.

45. **Moteff, J.,** Fission Product Decay Gamma Energy Spectrum, APEX-134, Aircraft Nuclear Propulsion Project, General Electric, June 1953.

46. **Bearden, J. A.,** X-ray wavelengths, in *Handbook of Chemistry and Physics,* 56th ed., Weast, R. C., Ed., CRC Press, Cleveland, 1976.

47. **Fine, S. and Hendee, C. F.,** X-Ray critical-absorption and emission energies in keV, *Nucleonics,* 13(3), 36, 1956.

48. **Wyard, S. J.,** Radioactive-source corrections for bremsstrahlung and scatter, *Nucleonics,* 13(7), July 1955.

49. **Marion, J. B.,** Nuclear constants and calibrations, in *American Institute of Physics Handbook,* 3rd ed., Gray, D. E., Ed., McGraw-Hill, New York, 1972.

50. **Evans, R. D.,** Gamma rays, in *American Institute of Physics Handbook,* 3rd ed., Gray, D. E., Ed., McGraw-Hill, New York, 1972.

51. **Loveland, W. D.,** Nuclear fission, in *American Institute of Physics Handbook,* 3rd ed., Gray, D. E., Ed., McGraw-Hill, New York, 1972.

52. **Bichsel, H.,** Passage of charged particles through matter, in *American Institute of Physics Handbook,* 3rd ed., Gray, D. E., Ed., McGraw-Hill, New York, 1972.

53. **Rossi, H. H.,** Microscopic energy distribution in irradiated matter, in *Radiation Dosimetry,* Vol. 1., 2nd ed., Attix, F. H. and Roesch, W. C., Eds., Academic Press, New York, 1968 543–593.

54. **Ellett, W. and Braby, L.,** The microdosimetry of 250 kVp and 65 kVp X-rays, 60 C_0 gamma rays, and tritium beta particles, *Radiat. Res.,* 51, 229, 1972.

55. **Kellerer, A. and Brenot, J.,** On the statistical evaluation of dose response functions, *Radiat. Environ. Biophys.,* 11, 1, 1974.

56. **Kellerer, A. and Chmelevsky, D.,** Criteria for the applicability of LET, *Radiat. Res.,* 63, 226, 1975.

57. **Kellerer, A. and Rossi, H. H.,** *Curr. Top. Radiat. Res. Q.,* 8, 85, 1972.

58. **Raju, M. R., Lyman, J. T., Brustad, T., and Tobias, C. A.,** Heavy charged-particle beams, in *Radiation Dosimetry,* Vol. 3, 2nd ed., Attix, F. H. and Tochilin, E., Eds., Academic Press, New York, 1969, 151–200.

59. **Voigt, A. F.**, Table of Fission Product Nuclides, UC-34C, Ames Laboratory, USERDA, Iowa State University, Ames, Iowa, 1976.
60. **Evans, R. D.**, X-ray and γ-ray interactions, in *Radiation Dosimetry,* Vol. 1, 2nd ed., Attix, F. H. and Roesch, W. C., Eds., Academic Press, New York, 1968.
61. **Hettinger, G. and Starfelt, N.**, Bremsstrahlung spectra from roentgen tubes, *Acta Radiol.,* 50, 381, 1958.
62. **Ehrlich, M.**, Reciprocity law for X-rays. I. Validity for high-intensity exposures in the negative region, *J. Opt. Soc. Am.,* 46, 797, 1956.
63. **Cormack, D. V., Till, J. E., Whitmore, G. F., and Johns, H. E.**, Measurement of continuous X-ray spectra with a scintillation spectrometer, *Br. J. Radiol.,* 28, 605, 1955.
64. **Johns, H. E., Fedoruk, S. O., Kornelsen, P. O., Epp, E. R., and Darby, E. K.**, Depth dose data, 150 kVp to 400 kVp, *Br. J. Radiol.,* 25, 542, 1952.
65. **Starfelt, N. and Koch, H. W.**, Differential cross section measurements of thin-target bremsstrahlung produced by 2.7- to 9.7-MeV electrons, *Phys. Rev.,* 102, 1958, 1956.
66. Protection against betatron-synchrotron radiation up to 100 MeV, *Natl. Bur. Stand. Handb.,* 55, 1954.
67. **Cormack, D. V., Griffith, T. J., and Johns, H. E.**, Measurement of the spectral distribution of scattered 400 kVp X-rays in a water phantom, *Br. J. Radiol.,* 30, 129, 1957.
68. **Bruce, W. R., Pearson, M. L., and Freedhoff, H. S.**, The linear energy transfer distributions resulting from primary and scattered X-rays and gamma rays with primary HVL's from 1.25 mm Cu to 11 mm Pb, *Radiat. Res.,* 19, 606, 1963.
69. **Rosenzweig, W. and Rossi, H. H.**, Determination of the quality of the absorbed dose delivered by monoenergetic neutrons, *Radiat. Res.,* 10, 532, 1959, and private communication.
70. **Schaefer, H. J.**, Local LET spectra in tissue for solar flare protons in space and for neutron-produced recoil protons, in *Symp. Biological Effects Neutron Proton Irradiations,* Upton, N.Y., Vol. 1, IAEA, Vienna, 1963, 297.
71. **Schaefer, H. J.**, LET spectrum and RBE of high energy protons, in *Proc. Symp. Protection Against Radiation Hazards in Space,* Gatlinburg, Tenn., Book 1, USAEC-TID-7652, USAEC, 1962, 393.
72. **Maccabee, H. O., Raju, M. R., and Tobias, C. A.**, Fluctuations of energy loss in semiconductor detectors, *IEEE Trans. Nucl. Sci.,* 13(3), 176, 1966.
73. **Raju, M. R.**, Heavy particle studies with silicon detectors, *Radiat. Res. Suppl.,* 7, 43, 1967.
74. **Weale, J. W., Benjamin, P. W., Kemshall, C. D., Paterson, W. I., and Redfearn, J.**, Neutron Spectrum Measurements in the Zero Power Fast Reactor VERA, in Proc. Int. Conf. Radiation Measurement in Nuclear Power, Berkeley Nuclear Laboratory, September 1966, p. 231.
75. **Dixon, W. R., Bielesch, A., and Geiger, K. W.**, Neutron spectrum of an actinium-beryllium source, *Can. J. Phys.,* 35, 699, 1957.
76. **Stewart, L.**, Neutron spectrum and absolute yield of a plutonium-beryllium source, *Phys. Rev.,* 98, 740, 1955.
77. **Broek, H. W. and Anderson, C. E.**, The stilbene scintillation crystal as a spectrometer for continuous fast-neutron spectra, *Rev. Sci. Instrum.,* 31, 1063, 1960.
78. **Medvecsky, L.**, Energy spectrum of neutron sources Be(α, n), *ATOMKI Kozl.,* 3, 2–3, 1961.
79. **Geiger, K. W. and Jarvis, C. J. D.**, Neutrons and gamma rays from Po^{210}-B^{10}(α, n) and Po^{210}-B^{11}(α, n) sources, *Can. J. Phys.,* 40, 33, 1962.
80. **Khabakhpashev, A. G.**, Neutron spectrum of a Po-α-O source, *Sov. J. At. Energy,* 1, 591, 1960 (English translation).
81. **Eggler, C. and Hughes, D. J.**, The Neutron Spectrum of a Radium-Beryllium Photo Source, Report ANL 4476, USAEC, 1950.
82. **Tochilin, E. and Alves, R. V.**, Neutron spectra from mock-fission sources, *Nucleonics,* 16(11), 145, 1958.
83. **Tochilin, E. and Kohler, G. D.**, Neutron beam characteristics from the University of California 60-in. cyclotron, *Health Phys.,* 1, 332, 1958.
84. **Nelms, A. T.**, *Natl. Bur. Stand. Circ.,* 512, 1953.

3.8. DOSIMETRY OF INTERNAL EMITTERS –
A GUIDE TO THE MIRD TECHNIQUE

Tuvia Schlesinger

3.8.1 Introduction

After its first systematic formulation (in 1948) by Marinelli et al.[1] the dosimetry of biologically distributed radionuclides did not undergo significant changes for about 20 years. Their approach was to separate the radiations involved in the dose calculations into the penetrating x and γ radiations and the nonpenetrating particulate (and very soft X-ray) categories. Their basic equations for calculating the doses from these two types of radiations were quickly accepted and widely used by physicists and physicians. Marinelli's theory has been reviewed and somewhat extended by Loevinger and his collaborators[2] in 1956 and has since been regarded as the "standard" method for the dosimetry of internally deposited radionuclides. This theory is well known, and we will not go into its details here. The cumulated dose D (rad) to an organ at a time t (hours) after administration of the radioisotope was determined by the familiar equation:

$$D = [2.134\overline{E}_\beta + 0.001\Gamma_t\rho\overline{g}] \, _0\!\int^t c(t)dt \text{ rad} \tag{3.8.1}$$

where $\overline{E}_\beta$(MeV) is the average effective energy of the nonpenetrating radiations (taking into account their fractional frequencies per disintegration), $\Gamma_t \frac{(rad\text{-}cm^2)}{mCi\text{-}hr}$ is the gamma ray dose rate constant of the radioisotope, $\rho \left(\frac{g}{cm^3}\right)$ is the density of the tissue, $\overline{g}$ (cm) is the average geometry factor, and $c(t) \frac{\mu Ci}{g}$ is a function that describes the variation of the concentration of the radionuclide in the organ with time t (hours) after administration.

Although used successfully for many years, the above method for internal dose calculations gradually appeared to be insufficient and not general enough. This feeling grew in the late 1960s with the rapid development of nuclear medicine and the introduction of many new radiopharmaceuticals containing radionuclides that show rather complex decay schemes involving internal conversion electron, Auger electron, and characteristic X-ray emissions. It appeared at this time, as pointed out by Smith,[3] that the separation of the radiations into the penetrating and nonpenetrating categories is often artificial and inappropriate, since, for many of the new radiopharmaceuticals, the line between these two kinds of radiations seems to be very vague, and the equations of the kind (Equation 3.8.1) to be solved were difficult to manipulate.

Therefore, a more general and convenient approach was needed. Based on the "absorbed fraction" concept first used by Ellett et al.,[4,5] Loevinger and Berman[6] introduced in 1968 a generalized formalism for internal dose calculations applicable to all radiations and radionuclides. The Medical Internal Radiation Dose (MIRD) Committee of the Society of Nuclear Medicine adopted this formalism as the basis of its work and issued a set of publications[7-13] under the name "nm/mird pamphlets" describing this formalism and presenting the necessary biological and physical information needed for internal dose calculations. These publications were reviewed and complemented in a recent new series by the same committee.[14,15] We will hence refer to the new method of internal dosimetry as the MIRD technique.

The purpose of this section is to familiarize the reader with the basic principles and applied details of the MIRD technique for internal dosimetry. We will skip most of the details of derivation of these techniques which are mainly of theoretical interest and present the material in a way applicable by the practicing radiopharmacist, physician, and health physicist. In its present modified form and with the use of the new nm/mird pamphlets,[14,15] the MIRD technique reduces the internal dose calculations in most cases to a few simple arithmetic operations, as explained in Section 3.8.3. To gain competence in internal dosimetry, however, one must be acquainted with the very basic aspects of the theory which are given in Section 3.8.2. To clarify the concepts and illustrate the techniques, numerical examples are given along with the theoretical presentation. The reader who is already familiar with the principles of the MIRD technique and is mainly interested in the use of the new simplified method[15] can skip Sections 3.8.1 and 3.8.2.1 to 3.8.2.11 and begin with Section 3.8.2.12. In any case, we urge the reader to have the nm/mird pamphlets in front of him while reading this "guide chapter" referring to the relevant pamphlet and paragraph corresponding to the material discussed.

3.8.2. Basic Quantities and Units
3.8.2.1. The Radiation Sources
The principal types of radiations involved in clinical nuclear medicine, radiopharmacy, and medical radiation protection (at which the MIRD dosimetry technique is aimed) are the primary α, β, and γ radiations, and the secondary internal conversion electrons, Auger electrons, and characteristic X-rays. These radiations were formally separated into the nonpenetrating group, e.g., electrons, positrons, α particles, and very soft (<10 KeV) X-rays and the penetrating radiations, e.g., the gamma and X-ray radiations.

3.8.2.2. The Activity – (A)
The activity of a radioactive material is a measure of the transformation rate of its radionuclides. The activity is measured in curie (Ci) units; 1 Ci is equal to 3.7×10^{10} disintegrations per second. In nuclear medicine, radiopharmacy, and radiation protection, the more frequently used units of activity are the microcurie (10^{-6} Ci, 3.7×10^4 disintegrations per second), or the millicurie (10^{-3} Ci, 3.7×10^7 disintegrations per second).

3.8.2.3. The Activity Concentration – (C)
The activity concentration is the activity per unit mass. It is measured in microcuries per gram (μCi/g) units.

3.8.2.4. The Energy
The energy of the individual particles and photons is measured in megaelectronvolt (MeV) units; 1 MeV is equal to 1.6×10^{-6} erg. Gamma-ray and characteristic X-ray photons, as well as internal conversion electrons and Auger electrons, are monoenergetic. Beta particles (electrons and positrons) are emitted with a continuous energy distribution from zero up to a maximum energy characteristic of the specific radionuclide. For dosimetry purposes, the average energy of these particles is used. Data on the energies of the gamma radiations, internal conversion electrons, characteristic X-rays, and the maximal and average β particle energies have been published by the Society of Nuclear Medicine in the nm/mird pamphlet No. 10.[14] Table 3.8-1 is an example of the data for [113m]In as it is presented in the above publication (Indium 113m is an isomeric state of [113]In decaying with the emission of a single photon of 0.3916 MeV and with a half-life of 99.4 min).

The input data table refers to the energy and fractional frequency of the radiation emitted in the course of the primary nuclear transformation. The nuclear data in the last column of the input data table refer to the parameters that determine the fractional frequencies,   (the mean number of photons per particles per disintegration) of the different internal conversion electrons related to the corresponding gamma rays. Thus, K/L = 5.30 means that 5.3 times more K conversion electrons than L conversion electrons

Table 3.8-1
DECAY SCHEME AND DATA FOR INTERNAL
DOSIMETRY OF INDIUM-113m[14]

INDIUM-113M
ISOMERIC LEVEL DECAY

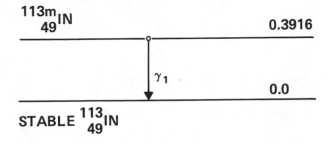

Input Data

(49 Indium-113m, half-life = 99.4 min, decay mode — isomeric level)

Transition	Mean number/ disinte- gration	Tran- sition energy (MeV)	Other nuclear data
Gamma 1	1.000	0.3916	M4, AK = 0.430 K/L = 5.30 K/M = 4.30

Output Data

(49 Indium-113m, half-life = 99.4 min, decay mode — isomeric level)

Radiation	Mean number/ disinte- gration n_i	Mean energy/ particle $\bar{E}_1$ (MeV)	Equi- librium dose constant Δ_i (g-rad/μCi-hr)
Gamma	0.6206	0.3916	0.5178
K Int con elect	0.2668	0.3637	0.2067
L Int con elect	0.0503	0.3877	0.0415
M Int con elect	0.0620	0.3910	0.0516
K Alpha-1 X-ray	0.1243	0.0242	0.0064
K Alpha-2 X-ray	0.0632	0.0240	0.0032
K Beta-1 X-ray	0.0324	0.0272	0.0018
K Beta-2 X-ray	0.0065	0.0279	0.0003
L X-rays	0.0325	0.0032	0.0002
KLL Auger elect	0.0274	0.0201	0.0011
KLX Auger elect	0.0110	0.0233	0.0005
LMM Auger elect	0.2715	0.0025	0.0014
HXY Auger elect	0.6846	0.0006	0.0009

Note: Reference — Nuclear Data B5, 195 (1971).

From nm/mird Pamphlet No. 10, Society of Nuclear Medicine, New York, 1975, 66.

are produced in the partial internal conversion of the 0.3916 MeV gamma-rays of ^{113m}In. Ak = 0.43 means that the ratio of the frequency per disintegration of the K conversion electrons to that of the unconverted (0.3916 MeV) gamma-rays is 0.43. The information presented in the last column of the input data table together with additional data from the reference quoted at the bottom of the table is used to precisely calculate the energies, E_i, and fractional frequencies, n_i, of the internal conversion electrons (INT. CON ELECT) and Auger electrons (KLL AUGER ELECT, KLX AUGER ELECT, etc.) associated with the corresponding gamma-ray. These are presented in the output data table. Details on the methods to calculate these fractional frequencies are given in Appendices B and C of nm/mird pamphlet 10.[16]

3.8.2.5. The Absorbed Dose — (D)

The absorbed dose (frequently referred to in brief terms as the dose) is the energy absorbed in a unit mass of material (in our case, human tissue). The unit of absorbed dose is the rad (abbreviation of radiation absorbed dose). One rad equals 100 erg/gram of material, e.g., if 10^4 erg is uniformly absorbed in an organ of 200-g mass, the dose will be $10^4/(100 \times 200) = 0.5$ rad, etc. The rad is a universal unit applicable to all types of radiations, i.e., α-, β-, γ- and X-rays as well as to other charged particles and neutrons. Due to the inhomogeneity of the biological material composing human tissues, the concentration of a radionuclide may vary in different parts of the same organ or tissue. Therefore, usually only the average dose to an organ as a whole is calculated.

The absorbed dose to an organ will depend on the activity of the radiochemical administered into the body, its physical state and chemical composition, the energy released per disintegration, the type of radiation emitted (particulate or electromagnetic), and the metabolic route of the radiochemical (i.e., the distribution of the radiochemical in the body, and the duration of its stay in a particular organ). These biochemical factors are the limiting factors today in computing reliable estimates of the absorbed dose, since for many radiochemicals and radiopharmaceuticals, they must be evaluated on the basis of very limited data from animals and patients.

3.8.2.6. The Absorbed Dose Rate —(R)*

The absorbed dose rate is the absorbed dose per unit time. For most clinical purposes, the unit of absorbed dose rate used is the rad per hour (rad/hr). For radiation protection dose calculations, the rad per day, rad per week, and rad per year are used frequently.

3.8.2.7. The Dose Equivalent — (DE)

In order to assess the biological effects of ionizing radiations mainly for radiation protection related issues, a special unit called the rem (rad equivalent man) has been defined. The introduction of this unit was necessary since a dose of one type of radiation may cause a much larger biological effect than the same dose of a different type of radiation. In particular, the effects observed with a given dose, in rads, of neutrons, alpha radiation, and other heavy charged particles are significantly enhanced in relation to effects of the same dose of gamma- and X-radiation. The equivalent dose in rem units is equal to the dose in rad units multiplied by the relative biological effectiveness (RBE) related to the microdistribution of the radiochemical in the tissue and the linear energy transfer of the radiation. For beta-, gamma-, and X-ray radiations, the RBE is one. For this reason, the rem unit is not used now in radiopharmaceutical dosimetry since this field presently involves only beta/gamma emitters. The reader interested in more details on the rem unit and in the relative biological effectiveness can find them in the literature[16] and in other sections of this handbook.

*Caution: Elsewhere, R is used as the symbol for the unit of exposure, the roentgen.

3.8.2.8. The Special Energy Unit – g-rad

A logical way to calculate the average dose to an organ is to determine first the total energy absorbed by the organ and divide this quantity by the mass of the organ. Since the absorbed dose is measured in rad units and the mass in grams, the unit of the total energy absorbed in the organ is the gram-rad (g-rad). One g-rad is equal to 100 erg. This amount of energy is referred to as the integral dose (not to be confused with the cumulated dose measured in rad units).

3.8.2.9. The Cumulated Activity – $\widetilde{A}$

The total dose to the body or an organ is determined by the dose rate integrated over the length of time the irradiation is maintained. The dose rate is not constant in time. The change in the dose rate with time is due to the physical decay of the radionuclides and the biological redistribution or excretion of the radiochemical from the body. The total dose to a tissue is determined by the energy absorbed in this tissue from the moment of intake of the activity to its total clearance from the body. If $A(t)$ is the function that describes the variation of the activity with time in a certain tissue, then the contributions of this activity to the cumulated radiation doses in this tissue and any other nearby tissue will be obviously proportional to the integral $\widetilde{A}(t) = {_0}\int^t A(t)dt$, where t is the time elapsed from administration. $\widetilde{A}(t)$ is called the "cumulated activity" and is measured in units of microcurie-hour (μCi-hr). The concept of cumulated activity is a very basic one in the dosimetry of internal emitters, and it is important to have a clear understanding of it. In Section 3.8.2.12.3, we will present a more detailed explanation of this quantity and demonstrate by numerical examples how it can be calcualted.

3.8.2.10. The Equilibrium Absorbed Dose Rate Constant – Δ

The dose to an organ from a given distribution of a radiochemical in the body is a function of many physical, geometrical, and biological factors, e.g., the types of the radiations emitted (particulate or electromagnetic or both); the initial uptake of the activity by the organs; the excretion rate (biological half-life); the size and shape of the organ for which the dose is calculated; and its geometrical position relative to organs containing the radioactivity.

To simplify and generalize dose calculations, these factors can be divided into two main groups: factors related to the energy emission process and factors related to the absorption of the energy and its transfer from site to site in the body. The second group is described and discussed in Section 3.8.2.11.

To evaluate the factors involved in the energy emission process, we will determine the total energy emitted by one unit of activity (1 μCi) of a certain radionuclide in a unit time (1 hr). We assume that in each disintegration, j types of radiation with average energies $\overline{E}_i$ (MeV) and fractional frequencies n_i (per disintegration) are produced. The energy emitted by the ith type of radiation in 1 hr will be given by

$$\Delta_i = 3.7 \times 10^4 \left(\frac{\text{disintegrations}}{\text{sec-}\mu\text{Ci}}\right) \times 3600 \left(\frac{\text{sec}}{\text{hr}}\right) \times n_i \left(\frac{\text{photons/particles}}{\text{disintegrations}}\right) \times \overline{E}_i \left(\frac{\text{MeV}}{\text{photon/particle}}\right) \times$$

$$1.6 \times 10^{-6} \left(\frac{\text{erg}}{\text{MeV}}\right) = 2.134 \times 10^2 \ n_i \overline{E}_i \ \text{erg/}\mu\text{Ci-hr} \tag{3.8-2}$$

and since 1 erg = 10^2 g-rad

$$\Delta_i = 2.134 n_i \overline{E}_i \ \frac{\text{g-rad}}{\mu\text{Ci-hr}} \tag{3.8-3}$$

The rate of energy released by 1 μCi of this radionuclide is the sum of the contributions from all the j types of radiations, i.e.,

$$\Delta = \sum_{i=1}^{j} \Delta_i = \sum_{i=1}^{j} 2.134 \, n_i \overline{E}_i \frac{g\text{-rad}}{\mu Ci\text{-hr}} \tag{3.8-4}$$

Δ_i and Δ are called the equilibrium absorbed dose rate constants. This name was given to Δ due to the fact that if the organ were infinite in mass or at least very large relative to the range of the radiations involved, we could state (assuming homogeneous distribution) that the energy absorbed per gram of tissue (the dose) is in equilibrium with the energy emitted here. In this case, the dose rate to the organ would be simply

$$R = \frac{\Sigma \Delta_i}{m} \, rad/\mu Ci\text{-hr} \tag{3.8-5}$$

where m is the mass of the organ in grams. Indeed, this is the case when the dose due to particulate radiation only is calculated.

The equilibrium absorbed dose rate constant is a quantity that, once determined for a radionuclide, can be used in any absorbed dose calculations involving this radionuclide. The main difficulty in determining Δ_i values is to gather adequate information on the energies $(\overline{E}_i)$ and relative frequencies (n_i) of the conversion electrons and characteristic X-rays, which, for many radionuclides, contribute a significant part of the dose. In the past, contradictory and inadequate data have been published in the literature regarding these parameters. A recent updated list of $\overline{E}_i$, n_i, and Δ_i values carefully calculated and adopted by the Medical Internal Radiation Dose Committee was published in September 1975 by the Society of Nuclear Medicine, as nm/mird pamphlet No. 10 mentioned above.[14] This publication includes complete information on approximately 120 radionuclides. Table 3.8-2 is an example of the information presented in the above publication for ^{18}F. We will use the data on this material in subsequent paragraphs to illustrate the MIRD dosimetry method; therefore, we will discuss some of the details related to this radionuclide here. The decay scheme in Table 3.8-2 shows that ^{18}F disintegrates with a frequency of 3% by electron capture and with a frequency of 97% by β^+. The input data table gives the relative frequencies and the transition energies of these decay modes. The electron capture transition energy has no consequence for the dose calculation, since this energy is carried away by a neutrino that is ejected from the nucleus during the electron capture process. The maximal energy of the β^+ particles in 0.633 MeV. The output data table shows those parameters that are of importance for the dosimetry — the fractional frequency n_i of the β^+ particles (0.97) and their average energy (0.2496 MeV). After slowing down in the tissue and delivering their kinetic energy, the β^+ particles undergo annihilation in which the mass of a positron together with the mass of a free electron in the tissue are transformed into two gamma photons of 0.511 MeV each. The fractional frequency of these photons is therefore $0.97 \times 2 = 1.94$. Using Equation 9.2-3 we can now calculate the Δ_i values for ^{18}F.

$$\Delta_i \text{ (positrons)} = 2.134 \times 0.97 \times 0.2496 = 0.5157 \frac{g\text{-rad}}{\mu Ci\text{-hr}} \tag{3.8-6}$$

$$\Delta_i \text{ (annihilation)} = 2.134 \times 1.94 \times 0.511 = 2.115 \frac{g\text{-rad}}{\mu Ci\text{-hr}} \tag{3.8-7}$$

These values are listed in the last column of the output data table.

3.8.2.11. The Absorbed Fraction — ϕ

Because of the finite dimensions of the human body and its organs, photons originated in a given organ may escape from it and deposit a part or all of their energy elsewhere. Therefore, not all the energy emitted by a radionuclide within an organ will be absorbed locally. Similarly, not all the energy absorbed in an organ necessarily originates from radiation sources located within its physical boundaries. To establish the relationship

Table 3.8-2
DOSIMETRY DATA FOR FLOURINE-18[14]

FLUORINE-18
ELECTRON CAPTURE AND
BETA-PLUS DECAY

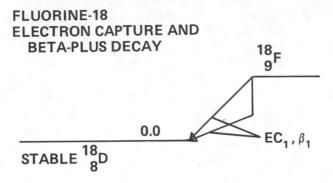

Input Data

(9 Fluorine-18, half-life = 109 min, decay modes — electron capture and beta plus.)

Transition	Mean number/ disintegration	Transition energy (MeV)	Other nuclear data
Electron capture 1	0.0300	1.6550	Allowed
Beta plus 1	0.9700	0.6330[a]	Allowed

[a]Endpoint Energy

Output Data

(9 Fluorine-18, half-life = 109 min, decay modes — electron capture and beta plus.)

Radiation	Mean number/ disintegration n_i	Mean energy/ particle $\overline{E}_i$ (MeV)	Equilibrium dose constant Δ_i (g-rad/ μCi-hr)
Beta plus 1	0.9700	0.2496	0.5157
Annih. radiation	1.9400	0.5110	2.1115

Note: Reference — Lederer, C. M. et al., Table of Isotopes, 6th De.

From nm/mird Pamphlet No. 10, Society of Nuclear Medicine, New York, 1975, 18.

between the amount of energy produced in an organ and the fraction of this energy that is absorbed within this organ and/or in any other organ in the body, the absorbed fraction concept is introduced. To do this, we distinguish between the source organ that contains the radionuclide and the target organ for which the absorbed dose is calculated.

The absorbed fraction ϕ is, by definition, the fraction of energy emitted by the radionuclide in the source organ that is deposited in the target organ; or in mathematical terms, the absorbed fraction ϕ is the energy absorbed in the target organ divided by the energy emitted in the source organ. From the definition, it is obvious that the absorbed fraction is a dimensionless quantity and that its value is always between zero and one. It is also evident that for particulate (non-penetrating radiations), the absorbed fraction will be one when the target organ coincides with the source organ and zero when target organ is different from the source organ. ϕ will be different for the same pair of target-source organs for gamma radiations of different energies. The absorbed fraction is a complicated function of the composition of the source and target organs, their physical dimensions and relative distances, the probabilities of photon interaction for the specific energies, etc. The most useful method to carry out calculations of the absorbed fraction values is the Monte Carlo technique. In a recent work, Snyder and his collaborators[11] used this technique with a mathematical nonhomogeneous phantom consisting of muscle, bone, and lung tissue (the "Standard Man") to generate absorbed fraction tables for about 400 source-target organ pairs covering 12 different energy ranges from 0 to 10 MeV. Their results have been published in the nm/mird pamphlet No. 5.[11] Table 3.8-3 is a typical example of the data presented in the above publication.[11] This table lists the absorbed fractions in different target organs for uniformly distributed radioactive material in the liver as the source organ. The absorbed fractions are given for different gamma ray energies together with their coefficients of variation ($\frac{100\sigma\phi}{\phi}$ or the percentage standard deviation).

From Table 3.8-3 we learn, for example, that the absorbed fraction in the liver (the target organ) for a homogeneously distributed gamma emitter in the liver itself (the source organ) is 0.157 for a gamma energy of 0.500 MeV and 0.132 for a gamma energy of 1.5 MeV; that is, 15.7 and 13.2% of the energy emitted in the liver is absorbed there for the two energies, respectively. If the target organ is the kidney and the source organ is still the liver, the same table tells us that the absorbed fraction in the kidney is 0.386% (0.386×10^{-2}) for the 0.500-MeV gamma rays, etc. The nm/mird pamphlet No. 5 includes many tables similar to Table 3.8-3 presenting the absorbed fractions for uniform sources in many other organs.[11] Additional tables are presented in Section 3.8.3.6 and Appendix A of this handbook.

3.8.2.12. Dose Calculations

The basic quantities and units introduced in Sections 3.8.2.1 to 3.8.2.11 enable us to calculate the dose to any organ provided that the distribution of the activity in the body is known.

3.8.2.12.1. Dose Rate Calculations – S Values

Let us begin with a target organ of mass m_T(g) and a source organ in which a unit activity (1 μCi) of a specific radionuclide is homogeneously distributed. We assume that the radionuclide emits ℓ types of radiations with known energies $\overline{E}_i$ (MeV) and fractional frequencies n_i, (per disintegration). The rate of production of energy by the ith type of radiation in the source organ is Δ_i $\frac{\text{g-rad}}{\mu\text{Ci-hr}}$. The absorbed fraction for photons of energy $\overline{E}_i$ for this particular source-target organ pair is $\phi_i(T \leftarrow S)$.

The energy deposited in the target organ in 1 hr due to the presence of unit activity of the radionuclide in the source organ is, therefore

$$S_{T \leftarrow S} = \sum_{i=1}^{\ell} \frac{\Delta_i \phi_i (T \leftarrow S)}{m_T} \quad \frac{\text{rad}}{\mu\text{Ci-hr}} \tag{3.8-8}$$

The quantity S defined by Equation 3.8-8 represents the absorbed dose rate in the target organ per unit activity in the source organ (and also the total absorbed dose (rad) in the target organ per unit cumulated activity μCi-h in the source organ).

To illustrate the meaning of $S_{T \leftarrow S}$, we present here a quantitative example using again the data for ^{18}F. To begin with, we assume the presence of 1 μCi of ^{18}F in the

Table 3.8-3

ABSORBED FRACTIONS AND COEFFICIENTS OF VARIATION, UNIFORM SOURCE IN LIVER

Photon energy, E (MeV)

Target organ	0.200 φ	0.200 $\frac{100\sigma_\phi}{\phi}$	0.500 φ	0.500 $\frac{100\sigma_\phi}{\phi}$	1.000 φ	1.000 $\frac{100\sigma_\phi}{\phi}$	1.500 φ	1.500 $\frac{100\sigma_\phi}{\phi}$	2.000 φ	2.000 $\frac{100\sigma_\phi}{\phi}$	4.000 φ	4.000 $\frac{100\sigma_\phi}{\phi}$
Adrenals	0.237E−03	16	0.198E−03	25	0.221E−03	23	0.235E−03	29	0.327E−03	26	0.265E−03	31
Bladder	0.389E−03	15	0.358E−03	17	0.521E−03	16	0.515E−03	17	0.735E−03	17	0.474E−03	20
GI (stom)	0.271E−02	5.6	0.280E−02	6.9	0.240E−02	8.4	0.254E−02	8.9	0.210E−02	9.6	0.150E−02	12
GI (SI)	0.100E−01	3.2	0.971E−02	3.7	0.925E−02	4.4	0.904E−02	4.6	0.798E−02	5.1	0.646E−02	5.9
GI (ULI)	0.387E−02	4.9	0.364E−02	5.9	0.320E−02	7.2	0.287E−02	7.8	0.340E−02	7.9	0.232E−02	10
GI (LLI)	0.211E−03	17	0.391E−03	19	0.268E−03	22	0.286E−03	24	0.334E−03	23	0.391E−03	25
Heart	0.570E−02	4.1	0.573E−02	4.9	0.501E−02	5.9	0.498E−02	6.3	0.392E−02	7.3	0.373E−02	8.0
Kidneys	0.390E−02	4.7	0.386E−02	5.8	0.335E−02	7.2	0.294E−02	8.0	0.301E−02	8.2	0.256E−02	9.6
Liver	0.158	0.79	0.157	0.93	0.144	1.1	0.132	1.2	0.122	1.3	0.101	1.5
Lungs	0.923E−02	3.0	0.838E−02	3.8	0.825E−02	4.5	0.696E−02	5.2	0.601E−02	5.8	0.568E−02	6.4
Marrow	0.133E−01	1.7	0.107E−01	2.2	0.935E−02	2.7	0.924E−02	2.8	0.888E−02	3.1	0.720E−02	3.6
Pancreas	0.102E−02	8.2	0.822E−03	12	0.864E−03	14	0.587E−03	18	0.765E−03	17	0.612E−03	19
Sk. (rib)	0.111E−01	2.7	0.867E−02	3.8	0.760E−02	4.7	0.748E−02	5	0.635E−02	5.6	0.527E−02	6.4
Sk. (pelvis)	0.216E−02	6.0	0.182E−02	7.7	0.171E−02	9.1	0.181E−02	9.7	0.207E−02	9.7	0.157E−02	12
Sk. (spine)	0.108E−01	3.1	0.857E−02	3.9	0.694E−02	4.8	0.670E−02	5.3	0.701E−02	5.5	0.521E−02	6.6
Sk. (skull)	0.140E−03	25	0.187E−03	27	0.262E−03	23	0.360E−03	23	0.420E−03	22	0.370E−03	24
Skeleton (total)	0.324E−01	1.7	0.260E−01	2.2	0.231E−01	2.6	0.229E−01	2.8	0.217E−01	3	0.179E−01	3.5
Skin	0.507E−02	3.8	0.561E−02	4.6	0.581E−02	5.2	0.567E−02	5.7	0.560E−02	5.9	0.543E−02	6.4
Spleen	0.645E−03	10	0.619E−03	13	0.633E−03	15	0.396E−03	21	0.336E−03	21	0.503E−03	20
Thyroid									0.238E−04	49		
Uterus	0.136E−03	22	0.130E−03	33	0.127E−03	30	0.729E−04	44	0.920E−04	47		
Trunk	0.413	0.47	0.404	0.51	0.377	0.58	0.351	0.64	0.331	0.69	0.276	0.81
Legs	0.716E−03	14	0.141E−02	10	0.208E−02	9.6	0.248E−02	9.0	0.298E−02	8.5	0.390E−02	7.8
Head	0.867E−03	11	0.106E−02	11	0.179E−02	10	0.225E−02	9.5	0.220E−02	9.7	0.211E−02	10
Total body	0.415	0.47	0.407	0.51	0.381	0.58	0.355	0.64	0.336	0.68	0.282	0.80

Note: The digits following the symbol E indicate the powers of ten by which each number is to be multiplied. A blank in the table indicates that the coefficient of variation was greater than 50%. Total body = head + trunk + legs.

From nm/mird pamphlet No. 5, Society of Nuclear Medicine, New York, 1974, 29.

liver, kidney, and spleen of a "standard man." Using Table 3.8-3 and the absorbed fractions from nm/mird pamphlet No. 5 for the kidney and spleen as source organs, we can determine the ϕ_i values (Table 3.8-4 for the β- and γ-radiations of ^{18}F (remembering that for the nonpenetrating β^+ particles ϕ (T ←S) = 1 when the source and target organs coincide and ϕ (T ← S) = 0 when these two are not identical). From Table 3.8-2 we have for ^{18}F

$$\Delta_i \text{ (positrons)} = 0.5157 \frac{\text{g-rad}}{\mu\text{Ci-hr}}$$

and

$$\Delta_i \text{ (annihilation)} = 2.1115 \frac{\text{g-rad}}{\mu\text{Ci-hr}}$$

Using Equation 3.8-8 and the value of 1809 g for the mass of the liver (mn/mird No. 11), we can now calculate the dose rate to the liver due to unit activity of ^{18}F in the liver, kidney, and spleen

$$S_{L \leftarrow L} = \frac{(0.5157 \times 1) + (2.1115 \times 0.157)}{1809} = 4.68 \times 10^{-4} \text{ rad}/\mu\text{Ci-hr}$$

$$S_{L \leftarrow K} = \frac{(0.5157 \times 0) + (2.1115 \times 0.0229)}{1809} = 2.7 \times 10^{-5} \text{ rad}/\mu\text{Ci-hr}$$

$$S_{L \leftarrow S} = \frac{(0.5157 \times 0) + (2.1115 \times 0.00566)}{1809} = 6.6 \times 10^{-6} \text{ rad}/\mu\text{Ci-hr}$$

In this way, the S values can be calculated for any radioisotope and source-target organ pair from the Δ_i values given in nm/mird pamphlet No. 10[14] and the ϕ_i values presented in nm/mird pamphlet No. 5.[11] The laborious task of calculating S values for most of the isotopes used in nuclear medicine has been recently completed by Snyder and his collaborators[15] from the Oak-Ridge National Laboratories and published in 1975 by the Society of Nuclear Medicine as nm/mird pamphlet No. 11. This publication gives a tabulation of S values for about 120 radionuclides for 400 different source-target organ pairs.[15] Table 3.8-5 lists a part of the S values for ^{18}F as they are presented in the mird pamphlet No. 11.[15] From the table, we can see that $S_{L \leftarrow L}$ (the liver dose rate for ^{18}F unit activity in the liver itself) is $4.7 \times 10^{-4} \frac{\text{rad}}{\mu\text{Ci-hr}}$ $S_{L \leftarrow K}$ (the liver dose rate for unit ^{18}F activity in the kidney) is $2.9 \times 10^{-5} \frac{\text{rad}}{\mu\text{Ci-hr}}$ and $S_{L \leftarrow S}$ (spleen to liver) is $7.8 \times 10^{-6} \frac{\text{rad}}{\mu\text{Ci-hr}}$). The reader can compare these values to the results for the same quantities calculated above. The slight differences for the spleen and the kidney are due to the more updated ϕ values used in nm/mird pamphlet No. 11 issued in 1975. (nm/mird No. 5 was published in 1969. Revised calculations of ϕ have been published by Snyder et al. in ICRP Report 23.[21])

Table 3.8-4
ABSORBED FRACTION VALUES FOR
F-18: THE LIVER AS TARGET ORGAN

Source organ	Radiation	$\phi(\text{T} \leftarrow \text{S})$
Liver	β^+	1
	Annihilation (0.511 MeV)	0.157
Kidney	β^+	0
	Annihilation	0.229×10^{-1}
Spleen	β^+	0
	Annihilation	0.566×10^{-2}

Table 3.6-3

S, ABSORBED DOSE PER UNIT CUMULATED ACTIVITY (RAD/μCi·HR, FLUORINE-18, HALF-LIFE 1.83 HR)

Target organs	Adrenals	Bladder contents	Stomach contents	Intestinal tract SI contents	Intestinal tract ULI contents	Intestinal tract LLI contents	Kidneys	Liver	Lungs	Spleen
Adrenals	4.0E−02	1.7E−06	1.5E−05	9.0E−06	7.3E−06	3.6E−06	9.0E−05	3.5E−05	1.8E−05	4.4E−05
Bladder wall	8.0E−07	1.8E−03	3.3E−06	2.5E−05	1.4E−05	4.1E−05	2.8E−06	2.5E−06	5.4E−07	9.9E−07
Bone (total)	9.9E−06	4.3E−06	4.4E−06	5.7E−06	5.3E−06	7.6E−06	7.2E−06	5.3E−06	7.3E−06	5.6E−06
GI (stom wall)	2.1E−05	2.4E−06	1.4E−03	2.5E−05	2.6E−05	1.3E−05	2.4E−05	1.4E−05	1.3E−05	7.1E−05
GI (SI)	7.0E−06	1.9E−05	1.9E−05	8.7E−04	1.2E−04	6.5E−05	2.0E−05	1.1E−05	2.0E−06	1.0E−05
GI (ULI wall)	7.6E−06	1.7E−05	2.4E−05	1.7E−04	1.5E−03	3.1E−05	2.1E−05	1.9E−05	2.8E−05	1.0E−05
GI (LLI wall)	2.6E−06	5.1E−05	9.2E−06	4.9E−05	2.2E−05	2.3E−03	6.4E−06	2.4E−06	5.7E−07	5.1E−06
Kidneys	8.3E−05	2.8E−06	2.4E−05	2.3E−05	2.0E−05	6.0E−06	2.2E−03	2.7E−05	7.6E−06	6.2E−05
Liver	3.5E−05	2.1E−06	1.5E−05	1.3E−05	1.8E−05	2.5E−06	2.9E−05	4.7E−04	1.7E−05	7.8E−06
Lungs	1.7E−05	3.6E−07	1.4E−05	2.4E−06	2.5E−06	7.5E−07	6.9E−06	1.7E−05	6.2E−04	1.6E−05
Marrow (red)	1.8E−05	9.1E−06	7.6E−06	1.8E−05	1.6E−05	2.3E−05	1.8E−05	7.8E−06	9.2E−06	8.2E−06
Oth tiss (musc)	1.1E−05	1.3E−05	1.0E−05	1.1E−05	1.1E−05	1.2E−05	1.0E−05	8.1E−06	9.7E−06	1.1E−05
Ovaries	4.0E−06	4.5E−05	3.7E−06	6.3E−05	9.2E−05	1.1E−04	8.4E−06	1.5E−06	1.1E−06	7.7E−06
Pancreas	6.1E−05	1.9E−06	1.3E−04	1.5E−05	1.3E−05	5.4E−06	4.8E−05	3.5E−05	2.0E−05	1.4E−04
Skin	5.0E−06	4.7E−06	4.3E−06	3.9E−06	4.0E−06	4.3E−06	4.8E−06	4.4E−06	5.0E−06	4.6E−06
Spleen	4.7E−05	1.7E−06	6.9E−05	1.1E−05	8.9E−06	6.9E−06	6.0E−05	7.3E−06	1.5E−05	3.8E−03
Testes	5.2E−07	3.7E−05	1.9E−07	2.6E−06	3.8E−06	1.5E−05	1.1E−06	8.5E−07	1.8E−07	7.0E−07
Thyroid	1.4E−06	7.0E−08	1.1E−06	2.9E−07	3.1E−07	1.3E−07	6.9E−07	1.3E−06	8.8E−06	1.1E−06
Uterus (nongrvd)	2.7E−06	1.1E−04	6.5E−06	6.3E−05	3.2E−05	4.0E−05	6.1E−06	3.2E−05	7.0E−07	3.4E−06
Total body	2.0E−05	1.4E−05	1.5E−05	2.0E−05	1.7E−05	1.8E−05	2.0E−05	2.0E−05	1.8E−05	2.0E−05

Note: Decay data revised – March 1972. Reference – MIRD Pamphlet No. 10. Date of issue – May 13, 1975.

From nm/mird pamphlet No. 11, Society of Nuclear Medicine, New York, 1975, 23–24.

3.8.2.12.2. Dose Rates for Practical Body Distributions

With the above calculations understood, we can now proceed to calculate the dose to an organ for practical body distributions. The S values described in the preceeding section tell us what the dose rate is from the source organ to the target organ when a unit activity is present in the source organ. In a practical case, however, we do not usually know the activity in the source organ. What we do know is the amount of activity that was administered; therefore, we will have to relate the doses in the different organs to the administered activity.

Whenever a radiochemical is introduced into the body (accidentally or deliberately), the actual amount of activity in an organ at a given time after administration will depend on the chemical form of the material, the means of administration, and the metabolic patterns of the individual human being involved (see Section 5.5). Furthermore, the relative amount of radionuclide present in each organ will be different at different times after administration. The variation of the activity with time will be determined by the biological excretion rates from the specific organ and the physical half-life of the radionuclide. To calculate the dose rate to a target organ at a time t, let use assume that a unit activity (1 μCi) of the radionuclide was administered and that after time t (hours) the fraction of this unit activity present in the jth source organ is f_j. The dose rate to the target organ due to this activity will be given by

$$R_{T \leftarrow j} = S_{T \leftarrow j} f_j \frac{rad}{\mu Ci_a\text{-hr}} \qquad (3.8\text{-}9)$$

where $S_{T \leftarrow j}$ is the S value for this specific target-source organ pair (the subscript a is added to the microcurie unit to indicate that this is the dose rate per microcurie administered). To illustrate this numerically, let us go back to our previous example using an actual body distribution of ^{18}F. Experimental studies with intravenously administered ^{18}F-5-Fluorouracil (a labeled antitumor agent) indicated the initial body distribution of the drug in animals.[17]

Table 3.8-6 tells us that a few minutes after administration, 31% of the activity of ^{18}F-5-Fluorouracil administered is found in the liver, 6.8% in the kidney, and 0.28% in the spleen. The initial dose rate to the liver from the activities in the liver, kidney, and spleen will, therefore, be (using Tables 3.8-5 and 3.8-6, and Equation 3.8-8).

$$R_{L \leftarrow L} = S_{L \leftarrow L} \; f_L = 4.7 \times 10^{-4} \times 0.31 = 1.457 \times 10^{-4} \; \frac{rad}{\mu Ci_a\text{-hr}}$$

$$R_{L \leftarrow K} = S_{L \leftarrow K} \; f_K = 2.9 \times 10^{-5} \times 0.068 = 1.972 \times 10^{-6} \; \frac{rad}{\mu Ci_a\text{-hr}}$$

$$R_{L \leftarrow S} = S_{L \leftarrow S} \; f_S = 7.8 \times 10^{-6} \times 0.0028 = 2.186 \times 10^{-8} \; \frac{rad}{\mu Ci_a\text{-hr}}$$

Table 3.8-6
INITIAL DISTRIBUTION OF
F^{18}-5-FLUOROURACIL
IN RATS[17]

Organ	f_{jo} Initial organ activity (μCi/μCi administered)
Liver	0.3100
Kidney	0.0680
Spleen	0.0028

3.8.2.12.3 Cumulative Activities and Doses

The radiation damage to the human tissue is ultimately proportional to the total absorbed dose and is almost independent of dose rate (for total doses below about 100 rad). We are, therefore, more interested in the total dose than the dose rate. To calculate the total doses from the dose rates, we must integrate over time. The dose to a target organ due to the jth source organ (from the time of administration to a time t[hours]) will be given by

$$D_{T \leftarrow S} = {}_0\int^t R dt = S_{T \leftarrow j} {}_0\int^t f_j(t) \, dt = S_{T \leftarrow j} \tilde{A}_j(t) \text{ rad} \qquad (3.8\text{-}10)$$

where

$$\tilde{A}_j(t) = {}_0\int^t f_j(t) \, dt \qquad (3.8\text{-}11)$$

$\tilde{A}_j(t)$ is called the cumulated activity in the jth organ and is measured in μCi-hr units (per microcurie administered). Note that $A_j(t)$ is equivalent to that amount of activity (microcurie) which, if present constantly in the source organ j, will deliver to the target organ, in 1 hr, the same dose that is practically delivered (from the time of administration to time t) by the amount of activity that actually reaches the source organ following the administration of a unit activity (1 μCi) to the body.

When an effective half-life of the radionuclide in the source organ can be established, and if the amount of time required for uptake in the source organ is very short, $\tilde{A}_j(t)$ is simply calculated

$$\tilde{A}_j(t) = {}_0\int^t f_{jo} \, e^{-\lambda_e t} \, dt \; \mu\text{Ci-hr}/\mu\text{Ci administered} \qquad (3.8\text{-}12)$$

where f_{jo} is the initial fraction of the administered activity in the jth organ and $\lambda_e = \frac{\ln 2}{T_e}$, T_e being the effective half-life of the material in the source organ. The integral in Equation (3.8-12) is readily evaluated to get

$$\tilde{A}_j(t) = 1.44 \, T_e f_{jo} \, (1 - e^{-\lambda_e t}) \; \mu\text{Ci-hr}/\mu\text{Ci administered} \qquad (3.8\text{-}13)$$

We are usually interested in estimating the cumulated dose to the target organ from the time of administration to total clearance from the body. To do this, we integrate from 0 to ∞ and call $\tilde{A}_j(\infty)$ simply $\tilde{A}_j$. From Equation 3.8-13 to is obvious that

$$\tilde{A}_j = 1.44 \, T_e \, f_{jo} \; \mu\text{Ci-hr}/\mu\text{Ci administered}$$

When an effective half-life in the source organ cannot be established and/or when the uptake time is significant, the evaluation of the integral $\int f_j(t) dt$ may be more complicated. Methods for calculating cumulated organ activities for 60 different radionuclides have been described elsewhere in this book[18] (see Section 6.3). In the absence of any estimates of the retention times of the material in the body, we may assume that the effective half-life in each organ is the same as the physical half-life. This will lead to an overestimation of doses, but we still remain on the safe side.

3.8.2.12.4. The Total Dose to the Target Organ

We have learned how to evaluate the dose to a target organ due to the activity in a source organ after administration of a unit activity of radionuclide to the body. The target organ is actually exposed to radiations from all source organs (including itself). The total dose to the target organ due to administration of a unit activity will therefore be

$$D_T = \sum_j S_{T \leftarrow j} \tilde{A}_j \text{ rad/}\mu\text{Ci administered} \qquad (3.8\text{-}14)$$

To complete the numerical example, we can now estimate the doses to the liver due to the activity in the liver, kidney, and spleen following the administration of 1 μCi ^{18}F-5-Fluorouracil. The distribution studies[17] showed that the effective half-lives of this compound in the above organs are as shown in Table 3.8-7.

To calculate the total dose to the liver due to the activity in itself and these other two organs, we generate the following data table using Tables 3.8-5 to 3.8-7. The dose to the liver due to the activity in the three organs can now be calculated by Equation 3.8-14.

$$D_{Liver} = S_{L \leftarrow L} \tilde{A}_L + S_{L \leftarrow K} \tilde{A}_K + S_{L \leftarrow S} \tilde{A}_S$$

$$= (4.7 \times 10^{-4} \times 0.232) + (2.9 \times 10^{-5} \times 0.0793) + (7.810^{-6} \times 0.0032) \text{ rad}$$

$$= 1.11 \times 10^{-4} \text{ rad (per } \mu\text{Ci administered)} \qquad (3.8\text{-}15)$$

In an actual diagnostic procedure with ^{18}F-5-Fluorouracil, an activity of the order of 1 mCi will be administered. The dose to the liver then will be of the order of $1.11 \times 10^{-4} \times 10^3 = 0.11$ rad $= 110$ mrad.

3.8.3. Calculating the Dose in Steps

To sum up the MIRD technique, we can now formulate the following sequence of steps to calculate the dose to each target organ following the administration of Q μCi of a radiochemical:

1. From estimation or detailed body distribution studies, evaluate the initial fractions f_{j0} of the administered activity in each organ of interest.

2. Estimate the effective half-lives T_{ej} of the material in each organ.

3. For each target organ, generate a table of f_{j0}, T_{ej}, $\tilde{A}_j$, and S_j values for the different source organs (see Table 3.8-8, above) using nm/mird pamphlet No. 11[15] for

Table 3.8-7
HALF-LIVES OF
[18]F-5-FLUOROURACIL
IN THE RAT[17]

Organ	Liver	Kidney	Spleen
Effective half-life (hr)	0.52	0.81	0.79

Table 3.8-8
DATA FOR DOSE CALCULATIONS FOR [18]F
IN THE LIVER, KIDNEY, AND SPLEEN

Liver as Target Organ

Organ j	f_{j0}	T_{ej}(hr)	$\tilde{A}_j(\mu\text{Ci·hr})$ $= 1.44 f_{j0} T_{ej}$	$S_{T \leftarrow j}(\frac{rad}{\mu Ci \cdot hr})$
Liver	0.31	0.52	0.232	4.7×10^{-4}
Kidney	0.068	0.81	0.0793	2.9×10^{-5}
Spleen	0.0028	0.79	0.0032	7.8×10^{-6}

the S values. (To calculate $\widetilde{A}_j$ in cases with more complex excretion patterns, see Section 3.8.2.12).

   4. Calculate the dose to each target organ per unit activity administered

$$D_T = \sum_j S_{T \leftarrow j} \widetilde{A}_j \frac{rad}{\mu C_{i_a}}$$

   5. Multiply each value in 4 by $Q(\mu Ci)$ to get the total doses to each organ.

3.8.4. Limitations

Although the mird publications provide us with a simple and elegant way to calculate radiation doses from biologically distributed radionuclides, we must remember the limitations inherent in the data used. The first and most obvious limitation is that the S tables published in nm/mird pamphlet No. 11[15] allow us to calculate only the dose averaged over a target organ per unit cumulated activity distributed uniformly in a source organ. When this is not the case, the use of these S values gives us only an approximation of the dose. Second and not less important, the S values in nm/mird pamphlet No. 11 are calculated for the "standard man." The masses of the organs of this man (as given in nm/mird pamphlet No. 11[15]) are the typical average values for a 70-kg man and, therefore, can hardly be used for dose calculations in, for example, a 5-year-old child or for a 120-kg patient, etc. There are also other less significant approximations and interpolations involved in the use of the S values. Some of these are explained in detail in the introduction to nm/mird pamphlet No. 11.[15]

3.8.5. Recent Developments

As explained in the last section, values of S and ϕ are needed for phantoms different in size from the "standard man." Recently, Poston and his group[19] developed phantoms for newborn, 1-, 5-, 10-, and 15-year-old humans. These will be used to generate S values applicable to dose calculations in such humans. A pregnant woman model used to provide ϕ values for the developing fetus during its early growth has been developed by Cloutier et al.[20] A comprehensive collection of anatomical, physiological, and analytical data has recently been published by Snyder et al.,[21] including variations in biological parameters with age. Adjustments to the MIRD dose calculations for various age, sex, or anatomical groups can make use of this biological compendium[21] together with the Monte Carlo calculational approach.

The problem of dose calculations with nonuniform activity distribution in the source organ has not yet been solved, although Blau[22] has recently discussed this problem for the particular case of some bone-seeking radionuclides.

The most difficult problem that still faces people engaged in dosimetry of internal emitters remains, however, the lack of good and reliable biological data on the distribution and retention of radiochemicals in the human body, especially in diseased patients. The solution of this problem will mark real progress in radiation dose estimates for biologically distributed radiochemicals.

ACKNOWLEDGMENTS

The author wishes to thank Dr. Walter Wolf, Director of the Radiopharmacy Program, U.S.C. School of Pharmacy for reading the manuscript and for his valuable remarks. The author acknowledges an IAEA fellowship for 1975/1976. This work was supported, in part, by Cancer Grants NCI-CA-14089 and CA-19438.

REFERENCES

1. **Marinelli, L. D., Quimby, E. H., and Hine, G. J.,** Dosage determination with radioactive isotopes. II. Practical considerations in therapy and protection, *Am. J. Roentgenol. Radium Ther.,* 59, 260–280, 1968; *Nucleonics,* 2(4), 56–66, 1968; 2(5), 44–49, 1968.
2. **Loevinger, R., Holt, J. G., and Hine, G. J.,** Internally administered radioisotopes, in *Radiation Dosimetry,* Hine, G. J. and Brownell, G. L., Eds., Academic Press, New York, 1956, chap. 17.
3. **Smith, E. M.,** Internal dose calculations for Tc – 99m, *J. Nucl. Med.,* 6, 231–251, 1965.
4. **Ellett, W. H., Callahan, A. B., and Brownell, G. L.,** Gamma-ray dosimetry of internal emitters. I. Monte Carlo calculations of absorbed dose from point sources, *Br. J. Radiol.,* 37, 45–64, 1964.
5. **Ellett, W. H., Callahan, A. B., and Brownell, G. L.,** Gamma-ray dosimetry of internal emitters. II. Monte Carlo calculations of absorbed dose from uniform sources, *Br. J. Radiol.,* 38, 541–544, 1965.
6. **Loevinger, R. and Berman, M.,** A formalism for calculation of absorbed dose from radionuclides, *Phys. Med. Biol.,* 13, 205–217, 1968.
7. **Loevinger, R. and Berman, M.,** A schema for absorbed dose calculations for biologically distributed radionuclides, *J. Nucl. Med.,* Suppl. No. 1, Pamphlet No. 1, 1968.
8. **Berger, M. J.,** Energy deposition in water by photons from point isotropic sources, *J. Nucl. Med.,* Suppl. No. 1, Pamphlet No. 2, 1968.
9. **Brownell, G. L., Ellett, W. H., and Reddy, A. R.,** Absorbed Fractions for Photon Dosimetry, *J. Nucl. Med.,* Suppl. No. 1, Pamphlet No. 3, 1968.
10. **Dillman, L. T.,** Radionuclide decay schemes and nuclear parameters for use in radiation dose estimation, *J. Nucl. Med.,* Suppl. No. 2, Pamphlet No. 4, 1969; Suppl. No. 4, Pamphlet No. 6, 1970.
11. **Snyder, W. J., Ford, M. R., Warner, G. G., and Fisher, H. L., Jr.,** Estimates of absorbed fractions for monoenergetic photon sources uniformly distributed in various organs of an heterogeneous phantom, *J. Nucl. Med.,* Suppl. No. 3, Pamphlet No. 5, 1969.
12. **Berger, M. J.,** Distribution of absorbed dose around point sources of electrons and beta particles in water and other media, *J. Nucl. Med.,* Suppl. No. 5, Pamphlet No. 7, 1971.
13. **Ellet, W. H. and Humes, R. M.,** Absorbed fractions for small volumes containing photon emitting radioactivity, *J. Nucl. Med.,* Suppl. No. 5, Pamphlet No. 8, 1971.
14. **Dillman, L. T. and Von der Lage, F. C.,** *Radionuclide Decay Schemes and Nuclear Parameters For Use in Radiation Dose Estimation,* nm/mird Pamphlet No. 10, Society of Nuclear Medicine, New York, 1975.
15. **Snyder, W. S., Ford, M. R., Warner, G. G., and Watson, S. B.,** *"S" Absorbed Dose Per Unit Cumulated Activity for Selected Radionuclides and Organs,* nm/mird Pamphlet No. 11, Society of Nuclear Medicine, New York, 1975.
16. **International Commission on Radiation Protection (ICRP),** *Recommendations of the ICRP,* ICRP Publication 9, Pergamon Press, Oxford, Oxford Pergamon 1966, 115–119.
17. **Shani, J. and Wolf, W.,** Notice of claimed investigational exemption for a new drug – ^{18}F-5-Fluorouracil, private communication, 1976.
18. **Bernard, S. R.,** *Handbook of Radiation Measurement and Protection,* Section 6.3, CRC Press, Cleveland, 1977, in press.
19. **Poston, J. W.,** The effect of body organ size on absorbed dose: there is no standard patient, in *Radiopharmaceutical Dosimetry,* Proc. Symp. at Oak Ridge, April 1976, Cloutier, R. J., Coffey, J. L., Snyder, W. S., and Watson, E. E., Eds., Bureau of Radiological Health Publication (FDA)76-8044, U.S. Government Printing Office, Washington, D.C., 1976.
20. **Cloutier, R. J., Watson, E. E., and Snyder, W. S.,** Dose to the fetus during the first three months from gamma sources in Maternal Organs, in *Radiopharmaceutical Dosimetry,* Proc. Symp. at Oak Ridge, April, 1976, Cloutier, R. J. et al., Eds., Bureau of Radiological Health Publication (FDA)76-8044, U.S. Government Printing Office, Washington, D.C., 1976.
21. **Snyder, W. S., Cook, M. J., Nasset, E. S., Karhausen, L. R., Howells, G. P., and Tipton, I. H.,** *Report of the Task Group on Reference Man,* Int. Comm. on Radiological Protection Report No. 23, Pergamon Press, Oxford, 1975.
22. **Blau, M.,** Problems of dose calculations for technetium-99m bone scanning agents, in *Radiopharmaceutical Dosimetry,* Proc. Symp. at Oak Ridge, April 1976, Cloutier, R. J. et al., Eds., Bureau of Radiological Health Publication (FDA)76-8044, U.S. Government Printing Office, Washington, D.C., 1976.

Chemical Data

4.1. RADIATION CHEMISTRY: AQUEOUS SOLUTIONS OF ORGANIC SOLUTES

Edward B. Sanders

4.1.1. Introduction

The interaction of ionizing radiation with biological systems is a function of a diverse number of parameters. In addition to the inherent complexities of a functioning biological unit, variables such as the type of radiation, the mechanism of the interaction of the ionizing radiation with a particular component of the biological system, and the biological endpoint under study must all be taken into account. For example, for inactivation of an isolated enzyme by radiation, there is an excellent correlation between size of the macromolecule and absorbed dose in rads,[1] which is in accord with the target theory proposed by Lea.[2] Alper[1] points out that, "Since the target volume is defined as that within which a single energy deposition event is sufficient for inactivation, it follows that a second event would represent 'wasted dose'; so it is predictable that, if simple target theory applies, effectiveness of radiation decreases as energy deposition events occur increasingly close together, i.e., as linear energy transfer (LET) increases." This prediction has indeed been confirmed.[3,3a] The observed relationship between relative biological effectiveness (RBE) and LET changes markedly, however, when an intact cell is exposed to ionizing radiation, in that high LET radiation is more effective than low LET radiation when cell death is utilized as the biological endpoint.[4,4a] The suggestion has been made by numerous investigators that this result is the consequence of repair mechanisms within the cell, i.e., that damage leading to rapid cell death is likely to occur only when two sites within the subcellular target, e.g., a nucleic acid, are simultaneously inactivated.[1]

When dose levels are utilized that are considerably below those necessary for complete inactivation of the biological system being irradiated, indirect effects rather than direct interaction with a target molecule assume considerable importance. Indirect effects are mediated through the interaction of ionizing radiation with water, the universal cellular medium, which undergoes well-understood radiolytic cleavage reactions to yield a number of highly reactive radical species. These radicals, in turn, react with cellular components via typical radical reaction pathways to produce lesions within the cell.

The study of the radiation chemistry of dilute aqueous solutions of biologically important molecules has been extensively used as a model to study the indirect effects of ionizing radiation on cellular structures. There is no doubt that the chemical changes that occur in such model systems are the result of the interaction of water radiolysis products with the solute.[5] As a consequence, the assumption that similar chemical alterations brought about by indirect effects are equally important in biological systems seems warranted. This section will discuss the radiolysis of water and the basic mechanisms for the interactions between the radicals thus produced and biomolecules and summarize some kinetic and product data for such interactions.

4.1.2. Radiolysis of Water

Radiolysis of liquid water produces two molecular and four radical species, namely, hydrogen, hydrogen peroxide, the hydrated electron (e^-aq), hydrogen atoms, hydroxyl radicals, and perhydroxyl radicals (Equation 4.1-1).

$$H_2O - \!\!\!\sim\!\!\!\sim\!\!\!\sim \to H_2 + H_2O_2 + e^-_{aq} + H\cdot + OH\cdot + HO_2\cdot \qquad (4.1\text{-}1)$$

The yields of these products are dependent on both the type of ionizing radiation and the pH of the aqueous solution. Radiochemical yields are shown in Table 4.1-1. Yields are given in G values where a G unit is defined as being equivalent to one molecule of the specific material formed or destroyed per 100 eV energy absorbed. The two molecular products formed are of lesser importance to further chemical reactions than the four radical products. The hydrogen molecule is totally unreactive and need not be discussed further. Although hydrogen peroxide is not inert, its rates of reaction are considerably slower in general than the radical products; consequently it is involved only in secondary reactions. The hydrated electron and the hydrogen atom are frequently referred to as "reducing radicals," since they frequently bring about the reduction of dissolved substances; the hydroxyl and perhydroxyl radicals and hydrogen peroxide are described as the "oxidizing products," since they tend to bring about oxidation.[15]

The actual mechanism by which water is radiolyzed can be described as follows. Initially a water molecule absorbs energy from interaction with ionizing radiation to produce an excited water molecule (Equation 4.1-2).

$$H_2O - \!\!\!\sim\!\!\!\sim\!\!\!\sim \to H_2O^* \qquad (4.1\text{-}2)$$

The excited water molecule can then decompose by two distinct pathways: dissociation to give a hydrogen atom and a hydroxyl radical (Equation 4.1-3) or loss of an electron to give e^- and H_2O^+ (Equation 4.1-4).

$$H_2O^* \to H\cdot + OH\cdot \qquad (4.1\text{-}3)$$

$$H_2O^* \to H_2O^+ + e^- \qquad (4.1\text{-}4)$$

The two products from Equation 4.1-4 react with water via diffusion-controlled processes to give e^-aq (Equation 4.1-5a) and H_3O^+ and $OH\cdot$ (Equation 4.1-5b).

Table 4.1-1
RADICAL AND MOLECULAR PRODUCT YIELDS IN IRRADIATED WATER

Radiation	pH	G_{-H_2O}	G_{H_2}	$G_{H_2O_2}$	$G_{e^-_{aq}}$	G_H	G_{OH}	G_{HO_2}	Ref
γ-Rays and fast electrons with energies in the range	0.46	4.45	0.40	0.78	0	3.65	2.90	0.008[a]	6,7
0.1 to 20 MeV	3-13	4.08	0.45	0.68	2.63	0.55	2.72	0.026[b]	8, 7
Tritium β-particles (E_{av} 5.7 keV)	1	3.97	0.53	0.97	0	2.91	2.0		9
32 MeV He	~7	3.01	0.96	1.00	0.72	0.42	0.91	0.05	10
12 MeV He	~7	2.84	1.11	1.08	0.42	0.27	0.54	0.07	10
Polonium α-particles (5.3 MeV)	0.46	3.62	1.57	1.45	0	0.60	0.50	0.11	10
^{10}B $(n,\alpha)^7$ Li recoil nuclei	0.46	3.55	1.65	1.55	0	0.25	0.45		12
Particle with infinitely high LET, extrapolated from results with accelerated ^{12}C and ^{14}N ions	0.46	~2.9	~1.45	~1.45	0	0	0		13

[a]   From Reference 14.
[b]   From Reference 10.

From Spinks, J. W. T. and Woods, R. J., *An Introduction to Radiation Chemistry,* 2nd ed., John Wiley & Sons, New York, 1976. With permission.

$$e^- + H_2O \rightarrow e^-_{aq} \tag{4.1-5a}$$

$$H_2O^+ + H_2O \rightarrow H_3O^+ + OH\cdot \tag{4.1-5b}$$

The hydrogen atom and hydroxyl radical can react further as shown in Equations 4.1-6 to 4.1-9.

$$2OH\cdot \rightarrow H_2O_2 \tag{4.1-6}$$

$$2H\cdot \rightarrow H_2 \tag{4.1-7}$$

$$H_2O_2 + H\cdot \rightarrow HO_2\cdot + H_2 \tag{4.1-8}$$

$$H_2O_2 + OH\cdot \rightarrow HO_2\cdot + H_2O \tag{4.1-9}$$

By combining Equations 4.1-3 to 4.1-5 and 4.1-7, it can be seen that the G value for destruction of H_2O is

$$G_{-H_2O} = 2G_{H_2} + G_H + G_{e^-_{aq}} \tag{4.1-10}$$

while combining Equations 4.1-4, 4.1-6, 4.1-8, and 4.1-9 it can be seen that G is also given by:

$$G_{-H_2O} = 2G_{H_2O_2} + G_{OH} + G_{HO_2} \tag{4.1-11}$$

At low pH, the hydrated electron is rapidly scavenged by hydronium ions (Equation 4.1-12) and is consequently reported as $H\cdot$ in Table 4.1-1 at pH below 1.

$$e^-_{aq} + H_3O^+ \rightarrow H\cdot + 2H_2O \tag{4.1-12}$$

At pH above 4, the perhydroxyl radical dissociates (Equation 4.1-13), while in very basic solutions (pH 11), the hydroxyl radical and hydrogen peroxide also dissociate[15] (Equations 4.1-14 and 4.1-15).

$$HO_2\cdot \rightleftharpoons O_2^- + H^+ \qquad \text{pK 4.88} \tag{4.1-13}$$

$$OH\cdot \rightleftharpoons O^- + H^+ \qquad \text{pK 11.9} \tag{4.1-14}$$

$$H_2O_2 \rightleftharpoons HO_2^- + H^+ \qquad \text{pK 11.6} \tag{4.1-15}$$

Of these pH dependent processes, only the dissociation of the perhydroxyl radical (Equation 4.1-13) is important at physiological pH. This process gives rise to the superoxide radical, O_2^-, the reactions of which have only recently come under intensive study.

4.1.3. Modes of Radical Reactions

Interaction of ionizing radiation with water gives rise to four highly reactive free radicals which, in turn, react with any solute that may be present. Attack of a radical on an organic substrate can occur by two main pathways, namely, abstraction or addition. Abstraction occurs when the radical abstracts an atom (usually hydrogen) from the substrate to leave an unpaired electron on the substrate (Equation 4.1-16),

$$RH + X\cdot \rightarrow XH + R\cdot \tag{4.1-16}$$

whereas addition comprises the addition of the radical to the substrate producing an unpaired electron once again in the substrate (Equation 4.1-17).

$$X\cdot + R-CH=CH_2 \rightarrow R-\overset{\cdot}{C}H-CH_2X \tag{4.1-17}$$

Addition can occur only in molecules that possess unsaturation, i.e, molecules that contain at least one π bond. The position of attack on a given organic substrate will not be random but will be determined by the energy of the radical produced, that is, the reaction will occur most rapidly at a position where the most stabilized radical will be formed. Radicals are highly stabilized by interaction with neighboring unsaturation, to a lesser extent by the presence of heteroatoms (e.g., oxygen and nitrogen) next to the radical center and to some extent by the presence of alkyl groups next to the radical center.

The intermediates produced by both abstraction and addition are themselves radicals and consequently will react further. In addition to abstraction and addition reactions already described, three additional modes of reaction are open to these secondary radicals: dimerization, the combination of two like or unlike radicals to give a neutral product containing both radicals (Equation 4.1-18);

$$R \cdot + R \cdot \rightarrow R{-}R \tag{4.1-18}$$

disproportionation, the combination of two like or unlike radicals to give two neutral molecules with different degrees of unsaturation (Equation 4.1-19);

$$R{-}\overset{\cdot}{C}H{-}CH_3 + R{-}\overset{\cdot}{C}H{-}CH_3 \rightarrow R{-}CH{=}CH_2 + R{-}CH_2{-}CH_3 \tag{4.1-19}$$

and fragmentation, the loss of a portion of the secondary radical to give a neutral molecule and a new radical* (Equation 4.1-20). As can be seen by inspection of Equation 4.1-20,

$$R{-}CH_2{-}CH_2 \cdot \rightarrow R \cdot + CH_2{=}CH_2 \tag{4.1-20}$$

it is the reverse of the addition process. An understanding of the mechanisms of radical reactions is central to an understanding of the radiolysis of aqueous solutions of biomolecules.

The presence of certain radical "scavengers" can significantly alter the course of the reaction of both the primary and secondary radicals. Scavengers are frequently used in in vitro radiation chemical studies to suppress the reaction of one or more primary radicals, e.g., the use of nitrous oxide, which scavenges hydrated electrons (e^-aq) and converts them to hydroxyl radicals. In vivo, however, the concentration of a potential scavenger will, in general, be too low to affect the outcome of the reactions in question, with the possible exception of oxygen. Oxygen, which exists as a triplet in its ground state, acts very much like a radical and adds rapidly to other radicals (Equation 4.1-21).

$$R \cdot + O_2 \rightarrow RO_2 \cdot \tag{4.1-21}$$

It specifically scavenges both H· and e^-_{aq} producing $HO_2 \cdot$ and O_2^-, thereby preventing their reaction with a given substrate.

4.1.4. Kinetics of Primary Water Radiolysis Products with Organic Substrates

The reaction rates of the hydrated electron and the hydrogen atom with simple organic compounds are presented in Table 4.1-2. Reaction of e^-_{aq} with aliphatic molecules generally tends to be slow. Rates of reaction of e^-_{aq} with saturated aliphatic compounds (ethanol, D-glucose, and diethyl ether) are quite low as would be expected, in that e^-_{aq} can undergo facile reaction only by addition to unsaturated linkages. Addition of e^-_{aq} to systems containing carbon-carbon π bonds is appreci-

---

*   It should be noted that complex organic radicals may rearrange to give radicals of altered structure preceding any of the reactions cited.

Table 4.1-2
RATE CONSTANTS FOR REACTIONS OF THE HYDRATED ELECTRON AND HYDROGEN ATOM

Solute	Reaction with e^-_{aq}		Reaction with H	
	Products	Rate constant[a] $(10^7 mol^{-1}sec^{-1})$	Products	Rate constant $(10^7 mol^{-1} sec^{-1})$
Acetaldehyde		350		3.4
Acetic acid	$CH_3COO^- + H$	8.4[b]	(pH 1)	0.02
	$CH_3CO\cdot + OH^-$	9.6[b]		
Acetate ion (pH 10)		<0.1	(pH 7)	0.027
Acetone	$(CH_3)_2\dot{C}-O^-$	590	H addition	0.04
			H abstraction	0.19
Acetonitrile	$CH_3CH=\dot{N} + OH^-$	2.5	$CH_3CH=N\cdot$	0.35
N-Acetylalanine (pH ~ 7)		1.0	(pH 1)	0.8
Acetylene		3500		
Acrylamide	$CH_3\dot{C}HCONH_2$	1800	$CH_3\dot{C}HCONH_2$	1800
Allyl alcohol		<0.1		230
Aniline		<2		180
Benzene	$C_6H_7 + OH^-$	1.3		53
Benzoic acid (pH 5.4)		3300		100
Benzoate ion (pH 5 to 14)		330	(pH 7)	87
Benzophenone	$(C_6H_5)_2C-O^-$	3000		
Benzyl alcohol		13		65
Benzyl chloride		550		
Bromoacetate ion (pH 10)	$Br^- + \cdot CH_2COO^-$	620 (pH 9)	$HBr + \cdot CH_2COO^-$	35
			$H_2 + Br\dot{C}HCOO^-$	<0.2
Bromobenzene		430		
p-Bromophenol		1200		
Chloroacetic acid (pH 1 to 1.5)	$Cl^- + \cdot CH_2COOH$	690	$HCl + \cdot CH_2COOH$	0.008
			$H_2 + Cl\dot{C}HCOOH$	0.018
Chloroacetate ion (pH 7 to 11)	$Cl^- + \cdot CH_2 COO^-$	120	$HCl + \cdot CH_2COO^-$	0.026
			$H_2 + Cl\dot{C}HCOO^-$	0.26
Chlorobenzene		50		
Chloroform	$Cl^- + \cdot CHCl_2$	3000		1.2
Cystamine (RSSR) (pH 7.3)	$RSSR^-$	4000		
Cysteine				
+ Ion (pH 1)		3000	$H_2 + \cdot SCH_2CH(NH_3^+)COOH$	250
Zwitterion (pH 5.5 to 7)		820	$H_2 + \cdot SCH_2CH(NH_3^+)COO^-$	100
	$HS^- + \cdot CH_2CH(NH_3^+)COO^-$		$H_2S + \cdot CH_2CH(NH_3^+)COO^-$	12
			H abstraction from C—H	5
− Ion (pH 11.6)		7.5		
Cystine (RSSR)				
+ Ion (pH 1)			$RSH + RS\cdot$	500
Zwitterion (pH 6.1)	$RSSR^-$	1300		>150
− Ion (pH 10.7 to 12)		300		
Diethyl ether		<1		4.7
Ethanol	$C_2H_5O^- + H$	<0.01		1.7
Ethyl acetate		5.9		0.06
Ethylamine (pH >11)	$CH_3\dot{C}H_2 + NH_3 + OH^-$			
+ Ion (pH <10)	$H + CH_3CH_2NH_2$	0.25		
Ethylene		<0.025		300
Fluoracetate ion (pH ~ 10)		0.12 (pH 7)	$H_2 + F\dot{C}HCOO^-$	0.04
Fluorobenzene		6.5		
Formaldehyde		<1		0.5
Formic acid	$HCOO^- + H$	15[b]	$H_2 + \cdot COOH$	0.11
	$H\dot{C}O + OH^-$		19[b]	

Table 4.1-2 (continued)
RATE CONSTANTS FOR REACTIONS OF THE HYDRATED ELECTRON AND HYDROGEN ATOM

Solute	Reaction with e^-_{aq}		Reaction with H	
	Product	Rate constants[a] (10^7 mol^{-1} sec^{-1})	Products	Rate constant (10^7 mol^{-1} sec^{-1})
Formate ion (pH 9 to 11)	$H\dot{C}O + OH^-$	~0.001		22
D-Glucose		~0.03		4
Glycine		.47	(pH 2)	
+ Ion (pH 3)				1.7
Zwitterion (pH 6.4 to 8.5)	$NH_3 + \cdot CH_2COO^-$	0.7	$H_2 + NH_3^+CH\dot{C}OO^-$	0.008
− Ion (pH 11)		0.18		
Glycylglycine zwitterion (pH 6.4)		25		15
Iodoacetate ion (pH 10)	$I^- + \cdot CH_2COO^-$	1200		
Iodobenzene		1200		
Iodoethane	$I^- + \cdot C_2H_5$	1500		
Methacrylate ion (pH 10)		840		
Methane		<1		
Methane thiol	$SH^- + \cdot CH_3$	1800		
Methanol	$CH_3O^- + H$	<0.001		0.16
Methanol radical	$\cdot CH_2OH \rightarrow CH_2OH^-$	<10		
Nitrobenzene	$C_6H_5NO_2^-$ ($\rightarrow C_6H_5NO_2H$)	3000	$\cdot C_6H_6NO_2$	170
Nitromethane (pH 0 to 6)	$CH_3NO_2^-$	2500	(pH 1)	4.4
− Ion (pH 12)	$CH_3NO_2^-$	660		
Oxalic acid (pH 1.3)	$OCCOOH + OH^-$	2500		0.04
−Monoanion (pH 2.8 to 4)		330		
−Dianion (pH 7 to 10)		3		
Phenol (pH 6.3 to 6.8)		1.8	(pH 7)	420
− Ion (pH ~11)		0.4		
DL-Phenylalanine zwitterion (pH ~ 7)		1300		12
2-Propanol			CH_3COHCH_3	5
Purine (pH 7.2)		1700		
Pyridine (pH 5.5 to 7.3)		300	(pH 7)	6.4
Pyrrole (pH 10.3)		0.06		
Ribose		<1		
Styrene		1300		
Tetrachloroethylene		1300		
Tetracyanoethylene		1500		
Tetranitromethane (pH 5.5 to 7)	$C(NO_2)_3^- + NO_2$	5000	(pH 7)	260
Thiourea		300		
Thiophene		6.5		
Thymine		1800		
Trichloroacetate ion (pH ~10)		1200		
Uracil		1100		300
Urea		0.028		

[a] Hydrated electron rate constants are for neutral or basic solutions and are taken from Reference 17; rate constants for hydrogen atom reactions are from References 18 or 19; values are generally known to ± 25%.

[b] From Reference 16.

From Spinks, J. W. T. and Woods, R. J., *An Introduction to Radiation Chemistry*, 2nd ed., John Wiley & Sons, New York, 1976. With permission.

ably faster. For example, rate constants for the reaction of e^-_{aq} with benzoic acid, benzophenone, nitrobenzene, thymine, and uracil range from 1100 to 3300×10^7 mol^{-1} sec^{-1}. Large rate constants are also observed for substrates containing heteroatoms, such as chlorine or sulfur, in that stabilization of the radical originally formed can be achieved via fragmentation leading to a stable halide atom or thiol radical. Rates of hydrogen atom reactions with organic substrates tend to be large only when a stabilized radical can be produced by abstraction. For instance, compounds, such as acrylamide, allyl alcohol, and cysteine react quite rapidly, giving rise to radicals stabilized by the unsaturated amide group and vinyl group and by a sulfur atom, respectively. Aromatic substrates with heteroatom-containing substitutents, such as benzoic acid, phenol, and nitrobenzene, react rapidly with H $\cdot$ via an addition mode.

Table 4.1-3 presents analogous data for the hydroxyl radical. As expected, the data are similar to the data for H $\cdot$, although the hydroxyl radical can be seen to be somewhat more reactive. Once again, abstraction reactions are rapid for those substrates where a stabilized radical can be formed, such as acrylamide, isobutanol, amines, and formaldehyde. Addition reactions occur facilely when carbon-carbon π bonds are present.

Table 4.1-4 lists data for the reaction of hydroxyl radicals with molecules of biological importance. The rate of reaction of OH $\cdot$ with nucleic acids and nucleic acid bases is extremely rapid, since these molecules are all cyclic structures with carbon-carbon π bonds and heteroatoms within the ring. As a consequence, addition reactions are allowed, and the resulting radical is stabilized by the nitrogens within the ring. Nonaromatic amino acids, such as glycine and L-alanine, tend to exhibit slower rates of reaction, and the mode of reaction is unquestionably abstraction. Aromatic amino acids, such as L-phenylalanine, L-tyrosine, and L-tryptophan, exhibit much more rapid rates of reaction due to both the ability of the aromatic ring to undergo addition reactions and the stabilization of a radical formed by abstraction imparted by the unsaturated aromatic system. Reaction of OH $\cdot$ with free sugars, such as D-glucose and D-ribose, tends to be slow. Conversion of a free sugar to a disaccharide increases the rate of reaction as can be seen for D-cellobiose and D-lactose.

The 1969 discovery of an enzyme, superoxide dismutase, which can function as a defense toward superoxide anion, has stimulated considerable recent interest in the interaction of superoxide anion, O_2^-, with biologically important molecules.[21] Recent work has indicated that O_2^- is inert to most simple biomolecules, such as formate, acetate, imidizole, pyruvate, succinate, etc. Rates of reaction with more complicated molecules, e.g., ferricytochrome C and ascorbate, were found to be 2.6×10^5 mol^{-1} sec^{-1} at pH 9.0 and 1.51×10^5 mol^{-1} sec^{-1} at pH 9.9, respectively.[22] Apparently, the rates of reaction of superoxide anion with biomolecules are considerably slower than other water radiolysis products. Nevertheless, O_2^- is known to be extremely powerful nucleophile,[23] a property not shared by the radical species already discussed. Further research may show this property to have considerable importance to the effect of ionizing radiation on biological systems.

4.1.5. Mechanisms for the Degradation of Aqueous Solutions of Biomolecules by Ionizing Radiation

4.1.5.1. Amino Acids

The radiolysis of amino acids has been extensively studied.[24,24a] Glycine (Substance 1), the simplest amino acid, serves as a useful model for reactions occurring at that carbon atom bonded to the amino group and the carboxyl group. Glycine exists primarily as its zwitterion (Substance 2) at physiological pH (Equation 4.1-22).

$$NH_2-CH_2-COOH \; (1) \rightleftharpoons \; {}^+NH_3-CH_2-COO^- \; (2) \qquad (4.1\text{-}22)$$

Table 4.1-3

RATE CONSTANTS FOR REACTIONS OF THE HYDROXYL RADICAL AND OXIDE ION

Reactant	Products or reaction type	Rate constant[a] (10^7 mol^{-1} sec^{-1})
Acetic acid (pH 1 to 2)	$\cdot CH_2COOH + H_2O$	1.65
Acetate ion (pH >6)	$\cdot CH_2COO^- + H_2O$	7.5
Acetone	$\cdot CH_2COCH_3 + H_2O$	8.5
Acetonitrile	$CH_3C(OH)=\dot{N}$ or $CH_3C=NOH$	0.55
N-Acetylalanine	Abstraction	46
Acrylamide	Addition	450
Aniline	Addition	790
Benzene	Addition	530
Benzoic acid (pH 3)	Addition	400
Benzoate ion (pH >6)	Addition	540 (<0.6)
Benzyl alcohol	Addition	840
Bromoacetate ion	$\cdot CHBrCOO^- + H_2O$	4.4
1-Butanol	Abstraction	390
2-Butanol	Abstraction	260
Isobutanol	Abstraction	340
t-Butanol	Abstraction	51
Chloroacetic acid	$\cdot CHClCOOH + H_2O$	5.6
p-Chlorobenzoate ion	Addition	500
Chlorobenzene	Addition	620
Chloroform	$\cdot CCl_3 + H_2O$	1.4
Cysteamine (pH 1.4 to 9)		1470
Cysteine (pH 1 to 2)	$\cdot SCH_2CH(NH_3^+)COOH + H_2O$	~900
Cystine (pH2) (RSSR)	$RSOH + RS\cdot$	480
Diethyl ether	$\cdot C_2H_4OC_2H_5 + H_2O$	235
Ethanol	$\cdot C_2H_4OH + H_2O$	180 (95)
Ethyl acetate	Abstraction	32
Ethylamine (pH > 11)	$H_2O + \cdot CH_3CH_2\dot{N}H$ and $CH_3\dot{C}HNH_2$	630
+ Ion (pH <10)	$H_2O + \cdot CH_2CH_2NH_2$	50
Ethylene	$HOCH_2CH_2\cdot$	180
Formaldehyde	$\cdot CHO + H_2O$	~200
Formic acid (pH 1)	$\cdot COOH + H_2O$	13
Formate ion (pH 6 to 11)	$\cdot COO^- + H_2O$	280
Glycine		
+ Ion (pH 3)	$NH_3^+CHCOOH + H_2O$	1.1
Zwitterion (pH ~ 7)	$NH_3^+\dot{C}HCOO^- + H_2O$	0.9
− Ion (pH >9.5)	Abstraction	260
Glycylglycine (pH ~ 6)	Abstraction	25
D-Glucose	Abstraction	190
Methane	$\cdot CH_3 + H_2O$	24
Methanol	$\cdot CH_2OH + H_2O$	84 (55)
Nitrobenzene	Addition	340
Nitromethane	Addition	<0.9
− Ion	Addition	850
p-Nitroso-*N,N*-dimethyl aniline (PNDA)	Addition	1250
Oxalic acid (pH 2)	$\cdot OOCCOOH + H_2O$	0.8
Dianion	$\cdot OOCCOO^- + OH^-$	0.95 (1.6)
Phenol (pH 6 to 9)	Addition	1200
Phenylalanine (pH ~ 7)		630
1-Propanol	$\cdot C_3H_6OH + H_2O$	265
Isopropanol	$CH_3\dot{C}(OH)CH_3 + H_2O$	200

Table 4.1-3 (continued)
RATE CONSTANTS FOR REACTIONS OF THE HYDROXYL RADICAL AND OXIDE ION

Reactant	Products or reaction type	Rate constant[a] $(10^7 \text{ mol}^{-1} \text{ sec}^{-1})$
Pyridine (pH 7)	Addition	250
Ribose	Abstraction	210
Thymine	Addition	530
Toluene	Addition	300
Uracil	Addition	630

[a]   Rate constants are mean values calculated from the data collected by Dorfman and Adams;[20] values are generally known to ±25%. Reaction products and rate constants in parentheses are for the oxide radical ion, O$^-$, and were determined at pH 11 or above. Unless pH ranges are specified, the remaining rate constants (for OH) should be applicable at all pH below 11. Abstraction is H-atom abstraction.

From Spinks, J. W. T. and Woods, R. J., *An Introduction to Radiation Chemistry,* 2nd ed., John Wiley & Sons, New York, 1976. With permission.

Addition of e^-_{aq} to Substance 2 gives a radical anion (Substance 3) which fragments primarily into ammonia and an acetyl radical (Substance 4) with fragmentation into H· and Substance 1 as a less important pathway (Equation 4.1-23).

$$e^-_{aq} + {}^+NH_3 - CH_2 - COO^- \; (2) \rightarrow [NH_3 - CH_2 - COO]^{\cdot -} \; (3)$$

$$k = 7 \times 10^6 \text{ mol}^{-1} \text{ sec}^{-1}$$

$$NH_3 + \cdot CH_2 COO^- (4) \qquad\qquad H\cdot + NH_2 - CH_2 - COO^- (1) \qquad (4.1\text{-}23)$$

Reaction of H· and OH· with Substance 2 gives the glycine radical (Substance 5) (Equations 4.1-24 and 4.1-25). In the absence of oxygen the radicals (Substances 4 and 5) interact by disproportionation to give acetate (Substance 6), the glycine zwitterion (Substance 2), and the iminoacetate zwitterion (Substance 7) (Equations 4.1-26 and 4.1-27).

$$H\cdot + {}^+NH_3 - CH_2 - COO^- (2) \rightarrow H_2 + {}^+NH_3 - \overset{\cdot}{C}H - COO^- (5) \qquad (4.1\text{-}24)$$

$$k = 8 \times 10^4 \text{ mol}^{-1} \text{ sec}^{-1}$$

$$OH\cdot + {}^+NH_3 - CH_2 - COO^- (2) \rightarrow H_2O + {}^+NH_3 - \overset{\cdot}{C}H - COO^- (5) \qquad (4.1\text{-}25)$$

$$k = 9 \times 10^6 \text{ mol}^{-1} \text{ sec}^{-1}$$

$${}^+NH_3 - CH - COO^- (5) + \cdot CH_2 - COO^- (4) \rightarrow {}^+NH_2 = CHCOO^- (7) + CH_3 COO^- (6) \qquad (4.1\text{-}26)$$

$$2\,{}^+NH_3 - \overset{\cdot}{C}H - COO^- (5) \rightarrow {}^+NH_2 = CH - COO^- (7) + {}^+NH_3 - CH_2 - COO^- (2) \qquad (4.1\text{-}27)$$

Iminoacetate (Substance 7) then undergoes hydrolysis to give glyoxyllic acid (Substance 8), the major product (Equation 4.1-28). Small amounts of aminosuccinic acid and diaminosuccinic acids are formed by dimerization pathways.

$$\overset{O}{\overset{\|}{}}$$
$${}^+NH_2 = CH - COO^- (7) + H_2O \rightarrow HC\overset{\|}{C}OO^- (8) + NH_4^+ \qquad (4.1\text{-}28)$$

Introduction of oxygen into the aqueous glycine system blocks Equations 4.1-23 and 4.1-24 by scavenging e^-_{aq} and H·, but is without effect on the hydroxyl radical. The initial radical (Substance 5) formed by abstraction of a hydrogen by a hydroxyl radical (see Equation 4.1-25) is completely scavenged by oxygen to give a perhy-

Table 4.1-4
RATE CONSTANTS OF THE REACTION
BETWEEN VARIOUS COMPOUNDS AND
HYDROXYL RADICAL

Compound	$k(OH\cdot + compound)$ $mol^{-1}sec^{-1}$
DNA	$7\pm1.7 \times 10^{10}$
RNA	$1.9\pm0.3 \times 10^{9}$
Thymine	$7.6\pm1.5 \times 10^{9}$
Thymidine	$2.7\pm0.5 \times 10^{9}$
Thymidylic acid	$6.9\pm0.2 \times 10^{8}$
Uracil	$4.2\pm0.8 \times 10^{9}$
Uridine	$2.4\pm0.5 \times 10^{9}$
2'-,3'-Uridylic acid	$1.9\pm0.5 \times 10^{9}$
5-Aminouracil	$8.8\pm1.7 \times 10^{9}$
Cytosine	$2.9\pm0.6 \times 10^{9}$
Cytidine	$2.5\pm0.4 \times 10^{9}$
Deoxycytidine·HCl	$6.4\pm2.7 \times 10^{9}$
Deoxycytidylic acid	$8.5\pm2.3 \times 10^{9}$
Adenine	$4.9\pm1.0 \times 10^{9}$
Hypoxanthine	$4.2\pm0.8 \times 10^{9}$
Adenosine	$2.8\pm0.3 \times 10^{9}$
Adenylic acid	$3.6\pm1.4 \times 10^{9}$
Deoxyadenosine	$2.0\pm0.4 \times 10^{9}$
Deoxyadenylic acid·NH$_4$	$2.0\pm0.7 \times 10^{9}$
Guanine	$4.1\pm0.3 \times 10^{9}$
Deoxyguanosine	$1.25\pm0.37 \times 10^{9}$
Deoxyguanylic acid·NH$_4$	$2.3\pm0.6 \times 10^{9}$
2-Deoxyribose	$8.8\pm1.9 \times 10^{8}$
Ribose	$1.36\pm0.45 \times 10^{8}$
Glucose-1-phosphate dipotassium salt	$4.2\pm0.7 \times 10^{7}$
Human serum albumin	$4.0\pm0.8 \times 10^{8}$
β-Phenylalanine	$3.7\pm0.7 \times 10^{9}$
Tyrosine	$1.2\pm0.35 \times 10^{10}$
Tryptophan	$4.3\pm0.9 \times 10^{10}$
Cysteine	$5.6\pm1.9 \times 10^{9}$
Cystine	$5.7\pm0.9 \times 10^{10}$
Alanine	$9.8\pm1.9 \times 10^{7}$
Glycine	$3.7\pm1.9 \times 10^{8}$
Benzene	$2.3\pm0.7 \times 10^{9}$
Phenol	$7.7\pm1.5 \times 10^{10}$
Indole	$4.7\pm0.8 \times 10^{10}$
Maize starch	$2.9\pm0.5 \times 10^{8}$
Waxy starch	$2.5\pm0.4 \times 10^{8}$
Glucose	$9.5\pm1.9 \times 10^{7}$
Cellobiose	$4.9\pm1.0 \times 10^{8}$
Lactose	$2.4\pm0.2 \times 10^{9}$

From Strazhevskaya, N. B., Ed., *Molecular Radiobiology,* Keter Publishing House, Jerusalem, 1975.[9] With permission.

droxyl radical (Substance 9), which breaks down to glyoxyllic acid (Substance 8) and ammonium ion (Equation 4.1-29). As a consequence in the presence of oxygen:

$$^+NH_3-\overset{\cdot}{C}H-COO^-\,(5) + O_2 \rightarrow (9) \rightarrow NH_4^+ + H\overset{\overset{\displaystyle O}{\|}}{C}COO^-\,(8) + C(NH_3) \simeq C(8) \simeq C_{OH} \quad (4.1\text{-}29)$$

G values of the products from the radiolysis of glycine are given in Table 4.1-5. Inspection of Equations 4.1-23 to 4.1-27 indicates that prediction of G values for the products in the absence of oxygen is quite complex, since 1. the ratio of the two

Table 4.1-5
RADIOLYSIS OF AQUEOUS GLYCINE SOLUTIONS

Product	G Values	
	Oxygen-saturated	Oxygen-free
Hydrogen	—	2.02
Hydrogen peroxide	3.6	0.01
Ammonia	4.3	3.97
Carbon dioxide	—	0.90
Formaldehyde	1.1	0.53
Formic acid	—	0.085
Methylamine	0.16	0.19
Acetic acid	—	1.20
Glyoxyllic acid	3.4	2.10
Aminosuccinic acid	—	0.25
Diaminosuccinic acid	—	0.08

From Strazhevskaya, N. B., Ed., *Molecular Radiobiology,* Keter Publishing House, Jerusalem, 1975.[9] With permission.

fragmentation reactions described by Equation 4.1-23 must be known and 2. product yields must be corrected for the amount of dimerization. The mechanism outlined for the cleavage of glycine must be applied to more complex amino acids with caution and is completely inadequate for amino acids containing aromatic rings or sulfur atoms.

4.1.5.2. Sugars

The radiation chemistry of even simple sugars is extraordinarily difficult to elucidate due to the fact that a diverse number of similar products are formed, all of which are quite difficult to isolate and identify. Disdaroglu et al.[25] have identified a large number of products formed in the radiolysis of aqueous solutions of D-glucose (Table 4.1-6). These results were obtained in solutions saturated with N_2O which converts e^-_{aq} into OH· as previously mentioned. The major products are formed by oxidation reactions with or without concomitant dehydration. Initiation via hydrogen abstraction at C-1 leads to sugar acids; whereas, initiation via abstraction at any other site yields sugars containing a second carbonyl group. The mechanism involved in formation of the observed products is exemplified by attack of OH· or H· at C-1 of D-glucose (Substance 10), which, because it leads to the most stable radical, is the major point of attack. Reaction is initiated by abstraction of a hydrogen atom to give a radical (Substance 11) (Equation 4.1-30).

$$(4.1\text{-}30)$$

(10)   (11)

Loss of a hydrogen at O-1 from Substance 11 leads to formation of D-gluconic acid as its lactone (Substance 12) (Equation 4.1-31).

Table 4.1-6
CO-60-γ-RADIOLYSIS OF
N₂O-SATURATED AQUEOUS
SOLUTIONS OF D-GLUCOSE
(10⁻²MOL⁻¹) INITIAL G
VALUES OF THE PRODUCTS
AT 25° C

Product	G Value
Gluconic acid	0.15
Glucosone	0.15
3-Ketoglucose	0.10
4-Ketoglucose	0.075
5-Ketoglucose	0.18
Gluco-hexodialdose	0.22
2-Deoxygluconic acid	0.95
5-Deoxy-4-ketoglucose	
5-Deoxygluconic acid	0.08
2-Deoxy-5-ketoglucose	
5-Deoxy-*xylo*-hexodialdose	
3-Deoxy-4-ketoglucose	0.25
3-Deoxyglucosone	
4-Deoxy-5-ketoglucose	
6-Deoxy-5-ketoglucose	0.05
Arabinose	0.01
Ribose	< 0.005
Xylose	< 0.005
2-Deoxyribose	0.04
3-Deoxypentulose	< 0.005
Erythrose	0.01
Threose	< 0.005
2-Butanone-1,4-diol	0.02
Dihydroxyacetone	0.03
Glucose consumption	5.6

From Dizdaroglu, M., Henneberg, D., Schomburg, G., and von Sonntag, C., *Z. Naturforsch.*, 30(b), 416, 1975. With permission.

$$(4.1\text{-}31)$$

This process results in oxidation of D- glucose (Substance 10) without accompanying dehydration. Radical 11, however, also dehydrates to a significant extent to give radical 13, which can abstract a hydrogen from another molecule of D-glucose (or disproportionate with radical 11) to give 2-deoxy-D-gluconic acid as its lactone (Structure 14) (Equation 4.1-32).

(4.1-32)

Lastly, radical 11 can fragment to give the ring-opened radical 15, which forms 5-deoxy-D-gluconic acid (Structure 16) via either abstraction or disproportionation (Equation 4.1-33).

(4.1-33)

Inspection of Table 4.1-6 shows that the sum total of all identified products is significantly below the amount of D-glucose radiolyzed indicating that either the G values quoted are understated or that additional products were formed but not detected. The methodology that by necessity was used to identify the radiolysis products — namely elution and gas chromatography — is subject to both errors. A recent report by Esterbauer et al.[26] describes the identification and product yields for acidic products from the radiolysis of He-saturated D-glucose solutions (Table 4.1-7). In this case, the acidic nature of the products was utilized, allowing their separation from the neutral components by ion exchange chromatography prior to gas chromatographic identification and quantitation. As can be seen, there are additional products not identified by Dizdaroglu.[25] Although the G values are lower than those in Table 4.1-6, the fact that no scavenger was used would decrease the values considerably.

Table 4.1-7
INITIAL G VALUES OF SUGAR ACIDS FROM HELIUM DEGASSED GAMMA-IRRADIATED 0.055 MOL^{-1} D-GLUCOSE

Sugar Acid	G Value
Glyceric Acid	0.03
2-Deoxytetronic acid	0.04
Tetronic acid	0.03
4-Deoxypentenoic acid	0.02
Deoxyketogluconic acid (1)	0.17
Deoxyketogluconic acid (2)	0.05
2-Deoxygluconic acid	0.62
Hexonic acid	0.03
Gluconic acid	0.20
Pentonic acid	< 0.01
5-Deoxygluconic acid	< 0.01

From Esterbauer, H., Schubert, J., Sanders, E. B., and Sweeley, C. C. *Z. Naturforsch.*, 32(b), 315, 1977. With permission.

4.1.5.3. Nucleic Acid Bases

The radiolysis of thymine (Substance 17) has recently been studied both in the presence and absence of oxygen.[27] The results are displayed in Table 4.1-8. Products formed and the mechanisms for their formation are fairly straightforward in oxygenated solution where both e^-_{aq} and $H \cdot$ are effectively scavenged. Hydroxyl radical can add to the 5 or 6 position of thymine (Substance 17), giving two hydroxylated radicals (Substances 18 and 19) (Equation 4.1-34).

(4.1-34)

Both radicals add oxygen to give the corresponding peroxy radicals (Substances 20 and 21), which are then converted to *cis-* and *trans-*thymine glycol (Substances 22 and 23) (Equation 4.1-35)

(4.1-35)

In addition, peroxy radical 20 can lose $HO_2 \cdot$ to give 5-hydroxy-5-methyl-barbituric acid (Substance 24) (Equation 4.1-36).

(4.1-36)

A small amount of thymine dimer is most probably formed via an initial abstraction step. The mechanism outlined predicts that G(products) $\simeq$ G(-thymine) $\simeq$ G_{OH}, which is, indeed, observed.

Results in the absence of oxygen are qualitatively similar when nitrous oxide is used as a scavenger. This is expected in that e^-_{aq} is still scavenged and the amount of $H \cdot$ formed is small compared to $OH \cdot$. However, when $0.01 M$ isopropanol is utilized as a scavenger removing both $H \cdot$ and $OH \cdot$, dihydrothymine (Substance 25) becomes the major product. The formation of Substance 25 can be explained by attack of e^-_{aq} on thymine (Substance 17) to then give the thymine radical anion

Table 4.1-8
YIELDS OF RADIOLYTIC PRODUCTS IN
IRRADIATED AQUEOUS THYMINE SOLUTIONS
AT NEUTRAL pH

Product	Yields (G values)[a]		
	A	B	C
Thymine dimer	0.13	0.26	0.22
cis-Thymine glycol	0.08	1.45	0.97
trans-Thymine glycol	0.00	0.81	0.54
5-Hydroxydihydrothymine	0.13	—	—
6-Hydroxydihydrothymine	0.17	0.13	—
Dihydrothymine	0.38	0.10	—
5-Hydroxymethyluracil	0.09	0.22	—
5-Hydroxy-5-methyl barbituric acid	—	0.11	0.16
Formylpyruvylurea	—	0.12	0.23
Acetylurea	0.03	—	—
Urea	—	0.19	—
Alcohol adduct	0.21	—	—
5-Hydroxyperoxydihydrothymine	—	—	0.33
Unknown	0.10	0.11	—
Total Products	1.32	3.50	2.45
-Thymine	1.42	3.92	2.70

[a]   Conditions: A, degassed, 0.10 M isopropanol
as scavenger; B, N_2O-saturated solution; C,
air-saturated, no scavengers.

From Infante, G. A., Jirathana, P., Fendler, J. H., and Fendler, E.
J., *J. Chem. Soc. Faraday Trans. 1*, 69, 1586, 1973.

(Substance 26). Protonation of Substance 26 followed by abstraction of a hydrogen
from isopropanol gives Substance 25 (Equation 4.1-37).

$$(4.1-37)$$

4.1.6. Radiation Chemistry of Biopolymers

Rate constants for reaction of DNA with the reactive species produced by radi-
olysis of water are $k_{(e^-_{aq} + DNA)} \geq 6 \times 10^7 mol^{-1}sec^{-1}$, $k_{(H \cdot + DNA)} = 5 \times 10^7 mol^{-1}sec^{-1}$,
and $k_{(OH \cdot + DNA)} = 6 \times 10^8 mol^{-1}sec^{-1}$.[28-30] These values are based on the nucleotide
concentration assuming that nucleotides in DNA react independently.[31] Optical
spectra of radicals produced by the reaction of $OH \cdot$ with DNA are virtually identical to spectra of the corresponding bases[31] indicating that the same radicals are

Table 4.1-9
RATE CONSTANTS FOR REACTIONS
OF HYDRATED ELECTRONS WITH
PROTEINS

Protein	pH	Rate constant (10^{10} mol^{-1}sec^{-1})	Ref.
Gelatin	6	6	32
	10	3	32
Ribonu-clease	5.6	2.7	32
	6.8	1.3	32
	8.35	0.6	32
Lysozyme	6.2	7.5	32
	—	5.2	33
	10.1	2.7	32
Egg albu-min	11.5	1.3	32
Catalase	—	0.4	34

From Meyers, L. S., Jr., *The Radiation Chemistry of Macromolecules,* Vol. 2, Dale, M., Ed., Academic Press, New York, 1973. With permission.

produced and that neither the pentose or phosphate units react appreciably with OH·. As a consequence, one might expect that the radiation chemistry of nucleic acids in water could be predicted significantly from the radiation chemistry of the component bases in aqueous solution.

In that proteins all contain significant proportions of highly reactive amino acids, it is not surprising that their rates of reaction with water radiolysis products are similar. Representative data for the reaction of e^-_{aq} with proteins are shown in Table 4.1-9. Values are quite pH dependent, decreasing as pH is increased. This is not unexpected since at higher pH, the protein will carry an increased negative charge due, for example, to the ionization of the carboxyl groups of glutamate and aspartate, which will decrease its ability to react with a negatively charged species. Rates of reaction of hydroxyl radicals with proteins are about the same as the rate of reaction of e^-_{aq}.[31] Rates of reaction of both OH· and e^-_{aq} are dependent on protein conformation.

Although no information has been obtained on actual product formation following irradiation of proteins in aqueous solution, an elegant experiment has shown that inactivation of lysozyme is due to the reaction of OH· with tryptophan.[35] Most of the radiation chemical studies of biopolymers has been carried out in the "dry" state. For a brief review of these studies, see Meyers[31] and references cited therein.

REFERENCES

1. **Alper, T.,** Biological effects of neutron irradiation, *Proc. Symp. Effects of Neutron Irradiation upon Cell Function,* International Atomic Energy Agency, Munich, October 1973.
2. **Lea, D. E.,** *Actions of Radiation on Living Cells,* Cambridge University Press, 1946.

3. Dolphin, G. W. and Hutchinson, F., The action of fast carbon and heavier ions on biological materials. I. The inactivation of dried enzymes, *Radiat. Res.*, 13, 403, 1960.

3a. Schambra, P. E. and Hutchinson, F., The action of fast heavy ions on biological material. II. Effects on T1 and ϕX-174 bacteriophage and double-strand and single-strand DNA, *Radiat. Res.*, 23, 514, 1964.

4. Alper, T., Moore, J. L., and Bewley, D. K., LET as a determinant of bacterial radiosensitivity, and its modification by anoxia and glycerol, *Radiat. Res.*, 32, 277, 1967.

4a. Munson, R. J., Neary, G. J., Bridges, B. A., and Preston, R. J., The sensitivity of *Escherichia coli* to ionizing particles of different LET's, *Int. J. Radiat. Biol.*, 13, 205, 1967.

5. Strazhevskaya, N. B., Ed., *Molecular Radiobiology*, Keter Publishing House, Jerusalem, 1975.

6. Buxton, G. V., Primary radical and molecular yields in aqueous solution; the effect of pH and solute concentration, *Radiat. Res. Rev.*, 1, 209, 1968.

7. Draganić, I. G. and Draganić, Z. D., *The Radiation Chemistry of Water*, Academic Press, New York, 1964.

8. Draganić, I. G., Nenadovic, M. T., and Draganić, Z. D., Radiolysis of $HCOOH + O_2$ at pH 1.3—13 and the yields of primary products in γ radiolysis of water, *J. Phys. Chem.*, 73, 2564, 1969.

9. Collinson, E., Dainton, F. S., and Kroh, J., Effects of linear energy transfer on the radiolysis of water and heavy water, *Nature*, 187, 475, 1960.

9a. Collinson, E., Dainton, F. S., and Kroh, J., The radiation chemistry of aqueous solutions. II. Radical and molecular yields for tritium β-particles, *Proc. R. Soc. London Ser. A*, 265, 422, 1962.

10. Appleby, A. and Schwarz, H. A., Radical and molecular yields in water irradiated by γ rays and heavy ions, *J. Phys. Chem.*, 73, 1937, 1969.

11. Lefort, M and Tarrago, X., Radiolysis of water by particles of high linear energy transfer. The primary chemical yields in aqueous acid solutions of ferrous sulfate and in mixtures of thallous and ceric ions, *J. Phys. Chem.*, 63, 833, 1959.

12. Lefort, M., Radiation chemistry, *Annu. Rev. Phys. Chem.*, 9, 123, 1958.

13. Imamura, M., Matsui, M., and Karasawa, T., Radiation chemical studies with cyclotron beams. II. Radiolysis of an aqueous ferrous ammonium sulfate solution with carbon- and nitrogen-ion radiations, *Bull. Chem. Soc. Jpn.*, 43, 2745, 1970.

14. Kuppermann, A., in *Actions Chimique et Biologiques des Radiations*, Vol. 5, Haissinsky, M., Ed., Masson et Cie, Paris, 1961.

14a. Kupperman, A., in *Radiation Research*, Silini, G., Ed., North Holland, Amsterdam, 1967.

15. Spinks, J. W. T. and Woods, R. J., *An Introduction to Radiation Chemistry*, 2nd ed., John Wiley & Sons, New York, 1976.

16. Mićić, O. I. and Marković, V., Rates of hydrated electron reactions with undissociated carboxylic acids, *Int. J. Radiat. Phys. Chem.*, 4, 43, 1972.

17. Anbar, M., Bambenek, M., and Ross, A. B., Selected Specific Rates of Reactions of Transients in Aqueous Solution. I. Hydrated Electron, NSRDS-NBS 43, U.S. Department of Commerce-National Bureau of Standards, Washington, D. C., 1973.

17a. Ross, A. B., Supplement to NSRDS-NBS 43, U.S. Department of Commerce-National Bureau of Standards, Washington, D. C., 1975.

18. Anbar, M. and Neta, P., A compilation of specific bimolecular rate constants for the reactions of hydrated electrons, hydrogen atoms and hydroxyl radicals with inorganic and organic compounds in aqueous solution, *Int. J. Appl. Radiat. Isot.*, 18, 493, 1967.

18a. Anbar, M., Farhataziz, A., and Ross, A. B., Selected Specific Rates of Reactions of Transients from Water in Aqueous Solution. II. Hydrogen Atom, NSRDS-NBS 51, U.S. Department of Commerce-National Bureau of Standards, Washington, D. C., 1975.

19. Neta, P., Fessenden, R. W., and Schuler, R. H., An electron spin resonance study of the rate constants for reaction of hydrogen atoms with organic compounds in aqueous solution, *J. Phys. Chem.*, 75, 1654, 1971.

20. Dorfman, L. M. and Adams, C. E., Reactivity of the Hydroxyl Radical in Aqueous Solution, NSRDS-NBS 46, U.S. Department of Commerce-National Bureau of Standards, Washington, D. C., 1973.

21. McCord, J. M. and Fridovich, I., Superoxide dismutase. An enzymatic function for erythrocuprein (hemocuprein), *J. Biol. Chem.*, 244, 6049, 1969.

22. Bielski, B. H. J. and Richter, H. W., A study of the superoxide radical chemistry by stopped-flow radiolysis and radiation induced oxygen consumption, *J. Am. Chem. Soc.*, 99, 3019, 1977.

23. Danen, W. C. and Warner, R. J., The remarkable nucleophilicity of superoxide anion radical. Rate constants for reaction of superoxide ion with aliphatic bromides, *Tetrahedron Lett.*, 989, 1977.

24. Garrison, W. M., Actions of ionizing radiations on nitrogen compounds in aqueous media, *Radiat. Res. Suppl.*, 4, 158, 1964.

24a. Garrison, W. M., Radiation-induced reactions of amino acids and peptides, *Radiat. Res. Rev.*, 3, 305, 1972.

25. **Dizdaroglu, M., Henneberg, D., Schomburg, G., and von Sonntag, C.,** Radiation Chemistry of Carbohydrates. VI. γ-Radiolysis of glucose in deoxygenated N$_2$O saturated aqueous solution, *Z. Naturforsch.,* 30(b), 416, 1975.
26. **Esterbauer, H., Schubert, J., Sanders, E. B., and Sweeley, C. C.,** Yields of 2-deoxy-D-gluconic, D-gluconic, and other sugar acids in gamma-irradiated aqueous solutions of D-glucose, *Z. Naturforsch.,* 32(b), 315, 1977.
27. **Infante, G. A., Jirathana, P., Fendler, J. H., and Fendler, E. J.,** Radiolysis of pyrimidines in aqueous solutions. I. Product formation in the interaction of e^-_{aq}, ·H, ·OH and Cl^-_2· with thymine, *J. Chem. Soc. Faraday Trans. 1,* 69, 1586, 1973.
28. **Scholes, G., Shaw, P., Wilson, R. L., and Ebert, M.,** in *Pulse Radiolysis,* Ebert, M., Keene, J. P., Swallow, A. J., and Baxendale, J. H., Eds., Academic Press, New York, 1965.
29. **Scholes, G. and Simić, M.,** Radiolysis of aqueous solutions of DNA and related substances, reaction of hydrogen atoms, *Biochim. Biophys. Acta,* 166, 255, 1968.
30. **Loman, H. and Ebert, M.,** Some pulse radiolytic observations on the reactivity of hydroxyl radicals with DNA and poly A, *Int. J. Radiat. Biol.,* 13, 549, 1967.
31. **Meyers, L. S., Jr.,** in *The Radiation Chemistry of Macromolecules,* Vol. 2, Dole, M., Ed., Academic Press, New York, 1973.
32. **Braams, R.,** Rate constants of hydrated electron reactions with peptides and proteins, *Radiat. Res.,* 31, 8, 1967.
33. **Davies, J. V., Ebert, M., and Shalek, R. J.,** The radiolysis of dilute solutions of lysozyme II. Pulse radiolysis studies with cysteine and oxygen, *Int. J. Radiat. Biol.,* 14, 19, 1968.
34. **Karmann, W.,** Investigations, by pulse radiolysis, of the reactivity of catalase towards hydrated electrons and free OH radicals, *Angew. Chem. Int. Ed. Engl.,* 5, 738, 1966.
35. **Aldrich, J. E., Cundall, R. B., Adams, G. E., and Wilson, R. L.,** Identification of essential residues in lysozyme: a pulse radiolysis method, *Nature,* 221, 1049, 1969.

4.2. CHEMICAL PROCEDURES FOR PREPARATION OF RADIONUCLIDE SOURCES

Imogene F. Sevin

4.2.1. General Methods of Preparing Sources

Because of the statistical nature of radiation; properties attributable to the source, such as scattering and absorption; and instrumental errors, such as background and geometric variations, resolving time, and aging of the detector; a primary standard is difficult to prepare. Elaborate equipment for the measurement of radiation and additional knowledge of the characteristics of the material being calibrated are needed. Once such a source is prepared, its radiation, in turn, may be used for the calibration of secondary standards.

Direct weighing of a sample would be the simplest method except for two factors. First, the amount of material required is often not weighable. Consider, for instance, that 1 μCi of carbon-14 weighs 0.224 μg. Second, since the sample must be uniformly distributed over the surface area of the backing material, this requires preparation from a liquid. If there is sufficient material to be weighed and if a suitable solvent that will, in addition, minimize sample loss, can be identified, then it may be possible to prepare the standard without a measurement of radioactivity. The specific activity of the solution must be known or calculable, and the material must be radiochemically pure.

To prepare a source, an aliquot is removed from a freshly prepared solution. This material is placed on a suitable backing and dried, and a cover, if necessary, is added. Micropipettes contribute an error of 2% or greater, and weighing of the aliquot is often more accurate. Additional steps such as solvent extraction for purification or electroplating for a more uniform source may also be conducted, provided that the chemical yield of each step is known. Since large quantities of radioactivity may be involved, proper procedures for safe handling must be strictly adhered to. Commercially prepared radioactive standards generally range from ± 1 to 10% in accuracy, as reported by the manufacturer.

4.2.2. General Considerations in Experiments Using Radioactivity

Most users of isotopes never have occasion to prepare a primary standard. However, in order to exploit the most beneficial characteristic of radioactivity, the sensitivity of the measurement with regard to the amount of material needed for analysis, several factors must be remembered to obtain meaningful results. First, the measurement of radioactivity will not demonstrate small differences between two numbers. Second, a greater degree of accuracy can be obtained in experiments designed to measure relative changes in count rates rather than in experiments where absolute activity must be determined. Third, most counting equipment have optimum detection ranges of approximately several thousand counts per minute, although this varies somewhat depending on individual instruments and the type of radiation being measured. Within the limitations of an experiment, appropriate adjustments in sample size and other variables should be made in order to obtain, in all samples, an optimum count rate.

A second feature of radioactive isotopes that has widespread use is their ability to tag molecules. In experiments of this type, the following conditions must be met. The reaction or its rate, depending on what is being measured, must not be altered significantly by isotopic effects. This is especially important in experiments using tritium since there is a threefold difference in atomic weight between tritium and hydrogen. In addition, attention must be paid to the position of the label on a molecule to ensure that the proper species in a reaction is being measured. Radioactive molecules may be made by either specific or universal labeling; in particular, if a universally labeled molecule is used, the possibility of labile groups must be considered. Lability can result in the tagging of an undesired molecule with a subsequent loss of meaningful results. Although many tagged molecules have a long shelf life, some are particularly sensitive to radiation effects, resulting in decomposition products of a completely different chemical structure.[1] As can be seen from the few examples given in Table 4.2-1, storage conditions can play a major role in the rate of decomposition. Manufacturer's specifications for storage and shelflife should be consulted and adhered to. Finally, many inorganic compounds, especially those in carrier-free solutions with high specific activity and consequent low concentration, are particularly susceptible to the formation of radiocolloids.[2-4] Acid pH,[5] highly purified water,[6] and avoidance of mechanical agitation[7] of such solutions will substantially reduce the possibility of errors introduced by radiocolloid formation and limit the amount of activity adsorbed on container surfaces.

When in vivo experiments are being considered, additional constraints must be satisfied. The chemical form of the injected material must not introduce spurious results because of incompatibility with normal reactions in the body. An example is plutonium, where the monomer reacts quite differently from polymeric forms.[8,9] Additionally, the route of administration must be appropriate to the desired results. Finally, the administered isotope should reach the proper site of translocation before additional experimental testing is performed. This is especially important when the radioactive material administered has a stable form occurring in the body.

4.2.3. Characteristics of the Measurement of Radiation

Regardless of the type of radiation being measured or the purpose of an experiment, several factors should be considered to improve results. These include the optimization of counting time, geometry, and scattering effects, the normalization of all counts to a standard value, and the initial use of material whose composition is appropriate to the experiment.

Table 4.2-1.
STABILITY OF LABELED COMPOUNDS

Compound	Relative stability	Protective measures
^{14}C	More stable than tritium; reducing sugars not stable	Use of scavengers; storage at $-20°C$; lower specific activity
Tritium	Some decomposition usual	Use of scavengers; storage above freezing or at $-196°C$; lower specific activity
^{32}P	Limited data	Storage as thin film
^{35}S	Variable in solutions	Storage as solid
^{131}I	Most compounds unstable	^{125}I is more stable

Derived from information provided in Bayly, R. J. and Evans, E. A., Storage and Stability of Compounds Labeled with Radioisotopes, Amersham/Searle Corp., Des Plaines, Ill.

Radiation measurements cannot be quantitated absolutely because of the statistical nature of radioactive decay and the background level, although uncertainties due to these factors can be minimized. Consider that the standard error of a net count rate decreases as

$$\sqrt{r_s/t_s + r_b/t_b} \qquad\qquad (4.2\text{-}1)$$

where r_s is the sample count rate, t_s is the sample counting time, r_b is the background count rate, and t_b is the background counting time. Ideally, background contributions could be minimized by a very large sample count rate, but in practice this factor is limited because of the finite resolving time of detectors or in the case of low-level measurements, because of the small amount of radioactivity to be measured. The optimum time to obtain a given precision in a fixed total counting time can be determined from the ratio

$$t_s/t_b = \sqrt{r_s/r_b} \qquad\qquad (4.2\text{-}2)$$

In practice, it is often possible to take a background count while preparing the other samples, so that the long counting time makes this contribution to the total counting error negligible.

The reproducibility of the geometry of a source is a second factor that must be considered. Except at a very close distance where the cross sectional area of the detector is a factor, the fraction of radiation reaching a detector follows the inverse square law. Thus, it is important to maintain a constant geometry during a series of measurements, especially in the region where the smallest change in distance makes the greatest change in the percentage of radiation reaching the detector.

If the count rate is sufficiently high, corrections must be included for the resolving time of the detector. This limitation is particularly severe for Geiger-Mueller detectors. An exact calculation for an individual instrument may be made by the paired sources method,[10] or correction factors for resolving time[11] may be consulted (see Figure 4.2-1).

For materials that are counted as solids, the effect of backscattering on count rate should be considered. As the atomic number and thickness of a backing material increase, count rate increases for the same amount of activity. Therefore, stainless steel and aluminum planchets should not be used interchangeably in the same experiment. Often, for standard sources, a protective covering, also used to eliminate undesired low-energy radiation, is added. This covering will alter the count rate of a source. To calibrate a source from a standard, the fraction of the counts actually emitted, compared to the theoretical activity for the standard, must be known; the calibration should be made on a source prepared on a "weightless" backing, such as a thin sheet of plastic foil. Backing and covering material also contribute to forescattering. Since the effects of forescattering are not easily distinguishable from the effects of air absorption, again, it becomes important to maintain identical geometry. When a souce is collimated, sidescattering provides an important contribution to the enhancement of count rate. Sidescatter effects, like backscatter, increase with the atomic number of the collimating material.

Finally, absorption of the sample itself will decrease the number of counts measured in all but infinitely thin sources. Consequently, it is important that the source be of uniform thickness. If a thin source is not possible, variability can be lessened by an infinitely thick source, in which the count rate is unaffected by small differences in individual sources. However, the count rate is severely lowered, so that this method is often not useful. Alpha emissions are especially affected by self-absorption, a sheet of paper effectively dropping their count rate to zero. Thus, electro-

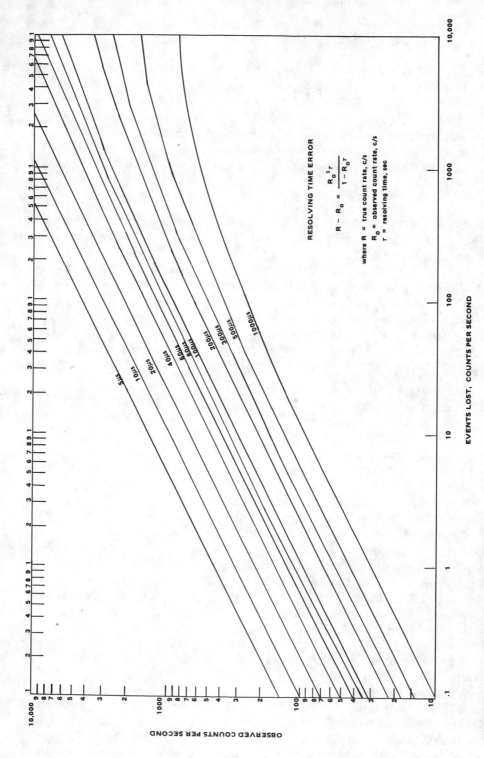

FIGURE F4.-1. Resolving time corrections for lost counts. (From Bureau of Radiological Health, Radiological Health Handbook, U.S. Department of Health, Education, and welfare, Public Health Service, Rockville, Md., 1970, 121.

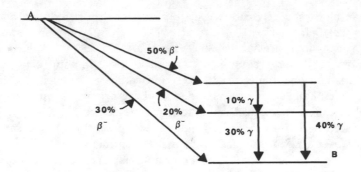

FIGURE 4-2.   Example decay scheme.

than 100% because of the presence of metastable states. Second, if one type of emission, either beta or gamma, is measured, even when scatter, geometry, and counter efficiency are taken into account, the absolute number of disintegrations per minute is not necessarily equal to the number of microcuries times 2.22×10^6 in the sample. Before the conversion can be made, the count rate must be divided by the fractional abundance (see Section 3.6).

When more than one type of radiation is present, a decision on which one to measure must be made. In general, the one with the highest count rate would be preferable, to improve sensitivity, accuracy, and precision. It must be remembered that while beta counters are fairly insensitive to gamma emissions, the presence of a similar number of gamma emissions, especially if they are of low energy, will contribute to the overall count rate. For gamma-ray spectroscopy, it is preferable to measure the highest energy emitted, unless it is of extremely low abundance. This obviates the necessity of subtractions from the count rate for backscatter and Compton effects from higher energy emissions. For mixed alpha and beta sources, measurement of the alpha particle is preferred in proportional counting, since counts will not be contributed by the beta fraction.

Occasionally, two isotopes of the same chemical species might be found together; more often, as the result of decay, radioactive daughters may be present. First, it must be considered whether or not the presence of the undesired isotope will affect the results of an experiment. For example, it would be extremely difficult to measure cobalt-57 in the presence of a significant amount of cobalt-60. If their presence is undesirable, daughters may be separated from the parent by chemical reaction. If the two isotopes are the same chemically, their separation is not possible.

As the half-life increases, so does the amount of material required to give a desirable count rate. In addition, samples of low specific activity may be encountered, so that at times the measurement of a very thick source is unavoidable. In this case, high-energy gamma rays are the least easily absorbed radiation, and given the option, it is preferable to detect the gamma emission. When the isotope being used has a very short half-life, time becomes a critical factor and separations or purification steps that would improve experimental accuracy and sensitivity or provide a high chemical yield often must be sacrificed. Many theoretically available isotopes have half-lives on the order of seconds to days, and it would appear that they would not be useful in most experiments. However, some of them are readily obtained by neutron activation or separation of a daughter and, thus, can be prepared in the laboratory. Many of these isotopes have widespread use in student experiments and in nuclear medicine, where their short half-lives greatly reduce the total radiation dose to the patient.

plating of an alpha source is preferred, to obtain the most uniform and thinnest source possible. Beta sources are less affected; gamma sources, particularly for higher energy emissions, are hardly affected at all.

In summary, to convert a count rate to disintegrations in a unit time, the following corrections must be made: the background rate is subtracted, the number of counts lost to resolving time error is added, and the net count rate is corrected for the fraction of counts theoretically able to reach the detector (geometry), for thickness (self-absorption), for scattering, and for air absorption. If the samples have been counted over a period of time, a normalization factor for changes in counter efficiency is then included. Finally, if the sample has decayed by a finite amount, the activity must be corrected to a standard time according to the equation

$$A = A_o e^{-\lambda t} \tag{4.2-3}$$

where A is the activity at time t, λ is the decay constant, and A_o is the initial activity. Corrections for geometry, scatter, and self-absorption can be omitted, if the sample is being compared to an identically prepared standard of known activity or if only a relative count rate is needed. A summary of data for counting efficiency corrections is given in Appendix I (Section 4.2.7) provided by the editor.

Several additional considerations are necessary to successfully complete an experiment. They are ease of handling, proper storage facilities, suitable specific activity, and commercial availability of the required radioisotopes. Specific activity is important, because it governs the amount of sample used during source preparation. Consider, for example, that in beta counting for the direct preparation of a source on a disc a 10- to 25- $\mu\ell$ aliquot is optimal; whereas, in liquid scintillation counting a 1-ml aliquot can often be used without severe quenching effects. For some well detectors used in gamma counting, very large volumes may be acceptable. For most experiments, it is necessary to have some assurance of chemical purity. Although this information is often provided by the manufacturer, if there is reason to believe that decomposition has occurred as the result of self-radiolysis, the source should be rechecked before an experiment has begun. Paper and thin layer chromatography are usually excellent methods for organic chemicals. Third, instruments vary in their performance from day to day. Thus, a check source should be counted each time an instrument is used, and a log of results kept. This will also ensure that any major malfunction will be observed immediately. Finally, some isotopes, including such commonly used ones as phosphorous-32, iodine-131, and sulfur-35, have half-lives that are sufficiently short that the results of experiments conducted over several days must be corrected for decay to a normalized value. Since the half-lives of these isotopes are too short to warrant their accurate calibration, mock standards are available for this purpose. Since the characteristics of the mock standard resemble those of the source over a limited range only, the recommendations of the manufacturer must be followed precisely.

4.2.4. Special Considerations

To determine the absolute amount of radioactivity present in a source, the decay scheme must be known. Fortunately, many commonly used isotopes decay by a single type of emission. However, some isotopes have alternate pathways or branching for decay. Gamma emitters can and often do proceed through metastable states as intermediates. Since the unit of measurement of radiation, the curie (Ci), is defined as 3.7×10^{10} disintegrations/sec, the relative abundance of the emission that is being measured must be known to convert the count rate to activity in microcuries. For example, consider the hypothetical isotope A decaying to B as shown in Figure 4.2-2. First, notice that the sum of all emissions for the decay of A is greater

4.2.5. Counting Equipment

A complete description of the principles upon which the various instruments are based may be obtained in any of several books on the subject.[12-14] The following discussion is limited to a brief description of the characteristics of each instrument for the measurement of radioactivity, i.e., types of radiation measured, their levels, background characteristics, and any other information necessary to provide the basis on which to choose an instrument appropriate for an experiment. This discussion pertains only to the measurement of sources as encountered in radiochemical analysis, and extrapolation to other situations is not warranted.

NaI(Tl) crystals are the most common type of detectors for gamma counting. In principle, energy from radioactive decay is fluoresced or scintillated from the crystal in the form of light which is converted into a signal and amplified through a photomultiplier. The crystal may be a flat disc or well, in which case a hole is bored into the crystal in order to improve the geometry. A well counter can have greater than 80% efficiency; a flat crystal, even theoretically, cannot exceed 50%. The scintillation counter can be operated in conjunction with a spectrometer, since the signal emitted is dependent on the energy of the gamma ray. However, a peak can be resolved to only about 30% full width at half maximum, and a complete spectrum at all energies, less than or equal to the gamma emission, is obtained because of Compton scatter and backscatter. The resolving time of the system is good and it rarely presents a problem in measurement. Gamma counters are not exceptionally sensitive to extremely low amounts of radioactivity, and background count rate is fairly high. Background is considerably higher for the lowest range of detection which is about 0 to 30 keV. The presence of higher energy gamma emissions makes the detection of lower energies extremely difficult. In this case, spectrum stripping is necessary. For the above two reasons, the gamma scintillation spectrometer is most effective in the detection of higher energy gamma rays.

Geiger-Mueller counters are capable of detecting alpha and beta particles although they cannot discriminate between the two. Resolving times are long, and the detectors lose a significant number of counts even in the range of 5,000 to 10,000 counts/min. Windows are relatively thick, so that counting efficiency for alphas and low-energy betas is low. Most gamma rays pass through the detector without interaction unless a metal shield is added to cover the tube. A classical method for radiation detection, the Geiger-Mueller counter, still has use in providing a simple and inexpensive detection method. Because the detector tube itself is rugged and easily replaced, Geiger-Mueller counters are often used in student laboratories and as probes for portable survey meters and monitors.

The preferred method for detection of low-energy beta emitters, tritium and carbon-14, is liquid scintillation counting.[15] The geometry in this detector is nearly 4π with a theoretical counting efficiency of nearly 100%. However, a substantial proportion of the energy from tritium decay is lost, so that the counting efficiency is about 40% for tritium. The contents of a scintillation cocktail, i.e., the solution into which the radioactive material is placed, vary somewhat,[15] but all contain a primary scintillator such as 2,5-diphenyloxazole (PPO) and a secondary scintillator, usually 1,4-bis-(5-phenyloxazol-2-yl)benzene (POPOP), which is used to shift the wavelength of the emitted light to an optimum range for detection. Ethanol may be present to aid in the suspension of aqueous samples. Toluene or a similar solvent with favorable properties of transfer of the energy from emissions to the scintillator is the main constituent, a factor that must be considered in the selection of counting vials. The use of polyethylene vials can be advantageous, but toluene will diffuse through them, an unacceptable event if the samples are to be stored for later use. Glass vials should always be of the borosilicate type with a low potassium content in order to minimize background counts contributed by potassium-40. The background count rate is in the range of 30 counts/min. The sample prepared for counting must be clear; if any opacity is observable, severe quenching will occur and the sample results will be in error and should be rejected. Since opacity is most often the result of too great a sample volume being placed in the cocktail, opacity can sometimes be corrected by adding extra scintillation solution.

Even though a liquid scintillation counter is capable of energy discrimination, beta emissions do not occur at a discrete energy; rather, a range of energies is found with the excess being carried off by simultaneous emission of neutrinos. As a consequence, it is difficult to measure two beta emitters present in the same solution, even if their maximum energies are extremely different. For similar energies, the task is impossible. Thus, the use of the liquid scintillation counter is easily extended to other beta emitters, especially sulfur-35 and phosphorus-32, but direct standardization by the use of two isotopes is not possible. There are two methods for standardization. One is done instrumentally by counting a source, such as radium, in different positions. This method is available only if the instrument is equipped with this feature. A second method is internal standardization. The sample is first counted, a known quantity of the radioactive substance is added, and the sample, recounted. The calculated number of disintegrations per minute for the added material, divided by the difference between the two measured count rates, represents the correction factor by which the original number must be multiplied, to obtain the actual number of counts contributed by the sample.

In recent years, the liquid scintillation counting technique has been extended to the counting of alpha particles.[16-20] Since alphas are monoenergetic and have high energies, this technique can be extremely successful even though alpha particles are readily absorbed. The advantage of this technique is that the source does not require the careful and tedious preparation required for other alpha counting techniques, since rather large amounts of extraneous materials can be tolerated. A disadvantage is that standardization of a sample cannot be carried through each step of an analysis by using two isotopes of the same element as when using a solid state detector. Thus to determine chemical yield, spiked controls must be run simultaneously and this method is less reliable, especially for biological materials.

The proportional counter may be used for measurement of alpha particles, even in the presence of beta emission, or for beta particles, if alpha decay does not occur. Although the proportional counter is similar in concept to the Geiger-Mueller counter, different electrode shapes make possible discrimination between alpha and beta particles and a much lower resolving time. Additional differences include a very thin, and therefore, fragile window and a continuous flow of gas to the detector. The background for beta counting averages about 30 cpm, but it can be lowered considerably with extensive shielding. Background in the alpha region is less than 1 count/min, so that the proportional counter is a very sensitive instrument for alpha detection. Proportional counters have been adapted to multiple sample changing devices and motorized drive mechanisms for scanning thin layer and paper chromatography strips, thus extending greatly their range of applicability.

Silicon and germanium surface barrier detectors represent the most sensitive method available to date, for the detection of extremely low quantities of alpha emitters. They are readily capable of detecting a fraction of a count per minute with negligible contributions from background. Thus, one of their most important applications in health physics is the measurement of plutonium-239 and americium-241 in urine, where a maximum permissible body burden may be represented by 1 to 2 disintegrations/min or less in a 24-hr sample. A lithium-drifted germanium detector is available for counting gamma emissions, but it is difficult to use and the sensitivity it offers is rarely necessary in studies involving radiological health; thus, it has received only limited use in this field.

The advantages of solid state detectors besides extreme sensitivity are good energy resolution,[13] which can result in separation of the spectra of two alpha emitters such as americium-241 and americium-243, even though their energies are nearly identical, and the ability to use an internal tracer throughout an experiment so that the exact yield for individual samples can be determined. However, the disadvantages are rather severe. First, a good vacuum system must be available, and considerable time must be spent in pumping down the system. Thus, it is impractical to use the solid state detector for short sampling times when the count rate is sufficiently high to be detectable using other instruments. Second, in a high vacuum, improperly prepared solids will be lifted off the sample, and some of these solids will be deposited on the gold surface of the detector, with consequent irreversible damage to the detector. Solid state detectors should not be used unless the sample being counted has been electroplated. Finally, in order to take full advantage of the exceptional spectral resolving powers of the detector, a multichannel analyzer with at least 1056 channels is preferable. In summary, surface barrier detectors should be used for extremely low activities, not when alternate detectors would do the same job.

4.2.6. Analytical Methods Adaptable to the Study of Radioisotopes

Certain techniques in analytical and biochemistry are particularly useful in the study of the effects of radioisotopes or in their separation from biological or environmental matrices. Ion exchange techniques, using both solid[14] and liquid exchangers, are especially useful. Liquid ion exchangers have some advantage over solid exchangers, because the exchange reactions are rapid and reliable. The liquid ion exchangers, bis-2-ethylhexyl hydrogen phosphate (HDEHP), trioctanolamine (TOA), and 2-thenoyl trifluoroacetone (TTA) have been used extensively.[21-27] Each liquid ion exchanger is most efficient for the extraction of metal ions at a specific pH. However, most are used at an acidic pH, an important consideration in the extraction of actinides where radiocolloid formation presents a problem. An example illustrates the power of this technique.[25] Plutonium and uranium can be exchanged to HDEHP-toluene in 8 N nitric acid; whereas, americium is extractable at pH 4 to 5. Plutonium and uranium can then be separated from a 2 N hydrochloric acid solution by TOA-xylene, since uranium is exchanged and plutonium is not.

Gel permeation chromatography (GPC) [28],[29] is readily adaptable to the use of isotopes in biological systems when binding of the radioactive species to a biological molecule is known or suspected. GPC has been especially useful in the identification of proteins that bind the isotopes plutonium [30],[31] and americium [32],[33] and has contributed greatly to the understanding of the biological properties of these elements. Since very little radioactivity, on a molar basis, is needed to be detectable, microanalytical GPC methods can be developed when extremely small amounts of material are available. However, the possibility of radiocolloid formation must be considered for metal ions, since for the study of biological systems, a pH near neutrality is preferred. Radiocolloids can form high molecular weight complexes so that failure to identify this problem can lead to anomalous results. At a minimum, a significant amount of the isotope may be lost by adsorption on the container walls.

A group of related techniques, electrophoresis and thin layer and paper chromatography are adaptable to work with isotopes.[10,14] In each case, components of a solution become physically separated, in electrophoresis because of differential migration in a charged field, and in chromatography because of differences in migration with a solvent front. Thus, a major use of these techniques is in separating the various components with which an isotope has reacted when it is added to a mixture.

These conditions are most often encountered in ascertaining purity after organic synthesis and in metabolism studies. Electrophoresis is especially useful in determining the possibility of binding an isotope to one or more of the components in a mixture of proteins such as serum. Procedures for tagged molecules do not differ from well-established techniques for other molecules, except that ultimately measurement is quantitated by counting the radioactivity.

Electroplating, which has been mentioned before, is especially useful for the preparation of alpha sources. Although this technique is extended to other uses as well, it is somewhat tedious and therefore has little use for isotope work, unless a source must be prepared with the greatest purity and precision possible. Electroplating has been used extensively in the preparation of americium-241[34,35] and plutonium-239[36].* sources of very low activity such as are encountered in radiological health.

Two additional techniques in common use are solvent extraction and precipitation of an isotope on a carrier.[10,14] Solvent extraction requires that the substance extracted be more soluble in the nonpolar phase than in water. Thus, the method is applicable to the extraction of many tagged organic molecules and to the removal of organic impurities from polar isotopes. Since carrier-free isotopes are present in solutions in very small quantities, they cannot be separated from impurities by precipitation. However, addition of several milligrams of a nonradioactive form of the isotope makes precipitation possible. Often, the precipitate is collected on a filter by suction filtration in order to prepare a source directly for counting. Chemical yields can be determined by weighing the recovered material. The method is not applicable to alpha counting, since a thick source is obtained.

Instruments that are not readily adaptable to radioisotope techniques include gas-liquid chromatography, nuclear magnetic resonance spectroscopy, mass spectroscopy, infrared spectroscopy, colorimetry, and ultraviolet spectrophotometry. In these instances, with the exception of gas-liquid chromatography, no additional sensitivity of the method can be attained by use of isotopes. For gas-liquid chromatography, no read-out devices to measure radioactivity are available. In mass spectroscopy, considerable confusion would be added into the interpretation of fragmentation patterns, since the measurement of mass would be affected by the different atomic weights for the radioactive and stable forms. In all cases, the possibility of irreversible contamination of at least some part of the equipment is high.

APPENDIX I

4.2.7. DETERMINING COUNTING EFFICIENCY FOR INTERNAL PROPORTIONAL COUNTERS**

Allen Brodsky

Since internal proportional counters do not detect all beta rays emitted by a sample, it is necessary to divide the net counting rate by an appropriate efficiency correction decimal to determine the total beta emission. This efficiency (E) is the prod-

---

* Data obtained from Setter, L. R., Goldin, A. S., and Nader, J. S., Radioactivity assay of water and industrial wastes with internal proportional counter, *Anal. Chem.*, 26, 1305, 1954.

** Section 4.2.7 is reprinted from Bureau of Radiological Health, Radiological Health Handbook, U.S. Department of Health, Education, and Welfare, Public Health Service, Rockville, Md., 1970, with their permission. For spectra of other beta emitters, see Section 3.7.

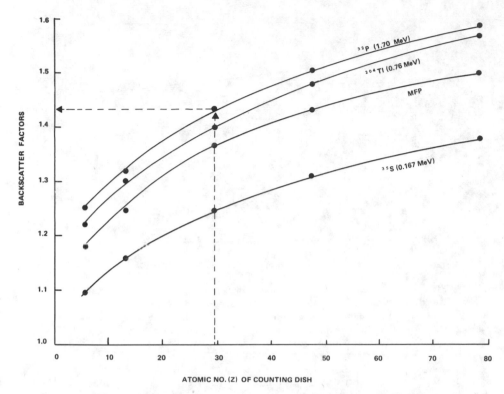

FIGURE 4-3. Backscatter factors vs. atomic number for beta emitters. (From Nader, J. S., Hagee, G. R., and Setter, L. R., (*Nucleonics,*) 12(6), 1954, as given in Bureau of Radiological Health, Radiological Health Handbook, U.S. Department of Health, Education, and Welfare, Public Health Service, Rockville, Md., 1970, 127. With permission.)

uct of three factors — geometry (G), backscatter (B), and self-absorption, generally expressed in terms of a transmission factor (T): $E = G \times B \times T$.

Geometry factor (G) — Not all radiation from a sample is emitted in the direction of the detector. The geometry factor accounts for the fraction emitted in the proper direction. For internal proportional counters with hemispherical chambers, this factor is 0.50.

Backscatter factor (B) — The backscattering of beta rays is a function of their energy and the atomic number (Z) of the counting dish. The following curves may be conveniently used for estimating this factor. To illustrate their use, consider a ^{32}P sample in a copper counting dish (Z = 29). From the appropriate value on the abscissa, draw a vertical line until it intersects the curve for ^{32}P. A horizontal line projected from the point of intersection to the ordinate reveals the resultant factor to be about 1.43. (Figure 4-3)

Self-absorption or transmission factor (T)* — A fraction of the beta particles emitted by a sample may be absorbed within the sample itself. This loss, which increases with sample thickness, is known as self-absorption. For counting purposes, it may be conveniently expressed in terms of a transmission factor, the fraction of the emitted beta particles not absorbed within a sample. The transmission factor (T) may be estimated using the curves given below. If, for example, a sample containing ^{32}P weighs 200 mg and is evenly distributed on a 2-in. diameter dish, then the aver-

---

* Data obtained from Setter, L. R., Goldin, A. S., and Nader, J. S., Radioactivity assay of water and industrial wastes with internal proportional counter, *Anal. Chem.*, 26, 1305, 1954.

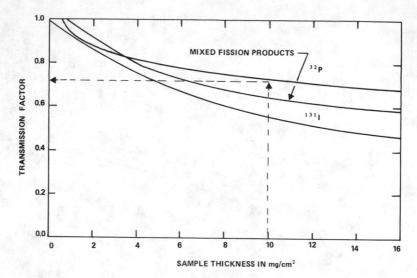

FIGURE 4-4.   Self-absorption vs. sample thickness for beta emitters. (Data obtained from Setter, L. R., Goldin, A. S., and Nader, J. S., *Anal. Chem.*, 26, 1305, 1954, as given in Bureau of Radiological Health, Radiological Health Handbook, U.S. Department of Health, Education, and Welfare, Public Health Service, Rockville, Md., 1970, 128. With permission.)

age sample thickness can be calculated to be 10 mg/cm². To estimate the factor, draw a vertical line from the appropriate value on the abscissa until it intersects the curve for ³²P. A horizontal line projected from the point of intersection to the ordinate reveals the resultant factor to be about 0.73. (Figure 4-4)

Overall counting efficiency (E)* — The efficiency correction decimal fraction for the previous example, in which a sample containing ³²P was counted, would be:

$$E = 0.50 \times 1.43 \times 0.73 = 0.52$$

If the net sample counting rate was 1000 counts/min, the disintegration rate could be calculated to be:

$$dpm = net\ cpm \div E = 1000\ cpm \div 0.52 = 1920\ dpm.$$

REFERENCES

1. **Bayly, R. J. and Evans, E. A.,** Storage and Stability of Compounds Labeled with Radioisotopes, Amersham/Searle Corp., Des Plaines, Ill., undated.
2. **Schweitzer, G. K. and Jackson, W. M.,** Studies in low concentration chemistry. I. The radiocolloidal properties of lanthanum-140, *J. Am. Chem. Soc.,* 74, 4178, 1952.
3. **Davydov, Y. P.,** Experimental determination of the degree of polymerization of hydrolyzed ions in solution, *Radiokhimiya,* 14, 142, 1972.
4. **Starik, I. E., Ampelogova, N. I., Barbanel, Y. A., Ginsburg, F. L., Ilmenhova, L. I., Groozovskaya, N. G., Skulskii, I. A., and Sheidina, L. D.,** Reply to Yu P. Davydov's article "The nature of colloids of radioactive elements", *Radiokhimiya,* 9, 105, 1967.

* Data obtained from Setter, L. R., Goldin, A. S., and Nader, J. S., Radioactivity assay of water and industrial wastes with internal proportional counter, *Anal. Chem.,* 26, 1305, 1954.

5. **Davydov, Y. P.,** The question of colloids and radiocolloids, *Radiokhimiya,* 14, 140, 1972.
6. **Shchebothovskii, V. N.,** A method of determining the sorption of microquantities of radioactive isotopes in the region of their hydrolysis, *Radiokhimiya,* 14, 635, 1972.
7. **Sheidina, L. D., Razovskaya, N. G., and Kovarskaya, E. N.,** A method of investigating the sorption of radioelements, *Radiokhimiya,* 13, 180, 1971.
8. **Lindenbaum, A. and Westfall, W.,** Colloidal properties of plutonium in dilute aqueous solution, *Int. J. Appl. Radiat. Isot.,* 16, 545, 1965.
9. **Schubert, J., Fried, J. F., Rosenthal, M. W., and Lindenbaum, A.,** Tissue distribution of monomeric and polymeric plutonium as modified by a chelating agent, *Radiat. Res.,* 15, 220, 1961.
10. **Chase, G. D. and Rabinowitz, J. L.,** *Principles of Radioisotope Methodology,* 3rd ed., Burgess Publishing, Minneapolis, 1970.
11. Bureau of Radiological Health, Radiological Health Handbook, U.S. Department of Health, Education, and Welfare, Public Health Service, Rockville, Md., 1970.
12. **Lapp, R. E. and Andrews, H. L.,** *Nuclear Radiation Physics,* 4th ed., Prentice-Hall, Englewood Cliffs, N. J., 1972.
13. **Friedlander, G., Kennedy, J. W., and Miller, J. M.,** *Nuclear and Radiochemistry,* 2nd ed., John Wiley & Sons, New York, 1966.
14. **Finston, H. L., Crouthamel, C. E., Heinrich, R. R., Seaman, W., and Guinn, V. P.,** Section D-6: radioactive methods, *Treatise on Analytical Chemistry,* Part 1, Vol. 9, Kolthoff, I. M., and Elving, P. J., Eds., Wiley-Interscience, New York, 1971.
15. **Turner, J. C.,** Sample Preparation for Liquid Scintillation Counting, Amersham/Searle Corp., 2000 Nuclear Drive, Des Plaines, Ill., 60018.
16. **Lindenbaum, A. and Lund, C. J.,** Alpha counting by liquid scintillation spectrometry: plutonium-239 in animal tissues, *Radiat. Res.,* 37, 131, 1969.
17. **Seidel, A. and Volf, V.,** Rapid determination of some trace transuranic elements in biological material by liquid scintillation counting, *Int. J. Appl. Radiat. Isot.,* 23, 1, 1972.
18. **Crawley, F. E. H., Humphries, E. R., and Stather, J. W.,** A comparison of 239-plutonium in soft tissues and skeleton of mice, rats, and hamsters, *Health Phys.,* 30, 491, 1976.
19. **Keough, R. F. and Powers, G. J.,** Determination of plutonium in biological materials by extraction and liquid scintillation counting, *Anal. Chem.,* 42, 419, 1970.
20. **Cross, P. and McBeth, G. W.,** Liquid scintillation alpha particle assay with energy and pulse shape discrimination, *Health Phys.,* 30, 306, 1976.
21. **Moore, F. L.,** Selective liquid-liquid extraction of beryllium (IV) with 2-thenoyltrifluoroacetone — xylene, *Anal. Chem.,* 38, 1872, 1966.
22. **Schneider, R. A.,** Analytical extraction of neptunium using tri-*iso*-octylamine and thenoyltrifluoroacetone, *Anal. Chem.,* 34, 522, 1962.
23. **Bruegner, F. W., Stover, B. J., and Atherton, D. R.,** Determination of plutonium in biological material by solvent extraction with primary amines, *Anal. Chem.,* 35, 1671, 1963.
24. **Moore, F. L. and Hudgens, J. E.,** Separation and determination of plutonium by liquid-liquid extraction, *Anal. Chem.,* 29, 1767, 1957.
25. **Butler, F. E., Boulogne, A. R., and Whitley, E. A.,** Bioassay and environmental analyses by liquid ion exchange, *Health Phys.,* 12, 927, 1966.
26. **Peppard, D. F., Mason, G. W., and Andrejasich, C. M.,** Applicability of certain organo-phosphorus extractants to environmental and biological radioactivity surveys, *J. Inorg. Nucl. Chem.,* 25, 1175, 1963.
27. **Peppard, D. F., Mason, G. W., Driscoll, W. J., Sironen, R. J.,** Acidic esters of orthophosphoric acid as selective extractants for metallic cations — tracer studies, *J. Inorg. Nucl. Chem.,* 7, 276, 1958.
28. **Anderson, D. M. W. and Stoddard, J. F.,** Some observations on molecular weight estimations by molecular seive chromatography, *Anal. Chim. Acta,* 34, 401, 1966.
29. **Sephadex Gel Filtration in Theory and Practice, Pharmacia Fine Chemicals, Inc., Piscataway, N. J.**
30. **Turner, G. A. and Taylor, D. M.,** The binding of plutonium to serum proteins *in vitro, Radiat. Res.,* 36, 22, 1968.
31. **Stevens, W., Bruegner, F. W., and Stover, B. J.,** *In vivo* studies on the interactions of Pu(IV) with blood constituents, *Radiat. Res.,* 33, 490, 1968.
32. **Boocock, G. and Popplewell, D. S.,** *In vitro* distribution of americium in human blood serum proteins, *Nature,* 210, 1283, 1966.
33. **Bruegner, F. W., Stevens, W., and Stover, B. J.,** Americium-241 in the blood: *in vivo* and *in vitro* observations, *Radiat. Res.,* 37, 349, 1969.
34. **Mitchell, R. F.,** Electrodeposition of actinide elements at tracer concentrations, *Anal. Chem.,* 32, 326, 1960.

35. **Donnan, M. Y. and Dukes, E. K.,** Carrier techniques for quantitative electrodeposition of actinides, *Anal. Chem.,* 36, 392, 1964.
36. **Harley, J. H., Ed.,** Manual of Safety Procedures, NYO-4700, Health and Safety Laboratory, U.S. Atomic Energy Commission, New York, 1967.

4.3. PRODUCTION AND TESTING OF RADIOPHARMACEUTICALS

Mitchell L. Borke

4.3.1. Production of Nuclides Used in Radiopharmaceuticals

Radiopharmaceuticals can be defined as (1) drugs that are radioactive and, because of this property, used exclusively for supplying information about normal and pathological processes in human organisms, or (2) compounds that are radioactive or that can be made radioactive after introduction into tissue, where its radiation is effective in destroying tissue.[1] The first group of compounds represents *diagnostic* and the latter *therapeutic radiopharmaceuticals.* It should be noted that neither the diagnostic nor therapeutic radiopharmaceuticals exert any pharmacological action.[2] Furthermore, radiopharmaceuticals differ from radiochemicals with respect to methods of synthesis, quality control, and intended use. Generally, synthetic methods used for preparation of radiochemicals are similar to those employed in the synthesis of nonradioactive compounds. This is not always true in the case of radiopharmaceuticals. Radiochemicals may be extremely pure both chemically and radiochemically; nevertheless, they cannot be used as radiopharmaceuticals. Finally, radiochemicals are not intended for the same use as radiopharmaceuticals. Therefore, it can be said that some radiopharmaceuticals are radiochemicals, but no radiochemicals are radiopharmaceuticals.

Because of the intended use of radiopharmaceuticals, they are subject to certain definite requirements; namely,[2]

1. They must have specific physical and chemical characteristics to be useful in diagnostic, therapeutic, and research procedure in order to provide information on translocation, deposition, and metabolism.
2. They must provide the sought information when used in minimal quantities in order to avoid undesirable biological and radiation effects.
3. Their chemical and physical toxicities as well as those of their metabolites and daughter nuclides must be known.
4. They must be radiochemically pure or almost pure, i.e., they must contain the radionuclide only in the desired chemical form.

Consequently, the usefulness of a radiopharmaceutical depends on the physical characteristics of the radionuclide and the chemical and biological behavior of the labeled material. Therefore, a radiopharmaceutical is characterized by the following four parameters:[3] biological behavior, type and rate of radioactive decay, detection characteristics, and production factors.

There are three sources of radionuclides used as radiopharmaceuticals per se or used in the preparation of radiopharmaceuticals, i.e., nuclear reactors, cyclotrons, and generators. Nuclear reactors can produce radionuclides through neutron capture reactions, e.g., (n, γ), (n, p), and (n, α), as well as through the (n, f) reaction. In the latter case, the radionuclides are encountered in the products of nuclear fission, e.g., ^{131}I, ^{137}Cs, ^{144}Ce, ^{90}Sr, and ^{133}Xe.

The (n, γ) reaction, which involves "thermal" neutrons as the bombarding particles, is the most important source of radionuclides. However, the (n, p) and (n, α), reactions have the advantage of providing products with high specific activities. The use of enriched stable isotopes as targets contributes to increased yields, particularly in those cases where the natural abundance of the target nuclide is low, and also minimizes undesirable side reactions due to activation of other stable isotopes be-

Table 4.3-1
NEUTRON-INDUCED REACTIONS

Type	Illustration
(n,γ)	$^{46}Ca(n,\gamma)^{47}Ca$
	$^{197}Au(n,\gamma)^{198}Au$
$(n,fission)$	$^{235}U(n,fission)^{131}I$
	$^{235}U(n,fission)^{90}Sr$
(n,p)	$^{32}S(n,p)^{32}P$
	$^{14}N(n,p)^{14}C$
(n,α)	$^{40}Ca(n,\alpha)^{37}Ar$
	$^{6}Li(n,\alpha)^{3}H$
(n,γ) decay	$^{124}Xe(n,\gamma)^{125}Xe\ EC\ ^{125}I$
	$^{198}Pt(n,\gamma)^{199}Pt\ \beta\ ^{199}Au$

From Baker, P.S., in *Radioactive Pharmaceuticals,* Andrews, G. A., Kniseley, R. M., Wagner, H. N., Eds., USAEC-Division of Technical Information, Oak Ridge, Tenn., 1966, 129. With permission.

sides the desired target. Table 4.3-1 presents types of commonly used neutron reactions and specific examples for each reaction type.[4]

Next to nuclear reactors, cyclotrons are the second important source of radionuclides. Cyclotron-produced radionuclides are neutron deficient and, consequently, have a number of advantages over those produced in a nuclear reactor. Many have relatively short halflives and are generally carrier-free.[5] Approximately 40 cyclotron-produced radionuclides that have been proposed for various procedures in nuclear medicine are listed by MacDonald[6] in table form, together with typical production reactions and 151 references.

Modern scanners and scintillation cameras have brought about a need for readily available radionuclides that provide a high photon flux but deliver a low radiation dose to the patient. It has been suggested by Wagner[7,8] that the concept of maximum information with minimum radiation exposure to the patient can be best satisfied by radionuclides whose effective halflives fall into the range of 0.693 times the time required for completion of the study. Therefore, radionuclides with $T_{1/2} \leqslant 1$ min could be very useful for dynamic studies. Not only would the radiation exposure to the patient be minimal, but the study could also be repeated within a few minutes, because the administered radioactivity would rapidly decay.[9]

A convenient source of short-lived radionuclides is a system in which a parent having an intermediate halflife decays to a short-lived daughter. The parent-daughter transient equilibrium mixture adsorbed on a suitable material, e.g., alumina, forms the heart of a radionuclide generator. Because of differences in chemical behavior of the daughter nuclide as compared to those of the parent, the former can be eluted at convenient time intervals. Radionuclide generator systems for production of short-lived radionuclides for use in nuclear medicine are listed in Table 4.3-2.[10]

Of the 27 generator systems listed in Table 4.3-2, three are under preliminary investigation, three have been used in animal studies, seven have undergone preliminary human studies, no status is listed for eleven systems, and only three (^{87m}Sr, ^{113m}In and ^{99m}Tc) have clinical applications.

4.3.2. Preparation and Synthesis of Radiopharmaceuticals

From a chemical viewpoint, radiopharmaceuticals can be divided into inorganic

Table 4.3-2
GENERATOR-PRODUCED RADIONUCLIDES FOR USE IN NUCLEAR MEDICINE

Daughter					Parent				
Isotope	$T_{1/2}$	Decay (%)	Photon MeV (%)	Status[1]	Isotope	$T_{1/2}$	Decay (%)	Photon MeV (%)	Production
^{28}Al	2.3 min	β^-(100)	1.78(100)		^{28}Mg	21.2 hr	β^-(100)	0.031(96), 0.40(30), 0.95(30), 1.35(70)	^{26}Mg(t,p), ^{26}Mg(^{4}He,2p)
^{38}Cl	37.3 min	β^-(100)	1.60(38), 2.17(47)		^{38}S	2.9 hr	β^-(100)	1.88(95)	^{37}Cl(^{4}He,3p)
^{47}Sc	3.4 days	β^-(100)	0.160(73)	C	^{47}Ca	4.5 days	β^-(100)	0.49(5), 0.815(5), 1.31(74)	^{46}Ca(n,γ)
^{52m}Mn	21.1 min	β^+(98), IT(2)	0.511(196), 1.43(100)		^{52}Fe	8.2 hr	β^+(56), EC(44)	0.165(100), 0.511(112)	^{50}Cr(^{4}He,2n)
^{62}Cu	9.8 min	β^+(97)	0.511(194)		^{62}Zn	9.1 hr	β^+(18), 0.51(47), 0.59(22), EC(82)	0.042(20)	^{63}Cu(p,2n)
^{68}Ga	68.3 min	β^+(88), EC(12)	0.511(176), 1.08(3.5)	C	^{68}Ge	275 days	EC(100)	—	^{66}Zn(^{4}He,2n)
^{72}As	1.1 days	β^+(75), EC	0.511(150), 0.63(8), 0.835(78)		^{72}Se	8.4 days	EC(100)	0.046(59)	^{75}As(d,5n), ^{70}Ge(^{4}He,2n)
^{81m}Kr	13 sec	IT(100)	0.190(65)	C	^{81}Rb	4.7 hr	β^+(13), EC(87)	0.253, 0.450, 0.511(26)	^{79}Bi(^{4}He,2n)
^{82}Rb	1.3 min	β^+(96), EC(4)	0.511(192), 0.777(9)	C	^{82}Sr	25 days	EC(100)	—	^{85}Rb(p,4n)
^{83m}Kr	1.86 hr	IT(100)	0.009(9)	B	^{83}Rb	83 days	EC	0.53(93), 0.79(0.9)	^{83}Br(^{4}He,2n)^{83}Sr
^{87m}Sr	2.83 hr	IT(99+)	0.388(80)	D	^{87}Y	3.3 days	β^+(<1), EC(99)	0.483	^{86}Sr(d,n), ^{87}Sr(p,n), ^{88}Sr(p,2n)
^{89}Y	16.1 sec	IT(100)	0.91(99)		^{89}Zr	3.3 days	β^+(22), EC(78)	0.511(44)	^{89}Y(p,n)
^{90m}Nb	24 sec	IT(100)	0.122(71)	A	^{90}Mo	5.7 hr	β^+(25), EC(75)	0.257(85), 0.445(9)	^{93}Nb(p,4n)

Table 4.3-2 (continued)
GENERATOR-PRODUCED RADIONUCLIDES FOR USE IN NUCLEAR MEDICINE

Isotope	$T_{1/2}$	Daughter Decay (%)	Photon MeV (%)	Status[1]	Parent Isotope	$T_{1/2}$	Decay (%)	Photon MeV (%)	Production
^{99m}Tc	6.0 hr	IT(100)	0.140(90)	D	^{99}Mo	2.78 days	β^-(100)	0.511(50) 0.181(7) 0.740(12) 0.780(4)	Fission, ^{98}Mo(n,γ)
^{109m}Ag	39.2 sec	IT(100)	0.088(5)	B	^{109}Cd	1.24 years	EC(100)	—	^{108}Cd(n,γ), ^{109}Ag(d,2n)
^{113m}In	1.66 hr	IT(100)	0.393(64)	D	^{113}Sn	115 days	EC(100)	0.225(1.8)	^{112}Sn(n,γ)
^{118}Sb	3.5 min	β^+(75) EC(25)	0.511(150) 1.23(3)		^{118}Te	6.0 days	EC(100)	—	^{121}Sb(d,5n)
^{122}I	3.5 min	β^+(65) EC	0.511(130) 0.564		^{122}Xe	20.1 hr	EC(100)	0.060 0.090 0.110	^{127}I(p,6n)
^{132}I	2.26 hr	β^-(100)	0.52(20) 0.67(144) 0.73(89) 0.95(22)	C	^{132}Te	3.24 days	β^-(100)	0.053(17) 0.230(90)	Fission
^{128}Cs	3.8 min	β^+(51) EC(49)	0.441(27) 0.511(102) 0.528	B	^{128}Ba	2.43 days	EC(100)	0.134 0.278	^{133}Cs(p,6n)
^{137m}Ba	2.55 min	IT(100)	0.662(89)	C	^{137}Cs	30.0 years	β^-(100)	—	Fission
^{134}La	6.8 min	β^+(62) EC(38)	0.511(124) 0.605(61)		^{134}Ce	3.0 days	EC(100)	—	^{139}La(p,6n)
^{140}Pr	3.4 min	β^+(50) EC(50)	0.511(100) 1.60(0.3)	A	^{140}Nd	3.3 days	EC	—	^{141}Pr(^{4}He,2n)
^{178}Ta	9.4 min	β^+(1) EC(99)	0.093 0.511		^{178}W	21.5 days	EC	—	^{181}Ta(p,4n)
^{183m}W	5.3 sec	IT(100)	0.108(19)		^{183}Ta	5.0 days	β^-(100)	0.246(33)	^{181}Ta(nn,γ)
^{191m}Ir	4.9 sec	IT(100)	0.129(25)	C	^{191}Os	15.0 days	β^-	—	^{190}Os(n,γ)
^{195m}Au	30.6 sec	IT(100)	0.261(77)	A	^{195m}Hg	1.67 days	IT(50) EC(50)	0.200(35) 0.261(20) 0.560(20)	^{197}Au(p,3n)

[1] A = Preliminary investigation; B = separation of daughter, animal studies; C = preliminary human studies; D = clinical applications.

preparations and organic compounds. Radioactive inorganic compounds can be prepared by uranium fission, e.g., ^{131}I, or by neutron bombardment of a suitable target nuclide, e.g., $^{32}S(n, p)^{32}P$. The produced radionuclide is purified, converted to a suitable chemical and physical form, e.g., sodium iodide ^{131}I solution, sodium phosphate ^{32}P solution, etc., stabilized, and either diluted with a suitable carrier or supplied in a carrier-free form.[2] It can also be used in the preparation of other inorganic radiopharmaceuticals. For example, sodium phosphate ^{32}P is mixed with a stoichiometric amount of chromium (III) nitrate, and the precipitate of chromic phosphate ^{32}P is washed, dried, and incinerated. The hard, sintered, insoluble mass is ground in a ball mill to less than 1 μm particle size, sterilized by autoclaving, and suspended in gelatin or 25% glucose.[11]

Organic compounds are usually tagged with radioactive isotopes using exchange reactions, radiochemical synthesis, biosynthesis, or trace iodination. Radioisotopes of iodine, e.g., ^{125}I and ^{131}I, have been introduced into organic molecules, e.g., sodium iodohippurate or iodobenzoic acids, by exchanging the radioiodine for ^{127}I.[12,13] Some exchange reactions proceed more readily in the presence of ICI.

Under proper conditions, tritium in the form of ^3H_2O, $^3H_2SO_4$, or $^3H_2$ replaces part of the hydrogen in a compound. However, the most important method of introducing tritium into an organic compound was reported by Wilzbach in 1957.[14] The unlabeled compound in finely powdered form is sealed in a tube with high specific-activity tritium gas and kept for several days. While the Wilzbach process is rapid and provides an easy method for preparation of tritiated organic compounds, it also has certain drawbacks. Tritium may add to double bonds rather than exchange with hydrogen, and the resulting compounds are different from those originally exposed to tritium gas and difficult to separate. Production of labeled impurities in addition to the desired compound necessitates strict purification procedures. Sometimes, amounts of those impurities are so small that neither their detection nor their removal is possible. Finally, dissolution of the labeled compound may cause cleavage of $O-^3H$ bonds and loss of tritium atoms which have to be removed from the material by repeated equilibration with a solvent, e.g., water or alcohol, that also must be removed.

The usual starting materials for radiochemical synthesis are $^{14}CO_2$ (obtained from $Ba^{14}CO_3$) for ^{14}C-tagged compounds and either ^3H_2O or tritium gas for ^3H-tagged compounds. Special texts have been published for the syntheses of both organic and inorganic radioisotopically labeled compounds.[15,15a,16]

Frequently it is either too difficult or not practical to prepare labeled compounds of biological origin by radiochemical synthesis. In those cases, it is advantageous to employ biosynthetic procedures. In biosynthetic procedures, a suitable form of the radionuclide is administered to a living organism, which synthesizes the desired labeled compound or derivative. Subsequently, the product is isolated and purified. Vitamin B_{12}, labeled with radiocobalt, and ^{75}Se-selenomethionine are examples of biosynthetically prepared radiopharmaceuticals.[17,18]

Radionuclides, incorporated into organic molecules by biosynthetic procedures, can be replaced with chemically similar or dissimilar radionuclides through exchange reactions.[19] Other reactions employed for replacement of groups by atoms, e.g., replacement of amino groups by halogens, or for chemical modification of compounds by incorporation of labeled reactive functional groups, have been described.[19—21]

In trace iodination, a minimum number of radioiodine atoms is introduced into a molecule, so that neither the chemical properties nor the biological behavior of the iodinated molecule changes significantly. For example, trace iodination of human serum albumin involves introduction of a gram atom of iodine per gram mole of protein. Primary sites of iodination are the tyrosyl moieties and, to a much lesser extent, the imidazole rings of the histidyl residues.[22] In the case of oleic acid and triolein, iodine saturates the double bonds of these compounds.

4.3.3. Quality Control of Radiopharmaceuticals

It was mentioned in the beginning of this section that one of the differences between radiochemicals and radiopharmaceuticals is the intended use of the latter group of compounds. Because radiopharmaceuticals are drugs they are subject to stringent quality control. The WHO Expert Committee on Specifications for Pharmaceutical Preparations[23] made the following statement about pharmaceutical quality control: "The suitability of drugs for their intended use is determined by (a) their efficacy weighed against safety to health according to label claim or as promoted or publicized and (b) their conformity to specifications regarding identity, strength, purity and other characteristics."

When a new drug is in the process of being developed, it is normally subject to the following steps: (1) laboratory studies of properties, toxicity, etc., (2) clinical tests, and (3) approval by a regulatory agency, e.g., the Food and Drug Administration (FDA). When the pharmaceutical is finally marketed, it becomes subject to quality control by the manufacturer, testing by the dispenser or the user, e.g., a radiopharmacist or a physician, and testing of the product and inspection of manufacturing facilities by FDA inspectors. These steps also apply to radiopharmaceuticals; however, in this case it may be useful to consider two different types of radiopharmaceuticals

1. Those supplied by a manufacturer in a form ready to use, i.e., the user only has to determine the dose and administer it to the patient
2. "In-house" preparations

The first group of radiopharmaceuticals poses no problem to the user, because it is the responsibility of the manufacturer to set up a proper quality control program. Nevertheless, upon receipt of a radiopharmaceutical the shipment should be subjected to physical quality control,[24] which first involves careful examination of the shipping container for contamination due to possible leakage. It is advisable to perform a "wipe test" on the container in order to check for external contamination. This precautionary measure will avert exposure of personnel to radiation, contamination of equipment, and possible loss of sterility. Then the container with the radiopharmaceutical should be assayed in a suitable ionization chamber to check if its activity corresponds to that stated on the label as well as on the purchase requisition. If there is no external contamination, containers with parenteral solutions should also be checked visually for any particulate matter or any altered appearance, such as change of color.

Quality control of the second type of radiopharmaceutical, the "in-house" preparation, is the responsibility of the radiopharmacist who prepares and dispenses the desired dosage form. Any quality control program, whether in a large manufacturing facility or in a radiopharmacy, should be based on guidelines specified by U.S.P. XIX and should include a check of radio-chemical purity, radionuclidic purity, sterility, and pyrogenicity.[25]

Radiochemical purity of a radiopharmaceutical refers to the fraction of the stated radionuclide present in the stated form. The appearance of radiochemical impurities may be due to decomposition caused by radiation or improper preparative procedures. Procedures used to ascertain radiochemical purity are described in individual monographs and include various chromatographic techniques, e.g., thin-layer, column or paper chromatography, and electrophoresis, as well as other suitable analytical techniques.[25] For example, radiochemical purity of a sodium chromate ^{51}Cr injection is determined by removing chromate ^{51}Cr as the lead salt, centrifuging, and checking the supernatant for residual radioactivity.[25]

Radionuclidic purity of a radiopharmaceutical refers to the proportion of radioactivity due to the desired radionuclide in the total radioactivity measured.[21] The presence of radionuclidic impurities may be the result of impurities in the target materials, various competing nuclear reactions, or differences in energies of the bombarding particles during production of the radionuclide. A radionuclide can be identified by its halflife, its mode of decay, and the energies of nuclear emissions.[25] U.S.P. XIX describes procedures for determination of halflife, for assays of beta- and gamma-emitting radionuclides as well as for identification of radionuclidic impurities.[25] The obtained data can be compared with those published in reference sources, such as References 26 and 27.

Sterility can be defined as absence of viable organisms, such as bacteria, fungi, and yeasts, that grow and are detectable in standard culture media.[19] Radiopharmaceuticals are not self-sterilizing because of their own radiation. Consequently, they must be sterilized by one of the accepted sterilization methods. Thermostable radiopharmaceuticals are steam sterilized. Those which are heat-labile, e.g., colloids, micro- and macro-aggregates, some organic molecules, and biochemicals, are sterilized by membrane filtration.[25] Because of the highly exacting nature of sterility tests, they should be conducted by personnel with training in microbiology or experience in rigid aseptic techniques, in an area free from dust and air-borne microorganisms.[25]

While the Pharmacopeia of the United States describes procedures for sterility tests that are suitable for revealing the presence of viable forms of bacteria, fungi, and yeasts, it also permits use of alternative procedures or procedural details provided that the obtained results are of equivalent reliability.[25] Since these guidelines are designed for nonradioactive pharmaceuticals and involve large volumes, they must be reduced accordingly in the case of radiopharmaceuticals because of the small volumes prepared and administered. However, in the case of long-lived radiopharmaceuticals that are prepared in larger volumes, no adjustment of U.S.P. procedures is necessary.[19]

Pyrogens are the products of growth and metabolism of microorganisms. When injected into animals or men, pyrogens cause a rise in body temperature, probably by acting on the temperature control center in the hypothalamus. Chemically, pyrogens are believed to be compound lipids attached to a carrier, possibly a polysaccharide or polypeptide. While many microorganisms produce pyrogens, Gram-negative bacilli are the source of the most potent ones.[28] Pyrogens do not affect all animals to the same degree; however, the fever response of rabbits most closely resembles that of man. Consequently, the U.S.P. pyrogen test utilizes rabbits as test animals.[25] However, the test is highly impractical for pyrogen testing of parenteral radiopharmaceuticals because of the small volumes involved and very frequently short halflives.

During the past several years, the *Limulus* amoebocyte lysate (LAL) test has been favorably accepted as a test for pyrogens and endotoxins. The test is based on the discovery of Bang, that the blood of the horseshoe crab (*Limulus polyphemus*) coagulates when the animal becomes infected.[29] This reaction was traced to clottable proteins in the amoebocyte, which gelled in the presence of very small quantities of endotoxin.[30] Presently, the lyophilized LAL reagent is available commercially in a kit form with reference endotoxin.[31-33] The reagent must produce a positive reaction with 0.1 ng of the reference endotoxin and a negative reaction with 0.0125 ng of the reference endotoxin.[34] The test has been demonstrated to be at least five times more sensitive to purified endotoxin than the rabbit test.[35] It is capable of measuring endotoxin at levels at which they were not always detected by the conventional rabbit test.

Although at the present time, the LAL test is not considered to be a replacement for the U.S.P. pyrogen test, sufficient data have been accumulated to justify its use as an in-process quality control test for endotoxin contamination.[31]

REFERENCES

1. **Soloway, A. H. and Davis, M. A.,** Survey of radiopharmaceuticals and their current status, *J. Pharm. Sci.,* 63, 647, 1974.
2. **Tubis, M.,** Radioactive tracers and radiopharmaceuticals, in *Nuclear Medicine,* Blahd, W. H., McGraw-Hill, New York, 1971, 143.
3. **Wagner, H. N. and Rhodes, B. A.,** The radiopharmaceutical, in *Principles of Nuclear Medicine,* Wagner, N. H., Jr., Ed., W. B. Saunders, Philadelphia, 1969, 259.
4. **Baker, P. S.,** Reactor-produced radionuclides, in *Radioactive Pharmaceuticals,* Andrews, G. A., Kniseley, R. M., and Wagner, H. N., Eds., USAEC-Division of Technical Information, Oak Ridge, Tenn., 1966, 129.
5. **Pinajian, J. J.,** ORNL 86 in. cyclotron, in *Radioactive Pharmaceuticals,* Andrews, G. A., Kniseley, R. M., and Wagner, H. N., Eds., USAEC-DTI, 1966, 143.
6. **MacDonald, N. S.,** Cyclotron-produced radionuclides, in *Radiopharmaceuticals,* Subramanian, G., Rhodes, B. A., Cooper, J. F., and Sodd, V. J., Eds., Society of Nuclear Medicine, New York, 1975, 165.
7. **Wagner, H. N. and Emmons, H.,** Characteristics of an ideal radiopharmaceutical, in *Radioactive Pharmaceuticals,* Andrews, G. A., Kniseley, R. M., and Wagner, H. N., Eds., USAEC-DTI, 1966, 1.
8. **Wagner, H. N.,** Present and future applications of radiopharmaceuticals from generator-produced radioisotopes, in *Radiopharmaceuticals* from Generator-produced radionuclides, IAEA, Vienna, 1970, 163.
9. **Yano, Y. and Anger, H. O.,** Ultrashort lived radioisotopes for visualizing blood vessels and organs, *J. Nucl. Med.,* 9, 2, 1968.
10. **Yano, Y.,** Radionuclide generators: current and future applications in nuclear medicine, in *Radiopharmaceuticals,* Subramanian, G., Rhodes, B. A., Cooper, J. F., and Sodd, V. J., Eds., Society of Nuclear Medicine, New York, 1975, 236.
11. **Morton, M. E.,** Colloidal chromic radiophosphate in high yields for radiotherapy, *Nucleonics,* 10, 92, 1952.
12. **Tubis, M., Pasnick, E., and Nordyke, R. A.,** Preparation and use of ^{131}I-labeled sodium hippurate in kidney function tests, *Proc. Soc. Exp. Biol. Med.,* 103, 497, 1960.
13. **Tubis, M., Endow, J. S., and Rawalay, S. S.,** The preparation and properties of ^{131}I-labeled iodobenzoic acids, *Int. J. Appl. Radiat. Isot.,* 15, 397, 1964.
14. **Wilzbach, J.,** Tritium-labeling by exposure of organic compounds to tritium gas, *J. Am. Chem. Soc.,* 79, 1013, 1957.
15. **Murray, A., III, and Williams, D. L.,** *Organic Synthesis with Isotopes,* Interscience, New York, 1958.
15a. **Evans, E. A.,** *Tritium and Its Compounds,* John Wiley & Sons, New York, 1972, 822 pp.
16. **Herber, R. H., Ed.,** *Inorganic Isotopic Syntheses,* W. A. Benjamin, New York, 1962.
17. **Chaiet, L., Rosenbloom, C., and Woodbury, D. T.,** Biosynthesis of radioactive vitamin B_{12} containing ^{60}Co, *Science,* 111, 601, 1950.
18. **Blau, M.,** Biosynthesis of ^{75}Se-selenomethionine and ^{75}Se-selenocystine, *Biochim. Biophys. Acta,* 49, 389, 1961.
19. **Tubis, M.,** Special iodinated compounds for biology and medicine, in *Radioactive Pharmaceuticals,* Andrews, G. A., Kniseley, R. M., and Wagner, H. N., Eds. USAEC-DTI, 1966, 281.

20. **Charlton, J. C.,** Problems characteristic of radioactive pharmaceuticals, in *Radioactive Pharmaceuticals,* Andrews, G. A., Kniseley, R. M., and Wagner, H. N., Eds., USAEC-DTI, 1966, 33.

21. **Oliver, G. C., Parker, B. M., Brasfield, D. A., and Parker, C. W.,** The measurement of digitoxin in human serum by radioimmunoassay *J. Clin. Invest.,* 47, 1035, 1968.

22. **Hughs, W. L.,** The chemistry of iodination, *Ann. N.Y. Acad. Sci.,* 70, 3, 1957.

23. **World Health Organization,** WHO Expert Committee on Specifications for Pharmaceutical Prepa-

24. **Tubis, M.,** Quality control of radiopharmaceuticals, in *Radiopharmacy,* Tubis, M. and Wolf, W., Eds., John Wiley & Sons, New York, 1976, 555.

25. *The Pharmacopeia of the United States,* 19th rev., Mack, Easton, Pa., 1975.

26. **Scientific tables, in** *Documenta Geigy,* Diem, K. and Lentner, C., Eds., Ciba-Giegy, Basel, 1973, 292.

27. **Heath, R. L.,** *Scintillation Spectrometry Gamma-Ray Spectrum Catalog,* Vol. 2, 2nd ed., USAEC IDO-16880-2, U. S. Clearing House of Federal Scientific and Technical Information, National Bureau of Standards, U. S. Department of Commerce, Springfield, Va., 1964.

28. **Avis, K. E. and Turnbull, R. T.,** Parenteral medications, in *Dispensing of Medication,* Martin, E. W., Ed., Mack, Easton, 1971, 994.

29. **Bang, F. B.,** A bacterial disease of *Limulus polyphemus, Bull. Johns Hopkins Hosp.,* 98, 325, 1956.

30. **Levin, J. and Bang, F. B.,** Clottable protein in *Limulus*: its localization and kinetics of its coagulation by endotoxin, *Thromb. Diath. Haemorrh.,* 19, 189, 1968.

31. **Pyrogent**^{TM} *Limulus* amebocyte lysate for in process endotoxin detection, informational bulletin,

32. *Limulus* amebocyte lysate, Pyrogent^{TM} for in process endotoxin detection test procedure, Mallinckrodt Chemical Works, St. Louis, April 1973.

33. **Pyrostat**^{TM} **pyrogen test, in** *Worthington's World,* Worthington Biochemical, Freehold, N.J., April 1976.

34. *Limulus* amebocyte lysate, additional standards, DHEW, Food and Drug Administration (21CFR Part 273), *Fed. Regist.,* 38 (180), 1973.

35. **Nickoloff, E. L.,** Quality Control for Commercial and Developmental Agents, paper presented at the Symposium on Chemistry of Radiopharmaceuticals, Hahnemann Medical College, Philadelphia, April 23, 1976.

Appendixes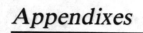

APPENDIX A.3.6

SUPPLEMENTARY DECAY SCHEME DATA

Allen Brodsky

The following data were taken from Reference 1,* which provides recent calculations supplementing those given in Section 3.6 for some of the radionuclides occurring in routine releases from nuclear fuel cycle facilities. Reference 1 presents such data for 240 radionuclides in this category. Not all of the nuclides covered in Section 3.6 are included in Reference 1, since the nuclides in Section 3.6 also include those of interest in medical institutions and other facilities using radioactive materials. However, some of the nuclide schemes included in the following table use later Oak Ridge calculations and have values that may be somewhat different from those given in Section 3.6. Users should cross-check references in order to ascertain whether values used for their applications are those with the desired degree of accuracy. Any further questions should be directed to the authors of the references cited who may be able to provide updated information or even information for nuclides not included in this handbook.

* Reference 1 is available from the National Technical Information Service, Department of Commerce, 5285 Port Royal Road, Springfield, Va. 22161. The address of the author is Dr. D. C. Kocher, Health and Safety Research Division, Oak Ridge National Laboratory, Oak Ridge, Tenn. 37830.

Table A.3.6-1
SUPPLEMENTARY DECAY SCHEME DATA

14C B- DECAY (5730 Y 40)   I(MIN) = 0.10%

Radiation Type	Energy (keV)	Intensity (%)	Δ(g-rad/μCi-h)
β- 1 max	156.478 9		
avg	49.470 3	100	0.105

18P B+ DECAY (109.74 M 4)   I(MIN) = 0.10%

Radiation Type	Energy (keV)	Intensity (%)	Δ(g-rad/μCi-h)
Auger-K	0.52	3.066 7	≈0
β+ 1 max	633.7 6		
avg	249.9 3	96.73 17	0.515

Maximum γ±-intensity = 193.46%

22NA B+ DECAY (2.602 Y 2)   I(MIN) = 0.10%

Radiation Type	Energy (keV)	Intensity (%)	Δ(g-rad/μCi-h)
Auger-K	0.82	9.20 4	0.0002
β+ 1 max	545.5 5		
avg	215.54 21	89.84 9	0.412

1 weak β's omitted:
Eβ(avg) = 834.8; ΣIβ = 0.06%

K X-ray	0.84	0.12 4	≈0
γ 1	1274.540 20	99.940 20	2.71

Maximum γ±-intensity = 179.80%

24NA B- DECAY (15.00 H 4)   I(MIN) = 0.10%

Radiation Type	Energy (keV)	Intensity (%)	Δ(g-rad/μCi-h)
β- 1 max	1390.6 7		
avg	554.1 4	99.935 4	1.18

1 weak β's omitted:
Eβ(avg) = 88.8; ΣIβ = 0.06%

γ 2	1368.53 5	99.99916 5	2.91
γ 3	2754.09 5	99.862 5	5.86

4 weak γ's omitted:
Eγ(avg) = 3823.6; ΣIγ = 0.06%

Table A.3.6-1 (continued)
SUPPLEMENTARY DECAY SCHEME DATA

32P B- DECAY (14.29 D 3)   I(MIN) = 0.10%

Radiation Type	Energy (keV)	Intensity (%)	Δ(g-rad/μCi-h)
β- 1 max	1710.3 6		
avg	694.8 3	100	1.48

54MN EC DECAY (312.5 D 5)   I(MIN) = 0.10%

Radiation Type	Energy (keV)	Intensity (%)	Δ(g-rad/μCi-h)
Auger-L	0.54	142.0 12	0.0016
Auger-K	4.78	63.9 7	0.0065
X-ray L	0.57	0.37 13	≈0
X-ray Kα2	5.40551	7.43 21	0.0009
X-ray Kα1	5.41472	14.7 4	0.0017
X-ray Kβ	6	2.94 10	0.0004
γ 1	834.827 21	99.9760 2	1.78

57CO EC DECAY (270.9 D 6)   I(MIN) = 0.10%

Radiation Type	Energy (keV)	Intensity (%)	Δ(g-rad/μCi-h)
Auger-L	0.67	249 3	0.0036
Auger-K	5.62	105.5 14	0.0126
ce-K- 1	7.303 3	69.6 4	0.0108
ce-L- 1	13.569 3	7.79 22	0.0023
ce-MNO- 1	14.322 3	1.15 7	0.0004
ce-K- 2	114.951 4	1.87 10	0.0046
ce-L- 2	121.217 3	0.183 10	0.0005
ce-K- 3	129.364 4	1.40 12	0.0039
ce-L- 3	135.630 3	0.140 12	0.0004
X-ray L	0.7	0.8 3	≈0
X-ray Kα2	6.39084	16.6 5	0.0023
X-ray Kα1	6.40384	32.8 8	0.0045
X-ray Kβ	7	6.63 21	0.0010
γ 1	14.4147 25	9.54 13	0.0029
γ 2	122.063 3	85.59 19	0.223
γ 3	136.476 3	10.61 18	0.0309
γ 9	692.00 3	0.160 5	0.0024

6 weak γ's omitted:
Eγ(avg) = 536.0; ΣIγ= 0.03%

Table A.3.6-1 (continued)
SUPPLEMENTARY DECAY SCHEME DATA

58CO EC DECAY (70.8 D 1)   I(MIN) = 0.10%

Radiation Type	Energy (keV)	Intensity (%)	Δ(g-rad/μCi-h)
Auger-L	0.67	116.5 13	0.0017
Auger-K	5.62	49.4 6	0.0059
β+ 1 max	474.7 14		
avg	201.1 6	14.90 14	0.0638
X-ray L	0.7	0.36 12	≈0
X-ray Kα₂	6.39084	7.79 21	0.0011
X-ray Kα₁	6.40384	15.4 4	0.0021
X-ray Kβ	7	3.10 10	0.0005
γ 1	810.750 20	99.450 10	1.72
γ 2	863.940 20	0.676 10	0.0124
γ 3	1674.73 4	0.517 10	0.0184

Maximum γ±-intensity = 29.80%

59FE β- DECAY (44.6 D 4)   I(MIN) = 0.10%

Radiation Type	Energy (keV)	Intensity (%)	Δ(g-rad/μCi-h)
β- 1 max	131.0 22		
avg	35.8 7	1.29 6	0.0010
β- 2 max	273.6 22		
avg	81.0 8	45.3 11	0.0782
β- 3 max	466.0 22		
avg	149.3 9	53.1 11	0.169
β- 4 max	1565.2 22		
avg	614.6 10	0.18 4	0.0024
total β-			
avg	117.6 10	100.0 16	0.250

1 weak β's omitted:
Eβ(avg) = 22.1; ΣIβ = 0.09%

γ 1	142.648 4	1.03 5	0.0031
γ 2	192.344 6	3.11 16	0.0127
γ 3	334.80 20	0.260 20	0.0019
γ 5	1099.224 25	56.5 10	1.32
γ 6	1291.56 3	43.2 10	1.19

2 weak γ's omitted:
Eγ(avg) = 1227.9; ΣIγ = 0.09%

60CO β- DECAY (5.271 Y 1)   I(MIN) = 0.10%

Radiation Type	Energy (keV)	Intensity (%)	Δ(g-rad/μCi-h)
β- 1 max	317.93 12		
avg	95.80 4	99.920 20	0.204

1 weak β's omitted:
Eβ(avg) = 625.9; ΣIβ = 0.08%

γ 3	1173.208 25	99.900 20	2.50
γ 4	1332.46 3	100	2.84

4 weak γ's omitted:
Eγ(avg) = 637.2; ΣIγ = 0.01%

64CU β+ DECAY (12.701 H 2)   I(MIN) = 0.10%
%(EC + β+) DECAY=62.8 4
SEE ALSO 64CU β- DECAY

Radiation Type	Energy (keV)	Intensity (%)	Δ(g-rad/μCi-h)
Auger-L	0.84	59.1 21	0.0011
Auger-K	6.54	23.3 12	0.0033
β+ 1 max	653.0 8		
avg	278.2 4	17.88 13	0.106
X-ray L	0.85	0.23 8	≈0
X-ray Kα₂	7.46089	4.9 4	0.0008
X-ray Kα₁	7.47815	9.6 7	0.0015
X-ray Kβ	8.26	1.96 14	0.0003
γ 1	1345.9 3	0.49 4	0.0140

Maximum γ±-intensity = 35.76%

Table A.3.6-1 (continued)
SUPPLEMENTARY DECAY SCHEME DATA

64CU B- DECAY (12.701 H 2)   I (MIN) = 0.10%
%B- DECAY=37.2 4
SEE ALSO 64CU B+ DECAY

Radiation Type	Energy (keV)	Intensity (%)	Δ(g-rad/µCi-h)
B- 1 max	577.7 16		
avg	190.1 7	37.2 4	0.151

65ZN EC DECAY (244.1 D 2)   I (MIN) = 0.10%

Radiation Type	Energy (keV)	Intensity (%)	Δ(g-rad/µCi-h)
Auger-L	0.92	126.7 18	0.0025
Auger-K	7	48.3 8	0.0072
B+ 1 max	329.9 11		
avg	143.0 5	1.416 18	0.0043
X-ray L	0.93	0.57 20	≈0
X-ray $K\alpha_2$	8.027830	11.5 3	0.0020
X-ray $K\alpha_1$	8.047780	22.6 5	0.0039
X-ray Kβ	9	4.61 13	0.0009
γ 3	1115.52 3	50.75 10	1.21

Maximum γ±-intensity = 2.83%

82BR B- DECAY (35.30 H 3)   I (MIN) = 0.10%

Radiation Type	Energy (keV)	Intensity (%)	Δ(g-rad/µCi-h)
β- 1 max	264.4 15		
avg	76.2 5	1.40 10	0.0023
β- 2 max	444.1 15		
avg	137.7 6	98.60 10	0.289
total β-			
avg	136.8 6	100.00 15	0.291
γ 1	92.184 8	0.75 5	0.0015
γ 4	137.40 20	0.14 3	0.0004
γ 6	221.45 3	2.25 9	0.0106
γ 7	273.45 3	0.79 5	0.0046
γ 12	554.320 20	70.6 9	0.834
γ 14	606.30 10	1.17 9	0.0151
γ 15	619.070 20	43.1 7	0.569
γ 17	698.330 20	27.9 6	0.416
γ 19	776.49 3	83.4 9	1.38
γ 20	827.81 3	24.2 4	0.426
γ 23	952.10 20	0.37 3	0.0074
γ 24	1007.57 9	1.31 4	0.0281
γ 25	1043.97 3	27.4 5	0.608
γ 27	1081.40 20	0.63 4	0.0144
γ 30	1317.47 5	26.9 5	0.756
γ 32	1426	0.11 5	0.0033
γ 33	1474.82 8	16.60 25	0.521
γ 34	1650.30 10	0.79 5	0.0279
γ 35	1779.60 20	0.117 17	0.0044

18 weak γ's omitted:
Eγ(avg) = 734.5; ΣIγ= 0.75%

Table A.3.6-1 (continued)
SUPPLEMENTARY DECAY SCHEME DATA

85KR IT DECAY (4.48 H 1)   I(MIN) = 0.10%
%IT DECAY=21.1 6
FEEDS 85KR (10.72 Y)
SEE ALSO 85KR B- DECAY (4.48 H)

Radiation Type	Energy (keV)	Intensity (%)	Δ(g-rad/ μCi-h)
Auger-L	1.5	7.6 5	0.0002
Auger-K	10.8	2.09 20	0.0005
ce-K- 1	290.544 20	5.92 21	0.0366
ce-L- 1	302.949 20	0.90 4	0.0058
ce-MNO- 1	304.582 20	0.295 9	0.0019
X-ray L	1.59	0.10 4	≈0
X-ray Kα₂	12.5980 20	1.11 7	0.0003
X-ray Kα₁	12.6490 20	2.16 13	0.0006
X-ray Kβ	14	0.55 4	0.0002
γ 1	304.870 20	14.0 5	0.0908

89SR B- DECAY (50.55 D 9)   I(MIN) = 0.10%

Radiation Type	Energy (keV)	Intensity (%)	Δ(g-rad/ μCi-h)
β- 1 max	1492 4		
avg	583.4 15	99.985 5	1.24

1 weak β's omitted:
Eβ(avg) = 188.0; ΣIβ = 0.02%

1 weak γ's omitted:
Eγ(avg) = 909.1; ΣIγ = 0.02%

90SR B- DECAY (28.5 Y 8)   I(MIN) = 0.10%
FEEDS 90Y (64.0 H)

Radiation Type	Energy (keV)	Intensity (%)	Δ(g-rad/ μCi-h)
β- 1 max	546.0 20		
avg	195.8 8	100	0.417

90Y IT DECAY (3.19 H 1)   I(MIN) = 0.10%
FEEDS 90Y (64.0 H)

Radiation Type	Energy (keV)	Intensity (%)	Δ(g-rad/ μCi-h)
Auger-L	2	11.8 7	0.0005
Auger-K	12.7	2.9 4	0.0008
ce-K- 1	185.47 3	2.65 15	0.0105
ce-L- 1	200.14 3	0.309 20	0.0013
ce-K- 2	462.49 4	7.29 13	0.0718
ce-L- 2	477.16 4	1.025 18	0.0104
ce-MNO- 2	479.14 4	0.28574	0.0029
X-ray L	2	0.27 9	≈0
X-ray Kα₂	14.88290 2	2.04 11	0.0006
X-ray Kα₁	14.95840 2	3.94 19	0.0013
X-ray Kβ	16.7	1.08 6	0.0004
γ 1	202.51 3	96.58 18	0.417
γ 2	479.53 4	90.71 7	0.927
γ 3	682	0.36 8	0.0053

Table A.3.6-1 (continued)
SUPPLEMENTARY DECAY SCHEME DATA

90Y B- DECAY (64.0 H 1)   I(MIN) = 0.10%

Radiation Type	Energy (keV)	Intensity (%)	Δ(g-rad/μCi-h)
β- 1 max	2276 3		
avg	931.0 12	99.984 3	1.98

1 weak β's omitted:
Σβ(avg)= 183.4; ΣIβ= 0.02%

91Y IT DECAY (49.71 M 4)   I(MIN) = 0.10%
FEEDS 91Y (58.51 D)

Radiation Type	Energy (keV)	Intensity (%)	Δ(g-rad/μCi-h)
Auger-L	2	4.9 3	0.0002
Auger-K	12.7	1.20 13	0.0003
ce-K- 1	540.53 5	4.17 8	0.0480
ce-L- 1	555.20 5	0.561 10	0.0066
ce-MNO- 1	557.18 5	0.18544	0.0022
X-ray L	2	0.11 4	≈0
X-ray Kα2	14.88290 2	0.86 5	0.0003
X-ray Kα1	14.95840 2	1.65 8	0.0005
X-ray Kβ	16.7	0.454 23	0.0002
γ 1	557.57 5	95.10 10	1.13

95ZR β- DECAY (63.98 D 6)   I(MIN) = 0.10%
% FEEDING TO 95NB (35.15 D)=99.30 6
% FEEDING TO 95NB (86.6 H)=0.70 6
SEE ALSO 95NB IT DECAY

Radiation Type	Energy (keV)	Intensity (%)	Δ(g-rad/μCi-h)
β- 1 max	365 4		
avg	108.8 13	55.0 3	0.127
β- 2 max	398 4		
avg	120.0 13	44.6 6	0.114
β- 3 max	887 4		
avg	326.9 15	0.70 6	0.0049
total β-			
avg	115.3 14	100.3 7	0.246
γ 2	724.23 4	44.5 6	0.687
γ 3	756.74 4	55.0 3	0.887

95NB IT DECAY (86.6 H 8)   I(MIN) = 0.10%
FEEDS 95NB (35.15 D)

Radiation Type	Energy (keV)	Intensity (%)	Δ(g-rad/μCi-h)
Auger-L	2.15	71 4	0.0032
Auger-K	14	14.7 19	0.0044
ce-K- 1	215.71 14	58.4 5	0.268
ce-L- 1	232.00 14	11.70 10	0.0578
ce-MNO- 1	234.23 14	3.86 3	0.0193
X-ray L	2.17	2.2 8	0.0001
X-ray Kα2	16.52100 2	12.6 6	0.0044
X-ray Kα1	16.61510 2	24.2 11	0.0085
X-ray Kβ	18.6	6.9 4	0.0028
γ 1	234.70 14	26.1 6	0.130

Table A.3.6-1 (continued)
SUPPLEMENTARY DECAY SCHEME DATA

99MO B− DECAY (66.02 H 1) I(MIN) = 0.10%

%FEEDING TO 99TC (2.13E5 Y)=13.2 2
%FEEDING TO 99TC (6.02 H)=86.8 9
SEE ALSO 99TC IT DECAY (6.02 H)

Radiation Type	Energy (keV)	Intensity (%)	Δ(g-rad/μCi-h)
Auger-L	2.17	6.2 5	0.0003
Auger-K	15.5	1.27 20	0.0004
ce-K- 2	19.5400 22	4.4 4	0.0018
ce-L- 2	37.5415 21	0.53 4	0.0004
ce-MNO- 2	40.0400 23	0.127 10	0.0001
ce-K- 3	119.465 3	0.56 5	0.0014
ce-K- 6	160.023 7	0.82 3	0.0028
ce-L- 6	178.024 7	0.123 5	0.0005
β- 1 max	215.2 10		
avg	59.9 3	0.113 5	0.0001
β- 2 max	352.9 10		
avg	104.3 4	0.142 6	0.0003
β- 3 max	436.4 10		
avg	133.1 4	17.6 5	0.0499
β- 4 max	847.9 10		
avg	289.7 4	1.33 6	0.0082
β- 5 max	1214.3 10		
avg	442.8 4	80.7 8	0.761
total β- avg	385.1 5	99.9 10	0.820

5 weak β's omitted:
Eβ(avg)= 184.6; ΣIβ = 0.06%

Radiation Type	Energy (keV)	Intensity (%)	Δ(g-rad/μCi-h)
X-ray L	2.42	0.29 10	≈0
X-ray Kα₂	18.2508 8	1.28 9	0.0005
X-ray Kα₁	18.3671 8	2.45 17	0.0010
X-ray Kβ	20.6	0.74 6	0.0003
γ 2	40.5840 20	1.33 10	0.0012
γ 3	140.509 3	5.7 5	0.0171
γ 6	181.067 7	6.52 19	0.0252
γ 9	366.439 14	1.37 6	0.0107
γ 21	739.48 4	13.0 4	0.204
γ 23	777.88 3	4.62 19	0.0766
γ 24	822.98 3	0.140 6	0.0024
γ 26	960.69 3	0.101 5	0.0021

21 weak γ's omitted:
Eγ(avg)= 504.4; ΣIγ = 0.15%

99TC IT DECAY ª (6.02 H 3) I(MIN) = 0.10%

FEEDS 99TC (2.13E5 Y)

Radiation Type	Energy (keV)	Intensity (%)	Δ(g-rad/μCi-h)
ce-M- 1	1.630 5	86.82 5	0.0030
ce-NOP- 1	2.106 5	12.177 7	0.0005
Auger-L	2.17	10.1 6	0.0005
Auger-K	15.5	2.1 3	0.0007
ce-K- 2	119.465 3	8.78 4	0.0224
ce-K- 3	121.59 3	0.71 4	0.0018
ce-L- 2	137.466 3	1.058 10	0.0031
ce-L- 3	139.59 3	0.219 11	0.0007
ce-M- 2	139.965 4	0.1920 1	0.0006
X-ray L	2.42	0.49 17	≈0
X-ray Kα₂	18.2508 8	2.12 10	0.0008
X-ray Kα₁	18.3671 8	4.05 18	0.0016
X-ray Kβ	20.6	1.22 6	0.0005
γ 2	140.509 3	88.9 3	0.266

2 weak γ's omitted:
Eγ(avg)= 142.6; ΣIγ= 0.02%

ª 99Tc is usually designated as 99mTc and is of prime importance in nuclear medicine

106RU B− DECAY (368.2 D 12) I(MIN) = 0.10%

FEEDS 106RH (29.9 S)

Radiation Type	Energy (keV)	Intensity (%)	Δ(g-rad/μCi-h)
β- 1 max	39.4 3		
avg	10.03 8	100	0.0214

Table A.3.6-1 (continued)
SUPPLEMENTARY DECAY SCHEME DATA

110AG IT DECAY (252.2 D 3)   I(MIN) = 0.10%
%IT DECAY=1.33 10
FEEDS 110AG B- DECAY (24.6 S)
SEE ALSO 110AG B- DECAY (250.8 D)

Radiation Type	Energy (keV)	Intensity (%)	Δ(g-rad/μCi-h)
ce-M- 1	0.56 10	1.11 9	≈0
ce-NOP- 1	1.18 10	0.215 17	≈0
Auger-L	2.6	1.11 8	≈0
Auger-K	18.5	0.140 24	≈0
ce-K- 2	90.97 5	0.83 7	0.0016
ce-L- 2	112.67 5	0.39 3	0.0009
X-ray Kα2	21.9903 3	0.196 16	≈0
X-ray Kα1	22.16290 1	0.37 3	0.0002
X-ray Kβ	24.9	0.119 10	≈0

110AG B- DECAY (24.6 S 2)   I(MIN) = 0.10%
%B- DECAY=99.70 6
SEE ALSO 110AG EC DECAY (24.6 S)

Radiation Type	Energy (keV)	Intensity (%)	Δ(g-rad/μCi-h)
β- 1 max	2234.9 20		
avg	894.0 10	4.4 3	0.0838
β- 2 max	2892.7 20		
avg	1199.3 10	95.2 3	2.43
total β- avg	1185.1 10	99.7 5	2.52

9 weak β's omitted:
Eβ(avg) = 407.0; ΣIβ = 0.09%

| γ 2 | 657.749 10 | 4.50 23 | 0.0630 |

12 weak γ's omitted:
Eγ(avg)=1046.0; ΣIγ = 0.10%

129I B- DECAY (1.57E7 Y 4)   I(MIN) = 0.10%

Radiation Type	Energy (keV)	Intensity (%)	Δ(g-rad/μCi-h)
Auger-L	3.43	73 4	0.0054
ce-K- 1	5.02 3	78.60 10	0.0084
Auger-K	24.6	8.7 16	0.0046
ce-L- 1	34.13 3	10.90 10	0.0079
ce-M- 1	38.44 3	2.210 10	0.0018
ce-NOP- 1	39.37 3	0.7280 2	0.0006
β- 1 max	149 4		
avg	40.1 12	100	0.0854
X-ray L	4.1	8.2 25	0.0007
X-ray Kα2	29.4580 10	19.9 6	0.0125
X-ray Kα1	29.7790 10	36.9 9	0.0234
X-ray Kβ	33.6	13.1 4	0.0094
γ 1	39.58 3	7.54 21	0.0064

Table A.3.6-1 (continued)
SUPPLEMENTARY DECAY SCHEME DATA

131I B- DECAY (8.04 D 1)   I(MIN) = 0.10%
%FEEDING TO 131XE (11.9 D)=1.086 13
SEE ALSO 131XE IT DECAY

Radiation Type	Energy (keV)	Intensity (%)	Δ(g-rad/ μCi-h)
Auger-L	3.43	5.0 3	0.0004
Auger-K	24.6	0.59 11	0.0003
ce-K- 1	45.622 10	3.53 9	0.0034
ce-L- 1	74.730 10	0.463 12	0.0007
ce-MNO- 1	79.041 10	0.1176 2	0.0002
ce-K- 7	249.737 11	0.248 6	0.0013
ce-K- 14	329.919 11	1.54 4	0.0108
ce-L- 14	359.027 11	0.244 6	0.0019
β- 1 max	247.9 6		
β- 1 avg	69.36 19	2.12 3	0.0031
β- 2 max	303.9 6		
β- 2 avg	86.95 20	0.627 9	0.0012
β- 3 max	333.8 6		
β- 3 avg	96.62 20	7.36 10	0.0151
β- 4 max	606.3 6		
β- 4 avg	191.58 23	89.3 11	0.364
β- 5 max	806.9 6		
β- 5 avg	283.25 23	0.420 20	0.0025
total β- avg	181.72 24	99.9 11	0.387

1 weak β's omitted:
Eβ (avg) = 200.2; ΣIβ = 0.07%

Radiation Type	Energy (keV)	Intensity (%)	Δ(g-rad/ μCi-h)
X-ray L	4.1	0.55 17	≈0
X-ray $K\alpha_2$	29.4580 10	1.35 5	0.0008
X-ray $K\alpha_1$	29.7790 10	2.50 8	0.0016
X-ray $K\beta$	33.6	0.89 3	0.0006
γ 1	80.183 10	2.62 5	0.0045
γ 4	177.210 10	0.265 4	0.0010
γ 7	284.298 11	6.05 8	0.0366
γ 12	325.781 11	0.251 6	0.0017
γ 14	364.480 11	81.2 11	0.630
γ 16	502.991 11	0.361 6	0.0039
γ 17	636.973 10	7.26 10	0.0985
γ 18	642.703 11	0.220 4	0.0030
γ 19	722.893 10	1.803 25	0.0278

10 weak γ's omitted:
Eγ (avg) = 329.4; ΣIγ = 0.23%

132I B- DECAY (2.30 H 3)   I(MIN) = 0.10%

Radiation Type	Energy (keV)	Intensity (%)	Δ(g-rad/ μCi-h)
Auger-L	3.43	0.55 4	≈0
ce-K- 24	488.09 9	0.138 17	0.0014
ce-K- 34	633.13 8	0.351 6	0.0047
ce-K- 41	738.04 8	0.190 6	0.0030
β- 1 max	226 20		
β- 1 avg	63 7	0.10 6	0.0001
β- 2 max	320 20		
β- 2 avg	92 7	0.283 22	0.0006
β- 3 max	353 20		
β- 3 avg	103 7	0.12 5	0.0003
β- 4 max	366 20		
β- 4 avg	107 7	0.10 3	0.0002
β- 5 max	425 20		
β- 5 avg	127 7	0.21 4	0.0006
β- 6 max	504 20		
β- 6 avg	154 8	0.21 6	0.0007
β- 7 max	522 20		
β- 7 avg	161 8	0.33 7	0.0011
β- 8 max	689 20		
β- 8 avg	223 8	0.79 9	0.0038
β- 9 max	741 20		
β- 9 avg	242 8	12.5 8	0.0644
β-10 max	740 20		
β-10 avg	242 8	1.96 10	0.0101
β-11 max	826 20		
β-11 avg	275 8	0.33 6	0.0019
β-12 max	910 20		
β-12 avg	309 8	3.55 14	0.0234
β-13 max	967 20		
β-13 avg	331 9	8.2 4	0.0578
β-14 max	991 20		
β-14 avg	342 9	2.87 15	0.0209
β-15 max	996 20		
β-15 avg	344 9	3.46 16	0.0254
β-16 max	1155 20		
β-16 avg	409 9	2.57 7	0.0224
β-17 max	1185 20		
β-17 avg	422 9	18.8 7	0.169
β-18 max	1229 20		
β-18 avg	440 9	0.82 15	0.0077
β-19 max	1393 20		
β-19 avg	510 9	0.10 3	0.0011

Table A.3.6-1 (continued)
SUPPLEMENTARY DECAY SCHEME DATA

132I B- DECAY   (2.30 H 3)   (Continued)

Radiation Type	Energy (keV)	Intensity (%)	Δ(g-rad/µCi-h)
β-20 max	1468 20		
avg	543 9	2.0 8	0.0231
β-21 max	1470 20		
avg	543 9	10.0 10	0.116
β-22 max	1617 20		
avg	608 9	12.6 7	0.163
β-23 max	2140 20		
avg	841 9	19.2 22	0.344
total β-			
avg	490 11	101 3	1.06

8 weak β's omitted:
Eβ(avg)= 134.1; ΣIβ= 0.32%

Radiation Type	Energy (keV)	Intensity (%)	Δ(g-rad/µCi-h)
X-ray Kα₂	29.4580 10	0.172 7	0.0001
X-ray Kα₁	29.7790 10	0.319 12	0.0002
X-ray Kβ	33.6	0.113 5	≈0
γ 2	147.20 10	0.237 20	0.0007
γ 3	183.3	0.15792	0.0006
γ 4	254.80 20	0.19 3	0.0010
γ 5	262.70 10	1.44 9	0.0081
γ 7	284.80 10	0.79 7	0.0048
γ 10	306.6 4	0.11 4	0.0007
γ 12	316.5 4	0.16 4	0.0011
γ 15	363.5 4	0.49 10	0.0038
γ 16	387.8 4	0.17 3	0.0014
γ 17	416.8 4	0.46 9	0.0041
γ 18	431.9 4	0.45 4	0.0042
γ 19	446.0 4	0.67 8	0.0064
γ 20	473.4 7	0.27 5	0.0027
γ 22	487.5 7	0.18 5	0.0018
γ 23	505.90 15	5.03 20	0.0542
γ 24	522.65 8	16.1 6	0.179
γ 25	535.5 4	0.52 8	0.0060
γ 26	547.10 20	1.25 9	0.0146
γ 30	620.8 10	0.3948	0.0052
γ 31	621.2 10	.5792 1	0.0209
γ 32	630.22 9	13.7 6	0.184
γ 33	650.60 20	2.66 20	0.0369
γ 34	667.69 8	98.70 10	1.40
γ 35	669.8 3	4.9 8	0.0704

132I B- DECAY   (2.30 H 3)   (Continued)

Radiation Type	Energy (keV)	Intensity (%)	Δ(g-rad/µCi-h)
γ 36	671.6 3	5.2 4	0.0748
γ 37	727	2.2 6	0.0336
γ 38	727.2 10	3.2 6	0.0489
γ 39	729.5 4	1.1 3	0.0169
γ 41	772.60 8	76.2 18	1.25
γ 42	780.2 3	1.23 6	0.0205
γ 43	784.5 4	0.42 5	0.0071
γ 45	809.80 20	2.9 3	0.0494
γ 46	812.20 20	5.6 5	0.0973
γ 47	863.30 20	0.59 5	0.0109
γ 48	876.80 20	1.08 5	0.0201
γ 50	910.30 20	0.92 5	0.0178
γ 51	927.3 3	0.44 8	0.0088
γ 53	954.55 9	18.1 6	0.367
γ 54	984.50 20	0.56 6	0.0118
γ 58	1034.70 20	0.57 5	0.0126
γ 64	1136.03 12	2.96 20	0.0716
γ 65	1138	0.2961	0.0072
γ 66	1143.40 20	1.33 10	0.0337
γ 67	1148.2 7	0.21 5	0.0051
γ 68	1173.20 20	1.09 10	0.0271
γ 71	1272.7 4	0.15 3	0.0040
γ 72	1290.7 3	1.14 6	0.0312
γ 73	1295.3 3	1.97 10	0.0545
γ 74	1298.2 5	0.89 10	0.0246
γ 76	1317.1 7	0.118 20	0.0033
γ 77	1372.07 13	2.47 10	0.0721
γ 78	1398.57 10	7.1 3	0.212
γ 80	1442.56 10	1.42 6	0.0437
γ 82	1476.80 20	0.138 20	0.0043
γ 94	1757.50 20	0.38 3	0.0140
γ 101	1921.08 12	1.18 10	0.0485
γ 103	2002.30 12	1.09 10	0.0463
γ 104	2086.82 15	0.25 4	0.0110
γ 105	2172.68 15	0.19 3	0.0087
γ 107	2223.17 15	0.118 20	0.0056
γ 110	2390.48 15	0.168 20	0.0085

59 weak γ's omitted:
Eγ(avg)=1073.6; ΣIγ= 2.23%

(Continued)

Table A.3.6-1 (continued)
SUPPLEMENTARY DECAY SCHEME DATA

133XE IT DECAY (2.188 D 8)   I(MIN) = 0.10%
PEEDS 133XE (5.245 D)

Radiation Type	Energy (keV)	Intensity (%)	Δ(g-rad/ μCi-h)
Auger-L	3.43	70 4	0.0051
Auger-K	24.6	7.0 13	0.0037
ce-K- 1	198.62 4	63.30 20	0.268
ce-L- 1	227.73 4	20.60 10	0.0999
ce-M- 1	232.04 4	4.560 10	0.0225
ce-NOP- 1	232.97 4	1.220 10	0.0061
X-ray L	4.1	7.8 24	0.0007
X-ray Kα₂	29.4580 10	16.0 5	0.0100
X-ray Kα₁	29.7790 10	29.7 8	0.0188
X-ray Kβ	33.6	10.6 3	0.0076
γ 1	233.18 4	10.3 3	0.0512

133XE B- DECAY (5.245 D 6)   I(MIN) = 0.10%

Radiation Type	Energy (keV)	Intensity (%)	Δ(g-rad/ μCi-h)
Auger-L	3.55	50 5	0.0038
Auger-K	25.5	5.7 8	0.0031
ce-K- 1	43.636 11	0.9 4	0.0008
ce-K- 2	45.012 5	53 5	0.0509
ce-L- 2	75.283 5	8.2 7	0.0131
ce-M- 2	79.780 5	1.68 15	0.0029
ce-NOP- 2	80.766 5	0.44 4	0.0007
β- 1 max	266 3		
avg	75.0 10	1.8 8	0.0029
β- 2 max	346 3		
avg	100.5 10	98.2 8	0.210
total β-			
avg	100.0 10	100.0 12	0.213
X-ray L	4.29	6.2 18	0.0006
X-ray Kα₂	30.6251 3	13.7 12	0.0090
X-ray Kα₁	30.9728 3	25.5 22	0.0168
X-ray Kβ	35	9.1 8	0.0068
γ 1	79.621 11	0.6 3	0.0010
γ 2	80.997 5	37 3	0.0632

4 weak γ's omitted:
Eγ (avg) = 177.8; ΣIγ = 0.07%

Table A.3.6-1 (continued)
SUPPLEMENTARY DECAY SCHEME DATA

137CS B- DECAY (30.17 Y 3)   I(MIN) = 0.10%
%FEEDING TO 137BA (2.522 M)=94.6 3
SEE ALSO 137BA IT DECAY

Radiation Type	Energy (keV)	Intensity (%)	Δ(g-rad/μCi-h)
β- 1 max	511.6 9		
avg	156.8 7	94.6 3	0.316
β- 2 max	1173.2 9		
avg	415.2 8	5.4 3	0.0478
total β-			
avg	170.8 8	100.0 5	0.364

140LA B- DECAY (40.22 H 2)   γ(MIN) = 0.10%

Radiation Type	Energy (keV)	Intensity (%)	Δ(g-rad/μCi-h)
Auger-L	4	1.53 11	0.0001
ce-L- 1	18.0512 5	0.25 3	≈0
Auger-K	28.4	0.15 5	≈0
ce-K- 3	28.473 6	0.18 5	0.0001
ce-K- 4	68.975 7	0.159 12	0.0002
ce-K- 5	90.678 8	0.227 15	0.0004
ce-K- 10	288.325 12	0.73 3	0.0045
ce-K- 15	446.586 19	0.415 8	0.0040
β- 1 max	1213.0 20		
avg	430.3 9	0.63 3	0.0058
β- 2 max	1238.8 20		
avg	441.1 9	10.8 4	0.101
β- 3 max	1244.4 20		
avg	443.5 9	5.31 22	0.0502
β- 4 max	1279.3 20		
avg	458.2 9	1.13 10	0.0110
β- 5 max	1296.2 20		
avg	465.3 9	5.4 3	0.0535
β- 6 max	1348.2 20		
avg	487.4 9	41.4 9	0.430
β- 7 max	1410.5 20		
avg	513.9 9	0.19 7	0.0021
β- 8 max	1412.1 20		
avg	514.7 9	4.84 21	0.0531
β- 9 max	1677.0 20		
avg	629.5 9	20.4 7	0.274

140LA B- DECAY (40.22 H 2)   (Continued)

Radiation Type	Energy (keV)	Intensity (%)	Δ(g-rad/μCi-h)
β-10 max	2164.1 20		
avg	846.2 9	10.0 9	0.180
total β-			
avg	544.0 10	100.3 16	1.16

5 weak β's omitted:
Eβ(avg)= 334.2; ΣIβ= 0.15%

Radiation Type	Energy (keV)	Intensity (%)	Δ(g-rad/μCi-h)
X-ray L	4.84	0.23 6	≈0
X-ray Kα2	34.27890 2	0.444 21	0.0003
X-ray Kα1	34.71970 2	0.81 4	0.0006
X-ray Kβ	39.3	0.303 15	0.0003
γ 4	109.418 7	0.201 15	0.0005
γ 5	131.121 8	0.48 3	0.0013
γ 6	173.550 11	0.124 20	0.0005
γ 7	241.966 12	0.39 3	0.0020
γ 8	266.551 14	0.47 3	0.0027
γ 10	328.768 12	18.5 6	0.130
γ 11	397.79 11	0.11 4	0.0009
γ 12	432.53 3	2.72 15	0.0251
γ 15	487.029 19	43.00 21	0.446
γ 17	751.83 8	4.20 20	0.0673
γ 18	815.85 7	22.5 7	0.390
γ 19	867.82 14	5.4 3	0.0989
γ 20	919.63 15	2.52 16	0.0494
γ 21	925.24 9	6.8 3	0.134
γ 23	951.4 4	0.53 3	0.0107
γ 24	1596.49 24	95.56 17	3.25
γ 26	2348.1 7	0.86 6	0.0430
γ 28	2521.7 5	3.44 18	0.185
γ 30	2547.5 6	0.105 7	0.0057

14 weak γ's omitted:
Eγ(avg)=1305.9; ΣIγ= 0.33%

141BA B- DECAY (18.27 M 7)   I(MIN) = 0.10%
FEEDS 141LA

Radiation Type	Energy (keV)	Intensity (%)	Δ(g-rad/μCi-h)
Auger-L	3.8	6.3 7	0.0005

Table A.3.6-1 (continued)
SUPPLEMENTARY DECAY SCHEME DATA

141BA B- DECAY (18.27 M 7) FEEDS 141LA   (Continued)

Radiation Type	Energy (keV)	Intensity (%)	Δ(g-rad/μCi-h)
Auger-K	27.4	0.67 20	0.0004
ce-K- 4	151.30 8	7.2 7	0.0231
ce-L- 4	183.95 8	0.95 9	0.0037
ce-M- 4	188.86 8	0.197 18	0.0008
β- 1 max	560 50	0.59 12	
avg	174 19		0.0022
β- 2 max	590 50	0.23 5	
avg	185 19		0.0009
β- 3 max	640 50	0.17 4	
avg	205 19		0.0007
β- 4 max	650 50	0.77 16	
avg	208 19		0.0034
β- 5 max	810 50	0.62 13	
avg	269 20		0.0036
β- 6 max	850 50	0.54 11	
avg	283 20		0.0033
β- 7 max	1100 50	2.2 3	
avg	386 21		0.0181
β- 8 max	1160 50	6.4 7	
avg	408 21		0.0556
β- 9 max	1190 50	2.5 3	
avg	420 21		0.0224
β- 10 max	1290 50	2.4 3	
avg	463 22		0.0237
β- 11 max	1400 50	2.2 3	
avg	511 22		0.0239
β- 12 max	1460 50	0.20 10	
avg	538 22		0.0023
β- 13 max	1530 50	6.3 7	
avg	566 22		0.0760
β- 14 max	1600 50	0.30 10	
avg	599 22		0.0038
β- 15 max	1840 50	0.30 10	
avg	703 23		0.0045
β- 16 max	1860 50	1.40 20	
avg	711 23		0.0212
β- 17 max	1960 50	3.5 4	
avg	758 23		0.0565
β- 18 max	2040 50	0.20 10	
avg	791 23		0.0034
β- 19 max	2100 50	12.0 20	
avg	819 23		0.209

141BA B- DECAY (18.27 M 7) FEEDS 141LA   (Continued)

Radiation Type	Energy (keV)	Intensity (%)	Δ(g-rad/μCi-h)
β- 20 max	2200 50	2.00 20	
avg	863 23		0.0368
β- 21 max	2200 50	0.50 10	
avg	866 23		0.0092
β- 22 max	2380 50	23 3	
avg	947 23		0.464
β-23 max	2560 50	18.0 20	
avg	1029 23		0.395
β-24 max	2840 50	12 3	
avg	1156 23		0.295
β-25 max	3030 50	5 5	
avg	1244 24		0.132
total β- avg	850 30	103 8	1.87

1 weak β's omitted:
Eβ(avg) = 770.0; Σiβ = 0.09%

Radiation Type	Energy (keV)	Intensity (%)	Δ(g-rad/μCi-h)
X-ray L	4.65	0.94 24	≈0
X-ray Kα	33.03410 2	1.85 18	0.0013
X-ray Kα	33.44180 2	3.4 4	0.0024
X-ray Kβ	37.8	1.25 12	0.0010
γ 1	112.94 9	0.93 9	0.0022
γ 2	162.96 12	0.44 5	0.0015
γ 3	180.50 8	0.49 5	0.0019
γ 4	190.22 8	46 4	0.186
γ 7	276.99 9	23.3 20	0.137
γ 9	304.24 8	25.2 22	0.163
γ 10	343.71 8	14.2 13	0.104
γ 11	349.35 20	0.29 6	0.0021
γ 12	364.38 10	0.58 6	0.0045
γ 13	381.31 22	0.115 25	0.0009
γ 14	389.78 9	1.32 12	0.0110
γ 16	457.58 8	4.8 5	0.0466
γ 17	462.15 8	4.8 5	0.0471
γ 18	467.26 8	5.5 5	0.0545
γ 20	522.19 18	0.43 7	0.0048
γ 21	524.20 20	0.40 7	0.0045
γ 22	527.42 13	0.38 5	0.0042
γ 26	572.09 19	0.25 4	0.0031
γ 27	572.09 19	0.25 4	0.0031
γ 29	599.28 19	0.23 4	0.0030

Table A.3.6-1 (continued)
SUPPLEMENTARY DECAY SCHEME DATA

141BA B- DECAY (18.27 M 7) FEEDS 141LA   (Continued)

Radiation Type	Energy (keV)	Intensity (%)	Δ(g-rad/μCi-h)
γ 30	608.91 18	0.24 4	0.0032
γ 31	625.23 8	3.3 3	0.0435
γ 32	636.05 20	0.28 5	0.0038
γ 33	641.38 16	0.36 6	0.0050
γ 34	647.88 8	5.6 5	0.0774
γ 35	670.04 24	0.18 4	0.0026
γ 36	674.2 10	0.11 11	0.0015
γ 37	685.7 6	0.13 6	0.0019
γ 38	687.8 7	0.10 5	0.0015
γ 39	698.5 4	0.28 12	0.0042
γ 40	700.0 5	0.21 12	0.0032
γ 41	704.80 14	0.30 4	0.0045
γ 42	739.10 8	4.3 4	0.0673
γ 45	762.2 4	0.14 4	0.0022
γ 46	778.2 5	0.11 4	0.0018
γ 49	826.34 19	0.33 5	0.0057
γ 50	831.72 9	1.51 14	0.0268
γ 51	832.6 8	0.16 10	0.0029
γ 53	867.9 4	0.15 4	0.0027
γ 54	876.29 8	3.4 3	0.0635
γ 55	880.6 3	0.20 5	0.0037
γ 57	908.8 6	0.12 5	0.0024
γ 58	929.47 10	0.69 7	0.0138
γ 59	943.25 12	0.73 8	0.0146
γ 61	981.63 13	0.78 9	0.0163
γ 62	996.6 4	0.12 4	0.0026
γ 63	1012.3 6	0.10 4	0.0022
γ 64	1034.49 24	0.29 5	0.0065
γ 66	1046.32 21	0.34 6	0.0077
γ 67	1094.0 3	0.22 5	0.0051
γ 68	1160.8 5	0.24 10	0.0059
γ 69	1160.84 9	0.92 11	0.0227

141PA B- DECAY (18.27 M 7) FEEDS 141LA   (Continued)

Radiation Type	Energy (keV)	Intensity (%)	Δ(g-rad/μCi-h)
γ 70	1197.47 8	4.6 4	0.117
γ 71	1224.79 16	0.41 6	0.0107
γ 72	1235.5 4	0.14 4	0.0038
γ 73	1264.20 14	0.82 9	0.0220
γ 74	1273.64 19	0.52 7	0.0140
γ 75	1278.24 16	0.66 9	0.0179
γ 76	1309.1 7	0.23 12	0.0065
γ 77	1311.2 3	0.60 15	0.0167
γ 78	1323.72 10	0.94 9	0.0266
γ 79	1345.27 25	0.22 4	0.0063
γ 80	1376.99 14	0.70 8	0.0206
γ 81	1405.59 25	0.27 5	0.0081
γ 82	1436.84 13	0.82 9	0.0252
γ 83	1458.56 14	0.68 8	0.0210
γ 84	1501.82 21	0.31 5	0.0099
γ 86	1550.55 19	0.31 5	0.0102
γ 87	1568.81 25	0.25 5	0.0085
γ 92	1653.95 12	0.75 8	0.0263
γ 93	1682.35 10	1.34 13	0.0480
γ 94	1713.23 22	0.17 3	0.0062
γ 96	1735.6 4	0.18 4	0.0068
γ 97	1740.83 21	0.31 5	0.0116
γ 98	1795.85 18	0.48 6	0.0185
γ 102	1912.7 4	0.129 25	0.0052
γ 104	1990.3 3	0.18 3	0.0078
γ 105	2026.56 23	0.38 6	0.0163
γ 107	2136.7 4	0.110 20	0.0050
γ 108	2164.7 3	0.16 4	0.0072
γ 112	2469.0 4	0.18 4	0.0097

31 weak γ's omitted:
Eγ(avg)=1158.1; ΣIγ= 2.13%

Table A.3.6-1 (continued)
SUPPLEMENTARY DECAY SCHEME DATA

141CE B- DECAY (32.50 D 7)   I(MIN) = 0.10%

Radiation Type	Energy (keV)	Intensity (%)	Δ(g-rad/μCi-h)
Auger-L	4	16.1 10	0.0014
Auger-K	29.4	1.6 5	0.0010
ce-K- 1	103.449 10	18.6 9	0.0410
ce-L- 1	138.605 10	2.57 11	0.0076
ce-M- 1	143.929 10	0.538 25	0.0016
ce-NOP- 1	145.135 10	0.148 7	0.0005
β- 1 max	434.8 15		
avg	129.7 6	70 3	0.193
β- 2 max	580.2 15		
avg	180.7 6	30 3	0.115
total β- avg	145.0 7	100 5	0.309
X-ray L	5	2.6 4	0.0003
X-ray Kα2	35.55020 2	4.8 3	0.0037
X-ray Kα1	36.02630 2	8.9 5	0.0068
X-ray Kβ	40.7	3.33 19	0.0029
γ 1	145.440 10	48.0 20	0.149

143PR B- DECAY (13.56 D 2)   I(MIN) = 0.10%

Radiation Type	Energy (keV)	Intensity (%)	Δ(g-rad/μCi-h)
β- 1 max	933.2 19		
avg	314.8 8	100	0.671

144CE B- DECAY (284.3 D 3)   I(MIN) = 0.10%
%FEEDING TO 144PR (17.28 M)=98.80 9
%FEEDING TO 144PR (7.2 M)=1.20 9
SEE ALSO 144PR IT DECAY

Radiation Type	Energy (keV)	Intensity (%)	Δ(g-rad/μCi-h)
Auger-L	4	11.3 7	0.0010
ce-K- 4	11.42 5	0.82 9	0.0002

144CE B- DECAY (284.3 D 3)   (Continued)
%FEEDING TO 144PR (17.28 M)=98.80 9
%FEEDING TO 144PR (7.2 M)=1.20 9
SEE ALSO 144PR IT DECAY

Radiation Type	Energy (keV)	Intensity (%)	Δ(g-rad/μCi-h)
ce-L- 1	26.74 3	0.93 11	0.0005
Auger-K	29.4	0.9 3	0.0006
ce-M- 1	32.06 3	0.196 22	0.0001
ce-L- 2	34.10 3	1.0 3	0.0007
ce-K- 7	38.13 3	3.5 3	0.0028
ce-M- 2	39.42 3	0.21 6	0.0002
ce-K- 8	44.5094 5	0.61 15	0.0006
ce-K- 4	46.58 5	0.114 12	0.0001
ce-L- 4	49.0 20	0.53 13	0.0006
ce-L- 7	73.29 3	0.48 5	0.0008
ce-M- 7	78.61 3	0.101 9	0.0002
ce-L- 8	79.6652 5	0.28 21	0.0005
ce-L- 9	84.2 20	0.22 17	0.0004
ce-K- 11	91.54 3	5.3 4	0.0104
ce-L- 11	126.70 3	0.73 5	0.0020
ce-M- 11	132.02 3	0.153 11	0.0004
β- 1 max	181.9 15		
avg	49.3 5	19.6 13	0.0206
β- 2 max	235.3 15		
avg	65.3 5	4.6 5	0.0064
β- 3 max	315.3 15		
avg	90.2 5	75.9 12	0.146
total β- avg	81.0 6	100.1 19	0.173
X-ray L	5	1.8 3	0.0002
γ 1	33.57 3	0.25 3	0.0002
X-ray Kα2	35.55020 2	2.80 16	0.0021
X-ray Kα1	36.02630 2	5.1 3	0.0039
X-ray Kβ	40.7	1.93 11	0.0017
γ 2	40.93 3	0.49 14	0.0004
γ 4	53.41 5	0.119 13	0.0001
γ 7	80.12 3	1.64 14	0.0028
γ 8	86.5	0.35 8	0.0006
γ 9	91.0 20	0.35 8	0.0007
γ 11	133.53 3	10.8 7	0.0307

5 weak γ's omitted:
E_γ(avg)= 80.4; ΣI_γ = 0.09%

Table A.3.6-1 (continued)
SUPPLEMENTARY DECAY SCHEME DATA

147PM B- DECAY (2.6234 Y 2)   I(MIN) = 0.10%
FEEDS 147SM

Radiation Type	Energy (keV)	Intensity (%)	Δ(g-rad/μCi-h)
β- 1 max	224.8 4		
avg	61.99 12	100	0.132

154EU B- DECAY (8.6 Y 1)   I(MIN) = 0.10%

Radiation Type	Energy (keV)	Intensity (%)	Δ(g-rad/μCi-h)
Auger-L	4.84	32.5 15	0.0034
Auger-K	34.9	1.8 6	0.0013
ce-K- 3	72.90 4	26.8 10	0.0416
ce-L- 3	114.76 4	16.8 6	0.0410
ce-M- 3	121.26 4	3.90 14	0.0101
ce-NOP- 3	122.76 4	1.09 4	0.3029
ce-K- 20	197.80 4	0.536 19	0.0023
ce-L- 20	239.66 4	0.149 6	0.0008
β- 1 max	255		
avg	71.1 11	27.7 8	0.0419
β- 2 max	314		
avg	89.3 12	0.78 3	0.0015
β- 3 max	329		
avg	94.1 12	0.150 6	0.0003
β- 4 max	357		
avg	103.2 12	1.37 5	0.0030
β- 5 max	415		
avg	122.3 12	0.128 5	0.0003
β- 6 max	443		
avg	131.8 12	0.300 17	0.0008
β- 7 max	556		
avg	171.0 13	0.119 6	0.0004
β- 8 max	577		
avg	178.4 13	36.5 12	0.139
β- 9 max	711		
avg	227.3 13	0.613 22	0.0030
β-10 max	723		
avg	231.8 14	0.244 12	0.0012
β-11 max	847		
avg	279.0 14	16.9 6	0.100

154EU B- DECAY (8.6 Y 1)   (Continued)

Radiation Type	Energy (keV)	Intensity (%)	Δ(g-rad/μCi-h)
β-12 max	978 4		
avg	330.5 14	2.1 6	0.0148
β-13 max	1159 4		
avg	403.5 15	0.27 7	0.0023
β-14 max	1604 4		
avg	590.6 15	0.50 18	0.0063
β-15 max	1851 4		
avg	698.3 16	12.1 20	0.180
total β-			
avg	232.7 20	100 3	0.496
	12 weak β's omitted:		
	Σβ(avg) = 130.9; ΣIβ = 0.26%		
X-ray L	6	7.6 9	0.0010
X-ray Kα₂	42.3089 3	7.3 4	0.0065
X-ray Kα₁	42.9962 3	13.1 6	0.0120
X-ray Kβ	48.7	5.17 24	0.0054
γ 3	123.14 4	40.5 13	0.106
γ 13	188.22 5	0.228 11	0.0009
γ 20	248.04 5	6.59 21	0.0348
γ 43	401.30 5	0.210 9	0.0018
γ 47	444.40 5	0.503 20	0.0048
γ 51	478.26 5	0.215 9	0.0022
γ 63	557.56 5	0.254 10	0.0030
γ 65	582.00 5	0.84 4	0.0104
γ 66	591.74 5	4.84 16	0.0610
γ 73	625.22 5	0.310 13	0.0041
γ 80	676.59 5	0.140 6	0.0020
γ 82	692.41 5	1.70 6	0.0251
γ 84	715.76 4	0.175 8	0.0027
γ 85	723.30 4	19.7 7	0.304
γ 87	756.87 5	4.34 14	0.0700
γ 92	815.55 5	0.465 18	0.0081
γ 94	845.39 5	0.550 22	0.0099
γ 95	850.64 5	0.230 9	0.0042
γ 96	873.19 5	11.5 4	0.214
γ 93	892.73 5	0.460 18	0.0087
γ 100	904.05 5	0.82 4	0.0159
γ 108	996.32 4	10.3 4	0.219
γ 109	1004.76 4	17.4 6	0.372
γ 117	1118.50 10	0.103 4	0.0025
γ 119	1128.40 10	0.267 11	0.0064
γ 121	1140.90 10	0.216 9	0.0052

Table A.3.6-1 (continued)
SUPPLEMENTARY DECAY SCHEME DATA

154EU B- DECAY (8.6 Y 1)   I(MIN) = 0.10%   (Continued)

Radiation Type	Energy (keV)	Intensity (%)	Δ(g-rad/ μCi-h)
γ 127	1241.60 20	0.130 5	0.0034
γ 128	1246.60 20	0.70 3	0.0185
γ 129	1274.45 9	35.5 11	0.964
γ 142	1494.60 20	0.65 3	0.0207
γ 148	1596.48 20	1.67 6	0.0568

123 weak γ's omitted:
Eγ(avq) = 722.5; ΣIγ = 1.72%

210PB B- DECAY (22.3 Y 2)   I(MIN) = 0.10%
%B- DECAY=99.9999980 6
FEEDS 210BI (5.012 D)
%A DECAY=0.0000020 6

Radiation Type	Energy (keV)	Intensity (%)	Δ(g-rad/ μCi-h)
Auger-L	8.15	33.5 25	0.0058
ce-L- 1	30.113 20	57.8 16	0.0371
ce-M- 1	42.501 20	13.6 4	0.0123
ce-NOP- 1	45.562 20	4.50 12	0.0044
β- 1 max	16.5 5		
avg	4.15 13	80 3	0.0071
β- 2 max	63.0 5		
avg	16.13 14	20 3	0.0069
total β-			
avg	6.55 18	100 4	0.0139
X-ray L	10.8	24.3 24	0.0056
γ 1	46.500 20	4.05 8	0.0040

210PO A DECAY (138.378 D 7)   I(MIN) = 0.10%

Radiation Type	Energy (keV)	Intensity (%)	Δ(g-rad/ μCi-h)
	5306.48 10	99.999930	11.30

222RN A DECAY (3.8235 D 3)   I(MIN) = 0.10%
FEEDS 218PO

Radiation Type	Energy (keV)	Intensity (%)	Δ(g-rad/ μCi-h)
α 1	5489.7 3	99.920 10	11.68

2 weak α's omitted:
Eα(avq)=4986.0; ΣIα = 0.08%

1 weak γ's omitted:
Eγ(avq) = 512.0; ΣIγ = 0.08%

223RA A DECAY (11.434 D 2)   I(MIN) = 0.10%
FEEDS 219RN

Radiation Type	Energy (keV)	Intensity (%)	Δ(g-rad/ μCi-h)
ce-NOP- 1	3.333 5	56.4	0.0040
ce-M- 2	5.528 5	9.5377	0.0011
ce-K- 6	8.32 12	0.20 3	≈0
Auger-L	8.7	28 3	0.0051
ce-NOP- 2	8.913 5	3.15	0.0006
ce-M- 3	9.938 5	9.6	0.0020
ce-NOP- 3	13.323 5	3.1872	0.0009
ce-L- 4	13.55 11	0.158 3	≈0
ce-K- 8	23.91 7	7.38 13	0.0038
ce-K- 11	45.80 5	12.6 3	0.0123
ce-K- 12	55.79 4	18.2 5	0.0216
ce-K- 13	60.22 5	2.06 6	0.0026
Auger-K	62.7	1.5 8	0.0020
ce-K- 18	81.27 12	0.251 21	0.0004
ce-L- 7	92.77 9	0.180 16	0.0004
ce-L- 9	104.26 8	1.38 3	0.0031
ce-M- 9	117.83 6	0.330 6	0.0008
ce-NOP- 9	121.21 6	0.1150 2	0.0003
ce-L- 11	126.15 6	2.34 5	0.0063
ce-L- 12	136.14	3.29 9	0.0095
ce-M- 11	139.72 4	0.557 16	0.0017
ce-L- 13	140.57 6	0.379 10	0.0011
ce-NOP-11	143.10 4	1.95 5	0.0059
ce-M- 12	149.71 3	0.783 20	0.0025
ce-NOP-12	153.09 3	0.272 7	0.0009

Table A.3.6-1 (continued)
SUPPLEMENTARY DECAY SCHEME DATA

223RA A DECAY (11.434 D 2) FEEDS 219RN

Radiation Type	Energy (keV)	Intensity (%)	Δ(g-rad/μCi-h)
ce-K- 28	171.01 4	9.1 3	0.0330
ce-K- 30	225.49 5	1.57 5	0.0075
ce-K- 33	239.92 7	1.01 4	0.0052
ce-L- 28	251.36 5	1.66 5	0.0089
ce-M- 28	264.93 3	0.392 11	0.0022
ce-NOP-28	268.31 3	0.135 4	0.0008
ce-K- 38	273.44 8	0.139 5	0.0008
ce-L- 30	305.84 6	0.285 9	0.0019
ce-L- 33	320.27 8	0.182 6	0.0012
ce-K- 44	346.54 6	0.197 10	0.0015
α 1	5287.3 10	0.15	0.0169
α 2	5338.7 10	0.13	0.0148
α 3	5365.6 10	0.13	0.0149
α 4	5433.6 5	2.27 20	0.263
α 5	5501.6 10	1.00 15	0.117
α 6	5540.0 10	9.2 4	1.08
α 7	5606.9 3	24.2 4	2.89
α 8	5716.4 3	52.5 8	6.39
α 9	5747.2 4	9.5 6	1.16
α 10	5857.5 10	0.32 4	0.0399
α 11	5871.6 10	0.85 4	0.106

15 weak α's omitted:
Eα(avg)=5348.2; ΣIα= 0.29%

Radiation Type	Energy (keV)	Intensity (%)	Δ(g-rad/μCi-h)
X-ray L	11.7	24 3	0.0061
X-ray Kα₂	81.070 20	15.0 4	0.0259
X-ray Kα₁	83.780 20	24.9 6	0.0445
X-ray Kβ	94.9	11.3 3	0.0228
γ 9	122.31 6	1.190 20	0.0031
γ 11	144.20 4	3.26 7	0.0100
γ 12	154.19 3	5.59 10	0.0184
γ 13	158.62 4	0.688 13	0.0023
γ 18	179.67 12	0.153 12	0.0006
γ 28	269.41 3	13.6 3	0.0780
γ 29	288.17 12	0.154 6	0.0009
γ 30	323.89 4	3.90 9	0.0269

(Continued)
223RA A DECAY (11.434 D 2) FEEDS 219RN

Radiation Type	Energy (keV)	Intensity (%)	Δ(g-rad/μCi-h)
γ 31	328.50 6	0.198 10	0.0014
γ 33	338.32 6	2.78 7	0.0200
γ 34	342.90 7	0.20 3	0.0015
γ 38	371.84 8	0.490 12	0.0039
γ 44	444.94 5	1.27 6	0.0120

38 weak γ's omitted:
Eγ(avg)= 337.1; ΣIγ = 0.90%

224RA A DECAY (3.66 D 4) FEEDS 220RN   T(MIN) = 0.10%

Radiation Type	Energy (keV)	Intensity (%)	Δ(g-rad/μCi-h)
Auger-L	8.7	0.44 10	≈0
ce-K- 1	142.60 10	0.44 13	0.0013
ce-L- 1	222.95 11	0.48 14	0.0023
ce-M- 1	236.52 10	0.13 4	0.0006
α 1	5449	4.9 4	0.569
α 2	5685.56 20	95.1 4	11.52

3 weak α's omitted:
Eα(avg)=5093.6; ΣIα= 0.02%

Radiation Type	Energy (keV)	Intensity (%)	Δ(g-rad/μCi-h)
X-ray L	11.7	0.39 9	≈0
X-ray Kα₂	81.070 20	0.12 4	0.0002
X-ray Kα₁	83.780 20	0.21 6	0.0004
γ 1	241.00 10	3.9 11	0.0200

3 weak γ's omitted:
Eγ(avg)= 440.1; ΣIγ = 0.02%

(Continued)

Table A.3.6-1 (continued)
SUPPLEMENTARY DECAY SCHEME DATA

225RA B- DECAY (14.8 D 2) FEEDS 225AC   I(MIN) = 0.10%

Radiation Type	Energy (keV)	Intensity (%)	Δ(g-rad/μCi-h)
Auger-L	9.28	14.2 15	0.0028
ce-L- 1	20.2 10	28.9 5	0.0124
ce-M- 1	35.0 10	7.13 13	0.0053
ce-NOP- 1	38.7 10	2.35 4	0.0019
β- 1 max	322 12		
avg	90 4	67.3	0.129
β- 2 max	362 12		
avg	103 4	32.7	0.0717
total β-			
avg	94 4	100	0.201
X-ray L	12.7	14.7 15	0.0040
γ 1	40.0 10	29	0.0247

225AC A DECAY (10.0 D 1) FEEDS 221FR   I(MIN) = 0.10%

Radiation Type	Energy (keV)	Intensity (%)	Δ(g-rad/μCi-h)
ce-K- 20	7.06 10	1.8 3	0.0003
ce-L- 1	7.26 11	1.67526	0.0003
Auger-L	8.9	17.7 21	0.0034
ce-L- 2	17.96 11	13.3 17	0.0051
ce-L- 3	19.86 11	6.4 9	0.0027
ce-M- 1	21.25 10	0.44719	0.0002
ce-M- 23	23.26 20	0.147	*0
ce-NOP- 1	24.75 10	0.147	*0
ce-M- 2	31.95 10	3.6 5	0.0024
ce-M- 3	33.85 10	1.71 23	0.0012
ce-NOP- 2	35.45 10	1.17 15	0.0009
ce-NOP- 3	37.35 10	0.57 8	0.0004
ce-L- 4	44.21 11	3.7 5	0.0034
ce-L- 5	45.56 11	0.7 3	0.0006
ce-K- 27	48.76 20	0.104 15	0.0001
ce-L- 6	50.96 11	0.25 5	0.0003
ce-L- 7	52.96 11	0.51 10	0.0006
ce-L- 8	54.86 11	0.530 10	0.0006
ce-L- 10	55.76 21	0.22 12	0.0003

225AC A DECAY (10.0 D 1) FEEDS 221FR   (Continued)

Radiation Type	Energy (keV)	Intensity (%)	Δ(g-rad/μCi-h)
ce-K- 29	56.06 20	1.03 12	0.0012
ce-L- 11	56.26 21	0.28 14	0.0003
ce-M- 4	58.20 10	0.87 11	0.0011
ce-M- 5	59.55 10	0.17 7	0.0002
ce-NOP- 4	61.7~ 10	0.31 4	0.0004
ce-L- 12	64.26 4	0.58 16	0.0008
ce-M- 7	66.95 10	0.14 3	0.0002
ce-L- 13	68.66 21	1.03 4	0.0015
ce-M- 8	68.85 10	0.1420 2	0.0002
ce-L- 14	75.96 11	0.34 6	0.0006
ce-L- 15	77.46 11	0.106 14	0.0002
ce-M- 12	78.248 5	0.14 4	0.0002
ce-L- 16	80.76 11	1.45 25	0.0025
ce-L- 17	81.06 11	0.29 16	0.0005
ce-M- 13	82.65 20	0.246 9	0.0004
ce-L- 20	89.56 11	0.62 10	0.0012
ce-K- 33	94.4 4	0.162 22	0.0003
ce-M- 16	94.75 10	0.35 6	0.0007
ce-NOP-16	98.25 10	0.123 21	0.0003
ce-NOP- 20	103.55 10	0.16 3	0.0003
ce-L- 29	138.56 21	0.23 3	0.0007
α 1	5286 3	0.230 10	0.0259
α 2	5443 3	0.140 10	0.0162
α 3	5553 4	0.10	0.0118
α 4	5579.0 20	1.20 10	0.143
α 5	5608 3	1.10 10	0.131
α 6	5637.0 20	4.35 20	0.522
α 7	5681.0 10	1.30 20	0.157
α 8	5723.0 20	3.2 4	0.390
α 10	5731.0 20	10	1.23
α 11	5790.6	8.6	1.06
α 12	5792.5	18.1 20	2.23
	5829.0 10	50.65	6.29

24 weak α's omitted:
Eα(avg)=5443.7. ΣIα= 0.39%

Radiation Type	Energy (keV)	Intensity (%)	Δ(g-rad/μCi-h)
X-ray L	12	17.0 20	0.0043
γ 9	62.85 10	0.42 5	0.0006
γ 12	73.70 10	0.48 24	0.0008
	82.9	0.15 4	0.0003
X-ray Kα2	83.230 20	0.93 10	0.0016

Table A.3.6-1 (continued)
SUPPLEMENTARY DECAY SCHEME DATA

225AC A DECAY (10.0 D 1) FEEDS 221FR   (Continued)

Radiation Type	Energy (keV)	Intensity (%)	Δ(g-rad/µCi-h)
X-ray Kα₁	86.100 20	1.53 16	0.0028
γ 13	87.30 20	0.308 10	0.0006
γ 14	94.60 10	0.130 20	0.0003
X-ray Kβ	97.5	0.70 8	0.0015
γ 16	99.40 10	0.60 10	0.0013
γ 17	99.70 10	3.5 19	0.0074
γ 20	108.20 10	0.25 4	0.0006
γ 21	111.4	0.32 5	0.0008
γ 25	138.2	0.20 10	0.0006
γ 26	144.70 20	0.13 3	0.0004
γ 27	149.90 20	0.73 10	0.0023
γ 28	153.7	0.17 3	0.0006
γ 29	157.20 20	0.35 4	0.0012
γ 32	187.70 20	0.55 5	0.0022
γ 33	195.5 4	0.150 20	0.0006
γ 35	216.2	0.34 10	0.0016
γ 36	224.5	0.1	0.0005
γ 39	253.6 3	0.13 3	0.0007
γ 43	453	0.11 4	0.0011

26 weak γ's omitted:
Eγ(avg) = 178.2; ΣIγ = 0.63%

226RA A DECAY (1600 Y 7) FEEDS 222RN   I(MIN) = 0.10%

Radiation Type	Energy (keV)	Intensity (%)	Δ(g-rad/µCi-h)
Auger-L	8.7	0.91 9	0.0002
ce-K- 1	87.59 5	0.633 13	0.0012
ce-L- 1	167.94 6	1.207 24	0.0043
ce-M- 1	181.51 4	0.321 7	0.0012
ce-NOP- 1	184.89 4	0.1118 2	0.0004
α 1	4601.9 5	5.55 5	0.544
α 2	4784.50 25	94.45 5	9.63
X-ray L	11.7	0.81 9	0.0002
X-ray Kα₂	81.070 20	0.180 5	0.0003
X-ray Kα₁	83.780 20	0.299 8	0.0005
X-ray Kβ	94.9	0.136 4	0.0003
γ 1	185.99 4	3.28 3	0.0130

227AC A DECAY (21.773 Y 3)   I(MIN) = 0.10%
%A DECAY=1.38 4
FEEDS 223FR
SEE ALSO 227AC B- DECAY

Radiation Type	Energy (keV)	Intensity (%)	Δ(g-rad/µCi-h)
ce-M- 1	8.048 5	0.414 24	≈0
ce-NOP- 1	11.547 5	0.138 8	≈0
α 1	4938.1 20	0.52 4	0.0552
α 2	4950.5 20	0.65 4	0.0684

17 weak α's omitted:
Eα(avg)=4839.7; ΣIα = 0.19%

32 weak γ's omitted:
Eγ(avg) = 116.4; ΣIγ = 0.10%

Table A.3.6-1 (continued)
SUPPLEMENTARY DECAY SCHEME DATA

227AC B- DECAY (21.773 Y 3)   I(MIN) = 0.10%
%B- DECAY= 98.62 4
FEEDS 227TH
SEE ALSO 227AC A DECAY

Radiation Type	Energy (keV)	Intensity (%)	Δ(g-rad/μCi-h)
ce-L- 3	4.03 20	0.74 15	≈0
ce-M- 1	4.12 10	33 7	0.0029
ce-NOP- 1	7.97 10	10.8 23	0.0018
Auger-L	9.48	0.36 8	0.0014
ce-M- 2	10.02 10	6.7 15	0.0007
ce-NOP- 2	13.87 10	2.2 5	≈0
ce-M- 3	19.32 20	0.18 4	
β- 1 max	19.2 20		
avg	4.8 5	10	0.0010
β- 2 max	34.4 20		
avg	8.7 6	35	0.0065
β- 3 max	43.7 20		
avg	11.1 6	54	0.0128
total β-			
avg	9.6 7	99	0.0203
X-ray L	13	0.39 9	0.0001

3 weak γ's omitted:
Bγ(avg)= 16.0; ΣIγ= 0.04%

227TH A DECAY (18.718 D 5)   I(MIN) = 0.10%
FEEDS 223RA

Radiation Type	Energy (keV)	Intensity (%)	Δ(g-rad/μCi-h)
ce-L- 5	1.03 10	1.2 5	≈0
ce-M- 2	3.18 20	2.6 6	0.0002
ce-NOP- 2	6.79 20	0.85 17	0.0001
Auger-L	9	41 9	0.0080
ce-K- 58	9.28 6	0.16 4	≈0
ce-L- 8	10.34 10	0.53 11	0.0001
ce-L- 9	10.67 3	41 15	0.0094
ce-L- 10	12.383 11	16.5	0.0042
ce-M- 5	15.45 10	0.30 11	≈0
	20.56 10	0.18 5	

227TH A DECAY (18.718 D 5)   (Continued)
FEEDS 223RA

Radiation Type	Energy (keV)	Intensity (%)	Δ(g-rad/μCi-h)
ce-M- 8	24.76 10	0.13 3	≈0
ce-L- 16	24.88 10	0.7 3	0.0003
ce-M- 9	25.09 3	11 4	0.0058
ce-L- 18	25.19 3	0.35 18	0.0002
ce-M- 10	26.798 11	4.1 13	0.0023
ce-NOP- 9	28.70 3	3.6 13	0.0022
ce-L- 20	29.06 10	2.3 6	0.0014
ce-NOP-10	30.412 11	1.3 5	0.0009
ce-L- 22	30.64 10	0.35 8	0.0002
ce-L- 23	30.96 10	4.5 10	0.0030
ce-L- 24	31.64 10	0.7 4	0.0005
ce-L- 26	34.9633 15	0.119 24	≈0
ce-L- 28	37.31 10	0.27 6	0.0002
ce-M- 16	39.30 10	0.17 9	0.0001
ce-L- 30	42.27 10	5.7 19	0.0052
ce-L- 33	43.5 3	0.10 3	≈0
ce-M- 20	43.48 10	0.62 14	0.0006
ce-M- 23	45.38 10	1.10 23	0.0011
ce-M- 24	46.06 10	0.17 9	0.0002
ce-NOP-20	47.09 10	0.21 5	0.0002
ce-NOP-23	48.99 10	0.37 8	0.0004
ce-L- 38	49.56 10	0.53 17	0.0006
ce-L- 40	53.66 10	0.5 4	0.0006
ce-L- 41	54.46 10	0.50 13	0.0006
ce-L- 30	56.69 10	1.6 5	0.0019
ce-NOP-30	60.30 10	0.56 18	0.0007
ce-L- 43	60.53 6	0.32 7	0.0004
ce-M- 38	63.98 10	0.14 5	0.0002
Auger-K	65.9	0.13 8	0.0002
ce-M- 40	68.08 10	0.14 11	0.0002
ce-M- 41	68.88 10	0.14 4	0.0002
ce-L- 45	74.76 6	0.14 3	0.0002
ce-L- 46	75.76 10	0.11 3	0.0002
ce-L- 50	81.06 10	0.60 14	0.0012
ce-L- 57	93.96 6	0.60 24	0.0010
ce-M- 50	95.48 10	0.16 4	0.0003
ce-K- 81	101.08 10	0.26 8	0.0006
ce-M- 57	108.38 6	0.16 7	0.0004
ce-K- 91	130.98 20	0.53 13	0.0015
ce-K- 92	132.08 8	0.55 12	0.0016
ce-K- 96	146.28 10	0.12 4	0.0004
ce-K- 97	146.48 20	0.13 4	0.0003
ce-K- 98	148.63 15	0.11 3	0.0003
ce-K-100	152.33 5	0.67 15	0.0022

Table A.3.6-1 (continued)
SUPPLEMENTARY DECAY SCHEME DATA

227TH A DECAY (18.718 D 5) FEEDS 223RA   (Continued)

Radiation Type	Energy (keV)	Intensity (%)	A(g-rad/μCi-h)
ce-K-106	169.08 10	0.21 18	0.0008
ce-K-113	182.23 6	0.89 19	0.0035
ce-K-117	192.68 10	0.27 6	0.0011
ce-K-123	200.52 13	0.57 14	0.0025
ce-K-123	208.74 13	0.26 6	0.0011
ce-L- 92	216.76 8	0.104 22	0.0005
ce-K-131	230.48 14	0.35 8	0.0017
ce-L-100	237.01 5	0.77 17	0.0039
ce-M-100	251.43 5	0.21 5	0.0011
ce-L-113	266.91 6	0.18 4	0.0010
ce-L-120	285.20 13	0.11 3	0.0007
α 1	5585.9 16	0.176 6	0.0209
α 2	5600.6 18	0.170 17	0.0203
α 3	5613.3 16	0.216 8	0.0258
α 4	5668.0 15	2.06 12	0.249
α 5	5693.0 16	1.50 10	0.182
α 6	5700.8 16	3.63 30	0.441
α 7	5709.0 16	8.2 3	0.997
α 8	5713.2 16	4.89 20	0.595
α 9	5757.06 15	20.3 10	2.49
α 10	5762.3 15	0.228 10	0.0280
α 11	5795.5 15	0.311 5	0.0384
α 12	5807.5 15	1.270 20	0.157
α 13	5866.6 15	2.42 10	0.302
α 14	5909.9 15	0.174 8	0.0219
α 15	5916.0 15	0.78 3	0.0983
α 16	5959.7 15	3.00 15	0.381
α 17	5977.92 10	23.4 10	2.98
α 18	6008.81 15	2.90 15	0.371
α 19	6038.21 15	24.5 10	3.15

25 weak α's omitted:
Eα(avg)=5563.3; ΣIα = 0.20%

γ 2	8.00 20	0.14 3	*0
X-ray L	12.3	41 9	0.0108
γ 5	20.27 10	0.20 8	*0
γ 9	29.91 3	0.10 4	*0
γ 15	43.80 10	0.23 6	0.0002

227TH A DECAY (18.718 D 5) FEEDS 223RA   (Continued)

Radiation Type	Energy (keV)	Intensity (%)	A(g-rad/μCi-h)
γ 22	49.88 10	0.65 14	0.0007
γ 23	50.20 10	8.5 18	0.0091
γ 32	62.50 20	0.24 6	0.0003
γ 43	79.77 6	2.1 5	0.0036
X-ray Kα2	85.430 10	1.48 12	0.0027
X-ray Kα1	88.470 10	2.44 19	0.0046
γ 45	94.00 6	1.4 3	0.0028
X-ray Kβ	100	1.12 9	0.0024
γ 57	113.20 6	0.15 6	0.0004
γ 58	113.20 6	0.56 13	0.0014
γ 59	117.20 7	0.17 4	0.0004
γ 66	141.30 7	0.13 3	0.0004
γ 80	204.30 10	0.23 6	0.0010
γ 81	205.00 10	0.15 5	0.0007
γ 82	206.10 10	0.23 5	0.0010
γ 84	210.65 10	1.13 24	0.0051
γ 91	234.90 20	0.45 11	0.0023
γ 92	236.00 8	11.2 24	0.0563
γ 96	250.20 10	0.37 12	0.0020
γ 97	250.40 20	0.13 4	0.0007
γ 98	252.55 15	0.11 3	0.0006
γ 99	254.70 20	0.80 19	0.0043
γ 100	256.25 5	6.8 15	0.0371
γ 101	262.80 10	0.100 23	0.0006
γ 106	273.00 10	0.49 11	0.0028
γ 110	281.40 7	0.16 4	0.0010
γ 113	286.15 6	1.6 4	0.0096
γ 117	296.60 10	0.43 10	0.0027
γ 118	299.90 10	2.0 5	0.0128
γ 119	300.30 20	0.20 9	0.0013
γ 120	304.44 13	1.05 25	0.0068
γ 123	312.66 13	0.48 11	0.0032
γ 125	314.86 11	0.46 11	0.0031
γ 130	329.82 10	2.8 6	0.0193
γ 131	334.40 14	1.00 23	0.0071
γ 133	342.46 6	0.35 9	0.0026
γ 136	350.50 10	0.110 25	0.0008

202 weak γ's omitted:
Eγ(avg) = 184.8; ΣIγ = 1.79%

Table A.3.6-1 (continued)
SUPPLEMENTARY DECAY SCHEME DATA

228RA B- DECAY (5.75 Y 3) FEEDS 228AC   I (MIN) = 0.10%

Radiation Type	Energy (keV)	Intensity (%)	Δ(g-rad/ μCi-h)
ce-M- 2	1.668 5	75	0.0027
ce-NOP- 2	5.401 5	25	0.0029
β- 1 max	38.9 10		
avg	9.9 3	100	0.0211†

228TH A DECAY (1.9131 Y 9) FEEDS 224RA   I (MIN) = 0.10%

Radiation Type	Energy (keV)	Intensity (%)	Δ(g-rad/ μCi-h)
Auger-L	9	9.6 23	0.0019
ce-L- 1	65.16 5	19 4	0.0265
ce-M- 1	79.58 5	5.2 11	0.0088
ce-NOP- 1	83.19 5	1.9 4	0.0033
α 1	5175	0.18	0.0198
α 2	5212	0.36	0.0400
α 3	5340.54 15	26.70 20	3.04
α 4	5423.33 22	72.7	8.40

1 weak α's omitted:
Eα(avg)=5139.0; ΣIα= 0.05%

X-ray L	12.3	9.6 23	0.0025
γ 1	84.40 5	1.2 3	0.0022
γ 2	131.62 20	0.114 19	0.0003
γ 5	215.94 8	0.278 16	0.0013

2 weak γ's omitted:
Eγ (avq) = 176.5; ΣIγ= 0.11%

229TH A DECAY (7.34E3 Y 16) FEEDS 225RA   I (MIN) = 0.10%

Radiation Type	Energy (keV)	Intensity (%)	Δ(g-rad/ μCi-h)
ce-K- 17	3.25 5	8.94 16	0.0006
ce-L- 3	6.153 20	44.8 8	0.0059
ce-M- 1	6.2780 15	7.5	0.0010
Auger-L	9	88 9	0.0171
ce-NOP- 1	9.8916 16	2.5	0.0005
ce-L- 4	11.06 10	9.94 18	0.0023
ce-L- 6	12.06 20	16.0 3	0.0041
ce-M- 2	12.54 3	17.7 3	0.0047
ce-NOP- 2	16.15 3	5.81 10	0.0020
ce-L- 7	18.56 10	3.16 6	0.0013
ce-M- 3	20.568 20	12.72 22	0.0056
ce-K- 18	20.58 10	6.03 11	0.0026
ce-L- 19	20.78 10	6.01 11	0.0027
ce-NOP- 3	23.52 3	8.16 15	0.0041
ce-L- 8	24.182 20	4.20 8	0.0022
ce-M- 4	25.48 10	2.57 5	0.0014
ce-M- 6	26.48 20	3.99 7	0.0023
ce-K- 20	28.05 5	1.92 4	0.0011
ce-K- 21	28.68 10	0.201 4	0.0001
ce-NOP- 4	29.09 10	0.846 15	0.0005
ce-NOP- 6	30.09 20	1.316 23	0.0008
ce-L- 23	31.79 7	0.298 6	0.0002
ce-M- 7	32.98 10	0.781 14	0.0005
ce-K- 24	33.11 6	6.48 12	0.0046
ce-L- 10	33.96 10	2.4 24	0.0017
ce-NOP- 7	36.59 10	0.258 5	0.0002
ce-L- 11	37.36 3	4.19 8	0.0033
ce-M- 8	37.94 3	2.07 4	0.0017
ce-K- 25	39.03 10	2.01 4	0.0017
ce-NOP- 8	41.55 3	0.694 12	0.0006
ce-K- 28	44.38 20	0.206 4	0.0002
ce-K- 29	46.3 3	0.15 20	0.0002
ce-K- 31	47.7 3	0.1012 1	0.0001
ce-M- 10	48.38 10	0.6 7	0.0006
ce-L- 12	48.94 7	1.110 20	0.0012
ce-L- 13	49.66 4	4.56 8	0.0048
ce-K- 32	50.48 7	2.50 5	0.0027

Table A.3.6-1 (continued)

SUPPLEMENTARY DECAY SCHEME DATA

229TH A DECAY (7.34E3 Y 16) FEEDS 225RA   (Continued)

Radiation Type	Energy (keV)	Intensity (%)	Δ(g-rad/μCi-h)
ce-M-11	51.78 3	1.002 18	0.0011
ce-NOP-10	51.99 10	0.23 25	0.0003
ce-K-33	52.56 4	4.08 7	0.0046
ce-K-34	54.58 3	0.200 4	0.0002
ce-NOP-11	55.39 3	0.355 7	0.0004
ce-L-14	55.96 7	13.92 25	0.0166
ce-M-12	63.36 7	0.279 5	0.0004
ce-M-13	64.08 4	1.232 22	0.0017
Auger-K	65.9	1.4 9	0.0020
ce-L-15	67.06 10	2.16 4	0.0031
ce-L-16	67.20 5	10.99 19	0.0157
ce-NOP-13	67.69 4	0.443 8	0.0006
ce-K-37	68.98 10	0.6 6	0.0009
ce-M-14	70.38 7	3.78 7	0.0057
ce-NOP-14	73.99 7	1.357 24	0.0021
ce-K-39	80.08 10	0.339 6	0.0006
ce-M-15	81.48 10	0.571 10	0.0010
ce-M-16	81.62 5	2.63 5	0.0046
ce-NOP-15	85.09 10	0.206 4	0.0004
ce-NOP-16	85.23 5	0.940 17	0.0017
ce-L-40	86.28 20	0.1006 1	0.0002
ce-L-17	87.93 5	1.87 4	0.0035
ce-K-41	89.71 6	8.53 15	0.0163
ce-M-17	102.35 5	0.455 8	0.0010
ce-L-18	105.26 10	1.122 20	0.0025
ce-L-19	105.46 10	1.122 20	0.0025
ce-NOP-17	105.96 5	0.163 3	0.0004
ce-K-44	107.05 10	4.93 9	0.0112
ce-L-20	112.73 5	0.358 7	0.0009
ce-K-45	114.18 20	0.203 4	0.0005
ce-L-23	116.47 7	1.74 3	0.0043
ce-L-24	117.79 6	1.212 21	0.0030
ce-M-18	119.88 10	0.268 5	0.0007
ce-M-19	119.88 10	0.374 7	0.0010
ce-L-25	123.71 10	0.473 9	0.0013
ce-M-23	130.89 7	0.288 6	0.0008
ce-M-24	132.21 6	0.172 3	0.0005
ce-NOP-23	134.50 3	0.464 8	0.0013
ce-L-32	135.16 7		

229TH A DECAY (7.34E3 Y 16) FEEDS 225RA   (Continued)

Radiation Type	Energy (keV)	Intensity (%)	Δ(g-rad/μCi-h)
ce-NOP-24	135.82 6	0.1030 1	0.0003
ce-L-33	137.24 4	0.757 14	0.0022
ce-K-47	138.7 3	0.15 18	0.0004
ce-M-32	149.58 7	0.111 2	0.0004
ce-M-33	151.66 4	0.181 4	0.0006
ce-L-37	153.66 10	0.224 12	0.0007
ce-L-38	160.56 20	0.215 5	0.0007
ce-L-39	164.76 10	0.1424 2	0.0005
ce-L-41	174.39 6	1.58 3	0.0059
ce-M-41	188.81 6	0.378 7	0.0015
ce-L-44	191.73 10	0.918 16	0.0037
ce-NOP-41	192.42 6	0.1336 2	0.0005
ce-M-44	206.15 10	0.219 4	0.0010
α 1	4688	0.15	0.0150
α 2	4761	0.63	0.0639
α 3	4797.2 12	1.27	0.130
α 4	4809	0.22	0.0225
α 5	4814.0 12	8.4	0.861
α 6	4833	0.29	0.0299
α 7	4837	4.8	0.495
α 8	4844.7 12	56.2	5.80
α 9	4861	0.18	0.0186
α 10	4900.4 12	10.8	1.13
α 11	4927	0.1	0.0115
α 12	4966.9 12	6.4	0.677
α 13	4977.9 12	3.2	0.339
α 14	5033	0.24	0.0257
α 15	5050	5.2	0.559
α 16	5052	1.6	0.172

12 weak α's omitted:
Eα(avq)=4764.3; ΣIα= 0.27%

X-ray L	12.3	88 9	0.0231
γ 2	17.36 3	0.17	~0
γ 6	31.30 20	8.6	0.0057
γ 7	37.80 10	2.8	0.0023

Table A.3.6-1 (continued)
SUPPLEMENTARY DECAY SCHEME DATA

229TH A DECAY (7.34E3 Y 16)
FEEDS 225RA
(Continued)

Radiation Type	Energy (keV)	Intensity (%)	Δ(g-rad/μCi-h)
γ 8	42.76 3	0.15	0.0001
γ 11	56.60 3	0.32	0.0004
γ 12	68.18 7	0.10	0.0001
γ 13	68.90 4	0.1	0.0002
γ 14	75.20 7	0.5	0.0008
X-ray Kα₂	85.430 10	15.5 4	0.0281
γ 15	86.30 10	0.24	0.0004
γ 16	86.44 5	2.9	0.0053
X-ray Kα₁	88.470 10	25.5 6	0.0480
X-ray Kβ		11.7 3	0.0248
γ 17	107.17 5	0.86	0.0020
γ 18	124.50 10	0.85	0.0023
γ 19	124.70 10	0.85	0.0023
γ 20	131.97 5	0.32	0.0009
γ 23	135.71 7	1	0.0029
γ 24	137.03 6	1.2	0.0035
γ 25	142.95 10	0.42	0.0013
γ 28	148.30 20	1.4	0.0044
γ 32	154.40 7	0.65	0.0021
γ 33	156.48 4	1	0.0037
γ 37	172.90 10	0.4	0.0015
γ 38	179.80 20	0.43	0.0016
γ 39	184.00 10	0.32	0.0013
γ 41	193.63 6	4.2	0.0173
γ 42	204.9 3	1.2	0.0052
γ 44	210.97 10	3	0.0139
γ 45	218.10 20	0.14	0.0007
γ 47	242.6 2	0.25 20	0.0013
γ 51	290.0 5	0.26	0.0016

22 weak γ's omitted:
Eγ(avg)= 151.1; ΣIγ= 0.56%

230TH A DECAY (7.7E4 Y 2)   I(MIN) = 0.10%
FEEDS 226RA
% SPONTANEOUS FISSION LE 5E-11

Radiation Type	Energy (keV)	Intensity (%)	Δ(g-rad/μCi-h)
Auger-L	9	8.5 11	0.0016
ce-L- 1	48.5633 15	17.0 14	0.0176
ce-M- 1	62.9780 15	4.6 4	0.0062
ce-NOP- 1	66.5916 16	1.66 14	0.0024
α 1	4473	0.15	0.0143
α 2	4617.5 15	23.4	2.30
α 3	4684.0 15	76.3	7.61

1 weak α's omitted:
Eα(avg)=4438.0; ΣIα= 0.03%

| X-ray L | 12.3 | 8.5 11 | 0.0022 |
| γ 1 | 67.8 | 0.38 3 | 0.0005 |

7 weak γ's omitted:
Eγ(avg)= 174.2; ΣIγ= 0.07%

231TH B- DECAY (25.52 H 1)   I(MIN) = 0.10%
FEEDS 231PA

Radiation Type	Energy (keV)	Intensity (%)	Δ(g-rad/μCi-h)
ce-M- 1	3.9331 16	0.40 4	≈0
ce-L- 5	5.535 20	56.7	0.0066
ce-NOP- 1	7.9129 18	0.133 14	≈0
Auger-L	9.68	78 11	0.0162

Table A.3.6-1 (continued)

SUPPLEMENTARY DECAY SCHEME DATA

231TH B- DECAY (25.52 H 1) FEEDS 231PA   (Continued)

Radiation Type	Energy (keV)	Intensity (%)	Δ(g-rad/µCi-h)
ce-M- 5	21.273 20	14.5 17	0.0066
ce-L- 6	21.70 20	0.136 16	≈0
ce-K- 27	23.07 5	0.4 4	0.0002
ce-NOP- 5	25.253 20	4.8 6	0.0026
ce-K- 30	33.40 10	0.101 12	≈0
ce-L- 8	37.495 11	65 10	0.0516
ce-L- 9	37.8954 18	35 4	0.0282
ce-L- 10	42.1954 18	0.22 16	0.0002
ce-L- 11	47.3954 18	0.34 4	0.0003
ce-K- 31	50.56 6	0.86 16	0.0009
ce-L- 12	51.71 3	0.134 16	0.0001
ce-M- 8	53.233 11	18 3	0.0202
ce-M- 9	53.6331 16	9.6 10	0.0110
ce-NOP- 8	57.213 11	6.6 10	0.0080
ce-NOP- 9	57.6129 18	3.6 4	0.0044
ce-L- 14	60.10 10	7.3 9	0.0094
ce-L- 15	61.00 10	2.9 4	0.0037
ce-L- 16	63.135 11	1.15 14	0.0016
ce-L- 17	68.84 5	0.152 25	0.0002
ce-M- 14	75.83 10	1.83 21	0.0029
ce-L- 15	76.73 10	0.70 9	0.0011
ce-L- 19	78.20 5	0.69 15	0.0012
ce-M- 16	78.873 11	0.28 4	0.0005
ce-NOP-14	79.81 10	0.67 8	0.0011
ce-NOP-15	80.71 10	0.26 4	0.0004
ce-M- 19	93.93 5	0.17 4	0.0003
ce-L- 27	114.57 5	0.18 5	0.0005
ce-L- 31	142.06 6	0.17 4	0.0005
β- 1 max	143.4		
avg	37.6 10	2.6 5	0.0021
β- 2 max	172.4		
avg	45.8 10	0.139 16	0.0001

231TH B- DECAY (25.52 H 1) FEEDS 231PA   (Continued)

Radiation Type	Energy (keV)	Intensity (%)	Δ(g-rad/µCi-h)
β- 3 max	206.4		
avg	55.7 10	16.1 18	0.0191
β- 4 max	216.4		
avg	58.4 11	2.4 3	0.0030
β- 5 max	306.4		
avg	85.1 11	78 7	0.141
β- 6 max	312.4		
avg	87.0 11	0.46 5	0.0009
total β- avg	78.4 12	100 8	0.167

3 weak β's omitted:
Eβ(avg) = 17.5; ΣIβ = 0.09%

Radiation Type	Energy (keV)	Intensity (%)	Δ(g-rad/µCi-h)
X-ray L	13.3	92 11	0.0261
γ 5	26.640 20	18.7 22	0.0106
γ 7	42.80 20	0.156 18	0.0001
γ 8	58.600 10	0.56 9	0.0007
γ 9	59	0.31 4	0.0004
γ 12	72.81 3	0.62 8	0.0010
γ 14	81.20 10	1.01 12	0.0018
γ 15	82.10 10	0.48 6	0.0008
γ 16	84.240 10	8.0 10	0.0143
γ 17	89.94 5	1.25 21	0.0024
X-ray Kα2	92.2870 20	0.39 12	0.0008
X-ray Kα1	95.8680 20	0.63 19	0.0013
γ 19	99.30 5	0.17 4	0.0004
γ 20	102.31 3	0.51 7	0.0011
X-ray Kβ	108	0.29 9	0.0007
γ 27	135.67 5	0.101 12	0.0003
γ 31	163.16 6	0.20 4	0.0007

34 weak γ's omitted:
Eγ(avg) = 149.0; ΣIγ = 0.46%

Table A.3.6-1 (continued)
SUPPLEMENTARY DECAY SCHEME DATA

232TH A DECAY (1.405E10 Y 6)   I(MIN) = 0.10%
FEEDS 228RA

Radiation Type	Energy (keV)	Intensity (%)	Δ(g-rad/μCi-h)
Auger-L	9	8.4 16	0.0016
ce-L- 1	39.8 10	17 3	0.0141
ce-M- 1	54.2 10	4.5 8	0.0052
ce-NOP- 1	57.8 10	1.6 3	0.0020
ce-L- 2	107 5	0.10 5	0.0002
α 1	3830	0.20 8	0.0163
α 2	3953	23 3	1.94
α 3	4010 5	77 3	6.56
X-ray L	12.3	8.4 16	0.0022
γ 1	59.0 10	0.19 3	0.0002

1 weak γ's omitted:
Eγ(avg)= 126.0; ΣIγ= 0.04%

232U A DECAY (71.7 Y 9)   I(MIN) = 0.10%
FEEDS 228TH
% SPONTANEOUS FISSION=1E-10

Radiation Type	Energy (keV)	Intensity (%)	Δ(g-rad/μCi-h)
Auger-L	9.48	11.6 13	0.0023
ce-L- 1	37.2279 5	23.9 5	0.0190
ce-M- 1	52.5177 3	6.57 12	0.0074
ce-NOP- 1	56.3705 4	2.41 5	0.0029
ce-L- 2	108.53 20	0.222 11	0.0005
α 1	5139.0 20	0.280 20	0.0306
α 2	5263.54 9	31.2 4	3.50
α 3	5320.30 14	68.6 4	7.77
X-ray L	13	12.6 13	0.0035
γ 1	57.7	0.2	0.0003

6 weak γ's omitted:
Eγ(avg)= 142.1; ΣIγ= 0.09%

233U A DECAY (1.585E5 Y 18)   I(MIN) = 0.10%
FEEDS 229TH

Radiation Type	Energy (keV)	Intensity (%)	Δ(g-rad/μCi-h)
ce-L- 2	8.53 10	1.93 11	0.0004
Auger-L	9.48	5.8 7	0.0012
ce-L- 1	21.93 10	9.0 5	0.0042
ce-M- 2	23.82 10	0.464 25	0.0002
ce-NOP- 2	27.67 10	0.153 8	≈0
ce-L- 6	34.22 10	0.76 4	0.0006
ce-L- 4	37.22 10	2.36 12	0.0019
ce-NOP- 4	41.07 10	0.78 4	0.0007
ce-M- 6	49.51 10	0.198 11	0.0002
ce-L- 13	51.41 5	0.134 7	0.0001
ce-L- 22	76.74 10	0.266 14	0.0004
α 1	4729	1.6	0.162
α 2	4754	0.163	0.0165
α 3	4782.9 12	13.23	1.35
α 4	4796	0.28	0.0286
α 5	4824.0 10	84.4	8.67

27 weak α's omitted:
Eα(avg)=4673.6; ΣIα= 0.28%

| X-ray L | 13 | 6.3 7 | 0.0017 |

94 weak γ's omitted:
Eγ(avg)= 110.9; ΣIγ= 0.26%

234TH B- DECAY (24.10 D)   I(MIN) = 0.10%
FEEDS 234PA (1.17 M)
%IT DECAY FOR 234PA (1.17 M)=0.13 3

Radiation Type	Energy (keV)	Intensity (%)	Δ(g-rad/μCi-h)
ce-L- 2	8.385 20	4.9 7	0.0009
Auger-L	9.68	8.3 11	0.0017
ce-M- 1	14.653 20	2.6 4	0.0008
ce-NOP- 1	18.633 20	0.85 11	0.0003
ce-M- 2	24.123 20	1.34 19	0.0007

Table A.3.6-1 (continued)
SUPPLEMENTARY DECAY SCHEME DATA

234TH B- DECAY (24.10 D)   (Continued)
FEEDS 234PA (1.17 M)
%IT DECAY FOR 234mPA (1.17 M) =0.13 3

Radiation Type	Energy (keV)	Intensity (%)	Δ(g-rad/μCi-h)
ce-NOP- 2	28.103 20	0.44 6	0.0003
ce-L- 4	41.755 20	0.40 9	0.0004
ce-L- 5	42.177 3	1.21 10	0.0011
ce-L- 7	52.815 20	0.33 3	0.0004
ce-M- 4	57.493 20	0.103 23	0.0001
ce-M- 5	57.915 3	0.298 25	0.0004
ce-NOP- 5	61.895 3	0.104 9	0.0001
ce-L- 12	71.262 6	10.9 10	0.0165
ce-L- 13	71.687 6	0.34 3	0.0005
ce-M- 12	87.000 6	2.62 23	0.0049
ce-NOP-12	90.980 6	0.97 9	0.0019
β- 1 max	80.3 21		
avg	20.7 6	3.8 5	0.0017
β- 2 max	89.7 21		
avg	23.2 6	0.49 5	0.0002
β- 3 max	100.3 21		
avg	26.0 6	5.5 6	0.0030
β- 4 max	100.7 21		
avg	26.1 6	17.6 16	0.0098
β- 5 max	193.1 21		
avg	51.8 6	72.5 20	0.0800
total β- avg	44.5 7	100 3	0.0948

1 weak β's omitted:
P.β (avg)= 15.4; ΣIβ = 0.08%

X-ray L	13.3	9.8 11	0.0028
γ 5	63.2820 20	3.9 4	0.0053
γ 12	92.367 5	2.57 23	0.0051
γ 13	92.792 5	3.0 3	0.0059
γ 17	112.81 5	0.249 21	0.0006

15 weak γ's omitted:
Eγ (avg)= 76.0; ΣIγ = 0.20%

234U A DECAY (2.445E5 Y 10)   I(MIN) = 0.10%
FEEDS 230TH
% SPONTANEOUS FISSION=1E-9

Radiation Type	Energy (keV)	Intensity (%)	Δ(g-rad/μCi-h)
Auger-L	9.48	9.6 13	0.0019
ce-L- 1	32.748 20	19.9 18	0.0139
ce-M- 1	48.038 20	5.5 5	0.0056
ce-NOP- 1	51.890 20	2.01 18	0.0022
ce-L- 2	100.428 20	0.139 14	0.0003
α 1	4604.7 20	0.24 3	0.0235
α 2	4723.7 20	27.4 15	2.76
α 3	4775.8 20	72.4 20	7.36
X-ray L	13	10.4 14	0.0029
γ 1	53.220 20	0.118 10	0.0001

5 weak γ's omitted:
Eγ (avg)= 121.4; ΣIγ = 0.04%

235U A DECAY (7.1E8 Y 3)   I(MIN) = 0.10%
FEEDS 231TH

Radiation Type	Energy (keV)	Intensity (%)	Δ(g-rad/μCi-h)
Auger-L	9.48	27 13	0.0055
ce-L- 2	11.12 14	18 17	0.0042
ce-L- 4	21.63 10	18 19	0.0084
ce-M- 2	26.41 14	1.6 16	0.0027
ce-NOP- 2	30.26 14	6.2 14	0.0010
ce-L- 5	31.2 4		0.0042
ce-K- 16	34.119 20	1.71 19	0.0012
ce-M- 4	36.92 10		0.0039
ce-NOP- 4	40.77 10	1.6 17	0.0014
ce-K- 17	41.29 3	0.20 19	0.0002

Table A.3.6-1 (continued)
SUPPLEMENTARY DECAY SCHEME DATA

235U A DECAY (7.1E8 Y 3) FEEDS 231TH

Radiation Type	Energy (keV)	Intensity (%)	Δ(g-rad/μCi-h)
ce-M- 5	46.5 4	1.7 4	0.0017
ce-NOP- 5	50.4 4	0.63 14	0.0007
ce-L- 7	52.23 20	4.15 8	0.0046
ce-K- 18	53.709 20	0.58 7	0.0007
ce-M- 7	67.52 20	1.133 20	0.0016
Auger-K	69.2	0.23 16	0.0003
ce-NOP- 7	71.37 20	0.419 8	0.0006
ce-K- 21	73.05 20	0.6 6	0.0010
ce-L- 10	75.6279 5	0.87 12	0.0014
ce-K- 22	76.069 20	4.83 9	0.0078
ce-M- 10	90.9177 3	0.24 3	0.0005
ce-K- 25	92.469 20	1.1 11	0.0022
ce-K- 26	95.659 20	0.33 4	0.0007
ce-L- 16	123.298 20	0.37 4	0.0010
ce-L- 18	142.888 20	0.120 13	0.0004
ce-L- 21	162.223 20	0.23 5	0.0008
ce-L- 22	165.248 20	0.977 17	0.0034
ce-M- 22	180.538 20	0.235 4	0.0009
ce-L- 25	181.648 20	0.38 5	0.0015
α 1	4145 6	0.90 20	0.0795
α 2	4209 4	5.7 6	0.511
α 3	4219 6	0.9 5	0.0809
α 4	4322 4	4.7 5	0.433
α 5	4358 4	17.0 20	1.58
α 6	4392 3	54 3	5.05
α 7	4411 5	2.10 20	0.197
α 8	4435 5	0.7 3	0.0661
α 9	4501 5	1.70 20	0.163
α 10	4555 3	4.5 5	0.437
α 11	4597 3	5.4 5	0.529
X-ray L	13	29 14	0.0082
γ	72.70 20	0.1	0.0052
X-ray Kα₂	89.9530 20	2.7 4	0.0052
X-ray Kα₁	93.3500 20	4.4 6	0.0088
X-ray Kβ	105	2.0 3	0.0046

235U A DECAY (7.1E8 Y 3) FEEDS 231TH (Continued)

Radiation Type	Energy (keV)	Intensity (%)	Δ(g-rad/μCi-h)
γ 11	109.25 5	1.40 20	0.0033
γ 15	140.80 8	0.21 3	0.0006
γ 16	143.770 20	10.5 11	0.0322
γ 18	163.360 20	4.8 5	0.0167
γ 21	182.70 20	0.42 9	0.0016
γ 22	185.720 20	54	0.214
γ 23	194.940 20	0.59 6	0.0024
γ 25	202.120 20	1.01 10	0.0043
γ 26	205.310 20	4.7 5	0.0206
γ 28	221.380 20	0.110 10	0.0005

41 weak γ's omitted:
Eγ(avg) = 192.9; ΣIγ = 0.78%

236U A DECAY (2.3415E7 Y 14) FEEDS 232TH
% SPONTANEOUS FISSION=1.2E-7   I(MIN) = 0.10%

Radiation Type	Energy (keV)	Intensity (%)	Δ(g-rad/μCi-h)
Auger-L	9.48	9.1 17	0.0018
ce-L- 1	28.897 9	19 3	0.0117
ce-M- 1	44.187 9	5.2 8	0.0049
ce-NOP- 1	48.039 9	1.7 3	0.0017
α 1	4332 8	0.260 10	0.0240
α 2	4445 5	26 4	2.46
α 3	4494 3	74 4	7.08
X-ray L	13	9.9 18	0.0027

2 weak γ's omitted:
Eγ(avg) = 51.0; ΣIγ = 0.08%

Table A.3.6-1 (continued)
SUPPLEMENTARY DECAY SCHEME DATA

236PU A DECAY (2.851 Y 8)   I(MIN) = 0.10%
FEEDS 232U
% SPONTANEOUS FISSION=8E-8

Radiation Type	Energy (keV)	Intensity (%)	Δ(g-rad/ μCi-h)
Auger-L	9.89	8.9 19	0.0019
ce-L- 1	25.84 10	20 3	0.0111
ce-M- 1	42.05 10	5.6 9	0.0050
ce-NOP- 1	46.16 10	1.8 3	0.0018
ce-L- 2	87.24 10	0.1190 2	0.0002
α 1	5614	0.18	0.0215
α 2	5720	30.9 5	3.76
α 3	5767	68.9 5	8.46
X-ray L	13.6	11.3 22	0.0033

7 weak γ's omitted:
Eγ(avg)= 66.2; ΣIγ= 0.08%

237U B- DECAY (6.75 D 1)   I(MIN) = 0.10%
FEEDS 237NP

Radiation Type	Energy (keV)	Intensity (%)	Δ(g-rad/ μCi-h)
ce-L- 2	3.921 10	7.7 8	0.0006
ce-M- 1	8.087 21	42 4	0.0072
Auger-L	10	42 7	0.0089
ce-L- 3	10.768 11	16 8	0.0036
ce-NOP- 1	12.309 20	13.9 14	0.0037
ce-M- 2	20.625 11	2.01 21	0.0009
ce-L- 5	20.996 20	4.2 6	0.0019
ce-NOP- 2	24.847 10	0.66 7	0.0004
ce-M- 3	27.472 12	3.9 19	0.0023
ce-L- 6	28.58 3	0.13 6	≈0
ce-NOP- 3	31.694 11	1.3 6	0.0009
ce-L- 7	37.116 15	13.8 14	0.0109
ce-M- 5	37.700 21	1.12 16	0.0009
ce-NOP- 5	41.922 20	0.37 6	0.0003
ce-L- 8	42.403 20	0.38 4	0.0003
ce-K- 10	45.93	0.392 14	0.0004
ce-M- 7	53.820 16	3.4 4	0.0039
ce-NOP- 7	58.042 15	1.20 13	0.0015
Auger-K	74.3	1.1 9	0.0018

237U B- DECAY (6.75 D 1)   FEEDS 237NP   (Continued)

Radiation Type	Energy (keV)	Intensity (%)	Δ(g-rad/ μCi-h)
ce-K- 11	89.33 4	54 4	0.102
ce-K- 13	115.72 6	0.123 9	0.0003
ce-L- 10	142.183 20	2.21 8	0.0067
ce-K- 14	148.86 6	0.60 4	0.0019
ce-M- 10	158.887 21	0.609 22	0.0021
ce-NOP-10	163.109 20	0.232 9	0.0008
ce-L- 11	185.578 23	11.56 14	0.0457
ce-M- 11	202.282 24	2.843 24	0.0122
ce-NOP-11	206.504 23	1.058 5	0.0047
ce-L- 14	245.11 4	0.191 16	0.0010
β- 1 max	148.2 11		
β- 1 avg	39.1 3	0.196 14	0.0002
β- 2 max	150.5 11		
β- 2 avg	39.8 3	0.268 14	0.0002
β- 3 max	186.7 11		
β- 3 avg	50.0 4	3.6 3	0.0038
β- 4 max	237.8 11		
β- 4 avg	64.7 4	56 5	0.0772
β- 5 max	251.6 11		
β- 5 avg	68.7 4	42 7	0.0615
total β- avg	65.7 4	102 9	0.143
γ 2	13.810 20	0.108 5	≈0
X-ray L	13.9	57 8	0.0170
γ 3	26.348 10	2.41 24	0.0014
γ 6	33.195 11	0.11 5	≈0
γ 7	51.01 3	0.22 10	0.0002
γ 8	59.543 15	36 4	0.0457
X-ray Kα2	64.830 20	1.25 13	0.0017
X-ray Kα1	97.08 4	15.9 11	0.0328
X-ray Kβ	101.07 4	25.7 17	0.0552
	114	12.0 9	0.0292
γ 10	164.610 20	1.97 6	0.0069
γ 11	208.005 23	23.3 4	0.103
γ 14	267.54 4	0.765 14	0.0044
γ 16	332.36 4	1.29 5	0.0091
γ 17	335.38 4	0.102 7	0.0007
γ 20	370.94 4	0.119 7	0.0009

8 weak γ's omitted:
Eγ(avg)= 239.8; ΣIγ= 0.14%

Table A.3.6-1 (continued)
SUPPLEMENTARY DECAY SCHEME DATA

237NP A DECAY (2.14E6 Y 1)   I (MIN) = 0.10%
FEEDS 233PA

Radiation Type	Energy (keV)	Intensity (%)	Δ(g-rad/μCi-h)
ce-K- 12	5.119 21	1.83 10	0.0002
ce-L- 2	8.270 20	22.9 13	0.0040
Auger-L	9.68	50 8	0.0104
ce-K- 14	21.679 21	0.3 3	0.0001
ce-K- 2	24.008 20	5.8 4	0.0030
ce-NOP- 2	27.988 20	1.92 11	0.0011
ce-K- 15	30.649 11	2.92 16	0.0019
ce-L- 4	36.007 20	55 6	0.0419
ce-K- 17	38.809 11	1.24 7	0.0010
ce-L- 5	41.7954 18	9 7	0.0084
ce-L- 6	49.8954 18	14 9	0.0143
ce-M- 4	51.745 20	15.0 15	0.0166
ce-NOP- 4	55.725 20	5.6 6	0.0066
ce-M- 5	57.5331 16	2.6 20	0.0032
ce-NOP- 5	61.5129 18	0.9 8	0.0012
ce-L- 7	65.381 11	1.77 10	0.0025
ce-M- 6	65.6331 16	4 3	0.0052
ce-NOP- 6	69.6129 18	1.3 10	0.0020
Auger-K	70.8	0.14 11	0.0002
ce-M- 7	81.119 11	0.432 25	0.0007
ce-NOP- 7	85.099 11	0.152 9	0.0003
ce-L- 10	85.20 20	0.72 4	0.0013
ce-L- 12	96.615 20	0.418 22	0.0009
ce-M- 10	100.93 20	0.198 10	0.0004
ce-M- 12	112.353 20	0.103 6	0.0002
ce-L- 14	113.175 20	0.15 4	0.0004
ce-L- 15	122.145 11	0.57 3	0.0015
ce-L- 17	130.305 11	0.279 15	0.0008
ce-M- 15	137.883 11	0.136 8	0.0004
α 1	4572.7 20	0.10	0.0097
α 2	4580.0 20	0.40 4	0.0390
α 3	4597.6 20	0.34 4	0.0333
α 4	4618 5	0.2	0.0197
α 5	4622 5	0.10	0.0098
α 6	4629 3	0.3	0.0296
α 7	4638.4 20	6.18 12	0.611
α 8	4658.1 20	0.7	0.0695
α 9	4663.0 20	3.32 10	0.330
α 10	4678 3	0.2	0.0199
α 11	4687 5	0.10	0.0100
α 12	4693.4 20	0.48 20	0.0480
α 13	4707 20	1.1 14	0.1

237NP A DECAY (2.14E6 Y 1)   (Continued)
FEEDS 233PA

Radiation Type	Energy (keV)	Intensity (%)	Δ(g-rad/μCi-h)
α 14	4716 3	0.10	0.0100
α 15	4740.3 20	0.10	0.0101
α 16	4764.7 20	8	0.812
α 17	4769.8 20	25 9	2.54
α 18	4777 5	0.9	0.0916
α 19	4787.5 15	47 9	4.79
α 20	4802.3 20	3	0.307
α 21	4816.3 20	2.5 4	0.256
α 22	4824 5	0.5	0.0514
α 23	4832 3	0.4	0.0412
α 24	4854 3	0.4	0.0414
α 25	4861.8 20	0.6	0.0621
α 26	4863 3	0.30 10	0.0311
α 27	4872.3 20	2.60 20	0.270

10 weak α's omitted:
Eα(avg)=4577.2: ΣIα= 0.56%

Radiation Type	Energy (keV)	Intensity (%)	Δ(g-rad/μCi-h)
X-ray L	13.3	59 9	0.0167
γ 2	29.375 20	9.8 5	0.0061
γ 3	46.46 10	0.140 20	0.0001
γ 4	57.112 20	0.42 4	0.0005
γ 5	62.9	0.2	0.0003
γ 6	71	0.5	0.0008
γ 7	86.486 10	13.1 7	0.0241
γ 8	88.07 10	0.160 20	0.0003
X-ray Kα₂	92.2870 20	1.82 11	0.0036
γ 9	94.66 10	0.84 8	0.0017
X-ray Kα₁	95.8680 20	2.96 18	0.0060
γ 10	106.30 20	0.105 5	0.0002
X-ray Kβ	108	1.37 9	0.0032
γ 12	117.720 20	0.184 9	0.0005
γ 15	143.250 10	0.473 24	0.0014
γ 16	149.5	0.10	0.0003
γ 17	151.410 10	0.263 13	0.0008
γ 18	155.250 20	0.105 5	0.0003
γ 28	194.97 20	0.210 20	0.0009
γ 35	212.330 20	0.184 9	0.0008

30 weak γ's omitted:
Eγ(avg) = 186.8: ΣIγ= 0.76%

Section A: General Scientific and Engineering Information

Table A.3.6-1 (continued)
SUPPLEMENTARY DECAY SCHEME DATA

238U A DECAY (4.468E9 Y 3) I(MIN) = 0.10%
FEEDS 234TH
% SPONTANEOUS FISSION=4.55E-5

Radiation Type	Energy (keV)	Intensity (%)	Δ(g-rad/μCi-h)
Auger-L	9.48	8.1 14	0.0016
ce-L- 1	29.08 6	16.8 22	0.0104
ce-M- 1	44.37 6	4.6 6	0.0043
ce-NOP- 1	48.22 6	1.50 20	0.0015
α 1	4042 5	0.23 7	0.0198
α 2	4150 5	23 4	2.03
α 3	4200 5	77 4	6.89
X-ray L	13	8.7 15	0.0024

1 weak γ's omitted:
Eγ(avq)= 49.6: ΣIγ= 0.07%

238PU A DECAY (87.75 Y 5) I(MIN) = 0.10%
FEEDS 234U
% SPONTANEOUS FISSION=1.7E-7

Radiation Type	Energy (keV)	Intensity (%)	Δ(g-rad/μCi-h)
Auger-L	9.89	9.1 13	0.0019
ce-L- 1	21.733 10	20.7 8	0.0096
ce-M- 1	37.942 10	5.69 21	0.0046
ce-NOP- 1	42.049 10	1.88 7	0.0017
α 1	5357.7	0.10 3	0.0114
α 2	5456.5 4	28.3 6	3.29
α 3	5499.21 20	71.6 6	8.39
X-ray L	13.6	11.6 14	0.0034

25 weak γ's omitted:
Eγ(avq)= 55.5; ΣIγ= 0.05%

239PU A DECAY (2.439E4 Y 2) I(MIN) = 0.10%
FEEDS 235U
% SPONTANEOUS FISSION=4.5E-10

Radiation Type	Energy (keV)	Intensity (%)	Δ(g-rad/μCi-h)
ce-M- 2	7.3920 4	14.3 6	0.0023
ce-L- 3	8.33 10	0.18 16	≈0
Auger-L	9.89	3.5 6	0.0007
ce-NOP- 2	11.4992 4	4.77 19	0.0012
ce-L- 4	16.93 3	2.8 5	0.0010
ce-L- 6	24.46 5	0.11 10	≈0
ce-L- 8	29.86 3	4.78 18	0.0030
ce-M- 4	33.14 3	0.76 14	0.0005
ce-NOP- 4	37.25 3	0.25 5	0.0002
ce-M- 8	46.07 3	1.32 5	0.0013
ce-NOP- 8	50.18 3	0.493 18	0.0005
α 1	5104.6 10	11.50 20	1.25
α 2	5142.9 8	15.10 20	1.65
α 3	5155.4 7	73.3 7	8.05

20 weak α's omitted:
Eα(avq)=5030.0: ΣIα= 0.14%

X-ray L	13.6	4.4 6	0.0013

146 weak γ's omitted:
Eγ(avq)= 113.2: ΣIγ= 0.05%

240U B- DECAY (14.1 H 2) I(MIN) = 0.10%
FEEDS 240NP (7.4 M)
%IT DECAY FOR 240NP (7.4 M)=0.113

Radiation Type	Energy (keV)	Intensity (%)	Δ(g-rad/μCi-h)
Auger-L	10	32 6	0.0069
ce-L- 2	21.67 7	76 10	0.0352
ce-M- 2	38.38 7	18.6 23	0.0152
ce-NOP- 2	42.60 7	6.1 8	0.0055
β- 1 max	360		
avg	102 16	100	0.217
X-ray L	13.9	44 7	0.0131
γ 2	44.10 7	1.69 20	0.0016

Table A.3.6-1 (continued)
SUPPLEMENTARY DECAY SCHEME DATA

240PU A DECAY (6537 Y 10) I(MIN) = 0.10%
FEEDS 236U
% SPONTANEOUS FISSION=4.9E-6

Radiation Type	Energy (keV)	Intensity (%)	Δ(g-rad/μCi-h)
Auger-L	9.89	8.7 12	0.0018
ce-L- 1	23.485 6	19.7 4	0.0098
ce-M- 1	39.694 6	5.40 12	0.0046
ce-NOP- 1	43.801 6	1.79 4	0.0017
α 1	5123.43 23	26.5 4	2.89
α 2	5168.30 15	73.4 8	8.08

3 weak α's omitted:
Eα(avg)=5010.4; ΣIα= 0.09%

X-ray L	13.6	11.0 12	0.0032

9 weak γ's omitted:
Eγ(avg)= 54.3; ΣIγ= 0.05%

241PU B- DECAY (14.4 Y 2) I(MIN) = 0.10%
%B- DECAY =99.99755 8
FEEDS 241AM
%A DECAY=0.00245 8

Radiation Type	Energy (keV)	Intensity (%)	Δ(g-rad/μCi-h)
β- 1 max	20.80 20		
avg	5.23 5	99.99755	0.0111

241AM A DECAY (433 Y 2) I(MIN) = 0.10%
FEEDS 237NP
% SPONTANEOUS FISSION=3.8E-10

Radiation Type	Energy (keV)	Intensity (%)	Δ(g-rad/μCi-h)
ce-L- 1	3.9182 14	8.3 8	0.0007
Auger-L	10	20 3	0.0043
ce-L- 3	10.773 10	15.0 3	0.0034
ce-M- 1	20.622 4	2.15 19	0.0009
ce-L- 5	20.993 20	9.2 9	0.0041
ce-NOP- 1	24.8443 13	0.71 7	0.0004
ce-M- 3	27.477 11	3.74 8	0.0022
ce-NOP- 3	31.699 10	1.23 8	0.0008
ce-L- 6	33.113 20	0.976 20	0.0007
ce-L- 7	37.1102 14	13.9 3	0.0110
ce-M- 5	37.697 21	2.43 24	0.0019
ce-NOP- 5	41.919 20	0.80 8	0.0007
ce-M- 6	49.817 21	0.259 6	0.0003
ce-M- 7	53.814 4	3.43 7	0.0039
ce-NOP- 7	58.0363 13	1.205 25	0.0015
ce-L- 13	76.50 3	0.23 4	0.0004
α 1	5387.0 10	1.5 3	0.172
α 2	5442.98 13	12.3 6	1.43
α 3	5485.74 12	85.6 10	10.00
α 4	5511.0 10	0.21 5	0.0247
α 5	5543.0 10	0.35 8	0.0413

23 weak α's omitted:
Eα(avg)=5301.6; ΣIα= 0.03%

X-ray L	13.9	28 3	0.0082
γ 3	26.3450 10	2.58 22	0.0014
γ 7	33.200 10	0.1125 1	=0
γ 7	59.5370 10	36.3 4	0.0460

101 weak γ's omitted:
Eγ(avg)= 72.8; ΣIγ= 0.17%

Table A.3.6-1 (continued)
SUPPLEMENTARY DECAY SCHEME DATA

242PU A DECAY (3.87E5 Y 2)   I(MIN) = 0.10%
FEEDS 238U
% SPONTANEOUS FISSION=5.50E-4 5

Radiation Type	Energy (keV)	Intensity (%)	Δ(q-rad/μCi-h)
Auger-L	9.89	6.8 10	0.0014
ce-L- 1	23.156 13	15.5 3	0.0076
ce-M- 1	39.365 13	4.27 8	0.0036
ce-NOP- 1	43.472 13	1.409 25	0.0013
α 1	4856.3 12	21	2.18
α 2	4900.6 12	78.9	8.24
X-ray L	13.6	8.7 10	0.0025

3 weak γ's omitted:
Eγ (avg) = 56.4; ΣIγ = 0.04%

242CM A DECAY (163.2 D 4)   I(MIN) = 0.10%
FEEDS 238PU
% SPONTANEOUS FISSION=6.8E-6

Radiation Type	Energy (keV)	Intensity (%)	Δ(q-rad/μCi-h)
Auger-L	10.3	7.7 14	0.0017
ce-L- 1	20.98 3	19.2 18	0.0086
ce-M- 1	38.15 3	5.3 5	0.0043
ce-NOP- 1	42.52 3	1.77 17	0.0016
α 1	6069.63 12	26.0 5	3.36
α 2	6112.92 8	74.0 5	9.64
X-ray L	14.3	11.5 16	0.0035

6 weak α's omitted:
Eα (avg) =5948.5; ΣIα = 0.04%

9 weak γ's omitted:
Eγ (avg) = 59.2; ΣIγ = 0.04%

243PU B- DECAY (4.956 H 3)   I(MIN) = 0.10%
FEEDS 243AM

Radiation Type	Energy (keV)	Intensity (%)	Δ(q-rad/μCi-h)
ce-L- 2	10.2271 20	0.52 5	0.0001
Auger-L	10.5	6.7 15	0.0015
ce-L- 3	18.03 20	0.77 10	0.0003
ce-L- 4	18.4 5	9.0 8	0.0035
ce-M- 2	27.880 8	0.144 13	≈0
ce-L- 5	30.2271 20	3 3	0.0019
ce-M- 3	35.68 20	0.194 25	0.0001
ce-M- 4	36.1 5	2.36 21	0.0018
ce-NOP- 4	40.6 5	0.78 7	0.0007
ce-M- 5	47.880 8	0.9 7	0.0009
ce-NOP- 5	52.3829 11	0.3 3	0.0004
ce-L- 8	60.23 20	3.7 4	0.0048
ce-L- 7	72.6 4	0.21 4	0.0003
ce-M- 7	77.88 20	0.93 9	0.0015
ce-NOP- 7	82.38 20	0.33 3	0.0006
ce-K- 14	256.7 3	0.32 4	0.0017
β- 1 max	50 4		
avg	12.6 10	0.23 5	≈0
β- 2 max	116 4		
avg	30.4 10	1.24 14	0.0008
β- 3 max	473 4		
avg	136.9 12	0.24 17	0.0007
β- 4 max	486 4		
avg	141.1 12	4 4	0.0132
β- 5 max	498 4		
avg	145.1 12	29 3	0.0896
β- 6 max	540 4		
avg	158.8 12	6 4	0.0203
β- 7 max	582 4		
avg	172.7 12	59 4	0.217
total β- avg	160.3 13	100 8	0.342
X-ray L	14.6	10.9 20	0.0034
γ 3	41.80 20	0.76 10	0.0007
γ 6	67	0.23 12	0.0003
γ 7	84.00 20	23.0 20	0.0412
X-ray Kα₁	106.52 6	0.147 19	0.0003
γ 10	109.30 20	0.161 22	0.0004
γ 13	356.4 3	0.131 17	0.0010
γ 14	381.7 3	0.55 7	0.0045

13 weak γ's omitted:
Eγ (avg) = 137.6; ΣIγ = 0.18%

Table A.3.6-1 (continued)
SUPPLEMENTARY DECAY SCHEME DATA

243AM α DECAY (7380 Y 40)   I (MIN) = 0.10%
FEEDS 239NP

Radiation Type	Energy (keV)	Intensity (%)	Δ(g-rad/μCi-h)
ce-L- 1	8.67 15	9.2 10	0.0017
Auger-L	10	12.1 18	0.0026
ce-L- 3	21.10 15	4.8 5	0.0022
ce-M- 1	25.38 15	2.23 23	0.0012
ce-NOP- 1	29.60 15	0.73 8	0.0005
ce-L- 5	32.9732 9	0.85 5	0.0006
ce-M- 3	37.81 15	1.21 11	0.0010
ce-NOP- 3	42.03 15	0.40 4	0.0004
ce-M- 5	49.677 4	0.230 12	0.0002
ce-L- 6	52.24 15	13.9 7	0.0154
ce-M- 6	68.95 15	3.42 17	0.0050
ce-NOP- 6	73.17 15	1.21 6	0.0019
ce-L- 8	76.0732 9	0.106 6	0.0002
α 1	5181.0 10	1	0.121
α 2	5233.5 10	10.6	1.18
α 3	5275.4 10	87.9	9.88
α 4	5321.0 10	0.12	0.0136
α 5	5350.0 10	0.16	0.0182

8 weak α's omitted:
Eα(avg)=5032.0; ΣIα= 0.02%

X-ray L	13.9	16.7 19	0.0050
γ 3	43.53 15	5.5 5	0.0051
γ 6	74.67 15	66 3	0.105
γ 7	86.79 15	0.34 4	0.0006
γ 9	117.60 15	0.55 9	0.0014
γ 10	142.18 15	0.125 15	0.0004

12 weak γ's omitted:
Eγ(avg)= 52.0; ΣIγ= 0.10%

244PU α DECAY (8.26E7 Y 9)   I (MIN) = 0.10%
%α DECAY=99.875 6
FEEDS 240U
% SPONTANEOUS FISSION=0.125 6

Radiation Type	Energy (keV)	Intensity (%)	Δ(g-rad/μCi-h)
Auger-L	9.89	6.2 9	0.0013
ce-L- 1	21.2 10	14.0 6	0.0063
ce-M- 1	37.5 10	3.85 17	0.0031
ce-NOP- 1	41.6 10	1.27 6	0.0011
α 1	4546.0 10	19.4 8	1.88
α 2	4589.0 10	80.5 8	7.87
X-ray L	13.6	7.8 9	0.0023

1 weak γ's omitted:
Eγ(avg)= 43.0; ΣIγ= 0.02%

Table A.3.6-1 (continued)
SUPPLEMENTARY DECAY SCHEME DATA

244CM A DECAY (18.11 Y 2)   I(MIN) = 0.10%
FEEDS 240PU
% SPONTANEOUS FISSION=1.347E-4 2

Radiation Type	Energy (keV)	Intensity (%)	Δ(g-rad/µCi-h)
Auger-L	10.3	7.0 11	0.0015
ce-L- 1	19.727 9	17.4 8	0.0073
ce-M- 1	36.891 9	4.84 21	0.0038
ce-NOP- 1	41.265 8	1.60 7	0.0014
α 1	5762.84 3	23.60 20	2.90
α 2	5804.96 5	76.40 20	9.45

5 weak α's omitted:
Eα(avg)=5633.0; ΣIα= 0.03%

| X-ray L | 14.3 | 10.5 12 | 0.0032 |

1 weak γ's omitted:
Eγ(avg)= 42.8; ΣIγ= 0.03%

252CF A DECAY (2.638 Y 10)   I(MIN) = 0.10%
%A DECAY=96.908 8
FEEDS 248CM
% SPONTANEOUS FISSION=3.092 8

Radiation Type	Energy (keV)	Intensity (%)	Δ(g-rad/µCi-h)
Auger-L	10.7	4.1 7	0.0009
ce-L- 1	18.876 25	11.2 5	0.0045
ce-M- 1	37.086 25	3.15 14	0.0025
ce-NOP- 1	41.735 25	1.04 5	0.0009
ce-L- 2	75.677	0.16 3	0.0003
α 1	5976.6	0.23 4	0.0296
α 2	6075.7 5	15.2 3	1.97
α 3	6118.3 5	81.6 3	10.63
X-ray L	15	7.3 8	0.0023

2 weak γ's omitted:
Eγ(avg)= 68.0; ΣIγ= 0.03%

REFERENCE

1. **Kocher, D. C.,** Nuclear Decay Data for Radionuclides Occurring in Routine Releases from Nuclear Fuel Cycle Facilities, ORNL/NUREG/TM-102, Oak Ridge National Laboratory, Oak Ridge, Tenn., August 1977.

APPENDIX 3.7.
SOME DATA ON NATURAL RADIATION

Alfred W. Klement, Jr.

Man is exposed to sources of radiation found in nature in varying degrees depending on his activities and location. Cosmic radiation entering the earth's atmosphere and crust is one source. Nuclear interactions of cosmic rays with matter produce radionuclides to which man is exposed. Other sources of natural radiation are elements found in the earth's crust that are composed of one or more radioisotopes.

This appendix is intended to provide summary data on natural sources of radiation. This subject is discussed more fully in Section 18 of the next Volume of this handbook. This appendix includes figures and tables from that section (adapted from References 1 and 2) and additional data that have been reported more recently.[3] Reference 3 provides a comprehensive review of the subject from which only some of the data are listed here. The figures and tables in this appendix were taken from the referenced reviews. There are several important, or seemingly important, considerations with regard to such reviews and brief tabulation of data. These reviews for the most part merely indicate a general range of data. Definitive estimates of natural (as well as man-made) radiation doses or levels of natural radioactivity at a particular site or area require statistical approaches utilizing sophisticated methods and instrumentation because variability is great in many situations from site to site and from time to time. However, for many situations, the health physicist is not greatly concerned with natural radiation, and these ranges or averages shown here would be adequate for comparative purposes.

The data reported here in summary form are often the result of compromises with regard to opinions on the validity of measurements and theories. While some data reported appear to be definitive, compilations as a result of compromise among groups of researchers may include data obtained by different methods and based on different theories of varying sophistication. Again, however, the data probably represent the most definitive and best available. Also, they have had the benefit of scientific discussion and argument. Differences among researchers are probably not great nor of much consequence to many health physicists.

The review references cited here provide bibliographies of studies of natural radiation and radioactivity. They are recommended for the serious student on the subject.

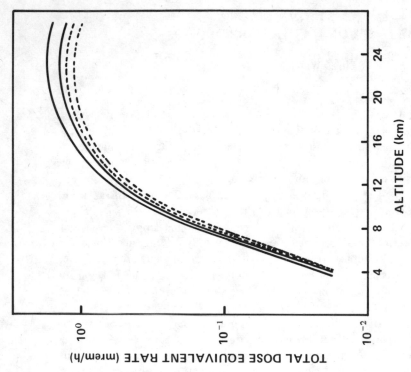

ABSORBED DOSE RATE IN AIR (mrad/y)

ALTITUDE (kilometers)

TISSUE DOSE EQUIVALENT RATE (mrem/y)

FIGURE A. 3.7.1. Long-term average dose from cosmic radiation. The charged particle absorbed dose rate in air or tissue is shown in the lower curve and the total dose equivalent rate (charged particles plus neutrons) is shown in the upper curve at 5 cm depth in a 30-cm thick slab of tissue. (From *Natural Background Radiation in the United States,* Report No. 45, National Council on Radiation Protection and Measurements, Washington, D. C., 1975. With permission.)

ALTITUDE (km)

TOTAL DOSE EQUIVALENT RATE (mrem/h)

FIGURE A.3.7.2. Total cosmic-ray dose equivalent rate at 5 cm depth in a 30-cm slab of tissue at latitude 55° N (–) and 43° N(- - - -) at solar minimum (upper curve) and solar maximum (lower curve). The quality factors for neutrons as a function of energy are included in the calculations. (From *Natural Background Radiation in the United States,* Report No. 45, National Council on Radiation Protection and Measurements, Washington, D. C., 1975. With permission.)

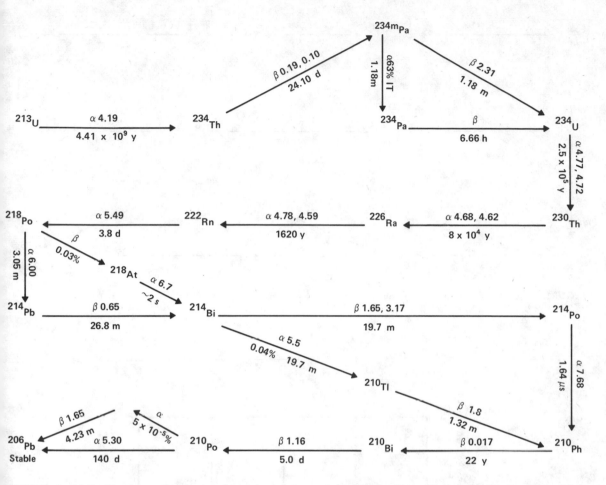

FIGURE A. 3.7.3. Uranium series decay scheme (α and β energies in MeV). (From Klement, A. W., Jr., *Radioactive Fallout, Soils, Plants, Foods, Man,* Fowler, E. B., Ed., Elsevier, Amsterdam, 1965, 113. With permission.)

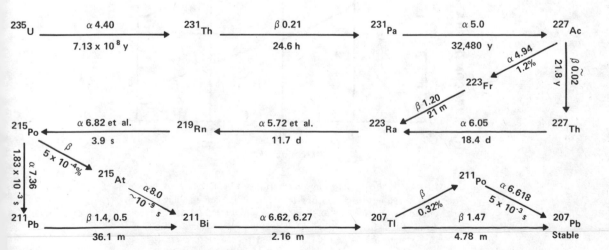

FIGURE A. 3.7.4. Actinium series decay scheme. (α and β energies in MeV). (From Klement, A. W., Jr., *Radioactive Fallout, Soils, Plants, Foods, Man,* Fowler, E. B., Ed., Elsevier, Amsterdam, 1965, 113. With permission.)

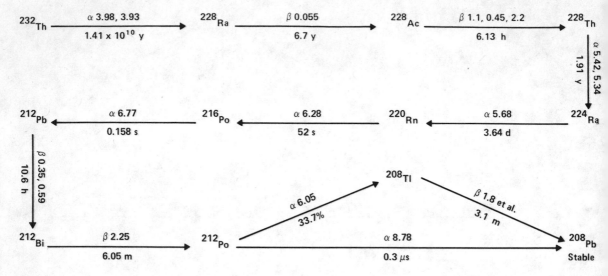

FIGURE A. 3.7.5.   Thorium series decay scheme.   (α and β energies in MeV).   (From Klement, A. W., Jr., *Radioactive Fallout, Soils, Plants, Foods, Man,* Fowler, E. B., Ed., Elsevier, Amsterdam, 1965, 113. With permission.)

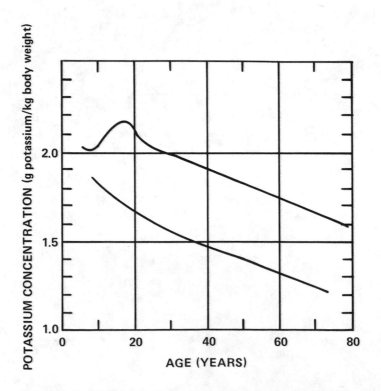

FIGURE A. 3.7.6.   Potassium concentration in the human body. The upper curve refers to men, the lower curve to women. The curves represent the results of 10,000 measurements (UNSCEAR, 1972). Note from Table 3.7.3 the values of isotopic abundance and 8.3 × 10^{-10} Ci/g element for which can be used to calculate average ^{40}K activities vs. age from this graph. (From *Natural Background Radiation in the United States,* Report No. 45, National Council on Radiation Protection and Measurements, Washington, D. C., 1975. With permission.)

Table A. 3.7.1
ESTIMATED ANNUAL COSMIC-RAY
WHOLE-BODY DOSES
(MREM/PERSON)

Political unit	Average annual dose	Political unit	Average annual dose
Alabama	40	New Jersey	40
Alaska	45	New Mexico	105
Arizona	60	New York	45
Arkansas	40	North Carolina	45
California	40	North Dakota	60
Colorado	120	Ohio	50
Connecticut	40	Oklahoma	50
Delaware	40	Oregon	50
Florida	35	Pennsylvania	45
Georgia	40	Rhode Island	40
Hawaii	30	South Carolina	40
Idaho	85	South Dakota	70
Illinois	45	Tennessee	45
Indiana	45	Texas	45
Iowa	50	Utah	115
Kansas	50	Vermont	50
Kentucky	45	Virginia	45
Louisiana	35	Washington	50
Maine	50	Wast Virginia	50
Maryland	40	Wisconsin	50
Massachusetts	40	Wyoming	130
Michigan	50	Canal Zone	30
Minnesota	55	Guam	35
Mississippi	40	Puerto Rico	30
Missouri	45	Samoa	30
Montana	90	Virgin Islands	30
Nebraska	75	District of Columbia	40
Nevada	85		
New Hampshire	45	Total United States	45

From Klement, A. W., Jr., Miller, C. R., Minx, R. P., and Shleien, B., *Estimates of Ionizing Radiation Doses in the United States 1960–2000,* Report ORP/CSD 72-1, U.S. Environmental Protection Agency, Washington, D. C., 1972, 7.

Table A. 3.7.2
MAJOR COSMIC-RAY-ACTIVATED RADIONUCLIDES

Nuclide	Half-life	Emission, decay	Energy of primary emission (MeV)	Production rate (atoms/cm²-sec)
^{3}H	12.26 years	β	0.0181	0.12–1.3
^{7}Be	53 days	ϵ, γ	–	0.021–0.035
^{10}Be	2.7×10^4 years	β	0.56	0.04–0.1
^{14}C	5760 years	β	0.156	2–2.6
^{22}Na	2.58 years	$\beta+, \epsilon, \gamma$	0.54	–
^{32}Si	~700 years	β	0.1 (to ^{32}P)	2×10^{-4}
^{32}P	14.3 days	β	1.71	1×10^{-4}
^{33}P	25 days	β	0.25	1×10^{-4}
^{35}S	86.7 days	β	0.168	2×10^{-4}
^{36}Cl	3×10^5 years	β, ϵ	0.71	–
^{39}Cl	55 min	β, γ	1.91, 2.18, 3.43	–

From Klement, A. W., Jr., *Radioactive Fallout, Soils, Plants, Foods, Man,* Fowler, E. B., Ed., Elsevier, Amsterdam, 1965, 113. With permission.

Table A. 3.7.3
PRIMORDIAL RADIONUCLIDES

Nuclide	Half-life (years)	% Isotopic abundance	Emissions, decay	Energy of primary emission (MeV)	Specific activity (Ci/g element)
^{40}K	1.3×10^{-9}	0.0118	β, γ, ϵ	1.32	8.3×10^{-10}
^{50}V	$\sim 6 \times 10^{-14}$	0.25	β, γ, ϵ	—	$\sim 2.8 \times 10^{-14}$
^{87}Rb	4.7×10^{10}	27.85	β	0.27	2.5×10^{-8}
^{115}In	6×10^{-14}	95.72	β	0.6	5.0×10^{-12}
^{138}La	1.1×10^{11}	0.089	β, γ, ϵ	0.205	2.1×10^{-12}
^{142}Ce	5×10^{-15}	11.07	α	1.5	5.7×10^{-14}
^{144}Nd	$\sim 5 \times 10^{-15}$	23.85	α	1.8	$\sim 1.2 \times 10^{-13}$
^{147}Sm	1.06×10^{11}	14.97	α	2.24	3.38×10^{-8}
^{148}Sm	1.2×10^{13}	11.24	α	2.14	2.24×10^{-11}
^{149}Sm	$\sim 4 \times 10^{-14}$	13.83	α	1.84	$\sim 8.2 \times 10^{-12}$
^{152}Gd	1.1×10^{-14}	0.200	α	2.15	4.15×10^{-12}
^{174}Hf	4.3×10^{-15}	0.18	α	2.5	8.4×10^{-14}
^{176}Lu	3.6×10^{10}	2.59	β, γ	0.42	1.47×10^{-10}
^{187}Re	7×10^{10}	62.93	β	<0.008	1.73×10^{-8}
^{190}Pt	7×10^{11}	0.0127	α	3.11	3.33×10^{-13}
^{192}Pt	$\sim 10^{15}$	0.78	α	2.6	$\sim 1.4 \times 10^{-14}$
^{204}Pb	1.4×10^{17}	1.48	α	2.6	1.83×10^{-16}
^{238}U (series[a])					
^{235}U (series[a])					
^{232}Th (series[a])					

[a] See Figures 3.7.3 to 3.7.5.

From Klement, A. W., Jr., *Radioactive Fallout, Soils, Plants, Foods, Man*, Fowler, E. B., Ed., Elsevier, Amsterdam, 1965, 113. With permission.

Table A. 3.7.4
NATURAL RADIONUCLIDES IN THE EARTH'S CRUST (μg/g)

Nuclide	Igneous rocks					Sedimentary rocks			Deep sea sediments	
			Granite							
	Ultra basic	Basaltic	High Ca	Low Ca	Syenites	Shales	Sand-stones	Carbo-nates	Carbo-nate	Clay
^{40}K	0.0047	0.98	2.97	4.96	5.66	3.14	1.26	0.32	0.34	2.95
^{50}V	0.096	0.60	0.21	0.10	0.072	0.31	0.048	0.048	0.048	0.29
^{87}Rb	0.056	8.4	31	47	31	39	17	0.8	2.8	31
^{115}In	0.01	0.21	0.0x	0.25	0.0x	0.1	0.x	0.x	0.x	1.4
^{138}La	0.000x	0.013	0.040	0.049	0.060	0.082	0.03	0.000x	0.009	0.102
^{142}Ce	0.0x	5.3	9.0	10	17.8	6.5	10	1.27	3.9	38.2
^{144}Nd	0.0x	4.8	7.9	8.8	15.5	5.7	8.8	1.1	3.3	33
^{147}Sm	0.0x	0.79	1.3	1.5	2.7	0.96	1.5	0.19	0.57	5.7
^{148}Sm	0.0x	0.60	0.99	1.1	1.0	0.72	1.1	0.15	0.43	4.3
^{149}Sm	0.0x	0.73	1.2	1.4	2.5	0.89	1.4	0.13	0.53	5.3
^{152}Gd	0.00x	0.011	0.018	0.02	0.036	0.013	0.02	0.0026	0.0076	0.076
^{174}Hf	0.0011	0.0036	0.041	0.0070	0.020	0.0050	0.0070	0.00054	0.00074	0.0074
^{176}Lu	0.00x	0.016	0.028	0.021	0.054	0.018	0.031	0.0052	0.013	0.12
^{204}Pb	0.015	0.09	0.22	0.28	0.18	0.30	0.10	0.12	0.12	1.2
Th	0.004	4	8.5	17	13	12	1.7	1.7	x	7
U	0.001	1	3	3	3	3.7	0.45	2.2	0.x	1.3

Note: x = an order of magnitude

From Klement, A. W., Jr., *Radioactive Fallout, Soils, Plants, Foods, Man*, Fowler, E. B., Ed., Elsevier, Amsterdam, 1965, 113. With permission.

Table A. 3.7.5
ESTIMATES OF NATURAL RADIONUCLIDES IN SOILS (pCi/g)[a,b,c]

Location	Radium	Thorium	Uranium	^{40}K
Cambridge Gault	1			
Dublin	5.2, 2.8			
U.S.	1.97	3.6—6.1		
U.S. Subsoils	1.52			
U.S.S.R. Central Mountains	0.28—0.95	0.25—1.50	0.3—6.0	
Poland		1.0—1.2	3.1—3.2	
Germany	0.15—1.30			
Colorado-New Mexico	1.2—2.0			
U.S.	80—800 mCi/mi²			
Germany-North America	39—780 mCi/mi²			
Germany (^{226}Ra)	433—1,660 mCi/mi²			
U.S.	0.9—1.5			0.0025—0.0075
Spain	0—237			

[a] Summarized from Klement.[1] This table merely shows the paucity of data, but more importantly that the variability is very wide and these data are of little use in health protection except in a gross way, so that detailed studies would be needed on a local or specific area basis to be of value where a problem in this regard appears significant.

[b] The values shown were mostly originally reported in units of grams per gram.

[c] It has been estimated that for U.S. soils, the average ^{40}K concentrations are about 21.3 Ci/mi².

Table A. 3.7.6
ESTIMATED AVERAGE CONCENTRATIONS OF PRIMORDIAL RADIONUCLIDES IN THE OCEANS AND SEDIMENTS

Nuclide	Seawater (g/l)	Surface sediments (g/g)	Nuclide	Seawater (g/l)	Surface sediments (g/g)
^{238}U	3.0×10^{-6}	1×10^{-6}	^{215}Po	$<8.1 \times 10^{-29}$	1.4×10^{-26}
^{235}U	2.1×10^{-8}	7.1×10^{-9}	^{214}Po	3.0×10^{-28}	1.1×10^{-27}
^{234}U	1.6×10^{-10}	8.1×10^{-11}	^{212}Po	1.2×10^{-22}	2.4×10^{-29}
^{234}Pa	1.4×10^{-19}	4.7×10^{-20}	^{211}Po	$<6.8 \times 10^{-20}$	1.2×10^{-26}
^{231}Pa	$<2 \times 10^{-12}$	1×10^{-11}	^{210}Po	2.2×10^{-17}	8.8×10^{-10}
^{234}Th	4.3×10^{-17}	1.4×10^{-17}	^{214}Bi	2.1×10^{-21}	8.8×10^{-20}
^{232}Th	$<2 \times 10^{-8}$	5.0×10^{-6}	^{212}Bi	2.2×10^{-22}	3.7×10
^{231}Th	8.6×10^{-20}	2.9×10^{-20}	^{211}Bi	$<5.6 \times 10^{-24}$	1.0×10^{-21}
^{230}Th	$<3 \times 10^{-13}$	2.0×10^{-10}	^{210}Bi	7.8×10^{-19}	3.1×10^{-17}
^{228}Th	4.0×10^{-18}	7×10^{-16}	^{214}Pb	2.9×10^{-21}	1.2×10^{-18}
^{227}Th	$<7.0 \times 10^{-20}$	1.3×10^{-17}	^{212}Pb	2.4×10^{-21}	3.9×10^{-19}
^{228}Ac	1.5×10^{-21}	2.4×10^{-19}	^{211}Pb	$<9.0 \times 10^{-23}$	1.6×10^{-20}
^{227}Ac	$<1 \times 10^{-15}$	5.9×10^{-15}	^{210}Pb	1.1×10^{-15}	4.5×10^{-14}
^{228}Ra	1.4×10^{-17}	2.3×10^{-15}	^{201}Pb	4.4×10^{-8}	—
^{226}Ra	1.0×10^{-13}	4.0×10^{-12}	^{208}Tl	4.1×10^{-12}	6.7×10^{-22}
^{224}Ra	2.1×10^{-20}	3.4×10^{-18}	^{207}Tl	$<1.2 \times 10^{-23}$	2.1×10^{-21}
^{223}Ra	$<4.4 \times 10^{-20}$	8.5×10^{-18}	^{142}Ce	4.4×10^{-4}	—
^{223}Fr	$<7.0 \times 10^{-24}$	1.4×10^{-21}	^{138}La	2.7×10^{-10}	—
^{220}Rn	6.3×10^{-19}	2.5×10^{-17}	^{87}Rb	3.3×10^{-5}	—
^{219}Rn	3.3×10^{-24}	5.4×10^{-22}	^{50}V	4.8×10^{-9}	—
^{218}Rn	$<1.7 \times 10^{-25}$	3.1×10^{-23}	^{40}K	4.2×10^{-5}	$0.8—4.5 \times 10^{-8}$
^{216}Po	3.4×10^{-22}	1.4×10^{-20}			
^{215}Po	1.0×10^{-27}	1.7×10^{-24}			

From Klement, A. W., Jr., *Radioactive Fallout, Soils, Plants, Foods, Man*, Fowler, E. B., Ed., Elsevier, Amsterdam, 1965, 113. With permission.

Table A. 3.7.7

ESTIMATED CONCENTRATIONS OF MAJOR COSMOGENIC RADIONUCLIDES IN THE OCEANS

Nuclide	Surface seawater (g/l)	Surface sediment (g/g)
^{35}S	$<1.8 \times 10^{-18}$	—
^{32}P	$<1.5 \times 10^{-18}$	—
^{33}P	$<3.1 \times 10^{-18}$	—
^{32}Si	5×10^{-19}	$0-2 \times 10^{-14}$
^{14}C	$2-3 \times 10^{-14}$	$0.1-1 \times 10^{-13}$
^{10}Be	$0.7-8 \times 10^{-17}$	$1-3 \times 10^{-13}$
^{7}Be	$<4.9 \times 10^{-17}$	—
^{3}H	$0.7-5 \times 10^{-16}$	—

From Klement, A. W., Jr., *Radioactive Fallout, Soils, Plants, Foods, Man*, Fowler, E. B., Ed., Elsevier, Amsterdam, 1965, 113. With permission.

Table A. 3.7.8

CONCENTRATION OF RADIUM IN FRESHWATERS[a,b] (pCi/l)[a]

Location	Concentrations
Various rivers	0.01-0.1
U.S.S.R. (normal)	1
U.S.S.R.	0.2-4
Germany (rivers)	0.07-0.84
U.S. (average)	0.1
U.S. (drinking water supplies)	0-6.5
Western U.S.	0.04-440
U.S.S.R. (deep spring brines)	up to 1000
Germany (springs)	up to 18.4
Maine-New Hampshire (maximum)	730 (^{226}Ra)
Illinois (maximum)	36 (^{226}Ra)
Illinois (sandstone wells)	up to 36.1 (^{226}Ra)
Spas	1×10^{5}
U.S.S.R.	$20-272 \times 10^{6}$ (^{224}Ra)

[a] Summarized from Klement.[1] This table merely shows the paucity of data, but more importantly that the variability is very wide and these data are of little use in health protection except in a gross way, so that detailed studies would be needed on a local or specific basis to be of value where a problem in this regard appeals significant.

[b] The values shown here were mostly originally reported in units of grams per gram.

Table A. 3.7.9

CONCENTRATIONS OF COSMOGENIC RADIONUCLIDES IN PRECIPITATION (ATOMS/ml)

Nuclide	Concentration
^{3}H	$5 \times 10^{2} -1 \times 10^{6}$
^{7}Be	$2 \times 10^{3} -6 \times 10^{3}$
^{14}C	14.7 dpm/g C
^{22}Na	~40
^{32}P	~40, 25
^{33}P	~30, 55
^{35}PS	~400, 316
^{39}Cl	2.5

From Klement, A. W., Jr., *Radioactive Fallout, Soils, Plants, Foods, Man*, Fowler, E. B., Ed., Elsevier, Amsterdam, 1965, 113. With permission.

Table A. 3. 7.10
CONCENTRATIONS OF RADIUM IN
DRINKING WATER (pCi/l)

Location	Concentration
U.S.	0–0.17
Joliet, Illinois	5.8
Chicago, Illinois	0.3(^{226}Ra)
Germany	0.02–0.62
London	0.1

From Klement, A. W., Jr., *Radioactive Fallout, Soils, Plants, Foods, Man,* Fowler, E. B., Ed., Elsevier, Amsterdam, 1965, 113. With permission.

Table A. 3.7.11
RADIUM-226 IN FOOD GROUPS FOR ESTIMATION OF
TOTAL DIETARY INTAKE[a,b]

Food groups	New York pCi ^{226}Ra/kg		New York pCi ^{226}Ra/year		Chicago pCi ^{226}Ra/kg		Chicago pCi ^{226}Ra/year		San Francisco pCi ^{226}Ra/kg		San Francisco pCi ^{226}Ra/year	
Whole wheat bread	3.2	1.2	36	13	3.5	2.9	38	33	2.8[b]	2.8	31[b]	31
White bread	3.2	1.5	118	54	3.3	2.0	123	76	2.9	2.5	106	94
Flour, white	2.7	1.7	117	74	2.4	2.0	106	84	1.34	0.83	58	36
Milk, liquid	0.25	0.24	56	54	0.24	0.22	53	49	0.22	0.20	49	44
Potatoes	2.0	2.5	90	113	1.4	0.77	64	35	1.0	2.0	46	91
Macaroni	2.1	1.8	6.2	5.3	1.6	1.9	4.8	5.8	1.2	1.7	3.5	5.1
Dried beans	6.1	3.2	18	9.7	7.0	2.5	21	7.5	2.3	4.1	6.8	12
Canned vegetables	2.2	0.54	44	11	1.8	1.1	37	22	0.91	1.0	18	20
Fresh vegetables	2.4	1.2	105	53	2.2	0.57	97	25	0.66	0.84	28	36
Root vegetables	3.4	2.3	57	39	2.0	1.8	34	31	2.6	2.4	44	40
Canned fruit	0.37	0.37	9.7	9.5	1.2	0.26	31	6.7	0.50	0.73	13	19
Fruit juices	1.6	0.49	31	9.3	0.68	0.86	13	16	0.71	0.62	14	12
Fresh fruit	1.5	2.8	99	192	1.4	0.57	92	39	0.91	0.65	62	44
Rice	1.5	1.0	4.6	3.1	0.70	0.37	2.1	1.1	0.63	0.80	1.9	2.4
Eggs	4.1	7.9	66	127	2.7	2.7[b]	44	44[b]	2.6	1.9	41	31
Fresh fish	1.2	0.68	9.6	5.4	0.71	1.0	5.7	8.3	0.80	1.2	6.4	9.2
Shellfish	1.2	1.1	1.2	1.1	2.5	1.7	2.5	1.7	2.0	1.0	2.0	1.0
Meat	0.44	0.47	32	34	0.45	0.64	32	47	0.81	0.55	59	40
Poultry	0.73	0.86	12	15	0.79	1.4	13	23	1.9	0.49	32	8.4

[a] Results of two samplings, 1960 to 1961.
[b] Only one sample, same value used for calculations.

From Hallden, N. A. and Fisenne, I. M., Fallout Program Quarterly Summary Report, Hardy, E. P., Jr., et al., Eds., Report HASL-113, U.S. Atomic Energy Commission, New York, 1961, 90.

Table A. 3.7.12
ESTIMATES OF POTASSIUM-40 CONTENT OF FOODS

Food groups	Estimated ^{40}K range pCi/kg food	Food groups	Estimated ^{40}K range pCi/kg food
Dairy Products	(770–1,510)	Vegetables	
Cows milk	1180	Roots	(910–4,450)
Cheese	800	Carrots (raw)	3400
Fats, Oils, Shortenings	(0–480)	Carrots (canned)	910
Butter	190	Sweet potatoes	4450
Margarine	480	White potatoes	3400
Fruits (fresh, raw)	(620–3,700)	Leaf or Stem	(1,080–6,500)
Assorted berries	(750–1,510)	Cabbage (raw)	1900
Apples	620	Lettuce	1150
Bananas	3520	Spinach (raw)	6500
Peaches	1320	Spinach (canned)	2140
Oranges	1400	Celery	2490
Cantaloupes	1900	Beans and Peas	(800–4,640)
Grain Products	(710–1,900)	Lima beans (raw)	4640
Wheat flour	710	Lima beans (green, canned)	1740
Rice (white)	1070	Snap beans (green, raw)	2490
White bread	1500	Peas (green, raw)	3070
Whole wheat bread	1900	Peas (green, canned)	800
Nuts	(3,500–6,400)	Other	
Brazil nuts	5600	Tomatoes (raw)	1900
Coconut (dry, sweet)	6400	Cucumbers (raw)	1900
Peanuts (roasted, salted)	5820	Corn, sweet (raw)	(1,990–3,070)
Meats (cooked)	(2,740–3,320)	Miscellaneous	
Seafoods (raw)	(925–4,540)	Sugar (cane)	4
Seafoods (cooked)	(925–2,520)	Sugar (brown)	1900
Eggs	840	Beer	390
		Cola beverages	430

From Klement, A. W., Jr., *Radioactive Fallout, Soils, Plants, Foods, Man,* Fowler, E. B., Ed., Elsevier, Amsterdam, 1965, 113. With permission.

Table A. 3.7.13
COMPARISON OF DIETARY ^{226}Ra, ^{40}K, AND ^{210}Pb BY INCOME AND AGE GROUPS

Sample groups	Sample weight Total kg	Sample weight kg/day	g Ca/ kg	pCi ^{226}Ra kg	pCi ^{226}Ra g Ca	pCi ^{226}Ra day	pCi ^{40}K kg	pCi ^{40}K day	pCi ^{210}Pb kg	pCi ^{210}Pb day
Middle-income teenage diet										
San Francisco, Calif.	47.3	3.38	0.51	0.4	0.8	1.4	1100	3700	0.8	2.7
Chicago, Ill.	46.5	3.32	0.59	0.8	1.4	2.7	1000	3300	1.1	3.7
New York, N.Y.	41.0	2.93	0.48	0.4	0.8	1.2	900	2600	0.7	2.0
Low-income teenage diet										
San Francisco, Calif.	46.0	3.28	0.51	0.7	1.4	2.3	900	2900	1.2	3.9
Chicago, Ill.	46.4	3.31	0.50	0.7	1.4	2.3	900	3000	0.9	3.0
New York, N.Y.	44.6	3.19	0.52	0.4	0.8	1.3	1100	3500	nd	nd
Middle-income adult diet										
San Francisco, Calif.	39.2	2.80	0.26	0.4	1.6	1.1				
Chicago, Ill.	42.2	3.01	0.28	nd	nd	nd				
New York, N.Y.	42.6	3.04	0.28	0.1	0.4	0.3				
Infant (1-year-old) diet										
San Francisco, Calif.	22.0	1.57	0.63	0.7	1.1	1.1				
Chicago, Ill.	22.6	1.62	0.59	nd	nd	nd				
New York, N.Y.	20.8	1.49	0.75	0.5	0.7	0.7				

From Michelson, I., Thompson, J. C., Jr., Hess, B. W., and Comar, C. L., *J. Nutr.,* 7, 371, 1962. With permission.

Table A. 3.7.14
SOME MEASURED OUTDOOR RADON-222 CONCENTRATIONS (pCi/m^3) IN THE UNITED STATES

Author	Location	Concentration range (mean)	
Moses et al. (1960)	Illinois	50–1,000	–
Glauberman and Breslin (1957)	New York	20–500	(130)
Wilkening (1959)	New Mexico	–	(240)
Fisenne and Harley (1974)	New York	40–230	(120)
George (1975)	New York	100–220	(170)[a]
			(210)[b]
Shleien (1963)	Ohio	170–1,040	(480)

[a] City.
[b] Rural.

From *Natural Background Radiation in the United States,* Report No. 45, National Council on Radiation Protection and Measurements, Washington, D.C., 1975. With permission.

Table A. 3.7.15
SOME OUTDOOR RADON CONCENTRATIONS OF RADON-222 (pCi/m^3) INFERRED FROM DAUGHTER MEASUREMENTS[a]

Author	Location	Concentration range (mean)	
Bradley and Pearson (1970)	Illinois	70–300	–
Golden (1968)	Florida	20–300	–
Lockhart (1964)	Little America		(2.5)
	South Pole		(0.5)
	Washington, D.C.		(122)
Yeates et al. (1972)	Massachusetts	<10–40	(20)
Lindeken (1966)	California		(90)
Fisenne and Harley (1974)	New York	15–200	(100)
Shleien (1963)	Ohio	70–850	(270)
Gold et al. (1964)	Ohio		(260)
Blifford et al. (1956)	Washington, D.C.		(47)
	Illinois		(25)
	California	2.5–10	(6)
	Tennessee		(17)
	Alaska		(3)
	Puerto Rico		(0.1)
	Washington		(2)

[a] Some authors assume radioactive equilibrium between radon and its daughters. This should not cause errors greater than about 20% for these particular data.

From *Natural Background Radiation in the United States,* Report No. 45, National Council on Radiation Protection and Measurements, Washington, D.C., 1975. With permission.

Table A. 3.7.16
GEOGRAPHIC VARIATION OF CONCENTRATIONS OF ^{210}Pb
AND ^{210}Po IN AIR

Author	Location	210 ^{210}Pb concentration (fCi/m^3)	^{210}Po concentration (fCi/m^3)
Poet et al. (1972)	Colorado	1–21	0.05–3
Shleien and Friend (1966)	Massachusetts	26[a]	–
Magno et al. (1970)	Illinois	26[a]	–
	Utah	26[a]	–
	Louisiana	21[a]	–
	Massachusetts	14[a]	–
	California	15[a]	–
	Alaska	8[a]	–
	Hawaii	5[a]	–
	Puerto Rico	8[a]	–
Patterson and Lockhart (1964)	Washington, D.C.	16[a]	–
	Florida	8[a]	–
	Hawaii	7[a]	–
Gold et al. (1964)	Ohio	8[a]	–

[a] Average over 1 year.

From *Natural Background Radiation in the United States,* Report No. 45, National Council on Radiation Protection and Measurements, Washington, D.C., 1975. With permission.

Table A. 3.7.17
^{226}Ra, ^{228}Ra AND ^{210}Pb CONCENTRATIONS
IN HUMAN BONE[a] (pCi/kg ASH)

Region	Number of cases	^{226}Ra concentration	^{228}Ra concentration	^{210}Pb concentration
Boston	77	14		
Houston	23	23		
Midwest[b]				
Low Ra water	128	16		105
High Ra water		37		
Midwest[c]	32	28	7	
New England	18	16		
New England	218	16		
New York	143	10		
New York		13		
Puerto Rico	42	5		
Puerto Rico		6		
San Francisco		11		
Wisconsin	75	12		

[a] UNSCEAR (1972).
[b] Holtzman (1963). These results include samples from people living in areas with high levels of ^{226}Ra in their drinking water (>0.5 pCi ^{226}Ra/l).
[c] Lucas et al. (1964). See note b.

From *Natural Background Radiation in the United States,* Report No. 45, National Council on Radiation Protection and Measurements, Washington, D.C., 1975. With permission.

Table A. 3.7.18

ESTIMATED ANNUAL EXTERNAL GAMMA WHOLE-BODY DOSES FROM NATURAL TERRESTRIAL RADIOACTIVITY (MREM/PERSON)

Political unit	Average annual dose	Political unit	Average annual dose
Alabama	70	New Jersey	60
Alaska	60[a]	New Mexico	70
Arizona	60[a]	New York	65
Arkansas	75	North Carolina	75
California	50	North Dakota	60[a]
Colorado	105	Ohio	65
Connecticut	60	Oklahoma	60
Delaware	60[a]	Oregon	60[a]
Florida	60[a]	Pennsylvania	55
Georgia	60[a]	Rhode Island	65
Hawaii	60[a]	South Carolina	70
Idaho	60[a]	South Dakota	115
Illinois	65	Tennessee	70
Indiana	55	Texas	30
Iowa	60	Utah	40
Kansas	60[a]	Vermont	45
Kentucky	60[a]	Virginia	55
Louisiana	40	Washington	60[a]
Maine	75	West Virginia	60[a]
Maryland	55	Wisconsin	55
Massachusetts	75	Wyoming	90
Michigan	60[a]	Canal Zone	60[a]
Minnesota	70	Guam	70
Mississippi	65	Puerto Rico	65
Missouri	60[a]	Samoa	60[a]
Montana	60[a]	Virgin Islands	60[a]
Nebraska	55	District of Columbia	55
Nevada	40	Others	60[a]
New Hampshire	65	Total U.S.	60

[a] Assumed to be equal to the U.S. average.

From Klement, A. W., Jr., Miller, C. R., Minx, R. P., and Shleien, B., *Estimates of Ionizing Radiation Doses in the United States. 1960–2000.* Report ORP/CSD 72-1, U.S. Environmental Protection Agency, Washington, D. C., 1972, 7.

Table A. 3.7.19

ANNUAL INTERNAL BETA AND GAMMA DOSE EQUIVALENT RATES (MREM/YEAR) IN TISSUE FROM INTERNALLY DEPOSITED NATURALLY OCCURRING RADIONUCLIDES[a]

Radionuclide	Soft tissues (gonads)	Cortical bone		Trabecular bone	
		Osteocytes	Haversian canals	Surfaces[b]	Marrow
^{3}H	~0.001	~0.001	~0.001	~0.001	~0.001
^{14}C	0.7	0.8	0.8	0.8	0.7
^{40}K	19	6	6	15	15
^{87}Rb	0.3	0.4	0.4	0.6	0.6
Total	20.0	7.2	7.2	16.4	16.3

[a] UNSCEAR (1972) gives the data as absorbed dose in tissue in mrad/year.
[b] Cells close to surfaces of bone trabeculae.

From *Natural Background Radiation in the United States*, Report No. 45, National Council on Radiation Protection and Measurements, Washington, D.C., 1975. With permission.

Table A. 3.7.20

ANNUAL ALPHA DOSE EQUIVALENT RATES (MREM/YEAR) FROM NATURALLY OCCURRING RADIONUCLIDES

	Concentration		Dose equivalent rates				
			Cortical bone			Trabecular bone	
Radionuclide	In air (pCi/m^3)	In bone[a] (pCi/kg)	Gonads	Osteocytes	Haversian canals[b]	Surfaces[c]	Marrow
$^{228-234}U$[d]	—	6.9	0.8	12.4	7.7	4.8	0.9
^{226}Ra[d]	—	7.8	0.2	16.4	10.2	6.6	1.2
^{228}Ra[d]	—	3.8	0.3	19.0	11.0	8.0	1.0
^{222}Rn[e]	150	—	0.4	0.2	0.2	0.4	0.4
^{220}Rn[e]	1	—	0.01	0.1	0.1	0.2	0.2
^{210}Po[d]	—	60	6	60	36	24	4.8
Total			8	110	65	44	8.5

[a] The alpha-emitting nuclides are assumed to be uniformly distributed in mineral bone, although this may not be the case (ICRP, 1968).
[b] Cells lining the Haversian canals.
[c] Cells close to surfaces of bone trabeculae (dose) averaged over the first 10 μm.
[d] Calculated by the method of Spiers (1968).
[e] Derived from UNSCEAR (1972).

From *Natural Background Radiation in the United States,* Report No. 45, National Council on Radiation Protection and Measurements, Washington, D.C. With permission.

Table A. 3.7.21

SUMMARY OF AVERAGE DOSE EQUIVALENT RATES (MREM/YEAR) FROM VARIOUS SOURCES OF NATURAL BACKGROUND RADIATION IN THE UNITED STATES

			Bone		
Source	Gonads	Lung	Surfaces	Marrow	GI tract
Cosmic radiation[a]	28	28	28	28	28
Cosmogenic radionuclides	0.7	0.7	0.8	0.7	0.7
External terrestrial[b]	26	26	26	26	26
Inhaled radionuclides[c]	—	100[d]	—	—	—
Radionuclides in the body[e]	27	24	60	24	24[f]
Rounded totals	80	180	120	80	80

[a] "Cosmic Radiation" includes 10% reduction to account for structural shielding.
[b] "External Terrestrial" includes 20% reduction for shielding by housing and 20% reduction for shielding by the body.
[c] Doses to organs other than lung included in "Radionuclides in the Body."
[d] Local dose equivalent rate to segmental bronchioles is 450 mrem/year.
[e] Excluding the cosmogenic contribution shown separately.
[f] This does not include any contribution from radionuclides in the gut contents.

From *Natural Background Radiation in the United States,* Report No. 45, National Council on Radiation Protection and Measurements, Washington, D.C., 1975. With permission.

Table A. 3.7.22
ESTIMATED AVERAGE ANNUAL INTERNAL RADIATION DOSES FROM NATURAL RADIOACTIVITY IN THE UNITED STATES

Radionuclide[a]	Whole-body	Annual doses (mrem/person) endosteal cells (bone)	Bone marrow
^{3}H	0.004	0.004	0.004
^{14}C	1.0	1.6	1.6
^{40}K	17	8	15
^{87}Rb	0.6	0.4	0.6
^{210}Po	3.0	21	3.0
^{222}Rn	3.0	3.0	3.0
^{226}Ra	—	6.1	0.3
^{228}Ra	—	7	0.3
Total	25	47	24

[a] Other natural radionuclides would contribute to doses but such a small fraction that they would not affect the totals within the accuracy of these estimates. As an example, doses from ^{3}H are shown here.

From Klement, A. W., Jr., Miller, C. R., Minx, R. P., and Shleien, B., *Estimates of Ionizing Radiation Doses in the United States, 1960–2000*, Report ORP/CSD, 72-1, U.S. Environmental Protection Agency, Washington, D.C., 1975, 7.

Table A. 3.7.23
ESTIMATED TOTAL ANNUAL AVERAGE WHOLE-BODY DOSES FROM NATURAL RADIATION IN THE UNITED STATES (MREM/PERSON)

Source	Annual doses
Cosmic rays	45
Terrestrial radiation	
External	60
Internal	25
Total	130

From Klement, A. W., Jr., Miller, C. R., Minx, R. P., and Shleien, B., *Estimates of Ionizing Radiation Doses in the United States 1960–2000*, Report ORP/CSD 72-1, U.S. Environmental Protection Agency, Washington, D.C., 1972, 7.

Table A. 3.7.24
ESTIMATED TOTAL ANNUAL WHOLE-BODY MAN-REM FROM NATURAL RADIATION IN THE UNITED STATES

Year	Population (millions)	Annual man-rem (millions)
1960	183	23.8
1970	205	26.6
1980	237	30.8
1990	277	36.0
2000	321	41.7

From Klement, A. W., Jr., Miller, C. R., Mixn, R. P., and Shleien, B., *Estimates of Ionizing Radiation Doses in the United States 1960–2000*, Report ORP/CSD 72-1, U.S. Environmental Protection Agency, Washington, D.C., 1972, 7.

REFERENCES

1. **Klement, A. W., Jr.,** Natural radionuclides in foods and food source materials, in *Radioactive Fallout, Soils, Plants, Foods, Man,* Fowler, E. B., Ed., Elsevier, Amsterdam, 1965, 113.
2. **Klement, A. W., Jr., Miller, C. R., Minx, R. P., and Shleien, B.,** *Estimates of Ionizing Radiation Doses in the United States 1960–2000,* Report ORP/CSD 72-1, U.S. Environmental Protection Agency, Washington, D.C., 1972, 7.
3. *Natural Background Radiation in the United States,* Report No. 45, National Council on Radiation Protection and Measurements, Washington, D.C., 1975.

APPENDIX A.3.8 SUPPLEMENTARY INFORMATION FOR ESTIMATING INTERNAL AND EXTERNAL EXPOSURE HAZARDS

Allen Brodsky

A.3.8.1. INTRODUCTION

This appendix contains supplementary information such as maximum permissible concentrations and external exposure rates for various radionuclides, pending publication of volumes in which more extensive data will be given for estimating internal and external radiation hazards. The data on permissible concentrations are based on dose limits to internal "critical organs," which are those organs likely to receive the maximum fractions of permissible dose rates, according to methods published by the International Commission on Radiological Protection (and the National Committee on Radiation Protection) in the "ICRP II Report", Health Physics 3, June 1960, or minor modifications of the methodology of the ICRP II report. The ICRP is currently recalculating annual limits of intake using new schematic representations of physiologic uptake and distribution of radionuclides and different constraints on the total doses and risks to various body organs. Also, various recent publications provide dose factors per unit intake of various radionuclides, based on variations of earilier ICRP models. Sections 3.6 and 3.8 of this volume, and Sections 5.5, 6.3, and 6.4 of Volume II will provide insight regarding more recent methods of calculating internal dose. (See also the recent report by Killough, G. C., Dunning, D. E., Jr., Pleasant, J. C., "INREM II: A Computer Implementation of Recent Models for Estimating the Dose Equivalent to Organs of Man from an Inhaled or Ingested Radionuclide," Oak Ridge National Laboratory Report No. ORNL/ NUREG/TM-84, NUREG/CR-0114, June 1968 available from National Technical Information Service, 5285 Port Royal Road, Springfield, Va. 22161.) Pending publication of further data, the values in this appendix can be considered within about an order of magnitude of those likely to appear for equivalent intake periods in future publications. Thus, the data in this appendix may still be useful for planning purposes in environmental health and health physics in which factors of safety of 50 or more may be reasonable to accept. In many cases where the new ICRP intake limit values will not be appreciably different from equivalent time integrated concentrations of the ICRP II report, the application of safety factors of 50 or more would provide consistency with the philosophy of maintaining exposures as low as reasonably achievable.

A.3.8.2. DATA AND METHODS FOR ESTIMATING RADIATION EXPOSURES FROM INTERNAL AND EXTERNAL RADIATION SOURCES*

There is now abundant literature on methods for estimating radiation exposures from radioactive sources external or internal to the body.[1-13] For the most precise calculations of radiation dose to various body organs, a specialist should be employed who can use the detailed physical calculations and correction factors—such as organ-shape factors—given in the references and in other literature. However, it is often useful and adequate to make rapid estimates of exposures that are generally accurate to within perhaps a factor of 2, particularly for radiation control purposes, evaluation of safety requirements for facilities, emergency planning, and evaluation of hazards of radionuclides dispersed to the environment. For such purposes, the ordinary biological variations between individual radiosensitivities usually require the application of additional safety factors of 10 or more, which overcome the ordinary errors of estimating organ exposures based on standard physical and biological constants.[9,10,14,15] It is usually only after the individual has received a known exposure under known physical geometry and circumstances, or a known intake of a given body content, that it is worthwhile, either for medical diagnosis or for treatment purposes, to estimate the dose to that particular individual as precisely as possible. Thus, only a summary of some of the general formulae and data useful for rapidly determining the general magnitudes of radiation exposures are given in this section.

The gamma and beta doses may be estimated from the following rules of thumb.[16]

Gamma Dose.

a) $1 \text{ r/hr} = \dfrac{5.6 \times 10^5}{E} = \text{photons/cm}^2/\text{sec}$, for E from 0.07 to 2.0 Mev, good

within ± 12 percent, assuming $W_{air} = 34$ ev/ion pair.
The above identity is equivalent to the equation:

$$I = \frac{5.64 \, C \sum_i E_i}{d^2}$$

where I = intensity in r/hr at d feet from a point source;
   C = curies of activity,
   $\sum_i E_i$ = sum of gamma energies in Mev per disintegration, if more
         than one gamma is emitted per disintegration,
   d = distance in feet.
Note: convert exposure in r to depth doses in rads, using tables in the first chapter of this section.

b) For any energy E in Mev and for its corresponding total minus Compton-scatter absorption coefficient ($\mu - \sigma_s$),

$$1 \text{ r} = \frac{7.1 \times 10^4}{(\mu - \sigma_s)E} \text{ photons/cm}^2$$

---

*   This section is reprinted from Wang, Y., Ed., *CRC Handbook of Radioactive Nuclides,* CRC Press, Cleveland, Ohio, 1969, 647—663.

Beta Surface Dose in Air.

In air of density 0.001293 g/cc,

$$1 \text{ rad/hour} = \frac{6.1 \times 10^5}{S_a} \text{ betas/cm}^2/\text{sec}$$

where S_a is the average number of ion pairs produced per centimeter of path in air. This equation holds within ± 6 percent for 0.01 to 2.0 Mev. Values of $(\mu - \sigma_s)$ are given in Table A.3.8-1, and the theoretical values of $(\mu - \sigma_s)$ for calculating external exposure rates for various photon energies may be taken from Table A.3.8-2.

Table A.3.8-1
SPECIFIC IONIZATION OF ELECTRONS[21]

Energy, Mev	Sa, ion pairs/cm	Range in Air	
		mg/cm²	cm
0.05	250	3.9	3.02
0.10	175	14.0	10.80
0.20	96	42.0	32.50
0.30	76	77.0	59.60
0.50	60	158.0	122.00
1.00	53	400.00	310.00
1.50	47	680.0	526.00

From Brodsky, A. and Beard, G. V., *A Compendium of Information for Use in Controlling Radiation Emergencies,* TID-8206, Rev. 1960. By permission of the U.S. Atomic Energy Commission, Washington, D.C.

Table A.3.8-2
THEORETICAL VALUES OF $(\mu - \sigma_s)$ AND OF r/hr AT 1 m FROM A 1-CURIE SOURCE

Energy, Mev	$(\mu - \sigma_s)_{air}$, cm²/g	r/hr at 1 m from 1- Ci Source
0.02	0.500	0.160
0.04	0.078	0.063
0.06	0.035	0.042
0.08	0.025	0.040
0.10	0.023	0.047
0.20	0.026	0.110
0.40	0.029	0.230
0.60	0.029	0.350
0.80	0.028	0.450
1.00	0.027	0.550
2.00	0.023	0.930
4.00	0.019	1.500
6.00	0.017	2.100
8.00	0.016	2.600
10.00	0.015	3.000
20.00	0.013	5.300
40.00	0.013	11.000
60.00	0.014	17.000
100.00	0.015	30.000

From Gray, *American Institute of Physics Handbook*, p. 8–257. Copyright 1957, by permission of McGraw-Hill Book Co., New York. With permission.

Internal doses for various nuclides taken into the body may be determined for various body organs by standard techniques given in the literature and by some of the equations presented in the two preceding chapters for single-intake situations. The total dose to the critical organ in 50 years per microcurie intake may be calculated, assuming instantaneous uptake in the critical organ from a single exposure of short duration (compared to the effective half-life of the radionuclide in the body), from the equation:

$$\text{Dose (rem/}\mu\text{Ci)} = \int_{t=0}^{t=50\times365} I_0 e^{-0.693t/T} \, dt = 1.44 \, I_0 T[1 - \exp(1.265 \times 10^4/T)], \quad (1)$$

where I_0 = initial dose rate to the critical organ per microcurie intake, or f_aR/qf_2 (with the standard symbols from Reference 3);

     q = body burden listed beside the corresponding critical organ in Table 1 of Reference 3;

     T = effective half-life in the body in days, from Table 12 of Reference 3;

     R = permissible dose rate for continuous exposure in rem per day fort he body organ concerned, obtained from Reference 3 as 0.1_7 rem/day for irradiation of the whole body, 0.08 rem/day for bone, 0.1_7 rem/day for the gonads, 0.6_7 rem/day for the thyroid and skin, and 0.3_7 rem/day for other parts of the body.

The expression in brackets in the above equation is essentially a factor of 1, except for the few bone-seeking radionuclides whose effective half-lives are not short compared to 50 years. For ^{90}Sr, with an effective half-life of 6.4×10^3 days, the factor in brackets becomes 0.861. For purposes of this chapter, the same relative dose from daughter products is assumed for single intake as for continuous exposure, where radionuclides that have radioactive daughters building up in the body are concerned. The contribution from daughters is thus taken into account by using the total body burdens, q, to give dose rates, R. These body burdens were calculated by the ICRP Committee, taking into account daughter products that build up in the body.

Since for materials insoluble in the lung an average of about 12½ percent of the material inhaled may remain in the lung, with a half-life of 120 days, the lung must also be taken into consideration as a possible critical organ. Deviations from this half-life in lung are described for various conditions,[10] and human data are in many cases still incomplete. However, the use of a 120-day half-life for lung exposure is reasonably conservative in the absence of other data. Rarely has lung clearance been more than an order of magnitude slower than 120 days, and it is often much less.[10] Single-intake doses based on the lung dose may be calculated from the equation

dose to the lung per microcurie inhaled (insoluble)

$$= \frac{\text{Permissible dose rate per week}}{(\text{MPC}_{air} \text{ based on continuous exposure to lung}) \times 1.4 \times 10^8 \text{ cc/week}}$$

$$= 2.14 \times 10^{-9}/\text{MPC}_{air} \text{ based on lung} \quad (2)$$

since the equilibrium dose rate from continuous exposure is the same as the average dose rate from a series of single intakes of the same total quantity of radioactivity per week for effective half-lives that are short compared to 50 years.

Similar equations may be written for ingested radionuclides taken into the body by mouth. The f_a of equations (1) and (2) and the air intake in cc's per week should be replaced by the respective f_w, representing the fractional amount of the ingested material reaching the critical organ, or the respective MPC_{water} multiplied by an intake of approximately 1.75×10^4 cc's of water per week.

Some of the specific data in regard to lung clearance rates may be obtained from the compilation of the ICRP Task Group on Lung Dynamics, given in Table A.3.8-3.[10] For approximate dose calculations, Tables A.3.8-4, A.3.8-5, and A.3.8-6, giving parameters for standard man and for the simplified lung model previously outlined by the ICRP, are still useful.

In Table A.3.8-7 some of the calculated maximum permissible body burdens of more frequently encountered radionuclides are given, based on the quantity of radioactivity (in microcuries) deposited in the organ of reference that delivers a radiation dose rate equal to the maximum permissible dose rate for continuous lifetime occupational exposure. Also represented are

Table A.3.8-3
BIOLOGICAL HALF-LIFE OF RADIOACTIVE SUBSTANCES IN THE LUNG[10]

Substance	Species	Biological Half-Life, days	Recommended Values for Man		Notes
			Single Exposure	Multiple Exposure	
Actinium (?)	Man	100–200	200		
Aluminum oxide (corundum)	Rats	100		100	Multiple
Antimony oxide	Rats	5	5		2-week study
Antimony chloride	Rats	94	94		
Arsenic trioxide	Man	16	16		
Barium sulfate	Rats	6 $27\frac{1}{2}$ and 10 10	10		
	Dogs	8			
Beryllium citrate	Rats	20–500 (?)	80		
Beryllium oxide	Rats	120 (?)	120		
Beryllium sulfate, carrier free	Rats	80	80		
Beryllium sulfate	Rats	80	80		
Calcium chloride, with carrier	Guinea pigs	18	18		
Cerium dioxide	Rats Dogs Rats	80 150 54	150		
Cerium (^{144}Pr) trichloride	Mice	58	60		
Cerium trichloride	Rats	63	60		Intratracheal injection
Cerous fluoride	Rats	200	200		
Cesium (?)	Man	89 109–149 140	90	150	Multiple Not only in lungs
Cesium chloride	Man	128			
Cesium sulfate	Man	78 (?)	80		Whole body (?),
Chromium sesquioxide	Dogs	>100	160		
Cobalt	Man	90			
Cobalt and cobalt oxide	Man	1.5–2.1 yrs	90	720	Multiple (?), mainly inhalation
Cobalt chloride	Mice	17	17		

Table A.3.8-3 (Continued)
BIOLOGICAL HALF-LIFE OF RADIOACTIVE SUBSTANCES IN THE LUNG[10]

Substance	Species	Biological Half-Life, days	Recommended Values for Man		Notes
			Single Exposure	Multiple Exposure	
Europium nitrate	Mice	28	28		
Ferric oxide	Man Dogs	70 63	70		
Indium sesquioxide	Rats	70		70	90-day exposure
Iodine	Mice, rats	<5 hrs	<5 hrs		
Iridium metal	Rats	14[a]	14[a]		
Iron powder	Rabbits	14	14		
Manganese (?)	Man	185	185		Not only in lungs (?)
Manganic oxide	Dogs Man	40 60	60		
Mercuric oxide	Dogs	33	33		
Mercuric sulfide	Rats	30	30		
Mercury vapor	Rats	20	20		
Plutonium dioxide	Mice Dogs Rats Dogs Rats	460 >1,400 150 >400 300 180	500		Excretion analysis
Plutonium nitrate	Rats Dogs Rats	12 40 38	38		
Polonium hydroxide	Dogs Rats Rabbits	29 22 18–30 35 30	30	30	10-day multiple excretion analysis
Praseodymium tri-chloride	Mice	24	24		
Protoactinium (Pa_2O_5 or $KPaO_3$)	Man	1,200 ± 600	1,500		
Radium sulfate	Man	180	180		
Ruthenium dioxide	Mice	230	230		
Silica (glass spheres)	Rats	180	180		

[a] Data from cases of human inhalation of iridium metal and other insoluble metals shows that the long-term biological half-life in humans will usually exceed 700 days Cool, D. A., Cool, W. S., Brodsky, A., and Eadie, G. G., Estimation of Long-Term Biological Elimination of Insoluble Iridium-192 from the Human Lung, submitted for publication, June 1978 (available from the editor).

Table A.3.8-3 (Continued)
BIOLOGICAL HALF-LIFE OF RADIOACTIVE SUBSTANCES IN THE LUNG[10]

Substance	Species	Biological Half-Life, days	Recommended Values for Man		Notes
			Single Exposure	Multiple Exposure	
Silicon dioxide	Rats	30	30		
	Rats (Winkelman)	56			
	Rats (Sprague-Dawley)	31	40	200	
	Rats	>180			Multiple
Silver iodide	Mice	<5 hrs	<5 hrs		Based on iodine
Stannic phosphate	Dogs	59	59		
Strontium sulfate	Mice	56	56		
Strontium titanate	Man	500	500		
Strontium chloride	Mice	9	9		
Thallium chloride	Rats	>40–80 (1,400?)	$^{60}/_{1,400}$		
Thallous nitrate	Rats	<2 hrs	<2 hrs		
Thorium dioxide	Rats	>400	500		Intratracheal administration
Uranium dioxide	Rats	270			Multiple
		135			
		141	150		17 days multiple
		169			20 days multiple
		239			58 days multiple
		289			140 days multiple
	Man	120–150			
	Dogs	180			
Uranium dioxide, natural	Dogs	380			Multiple
Uranium dioxide, enriched	Man	300			Multiple
Uranium octoxide	Man	380			
		>120	120	380	
	Rats	300			
Uranium octoxide, fume	Dogs	120	120	380	
Uranyl nitrate	Rats	<1	<1		
Yttrium chloride, with carrier	Guinea pigs	23			Excretion analysis
Yttrium chloride, carrier free	Guinea pigs	12	23		Excretion analysis

Table A.3.8-3 (Continued)
BIOLOGICAL HALF-LIFE OF RADIOACTIVE SUBSTANCES IN THE LUNG[10]

Substance	Species	Biological Half-Life, days	Recommended Values for Man		Notes
			Single Exposure	Multiple Exposure	
Yttrium chloride	Guinea pigs Mice	21 20			Lung analysis
Zinc phosphate	Dogs	15, 160	160		
Zirconium-niobium	Man	35			
Zirconium oxychloride, carrier free	Guinea pigs	23 30			Excretion analysis Lung analysis
Zirconium oxychloride, with carrier	Guinea pigs	23	35		Excretion analysis
Zirconium oxychloride	Guinea pigs	18			Lung analysis
Zirconium oxalate	Guinea pigs	12	12		Lung analysis

From Report of the ICRP Task Group on Lung Dynamics. *Health Phys.*, 12, 173—207, 1966. Copyright Pergamon Press, New York. With permission.

Table A.3.8-4
CHARACTERISTICS OF THE STANDARD MAN[16]*

Water intake in food	1,000 cc/day
Water intake in fluids	1,200 cc/day
Oxidation	300 cc/day
Total water intake per day	2,500 cc/day
Water excretion from lungs	300 cc/day
Water excretion by feces	200 cc day
Water excretion by urine	1,400 cc/day
Water excretion by sweat	600 cc/day
Total water excretion per day	2,500 cc/day
Air inhaled during 8-hour work day	10^7 cc/day
Air inhaled during 16 hours not at work	10^7 cc/day
Total air inhaled per day	2×10^7 cc/day
Vital capacity of lungs	
Men	3 to 4 liters
Women	2 to 3 liters
Total surface area of respiratory tract	70 m²
Total water in body	43 kg
Average life span of man	70 years
Occupational exposure of man	8 hours per day
	40 hours per week
	50 weeks per year
Minimum thickness of epidermis	0.07 mm (7 mg/cm²)
Depth of blood-forming organs	5 cm

* See Section 5 for more recent data on reference man. The data in this earlier ICRP II table (*Health Phys.*, 3, June 1960) should only be used for more rapid, approximate calculations now that it has been superseded.

From Brodsky, A. and Beard, G. V., *A Compendium of Information for Use in Controlling Radiation Emergencies, TID-8206* (Rev.). By permission of the U.S. Atomic Energy Commission, Washington, D.C.

Table A.3.8-5
MASSES OF CRITICAL ORGANS[16]

Organ	Mass, g
Total body	70,000
Muscles	30,000
Fat	10,000
Bone	7,000
Blood	5,400
Skin	2,000
Liver	1,700
Lungs	1,000
Kidneys	300
Spleen	150
Thyroid	20

From: Brodsky, A. and Beard, G. V., *A Compendium of information for Use in Controlling Radiation Emergencies, TID–8206* (Rev.). By permission of the U.S. Atomic Energy Commission, Washington, D.C.

maximum permissible concentrations of radionuclides of air and water for organ exposures of 40 hours per week, based on different body organs, and the single-intake doses per microcurie taken into the body by ingestion or by injections for particular radionuclides. In the case of the permissible body burden and concentrations, the more restrictive (smaller) values indicated for each respective nuclide should be used in order to limit the exposure to the critical organ (i.e., the part of the body exposed to the highest relative multiple of its permissible dose rate limit).

More detailed methods of estimating exposures, as applicable to various body organs, are presented in the references, particularly for radionuclides in bone;[11,17] a more detailed treatment of bone dosimetry would be beyond the scope of this handbook. Tables A.3.8-8 and A.3.8-9 present the specific dose rates from internal and external exposures to various nuclides, and the gamma-radiation levels from radioactive materials spread over wide land areas, for use in estimating exposures to populations from environmental radioactivity.[18]

Table A.3.8-6
PARTICULATE RETENTION IN THE RESPIRATORY TRACT[16]

Distribution	Readily Soluble Compounds, %	Other Compounds, %
Exhaled	25	25
Deposited in upper respiratory passages and swallowed	50	50
Deposited in the lungs (lower respiratory passages)	25 (this is taken up into the body)	25*

* Half of this is eliminated from the lungs and swallowed in the first 24 hours, making a total of 62½ percent swallowed. The remaining 12½ percent is retained in the lungs with a half-life of 120 days; it is assumed that this portion is taken up into body fluids.

From: Brodsky, A. and Beard, G. V., *A Compendium of Information for Use in Controlling Radiation Emergencies, TID–8206* (Rev.). By permission of the U.S. Atomic Energy Commission, Washington, D.C.

Table A.3.8-7

MPBB AND MPC VALUES FOR SOME RADIOISOTOPES[11]

Radionuclides and Type of Decay	Organ of Reference*	Maximum Permissible Burden in Total Body (q), μCi	Maximum Permissible Concentrations for a 40-Hour Week		Dose per μCi Administered†	
			MPC_{water}, μCi/cm³	MPC_{air}, μCi/cm³	By Ingestion, rem	By Injection, rem
^3H (HTO or ^3H_2O) β^-, soluble	Body tissue	1×10^3	0.1	5×10^{-6}	2.1×10^{-4}	2.1×10^{-4}
	Total body	2×10^3	0.2	8×10^{-6}	1.3×10^{-4}	1.3×10^{-4}
$^3H_2$, submersion	Skin			2×10^{-3}		
^{14}C (CO_2), soluble β^-	Fat	300	0.02	4×10^{-6}	2.4×10^{-3}	2.4×10^{-3}
	Total body	400	0.03	5×10^{-6}	5.7×10^{-4}	5.7×10^{-4}
	Bone	400	0.04	6×10^{-6}	5.7×10^{-4}	5.7×10^{-4}
^{14}C, submersion	Total body			5×10^{-5}		
^{22}Na, soluble β^+, γ	Total body	10	1×10^{-3}	2×10^{-7}	1.9×10^{-2}	1.9×10^{-2}
	G.I. (LLI)		0.01	2×10^{-6}	3.4×10^{-3}	
^{22}Na, insoluble	Lung			9×10^{-9}		
	G.I. (LLI)		9×10^{-4}	2×10^{-7}	6.8×10^{-2}	
^{24}Na, soluble β^-, γ	G.I. (SI)	7	6×10^{-3}	1×10^{-6}	6.2×10^{-3}	1.7×10^{-3}
	Total body		0.01	2×10^{-6}	1.7×10^{-3}	
^{24}Na, insoluble	G.I. (LLI)		8×10^{-4}	1×10^{-7}	4.8×10^{-2}	
	Lung			8×10^{-7}		
^{32}P, soluble β	Bone	6	5×10^{-4}	7×10^{-8}	3.8×10^{-2}	5.1×10^{-2}
	Total body	30	3×10^{-3}	4×10^{-7}	7.4×10^{-3}	9.8×10^{-3}
	G.I. (LLI)		3×10^{-3}	6×10^{-7}	2.1×10^{-2}	
	Liver	50	5×10^{-3}	6×10^{-7}	1.2×10^{-2}	1.7×10^{-2}
	Brain	300	0.02	3×10^{-6}	2.4×10^{-3}	3.2×10^{-3}
^{32}P, insoluble	Lung			8×10^{-8}	8.4×10^{-2}	
	G.I. (LLI)		7×10^{-4}	1×10^{-7}		
^{38}Cl, soluble β^-, γ	G.I. (S)	9	0.01	3×10^{-4}	4.9×10^{-3}	6.3×10^{-5}
	Total body		0.3	4×10^{-5}	6.3×10^{-5}	

Table A.3.8-7 (Continued)

MPBB AND MPC VALUES FOR SOME RADIOISOTOPES[11]

Radionuclides and Type of Decay	Organ of Reference*	Maximum Permissible Burden in Total Body (q), μCi	Maximum Permissible Concentrations for a 40-Hour Week		Dose per μCi Administered†	
			MPC_{water}, μCi/cm³	MPC_{air}, μCi/cm³	By Ingestion, rem	By Injection, rem
³⁸Cl, insoluble	G.I. (S)		0.01	2×10^{-6}	4.9×10^{-3}	
	Lung			1×10^{-5}		
⁴²K, soluble β⁻, γ	G.I. (S)		9×10^{-3}	2×10^{-6}	6.2×10^{-3}	
	Total body	10	0.02	3×10^{-6}	8.8×10^{-4}	8.8×10^{-4}
	Brain	20	0.04	6×10^{-6}	1.5×10^{-3}	1.5×10^{-3}
	Spleen	20	0.04	6×10^{-6}	1.5×10^{-3}	1.5×10^{-3}
	Muscle	20	0.04	6×10^{-6}	1.3×10^{-3}	1.3×10^{-3}
	Liver	50	0.08	1×10^{-5}	6.8×10^{-4}	6.8×10^{-4}
⁴²K, insoluble	G.I. (LLI)		6×10^{-4}	1×10^{-7}	5.9×10^{-2}	
	Lung			9×10^{-7}		
⁴⁵Ca, soluble β⁻, γ	Bone	30	3×10^{-4}	3×10^{-8}	7.9×10^{-2}	1.3×10^{-1}
	Total body	200	2×10^{-3}	3×10^{-7}	8.8×10^{-3}	1.5×10^{-2}
	G.I. (LLI)		0.01	3×10^{-6}	4.4×10^{-3}	
⁴⁵Ca, insoluble	Lung			1×10^{-7}		
	G.I. (LLI)		5×10^{-3}	9×10^{-7}	1.1×10^{-2}	
⁴⁷Ca, soluble β⁻, γ	Bone	5	1×10^{-3}	2×10^{-7}	1.9×10^{-2}	3.1×10^{-2}
	G.I. (LLI)		2×10^{-3}	5×10^{-7}	3.0×10^{-2}	
	Total body	10	4×10^{-3}	5×10^{-7}	4.3×10^{-3}	7.2×10^{-3}
⁴⁷Ca, insoluble	G.I. (LLI)		1×10^{-3}	2×10^{-7}	7.6×10^{-3}	
	Lung			2×10^{-7}		
⁵⁹Fe, soluble β⁻, γ	G.I. (LLI)		2×10^{-3}	4×10^{-7}	3.3×10^{-2}	1.4×10^{-1}
	Spleen	20	4×10^{-3}	1×10^{-7}	1.4×10^{-2}	3.6×10^{-2}
	Total body	20	5×10^{-3}	2×10^{-7}	3.6×10^{-3}	9.9×10^{-2}
	Liver	30	6×10^{-3}	2×10^{-7}	9.9×10^{-3}	
	Lung	100	0.02	8×10^{-7}		
	Bone	100	0.03	1×10^{-6}	1.3×10^{-3}	1.3×10^{-2}

Table A.3.8-7 (Continued)
MPBB AND MPC VALUES FOR SOME RADIOISOTOPES[11]

Radionuclides and Type of Decay	Organ of Reference*	Maximum Permissible Burden in Total Body (q), μCi	Maximum Permissible Concentrations for a 40-Hour Week		Dose per μCi Administered†	
			MPC_{water}, μCi/cm³	MPC_{air}, μCi/cm³	By Ingestion, rem	By Injection, rem
^{59}Fe, insoluble	Lung			5×10^{-8}		
	G.I. (LLI)		2×10^{-3}	3×10^{-7}	3.7×10^{-2}	
^{85}Sr, soluble ϵ, γ	Total body	60	3×10^{-3}	2×10^{-7}	6.8×10^{-3}	2.3×10^{-2}
	Bone	70	4×10^{-3}	4×10^{-7}	1.3×10^{-2}	4.4×10^{-2}
	G.I. (LLI)		7×10^{-3}	2×10^{-6}	8.1×10^{-3}	
^{85}Sr, insoluble	Lung			1×10^{-7}		
	G.I. (LLI)		5×10^{-3}	9×10^{-7}	1.2×10^{-2}	
^{90}Sr, soluble β^-	Bone	2	1×10^{-5}	1×10^{-9}		
	Total body	3	2×10^{-5}	2×10^{-9}		
	G.I. (LLI)		1×10^{-3}	3×10^{-7}		
^{90}Sr, insoluble	Lung			5×10^{-9}		
	G.I. (LLI)		1×10^{-3}	2×10^{-7}		
^{131}I, soluble β^-, γ, e^-	Thyroid	0.7	6×10^{-5}	9×10^{-9}	1.9	1.9
	Total body	50	5×10^{-3}	8×10^{-7}	3.5×10^{-3}	3.5×10^{-3}
	G.I. (LLI)		0.03	7×10^{-6}	1.5×10^{-3}	
^{131}I, insoluble	G.I. (LLI)		2×10^{-3}	3×10^{-7}	2.9×10^{-2}	
	Lung			3×10^{-7}		
^{132}I, soluble β^-, γ, e^-	Thyroid	0.3	2×10^{-3}	2×10^{-7}	2.0×10^{-2}	2.0×10^{-2}
	G.I. (SI)		0.01	3×10^{-6}	3.7×10^{-3}	
	Total body	10	0.1	2×10^{-5}	1.7×10^{-4}	1.7×10^{-4}
^{132}I, insoluble	G.I. (ULI)		5×10^{-3}	9×10^{-7}	4.1×10^{-3}	
	Lung			7×10^{-6}		
^{137}Cs, soluble β^-, γ, e^-	Total body	30	4×10^{-4}	6×10^{-8}		
	Liver	40	5×10^{-4}	8×10^{-8}		
	Spleen	50	6×10^{-4}	9×10^{-8}		
	Muscle	50	7×10^{-4}	1×10^{-7}		
	Bone	100	1×10^{-3}	2×10^{-7}		
	Kidney	100	1×10^{-3}	2×10^{-7}		
	Lung	300	5×10^{-3}	6×10^{-7}		
	G.I. (SI)		0.02	5×10^{-6}		

Table A.3.8-7 (Continued)
MPBB AND MPC VALUES FOR SOME RADIOISOTOPES[11]

Radionuclides and Types of Decay	Organ of Reference*	Maximum Permissible Burden in Total Body (q), μCi	Maximum Permissible Concentrations for a 40-Hour Week		Dose per μCi Administered†	
			MPC_{water}, μCi/cm³	MPC_{air}, μCi/cm³	By Ingestion, rem	By Injection, rem
^{137}Cs, insoluble	Lung			1×10^{-8}		
	G.I. (LLI)		1×10^{-3}	2×10^{-7}		
^{226}Ra, soluble α, β^-, γ	Bone	0.1	4×10^{-7}	3×10^{-11}		
	Total body	0.2	6×10^{-7}	5×10^{-11}		
	G.I. (LLI)		1×10^{-3}	3×10^{-7}		
^{226}Ra, insoluble	G.I. (LLI)		9×10^{-4}	2×10^{-7}		
^{235}U, soluble α, β^-, γ	G.I. (LLI)		8×10^{-4}	2×10^{-7}		
	Kidney	0.03	1×10^{-4}	5×10^{-10}		
	Bone	0.06	1×10^{-4}	6×10^{-10}		
	Total body	0.4	4×10^{-4}	2×10^{-9}		
^{235}U, insoluble	Lung			1×10^{-10}		
	G.I. (LLI)		8×10^{-4}	1×10^{-7}		
^{239}Pu, soluble α, γ	Bone	0.04	1×10^{-4}	2×10^{-12}		
	Liver	0.4	5×10^{-4}	7×10^{-12}		
	Kidney	0.5	7×10^{-4}	9×10^{-12}		
	G.I. (LLI)		8×10^{-4}	2×10^{-7}		
	Total body	0.4	1×10^{-3}	1×10^{-11}		
^{239}Pu, insoluble	Lung			4×10^{-11}		
	G.I. (LLI)		8×10^{-4}	2×10^{-7}		
^{241}Am, soluble α, γ	Kidney	0.1	1×10^{-4}	6×10^{-12}		
	Bone	0.05	1×10^{-4}	6×10^{-12}		
	Liver	0.4	2×10^{-4}	9×10^{-12}		
	Total body	0.3	4×10^{-4}	2×10^{-11}		
	G.I. (LLI)		8×10^{-4}	2×10^{-7}		
^{241}Am, insoluble	Lung			1×10^{-10}		
	G.I. (LLI)		8×10^{-4}	1×10^{-7}		

The abbreviations G.I., S, SI, ULI, and LLI refer to gastrointestinal tract, stomach, small intestine, upper large intestine, and lower large intestine respectively; critical organs are listed first.

* Data from the Report of ICRP Committee 2.

† Data from Vennart and Minski.

From Spiers, F. W., Radioisotopes in the Human Body: Physical and Biological Aspects, pp. 320–323. Copyright 1968, Academic Press, New York. With permission.

Table A.3.8-8

SPECIFIC DOSE RATES FROM INTERNAL AND EXTERNAL EXPOSURE TO VARIOUS RADIONUCLIDES[18]

Radionuclide	Maximum Dose to Any Body Organ, rem per μCi inhaled	Maximum External Dose Rate from Point Source, r/hr at 1 meter per μCi
^{3}H (as HTO)	0.00038	negligible
^{14}C as (CO_2)	0.00052	$<1.00 \times 10^{-8}$
^{99}Tc	0.0030	$<1.00 \times 10^{-9}$
^{55}Fe	0.0069	$<2.00 \times 10^{-6}$
^{198}Au	0.021	2.50×10^{-7}
^{35}S	0.021	$<1.00 \times 10^{-8}$
^{60}Co	0.021	1.30×10^{-6}
^{140}La	0.035	9.50×10^{-7}
^{59}Fe	0.037	6.50×10^{-7}
Mixed gross fission products, 180-day irradiation time	0.040	$<7.00 \times 10^{-7}$
^{192}Ir	0.047	5.10×10^{-7}
^{137}Cs	0.058	3.60×10^{-7}
^{140}Ba	0.11	1.54×10^{-6}
^{147}Pm	0.17	$<1.00 \times 10^{-8}$
^{32}P	0.17	$<1.00 \times 10^{-8}$
^{170}Tm	0.23	4.00×10^{-9}
^{45}Ca	0.35	$<1.00 \times 10^{-8}$
^{144}Ce	1.09	2.00×10^{-7}
^{131}I	1.25	2.50×10^{-7}
^{210}Po	11	$<1.00 \times 10^{-8}$
^{90}Sr	38	$<1.00 \times 10^{-8}$
^{233}U (with 20 ppm ^{232}U)	56	2.00×10^{-10}
^{242}Cm	57	$<1.00 \times 10^{-8}$
^{238}U	80 (but very low specific activity)	$<2.00 \times 10^{-9}$
^{235}U $+ 1\%$ ^{234}U	88	$<2.00 \times 10^{-9}$
U, natural	103 (but low specific activity)	$<2.00 \times 10^{-9}$
^{241}Am	2,280	3.90×10^{-8}
^{230}Th	5,320	9.00×10^{-9}
Th, natural	6,600 (but low specific activity)	$<2.00 \times 10^{-10}$
^{238}Pu	6,850	$<2.00 \times 10^{-8}$
^{239}Pu	7,370	$<1.00 \times 10^{-11}$
^{240}Pu	7,400	$<1.00 \times 10^{-9}$
^{85}Kr	see Reference 3	1.90×10^{-9}

From: Brodsky, A., Balancing Benefit versus Risk in Control of Consumer Items Containing Radioactive Material. *Amer. J Public Health*, 55(12):1971—1992, 1965.

Neutron dosimetry is particularly complicated, since it involves knowledge of the spectra of neutron radiation to which the individual is exposed as well as of the detailed absorption and scattering cross sections for all body elements as the functions of neutron energy.[7,19] Table A.3.8-10 from the *NBS Handbook 75* presents maximum permissible neutron fluxes in neutron/cm^2/second required to deliver 5 rems in a 2,000-hour work year. The assumed relative biological effectiveness of neutrons, as well as the flux, is given as a function of neutron energy. These values have been calculated for the multiple-collision tissue dose at the maximum of the depth dose curves,[7,19] taking into account the known neutron interactions, including secondary gamma radiation produced within the body. Snyder and Neufeld[20] have shown that the build-up ratio (maximum multiple-collision dose divided by first-collision dose) is 1.6 $\pm$ 0.2 for fast-neutron energies from 0.5 to 10 Mev.

Table A.3.8-9
GAMMA-RADIATION LEVELS FROM RADIOACTIVE MATERIALS SPREAD OVER WIDE LAND AREAS[18]

Radionuclides	Physical Half-Life	Activity in Curies per Square Mile to Produce 0.001 Roentgens per Hour at Three Feet Above Ground	
		Smooth Surface	Rough Soil
^{24}Na	15.1 hours	120	
^{59}Fe	45.0 days	340	
^{60}Co	5.2 years	168	
^{95}Zr–^{95}Nb	65.0 days 37.0 days	—	250
^{103}Ru	40.0 days	—	360
^{106}Ru	1.0 year	—	770
^{131}I	8.1 days	884	
^{137}Cs	27.0 years	620	290
^{140}Ba-^{140}La	12.8 days	144	118
^{144}Ce-^{144}Pr	285.0 days	1,100	5,900
^{170}Tm	129.0 days	53,000	
^{192}Ir	74.0 days	434	
^{226}Ra + daughters	1,620.0 years	263	
^{239}Pu	24,000.0 years	1,310	
^{241}Am	470.0 years	5,660	

From: Brodsky, A., Balancing Benefit versus Risk in the Control of Consumer Items Containing Radioactive Material. *Amer. J. Public Health*, 55(12): 1971—1992, 1965.

Table A.3.8-10
AVERAGE YEARLY MAXIMUM PERMISSIBLE NEUTRON FLUX[19]

Neutron Energy, Mev	RBE and Flux	Flux, $ncm^{-2} sec^{-1}$
Thermal	3.0	670
0.0001	2.0	500
0.005	2.5	570
0.02	5.0	280
0.1	8.0	80
0.5	10.0	30
1.0	10.5	18
2.5	8.0	20
5.0	7.0	18
7.5	7.0	17
10 *	6.5	17
10 to 30		10*

* Suggested limit.

The calculation of dose from implanted radium needles or other implanted radionuclides is also discussed in detail in the literature on radiological physics.[8,13] Basically, the dose rates near multiple-needle arrays have been calculated from the fundamental equation for the dose rate at any point near a single linear source by adding the dose rate contributed by each needle at each point in tissue. In Johns[8] this dose rate is expressed by the equation

$$I \cong \frac{K\rho}{h} (\theta_2 - \theta_1)e^{-\mu d}$$

where I = the dose rate, in r/hour, at point P in Figure A.3.8-1;

K = the specific dose-rate constant for radium, 8.26 ± 0.05 R/hr/mg at 1 cm from a ^{226}Ra source filtered by 0.5 mm Pt;

ρ = the linear density of activity, in mg Ra/cm of length;

h = the distance from the source, in cm, in a perpendicular to the center line;

θ_2 and θ_1 = the angles that the perpendicular makes with the lines drawn from P to the far and near ends of the active length of the source (as shown in Figure A.3.8-1);

μ = the average absorption coefficient of Pt, in cm^{-1};

d = the average path length that a Ra gamma must traverse through the Pt filter before it escapes the needle in the direction of the point P.

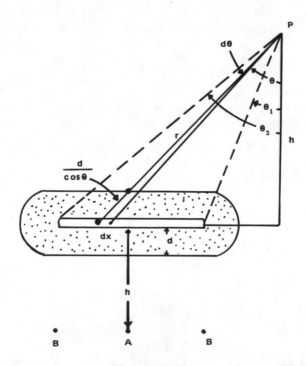

FIGURE A. 3.8-1. Diagram Illustrating How the Dose May Be Calculated at Points Near a Linear Source. (From: Johns, H. E., *The Physics of Radiology*, 2nd ed., Charles C. Thomas, Springfield, Illinois, 1964, 488. With permission.)

More precise calculations have been carried out integrating over the variable self-absorption path, and corrections are available for various thicknesses of Pt.[21] Table A.3.8-11 gives the mg-hours to deliver 1000 rad at a point h centimeters from the center of a linear source of active length L for filtration of 1.0 and 0.5 mm of Pt. This table was taken from Johns,[8] who obtained these values in rads by converting the original data of Meredith[22] by multiplying the values by 1.055 to correct for the newer measurements on K for Ra and a conversion from roentgens to rads.

Table A.3.8-11
LINEAR SOURCES[s]

Active Length, L	Treatment Distance, h cm						
	0.5	0.75	1.0	1.5	2.0	2.5	3.0

Filtration 1.0 mm Pt

Active Length, L	0.5	0.75	1.0	1.5	2.0	2.5	3.0
0.0	35	78	139	314	557	870	1253
0.5	41	82	142	319	562	875	1257
1.0	50	90	151	327	571	882	1266
1.5	58	102	165	338	585	896	1282
2.0	69	118	181	359	604	915	1302
2.5	79	135	200	381	628	942	1329
3.0	92	151	223	407	656	975	1361
3.5	106	170	245	438	690	1011	1397
4.0	118	187	268	467	726	1050	1438
5.0	145	225	317	532	805	1136	1532
6.0	171	263	366	601	891	1231	1638
7.0	197	302	417	673	981	1336	1754
8.0	225	343	468	748	1075	1447	1874
9.0	251	384	522	823	1171	1560	1998
10.0	280	424	576	900	1266	1677	2131
12.0	334	506	683	1056	1467	1916	2411
14.0	390	589	791	1214	1669	2163	2701
16.0	445	672	901	1374	1878	2415	2996
18.0	501	755	1011	1536	2087	2669	3298
20.0	557	839	1118	1699	2298	2929	3603

Filtration 0.5 mm Pt

Active Length, L	0.5	0.75	1.0	1.5	2.0	2.5	3.0
0.0	32	70	126	283	502	785	1130
0.5	35	74	128	287	504	788	1132
1.0	40	81	134	292	511	794	1142
1.5	50	91	146	303	524	806	1155
2.0	58	103	161	318	541	823	1175
2.5	68	118	177	338	564	844	1200
3.0	78	132	194	362	589	869	1227
3.5	89	148	214	388	615	900	1258
4.0	99	161	232	414	645	935	1289
5.0	122	193	273	467	712	1010	1365
6.0	146	226	313	525	783	1092	1454
7.0	168	257	356	583	857	1180	1551
8.0	190	289	401	646	936	1270	1654
9.0	211	324	444	708	1016	1363	1761
10.0	233	356	489	771	1099	1460	1872
12.0	277	423	576	900	1262	1660	2104
14.0	323	592	666	1030	1430	1866	2342
16.0	368	561	754	1161	1602	2079	2587
18.0	416	629	846	1291	1776	2290	2835
20.0	461	698	936	1426	1951	2502	3091

Table A.3.8-12 gives the maximum permissible concentrations of radionuclides in air and water for occupational exposure (columns A), and for relief of unrestricted areas (columns B), as reproduced in the *CRC Handbook of Radioactive Nuclides* from the 1966 Federal Regulations. This table is again used here since many present facilities are designed and operated based on these values. Thus, they still are of use for some purposes, and contain reasonable factors of safety. However, later versions should be consulted for some minor changes in official values. Updated tables will be included in future volumes.

Table A.3.8-13 contains some older, single-intake dose values and additional information on external exposure rates from various nuclides. These values are still useful as a single-page table of data for rapid hazard assessment under emergency conditions. However, for official use in estimating doses for permissible exposures more precisely, the most recent methods or references should be consulted, as discussed in other sections of this handbook.

Table A.3.8-12
MAXIMUM PERMISSIBLE CONCENTRATIONS OF RADIONUCLIDES IN AIR AND WATER*   (See Notes at End of Table)

Element	Atomic Number	Isotope*	A		B	
			In Air, μCi/ml	In Water, μCi/ml	In Air, μCi/ml	In Water, μCi/ml
Actinium	89	^{227}Ac				
		S	2×10^{-12}	6×10^{-5}	8×10^{-14}	2×10^{-6}
		I	3×10^{-11}	9×10^{-3}	9×10^{-13}	3×10^{-4}
		^{228}Ac				
		S	8×10^{-8}	3×10^{-3}	3×10^{-9}	9×10^{-5}
		I	2×10^{-8}	3×10^{-3}	6×10^{-10}	9×10^{-5}
Americium	95	^{241}Am				
		S	6×10^{-12}	1×10^{-4}	2×10^{-13}	4×10^{-6}
		I	1×10^{-10}	8×10^{-4}	4×10^{-12}	2×10^{-5}
		^{242m}Am				
		S	6×10^{-12}	1×10^{-4}	2×10^{-13}	4×10^{-6}
		I	3×10^{-10}	3×10^{-3}	9×10^{-12}	9×10^{-5}
		^{242}Am				
		S	4×10^{-8}	4×10^{-3}	1×10^{-9}	1×10^{-4}
		I†	5×10^{-8}	4×10^{-3}	2×10^{-9}	1×10^{-4}
		^{243}Am				
		S	6×10^{-12}	1×10^{-4}	2×10^{-13}	4×10^{-6}
		I	1×10^{-10}	8×10^{-4}	4×10^{-12}	3×10^{-5}
		^{244}Am†				
		S	4×10^{-6}	1×10^{-1}	1×10^{-7}	5×10^{-3}
		I	2×10^{-5}	1×10^{-1}	8×10^{-7}	5×10^{-3}
Antimony	51	^{122}Sb				
		S	2×10^{-7}	8×10^{-4}	6×10^{-9}	3×10^{-5}
		I	1×10^{-7}	8×10^{-4}	5×10^{-9}	3×10^{-5}
		^{124}Sb				
		S	2×10^{-7}	7×10^{-4}	5×10^{-9}	2×10^{-5}
		I	2×10^{-8}	7×10^{-4}	7×10^{-10}	2×10^{-5}
		^{125}Sb				
		S	5×10^{-7}	3×10^{-3}	2×10^{-8}	1×10^{-4}
		I	3×10^{-8}	3×10^{-3}	9×10^{-10}	1×10^{-4}
Argon	18	^{37}Ar				
		Sub	6×10^{-3}	—	1×10^{-4}	—
		^{41}Ar				
		Sub	2×10^{-6}	—	4×10^{-8}	—
Arsenic	33	^{73}As				
		S	2×10^{-6}	1×10^{-2}	7×10^{-8}	5×10^{-4}
		I	4×10^{-7}	1×10^{-2}	1×10^{-8}	5×10^{-4}
		^{74}As				
		S	3×10^{-7}	2×10^{-3}	1×10^{-8}	5×10^{-5}
		I	1×10^{-7}	2×10^{-3}	4×10^{-9}	5×10^{-5}
		^{76}As				
		S	1×10^{-7}	6×10^{-4}	4×10^{-9}	2×10^{-5}
		I	1×10^{-7}	6×10^{-4}	3×10^{-9}	2×10^{-5}
		^{77}As				
		S	5×10^{-7}	2×10^{-3}	2×10^{-8}	8×10^{-5}
		I	4×10^{-7}	2×10^{-3}	1×10^{-8}	8×10^{-5}

* Reprinted from Wang, Y., Ed., *Handbook of Radioactive Nuclides,* CRC Press, Cleveland, Ohio, 1969, 616. With permission. This table was taken from Title 10, Code of Federal Regulations, Part 20 (10 CFR 20) Revised August 9, 1966.

Table A.3.8-12 (Continued)
MAXIMUM PERMISSIBLE CONCENTRATIONS OF RADIONUCLIDES IN AIR AND WATER   (See Notes at End of Table)

Element	Atomic Number	Isotope*	A		B	
			In Air, μCi/ml	In Water, μCi/ml	In Air, μCi/ml	In Water, μCi/ml
Astatine	85	^{211}At				
		S	7×10^{-9}	5×10^{-5}	2×10^{-10}	2×10^{-6}
		I	3×10^{-8}	2×10^{-3}	1×10^{-9}	7×10^{-5}
Barium	56	^{131}Ba				
		S	1×10^{-6}	5×10^{-3}	4×10^{-8}	2×10^{-4}
		I	4×10^{-7}	5×10^{-3}	1×10^{-8}	2×10^{-4}
		^{140}Ba				
		S	1×10^{-7}	8×10^{-4}	4×10^{-9}	3×10^{-5}
		I	4×10^{-8}	7×10^{-4}	1×10^{-9}	2×10^{-5}
Berkelium	97	^{249}Bk				
		S	9×10^{-10}	2×10^{-2}	3×10^{-11}	6×10^{-4}
		I	1×10^{-7}	2×10^{-2}	4×10^{-9}	6×10^{-4}
		^{250}Bk†				
		S	1×10^{-7}	6×10^{-3}	5×10^{-9}	2×10^{-4}
		I	1×10^{-6}	6×10^{-3}	4×10^{-8}	2×10^{-4}
Beryllium	4	^{7}Be				
		S	6×10^{-6}	5×10^{-2}	2×10^{-7}	2×10^{-3}
		I	1×10^{-6}	5×10^{-2}	4×10^{-8}	2×10^{-3}
Bismuth	83	^{206}Bi				
		S	2×10^{-7}	1×10^{-3}	6×10^{-9}	4×10^{-5}
		I	1×10^{-7}	1×10^{-3}	5×10^{-9}	4×10^{-5}
		^{207}Bi				
		S	2×10^{-7}	2×10^{-3}	6×10^{-9}	6×10^{-5}
		I	1×10^{-8}	2×10^{-3}	5×10^{-10}	6×10^{-5}
		^{210}Bi				
		S	6×10^{-9}	1×10^{-3}	2×10^{-10}	4×10^{-5}
		I	6×10^{-9}	1×10^{-3}	2×10^{-10}	4×10^{-5}
		^{212}Bi				
		S	1×10^{-7}	1×10^{-2}	3×10^{-9}	4×10^{-4}
		I	2×10^{-7}	1×10^{-2}	7×10^{-9}	4×10^{-4}
Bromine	35	^{82}Br				
		S	1×10^{-6}	8×10^{-3}	4×10^{-8}	3×10^{-4}
		I	2×10^{-7}	1×10^{-3}	6×10^{-9}	4×10^{-5}
Cadmium	48	^{109}Cd				
		S	5×10^{-8}	5×10^{-3}	2×10^{-9}	2×10^{-4}
		I	7×10^{-8}	5×10^{-3}	3×10^{-9}	2×10^{-4}
		^{115m}Cd				
		S	4×10^{-8}	7×10^{-4}	1×10^{-9}	3×10^{-5}
		I	4×10^{-8}	7×10^{-4}	1×10^{-9}	3×10^{-5}
		^{115}Cd				
		S	2×10^{-7}	1×10^{-3}	8×10^{-9}	3×10^{-5}
		I	2×10^{-7}	1×10^{-3}	6×10^{-9}	4×10^{-5}
Calcium	20	^{45}Ca				
		S	3×10^{-8}	3×10^{-4}	1×10^{-9}	9×10^{-6}
		I	1×10^{-7}	5×10^{-3}	4×10^{-9}	2×10^{-4}
		^{47}Ca				
		S	2×10^{-7}	1×10^{-3}	6×10^{-9}	5×10^{-5}
		I	2×10^{-7}	1×10^{-3}	6×10^{-9}	3×10^{-5}

Table A.3.8-12 (Continued)
MAXIMUM PERMISSIBLE CONCENTRATIONS OF RADIONUCLIDES IN AIR AND WATER   (See Notes at End of Table)

Element	Atomic Number	Isotope*	A		B	
			In Air, μCi/ml	In Water, μCi/ml	In Air, μCi/ml	In Water, μCi/ml
Californium	98	^{249}Cf				
		S	2×10^{-12}	1×10^{-4}	5×10^{-14}	4×10^{-6}
		I	1×10^{-10}	7×10^{-4}	3×10^{-12}	2×10^{-5}
		^{250}Cf				
		S	5×10^{-12}	4×10^{-4}	2×10^{-13}	1×10^{-5}
		I	1×10^{-10}	7×10^{-4}	3×10^{-12}	3×10^{-5}
		^{251}Cf †				
		S	2×10^{-12}	1×10^{-4}	6×10^{-14}	4×10^{-6}
		I	1×10^{-10}	8×10^{-4}	3×10^{-12}	3×10^{-5}
		^{252}Cf				
		S	2×10^{-11}	7×10^{-4}	7×10^{-13}	2×10^{-5}
		I	1×10^{-10}	7×10^{-4}	4×10^{-12}	2×10^{-5}
		^{253}Cf				
		S	8×10^{-10}	4×10^{-3}	3×10^{-11}	1×10^{-4}
		I	8×10^{-10}	4×10^{-3}	3×10^{-11}	1×10^{-4}
		^{254}Cf †				
		S	5×10^{-12}	4×10^{-6}	2×10^{-13}	1×10^{-7}
		I	5×10^{-12}	4×10^{-6}	2×10^{-13}	1×10^{-7}
Carbon	6	^{14}C				
		S	4×10^{-6}	2×10^{-2}	1×10^{-7}	8×10^{-4}
		(CO_2) Sub	5×10^{-5}	—	1×10^{-6}	—
Cerium	58	^{141}Ce				
		S	4×10^{-7}	3×10^{-3}	2×10^{-8}	9×10^{-5}
		I	2×10^{-7}	3×10^{-3}	5×10^{-9}	9×10^{-5}
		^{143}Ce				
		S	3×10^{-7}	1×10^{-3}	9×10^{-9}	4×10^{-5}
		I	2×10^{-7}	1×10^{-3}	7×10^{-9}	4×10^{-5}
		^{144}Ce				
		S	1×10^{-8}	3×10^{-4}	3×10^{-10}	1×10^{-5}
		I	6×10^{-9}	3×10^{-4}	2×10^{-10}	1×10^{-5}
Cesium	55	^{131}Cs				
		S	1×10^{-5}	7×10^{-2}	4×10^{-7}	2×10^{-3}
		I	3×10^{-6}	3×10^{-2}	1×10^{-7}	9×10^{-4}
		^{134m}Cs				
		S	4×10^{-5}	2×10^{-1}	1×10^{-6}	6×10^{-3}
		I	6×10^{-6}	3×10^{-2}	2×10^{-7}	1×10^{-3}
		^{134}Cs				
		S	4×10^{-8}	3×10^{-4}	1×10^{-9}	9×10^{-6}
		I	1×10^{-8}	1×10^{-3}	4×10^{-10}	4×10^{-5}
		^{135}Cs				
		S	5×10^{-7}	3×10^{-3}	2×10^{-8}	1×10^{-4}
		I	9×10^{-8}	7×10^{-3}	3×10^{-9}	2×10^{-4}
		^{136}Cs				
		S	4×10^{-7}	2×10^{-3}	1×10^{-8}	9×10^{-5}
		I	2×10^{-7}	2×10^{-3}	6×10^{-9}	6×10^{-5}
		^{137}Cs				
		S	6×10^{-8}	4×10^{-4}	2×10^{-9}	2×10^{-5}
		I	1×10^{-8}	1×10^{-3}	5×10^{-10}	4×10^{-5}

Table A.3.8-12 (Continued)
MAXIMUM PERMISSIBLE CONCENTRATIONS OF RADIONUCLIDES IN
AIR AND WATER   (See Notes at End of Table)

Element	Atomic Number	Isotope*	A		B	
			In Air, μCi/ml	In Water, μCi/ml	In Air, μCi/ml	In Water, μCi/ml
Chlorine	17	^{36}Cl				
		S	4×10^{-7}	2×10^{-3}	1×10^{-8}	8×10^{-5}
		I	2×10^{-8}	2×10^{-3}	8×10^{-10}	6×10^{-5}
		^{38}Cl				
		S	3×10^{-6}	1×10^{-2}	9×10^{-8}	4×10^{-4}
		I	2×10^{-6}	1×10^{-2}	7×10^{-8}	4×10^{-4}
Chromium	24	^{51}Cr				
		S	1×10^{-5}	5×10^{-2}	4×10^{-7}	2×10^{-3}
		I	2×10^{-6}	5×10^{-2}	8×10^{-8}	2×10^{-3}
Cobalt	27	^{57}Co				
		S	3×10^{-6}	2×10^{-2}	1×10^{-7}	5×10^{-4}
		I	2×10^{-7}	1×10^{-2}	6×10^{-9}	4×10^{-4}
		^{58m}Co				
		S	2×10^{-5}	8×10^{-2}	6×10^{-7}	3×10^{-3}
		I	9×10^{-6}	6×10^{-2}	3×10^{-7}	2×10^{-3}
		^{58}Co				
		S	8×10^{-7}	4×10^{-3}	3×10^{-8}	1×10^{-4}
		I	5×10^{-8}	3×10^{-3}	2×10^{-9}	9×10^{-5}
		^{60}Co				
		S	3×10^{-7}	1×10^{-3}	1×10^{-8}	5×10^{-5}
		I	9×10^{-9}	1×10^{-3}	3×10^{-10}	3×10^{-5}
Copper	29	^{64}Cu				
		S	2×10^{-6}	1×10^{-2}	7×10^{-8}	3×10^{-4}
		I	1×10^{-6}	6×10^{-3}	4×10^{-8}	2×10^{-4}
Curium	96	^{242}Cm				
		S	1×10^{-10}	7×10^{-4}	4×10^{-12}	2×10^{-5}
		I	2×10^{-10}	7×10^{-4}	6×10^{-12}	3×10^{-5}
		^{243}Cm				
		S	6×10^{-12}	1×10^{-4}	2×10^{-13}	5×10^{-6}
		I	1×10^{-10}	7×10^{-4}	3×10^{-12}	2×10^{-5}
		$^{244\alpha}$Cm				
		S	9×10^{-12}	2×10^{-4}	3×10^{-13}	7×10^{-6}
		I	1×10^{-10}	8×10^{-4}	3×10^{-12}	3×10^{-5}
		^{245}Cm				
		S	5×10^{-12}	1×10^{-4}	2×10^{-13}	4×10^{-6}
		I	1×10^{-10}	8×10^{-4}	4×10^{-12}	3×10^{-5}
		^{246}Cm				
		S	5×10^{-12}	1×10^{-4}	2×10^{-13}	4×10^{-6}
		I	1×10^{-10}	8×10^{-4}	4×10^{-12}	3×10^{-5}
		^{247}Cm				
		S	5×10^{-12}	1×10^{-4}	2×10^{-13}	4×10^{-6}
		I	1×10^{-10}	6×10^{-4}	4×10^{-12}	2×10^{-5}
		^{248}Cm†				
		S	6×10^{-13}	1×10^{-5}	2×10^{-14}	4×10^{-7}
		I	1×10^{-11}	4×10^{-5}	4×10^{-13}	1×10^{-6}
		^{249}Cm				
		S	1×10^{-5}	6×10^{-2}	4×10^{-7}	2×10^{-3}
		I	1×10^{-5}	6×10^{-2}	4×10^{-7}	2×10^{-3}

Table A.3.8-12 (Continued)
MAXIMUM PERMISSIBLE CONCENTRATIONS OF RADIONUCLIDES IN
AIR AND WATER   (See Notes at End of Table)

Element	Atomic Number	Isotope*	A		B	
			In Air, μCi/ml	In Water, μCi/ml	In Air, μCi/ml	In Water, μCi/ml
Dysprosium	66	^{165}Dy				
		S	3×10^{-6}	1×10^{-2}	9×10^{-8}	4×10^{-4}
		I	2×10^{-6}	1×10^{-2}	7×10^{-8}	4×10^{-4}
		^{166}Dy†				
		S	2×10^{-7}	1×10^{-3}	8×10^{-9}	4×10^{-5}
		I	2×10^{-7}	1×10^{-3}	7×10^{-9}	4×10^{-5}
Einsteinium	99	^{253}Es				
		S	8×10^{-10}	7×10^{-4}	3×10^{-11}	2×10^{-5}
		I	6×10^{-10}	7×10^{-4}	2×10^{-11}	2×10^{-5}
		^{254m}Es				
		S	5×10^{-9}	5×10^{-4}	2×10^{-10}	2×10^{-5}
		I	6×10^{-9}	5×10^{-4}	2×10^{-10}	2×10^{-5}
		^{254}Es				
		S	2×10^{-11}	4×10^{-4}	6×10^{-13}	1×10^{-5}
		I	1×10^{-10}	4×10^{-4}	4×10^{-12}	1×10^{-5}
		^{255}Es				
		S	5×10^{-10}	8×10^{-4}	2×10^{-11}	3×10^{-5}
		I	4×10^{-10}	8×10^{-4}	1×10^{-11}	3×10^{-5}
Erbium	68	^{169}Er				
		S	6×10^{-7}	3×10^{-3}	2×10^{-8}	9×10^{-5}
		I	4×10^{-7}	3×10^{-3}	1×10^{-8}	9×10^{-5}
		^{171}Er				
		S	7×10^{-7}	3×10^{-3}	2×10^{-8}	1×10^{-4}
		I	6×10^{-7}	3×10^{-3}	2×10^{-8}	1×10^{-4}
Europium	63	^{152}Eu (T/2 = 9.2 hrs)				
		S	4×10^{-7}	2×10^{-3}	1×10^{-8}	6×10^{-5}
		I	3×10^{-7}	2×10^{-3}	1×10^{-8}	6×10^{-5}
		^{152}Eu (T/2 = 13 yrs)				
		S	1×10^{-8}	2×10^{-3}	4×10^{-10}	8×10^{-5}
		I	2×10^{-8}	2×10^{-3}	6×10^{-10}	8×10^{-5}
		^{154}Eu				
		S	4×10^{-9}	6×10^{-4}	1×10^{-10}	2×10^{-5}
		I	7×10^{-9}	6×10^{-4}	2×10^{-10}	2×10^{-5}
		^{155}Eu				
		S	9×10^{-8}	6×10^{-3}	3×10^{-9}	2×10^{-4}
		I	7×10^{-8}	6×10^{-3}	3×10^{-9}	2×10^{-4}
Fermium†	100	^{254}Fm				
		S	6×10^{-8}	4×10^{-3}	2×10^{-9}	1×10^{-4}
		I	7×10^{-8}	4×10^{-3}	2×10^{-9}	1×10^{-4}
		^{255}Fm				
		S	2×10^{-8}	1×10^{-3}	6×10^{-10}	3×10^{-5}
		I	1×10^{-8}	1×10^{-3}	4×10^{-10}	3×10^{-5}
		^{256}Fm				
		S	3×10^{-9}	3×10^{-5}	1×10^{-10}	9×10^{-7}
		I	2×10^{-9}	3×10^{-5}	6×10^{-11}	9×10^{-7}

Table A.3.8-12 (Continued)
MAXIMUM PERMISSIBLE CONCENTRATIONS OF RADIONUCLIDES IN AIR AND WATER   (See Notes at End of Table)

Element	Atomic Number	Isotope*	A		B	
			In Air, μCi/ml	In Water, μCi/ml	In Air, μCi/ml	In Water, μCi/ml
Fluorine	9	^{18}F				
		S	5×10^{-6}	2×10^{-2}	2×10^{-7}	8×10^{-4}
		I	3×10^{-6}	1×10^{-2}	9×10^{-8}	5×10^{-4}
Gadolinium	64	^{153}Gd				
		S	2×10^{-7}	6×10^{-3}	8×10^{-9}	2×10^{-4}
		I	9×10^{-8}	6×10^{-3}	3×10^{-9}	2×10^{-4}
		^{159}Gd				
		S	5×10^{-7}	2×10^{-3}	2×10^{-8}	8×10^{-5}
		I	4×10^{-7}	2×10^{-3}	1×10^{-8}	8×10^{-5}
Gallium	31	^{72}Ga				
		S	2×10^{-7}	1×10^{-3}	8×10^{-9}	4×10^{-5}
		I	2×10^{-7}	1×10^{-3}	6×10^{-9}	4×10^{-5}
Germanium	32	^{71}Ge				
		S	1×10^{-5}	5×10^{-2}	4×10^{-7}	2×10^{-3}
		I	6×10^{-6}	5×10^{-2}	2×10^{-7}	2×10^{-3}
Gold	79	^{196}Au				
		S	1×10^{-6}	5×10^{-3}	4×10^{-8}	2×10^{-4}
		I	6×10^{-7}	4×10^{-3}	2×10^{-8}	1×10^{-4}
		^{198}Au				
		S	3×10^{-7}	2×10^{-3}	1×10^{-8}	5×10^{-5}
		I	2×10^{-7}	1×10^{-3}	8×10^{-9}	5×10^{-5}
		^{199}Au				
		S	1×10^{-6}	5×10^{-3}	4×10^{-8}	2×10^{-4}
		I	8×10^{-7}	4×10^{-3}	3×10^{-8}	2×10^{-4}
Hafnium	72	^{181}Hf				
		S	4×10^{-8}	2×10^{-3}	1×10^{-9}	7×10^{-5}
		I	7×10^{-8}	2×10^{-3}	3×10^{-9}	7×10^{-5}
Holmium	67	^{166}Ho				
		S	2×10^{-7}	9×10^{-4}	7×10^{-9}	3×10^{-5}
		I	2×10^{-7}	9×10^{-4}	6×10^{-9}	3×10^{-5}
Hydrogen	1	^{3}H				
		S	5×10^{-6}	1×10^{-1}	2×10^{-7}	3×10^{-3}
		I†	5×10^{-6}	1×10^{-1}	2×10^{-7}	3×10^{-3}
		Sub	2×10^{-3}	—	4×10^{-5}	—
Indium	49	^{113m}In				
		S	8×10^{-6}	4×10^{-2}	3×10^{-7}	1×10^{-3}
		I	7×10^{-6}	4×10^{-2}	2×10^{-7}	1×10^{-3}
		^{114m}In				
		S	1×10^{-7}	5×10^{-4}	4×10^{-9}	2×10^{-5}
		I	2×10^{-8}	5×10^{-4}	7×10^{-10}	2×10^{-5}
		^{115m}In				
		S	2×10^{-6}	1×10^{-2}	8×10^{-8}	4×10^{-4}
		I	2×10^{-6}	1×10^{-2}	6×10^{-8}	4×10^{-4}
		^{115}In				
		S	2×10^{-7}	3×10^{-3}	9×10^{-9}	9×10^{-5}
		I	3×10^{-8}	3×10^{-3}	1×10^{-9}	9×10^{-5}

Table A.3.8-12 (Continued)
MAXIMUM PERMISSIBLE CONCENTRATIONS OF RADIONUCLIDES IN
AIR AND WATER   (See Notes at End of Table)

Element	Atomic Number	Isotope*	A		B	
			In Air, μCi/ml	In Water, μCi/ml	In Air, μCi/ml	In Water, μCi/ml
Iodine	53	^{125}I†				
		S	5×10^{-9}	4×10^{-5}	8×10^{-11}	2×10^{-7}
		I	2×10^{-7}	6×10^{-3}	6×10^{-9}	2×10^{-4}
		^{126}I††				
		S	8×10^{-9}	5×10^{-5}	9×10^{-11}	3×10^{-7}
		I	3×10^{-7}	3×10^{-3}	1×10^{-8}	9×10^{-5}
		^{129}I				
		S	2×10^{-9}	1×10^{-5}	2×10^{-11}	6×10^{-8}
		I	7×10^{-8}	6×10^{-3}	2×10^{-9}	2×10^{-4}
		^{131}I				
		S	9×10^{-9}	6×10^{-5}	1×10^{-10}	3×10^{-7}
		I	3×10^{-7}	2×10^{-3}	1×10^{-8}	6×10^{-5}
		^{132}I				
		S	2×10^{-7}	2×10^{-3}	3×10^{-9}	8×10^{-6}
		I	9×10^{-7}	5×10^{-3}	3×10^{-8}	2×10^{-4}
		^{133}I				
		S	3×10^{-8}	2×10^{-4}	4×10^{-10}	1×10^{-6}
		I	2×10^{-7}	1×10^{-3}	7×10^{-9}	4×10^{-5}
		^{134}I				
		S	5×10^{-7}	4×10^{-3}	6×10^{-9}	2×10^{-5}
		I	3×10^{-6}	2×10^{-2}	1×10^{-7}	6×10^{-4}
		^{135}I				
		S	1×10^{-7}	7×10^{-4}	1×10^{-9}	4×10^{-6}
		I	4×10^{-7}	2×10^{-3}	1×10^{-8}	7×10^{-5}
Iridium	77	^{190}Ir				
		S	1×10^{-6}	6×10^{-3}	4×10^{-8}	2×10^{-4}
		I	4×10^{-7}	5×10^{-3}	1×10^{-8}	2×10^{-4}
		^{192}Ir				
		S	1×10^{-7}	1×10^{-3}	4×10^{-9}	4×10^{-5}
		I	3×10^{-8}	1×10^{-3}	9×10^{-10}	4×10^{-5}
		^{194}Ir				
		S	2×10^{-7}	1×10^{-3}	8×10^{-9}	3×10^{-5}
		I	2×10^{-7}	9×10^{-4}	5×10^{-9}	3×10^{-5}
Iron	26	^{55}Fe				
		S	9×10^{-7}	2×10^{-2}	3×10^{-8}	8×10^{-4}
		I	1×10^{-6}	7×10^{-2}	3×10^{-8}	2×10^{-3}
		^{59}Fe				
		S	1×10^{-7}	2×10^{-3}	5×10^{-9}	6×10^{-5}
		I	5×10^{-8}	2×10^{-3}	2×10^{-9}	5×10^{-5}
Krypton	36	^{85m}Kr				
		Sub	6×10^{-6}	—	1×10^{-7}	—
		^{85}Kr				
		Sub	1×10^{-5}	—	3×10^{-7}	—
		^{87}Kr				
		Sub	1×10^{-6}	—	2×10^{-8}	—

Table A.3.8-12 (Continued)
MAXIMUM PERMISSIBLE CONCENTRATIONS OF RADIONUCLIDES IN AIR AND WATER   (See Notes at End of Table)

Element	Atomic Number	Isotope*	A		B	
			In Air, μCi/ml	In Water, μCi/ml	In Air, μCi/ml	In Water, μCi/ml
Krypton (*cont.*)	36	^{88}Kr$^+$ Sub	1×10^{-6}	—	2×10^{-8}	—
Lanthanum	57	^{140}La				
		S	2×10^{-7}	7×10^{-4}	5×10^{-9}	2×10^{-5}
		I	1×10^{-7}	7×10^{-4}	4×10^{-9}	2×10^{-5}
Lead	82	^{203}Pb				
		S	3×10^{-6}	1×10^{-2}	9×10^{-8}	4×10^{-4}
		I	2×10^{-6}	1×10^{-2}	6×10^{-8}	4×10^{-4}
		^{210}Pb				
		S	1×10^{-10}	4×10^{-6}	4×10^{-12}	1×10^{-7}
		I	2×10^{-10}	5×10^{-3}	8×10^{-12}	2×10^{-4}
		^{212}Pb				
		S	2×10^{-8}	6×10^{-4}	6×10^{-10}	2×10^{-5}
		I	2×10^{-8}	5×10^{-4}	7×10^{-10}	2×10^{-5}
Lutatium	71	^{177}Lu				
		S	6×10^{-7}	3×10^{-3}	2×10^{-8}	1×10^{-4}
		I	5×10^{-7}	3×10^{-3}	2×10^{-8}	1×10^{-4}
Manganese	25	^{52}Mn				
		S	2×10^{-7}	1×10^{-3}	7×10^{-9}	3×10^{-5}
		I	1×10^{-7}	9×10^{-4}	5×10^{-9}	3×10^{-5}
		^{54}Mn				
		S	4×10^{-7}	4×10^{-3}	1×10^{-9}	1×10^{-4}
		I	4×10^{-8}	3×10^{-3}	1×10^{-9}	1×10^{-4}
		^{56}Mn				
		S	8×10^{-7}	4×10^{-3}	3×10^{-8}	1×10^{-4}
		I	5×10^{-7}	3×10^{-3}	2×10^{-8}	1×10^{-4}
Mercury	80	^{197m}Hg				
		S	7×10^{-7}	6×10^{-3}	3×10^{-8}	2×10^{-4}
		I	8×10^{-7}	5×10^{-3}	3×10^{-8}	2×10^{-4}
		^{197}Hg				
		S	1×10^{-6}	9×10^{-3}	4×10^{-8}	3×10^{-4}
		I	3×10^{-6}	1×10^{-2}	9×10^{-8}	5×10^{-4}
		^{203}Hg				
		S	7×10^{-8}	5×10^{-4}	2×10^{-9}	2×10^{-5}
		I	1×10^{-7}	3×10^{-3}	4×10^{-9}	1×10^{-4}
Molybdenum	42	^{99}Mo				
		S	7×10^{-7}	5×10^{-3}	3×10^{-8}	2×10^{-4}
		I	2×10^{-7}	1×10^{-3}	7×10^{-9}	4×10^{-5}
Neodymium	60	^{144}Nd				
		S	8×10^{-11}	2×10^{-3}	3×10^{-12}	7×10^{-5}
		I	3×10^{-10}	2×10^{-3}	1×10^{-11}	8×10^{-5}
		^{147}Nd				
		S	4×10^{-7}	2×10^{-3}	1×10^{-8}	6×10^{-5}
		I	2×10^{-7}	2×10^{-3}	8×10^{-9}	6×10^{-5}

Table A.3.8-12 (Continued)
MAXIMUM PERMISSIBLE CONCENTRATIONS OF RADIONUCLIDES IN AIR AND WATER (See Notes at End of Table)

Element	Atomic Number	Isotope*	A		B	
			In Air, μCi/ml	In Water, μCi/ml	In Air, μCi/ml	In Water, μCi/ml
Neodymium (*cont.*)	60	^{149}Nd				
		S	2×10^{-6}	8×10^{-3}	6×10^{-8}	3×10^{-4}
		I	1×10^{-6}	8×10^{-3}	5×10^{-8}	3×10^{-4}
Neptunium	93	^{237}Np				
		S	4×10^{-12}	9×10^{-5}	1×10^{-13}	3×10^{-6}
		I	1×10^{-10}	9×10^{-4}	4×10^{-12}	3×10^{-5}
		^{239}Np				
		S	8×10^{-7}	4×10^{-3}	3×10^{-8}	1×10^{-4}
		I	7×10^{-7}	4×10^{-3}	2×10^{-8}	1×10^{-4}
Nickel	28	^{59}Ni				
		S	5×10^{-7}	6×10^{-3}	2×10^{-8}	2×10^{-4}
		I	8×10^{-7}	6×10^{-2}	3×10^{-8}	2×10^{-3}
		^{63}Ni				
		S	6×10^{-8}	8×10^{-4}	2×10^{-9}	3×10^{-5}
		I	3×10^{-7}	2×10^{-2}	1×10^{-8}	7×10^{-4}
		^{65}Ni				
		S	9×10^{-7}	4×10^{-3}	3×10^{-8}	1×10^{-4}
		I	5×10^{-7}	3×10^{-3}	2×10^{-8}	1×10^{-4}
Niobium (Columbium)	41	^{93m}Nb				
		S	1×10^{-7}	1×10^{-2}	4×10^{-9}	4×10^{-4}
		I	2×10^{-7}	1×10^{-2}	5×10^{-9}	4×10^{-4}
		^{95}Nb				
		S	5×10^{-7}	3×10^{-3}	2×10^{-8}	1×10^{-4}
		I	1×10^{-7}	3×10^{-3}	3×10^{-9}	1×10^{-4}
		^{97}Nb				
		S	6×10^{-6}	3×10^{-2}	2×10^{-7}	9×10^{-4}
		I	5×10^{-6}	3×10^{-2}	2×10^{-7}	9×10^{-4}
Osmium	76	^{185}Os				
		S	5×10^{-7}	2×10^{-3}	2×10^{-8}	7×10^{-5}
		I	5×10^{-8}	2×10^{-3}	2×10^{-9}	7×10^{-5}
		^{191m}Os				
		S	2×10^{-5}	7×10^{-2}	6×10^{-7}	3×10^{-3}
		I	9×10^{-6}	7×10^{-2}	3×10^{-7}	2×10^{-3}
		^{191}Os				
		S	1×10^{-6}	5×10^{-3}	4×10^{-8}	2×10^{-4}
		I	4×10^{-7}	5×10^{-3}	1×10^{-8}	2×10^{-4}
		^{193}Os				
		S	4×10^{-7}	2×10^{-3}	1×10^{-8}	6×10^{-5}
		I	3×10^{-7}	2×10^{-3}	9×10^{-9}	5×10^{-5}
Palladium	46	^{103}Pd				
		S	1×10^{-6}	1×10^{-2}	5×10^{-8}	3×10^{-4}
		I	7×10^{-7}	8×10^{-3}	3×10^{-8}	3×10^{-4}
		^{109}Pd				
		S	6×10^{-7}	3×10^{-3}	2×10^{-8}	9×10^{-5}
		I	4×10^{-7}	2×10^{-3}	1×10^{-8}	7×10^{-5}

Table A.3.8-12 (Continued)
MAXIMUM PERMISSIBLE CONCENTRATIONS OF RADIONUCLIDES IN
AIR AND WATER   (See Notes at End of Table)

Element	Atomic Number	Isotope*	A		B	
			In Air, μCi/ml	In Water, μCi/ml	In Air, μCi/ml	In Water, μCi/ml
Phosphorus	15	^{32}P				
		S	7×10^{-8}	5×10^{-4}	2×10^{-9}	2×10^{-5}
		I	8×10^{-8}	7×10^{-4}	3×10^{-9}	2×10^{-5}
Platinum	78	^{191}Pt				
		S	8×10^{-7}	4×10^{-3}	3×10^{-8}	1×10^{-4}
		I	6×10^{-7}	3×10^{-3}	2×10^{-8}	1×10^{-4}
		^{193m}Pt				
		S	7×10^{-6}	3×10^{-2}	2×10^{-7}	1×10^{-3}
		I	5×10^{-6}	3×10^{-2}	2×10^{-7}	1×10^{-3}
		^{197m}Pt				
		S	6×10^{-6}	3×10^{-2}	2×10^{-7}	1×10^{-3}
		I	5×10^{-6}	3×10^{-2}	2×10^{-7}	9×10^{-4}
		^{197}Pt				
		S	8×10^{-7}	4×10^{-3}	3×10^{-8}	1×10^{-4}
		I	6×10^{-7}	3×10^{-3}	2×10^{-8}	1×10^{-4}
Plutonium	94	^{238}Pu				
		S	2×10^{-12}	1×10^{-4}	7×10^{-14}	5×10^{-6}
		I	3×10^{-11}	8×10^{-4}	1×10^{-12}	3×10^{-5}
		^{239}Pu				
		S	2×10^{-12}	1×10^{-4}	6×10^{-14}	5×10^{-6}
		I	4×10^{-11}	8×10^{-4}	1×10^{-12}	3×10^{-5}
		^{240}Pu				
		S	2×10^{-12}	1×10^{-4}	6×10^{-14}	5×10^{-6}
		I	4×10^{-11}	8×10^{-4}	1×10^{-12}	3×10^{-5}
		^{241}Pu				
		S	9×10^{-11}	7×10^{-3}	3×10^{-12}	2×10^{-4}
		I	4×10^{-8}	4×10^{-2}	1×10^{-9}	1×10^{-3}
		^{242}Pu				
		S	2×10^{-12}	1×10^{-4}	6×10^{-14}	5×10^{-6}
		I	4×10^{-11}	9×10^{-4}	1×10^{-12}	3×10^{-5}
		^{243}Pu				
		S	2×10^{-6}	1×10^{-2}	6×10^{-8}	3×10^{-4}
		I	2×10^{-6}	1×10^{-2}	8×10^{-8}	3×10^{-4}
		^{244}Pu†				
		S	2×10^{-12}	1×10^{-4}	6×10^{-14}	4×10^{-6}
		I	3×10^{-11}	3×10^{-4}	1×10^{-12}	1×10^{-5}
Polonium	84	^{210}Po				
		S	5×10^{-10}	2×10^{-5}	2×10^{-11}	7×10^{-7}
		I	2×10^{-10}	8×10^{-4}	7×10^{-12}	3×10^{-5}
Potassium	19	^{42}K				
		S	2×10^{-6}	9×10^{-3}	7×10^{-8}	3×10^{-4}
		I	1×10^{-7}	6×10^{-4}	4×10^{-9}	2×10^{-5}
Praseodymium	59	^{142}Pr				
		S	2×10^{-7}	9×10^{-4}	7×10^{-9}	3×10^{-5}
		I	2×10^{-7}	9×10^{-4}	5×10^{-9}	3×10^{-5}
		^{143}Pr				
		S	3×10^{-7}	1×10^{-3}	1×10^{-8}	5×10^{-5}
		I	2×10^{-7}	1×10^{-3}	6×10^{-9}	5×10^{-5}

Table A.3.8-12 (Continued)
MAXIMUM PERMISSIBLE CONCENTRATIONS OF RADIONUCLIDES IN AIR AND WATER   (See Notes at End of Table)

Element	Atomic Number	Isotope*	A		B	
			In Air, μCi/ml	In Water, μCi/ml	In Air, μCi/ml	In Water, μCi/ml
Promethium	61	^{147}Pm				
		S	6×10^{-8}	6×10^{-3}	2×10^{-9}	2×10^{-4}
		I	1×10^{-7}	6×10^{-3}	3×10^{-9}	2×10^{-4}
		^{149}Pm				
		S	3×10^{-7}	1×10^{-3}	1×10^{-8}	4×10^{-5}
		I	2×10^{-7}	1×10^{-3}	8×10^{-9}	4×10^{-5}
Protoactinium	91	^{230}Pa				
		S	2×10^{-9}	7×10^{-3}	6×10^{-11}	2×10^{-4}
		I	8×10^{-10}	7×10^{-3}	3×10^{-11}	2×10^{-4}
		^{231}Pa				
		S	1×10^{-12}	3×10^{-5}	4×10^{-14}	9×10^{-7}
		I	1×10^{-10}	8×10^{-4}	4×10^{-12}	2×10^{-5}
		^{233}Pa				
		S	6×10^{-7}	4×10^{-3}	2×10^{-8}	1×10^{-4}
		I††	2×10^{-7}	3×10^{-3}	6×10^{-9}	1×10^{-4}
Radium	88	^{223}Ra				
		S	2×10^{-9}	2×10^{-5}	6×10^{-11}	7×10^{-7}
		I	2×10^{-10}	1×10^{-4}	8×10^{-12}	4×10^{-6}
		^{224}Ra				
		S	5×10^{-9}	7×10^{-5}	2×10^{-10}	2×10^{-6}
		I	7×10^{-10}	2×10^{-4}	2×10^{-11}	5×10^{-6}
		^{226}Ra				
		S	3×10^{-11}	4×10^{-7}	3×10^{-12}	3×10^{-8}
		I	5×10^{-11}	9×10^{-4}	2×10^{-12}	3×10^{-5}
		^{228}Ra				
		S	7×10^{-11}	8×10^{-7}	2×10^{-12}	3×10^{-8}
		I	4×10^{-11}	7×10^{-4}	1×10^{-12}	3×10^{-5}
Radon	86	^{220}Rn				
		S	3×10^{-7}	—	1×10^{-8}	—
		I	—	—	—	—
		^{222}Rn				
		S	1×10^{-7}	—	3×10^{-9}	—
		I	—	—	—	—
Rhenium	75	^{183}Re				
		S	3×10^{-6}	2×10^{-2}	9×10^{-8}	6×10^{-4}
		I	2×10^{-7}	8×10^{-3}	5×10^{-9}	3×10^{-4}
		^{186}Re				
		S	6×10^{-7}	3×10^{-3}	2×10^{-8}	9×10^{-5}
		I	2×10^{-7}	1×10^{-3}	8×10^{-9}	5×10^{-5}
		^{187}Re				
		S	9×10^{-6}	7×10^{-2}	3×10^{-7}	3×10^{-3}
		I	5×10^{-7}	4×10^{-2}	2×10^{-8}	2×10^{-3}
		^{188}Re				
		S	4×10^{-7}	2×10^{-3}	1×10^{-8}	6×10^{-5}
		I	2×10^{-7}	9×10^{-4}	6×10^{-9}	3×10^{-5}
Rhodium	45	^{103m}Rh				
		S	8×10^{-5}	4×10^{-1}	3×10^{-6}	1×10^{-2}
		I	6×10^{-5}	3×10^{-1}	2×10^{-6}	1×10^{-2}

Table A.3.8-12 (Continued)
MAXIMUM PERMISSIBLE CONCENTRATIONS OF RADIONUCLIDES IN AIR AND WATER   (See Notes at End of Table)

Element	Atomic Number	Isotope*	A		B	
			In Air, μCi/ml	In Water, μCi/ml	In Air, μCi/ml	In Water, μCi/ml
Rhodium (*cont.*)	45	^{105}Rh				
		S	8×10^{-7}	4×10^{-3}	3×10^{-8}	1×10^{-4}
		I	5×10^{-7}	3×10^{-3}	2×10^{-8}	1×10^{-4}
Rubidium	37	^{86}Rb				
		S	3×10^{-7}	2×10^{-3}	1×10^{-8}	7×10^{-5}
		I	7×10^{-8}	7×10^{-4}	2×10^{-9}	2×10^{-5}
		^{87}Rb				
		S	5×10^{-7}	3×10^{-3}	2×10^{-8}	1×10^{-4}
		I	7×10^{-8}	5×10^{-3}	2×10^{-9}	2×10^{-4}
Ruthenium	44	^{97}Ru				
		S	2×10^{-6}	1×10^{-2}	8×10^{-8}	4×10^{-4}
		I	2×10^{-6}	1×10^{-2}	6×10^{-8}	3×10^{-4}
		^{103}Ru				
		S	5×10^{-7}	2×10^{-3}	2×10^{-8}	8×10^{-5}
		I	8×10^{-8}	2×10^{-3}	3×10^{-9}	8×10^{-5}
		^{105}Ru				
		S	7×10^{-7}	3×10^{-3}	2×10^{-8}	1×10^{-4}
		I	5×10^{-7}	3×10^{-3}	2×10^{-8}	1×10^{-4}
		^{106}Ru				
		S	8×10^{-8}	4×10^{-4}	3×10^{-9}	1×10^{-5}
		I	6×10^{-9}	3×10^{-4}	2×10^{-10}	1×10^{-5}
Samarium	62	^{147}Sm				
		S	7×10^{-11}	2×10^{-3}	2×10^{-12}	6×10^{-5}
		I	3×10^{-10}	2×10^{-3}	9×10^{-12}	7×10^{-5}
		^{151}Sm				
		S	6×10^{-8}	1×10^{-2}	2×10^{-9}	4×10^{-4}
		I	1×10^{-7}	1×10^{-2}	5×10^{-9}	4×10^{-4}
		^{153}Sm				
		S	5×10^{-7}	2×10^{-3}	2×10^{-8}	8×10^{-5}
		I	4×10^{-7}	2×10^{-3}	1×10^{-8}	8×10^{-5}
Scandium	21	^{46}Sc				
		S	2×10^{-7}	1×10^{-3}	8×10^{-9}	4×10^{-5}
		I	2×10^{-8}	1×10^{-3}	8×10^{-10}	4×10^{-5}
		^{47}Sc				
		S	6×10^{-7}	3×10^{-3}	2×10^{-8}	9×10^{-5}
		I	5×10^{-7}	3×10^{-3}	2×10^{-8}	9×10^{-5}
		^{48}Sc				
		S	2×10^{-7}	8×10^{-4}	6×10^{-9}	3×10^{-5}
		I	1×10^{-7}	8×10^{-4}	5×10^{-9}	3×10^{-5}
Selenium	34	^{75}Se				
		S	1×10^{-6}	9×10^{-3}	4×10^{-8}	3×10^{-4}
		I	1×10^{-7}	8×10^{-3}	4×10^{-9}	3×10^{-4}
Silicon	14	^{31}Si				
		S	6×10^{-6}	3×10^{-2}	2×10^{-7}	9×10^{-4}
		I	1×10^{-6}	6×10^{-3}	3×10^{-8}	2×10^{-4}
Silver	47	^{105}Ag				
		S	6×10^{-7}	3×10^{-3}	2×10^{-8}	1×10^{-4}
		I	8×10^{-8}	3×10^{-3}	3×10^{-9}	1×10^{-4}

Table A.3.8-12 (Continued)
MAXIMUM PERMISSIBLE CONCENTRATIONS OF RADIONUCLIDES IN AIR AND WATER   (See Notes at End of Table)

Element	Atomic Number	Isotope*	A		B	
			In Air, μCi/ml	In Water, μCi/ml	In Air, μCi/ml	In Water, μCi/ml
Silver (*cont.*)	47	^{110m}Ag				
		S	2×10^{-7}	9×10^{-4}	7×10^{-9}	3×10^{-5}
		I	1×10^{-8}	9×10^{-4}	3×10^{-10}	3×10^{-5}
		^{111}Ag				
		S	3×10^{-7}	1×10^{-3}	1×10^{-8}	4×10^{-5}
		I	2×10^{-7}	1×10^{-3}	8×10^{-9}	4×10^{-5}
Sodium	11	^{22}Na				
		S	2×10^{-7}	1×10^{-3}	6×10^{-9}	4×10^{-5}
		I	9×10^{-9}	9×10^{-4}	3×10^{-10}	3×10^{-5}
		^{*24}Na				
		S	1×10^{-6}	6×10^{-3}	4×10^{-8}	2×10^{-4}
		I††	1×10^{-7}	8×10^{-4}	5×10^{-9}	3×10^{-5}
Strontium	38	^{85m}Sr				
		S	4×10^{-5}	2×10^{-1}	1×10^{-6}	7×10^{-3}
		I	3×10^{-5}	2×10^{-1}	1×10^{-6}	7×10^{-3}
		^{85}Sr				
		S	2×10^{-7}	3×10^{-3}	8×10^{-9}	1×10^{-4}
		I	1×10^{-7}	5×10^{-3}	4×10^{-9}	2×10^{-4}
		^{89}Sr				
		S	3×10^{-8}	3×10^{-4}	3×10^{-10}	3×10^{-6}
		I	4×10^{-8}	8×10^{-4}	1×10^{-9}	3×10^{-5}
		^{90}Sr				
		S	1×10^{-9}‡	1×10^{-5}	3×10^{-11}	3×10^{-7}
		I	5×10^{-9}	1×10^{-3}	2×10^{-10}	4×10^{-5}
		^{91}Sr				
		S	4×10^{-7}	2×10^{-3}	2×10^{-8}	7×10^{-5}
		I	3×10^{-7}	1×10^{-3}	9×10^{-9}	5×10^{-5}
		^{92}Sr				
		S	4×10^{-7}	2×10^{-3}	2×10^{-8}	7×10^{-5}
		I	3×10^{-7}	2×10^{-3}	1×10^{-8}	6×10^{-5}
Sulfur	16	^{35}S				
		S	3×10^{-7}	2×10^{-3}	9×10^{-9}	6×10^{-5}
		I	3×10^{-7}	8×10^{-3}	9×10^{-9}	3×10^{-4}
Tantalum	73	^{182}Ta				
		S	4×10^{-8}	1×10^{-3}	1×10^{-9}	4×10^{-5}
		I	2×10^{-8}	1×10^{-3}	7×10^{-10}	4×10^{-5}
Technetium	43	^{96m}Tc				
		S	8×10^{-5}	4×10^{-1}	3×10^{-6}	1×10^{-2}
		I	3×10^{-5}	3×10^{-1}	1×10^{-6}	1×10^{-2}
		^{96}Tc				
		S	6×10^{-7}	3×10^{-3}	2×10^{-8}	1×10^{-4}
		I	2×10^{-7}	1×10^{-3}	8×10^{-9}	5×10^{-5}
		^{97m}Tc				
		S	2×10^{-6}	1×10^{-2}	8×10^{-8}	4×10^{-4}
		I	2×10^{-7}	5×10^{-3}	5×10^{-9}	2×10^{-4}
		^{97}Tc				
		S	1×10^{-5}	5×10^{-2}	4×10^{-7}	2×10^{-3}
		I	3×10^{-7}	2×10^{-2}	1×10^{-8}	8×10^{-4}
		^{99m}Tc				
		S	4×10^{-5}	2×10^{-1}	1×10^{-6}	6×10^{-3}

Table A.3.8-12 (Continued)
MAXIMUM PERMISSIBLE CONCENTRATIONS OF RADIONUCLIDES IN
AIR AND WATER   (See Notes at End of Table)

Element	Atomic Number	Isotope*	A		B	
			In Air, μCi/Ml	In Water, μCi/ml	In Air, μCi/ml	In Water, μCi/ml
Technetium (*cont.*)	43	^{99m}Tc				
		I	1×10^{-5}	8×10^{-2}	5×10^{-7}	3×10^{-3}
		^{99}Tc				
		S	2×10^{-6}	1×10^{-2}	7×10^{-8}	3×10^{-4}
		I	6×10^{-8}	5×10^{-3}	2×10^{-9}	3×10^{-4}
Tellurium	52	^{125m}Te				
		S	4×10^{-7}	5×10^{-3}	1×10^{-8}	2×10^{-4}
		I	1×10^{-7}	3×10^{-3}	4×10^{-9}	1×10^{-4}
		^{127m}Te				
		S	1×10^{-7}	2×10^{-3}	5×10^{-9}	6×10^{-5}
		I	4×10^{-8}	2×10^{-3}	1×10^{-9}	5×10^{-5}
		^{127}Te				
		S	2×10^{-6}	8×10^{-3}	6×10^{-8}	3×10^{-4}
		I	9×10^{-7}	5×10^{-3}	3×10^{-8}	2×10^{-4}
		^{129m}Te				
		S	8×10^{-8}	1×10^{-3}	3×10^{-9}	3×10^{-5}
		I	3×10^{-8}	6×10^{-4}	1×10^{-9}	2×10^{-5}
		^{129}Te				
		S	5×10^{-6}	2×10^{-2}	2×10^{-7}	8×10^{-4}
		I	4×10^{-6}	2×10^{-2}	1×10^{-7}	8×10^{-4}
		^{131m}Te				
		S	4×10^{-7}	2×10^{-3}	1×10^{-8}	6×10^{-5}
		I	2×10^{-7}	1×10^{-3}	6×10^{-9}	4×10^{-5}
		^{132}Te				
		S	2×10^{-7}	9×10^{-4}	7×10^{-9}	3×10^{-5}
		I	1×10^{-7}	6×10^{-4}	4×10^{-9}	2×10^{-5}
Terbium	65	^{160}Tb				
		S	1×10^{-7}	1×10^{-3}	3×10^{-9}	4×10^{-5}
		I	3×10^{-8}	1×10^{-3}	1×10^{-9}	4×10^{-5}
Thallium	81	^{200}Tl				
		S	3×10^{-6}	1×10^{-2}	9×10^{-8}	4×10^{-4}
		I	1×10^{-6}	7×10^{-3}	4×10^{-8}	2×10^{-4}
		^{201}Tl				
		S	2×10^{-6}	9×10^{-3}	7×10^{-8}	3×10^{-4}
		I	9×10^{-7}	5×10^{-3}	3×10^{-8}	2×10^{-4}
		^{202}Tl				
		S	8×10^{-7}	4×10^{-3}	3×10^{-8}	1×10^{-4}
		I	2×10^{-7}	2×10^{-3}	8×10^{-9}	7×10^{-5}
		^{204}Tl				
		S	6×10^{-7}	3×10^{-3}	2×10^{-8}	1×10^{-4}
		I	3×10^{-8}	2×10^{-3}	9×10^{-10}	6×10^{-5}
Thorium	90	^{228}Th				
		S	9×10^{-12}	2×10^{-4}	3×10^{-13}	7×10^{-6}
		I	6×10^{-12}	4×10^{-4}	2×10^{-13}	1×10^{-5}
		^{230}Th				
		S	2×10^{-12}	5×10^{-5}	8×10^{-14}	2×10^{-6}
		I	1×10^{-11}	9×10^{-4}	3×10^{-13}	3×10^{-5}

Table A.3.8-12 (Continued)
MAXIMUM PERMISSIBLE CONCENTRATIONS OF RADIONUCLIDES IN
AIR AND WATER   (See Notes at End of Table)

Element	Atomic Number	Isotope*	A		B	
			In Air, $\mu Ci/ml$	In Water, $\mu Ci/ml$	In Air, $\mu Ci/ml$	In Water, $\mu Ci/ml$
Thorium (*cont.*)	90	^{232}Th				
		S	3×10^{-11}	5×10^{-5}	1×10^{-12}	2×10^{-6}
		I	3×10^{-11}	1×10^{-3}	1×10^{-12}	4×10^{-5}
		$^{natural}Th$				
		S	3×10^{-11}	3×10^{-5}	1×10^{-12}	1×10^{-6}
		I	3×10^{-11}	3×10^{-4}	1×10^{-12}	1×10^{-5}
		^{234}Th				
		S	6×10^{-8}	5×10^{-4}	2×10^{-9}	2×10^{-5}
		I	3×10^{-8}	5×10^{-4}	1×10^{-9}	2×10^{-5}
Thulium	69	^{170}Tm				
		S	4×10^{-8}	1×10^{-3}	1×10^{-9}	5×10^{-5}
		I	3×10^{-8}	1×10^{-3}	1×10^{-9}	5×10^{-5}
		^{171}Tm				
		S	1×10^{-7}	1×10^{-2}	4×10^{-9}	5×10^{-4}
		I	2×10^{-7}	1×10^{-2}	8×10^{-9}	5×10^{-4}
Tin	50	^{113}Sn				
		S	4×10^{-7}	2×10^{-3}	1×10^{-8}	9×10^{-5}
		I	5×10^{-8}	2×10^{-3}	2×10^{-9}	8×10^{-5}
		^{125}Sn				
		S	1×10^{-7}	5×10^{-4}	4×10^{-9}	2×10^{-5}
		I	8×10^{-8}	5×10^{-4}	3×10^{-9}	2×10^{-5}
Tungsten (Wolfram)	74	^{181}W				
		S	2×10^{-6}	1×10^{-2}	8×10^{-8}	4×10^{-4}
		I	1×10^{-7}	1×10^{-2}	4×10^{-9}	3×10^{-4}
		^{185}W				
		S	8×10^{-7}	4×10^{-3}	3×10^{-8}	1×10^{-4}
		I	1×10^{-7}	3×10^{-3}	4×10^{-9}	1×10^{-4}
		^{187}W				
		S	4×10^{-7}	2×10^{-3}	2×10^{-8}	7×10^{-5}
		I	3×10^{-7}	2×10^{-3}	1×10^{-8}	6×10^{-5}
Uranium	92	^{230}U				
		S	3×10^{-10}	1×10^{-4}	1×10^{-11}	5×10^{-6}
		I	1×10^{-10}	1×10^{-4}	4×10^{-12}	5×10^{-6}
		^{232}U				
		S	1×10^{-10}	8×10^{-4}	3×10^{-12}	3×10^{-5}
		I	3×10^{-11}	8×10^{-4}	9×10^{-13}	3×10^{-5}
		^{233}U				
		S	5×10^{-10}	9×10^{-4}	2×10^{-11}	3×10^{-5}
		I	1×10^{-10}	9×10^{-4}	4×10^{-12}	3×10^{-5}
		^{234}U				
		S	6×10^{-10}	9×10^{-4}	2×10^{-11}	3×10^{-5}
		I	1×10^{-10}	9×10^{-4}	4×10^{-12}	3×10^{-5}
		^{235}U				
		S	5×10^{-10}	8×10^{-4}	2×10^{-11}	3×10^{-5}
		I	1×10^{-10}	8×10^{-4}	4×10^{-12}	3×10^{-5}
		^{236}U				
		S	6×10^{-10}	1×10^{-3}	2×10^{-11}	3×10^{-5}
		I	1×10^{-10}	1×10^{-3}	4×10^{-12}	3×10^{-5}

Table A.3.8-12 (Continued)
MAXIMUM PERMISSIBLE CONCENTRATIONS OF RADIONUCLIDES IN AIR AND WATER (See Notes at End of Table)

Element	Atomic Number	Isotope*	A		B	
			In Air, μCi/ml	In Water, μCi/ml	In Air, μCi/ml	In Water, μCi/ml
Uranium (*cont.*)	92	^{238}U				
		S	7×10^{-11}	1×10^{-3}	3×10^{-12}	4×10^{-5}
		I	1×10^{-10}	1×10^{-3}	5×10^{-12}	4×10^{-5}
		$^{240}U\dagger$				
		S	2×10^{-7}	1×10^{-3}	8×10^{-9}	3×10^{-5}
		I	2×10^{-7}	1×10^{-3}	6×10^{-9}	3×10^{-5}
		$^{natural}U$				
		S	7×10^{-11}	5×10^{-4}	3×10^{-12}	2×10^{-5}
		I	6×10^{-11}	5×10^{-4}	2×10^{-12}	2×10^{-5}
Vanadium	23	^{48}V				
		S	2×10^{-7}	9×10^{-4}	6×10^{-9}	3×10^{-5}
		I	6×10^{-8}	8×10^{-4}	2×10^{-9}	3×10^{-5}
Xenon	54	^{131m}Xe				
		Sub	2×10^{-5}	—	4×10^{-7}	—
		^{133}Xe				
		Sub	1×10^{-5}	—	3×10^{-7}	—
		$^{133m}Xe\dagger$				
		Sub	1×10^{-6}	—	3×10^{-7}	—
		^{135}Xe				
		Sub	4×10^{-6}	—	1×10^{-7}	—
Ytterbium	70	^{175}Yb				
		S	7×10^{-7}	3×10^{-3}	2×10^{-8}	1×10^{-4}
		I	6×10^{-7}	3×10^{-3}	2×10^{-8}	1×10^{-4}
Yttrium	39	^{90}Y				
		S	1×10^{-7}	6×10^{-4}	4×10^{-9}	2×10^{-5}
		I	1×10^{-7}	6×10^{-4}	3×10^{-9}	2×10^{-5}
		^{91m}Y				
		S	2×10^{-5}	1×10^{-1}	8×10^{-7}	3×10^{-3}
		I	2×10^{-5}	1×10^{-1}	6×10^{-7}	3×10^{-3}
		^{91}Y				
		S	4×10^{-8}	8×10^{-4}	1×10^{-9}	3×10^{-5}
		I	3×10^{-8}	8×10^{-4}	1×10^{-9}	3×10^{-5}
		^{92}Y				
		S	4×10^{-7}	2×10^{-3}	1×10^{-8}	6×10^{-5}
		I	3×10^{-7}	2×10^{-3}	1×10^{-8}	6×10^{-5}
		^{93}Y				
		S	2×10^{-7}	8×10^{-4}	6×10^{-9}	3×10^{-5}
		I	1×10^{-7}	8×10^{-4}	5×10^{-9}	3×10^{-5}
Zinc	30	^{65}Zn				
		S	1×10^{-7}	3×10^{-3}	4×10^{-9}	1×10^{-4}
		I	6×10^{-8}	5×10^{-3}	2×10^{-9}	2×10^{-4}
		^{69m}Zn				
		S	4×10^{-7}	2×10^{-3}	1×10^{-8}	7×10^{-5}
		I	3×10^{-7}	2×10^{-3}	1×10^{-8}	6×10^{-5}
		^{69}Zn				
		S	7×10^{-6}	5×10^{-2}	2×10^{-7}	2×10^{-3}
		I	9×10^{-6}	5×10^{-2}	3×10^{-7}	2×10^{-3}

Table A.3.8-12 (Continued)
MAXIMUM PERMISSIBLE CONCENTRATIONS OF RADIONUCLIDES IN AIR AND WATER   (See Notes at End of Table)

Element	Atomic Number	Isotope*	A		B	
			In Air, μCi/ml	In Water, μCi/ml	In Air, μCi/ml	In Water, μCi/ml
Zirconium	40	^{93}Zr				
		S	1×10^{-7}	2×10^{-2}	4×10^{-9}	8×10^{-4}
		I	3×10^{-7}	2×10^{-2}	1×10^{-8}	8×10^{-4}
		^{95}Zr				
		S	1×10^{-7}	2×10^{-3}	4×10^{-9}	6×10^{-5}
		I	3×10^{-8}	2×10^{-3}	1×10^{-9}	6×10^{-5}
		^{97}Zr				
		S	1×10^{-7}	5×10^{-4}	4×10^{-9}	2×10^{-5}
		I	9×10^{-8}	5×10^{-4}	3×10^{-9}	2×10^{-5}
Any single radionuclide not listed above with decay mode other than alpha emission or spontaneous fission and with radioactive half-life less than 2 hours† Sub			1×10^{-6}	—	3×10^{-9}	—
Any single radionuclide not listed above with decay mode other than alpha emission or spontaneous fission and with radioactive half-life greater than 2 hours†			3×10^{-9}	9×10^{-5}	1×10^{-10}	3×10^{-6}
Any single radionuclide not listed above that decays by alpha emission or spontaneous fission†			6×10^{-13}	4×10^{-7}	2×10^{-14}	3×10^{-8}

* S = soluble; I = insoluble; "Sub" means that values given are for submersion in a semispherical infinite cloud of airborne material.‡

† Added 30 FR 15801.

‡ Revised 30 FR 15801.

†† Revised 29 FR 14434.

NOTE: In any case where there is a mixture in air or water of more than one radionuclide, the limiting values for purposes of this table should be derived as follows:

1. If the identity and concentration of each radionuclide in the mixture are known, the limiting values should be derived as follows: determine for each radionuclide in the mixture the ratio between the quantity present in the mixture and the limit otherwise established in this table for the specific radionuclide when not in a mixture. The sum of such ratios for all the radionuclides in the mixture may not exceed 1 (i.e., unity).

Example: If radionuclides A, B, and C are present in concentrations C_A, C_B, and C_C, and if the applicable MPC's are MPC_A, MPC_B, and MPC_C respectively, then the concentrations shall be limited so that the following relationship exists:

$$\frac{C_A}{MPC_A} + \frac{C_B}{MPC_B} + \frac{C_C}{MPC_C} \leqq 1$$

2. ‡ If either the identity or the concentration of any radionuclide in the mixture is not known, the limiting values for purposes of this table shall be:
   a) for purposes of Section A, Column 1—6×10^{-13};
   b) for purposes of Section A, Column 2—4×10^{-7};
   c) for purposes of Section B, Column 1—2×10^{-14};
   d) for purposes of Section B, Column 2—3×10^{-8}.

3. (26 FR 11046) If any of the conditions specified below are met, the corresponding values specified below may be used in lieu of those specified in paragraph 2 above:
   a) if the identity of each radionuclide in the mixture is known, but the concentration of one or more of the radionuclides in the mixture is not known, the concentration limit for the mixture is the limit specified in this table for the radionuclide in the mixture having the lowest concentration limit; or
   b) if the identity of each radionuclide in the mixture is not known, but it is known that certain radionuclides specified in this table are not present in the mixture, the concentration limit for the mixture is the lowest concentration limit specified in this table for any radionuclide that is not known to be absent from the mixture; or
   c) (30 FR 15801).

Table A.3.8-12 (Continued)
MAXIMUM PERMISSIBLE CONCENTRATIONS OF RADIONUCLIDES IN AIR AND WATER

Element	Atomic Number	Isotope*	A		B	
			In Air, μCi/ml	In Water, μCi/ml	In Air, μCi/ml	In Water, μCi/ml
If it is known that ^{90}Sr, ^{125}I, ^{126}I, ^{129}I, ^{131}I, (^{133}I, Section B only), ^{210}Pb, ^{210}Po, ^{211}At, ^{223}Ra, ^{224}Ra, ^{226}Ra, ^{227}Ac, ^{228}Ra, ^{230}Th, ^{231}Pa, ^{232}Th, ^{natural}Th, ^{248}Cm, ^{254}Cf, and ^{256}Fm are not present			—	9×10^{-5}	—	3×10^{-6}
If it is known that ^{90}Sr, ^{125}I, ^{126}I, ^{129}I, (^{131}I, ^{133}I, Section B only), ^{210}Pb, ^{210}Po, ^{223}Ra, ^{226}Ra, ^{228}Ra, ^{231}Pa, ^{natural}Th, ^{248}Cm, ^{254}Cf, and ^{256}Fm are not present			—	6×10^{-5}	—	2×10^{-6}
If it is known that ^{90}Sr, ^{129}I, (^{125}I, ^{126}I, ^{131}I, Section B only), ^{210}Pb, ^{226}Ra, ^{228}Ra, ^{248}Cm, and ^{254}Cf are not present			—	2×10^{-5}	—	6×10^{-7}
If it is known that (^{129}I, Section B only), ^{226}Ra, and ^{228}Ra are not present			—	3×10^{-6}	—	1×10^{-7}
If it is known that alpha emitters and ^{90}Sr, ^{129}I, ^{210}Pb, ^{227}Ac, ^{228}Ra, ^{230}Pa, ^{241}Pu, and ^{249}Bk are not present			3×10^{-9}	—	1×10^{-10}	—
If it is known that alpha emitters and ^{210}Pb, ^{227}Ac, ^{228}Ra, and ^{241}Pu are not present			3×10^{-10}	—	1×10^{-11}	—
If it is known that alpha emitters and ^{227}Ac are not present			3×10^{-11}	—	1×10^{-12}	—
If it is known that ^{227}Ac, ^{230}Th, ^{231}Pa, ^{238}Pu, ^{239}Pu, ^{240}Pu, ^{242}Pu, ^{244}Pu, ^{248}Cm, ^{249}Cf, and ^{251}Cf are not present			3×10^{-12}	—	1×10^{-13}	—

4. (25 FR 13952) If the mixture of radionuclides consists of uranium and its daughter products in ore dust prior to chemical processing of the uranium ore, the values specified below may be used in lieu of those determined in accordance with paragraph 1 above or those specified in paragraphs 2 and 3 above:

    a) for purposes of Section A, Column 1—1×10^{-10} μCi/ml gross alpha activity; or 2.5×10^{-11} μCi/ml natural uranium; or 75 μg/m^3 air natural uranium;

    b) for purposes of Section B, Column 1—3×10^{-12} μCi/ml gross alpha activity; or 8×10^{-13} μCi/ml natural uranium; or 3 μg/m^3 air natural uranium.

5. (26 FR 11046) For purposes of this note, a radionuclide may be considered as not present in a mixture if a) the ratio of the concentration of that radionuclide in the mixture (C_A) to the concentration limit for that radionuclide specified in Section B of this table (MPC_A) does not exceed 1/10 (i.e., $C_A/MPC_A \leq 1/10$), and b) the sum of such ratios for all the radionuclides considered as not present in the mixture does not exceed 1/4 (i.e., $C_A/MPC_A + C_B/MPC_B + \cdots \leq 1/4$).

Reprinted by permission of the U.S. Atomic Energy Commission from Title 10, *Code of Federal Regulations*, Part 20, Standards for Protection Against Radiation. Revised August 1966.

Table A.3.6—13

SINGLE INTAKE THAT WILL GIVE 5 REM TO CRITICAL ORGAN WITHIN 1 YEAR°

Isotope	Radiological half life	Inhaling soluble material (μc)	Inhaling soluble material Critical organ	Injecting soluble material (μc)	Injecting soluble material Critical organ	Inhaling insoluble (based on lung dose) (μc)	Inhaling insoluble (based on GI tract dose) (μc)	Activity of point source giving 100 mr/hr at 1 cm^b (μc)
H³ (HTO or H₂O)	12.5 years	5.42×10^4	T. body	4.14×10^4	T. body	8.6×10^2	2.0×10^5	>1000
Be⁷	53.5 days	1.27×10^4	Bone	3.18×10^3	Bone	3.86×10^2	7.0×10^3	315
C¹⁴ (CO₂)	5568 years	1.02×10^3	Fat	7.34×10^2	Fat	9.56×10	2.33×10^4	>1000
F¹⁸	110 min	2.13×10^5	Bone	5.10×10^4	Bone	8.92×10^4	2.83×10^4	1000
Na²⁴	15.1 hr	4.8×10^3	T. body	2.86×10^3	T. body	3.86×10^2	9.50×10^2	5.3^c
P³²	25 days	2.55×10^2	Bone	1.65×10^2	Bone	6.06×10	3.17×10^2	>1000
S³⁵	88 days	1.81×10^3	Skin	9.54×10^2	Skin	2.01×10^2	2.67×10^3	>1000
Cl³⁶	4.4×10^5 years	1.27×10^3	T. body	9.54×10^2	T. body	2.01×10	4.86×10^3	400^c
K⁴²	12.5 hr	5.42×10^3	Muscle	3.18×10^3	Muscle	7.97×10^2	1.63×10^3	51^d
Ca⁴⁵	180 days	2.13×10	Bone	1.46×10	Bone	9.56×10	6.66×10^3	>1000
Sc⁴⁶	84 days	2.13×10^2	Spleen	5.08×10	Spleen	8.29	1.53×10^2	9.8^d
		1.50×10^2	Liver	2.93×10	Liver			
		2.32×10^3	Spleen	4.46×10^2	Spleen			
Sc⁴⁷	3.40 days	1.69×10^3	Liver	3.18×10^2	Liver	7.97×10^2	3.50×10^2	189^e
		1.37×10^3	Spleen	3.18×10^2	Spleen			
Sc⁴⁸	44 hr	7.64×10^2	Liver	1.49×10^2	Liver	2.01×10^2	1.52×10^2	5.4
V⁴⁸	16 days	1.37×10^3	Bone	3.50×10^2	Bone	2.13×10	1.18×10^2	6.1^d
Cr⁵¹	28 days	3.18×10^4	Kidneys	9.23×10^3	Kidneys	2.13×10^3	8.34×10^3	13.7^c
		1.91×10^4	Kidneys	2.20×10^3	Kidneys			
Mn⁵⁶	2.6 hr	2.64×10^4	Liver	6.04×10^3	Liver	6.70×10^3	1.330×10^3	10.6^d
Fe⁵⁵	2.6 years	1.59×10^3	Blood	1.02×10^3	Blood	9.24×10^2	5.16×10^4	No γ (5μc for 6.5 kev X-ray)
Fe⁵⁹	45 days	3.82×10	Blood	2.51×10	Blood	2.01×10	1.13×10^3	15.3^{d,e,f}
Co⁶⁰	5.2 years	2.8×10^3	Liver	9.55×10^2	Liver	3.51	1.83×10^2	7.6^f
Ni⁵⁹	10⁵ years	4.14×10^4	Liver	1.43×10^3	Liver	1.02×10^2	1.57×10^3	>1000
Cu⁶⁴	12.9 hr	1.37×10^4	Liver	4.45×10^3	Liver	5.74×10^3	1.83×10^3	88^f
Zn⁶⁵	245 days	5.09×10^3	Bone	1.53×10^3	Bone	2.13×10	8.18×10^2	12.3^c
Ga⁷²	14 hr	3.82×10^3	Bone	7.64×10^2	Bone	5.42×10^2	1.83×10^2	6.4^d
Ge⁷¹	11 days	1.05×10^5	Kidneys	2.61×10^4	Kidneys	5.42×10^3	6.50×10^3	91^e
As⁷⁶	26 hr	7.00×10^3	Kidneys	1.69×10^3	Kidneys	4.15×10^2	9.00×10	45^d
Rb⁸⁶	19 days	1.05×10^3	Muscle	7.95×10^2	Muscle	4.46×10	2.17×10^3	172^c
Sr⁸⁹	51 days	1.49×10	Bone	8.27	Bone	2.65×10	2.84×10^2	>1000
Sr⁹⁰ + Y⁹⁰	28 years	1.81	Bone	9.54×10^{-1}	Bone	5.10	3.50×10^2	>1000

Table A.3.8-13 (Continued)
SINGLE INTAKE THAT WILL GIVE 5 REM TO CRITICAL ORGAN WITHIN 1 YEAR[a]

Isotope	Radiological half life	Inhaling soluble material (μc)	Inhaling soluble material Critical organ	Injecting soluble material (μc)	Injecting soluble material Critical organ	Inhaling insoluble (based on lung dose) (μc)	Inhaling insoluble (based on GI tract dose) (μc)	Activity of point source giving 100 mr/hr at 1 cm[b] (μc)
Y[91]	58 days	2.32×10	Bone	6.04	Bone	2.45×10	1.10×10^2	>1000
Zr[95] + Nb[95]	65 days	1.69×10^2	Bone	4.46×10	Bone	1.37×10	2.67×10^2	24.7
Nb[95]	35 days	5.74×10^2	Bone	2.73×10^2	Bone	3.82×10	7.67×10^2	23.4
Mo[99]	66 hours	1.27×10^6	Bone	8.28×10^5	Bone	3.82×10^2	1.20×10^3	71.4[c]
Tc[96]	4.2 days	6.68×10^3	Kidneys	3.82×10^3	Kidneys	8.92×10	4.00×10^2	7.2
Ru[106] + Rh[106]	1.0 years — 30 sec	7.64×10	Kidneys	1.910×10	Kidneys	4.15	4.50×10	74
Rh[105]	37 hr	2.23×10^2	Kidneys	8.28×10^2	Kidneys	8.28×10^2	3.67×10^2	1100
Pd[103] + Rh[103]*	17 days	1.59×10^3	Kidneys	6.36×10^2	Kidneys	4.78×10^2	2.00×10^3	330[e]
Ag[105]	40 days	2.77×10^4	Liver	9.23×10^3	Liver	9.56	1.57×10^2	16.5
Ag[111]	7.6 days	7.64×10^4	Liver	2.45×10^4	Liver	2.13×10^2	1.83×10^2	600
Cd[109] + Ag[109]*	470 days	1.91×10^2	Liver	4.77×10	Liver	8.60×10	2.17×10^3	205
Sn[113]	119 days	1.59×10^3	Bone	4.14×10^2	Bone	3.09×10	7.34×10^2	46
Te[127]	105 days — 9.4 hr	2.77×10^2	Kidneys	9.23×10	Kidneys	2.870×10	3.00×10^2	890
Te[129]	33 days — 72 min	1.18×10^2	Kidneys	3.82×10	Kidneys	1.50×10	1.02×10^2	77
I[131]	8.1 days	5.42	Thyroid	4.14	Thyroid	1.69×10^2	4.84×10^3	40[e]
Cs[137] + Ba[137]*	27 years	5.73×10^2	Muscle	4.45×10^2	Muscle	9.24	3.67×10^3	28[f]
Ba[140] + La[140]	13 days	4.77×10	Bone	1.30×10	Bone	2.45×10	1.17×10^2	6.5[h]
La[140]	40 hr	1.05×10^3	Bone	2.55×10^2	Bone	2.23×10^2	1.10×10^2	10.5
Ce[144] + Pr[144]	285 days — 17 min	5.42	Bone	1.37	Bone	4.78	4.84×10	
Pr[143]	14 days	4.46×10^2	Bone	1.08×10^2	Bone	1.50×10^2	2.17×10^2	>1000
Pm[147]	2.6 years	1.81×10^2	Bone	4.78×10	Bone	8.30×10	9.34×10^2	>1000
Sm[151]	93 years	3.82×10^2	Bone	9.24×10	Bone	2.55×10^2	3.17×10^3	>1000
Eu[154]	16 years	1.91×10	Bone	5.09	Bone	6.37	1.83×10^2	14.3[f]
Ho[166]	27 hr	2.13×10^3	Bone	6.04×10^2	Bone	7.66×10	1.67×10^2	79
Tm[171]	129 days	2.86×10	Bone	7.32	Bone	2.77×10^2	2.00×10^2	2500
Lu[177]	6.8 days	3.09×10^3	Bone	7.64×10^2	Bone	5.10×10^2	4.67×10^2	690
Ta[182]	115 days	5.73×10	Liver	1.47×10	Liver	1.18×10	2.00×10^2	16.7[e,i]
W[181]	145 days	1.05×10^4	Bone	3.18×10^3	Bone	7.97	2.67×10^2	>1000
		2.33×10^4	Thyroid	1.21×10^4	Thyroid			
Re[183]	71 days	7.63×10^4	Skin	1.11×10^4	Skin	1.50×10	8.34×10^2	69
		2.33×10^3	Kidneys	7.96×10^2	Kidneys	2.77×10^2	1.35×10^3	10.9

Table A.3.8-13 (Continued)
SINGLE INTAKE THAT WILL GIVE 5 REM TO CRITICAL ORGAN WITHIN 1 YEAR[a]

Isotope	Radiological half life	Inhaling soluble material (μc)	Critical organ	Injecting soluble material (μc)	Critical organ	Inhaling insoluble (based on lung dose) (μc)	Inhaling insoluble (based on GI tract dose) (μc)	Activity of point source giving 100 mr/hr at 1 cm[b] (μc)
Ir^{190}	11 days — 3.2 hr.	2.45×10^3	Spleen	1.81×10^2	Spleen			
Ir^{192}	74 days	1.27×10^2	Kidneys	4.79×10	Kidneys	1.18×10	2.00×10^2	19.6[f]
Pt^{191}	3.0 days	7.96×10	Spleen	5.74	Spleen	1.59×10^2	3.00×10^2	33 (?) decay scheme uncertain
Pt^{193}	1.8×10^5 days	5.73×10^2	Kidneys	1.62×10^2	Kidneys	1.28×10^2	3.50×10^2	417
		4.46×10^2	Liver	1.37×10^2	Liver			
Au^{196}	5.6 days	4.46×10^2	Kidneys	1.30×10^2	Kidneys	2.77×10^2	8.50×10^2	50.8
		4.46×10^2	Liver	1.05×10^2	Liver			
Au^{188}	2.7 days	2.87×10^2	Kidneys	7.97×10	Kidneys	3.82×10^2	2.34×10^2	40[c]
		9.56×10^2	Liver	2.36×10^2	Liver			
Au^{199}	3.1 days	6.69×10^2	Kidneys	1.99×10^2	Kidneys	7.96×10^2	6.66×10^2	113
			Liver		Liver			
Tl^{200}	26 hr	6.69×10^3	Muscle	2.58×10^4	Muscle	4.46×10^2	5.17×10^2	45.5
Tl^{201}	72 hr	1.91×10^4	Muscle	7.96×10^4	Muscle	1.18×10^3	3.67×10^3	108 (?)
Tl^{202}	12 days	5.42×10^3	Muscle	2.29×10^4	Muscle	2.33×10^2	2.00×10^3	40.8
Tl^{204}	3.6 years	1.91×10^3	Muscle	7.96×10^3	Muscle	2.14×10	4.50×10^2	No γ (100, K-X-ray)
Pb^{203}	52 hr	2.01×10^3	Bone	6.06×10^3	Bone	5.42×10^2	7.67×10^2	61.7
Pb^{210} + dr	19 years	8.28×10^{-2}	Bone	2.71×10^{-2}	Bone	9.23×10^{-2}	6.50×10^2	449
Po^{210}	138 days	1.18	Spleen	2.29×10^{-1}	Spleen	1.50×10^{-1}	1.07	>1000
At^{211} + dr	7.5 hr — 8.0 years	1.69	Thyroid	8.92×10^{-1}	Thyroid	3.50×10	4.00×10	>1000
Ra^{226} + 55% dr	1600 years	4.46×10^{-1}	Bone	1.56×10^{-1}	Bone	2.01×10^{-2}	8.5×10^{-1} (incld. all dr)	11.9[f]
Ac^{227} + dr	21.2 years	1.69×10^{-2}	Bone	4.14×10^{-3}	Bone	1.50×10^{-2}	1.83×10	29.8
Th (nat)	1.4×10^{10} years	4.78×10^{-3}	Bone	1.27×10^{-3}	Bone	1.37×10^{-2}	1.53 (Th^{232} + dr)	472
Th^{234} + Pa^{234}	24 days	2.35×10	Bone	5.73	Bone	3.09×10	7.34×10	12
U (nat)	4.5×10^9 years	8.92×10^{-2}	Kidneys	2.16×10^{-2}	Kidneys	5.42×10^{-2}	1.7 (based on U^{238})	498

Table A.3.8-13 (Continued)
SINGLE INTAKE THAT WILL GIVE 5 REM TO CRITICAL ORGAN WITHIN 1 YEAR*

Isotope	Radiological half life	Inhaling soluble material (μc)	Critical organ	Injecting soluble material (μc)	Critical organ	Inhaling insoluble (based on lung dose) (μc)	Inhaling insoluble (based on GI tract dose) (μc)	Activity of point source giving 100 mr/hr at 1 cm[b] (μc)
U^{233}	1.6×10^5 years	2.23×10^{-1}	Bone	3.50×10^{-2}	Bone	1.02×10^{-1}	1.25	2800
Pu^{239}	2.4×10^4 years	3.82×10^{-2}	Bone	9.86×10^{-3}	Bone	9.56×10^{-2}	1.18	59
Am^{241}	458 years	1.18×10^{-1}	Bone	3.06×10^{-2}	Bone	9.24×10^{-2}	1.12	256
Cm^{242}	162 days	2.13×10^{-1}	Bone	5.42×10^{-2}	Bone	1.27×10^{-1}	1.00	>1000

* The daughter products in these cases are isomers in an excited state.

a Calculated from tables by Morgan, K. Z., Snyder, W. S., and Ford, M. R., *Maximum Permissible Concentration of Radioisotopes in Air and Water for Short Period Exposure*, 1955 Geneva Conference paper, dated June 2, 1958.

b Intensities are not corrected for self-absorption by the source or its container.

c Slack, L. and Way, K., *Radiations from Radioactive Atoms in Frequent Use*, USAEC Report M-6965, National Research Council, February 1959.

d Marinello, L. D., Quimby, E. H., and Hine, G. J., *Am. J. Roetgenol. Radium Therapy*, February 1948.

e All values not referenced were calculated by A. Brodsky from tables of nuclear data by Strominger, D., Hollander, J. M., and Seaborg, G. T., *Revs. Modern Phys.*, 30, No. 2, Part II, April 1958. Low energy (below 10 key) X-rays and beta rays were neglected. Position annihilation radiation was included and adjustments were made for variation of absorption coefficient with energy from 10 to 60 key. From 60 key to 3 Mev, the approximate relation, mr/hr at 1 cm per microcurie = 5.58 E, where E is the average gamma energy in Mev per disintegration, was used.

f Kinsman, S. B., Ed., *Radiological Health Handbook*, Report PB-121784, p. 139, Robert A. Taft Sanitary Engineering Center, Cincinnati. January 1957.

g Price, Horton, and Spinney, *Radiation Shielding*, Pergamon Press, 1957.

h Brodsky, A., and Vogt, R., unpublished report, NRL, 1952.

i National Bureau of Standards Handbook 54, p. 42.

From Brodsky, A. and Beard, G. V., A Compendium of Information for Use in Controlling Radiation Emergencies, TID-8206(Rev), U.S. Atomic Energy Commission, Washington, D.C., 1960. (Available from the National Tech. Info. Service, Springfield, Virginia).

REFERENCES

1. International Commission on Radiological Units and Measurements, Clinical Dosimetry—Recommendations of the ICRU. *NBS Handbook 87*, p. 38. U.S. Government Printing Office, Washington, D.C., 1963.
2. Roesch, W. C. and Attix, F. H. *Radiation Dosimetry*, 2nd ed., Vol. 1, Ch. 1, Attix, F. H. and Roesch, W. C., eds. Academic Press, New York, 1968.
3. International Commission on Radiological Protection, Report of Committee 2. *Health Phys.*, 3: 1–380, 1960.
4. Attix, F. H. and Roesch, W. C., eds., *Radiation Dosimetry*, 2nd ed., Vol. 2. Academic Press, New York, 1968.
5. Hine, G. J. and Brownell, G. L., eds. *Radiation Dosimetry*, 1st ed. Academic Press, New York, 1956.
6. Morgan, K. Z. and Turner, J. E., eds., *Principles of Radiation Protection—A Textbook of Health Physics.* John Wiley and Sons, New York, 1967.
7. National Committee on Radiation Protection and Measurements, Measurement of Neutron Flux, and Spectra for Physical and Biological Applications, Recommendations of the NCRP. *NBS Handbook 72.* U.S. Government Printing Office, Washington, D. C., 1960.
8. Johns, H. E., *The Physics of Radiology*, 2nd ed. Charles C Thomas, Springfield, Illinois, 1964.
9. Society for Nuclear Medicine, MIRD Report. *J. Nucl. Med.*, Suppl. 1: 5–39, 1968.
10. International Commission on Radiological Protection, Report of the ICRP Task Group on Lung Dynamics. *Health Phys.*, 12: 173–207, 1966.
11. Spiers, F. W., *Radioisotopes in the Human Body: Physical and Biological Aspects.* Academic Press, New York, 1968.
12. International Atomic Energy Agency, Clinical Uses of Whole-Body Counting. *Proceedings of a Panel on the Clinical Uses of Whole-Body Counters.* International Atomic Energy Agency, Vienna, Austria, 1966.
13. Glasser, O., Quimby, E. H., Taylor, L. S., Weatherwax, J. L., and Morgan, R. H., *Physical Foundations of Radiology.* Hoeber Medical Division, Harper and Row, New York, 1963.
14. International Commission on Radiological Protection, Report of Committee 2. *Health Phys.*, 3: 1–380, 1960.
15. Fitzgerald, J. J., Brownell, G. L., and Mahoney, F. J., *Mathematical Theory of Radiation Dosimetry.* Gordon and Breach, New York, 1967.
16. Brodsky, A. and Beard, G. V., compilers and eds., *A Compendium of Information for Use in Controlling Radiation Emergencies, TID-8206* (Rev.). U.S. Atomic Energy Commission, Washington, D.C., 1960.
17. Saenger, E. L., *Medical Aspects of Radiation Accidents.* U.S. Government Printing Office, Washington, D.C., 1963.
18. Brodsky, A., Balancing Benefit versus Risk in the Control of Consumer Items Containing Radioactive Material. *Amer. J. Public Health*, 55(12): 1971–1992, 1965.
19. National Committee on Radiation Protection and Measurements, Measurement of Absorbed Dose of Neutrons, and of Mixtures of Neutrons and Gamma Rays. *NBS Handbook 75.* U.S. Government Printing Office, Washington, D.C., 1961.
20. Snyder, W. S. and Neufeld, J., Calculation of Depth Dose Curves in Tissue for Broad Beams of Fast Neutrons. *Brit. J. Radiol.*, 28: 342, 1955.
21. International Commission on Radiological Units and Measurements, Report of the ICRU. *NBS Handbook 62.* U.S. Government Printing Office, Washington, D.C., 1956.
22. Meredith, W. J., ed., *Radium Dosage, the Manchester Systems.* Williams and Wilkins, Baltimore, 1949.

Index

INDEX

A

N

Sulfur-37, gamma ray energy, 366
Sulfuric acid solution
conversion factors and densities, 252
fractions of elements in, 253
mass energy-transfer and -absorption coefficients, 237

T

Tantalum
maximum permissible concentrations in air and water, 657
X-ray critical-absorption and emission energies, 373
X-ray wavelengths, 383
Tantalum-178, production for medical use, 564
Tantalum-182, single intake giving 5 rem to critical organ within 1 year, 664
Technetium
Fission-product chains, 407
maximum permissible concentrations in air and water, 657
X-ray critical-absorption and emission energies, 372
X-ray wavelengths, 378
Technetium-94, gamma ray energy, 366
Technetium-96, single intake giving 5 rem to critical organ within 1 year, 664
Technetium-99
decay scheme, 319, 580
exposure to, specific dose rates, 640
Technetium-99m
isomeric level decay, 290
production for medical use, 564
Technetium-101, resonance detectors for measurement of neutron slowing down spectrum, 404
Technetium-104, fission yields, 409
Teflon, mass attenuation coefficients for low-energy X-rays, 241
Tellurium
fission-product chains, 407, 408
maximum permissible concentrations in air and water, 658
X-ray critical-absorption and emission energies, 372
X-ray wavelengths, 379
Tellurium-129, gamma ray energy, 364
Tellurium-127, single intake giving 5 rem to critical organ within 1 year, 664
Tellurium-129, single intake giving 5 rem to critical organ within 1 year, 664
Tellurium-131m, fission yields, 409
Tellurium-132, fission yields, 409
Tellurium-133, fission yields, 410
Tellurium-133m, fission yields, 410
Tellurium-134, fission yields, 409
Terbium
maximum permissible concentrations in air and water, 658
X-ray critical-absorption and emission energies, 373
X-ray wavelengths, 381
Terbium-155, gamma ray energy, 362, 363
Terbium-161, gamma ray energy, 364, 365
Terrestrial radioactivity, natural, estimated annual external gamma whole-body doses from, 623

Thallium
maximum permissible concentrations in air and water, 658
X-ray critical-absorption and emission energies, 373
X-ray wavelengths, 386
Thallium-100, single intake giving 5 rem to critical organ within 1 year, 665
Thallium-101, single intake giving 5 rem to critical organ within 1 year, 665
Thallium-102, single intake giving 5 rem to critical organ within 1 year, 665
Thallium-104, single intake giving 5 rem to critical organ within 1 year, 665
Thallium-208, beta-minus decay, 307, 308
Thallium chloride, biological half-life in lung, 633
Thallous nitrate, biological half-life in lung, 633
Thorium
alpha-ray spectrum, 358
decay schemes, 414
in soils, 617
maximum permissible concentrations in air and water, 658
transition of ThC^a to ThC^{11}, 415
X-ray critical-absorption and emission energies, 373
X-ray wavelengths, 388
Thorium-227
decay scheme data, 594, 595
gamma ray energy, 366
Thorium-228
concentrations in oceans and sediments, 617
decay scheme data, 596
Thorium-229, decay scheme data, 596, 597, 598
Thorium-230
concentrations in oceans and sediments, 617
decay scheme data, 598
exposure to, specific dose rates, 640
in absence of, maximum permissible concentrations of radionuclides in air and water, 662
neutron energy spectra, 393
Thorium-231
concentratios in oceans and sediments, 617
decay scheme data, 598, 599
gamma ray energy, 362
Thorium-232
concentrations in oceans and sediments, 617
decay scheme data, 600, 614
emissions, 616
energy variation for photon-induced fission, 503
half-life, 616
in absence of, maximum permissible concentrations of radionuclides in air and water, 662
isotopic abundance, 616
specific activity, 616
Thorium-233, gamma ray energy, 365
Thorium-234
concentrations in oceans and sediments, 617
decay scheme data, 600, 601
gamma ray energy, 365
single intake giving 5 rem to critical organ within 1 year, 665
Thorium-237, concentrations in oceans and sediments, 617
Thorium dioxide, biological half-life in lung, 633

U

V

W

X